Adaptive Filter Theory

PRENTICE-HALL INFORMATION AND SYSTEM SCIENCES SERIES

Thomas Kailath, *Editor*

ANDERSON & MOORE	*Optimal Filtering*
ÅSTRÖM & WITTENMARK	*Computer-Controlled Systems: Theory and Design*
GOODWIN & SIN	*Adaptive Filtering, Prediction, and Control*
GRAY & DAVISSON	*Random Processes: A Mathematical Approach for Engineers*
HAYKIN	*Adaptive Filter Theory*
KAILATH	*Linear Systems*
KUNG, WHITEHOUSE, & KAILATH, EDS.	*VLSI and Modern Signal Processing*
MACOVSKI	*Medical Imaging Systems*
MELSA & SAGE	*An Introduction to Probability and Stochastic Processes*
SPILKER	*Digital Communications by Satellite*
WILLIAMS	*Designing Digital Filters*

ADAPTIVE FILTER THEORY

SIMON HAYKIN
McMaster University

PRENTICE-HALL, *Englewood Cliffs, New Jersey 07632*

Library of Congress Cataloging-in-Publication Data

HAYKIN, SIMON S.   (date)
     Adaptive filter theory.

     Bibliography: p.
     Includes index.
     1. Adaptive filters.   I. Title.
TK7872.F5H368 1986   621.3815′324   85-19210
ISBN 0-13-004052-5

*Editorial/production supervision
   and interior design: Tracey L. Orbine*
Cover design: Wanda Lubelska Design
Manufacturing buyer: Rhett Conklin

Printed in the United States of America

10 9 8 7 6 5 4 3 2 1

ISBN 0-13-004052-5 025

Prentice-Hall International (UK) Limited, *London*
Prentice-Hall of Australia Pty. Limited, *Sydney*
Prentice-Hall Canada Inc., *Toronto*
Prentice-Hall Hispanoamericana, S.A., *Mexico*
Prentice-Hall of India Private Limited, *New Delhi*
Prentice-Hall of Japan, Inc., *Tokyo*
Prentice-Hall of Southeast Asia Pte. Ltd., *Singapore*
Editora Prentice-Hall do Brasil, Ltda., *Rio de Janeiro*
Whitehall Books Limited, *Wellington, New Zealand*

*To the many researchers
whose contributions have made it possible
to write this book*

Contents

Preface   xiii

1 *INTRODUCTION*   1

    1.1   The Linear Filtering Problem, 1
    1.2   Adaptive Filters, 3
    1.3   Approaches to the Development of Adaptive Filter Theory, 4
    1.4   Applications of Adaptive Filters, 7
    1.5   Some Historical Notes, 31

2 *STATIONARY DISCRETE-TIME STOCHASTIC PROCESSES*   44

    2.1   Introduction, 44
    2.2   Partial Characterization of a Discrete-time Stochastic Process, 45
    2.3   Properties of the Correlation Matrix, 48
    2.4   The Eigenvalue Problem, 53
    2.5   Properties of Eigenvalues and Eigenvectors, 54
    2.6   Power Spectral Density, 62
    2.7   Autoregressive Models, 67
    2.8   Computer Experiment: Autoregressive Process of Order 2, 76
    2.9   Other Stationary Stochastic Models, 86
    2.10   Selecting the Order of the Model, 87

2.11   The Innovations Process and Innovations Representation,   90
       Problems,   94

3   WIENER FILTER THEORY   100

3.1   Introduction,   100
3.2   Statement of the Optimum Filtering Problem,   101
3.3   Error-Performance Surface,   102
3.4   Differentiation with Respect to a Vector,   105
3.5   The Normal Equation,   107
3.6   Principle of Orthogonality,   109
3.7   Minimum Mean-squared Error,   110
3.8   Numerical Example,   111
3.9   Canonical Form of Error-performance Surface,   117
      Problems,   118

4   LINEAR PREDICTION   122

4.1   Introduction,   122
4.2   Forward Linear Prediction,   123
4.3   Forward Prediction-error Filter,   127
4.4   Backward Linear Prediction,   131
4.5   Backward Prediction-error Filter,   135
4.6   The Levinson–Durbin Recursion,   137
4.7   Relation among the Autocorrelation Function and the Reflection
      Coefficients,   147
4.8   Transfer Function of a Forward Prediction-error Filter,   149
4.9   Transfer Function of a Backward Prediction-error Filter,   152
4.10 The Schur-Cohn Test,   156
4.11 Whitening Property of Prediction-error Filters,   158
4.12 Eigenvector Representations of Prediction-error Filters,   161
4.13 Lattice Predictors,   162
4.14 Correlation Properties of Lattice Filters,   167
4.15 Generation of the Sequence of Backward Prediction Errors Viewed as
      a Gram–Schmidt Orthogonalization Process,   173
4.16 Equivalence of the Input Vector and the Backward Prediction-error
      Vector from an Information–theoretic Viewpoint,   178
4.17 Equivalence of the Positive Definiteness of the Correlation Matrix of
      the Tap Inputs and the Minimum–phase Property of the Prediction-
      error Filter,   180
4.18 Joint–process Estimation,   182

4.19   Inverse Filtering Using the Lattice Structure,   184

4.20   The Burg Formula,   185

Problems,   188

5 ADAPTIVE TRANSVERSAL FILTERS USING GRADIENT-VECTOR ESTIMATION

194

5.1   Introduction,   194

5.2   Structure of the Adaptive Filter,   195

5.3   The Method of Steepest Descent,   197

5.4   Stability of the Steepest-descent Algorithm,   199

5.5   The Mean-squared Error,   203

5.6   Example: Effects of Eigenvalue Spread and Step-size Parameter on the Steepest-descent Algorithm,   204

5.7   The Least-mean-square Adaptation Algorithm,   216

5.8   The Fundamental Assumption,   218

5.9   Average Tap-weight Vector,   219

5.10   Weight-error Correlation Matrix,   221

5.11   Average Mean-squared Error,   227

5.12   Properties of the Transient Behavior of the Average Mean-squared Error,   232

5.13   Summary of the LMS Algorithm,   237

5.14   Discussion,   238

5.15   Computer Experiment on Adaptive Prediction,   240

5.16   Computer Experiment on Adaptive Equalization,   244

5.17   Operation of the LMS Algorithm in a Nonstationary Environment, 251

5.18   Digital Implementation of the LMS Algorithm,   255

Problems,   259

6 KALMAN FILTER THEORY AND ITS APPLICATION TO ADAPTIVE TRANSVERSAL FILTERS

269

6.1   Introduction,   269

6.2   Recursive Minimum Mean-square Estimation for Scalar Random Variables,   270

6.3   Statement of the Kalman Filtering Problem,   273

6.4   The Innovations Process,   275

6.5   Estimation of the State Using the Innovations Process,   277

6.6   Filtering,   282

6.7   Summary and Discussion,   285

6.8   The Kalman Algorithm Applied to Adaptive Transversal Filters with
         Stationary Inputs, 287

6.9   The Kalman Algorithm Applied to Adaptive Transversal Filters with
         Nonstationary Inputs, 296

6.10   Square-root Kalman Filtering Algorithm, 300
         Problems, 301

7 METHOD OF LEAST SQUARES   307

7.1   Introduction, 307
7.2   Statement of the Linear Least-squares Estimation Problem, 308
7.3   Windowing of the Data, 310
7.4   The Deterministic Normal Equation, 312
7.5   The Principle of Orthogonality, 314
7.6   Uniqueness Theorem, 315
7.7   Minimum Sum of Error Squares, 316
7.8   Reformulation of the Deterministic Normal Equation in Terms of
         Correlation Functions, 317
7.9   Properties of Least-squares Estimates, 319
7.10   The Linear Prediction Problem, 324
7.11   The Forward–Backward Linear Prediction Method, 325
7.12   Singular-value Decomposition, 331
7.13   Estimation of Sine Waves in the Presence of Additive Noise, 341
7.14   Computer Experiment, 348
7.15   Two Other Eigendecomposition-based Methods for the Estimation of
         Sine Waves in Noise, 368
7.16   Discussion, 375
         Problems, 377

8 ADAPTIVE TRANSVERSAL FILTERS USING RECURSIVE LEAST SQUARES   381

8.1   Introduction, 381
8.2   Some Preliminaries, 382
8.3   The Matrix Inversion Lemma, 385
8.4   The Exponentially Weighted Recursive Least-squares Algorithm, 385
8.5   Update Recursion for the Sum of Weighted Error Squares, 389
8.6   Convergence Analysis of the RLS Algorithm, 390
8.7   Computer Experiment on Adaptive Prediction, 396
8.8   Computer Experiment on Adaptive Equalization, 398
8.9   Operation of the RLS Algorithm in a Nonstationary Environment,
         403

8.10   Relationship between the RLS Algorithm and Kalman Filter Theory, 407

8.11   Discussion,   407

8.12   Background Theory for Fast Algorithms,   409

8.13   Fast Transversal Filters (FTF) Algorithm,   420

8.14   Exact Initialization of the FTF Algorithm Using Zero Initial Condition, 425

8.15   Exact Initialization of the FTF Algorithm for Arbitrary Initial Condition,   433

8.16   Computer Experiment on System Identification,   436

8.17   Rescue Devices,   441

       Problems,   444

9   RECURSIVE LEAST-SQUARES LATTICE FILTERS

451

9.1   Introduction,   451

9.2   Some Preliminaries,   453

9.3   Order-update Recursions,   453

9.4   Time-update Recursion,   462

9.5   Summary of the LSL Algorithm and Discussion,   464

9.6   Initialization of the Recursive LSL Algorithm,   466

9.7   Computer Experiment on Adaptive Prediction Using the Recursive LSL Algorithm,   467

9.8   Joint-process Estimation,   469

9.9   Computer Experiment on Adaptive Equalization Using the Recursive LSL Algorithm,   478

9.10 Recursive LSL Algorithm Using A Priori Estimation Errors,   480

9.11 Modified Forms of the Recursive LSL Algorithms,   482

9.12 Discussion,   485

       Problems,   490

10   RECURSIVE LEAST-SQUARES ESTIMATION USING SYSTOLIC ARRAYS

494

10.1 Introduction,   494

10.2 Some Preliminaries,   495

10.3 Givens Rotation,   497

10.4 Solution of the Linear Least-squares Problem Using QR-Decomposition,   500

10.5 Solution of the Recursive Least-squares Problem Using a Sequence of Givens Rotations,   502

10.6 Systolic Array Implementation I,   512

10.7   Computer Experiment on Adaptive Equalization,   517

10.8   Recursive Least-squares Minimization without Explicit Computation of
         the Weight Vector,   522

10.9   Systolic Array Implementation II,   526

10.10   Computer Experiment on Adaptive Equalization (Continued),   528

10.11   Application of Systolic Array II to Adaptive Beamforming,   530

10.12   Another Systolic Array for Implementing the Minimum-variance
         Distortionless Response,   531

10.13   Computer Experiment on Adaptive Beamforming,   534

10.14   Discussion,   536

         Problems,   538

A   **MAXIMUM-LIKELIHOOD ESTIMATION**   543

B   **MAXIMUM-ENTROPY METHOD**   547

C   **z-TRANSFORM**   553

D   **METHOD OF LAGRANGE MULTIPLIERS**   557

     GLOSSARY   561

         Text Conventions,   561
         Abbreviations,   564
         Principal Symbols,   564

     REFERENCES AND BIBLIOGRAPHY   569

     INDEX   586

Preface

The subject of adaptive filters has matured to the point where it now constitutes an important part of statistical signal processing. Whenever there is a requirement to process signals that result from operation in an environment of unknown statistics, the use of an adaptive filter offers an attractive solution to the problem as it usually provides a significant improvement in performance over the use of a fixed filter designed by conventional methods. Furthermore, the use of adaptive filters provides new signal processing capabilities that would not be possible otherwise. We thus find that adaptive filters are successfully applied in such diverse fields as communications, control, radar, sonar, and seismology.

The principal aim of the book is twofold: (1) to develop the mathematical theory of various realizations of adaptive filters with finite impulse response, and (2) to illustrate the theory with examples and computer experiments. There is no unique solution to the adaptive filtering problem. Rather, we have a "kit of tools" represented by a variety of recursive algorithms, each of which offers desirable features of its own. Historical notes are added in the book in the hope that they provide a source of motivation for the interested reader to plough through the rich history of the subject. Each chapter of the book, except for Chapter 1, ends with problems that are designed to extend the theory and to challenge the readers to undertake computer experiments of their own.

In Chapter 1, we discuss, in general terms, the operation of adaptive filters and their practical applications. We end this chapter with some historical notes on the subject and related issues.

In Chapter 2, we consider the partial characterization of stationary discrete-time stochastic processes. In particular, we describe the properties of the correlation matrix of such a process, the related issue of eigenvalues and eigenvectors, and the power spectral density of the process. We describe, in some detail, the autoregressive model of a stationary process, and briefly consider other stochastic models. We consider the issue of selecting the order of the stochastic model. We end the chapter by considering the innovations representation of stochastic processes with rational power spectra.

In Chapter 3, we address the classical Wiener filter for the optimum extraction of a signal of interest that is corrupted by additive noise. We derive the well-known principle of orthogonality. We use the eigenvalue–eigenvector decomposition of the correlation matrix to develop a canonical description of the error–performance surface that results when operating a transversal filter in a stationary environment.

In Chapter 4, we apply the Wiener filter theory to develop a detailed understanding of the linear prediction problem. We consider the two basic forms of the problem: forward linear prediction and backward linear prediction. We use this theory to derive the Levinson–Durbin recursion, an important by-product of which is the set of reflection coefficients that bears a one-to-one correspondence with the autocorrelation sequence of the process. We discuss fundamental relationships that exist between forward and backward linear predictors. We derive the lattice predictor and discuss, in detail, its various properties and their implications. We consider the use of a lattice predictor as the structural basis of a joint-process estimator whereby an effective solution to the optimum filtering problem is obtained. We consider the use of the lattice structure as an inverse filter for generating an autoregressive process. We finish the chapter by deriving the Burg formula for the design of lattice predictors when operating in a stationary environment.

Thus, Chapters 2 through 4 provide the prerequisite material for the development of adaptive filter theory in subsequent chapters of the book.

In Chapter 5, we use the classical method of steepest descent to derive the widely used least mean-square (LMS) algorithm for adaptive transversal filters. The LMS algorithm is very simple to implement and yet quite effective in its own way. We study in detail the convergence properties of the LMS algorithm and thereby establish the conditions for its stability. We also consider the operation of the LMS algorithm in a nonstationary environment, and the effect of roundoff errors on its operation.

In Chapter 6, we use the innovations approach to derive the Kalman filter. Next, we apply this theory to the adaptive filtering problem in stationary and nonstationary environments. We briefly discuss the square-root version of the Kalman algorithm to overcome numerical instabilities.

Chapters 5 and 6 present two different approaches to the development of adaptive transversal filters. In both cases, we derive adaptive filtering algorithms by modifying the theory of linear filters that are optimized for stochastic inputs. In the rest of the book, Chapters 7 through 10, we use the classical method of least squares to derive another class of data-adaptive filtering algorithms starting from the input data directly.

In Chapter 7, we derive the least-squares version of the transversal filter, which represents the deterministic counterpart of the Wiener filter. We reformulate the principle of orthogonality in a time-averaged sense. We discuss the uniqueness theorem that establishes the condition for the least-squares transversal filter as a unique solution. We describe the properties of the least-squares estimator. We use the theory to solve the forward–backward linear prediction problem. We develop, in detail, the singular–value decomposition, which has many appealing analogies with the eigenvalue–eigenvector decomposition of a Hermitian matrix (e.g., correlation matrix). We discuss different algorithms, based on eigenvalue decomposition of the correlation matrix, for estimating sine waves in the presence of additive noise.

In Chapter 8, we use the matrix inversion lemma to derive the recursive least squares (RLS) algorithm for adaptive transversal filters. We study the convergence properties of this algorithm and consider its relationship to the Kalman algorithm. The RLS algorithm offers superior performance to the simple LMS algorithm at the cost of increased computational complexity. In the rest of the chapter, we develop the background theory of fast algorithms and apply this theory to derive a powerful algorithm, the fast transversal filters (FTF) algorithm, which retains the desirable features of the RLS algorithm, and yet its computational complexity is comparable to that of the LMS algorithm. We discuss the use of rescue devices for the FTF algorithm.

In Chapter 9, we derive the recursive least-squares lattice (LSL) algorithm for solving the adaptive filtering problem. The algorithm can be used for adaptive prediction or adaptive joint-process estimation. The LSL algorithm is robust and simple to implement. Moreover, its modular architecture makes it highly suited for very large scale integration (VLSI). We discuss two basic forms of the recursive LSL algorithm, one based on a posteriori estimation errors, and the other based on a priori estimation errors. We consider an error feedback modification of the latter algorithm that makes it less sensitive to roundoff errors. We conclude the chapter by summarizing the virtues and limitations of the RLS, FTF and LSL algorithms.

In Chapter 10, we develop yet another method based on the QR-decomposition for the recursive solution of the linear least-squares problem. This new approach differs from the adaptive filters considered previously in that it uses an open-loop structure. We consider two versions of the recursive QRD-LS algorithm, one of which yields on-the-fly the least-squares value of the weight vector, and the other provides recursive minimization without explicit computation of the least-squares weight vector. We consider the implementations of both versions in the form of systolic arrays whose highly concurrent architecture makes them well-suited for VLSI implementation. We consider the use of systolic arrays for solving a constrained optimization problem. We conclude the chapter by comparing the different systolic array-based adaptive filters.

The book includes four appendices on the maximum-likelihood estimation, the maximum-entropy method, the z-transform method, and the method of Lagrange multipliers to provide background material.

To help the reader, a Glossary for the book is included. It consists of a list of definitions, notations and conventions, a list of abbreviations, and a list of principal symbols used in the book.

Another item of interest is the Bibliography and References included at the end of the book, before the Index. All publications referred to in the text are compiled in this list. Each reference so made is identified in the text by the name(s) of the author(s) and the year of publication. The list also includes many other references that have been added for completeness.

The book is written at a level suitable for use in graduate courses on adaptive signal processing. In this context, the following material is available from the publisher as aids to the teacher of the course:

1. Solutions manual that presents detailed solutions to all of the problems in the book.
2. Master transparencies for all the figures and summaries of algorithms used in the book.

It is hoped that the book will also be useful to engineers in industry working on problems relating to the theory and application of adaptive filters.

ACKNOWLEDGMENTS

I wish to thank Dr. T. Kailath, editor of the series, for his encouragement and for his many useful comments. I would like to express my deep gratitude to Dr. John Cioffi for reviewing two early versions of the manuscript of the book and for providing me with many critical inputs.

I would like to express my gratitude to Qi-tu Zhang for many stimulating discussions on adaptive filters and related issues.

I am most indebted to the many graduate students who have taken my graduate course on Adaptive Systems at McMaster University during the past 5 years. Their critical inputs and comments over the years have helped shape the book into its present form.

I am most grateful to Dr. W.F. Gabriel for supplying me with the abstracts of papers accepted for publication in the March 1986 Special Issue of IEEE Transactions on Antennas and Propagation, helping to make the bibliography at the end of the book that much more up-to-date.

I am most thankful to W. Stewien, G. Macmillan, J.R. Raol, S. Kumar, H. Murthy, S. Daijavad, A. Macikunas, Y.P. Lee, and B. Dahanayake for their work on the computer experiments described in various chapters of the book. I also wish to record my appreciation to G.P. Madhavan and D. MacHattie for their comments on different parts of the book.

I wish to thank my secretary, Jill Monk, for typing many versions of the manuscript.

I am most indebted to Tracey Orbine and other staff members of Prentice-Hall for their help in the production of the book.

SIMON HAYKIN

1

Introduction

1.1 THE LINEAR FILTERING PROBLEM

The term *filter* is often used to describe a device in the form of a piece of physical hardware or computer software that is applied to a set of noisy data in order to extract information about a prescribed quantity of interest. The noise may arise from a variety of sources. For example, the data may have been derived by means of noisy sensors or may represent a useful signal component that has been corrupted by transmission through a communication channel. In any event, we may use a filter to perform three basic information-processing operations:

1. *Filtering*, which means the extraction of information about a quantity of interest at time t by using data measured up to and including time t.

2. *Smoothing*, which differs from filtering in that information about the quantity of interest need not be available at time t, and data measured later than time t can be used in obtaining this information. This means that in the case of smoothing there is a delay in producing the result of interest. Since in the smoothing process we are able to use data obtained not only up to time t,

but also data obtained after time t, we would expect it to be more accurate in some sense than the filtering process.

3. *Prediction*, which is the forecasting side of information processing. The aim here is to derive information about what the quantity of interest will be like at some time $t + \tau$ in the future, for some $\tau > 0$, by using data measured up to and including time t.

We say that the filter is *linear* if the filtered, smoothed, or predicted quantity of interest at the output of the device is a *linear function of the observations applied to the filter input*.

In the statistical approach to the solution of the *linear filtering problem* as classified above, we assume the availability of certain statistical parameters (i.e., *mean and correlation functions*) of the useful signal and unwanted additive noise, and the requirement is to design a linear filter with the noisy data as input so as to minimize the effects of noise at the filter output according to some statistical criterion. The useful approach to this filter-optimization problem is to minimize the mean-square value of the *error signal* that is defined as the difference between some desired response and the actual filter output. For stationary inputs, the resulting solution is *commonly* known as the *Wiener filter*, which is said to be *optimum in the mean-square sense*.

The Wiener filter is inadequate for dealing with situations in which *nonstationarity* of the signal and/or noise is intrinsic to the problem. In such situations, the optimum filter has to assume a *time-varying* form. A highly successful solution to this more difficult problem is found in the *Kalman filter*, a powerful device with a wide variety of engineering applications. Indeed, the Wiener filter may be viewed as a special case of the Kalman filter.

Linear filter theory, encompassing both Wiener and Kalman filters, has been developed fully in the literature for *continuous-time* as well as *discrete-time* signals. However, for technical reasons influenced by the wide availability of digital computers and the ever-increasing use of digital signal-processing devices, we find in practice that the discrete-time representation is often the preferred method. Accordingly, in subsequent chapters, we will only consider the discrete-time version of Wiener and Kalman filters; Wiener filters are considered in Chapter 3 and Kalman filters in Chapter 6. In this method of representation, the input and output signals, as well as the characteristics of the filters themselves, are all defined at discrete instants of time. In any case, a continuous-time signal may always be represented by a *sequence of samples* that are derived by observing the signal at uniformly spaced instants of time. No loss of information is incurred during this conversion process provided, of course, we satisfy the well-known *sampling theorem*, according to which the sampling rate has to be equal to or greater than twice the highest frequency component of the continuous-time signal. We may thus represent a continuous-time signal $u(t)$ by the sequence $\{u(n)\}$, $n = 0, \pm 1, = \pm 2, \ldots$, where for convenience we have normalized the sampling period to unity.

1.2 ADAPTIVE FILTERS

The design of a Wiener filter requires *a priori* information about the statistics of the data to be processed.   The filter is optimum only when the statistical characteristics of the input data match the a priori information on which the design of the filter is based.   When this information is not known completely, however, it may not be possible to design the Wiener filter or else the design may no longer be optimum.   A straightforward approach that we may use in such situations is the "estimate and plug" procedure.   This is a two-stage process whereby the filter first "estimates" the statistical parameters of the relevant signals and then "plugs" the results so obtained into a *nonrecursive* formula for computing the filter parameters.   For *real-time* operation, this procedure has the disadvantage of requiring excessively elaborate and costly hardware.   A more efficient method is to use an *adaptive filter*.   By such a device we mean one that is *self-designing* in that the adaptive filter relies for its operation on a *recursive algorithm*, which makes it possible for the filter to perform satisfactorily in an environment where complete knowledge of the relevant signal characteristics is not available.   The algorithm starts from some predetermined set of *initial conditions*, representing complete ignorance about the environment.   Yet, in a stationary environment, we find that after successive iterations of the algorithm it *converges* to the optimum Wiener solution in some statistical sense.   In a nonstationary environment, the algorithm offers a *tracking* capability, whereby it can track time variations in the statistics of the input data, provided that the variations are sufficiently slow.

As a direct consequence of the application of a recursive algorithm, whereby the parameters of an adaptive filter are updated from one iteration to the next, the parameters become *data dependent*.   This, therefore, means that an adaptive filter is a *nonlinear device*.

In another context, an adaptive filter is often referred to as linear in the sense that the estimate of a quantity of interest is obtained adaptively (at the output of the filter) as a *linear combination of the available set of observations applied to the filter input*.

A wide variety of recursive algorithms have been developed in the literature for the operation of adaptive filters.   In the final analysis, the choice of one algorithm over another is determined by various factors:

1. *Rate of convergence*: This is defined as the number of iterations required for the algorithm, in response to stationary inputs, to converge "close enough" to the optimum Wiener solution in the mean-square sense.   A fast rate of convergence allows the algorithm to adapt rapidly to a stationary environment of unknown statistics.   Furthermore, it enables the algorithm to *track* statistical variations when operating in a nonstationary environment.

2. *Misadjustment*: For an algorithm of interest, this parameter provides a quantitative measure of the amount by which the final value of the mean-squared

error, averaged over an ensemble of adaptive filters, deviates from the minimum mean-squared error that is produced by the Wiener filter.

3. *Robustness*: This refers to the ability of the algorithm to operate satisfactorily with ill conditioned input data.

4. *Computational requirements*: Here the issues of concern include (a) the number of operations (i.e., multiplications, divisions and additions/subtractions) required to make one complete iteration of the algorithm, (b) the size of memory locations required to store the data and the program, and (c) the investment required to program the algorithm on a computer.

5. *Structure*: This refers to the structure of information flow in the algorithm, determining the manner in which it is implemented in hard-ware form. For example, an algorithm whose structure exhibits high modularity, parallelism or concurrency is well-suited for implementation using very-large scale integration (VLSI)[1].

6. *Numerical properties*: When an algorithm is implemented numerically, inaccuracies are produced due to round-off noise and representation errors in the computer. There are two issues of concern, namely, the manner in which error introduced at an arbitrary point in the algorithm *propagates* to future time instants, and the effect and amplification of round-off noise on the output. For example, certain algorithms are known to be unstable with respect to such errors, which makes them unsuitable for continuous adaptation, unless some special *rescue* devices are incorporated.

1.3 APPROACHES TO THE DEVELOPMENT OF ADAPTIVE FILTER THEORY

There is no unique solution to the adaptive filtering problem. Rather, we have a "kit of tools" represented by a variety of recursive algorithms, each of which offers desirable features of its own. Basically, we may identify three distinct methods for deriving recursive algorithms for the operation of adaptive filters, as discussed next.

(a) Approach Based on Wiener Filter Theory. We use a *tapped-delay-line* or *transversal filter*[2] as the structural basis for implementing the adaptive filter.

---

[1]VLSI technology favors the implementation of algorithms that possess high modularity, parallelism or concurrency. We say a structure is *modular* when it consists of similar stages connected in cascade. By *parallelism* we mean a large number of operations being performed side-by-side. By *concurrency*, we mean a large number of *similar* computations being performed at the same time.

[2]The transversal filter was first described by Kallmann as a continuous-time device whose output is formed as a linear combination of voltages taken from uniformly spaced taps in a nondispersive delay line (Kallmann, 1940). In recent years, the transversal filter has been implemented using digital circuitry, charge-coupled devices, or surface-acoustic wave devices. Owing to its versatility and ease of implementation, the transversal filter has emerged as an essential signal-processing structure in a wide variety of applications.

The finite *impulse response* of such a filter is defined by a set of *tap weights*. For the case of stationary inputs, the *mean-squared error* (i.e., the mean-square value of the difference between the desired response and the transversal filter output) is precisely a second-order function of the tap weights in the transversal filter. The dependence of the mean-squared error on the unknown tap weights may be viewed to be in the form of a *multidimensional paraboloid* (i.e., punch bowl) with a uniquely defined bottom or *minimum point*. We refer to this paraboloid as the *error performance surface*. The tap weights corresponding to the minimum point of the surface define the optimum Wiener solution.

To develop a recursive algorithm for updating the tap weights of the adaptive transversal filter, we proceed in two stages. We first modify the *normal equation* (i.e., the matrix equation defining the optimum Wiener solution) through the use of the *method of steepest descent*, a well-known technique in optimization theory. This modification requires the use of a *gradient vector*, the value of which depends on two parameters: the *correlation matrix* of the tap inputs in the transversal filter and the *cross-correlation vector* between the desired response and the same tap inputs. Next, we use instantaneous values for these correlations so as to derive an *estimate* for the gradient vector. The resulting algorithm is widely known as the *least-mean square (LMS) algorithm*. This algorithm is simple and yet capable of achieving satisfactory performance under the right conditions. Its major limitations are a relatively slow rate of convergence and a sensitivity to variations in the *eigenvalue spread*, defined as the ratio of the maximum to minimum *eigenvalues* of the correlation matrix of the tap inputs.

In a nonstationary environment, the orientation of the error-performance surface varies continuously with time. In this case, the LMS algorithm has the added task of continually *tracking* the bottom of the error-performance surface. Indeed, tracking will occur provided that the input data vary slowly compared to the *learning rate* of the LMS algorithm.

In Chapter 5, we derive the LMS algorithm and study its behavior in a stationary or nonstationary environment.

(b) Approach Based on Kalman Filter Theory. Depending on whether the requirement is to operate in a stationary or nonstationary environment, we may exploit the Kalman filter as the basis for deriving an adaptive filtering algorithm appropriate to the situation.

The Kalman filtering problem for a linear dynamic system is formulated in terms of two basic equations: the *plant equation* that describes the dynamics of the system in terms of the *state vector*, and the *measurement equation* that describes *measurement errors* incurred in the system. The solution to the problem is expressed as a set of time-update recursions that are expressed in matrix form. To apply these recursions to solve the adaptive filtering problem, however, the theory requires that we postulate a *model* of the optimum operating conditions, which serves as the frame of reference for the Kalman filter to *track*. With a transversal filter used to provide the structural basis for the adaptive filter, we may identify the tap-weight vector in the filter as the state vector. Thus, for a stationary environment, we use a *fixed-state model* in which the tap weight or state vector of the model assumes a constant value. For a nonstationary environment, we use a *noisy-state model* in which, for example, the tap weight or state vector of the model executes a *random walk* around some mean value. Thus, by adopting one or the other of these idealized state models and also by making some other identifications, we may use the recursive solution to the Kalman filtering problem to derive different recursive algorithms for updating the tap-weight vector of the adaptive transversal filter. These algorithms are powerful in that usually they can provide a much faster rate of convergence than that attainable by the LMS algorithm. They are *robust* in that their rate of convergence is essentially insensitive to the eigenvalue-spread problem. Moreover, we may structure the Kalman algorithm to deal with a stationary or nonstationary environment. The basic limitation of these algorithms, however, is their computational complexity, which is a direct consequence of the matrix formulation of the solution to the Kalman filtering problem.

The development of Kalman filter theory and its use to derive recursive algorithms for adaptive transversal filters are discussed in Chapter 6.

(c) Method of Least Squares. The two procedures, described under (a) and (b), for deriving adaptive filtering algorithms follow from the Wiener filter and Kalman filter, respectively. The theory of both filters is based on statistical concepts. The approach based on the classical *method of least squares* differs from these two in that it is *deterministic* in its formulation right from the start. According to the method of least squares, we minimize an index of performance that consists of the *sum of weighted error squares*, where the *error* or *residual* is defined as the difference between some desired response and the actual filter output. Depending on the structure used for implementing the adaptive filter, we may identify three basically different classes of adaptive filtering alagorithms that originate from the method of least squares:

1. *Recursive least-squares algorithm:* As with adaptive filtering algorithms derived from the Wiener and Kalman filters, the *recursive least-squares (RLS) algorithm* also assumes the use of a transversal filter as the structural basis of the adaptive filter. The derivation of the algorithm relies on a basic result in linear algebra known as the *matrix-inversion lemma*. The RLS algorithm bears a close relationship to the Kalman algorithm for adaptive transversal filters. As such, the RLS algorithm enjoys the same virtues and suffers from the same limitation (computational complexity) as the Kalman algorithm. By exploiting certain properties that arise in the case of serialized data, various schemes have been developed to overcome the computational complexity of the RLS algorithm. One scheme is known as the *fast RLS algorithm*. Another particularly useful scheme is known as the *fast transversal filters (FTF) algorithm* that uses a parallel combination of 4 transversal filters. These *fast* algorithms effectively retain the advantages of the oridinary RLS algorithm, and yet their computational complexity is reduced to a level comparable to that of the simple LMS algorithm. We consider the method

of least squares in Chapter 7. The derivations of both the RLS algorithm and the FTF algorithm are presented in Chapter 8. The derivation of the fast RLS algorithm is presented as a problem at the end of Chapter 8.

2. *Least-squares lattice algorithm:* In this formulation of the recursive least-squares problem, the issue of computational complexity is resolved by using a *multistage lattice predictor* as the structural basis for implementing the adaptive filter. This predictor consists of a cascade of stages, each in the form of a *lattice*; hence, the name. An important property of the multistage lattice predictor is that its individual stages are *decoupled* from each other in a time-averaged sense. This property is exploited in the derivation of the recursive *least-squares lattice* (LSL) algorithm that involves both time– and order–updates. The algorithm is rapidly convergent, robust, and computationally efficient. Moreover, the highly pipelined, modular structure of the multistage lattice predictor and associated parts makes the recursive LSL algorithm well suited for implementation on a single silicon chip using very large scale integration (VLSI) technology. The derivation of the recursive LSL algorithm is presented in Chapter 9.

3. *QR decomposition-least squares algorithm:* This algorithm consists of an *iterative open-loop configuration*, and as such it differs from all the other adpative filtering algorithms (discussed above) that involve the use of iterative closed-loop configurations. The configuration consists of a two-stage process. The first stage involves an *orthogonal triangularization process* that is achieved by applying a special form of the *QR decomposition method* directly to the input data matrix in a recursive fashion. As new input data enter the computation, the recursive procedure maintains a linear transformation of the input data matrix into upper triangular form. At the end of the entire recursion, this special structure of the transformed data matrix is exploited to compute the least-squares weight vector, a computation that constitutes the second stage of the QR decomposition-least squares (QRD-LS) algorithm. This algorithm is stable, robust, rapidly convergent, and computationally efficient. Futhermore, it may be implemented using *systolic arrays*, which represent a highly efficient, dedicated structure. In particular, the systolic array architecture offers the desirable features of modularity, local interconnections, and highly pipelined and sychronized parallel processing. All these properties make systolic arrays particularly well suited for VLSI implementation. Two versions of the QRD-LS algorithm, including their implementations in systolic form, are discussed in Chapter 10.

In Table 1.1, we have summarized the origins of the various adaptive filtering algorithms briefly described in the preceding and the basic structures used for their implementation.

1.4 APPLICATIONS OF ADAPTIVE FILTERS

The desirable features of an adaptive filter, that is, the ability to operate effectively in an unknown environment and also track time variations of input statistics, make

TABLE 1.1 CLASSIFICATION OF ADAPTIVE FILTERING ALGORITHMS

Origin of the algorithm	Algorithm	Structural basis
Wiener filter	Least mean square (LMS)	Transversal filter
Kalman filter	Kalman algorithm assuming: (a) fixed-state model (b) random-walk state model	Transversal filter
Method of least squares	Recursive least squares (RLS) Fast RLS FTF	Transversal filters
	Recursive least-squares lattice (LSL)	Multistage lattice predictor
	Recursive QR decomposition-least squares (QRD-LS)	Systolic arrays

it a powerful device. It has been successfully applied in such diverse fields as communications, control, radar, sonar, seismology, image processing and pattern recognition. In this section, we describe applications of the adaptive filter in some of these areas, which clearly demonstrate its practical usefulness.[3]

For the applications considered here, we assume that the input data are in *baseband* form.[4] As such, the data may be *real valued* or *complex valued*, depending on the nature of the pertinent application. Both situations are considered in this section. Nevertheless, the theory presented in subsequent chapters of the book will be developed for the general case of complex valued data.

(a) System Identification. Suppose we have an unknown dynamic system, with a set of discrete-time measurements defining variation of the output signal of the system in response to a known stationary signal applied to the system input. We assume that the system is time invariant and linear. The requirement is to develop a *model* for this system in the form of a tranversal filter consisting of a set of delay-line elements (each of which is represented by the unit-delay operator z^{-1}) and a corresponding set of adjustable coefficients, which are interconnected

[3]For more detailed discussions of practical applications of adaptive filter theory, see the books by Cowan and Grant, 1985; Goodwin and Payne, 1977; Haykin, 1984; Hudson, 1981; Jayant and Noll, 1984; Ljung and Söderström, 1983; Proakis, 1983; Widrow and Stearns, 1985.

[4]The term *baseband* is used to designate the band of frequencies representing the original signal as received by the source of information. Typically, the signal at the receiver input consists of the original signal *modulated* on to a carrier wave. To obtain the baseband model of the receiver, the receiver signal is translated in frequency in such a way that the effect of the carrier is removed, and yet the information content of the original signal (in both amplitude and phase) is fully preserved. The filtering operations in the receiver are transformed in a corresponding way. For a discussion of the baseband representation of signals and systems, see Haykin (1983).

in the manner shown in Fig. 1.1. At time n, the available signal consists of a set of samples $u(n), u(n-1), \ldots, u(n-M+1)$. These samples are multiplied by a corresponding set of adjustable tap weights $w_1(n), w_2(n), \ldots, w_M(n)$, to produce an output signal denoted by $y(n)$. Let the actual output of the unknown system be denoted by $d(n)$. The adaptive filter output $y(n)$ is compared with the unknown system output $d(n)$ to produce an *estimation error $e(n)$*, as the difference between them. When the input data are real valued, assumed to be the case in Fig. 1.1, the tap weights of the transversal filter are likewise real valued.

Typically, at time n, the estimation error $e(n)$ is nonzero, implying that the model deviates from the unknown system. In an attempt to account for this deviation, the estimation error $e(n)$ is used as the input to an *adaptive control algorithm*, whereby it controls the *corrections* applied to the individual tap weights in the transversal filter. As a result, the tap weights of the filter assume a new set of values for use on the next iteration. Thus, at time $n+1$, a new filter output is produced, and with it a new value for the estimation error. The operation described is then repeated. This process is continued for a sufficiently large number of iterations (starting from time $n=0$), until the deviation of the model from the unknown dynamic system, measured by the estimation error $e(n)$, becomes sufficiently small in some statistical sense.

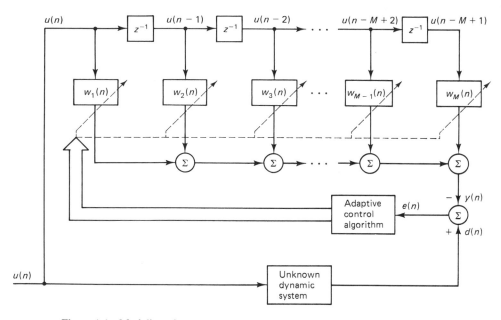

Figure 1.1 Modeling of an unknown dymanic system by a tapped-delay-line filter. The input data are assumed to be real valued; the tap weights are likewise real valued. From *Introduction to Adaptive Filters* by Simon Haykin. © 1984 by Macmillan Publishing Company, a division of Macmillan, Inc. Reproduced by permission of the publisher.

When the unknown dynamic system is time varying, the resulting system output (and therefore the desired response presented to the adaptive tapped-delay-line filter) becomes nonstationary. Correspondingly, the orientation of the error performance surface varies with time. In this case, the adaptive algorithm used to adjust the coefficients of the tapped-delay-line filter has the added task of continually tracking the statistical variations in the system.

(b) Adaptive Equalization for Data Transmission. During the last 20 years, a considerable effort has been devoted to the study of data-transmission systems that utilize the available channel bandwidth efficiently. The objective here is to design the system so as to accommodate the highest possible rate of data transmission, subject to a specified reliability that is usually measured in terms of the error rate or average probability of symbol error. The transmission of digital data through a linear communication channel is limited by two factors:

1. *Intersymbol interference (ISI):* This is caused by the dispersion of the transmitted pulse shape, which, in turn, results from deviations of the channel frequency response from the ideal characteristics of constant amplitude and linear phase (i.e., constant delay).

2. *Additive thermal noise:* This is generated at the front end of the receiver.

For bandwidth-limited channels (e.g., voice telephone channels), we usually find that intersymbol interference is the chief determining factor in the design of high-data-rate transmission systems.

Figure 1.2 shows the equivalent baseband model of a binary *pulse-amplitude modulation (PAM) system.* The signal applied to the input of the transmitter part of the system consists of a *binary data sequence* $\{b_k\}$, in which the binary symbol b_k consists of *1* or *0*. This sequence is applied to a pulse generator, the output of which is filtered first in the transmitter, then by the channel, and finally in the receiver. Let $\{u(k)\}$ denote the sampled output of the receiving filter in Fig. 1.2. Let a scaling factor a_k be defined by

$$a_k = \begin{cases} +1, & \text{if the input bit } b_k \text{ consists of symbol } 1 \\ -1, & \text{if the input bit } b_k \text{ consists of symbol } 0 \end{cases}$$

Then, in the absence of noise, we may express $\{u(k)\}$ as

$$\begin{aligned} u(k) &= \sum_n a_n p(k - n) \\ &= a_k p(0) + \sum_{\substack{n \\ n \neq k}} a_n p(k - n) \end{aligned} \tag{1.1}$$

where $\{p(n)\}$ is the sampled version of the impulse response of the cascade connection of the transmitting filter, the channel, and the receiving filter. The first term on the right side of Eq. (1.1) defines the desired symbol, whereas the remaining

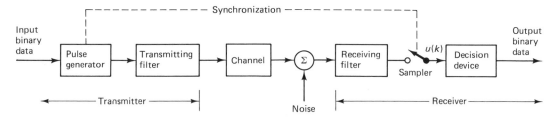

Figure 1.2 Block diagram of a baseband data transmission system (without equalization). From *Introduction to Adaptive Filters* by Simon Haykin. © 1984 by Macmillan Publishing Company, a division of Macmillan, Inc. Reproduced by permission of the publisher.

series represents the intersymbol interference caused by the combined action of the transmitting filter, the channel, and the receiving filter. This intersymbol interference, if left unchecked, can result in erroneous decisions when the sampled signal at the receiving filter output is compared with some preassigned threshold by means of a decision device.

To overcome the intersymbol interference problem, control of the time-sampled function $p(n)$ is required. In principle, if the characteristics of the channel are known precisely, then it is virtually always possible to design a pair of transmitting and receiving filters that will make the effect of intersymbol interference (at sampling times) arbitrarily small, and at the same time limit the effect of additive receiver noise by minimizing the average probability of symbol error. In practice, however, we find that a channel is random in the sense that it is one of an ensemble of possible channels. Accordingly, the use of a fixed pair of transmitting and receiving filters, designed on the basis of average channel characteristics, may not adequately reduce intersymbol interference. This suggests the need for an adaptive equalizer that provides precise control over the time response of the channel. *Equalizer* is a term that is used to describe a filter employed on telephone channels to flatten the amplitude and delay characteristics of the channel.

Among the basic philosophies for equalization of data-transmission systems are preequalization at the transmitter and postequalization at the receiver. Since the former technique requires the use of a feedback path, we will only consider equalization at the receiver, where the adaptive equalizer is placed after the receiving filter-sampler combination as in Fig. 1.2. The input into the decision device is thus modified by adding the adaptive equalizer.

Figure 1.3 shows the block diagram of an adaptive equalizer, the operation of which involves a *training mode* followed by a *tracking mode.* During the training mode, a known test signal is transmitted to probe the channel. A widely used test signal consists of a *maximal-length shift register* or *pseudonoise* (PN) sequence with a broad, even spectrum. The test signal must obviously be at least as long as the equalizer in order to make sure that the transmitted signal spectrum is adequately dense in the bandwidth of the channel to be equalized. By generating a synchronized version of the test signal in the receiver, the adaptive equalizer is supplied with a desired response. The equalizer output is subtracted from this desired

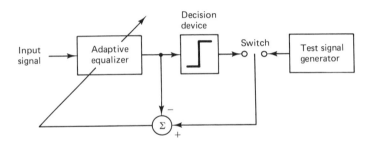

Figure 1.3 Adaptive equalizer with a decision-directed mode of operation. From *Introduction to Adaptive Filters* by Simon Haykin. © 1984 by Macmillan Publishing Company, a division of Macmillan, Inc. Reproduced by permission of the publisher.

response to produce an estimation error, which is in turn used to adaptively adjust the coefficients of the equalizer to their optimum values. The most popular class of adaptive algorithms used for this adjustment involves updating of each coefficient of the equalizer during each symbol period, starting from prescribed initial values.

When the initial training period is completed, the coefficients of the adaptive equalizer may be continually adjusted in a *decision-directed mode*. In this mode, the estimation error is derived from the final (not necessarily correct) receiver estimate of the transmitted sequence. The receiver estimate is obtained by applying the adaptive equalizer output to a decision device, as shown in Fig. 1.3. In normal operation, the receiver decisions are correct with a high probability, so the receiver estimate is correct often enough to allow the adaptive equalizer to maintain proper adjustment of its coefficients. Another attractive feature of a decision-directed adaptive equalizer is the fact that it can track slow variations in the channel characteristics or perturbations in the receiver front end, such as slow jitter in the sampler phase.

(c) Digital Representation of Speech.   During the last 20 years, there has been ever-increasing use of digital methods for the efficient encoding and transmission of speech. Two major factors are responsible for this trend. First, digital methods make it possible to control the degrading effects of noise and interference picked up during the course of transmission, thereby significantly improving the reliability of the system. Second, the revolution in digital technology (culminating in very large scale integration [VLSI]) has continually reduced both the cost and size of the hardware.

The coders used for the digital representation of speech signals fall into two broad classes: *source coders* and *wave-form coders*. Source coders are *model dependent* in that they use a priori knowledge about how the speech signal is generated at the source. Source coders for speech are generally referred to as *vocoders* (a contraction for voice coders). They can operate at 4.8 kb/s or below; however, they provide a synthetic quality, with the speech signal having lost substantial naturalness. Wave-form coders, on the other hand, essentially strive for

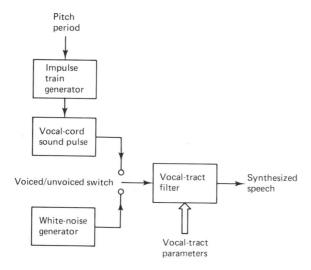

Figure 1.4 Block diagram of simplified model for the speech production process. From *Introduction to Adaptive Filters* by Simon Haykin. © 1984 by Macmillan Publishing Company, a division of Macmillan, Inc. Reproduced by permission of the publisher.

facsimile reproduction of the speech wave form. In principle, these coders are *signal independent*. They may be designed to provide telephone-toll quality for speech at coding rates as low as 16 kb/s.

Model of the Speech Production Process.   Figure 1.4 shows a simplified block diagram of the classical model for the speech production process. It assumes that the sound-generating mechanism (i.e., the source of excitation) is linearly separable from the intelligence-modulating, vocal-tract filter. The precise form of the excitation depends on whether the speech sound is *voiced* or *unvoiced*, as described next:

1. A voiced speech sound (such as[5] /i/ in eve) is generated from quasi-periodic vocal-cord sound. In the model of Fig. 1.4, the impulse-train generator produces a sequence of impulses (i.e., very short pulses), which are spaced by a fundamental period equal to the *pitch period*. This signal, in turn, excites a linear filter whose impulse response equals the vocal-cord sound pulse.

2. An unvoiced speech sound (such as /f/ in fish) is generated from random sound produced by turbulent airflow. In this case the excitation consists simply of a *white* (i.e., broad spectrum) noise source. The probability distribution of the noise samples does not appear to be critical.

The frequency response of the vocal-tract filter for unvoiced speech or that of the vocal tract multiplied by the spectrum of the vocal-cord sound pulses determines the short-time spectral envelope of the speech signal.

---

[5]The symbol / / is used to denote the *phoneme*, a basic linguistic unit.

Linear Predictive Coding.   The method of *linear predictive coding* (LPC) is an example of source coding.   This method is important because it provides not only a powerful technique for the digital transmission of speech at low bit rates but also accurate estimates of basic speech parameters.

The development of LPC relies on the model of Fig. 1.4 for the speech-production process.   The frequency response of the vocal tract for unvoiced speech or that of the vocal tract multiplied by the spectrum of the vocal sound pulse for voiced speech is described by the transfer function

$$H(z) = \frac{G}{1 + \sum_{k=1}^{M} a_k z^{-k}} \tag{1.2}$$

where G is a gain parameter and z^{-1} is the unit-delay operator.   The form of excitation applied to this filter is changed by switching between voiced and unvoiced sounds.   Thus, the filter with transfer function $H(z)$ is excited by a sequence of impulses to generate voiced sounds or a white-noise sequence to generate unvoiced sounds.   In this application, the input data are real valued; hence the filter coefficients, a_k, are likewise real valued.

In linear predictive coding, as the name implies, linear prediction is used to estimate the speech parameters.   Given a set of past samples of a speech signal, $u(n - 1), u(n - 2), \ldots, u(n - M)$, a linear prediction of $u(n)$, the present sample value of the signal, is defined by

$$\hat{u}(n) = \sum_{k=1}^{M} \hat{w}_k u(n - k) \tag{1.3}$$

The predictor coefficients, $\hat{w}_1, \hat{w}_2, \ldots, \hat{w}_M$, are optimized by minimizing the mean-square value of the prediction error, $e(n)$, defined as the difference between $u(n)$ and $\hat{u}(n)$.   The use of the minimum-mean-squared-error criterion for optimizing the predictor may be justified for two basic reasons:

1. If the speech signal satisfies the model described by Eq. (1.2), and if the mean-square value of the error signal $e(n)$ is minimized, then we find that $e(n)$ equals the excitation $u(n)$ multiplied by the gain parameter G in the model of Fig. 1.4, and $a_k = -\hat{w}_k, k = 1, 2, \ldots, M$.[6] Thus, the estimation error $e(n)$ consists of quasi-periodic pulses in the case of voiced sounds or a white-noise sequence in the case of unvoiced sounds.   In either case, the estimation error $e(n)$ would be small most of the time.

2. The use of the minimum-mean-squared-error criterion leads to tractable mathematics.

Figure 1.5 shows the block diagram of an LPC vocoder.   It consists of a transmitter and a receiver.   The transmitter first applies a *window* (typically 10 to

---

[6]The relationship between the set of predictor coefficients, $\{\hat{w}_k\}$, and the set of all-pole filter coefficients, $\{a_k\}$, is derived in Chapter 4.

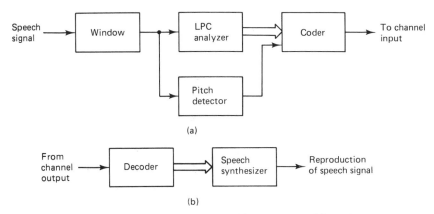

Figure 1.5 Block diagram of LPC vocoder: (a) transmitter, (b) receiver. From *Introduction to Adaptive Filters* by Simon Haykin. © 1984 by Macmillan Publishing Company, a division of Macmillan, Inc. Reproduced by permission of the publisher.

30 ms long) to the input speech signal, thereby identifying a block of speech samples for processing. This window is short enough for the vocal-tract shape to be nearly stationary, so the parameters of the speech-production model in Fig. 1.4 may be treated as essentially constant for the duration of the window. The transmitter then analyzes the input speech signal in an adaptive manner, block by block, by performing a linear prediction and pitch detection. Finally, it codes the parameters (1) the set of predictor coefficients, (2) the pitch period, (3) the gain parameter, and (4) the voiced–unvoiced parameter, for transmission over the channel. The receiver performs the inverse operations, by first decoding the incoming parameters. In particular, it computes the values of the predictor coefficients, the pitch period, and the gain parameter, and determines whether the segment of interest represents voiced or unvoiced sound. Finally, the receiver uses these parameters to synthesize the speech signal by utilizing the model of Fig. 1.4.

Wave-form Coding.   In wave-form coding the operations performed on the speech signal are designed to preserve the shape of the signal. Specifically, the operations include *sampling* (time discretization) and *quantization* (amplitude discretization). The rationale for sampling follows from a basic property of all speech signals: they are bandlimited. This means that a speech signal can be sampled in time at a finite rate in accordance with the sampling theorem. For example, commercial telephone networks designed to transmit speech signals occupy a bandwidth from 200 to 3200 Hz. To satisfy the sampling theorem, a conservative sampling rate of 8 kHz is commonly used in practice. Quantization is justified on the following grounds. Although a speech signal has a continuous range of amplitudes (and therefore its samples also have a continuous amplitude range), nevertheless, it is not necessary to transmit the exact amplitudes of the samples. Basically, the human ear (as ultimate receiver) can only detect finite amplitude differences. Examples of wave-form coding include *pulse-code modulation* (PCM) and *differential pulse-code modulation* (DPCM). In PCM, as used in telephony, the speech signal (after low-pass filtering) is sampled at the rate of 8 kHz, nonlinearly (e.g., logarithmically) quantized, and then coded into 8-bit words, as in Fig. 1.6(a). The result is a good signal-to-quantization-noise ratio over a wide dynamic range of input signal levels. This method requires a bit rate of 64 kb/s. DPCM involves the use of a predictor as in Fig. 1.6(b). The predictor is designed to exploit the correlation that exists between adjacent samples of the speech signal, in order to realize a reduction in the number of bits required for the transmission of each sample of the speech signal and yet maintain a prescribed quality of performance. This is achieved by quantizing and then coding the prediction error that results from the subtraction of the predictor output from the input signal. If the prediction is optimized, the variance of the prediction error will be significantly smaller than

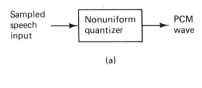

(a)

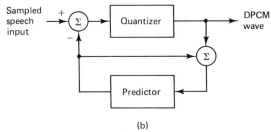

(b)

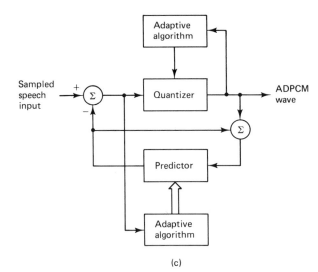

(c)

Figure 1.6 Wave-form coders: (a) PCM, (b) DPCM, (c) ADPCM. From *Introduction to Adaptive Filters* by Simon Haykin. © 1984 by Macmillan Publishing Company, a division of Macmillan, Inc. Reproduced by permission of the publisher.

White noise $\{v(n)\}$ → Discrete-time linear filter → Autoregressive process $\{u(n)\}$

Figure 1.7 Stochastic model.

that of the input signal, so a quantizer with a given number of levels can be adjusted to produce a quantizing error with a smaller variance than would be possible if the input signal were quantized directly as in a standard PCM system. Equivalently, for a quantizing error of prescribed variance, DPCM requires a smaller number of quantizing levels (and therefore a smaller bit rate) than PCM.

Differential pulse-code modulation uses a fixed quantizer and a fixed predictor. A further reduction in the transmission rate can be achieved by using an adaptive quantizer together with an adaptive predictor of sufficiently high order, as in Fig. 1.6(c). This type of wave-form coding is called *adaptive differential pulse-code modulation* (ADPCM), where "A" denotes adaptation of both quantizer and predictor algorithms. An adaptive predictor is used in order to account for the nonstationary nature of speech signals. ADPCM can digitize speech with toll quality (8-bit log PCM quality) at 32 kb/s. It can realize this level of quality with a 4-bit quantizer.

(d) Adaptive Autoregressive Spectrum Analysis. The *power spectrum* provides a quantitative measure of the second-order statistics of a discrete-time stochastic process as a function of frequency. In *parametric spectrum analysis*, we evaluate the power spectrum of the process by assuming a *model* for the process. In particular, the process is modeled as the output of a linear filter that is excited by a *white-noise process*, as in Fig. 1.7. By definition, a white-noise process has a constant power spectrum. A model that is of practical utility is the *autoregressive (AR) model*, in which the transfer function of the filter is assumed to consist of poles only. Let this transfer function be denoted by[7]

$$H(e^{j\omega}) = \frac{1}{1 + a_1^* e^{-j\omega} + \cdots + a_M^* e^{-jM\omega}}$$

$$= \frac{1}{1 + \sum_{k=1}^{M} a_k^* e^{-jk\omega}} \tag{1.4}$$

where the a_k are called the *autogressive (AR) parameters*, and M is the *model order*. Let σ_v^2 denote the constant power spectrum of the white-noise process $\{v(n)\}$ applied to the filter input. Accordingly, the power spectrum of the filter output $\{u(n)\}$ equals

$$S_{AR}(\omega) = \sigma_v^2 |H(e^{j\omega})|^2 \tag{1.5}$$

We refer to $S_{AR}(\omega)$ as the *autoregressive (AR) power spectrum*. Equation (1.5) assumes that the AR process $\{u(n)\}$ is stationary, in which case the AR parameters

---

[7]The statistical characterization of autoregressive models is considered in Chapter 2.

themselves assume constant values. For complex valued data, assumed to be the case here, the AR parameters are likewise complex valued.

If, however, the AR process is nonstationary, the model becomes time varying. Correspondingly, the model parameters become time dependent, as shown by $a_1(n)$, $a_2(n)$, . . . , $a_M(n)$. In this case, we express the power spectrum of the nonstationary AR process as

$$S_{\text{AR}}(\omega, n) = \frac{\sigma_v^2}{|1 + \sum_{k=1}^{M} a_k^*(n) e^{-jk\omega}|^2} \tag{1.6}$$

We may determine the AR parameters of the nonstationary model by applying the process $\{u(n)\}$ to an *adaptive prediction-error filter*, as indicated in Fig. 1.8. The filter consists of a transversal filter with adjustable tap weights. In the adpative scheme of Fig. 1.8, the prediction error produced at the output of the filter is used to control the adjustments applied to the tap weights of the filter.

The *adaptive AR model* provides a practical means for measuring the *instantaneous frequency* of a frequency-modulated process. In particular, we may do this by measuring the frequency at which the AR power spectrum $S_{\text{AR}}(\omega, n)$ attains its peak value for varying time n.

(e) Adaptive detection of a signal in noise of unknown statistics. The *detection problem*, that is, the problem of detecting a signal in noise, may be viewed as one of *hypothesis testing* with deep roots in *statistical decision theory*.[8] In the statistical formulation of hypothesis testing, there are two criteria of most interest: the *Bayes criterion* and the *Neyman–Pearson criterion*. In the Bayes test, we minimize the *average cost* or *risk* of the experiment of interest, which incorporates two sets of parameters: (1) *a priori probabilities* that represent the observer's information about the source of information before the experiment is conducted, and (2) a set of *costs* assigned to the various possible courses of action. As such, the Bayes criterion is directly applicable to digital communications. In the Neyman–Pearson test, on the other hand, we maximize the *probability of detection* subject to the constraint that the *probability of false alarm* does *not* exceed some preassigned value. Accordingly, the Neyman–Pearson criterion is directly applicable to radar or sonar. An idea of fundamental importance that emerges in hypothesis testing is that, for a Bayes criterion or Neyman–Pearson criterion, the optimum test consists of two distinct operations: (1) processing the observed data to compute a parameter called the *likelihood ratio*, and (2) comparing the likelihood ratio with a *threshold* to make a *decision* in favor of one of the two hypotheses. The choice of one criterion or the other merely affects the value assigned to the threshold. Let H_1 denote the hypothesis that the observed data consist of noise alone and H_2 denote the hypothesis that the data consist of signal plus noise. The likelihood ratio is defined as the ratio of two conditional probability density func-

---

[8]For a detailed discussion of hypothesis testing and its application to detection theory, see Van Trees (1968).

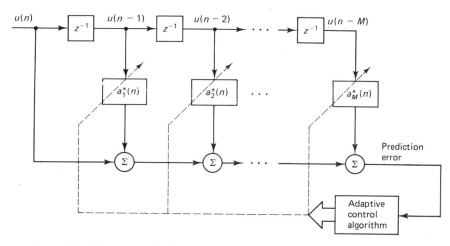

Figure 1.8 Adaptive prediction-error filter. The input data are complex valued; the tap weights are likewise complex valued.

tions, the numerator assuming that hypothesis H_2 is true and the denominator assuming that hypothesis H_1 is true. If the likelihood ratio exceeds the threshold, the decision is made in favor of hypothesis H_2; otherwise, the decision is made in favor of hypothesis H_1.

In simple binary hypothesis testing, it is assumed that the signal is known, and the noise is both white and Gaussian. In this case, the likelihood ratio test yields a *matched filter* (matched in the sense that its impulse response equals the time-reversed version of the known signal). When the additive noise is a *colored Gaussian noise* of known mean and correlation matrix, the likelihood ratio test yields a filter that consists of two sections: a *whitening filter* that transforms the colored noise component at the input into a white Gaussian noise process, and a *matched filter* that is matched to the new version of the known signal as modified by the whitening filter.

However, in some important operational environments such as *radar* and *active sonar*, the correlation matrix of noise in the observed data is *unknown*. In situations of this kind, the use of adaptivity offers an attractive approach to solve the detection problem.

The purpose of a *surveillance radar* is to *detect* moving targets (e.g., aircraft) within some prescribed range from the radar site under different environmental conditions. For example, in an airport surveillance radar, the system uses a sharply focused *fan beam*, rotated slowly in the horizontal plane while transmitting short radio pulses and listening for echoes of these pulses (Skolnik, 1982). The width of the fan beam determines the limit of resolution, or *resolution cell*, in *azimuth*. In typical radars of this type, between 10 and 20 pulses may pass through the cell before the antenna has scanned past the cell. Any object in the cell may return an echo from each of these pulses, forming a *time series* of samples representative of the object. If the radar receiver detects these pulses *coherently*, the time series contains both amplitude and phase information, the amplitude being related to the size and reflectivity of the object (radar cross section) and the phase being related to the rate of change in the range of the object (Doppler shift). This time series of samples, taken at a single range, form the basis for many radar signal-processing systems. The need for some type of processing of the radar signal is dictated by the fact that the desired echoes from a target (e.g., aircraft) are not the only signals detected. The additional *unwanted* signals, known as *clutter*, come from a number of sources (e.g., trees, buildings, weather disturbances, etc.) within the radar environment and may easily obscure the unprocessed target signal. The target detection problem is further complicated by the fact that, ordinarily, it is not possible to obtain a complete statistical description of the clutter before the radar is deployed in the field. Naturally, one other factor that also has to be taken into account is the additive noise generated at the front end of the receiver. Thus, the problem to be solved in a surveillance radar environment is to detect a target echo in the combined presence of clutter of unknown statistics and receiver noise.

The detection problem in active sonar is similar to that in radar. In active sonar, electromechanical transducers, known as *hydrophones*, are used to transmit an acoustic signal that travels through the ocean. When the signal is reflected from a target (e.g., enemy vessel), the echo is picked up by the hydrophones to provide the sonar receiver with a basis for detecting the target (Knight and others, 1981). The detection problem is compounded by several factors. First, the target echo is weakened by loss in signal strength due to propagation through the ocean and by reflection loss at the target. Second, the target echo is obscured by *reverberation* that results from reflections of the transmitted signal from such scatterers as the sea surface, the sea bottom, biologies, and inhomogeneities within the ocean volume. Third, there is receiver noise, as always, to be considered. Among these factors, however, it is reverberation that represents a major limiting factor in active sonar.

Typically, receiver noise is modeled as *white Gaussian noise of zero mean and known power spectrum*, and the clutter in radar and the reverberation in sonar are modeled as *colored Gaussian noise of unknown statistics*.

Innovations-based Detection Algorithm. Consider then the detection problem described by the model:

$$\text{hypothesis } H_2: \quad u(n) = s(n) + c(n) + v(n), \quad n = 1, 2, \ldots, N \quad (1.7)$$

$$\text{hypothesis } H_1: \quad u(n) = c(n) + v(n), \quad n = 1, 2, \ldots, N$$

where $\{u(n)\}$ denotes the observed data, $\{s(n)\}$ denotes a known signal or one taken from a Gaussian ensemble, $\{c(n)\}$ denotes a colored Gaussian noise process, and $\{v(n)\}$ denotes a white Gaussian noise process. Given the observed data $\{u(n)\}$, the problem is to decide whether the signal is present (hypothesis H_2) or not (hypothesis H_1). In radar, $\{c(n)\}$ refers to clutter, while in sonar it refers to reverberation; in both cases, $\{s(n)\}$ refers to the target echo.

Suppose that the observed data $\{u(n)\}$ are applied to two whitening filters

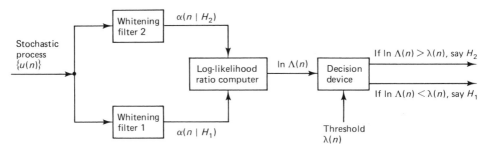

Figure 1.9 Whitening formulation of detector.

sharing a common input, as in Fig. 1.9. The filter in the upper path is designed to whiten the observed data $\{u(n)\}$, assuming that hypothesis H_2 is true. Let $\{\alpha(n|H_2)\}$ denote the conditional *innovations* process produced at the output of this whitening filter. The term *innovation* refers to new information. The filter in the lower path is designed to whiten the observed data $\{u(n)\}$, assuming that hypothesis H_1 is true. Let $\{\alpha(n)|H_1\}$ denote the innovations process produced at the output of this second whitening filter. For a finite observation length N and a stationary input process $\{u(n)\}$, these two whitening filters are generally time varying. With the colored noise process $\{c(n)\}$ and the white noise process $\{v(n)\}$ assumed to be both Gaussian, it follows that the conditional innovations processes $\{\alpha(n|H_2)\}$ and $\{\alpha(n|H_1)\}$ are both Gaussian. Since, by definition, an innovations process is white, we find that both of these innovations processes consist of statistically independent Gaussian random variables. Accordingly, we may express the log-likelihood ratio for the detection problem of Eq. (1.7) as follows:[9]

$$\ln\Lambda(N) = \frac{1}{2}\sum_{n=1}^{N}\ln\left[\frac{\sigma^2(n|H_1)}{\sigma^2(n|H_2)}\right] + \frac{|\alpha(n|H_1)|^2}{\sigma^2(n|H_1)} - \frac{|\alpha(n|H_2)|^2}{\sigma^2(n|H_2)} \qquad (1.8)$$

where ln denotes the natural logarithm, and the variance

$$\sigma^2(n|H_i) = E[|\alpha(n|H_i)|^2], \qquad i = 1, 2$$

The time dependence of this variance results from the time-varying character of the whitening filters. The decision-making process is therefore described by

$$\ln\Lambda(n) \underset{H_1}{\overset{H_2}{\gtrless}} \lambda(n) \qquad (1.9)$$

when $\lambda(n)$ is the threshold. In Eq. (1.8), the input data are complex valued; hence, the requirement for evaluating the squared amplitudes of the conditional innovations.

The detection algorithm described by Eqs. (1.8) and (1.9) is called the *innovations-based detection algorithm* (IBDA) in recognition of the use of the innovations process in its derivation. The algorithm is powerful in that it includes

[9]For a derivation of Eq. (1.8), see Problem 7 in Chapter 6.

many well-known results as special cases. For example, the algorithm may be reduced to the same form as the matched filter for a known signal in additive white Gaussian noise or that for a known signal in additive colored Gaussian noise.

To apply the innovations-based detection algorithm, we merely require knowledge of the conditional innovation $\alpha(n|H_i)$ and its variance $\alpha^2(n|H_i)$ for each of the two hypotheses H_1 and H_2. The practical utility of this algorithm is therefore limited only by the accuracy with which we are able to measure these parameters.

Also, it is noteworthy that although this algorithm is optimum only in the case when the colored-noise process $\{c(n)\}$ is Gaussian, nevertheless, use of the algorithm appears to be reasonable in the more general case when this process is non-Gaussian.

Empirical evidence suggests that radar clutter or sonar reverbration may be modeled as an autoregressive process of low order (Haykin and others, 1982; Hansen, 1984). The addition of a target signal to such a process may be accounted for by increasing the order of the autoregressive model. Accordingly, in a clutter-dominated radar environment or reverberation-dominated sonar environment, we may use adaptive prediction-error filters (whose orders are chosen to equal the orders of the pertinent autoregressive inputs) to perform the required whitening operations in an adaptive manner.

(f) Echo cancellation. In telephone connections that involve the use of both four- and two-wire transmissions, an echo is generated at the hybrid that connects a four- to a two-wire transmission. When the telephone call is made over a long distance (e.g., using geostationary satellites), an echo represents an impairment that can be as annoying subjectively as the more obvious impairments of low volume and noise. Figure 1.10 shows a satellite circuit with no echo protection. The hybrids at both ends of the circuit convert the two-wire transmissions used on customer loops and metallic trunks to the four-wire transmission needed for carrier circuits. Due to the high altitude of the satellite, a delay of 270 ms occurs in each four-wire path. Ideally, when person A on the left speaks, his speech should follow the upper transmission path to the hybrid on the right and from there be directed to the two-wire circuit. In practice, however, not all the

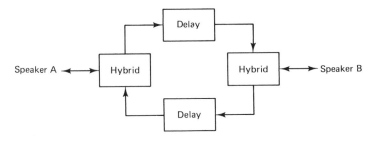

Figure 1.10 Satellite circuit with no echo protection. From *Introduction to Adaptive Filters* by Simon Haykin. © 1984 by Macmillan Publishing Company, a division of Macmillan, Inc. Reproduced by permission of the publisher.

speech energy is directed to this two-wire circuit, with the result that some is returned along the lower four-wire path to be heard by the person on the left as an echo that is delayed by 540 ms.

To overcome this problem, echo cancellers are installed in the network in pairs, as illustrated in Fig. 1.11(a). The cancellation is achieved by making an estimate of the echo and subtracting it from the return signal. The underlying assumption here is that the echo return path, from the point where the canceller bridges to the point where the echo estimate is subtracted, is linear and time invariant.

Thus, referring to the single canceller in Fig. 1.11(b) for definitions, the return signal at time n may be expressed as

$$d(n) = \sum_{k=0}^{\infty} h_k u(n - k) + v(n)$$

where $u(n)$, $u(n - 1)$, . . . , are samples of the far-end speech (from speaker A), $v(n)$ is the near-end speech (from speaker B) plus any additive noise at time n, and $\{h_k\}$ is the impulse response of the echo path. The echo canceller, consisting

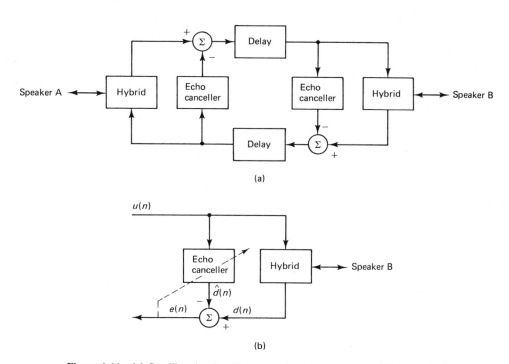

(a)

(b)

Figure 1.11 (a) Satellite circuit with a pair of echo suppressors; (b) signal definitions. From *Introduction to Adaptive Filters* by Simon Haykin. © 1984 by Macmillan Publishing Company, a division of Macmillan, Inc. Reproduced by permission of the publisher.

of some form of an adaptive filter, makes an estimate of this impulse response. Correspondingly, an estimate $\hat{d}(n)$ of the echo itself is produced at the output of the canceller. The estimation error

$$e(n) = d(n) - \hat{d}(n)$$

is, in turn, used to adaptively control the canceller coefficients. Thus, after a small number of iterations, the effect of the echo is minimized in some sense.

(g) Adaptive line enhancer. For our next example on adaptivity, we consider the *adaptive line enhancer* (ALE). This device can be used to detect a low-level sine wave embedded in a background of additive noise with a broadband spectrum.

As illustrated in Fig. 1.12, the ALE consists of a delay element and a linear predictor. The predictor output $y(n)$ is subtracted from the input signal $u(n)$ to produce the estimation error $e(n)$. This estimation error is, in turn, used to adaptively control the coefficients of the predictor. The predictor input equals $u(n - \Delta)$, the original input signal $u(n)$ delayed by Δ seconds, where Δ is equal to or greater than the sample period. The main function of the delay parameter Δ is to remove correlation that may exist between the noise component in the original input signal $u(n)$ and the noise component in the delayed predictor input $u(n - \Delta)$. For this reason, the delay parameter Δ is called the *decorrelation parameter* of the ALE. An ALE may thus be viewed as an adaptive filter that is designed to suppress broadband components (e.g., white noise) contained in the input, while at the same time passing narrowband components (e.g., sine waves) with little attenuation. In other words, it can be used to enhance the presence of sine waves (whose spectrum consists of harmonic lines) in an adaptive manner; hence, the name.

(h) Adaptive beamforming. The applications considered above pertain to adaptive signal processing problems in the *time-domain*. We now describe a *spatial* form of adaptive signal processing that finds applications in radar and sonar.

In the particular type of spatial filtering of interest to us in this book, a number

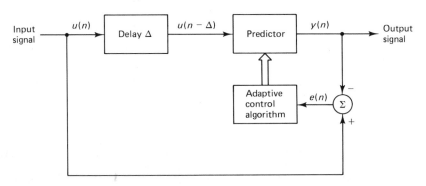

Figure 1.12 Adaptive line enhancer.

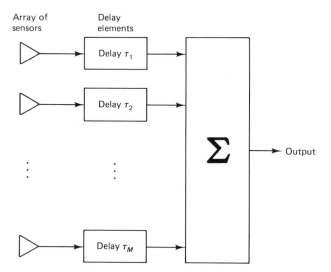

Figure 1.13 Delay–and–sum beam-former.

of independent *sensors* are placed at different points in space to "listen" to the received signal. In effect, the sensors provide a means of *sampling* the received signal *in space*. The set of sensor outputs collected at a particular instant of time constitutes a *snapshot*. Thus, a snapshot of data in spatial filtering (for the case when the sensors lie on a straight line) plays a role analogous to that of a set of consecutive tap inputs that exist in a transversal filter at a particular instant of time.[10]

In radar, the sensors consist of antenna elements (e.g., dipoles, horns, slotted waveguides) that respond to incident electromagnetic waves. In sonar, the sensors consist of hydrophones designed to respond to acoustic waves. In any event, spatial filtering is used in these systems to distinguish between the spatial properties of signal and noise.

In a primitive type of spatial filtering, known as the *delay-and-sum beam-former*, the various sensor outputs are delayed (by appropriate amounts to align signal components coming from the direction of a target) and then summed, as in Fig. 1.13. Thus, for a single target, the average power at the output of the delay-and-sum beamformer is maximized when it is steered towards the target. A major limitation of the delay-and-sum beamformer, however, is that it has no provisions for dealing with sources of *interference*.

In order to enable a beamformer to respond to an unknown interference environment, it has to be made *adaptive* in such a way that it places *nulls* in the direction(s) of the source(s) of interference automatically and in real time. By so doing, the output signal-to-noise ratio of the system is increased, and the *directional response* of the system is thereby improved. Below, we consider two examples of

---

[10]For a discussion of the analogies between time- and space-domain forms of signal processing, see Bracewell (1978).

adaptive beamformers that are well-suited for use with narrow-band signals in radar and sonar systems.

   Adaptive Beamformer with Minimum-Variance Distortionless Response.   Consider an adaptive beamformer that uses a linear array of M identical sensors, as in Fig. 1.14. The individual sensor outputs, assumed to be in *baseband* form, are weighted and then summed. The beamformer has to satisfy two requirements: (1) a *steering* capability whereby the target signal is always protected, and (2) the effects of sources of interference are minimized. One method of providing for these two requirements is to minimize the variance (i.e., average power) of the

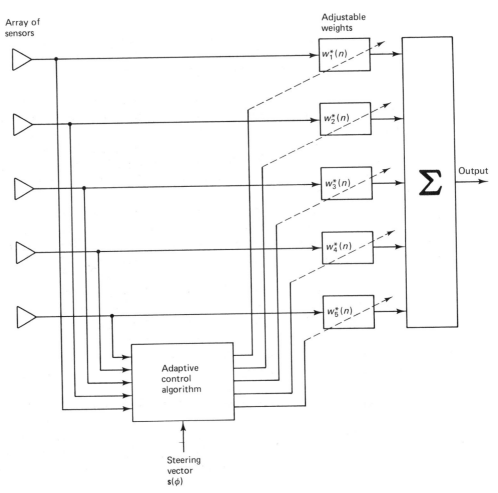

Figure 1.14   Adaptive MVDR beamformer illustrated for an array of 5 sensors. The sensor outputs (in baseband form) are complex valued; hence, the weights are complex valued.

beamformer output, subject to the *constraint* that, during the process of adaptation, the weights satisfy the condition:

$$\mathbf{w}^H(n)\, \mathbf{s}(\phi) = 1 \qquad \text{for all } n, \text{ and } \phi = \phi_t \qquad (1.10)$$

where $\mathbf{w}(n)$ is the M-by-1 weight vector and $\mathbf{s}(\phi)$ is an M-by-1 *steering vector*. The superscript H denotes Hermitian transposition (i.e., transposition combined with complex conjugation). In this application, the baseband data are complex valued; hence, the need for complex conjugation. The value of angle $\phi = \phi_t$ is determined by the direction of the target. The angle ϕ is itself measured with sensor 1 (at the top end of the array) treated as the point of reference.

The dependence of vector $\mathbf{s}(\phi)$ on the angle ϕ is defined by

$$\mathbf{s}^T(\phi) = [1, e^{-j\phi}, \ldots, e^{-j(M-1)\phi}]$$

The angle ϕ itself is related to incidence angle θ of a plane wave, measured with respect to the normal to the linear array, as follows[11]

$$\phi = \frac{2\pi d}{\lambda} \sin\theta \qquad (1.11)$$

where d is the spacing between adjacent sensors of the array, and λ is the wavelength (see Fig. 1.15). The incidence angle θ lies inside the range $-\pi/2$, to $\pi/2$. The permissible values that the angle ϕ may assume lie inside the range $-\pi$ to π. This means that we must choose the spacing $d \leq \lambda/2$, so that there is a one-to-one correspondence between the values of θ and ϕ without ambiguity. The condition $d \leq \lambda/2$ may be viewed as the spatial analog of the sampling theorem.

The imposition of the *signal-protection constraint* in Eq. (1.10) ensures that, for a prescribed look direction, the response of the array is maintained constant (i.e., equal to one), no matter what values are assigned to the weights. An algorithm that minimizes the variance of the beamformer output, subject to this constraint, is therefore referred to as the *minimum-variance distortionless response* (MVDR) algorithm. Note that the number of degrees of freedom of the adaptive beamformer is reduced to $M-2$ by the imposition of the constraint. The number of independent nulls it can produce is therefore limited to $M-2$. In Chapter 10, we describe a highly efficient structure, involving the use of systolic arrays, for the implementation of such an algorithm.

Adaptive Combiner with Fixed Beams.   Another method of designing an adaptive beamformer with provisions for main beam constraints is illustrated in the block diagram of Fig. 1.16. The system consists of a *linear array* of M identical sensors, an *orthogonal multiple-beam forming network*, and a *multiple side-lobe canceller*.

---

[11]When a plane wave impinges on a linear array as in Fig. 1.15, there is a spatial delay of $d \sin\theta$ between the signals received at any pair of adjacent sensors. With a wavelength of λ, this spatial delay is translated into an electrical angular difference defined by $\phi = 2\pi(d \sin\theta/\lambda)$.

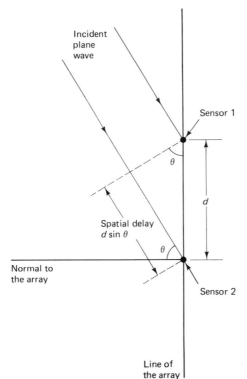

Figure 1.15 The spatial delay incurred when a plane wave impinges on a linear array.

The beam-forming network is designed to generate a set of *orthogonal* beams. The multiple outputs of the beam-forming network are referred to as *beam ports*. Assume that the sensor outputs are equally weighted and have a *uniform* phase. Under this condition, the response of the array produced by an incident plane wave arriving at the array along direction θ, measured with respect to the normal to the array, is given by

$$A(\phi,\alpha) = \sum_{n=-N}^{N} e^{jn\phi} e^{-jn\alpha} \qquad (1.12)$$

where $M = (2N + 1)$ is the total number of sensors in the array, with the sensor at the mid−point of the array treated as the point of reference. The angle ϕ is related to θ by Eq. (1.11), and α is a constant called the *uniform phase factor*. The quantity $A(\phi,\alpha)$ is called the *array pattern*. For $d = \lambda/2$, we find from Eq. (1.11) that

$$\phi = \pi \sin \theta$$

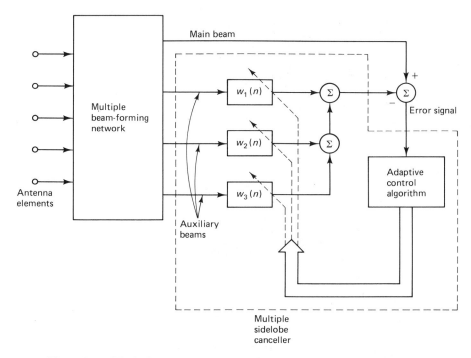

Figure 1.16 Block diagram of adaptive combiner with fixed beams; owing to the symmetric nature of the multiple beam–forming network, final values of the weights are real valued.

Summing the geometric series in Eq. (1.12), we may express the array pattern as

$$A(\phi,\alpha) = \frac{\sin\left[\frac{1}{2}(2N + 1)(\phi - \alpha)\right]}{\sin\left[\frac{1}{2}(\phi - \alpha)\right]} \qquad (1.13)$$

By assigning different values to α, the main beam of the antenna is thus scanned across the range $-\pi \leq \phi \leq \pi$. To generate an orthogonal set of beams, equal to $2N$ in number, we assign the following discrete values to the uniform phase factor

$$\alpha = \frac{\pi}{2N + 1}\, k, \qquad k = \pm 1, \pm 3, \ldots, \pm 2N - 1$$

Figure 1.17 illustrates the variations of the magnitude of the array pattern $A(\phi,\alpha)$ with ϕ for the case of $2N + 1 = 5$ elements and $\alpha = \pm\pi/5, \pm 3\pi/5$. Note that owing to the symmetric nature of the beamformer, the final values of the weights are real valued.

The orthogonal beams generated by the beam-forming network represent $2N$

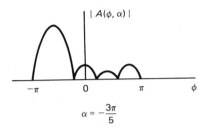

$$\alpha = -\frac{3\pi}{5}$$

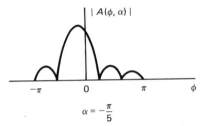

$$\alpha = -\frac{\pi}{5}$$

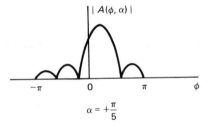

$$\alpha = +\frac{\pi}{5}$$

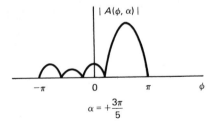

$$\alpha = +\frac{3\pi}{5}$$

Figure 1.17   Variations of the magnitude of the array pattern $A(\phi, a)$ with ϕ and α.

independent *look directions*, one per beam. Depending on the target direction of interest, a particular beam in the set is identified as the *main beam* and the remainder are viewed as *auxiliary beams*. We note from Fig. 1.17 that each of the auxiliary beams has a *null in the look direction of the main beam*. The auxiliary beams are adaptively weighted by the multiple sidelobe canceller so as to form a cancellation beam that is subtracted from the main beam. The resulting estimation error is fed back to the multiple sidelobe canceller so as to control the corrections applied to its adjustable weights.

Since all the auxiliary beams have nulls in the look direction of the main beam, and the main beam is excluded from the multiple sidelobe canceller, the overall output of the adaptive beamformer is constrained to have a constant response in the look direction of the main beam (i.e., along the direction of the target). Moreover, with $(2N-1)$ degrees of freedom (i.e., the number of available auxiliary beams) the system is capable of placing up to $(2N-1)$ nulls along the (unknown) directions of independent interferences.

Note that with an array of $(2N+1)$ sensors, we may produce a beam−forming network with $(2N+1)$ orthogonal beam parts by assigning the uniform phase factor the following set of values:

$$\alpha = \frac{k\pi}{2N+1}, \quad k = 0, \pm2, \ldots, \pm2N$$

In this case, small fraction of the main lobe of the beam port at either end lies in the nonvisible region. Nevertheless, with one of the beam ports providing the main beam and the remaining $2N$ ports providing the auxiliary beams, the adaptive beamformer is now capable of producing up to $2N$ independent nulls.

1.5 SOME HISTORICAL NOTES

> To understand a science it is necessary to know its history.
>   Auguste Comte

We complete this introductory chapter by presenting a brief historical review of developments in three areas that are closely related insofar as the subject matter of this book is concerned. The areas are linear estimation theory, adaptive filtering algorithms, and signal-processing applications of adaptive filters. It should, however, be emphasized that our interest in adaptive filters is restricted to those with a finite impulse response. The reader is referred to the paper by Johnson (1984) and to Treichler (1985) for tutorial presentations of the current status of adaptive infinite-impulse-response (IIR) filters and related open issues, and to the paper by Landau (1984) that presents a feedback system approach to the adaptive IIR filtering problem.

(a) Linear estimation theory.[12]   The earliest stimulus for the development of estimation theory was apparently provided by astronomical studies in which the motion of planets and comets was studied using telescopic measurement data. The

---

[12]The notes presented on linear estimation are influenced by the following review papers: Sorenson, 1970; Kailath, 1974; Makhoul, 1975.

beginnings of a "theory" of estimation in which attempts are made to minimize various functions of errors can be attributed to Galileo Galilei in 1632. However, the origin of linear estimation theory is credited to Gauss who, at the age of 18 in 1795, invented the *method of least squares* to study the motion of heavenly bodies (Gauss, 1809). Nevertheless, in the early nineteenth century, there was considerable controversy regarding the actual inventor of the method of least squares. The controversy arose because Gauss did not publish his discovery in 1795. Rather, it was first published by Legendre in 1805, who independently invented the method (Legendre, 1810).

The first studies of minimum mean-square estimation in stochastic processes were made by Kolmogorov, Krein, and Wiener during the late 1930s and early 1940s (Kolmogorov, 1939; Krein, 1945; Wiener, 1949). The work of Kolmogorov and Krein were independent of Wiener's, and while there was some overlap in the results, their aims were rather different. There were many conceptual differences (as one would expect after 140 years) between Gauss's problem and the problem treated by Kolmogorov, Krein, and Wiener.

Kolmogorov, inspired by some early work of Wold on discrete-time stationary processes (Wold, 1938), developed a comprehensive treatment of the linear prediction problem for discrete-time stochastic processes. Krein noted the relationship of Kolmogorov's results to some early work by Szegö on orthogonal polynomials (Szegö, 1939; Grenander and Szegö, 1958) and extended the results to continuous time by clever use of a bilinear transformation.

Wiener, independently, formulated the continuous-time linear prediction problem and derived an explicit formula for the optimum predictor. Wiener also considred the "filtering" problem of estimating a process corrupted by an additive "noise" process. The explicit formula for the optimum estimate required the solution of an integral equation known as the *Wiener–Hopf equation* (Wiener and Hopf, 1931).

In 1947, Levinson formulated the Wiener filtering problem in discrete time. In the case of discrete-time signals, the Wiener–Hopf equation takes on a matrix form that is identical to the *normal equation*:

$$\mathbf{R}\mathbf{w}_o = \mathbf{p}$$

where $\mathbf{w}_o$ is the tap-weight vector of the optimum Wiener filter structured in the form of a transversal filter, $\mathbf{R}$ is the correlation matrix of the tap inputs, and $\mathbf{p}$ is the cross-correlation vector between the top inputs and the desired response. For stationary inputs, the correlation matrix $\mathbf{R}$ assumes a special structure known as *Toeplitz*, so named after the mathematician O. Toeplitz. By exploiting the properties of a Toeplitz matrix, Levinson derived an elegant recursive procedure for solving the normal equation (Levinson, 1947). In 1960, Durbin rediscovered Levinson's recursive procedure as a scheme for recursive fitting of autoregressive models to scalar time-series data (Durbin, 1960). The problem considered by Durbin is a special case of the normal equation in that the column vector $\mathbf{p}$ comprises the same elements found in the correlation matrix $\mathbf{R}$. In 1963, Whittle showed there is a close relationship between the Levinson–Durbin recursion and that for Szegö's orthogonal polynomials and also derived a multivariate generalization of the Levinson–Durbin recursion (Whittle, 1963).

Wiener and Kolmogorov assumed an infinite amount of data and assumed the stochastic processes to be stationary. During the 1950s, some generalizations of the Wiener–Kolmogorov filter theory were made by various authors to cover the estimation of stationary processes given only for a finite observation interval and to cover the estimation of nonstationary processes. However, there were dissatisfactions with the most significant of the results of this period because they were rather complicated, difficult to update with increases in the observations interval, and difficult to modify for the vector case. These last two difficulties became particularly evident in the late 1950s in the problem of determining satellite orbits. In this application, there were generally vector observations of some combinations of position and velocity, and also there were large amounts of data sequentially accumulated with each pass of the satellite over a tracking station. Swerling was one of the first to tackle this problem by presenting some useful recursive algorithms (Swerling, 1958). For different reasons, Kalman independently developed a somewhat more restricted algorithm than Swerling's, but it was an algorithm that seemed particularly matched to the dynamical estimation problems that were brought by the advent of the space age (Kalman, 1960). After Kalman had published his paper and it had attained considerable fame, Swerling wrote a letter claiming priority for the Kalman filter equations (Swerling, 1963). However, history shows that Swerling's plea has fallen on deaf ears. It is ironic that orbit determination problems provided the stimulus for both Gauss's method of least squares and the Kalman filter and that there were squabbles concerning their inventors.

Kalman's original formulation of the linear filtering problem was derived for discrete-time processes. The continuous-time filter was derived by Kalman in his subsequent collaboration with Bucy; this solution is sometimes referred to as the *Kalman–Bucy filter* (Kalman and Bucy, 1961).

In a series of stimulating papers, Kailath reformulated the solution to the linear filtering problem by using the *innovations* approach (Kailath, 1968; Kailath and Frost, 1968; Kailath, 1970; Kailath and Geesey, 1973). In this approach, a stochastic process $\{u(n)\}$ is represented as the output of a causal and causally invertible filter driven by a white noise process $\{v(n)\}$. The requirement that the filter be causally invertible ensures that the white noise process $\{v(n)\}$, termed the *innovations process*, is probabilistically equivalent to the original process $\{u(n)\}$. This probabilistic equivalence is the reason for calling the white noise process $\{v(n)\}$ the "innovations process." Innovation denotes "newness", and this quality is represented by the whiteness of the process $\{v(n)\}$ where any redundant information in the form of correlation in the process $\{u(n)\}$ has been removed. Hence, only new information is retained in the innovations process $\{v(n)\}$. According to Kailath, the name "innovations process" was apparently first used by Wiener and Masani in the mid-fifties (Kailath, 1974).

(b) Adaptive filtering algorithms.   The earliest work on adaptive filters may be traced back to the late 1950s, during which time a number of researchers were working independently on different applications of adaptive filters.   From this early work, the *least-mean-square (LMS) algorithm* emerged as a simple algorithm for the operation of adaptive transversal filters.   The LMS algorithm was devised by Widrow and Hoff in 1959 in their study of a pattern recognition scheme known as the adaptive linear threshold logic element (Widrow and Hoff, 1960; Widrow, 1970).   The LMS algorithm is a stochastic gradient algorithm in that it iterates each tap weight in the transversal filter in the direction of the gradient of the squared amplitude of an error signal with respect to that tap weight.   As such, the LMS algorithm is closely related to the concept of *stochastic approximation* developed by Robbins and Monroe in statistics for solving certain sequential parameter estimation problems (Robbins and Monroe, 1951).   The primary difference between them is that the LMS algorithm uses a fixed step-size parameter to control the correction applied to the tap weight from one iteration to the next, whereas in stochastic approximation methods the step-size parameter is made inversely proportional to time n or to a power of n.

Another major contribution to the development of adaptive filtering algorithms was made by Godard in 1974.   He used Kalman filter theory to propose a new class of adaptive filtering algorithms for obtaining rapid convergence of the tap weights of a transversal filter to their optimum settings (Godard, 1974).   Although, prior to this date, several investigators had applied Kalman filter theory to solve the adaptive filtering problem, Godard's approach is widely accepted as the most successful.   This algorithm is referred to in the literature as the *Kalman algorithm* or *Godard algorithm*; we will use the former terminology.

The Kalman algorithm is closely related to the recursive least-squares (RLS) algorithm that follows from the method of least squares.   The RLS algorithm has been derived independently by several investigators.   However, the original reference on the RLS algorithm appears to be Plackett (1950).

As mentioned in Section 1.3, the Kalman or RLS algorithm usually provides a much faster rate of convergence than the LMS algorithm at the expense of increased computational complexity.   The desire to reduce computational complexity to a level comparable to that of the simple LMS algorithm prompted the search for computationally efficient RLS algorithms.   Various forms of such algorithms have been introduced in the literature.   In particular, mention should be made of two classes of computationally efficient RLS algorithms: one involves the use of transversal filters (Falconer and Ljung, 1978; Cioffi and Kailath, 1984), and the other the use of lattice predictors (Morf, Vieira, and Lee, 1977).   In one form or another, the development of these fast algorithms can be traced back to results that were derived by Morf in 1974 for solving the deterministic dual of the stochastic problem solved by Levinson for stationary inputs (Morf, 1974).

In 1981, Gentleman and Kung introduced yet another highly efficient algorithm for solving the linear least-squares problem by using an iterative open-loop two-stage process, a procedure well suited for implementation using systolic arrays

(Gentleman and Kung, 1981).   This algorithm, called the recursive QR decomposition-least squares algorithm, involves orthogonal triangularization of the input data matrix through direct application of a special form of QR decomposition known as Givens rotations (Givens, 1958).   In 1983, McWhirter described a modification of this algorithm by avoiding computation of the least-squares weight vector (McWhirter, 1983).   For applications that do not require explicit computation of the least-squares weight vector, the modification made by McWhirter offers an adaptive filtering algorithm that is attractive.

(c) Adaptive Signal-Processing Applications.

(1) Adaptive Equalization.   Until the early 1960s, the equalization of telephone channels to combat the degrading effects of intersymbol interference on data transmission was performed by using either fixed equalizers (resulting in a performance loss) or equalizers whose parameters were adjusted manually (a rather cumbersome procedure).   In 1965, Lucky made a major breakthrough in the equalization problem by proposing a *zero-forcing algorithm* for automatically adjusting the tap weights of a transversal equalizer (Lucky, 1965).   A distinguishing feature of the work by Lucky was the use of a *mini-max* type of performance criterion. In particular, he used a performance index called *peak distortion*, which is directly related to the maximum value of intersymbol interference that can occur.   The tap weights in the equalizer are adjusted to minimize the peak distortion.   This has the effect of *forcing* the intersymbol interference due to those adjacent pulses that are contained in the transversal equalizer to become *zero*; hence, the name of the algorithm.   A sufficient, but not necessary, condition for optimality of the zeroforcing algorithm is that the *initial distortion* (the distortion that exists at the equalizer input) be less than unity.   In a subsequent paper published in 1966, Lucky extended the use of the zero-forcing algorithm to the tracking mode of operation (Lucky, 1966).   In 1965, DiToro independently used adaptive equalization for combatting the effect of intersymbol interference on data transmitted over high-frequency links (DiToro, 1965).

The pioneering work by Lucky inspired many other significant contributions to different aspects of the adaptive equalization problem in one way or another. These contributions include the papers by Gersho, Proakis and Miller, Ungerboeck, Godard, Falconer and Ljung, and Satorius and Pack.   In 1969, Gersho, and Proakis and Miller, independently reformulated the adaptive equalization problem using a mean-square error criterion (Gersho 1969; Proakis and Miller, 1969).   In 1972, Ungerboeck presented a detailed mathematical analysis of the convergence properties of an adaptive transversal equalizer using the LMS algorithm (Ungerboeck, 1972).   In 1974, Godard used Kalman filter theory to derive a powerful algorithm for adjusting the tap weights of a transversal equalizer (Godard, 1974).   In 1978, Falconer and Ljung presented a modification of this algorithm that simplified its computational complexity to a level comparable to that of the simple LMS algorithm (Falconer and Ljung, 1978).   Satorius and Alexander in 1979 and Satorius and

Pack in 1981 demonstrated the power of lattice-based algorithms for adaptive equalization of dispersive channels (Satorius and Alexander, 1979; Satorius and Pack, 1981).

This brief historical review pertains to the use of adaptive equalizers for *linear synchronous receivers*; by "synchronous" we mean that the equalizer in the receiver has its taps spaced at the reciprocal of the symbol rate. Even though our interest in adaptive equalizers is restricted to this class of receivers, nevertheless, such a historical review would be incomplete without some mention of fractionally spaced equalizers and decision-feedback equalizers.

In a *fractionally spaced equalizer* (FSE), the equalizer taps are spaced closer than the reciprocal of the symbol rate. An FSE has the capability of compensating for delay distortion much more effectively than a conventional synchronous equalizer. Another advantage of the FSE is the fact that data transmission may begin with an arbitrary sampling phase. However, mathematical analysis of the FSE is much more complicated than for a conventional synchronous equalizer. It appears that early work on the FSE was initiated by Brady (1970). Other contributions to the subject include subsequent work by Ungerboeck (1976) and Gitlin and Weinstein (1981).

A *decision-feedback equalizer* consists of a feed-forward section and a feedback section connected as shown in Fig. 1.18. The feed-forward section itself consists of a transversal filter whose taps are spaced at the reciprocal of the symbol rate. The data sequence to be equalized is applied to the input of this section. The feedback section consists of another transversal filter whose taps are also spaced at the reciprocal of the symbol rate. The input applied to the feedback section is made up of decisions on previously detected symbols. The function of the feedback section is to subtract out that portion of intersymbol interference produced by previously detected symbols from the estimates of future symbols. This cancellation is an old idea known as the *bootstrap technique*. A decision-feedback equalizer yields good performance in the presence of severe intersymbol interference as experienced in fading radio channels, for example. The first report on decision-feedback equalization was published by Austin (1967), and the optimization of the decision-feedback receiver for minimum mean-squared error was first accomplished by Monsen (1971).

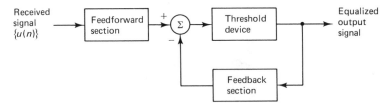

Figure 1.18 Block diagram of decision feedback equalizer. From *Introduction to Adaptive Filters* by Simon Haykin. © 1984 by Macmillan Publishing Company, a division of Macmillan, Inc. Reproduced by permission of the publisher.

For review material on adaptive equalization, see the papers by Lucky (1973), Proakis (1975), Qureshi (1982), and Belfiore and Park (1979).

(2) Coding of Speech.   In 1966, Saito and Itakura used a *maximum likelihood*[13] approach for the application of prediction to speech (Saito and Itakura, 1966). A standard assumption in the application of the maximum likelihood principle is that the input process is Gaussian. Under this condition, the exact application of the maximum likelihood principle yields a set of nonlinear equations for the parameters of the predictor. To overcome this difficulty, Itakura and Saito utilized approximations based on the assumption that the number of available data points greatly exceeds the prediction order. The use of this assumption makes the result obtained from the maximum likelihood principle assume an approximate form that is the same as the *autocorrelation method*[14] of linear prediction. The application of the maximum likelihood principle is justified on the assumption that speech is a stationary Gaussian process, which seems reasonable in the case of unvoiced sounds.

In 1970, Atal presented the first use of the term "linear prediction" for speech analysis (Atal, 1970). Details of this new approach, linear predictive coding (LPC), to speech analysis and synthesis were published by Atal and Hanauer in 1971, in which the speech wave form is represented directly in terms of time-varying parameters related to the transfer function of the vocal tract and the characteristics of the excitation (Atal and Hanauer, 1971). The predictor coefficients are determined by minimizing the mean-squared error, with the error defined as the difference between the actual and predicted values of the speech samples. In the work by Atal and Hanauer, the speech wave was sampled at 10 kHz and then analyzed by predicting the present speech sample as a linear combination of the 12 previous samples. Thus, 15 parameters (the 12 parameters of the predictor, the pitch period, a binary parameter indicating whether the speech is voiced or unvoiced, and the root mean square [rms] value of the speech samples) were used to describe the speech analyzer. For the speech synthesizer, an all-pole filter was used, with a sequence of quasi-periodic pulses or a white-noise source providing the excitation.

Another significant contribution to the linear prediction of speech was made in 1972 by Itakura and Saito; they used partial correlation techniques to develop a new structure, the lattice, for formulating the linear prediction problem[15] (Itakura and Saito, 1972). The parameters that characterize the lattice predictor are called *reflection coefficients* or *partial correlation (PARCOR) coefficients*. Although by that time the essence of the lattice structure had been considered by several other investigators, the invention of the lattice predicitor is credited to Saito and Itakura. In 1973, Wakita showed that the filtering actions of the lattice predictor model and

---

[13]For a discussion of the maximum likelihood principle, see Appendix A. See also Problem 19 of Chapter 2.

[14]The autocorrelation method of linear prediction is considered in Chapter 7.

[15]According to Markel and Gray (1976), the work of Itakura and Saito in Japan on the PARCOR formulation of linear prediction had been presented in 1969.

an acoustic tube model of speech are identical, with the reflection coefficients in the acoustic tube model as common factors (Wakita, 1973). This discovery made possible the extraction of the reflection coefficients by the use of a lattice predictor.

Early designs of a lattice predictor were based on a *block processing* approach (Burg, 1967). In 1981, Makhoul and Cossell used an *adaptive* approach for designing the lattice predictor for applications in speech analysis and synthesis (Makhoul and Cossell, 1981). They showed that the covergence of the adaptive lattice predictor is fast enough for its performance to equal that of the optimal (but more expensive) adaptive autocorrelation method.

This historical review on speech coding relates to LPC vocoders. We next present a historical review of the adaptive predictive coding of speech, starting with ordinary pulse-code modulation (PCM).

PCM was invented in 1937 by Reeves (1975). This was followed by the invention of differential pulse-code modulation (DPCM) by Cutler (1952). The early use of DPCM for the predictive coding of speech signals was limited to linear predictors with *fixed* parameters (McDonald, 1966). However, due to the nonstationary nature of speech signals, a fixed predictor cannot predict the signal values efficiently at all times. In order to respond to the nonstationary characteristics of speech signals, the predictor has to be adaptive (Atal and Schroeder, 1967). In 1970, Atal and Schroeder described a sophisticated scheme for adaptive predictive coding of speech (Atal and Schroeder, 1970). The scheme recognizes that there are two main causes of redundancy in speech (Schroeder, 1966): (1) quasi-periodicity during voiced segments, and (2) lack of flatness of the short-time spectral envelope. Thus, the predictor is designed to remove signal redundancy in two stages. The first stage of the predictor removes the quasi-periodic nature of the signal. The second stage removes formant information from the spectral envelope. The scheme achieves dramatic reductions in bit rate at the expense of a significant increase in circuit complexity. Atal and Schroeder (1970) report that the scheme can transmit speech at 10 kb/s, which is several times less than the bit rate required for logarithmic-PCM encoding with comparable speech quality.

For review papers on speech coding, see Gold (1977), Flanagan and others (1979), Gibson (1980) and Schroeder (1985). See also the books by Flanagan (1972), Markel and Gray (1976), Rabiner and Schafer (1978), Jayant and Noll (1984).

(3) Adaptive Autoregressive Spectrum Analysis. At the turn of the twentieth century, Schuster introduced the *periodogram* for analyzing the power spectrum[16] of a time series (Schuster, 1898). The periodogram is defined as the squared amplitude of the discrete Fourier transform of the time series. The periodogram was originally used by Schuster to detect and estimate the amplitude of a sine wave of known frequency that is buried in noise. Until the work of Yule in 1927, the periodogram was the only numerical method available for spectrum analysis. How-

---

[16]For a fascinating historical account of the concept of power spectrum, its origin and its estimation, see Robinson (1982).

ever, the periodogram suffers from the limitation that when it is applied to empirical time series observed in nature the results obtained are very erratic. This led Yule to introduce a new approach based on the concept of a *finite parameter model* for a stationary stochastic process in his investigation of the periodicities in time series with special reference to Wolfer's sunspot number (Yule, 1927). Yule, in effect, created a stochastic feedback model in which the present sample value of the time series is assumed to consist of a linear combination of past sample values plus an error term. This model is called an *autoregressive model* in that a sample of the time series regresses on its own past values, and the method of spectrum analysis based on such a model is accordingly called autoregressive spectrum analysis. The name "autoregressive" was coined by Wold in his 1938 thesis (Wold, 1938).

Interest in the autoregressive method was reinitiated by Burg in 1967 with the presentation of a paper that was to subsequently shake the foundations of spectrum estimation (Burg, 1967, 1975). Burg introduced the term *maximum-entropy method* to describe an algorithmic approach for estimating the power spectrum directly from the available time series. The idea behind the maximum-entropy method is to extrapolate the autocorrelation function of the time series in such a way that the *entropy* of the corresponding probability density function is maximized at each step of the extrapolation.[17] In 1971, Van den Bos showed that the maximum-entropy method is equivalent to least-squares fitting of an autoregressive model to the known autocorrelation sequence (Van den Bos, 1971).

In 1975, Griffiths extended the utility of the autoregressive model to deal with nonstationary stochastic processes by making the model adaptive through the incorporation of the LMS algorithm (Griffiths, 1975). In such a situation, the estimated power spectrum not only depends on frequency but also time. He successfully applied the adaptive autoregressive model to measure the instantaneous frequency of sine wave with rapid frequency-modulation (FM) variations (Griffiths, 1975). The adaptive autoregressive model has also been applied to the spectrum analysis of natural seismic events (Griffiths and Prieto-Diaz, 1977; Jurkevics and Ulrych, 1978).

Another application of autoregressive spectrum analysis that has attracted a great deal of attention in the literature is the estimation of the frequencies of superimposed sine waves that are corrupted by additive white noise.[18] The origin of the problem can be traced back to 1795, the year Gaspard Riche, Baron de Prony, published his work on the fitting of superimposed exponentials to (noiseless) data (Prony, 1795). A useful solution to the problem for noisy data was developed independently by Ulrych and Clayton and Nuttal, using the method of forward-backward linear prediction (formulated in terms of least squares) to evaluate the

---

[17]For a derivation of the maximum-entropy method, see Appendix B.

[18]For a detailed historical account of the superimposed signals problem, see Wax (1985). In this dissertation, an optimal solution to the joint estimation of the (unknown) number of signals, and their parameters is derived. The signals of interest represent sine waves of unknown frequencies in harmonic analysis or plane waves of unknown directions in its spatial counterpart.

autoregressive parameters (Ulrych and Clayton, 1976; Nuttall, 1976). To reduce the effects of noise, we may modify the method of least squares by manipulating the eigenvalue decomposition of the pertinent correlation matrix. The modified forward-backward linear prediction method (Tufts and Kumaresan, 1982), the minimum-norm method (Kumaresan and Tufts, 1983), and the MUSIC algorithm (Schmidt, 1981) exploit such an approach in different ways. These algorithms are discussed in Chapter 7.

For review material on autoregressive spectrum analysis (and other parametric spectrum analysis techniques), see the tutorial paper by Kay and Marple (1981), the collection book of papers by Childers (1978), and the book edited by Haykin (1983).

(4) Adaptive Noise Cancellation Schemes. The initial work on adaptive echo cancellers started around 1965. It appears that Kelly of Bell Telephone Laboratories was the first to propose the use of an adaptive filter for echo cancellation, with the speech signal itself utilized in performing the adaptation; Kelly's contribution is recognized in the paper by Sondhi (1967). This invention and its refinement are described in the patents by Kelly and Logan (1970) and Sondhi (1970).

The adaptive line enhancer was originated by Widrow and his co-workers at Stanford University. An early version of this device was built in 1965 to cancel 60-Hz interference at the output of an electrocardiographic amplifier and recorder. This work is described in the paper by Widrow and others (1975). The adaptive line enhancer and its application as an adaptive detector are patented by McCool and others (1980).

The adaptive echo canceller and the adaptive line enhancer, although intended for different applications, may be viewed as examples of the *adaptive noise canceller* discussed by Widrow and others (1975). This scheme operates on the outputs of two sensors: a *primary sensor* that supplies a desired signal of interest buried in noise, and a *reference sensor* that supplies noise alone, as illustrated in Fig. 1.19. It is assumed that (1) the signal and noise at the output of the primary sensor are uncorrelated, and (2) the noise at the output of the reference sensor is correlated with the noise component of the primary sensor output.

The adaptive noise canceller consists of an adaptive filter that operates on the reference sensor output to produce an *estimate* of the noise, which is subtracted from the primary sensor output. The overall output of the canceller is used to control the adjustments applied to the tap weights in the adaptive filter. The adaptive canceller tends to minimize the mean-square value of the overall output, thereby causing the output to be the best estimate of the desired signal in the minimum-mean-square sense.

(5) Adaptive Signal Detection. The first study of adaptive signal detection was apparently made by Glaser in 1960, with particular reference to the reception of a pulse signal of fixed periodic wave form, but unknown at the receiver (Glaser, 1961). The technique described by Glaser allows a matched filter for an unknown

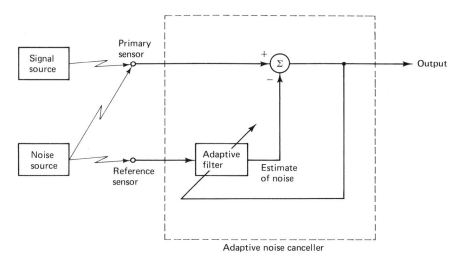

Figure 1.19   Concept of adaptive noise cancellation.

periodic signal to be synthesized using information obtained from measurement of the signal and noise.

In the late 1960s, Kailath reformulated the classical approach for the detection of a signal in additive noise by developing a new procedure, based on the innovations representation of a stochastic process, for solving large classes of Gaussian and non-Gaussian detection problems (Kailath, 1969, 1970).

Interest in adaptive signal detection was reawakened in the late 1970s with successful applications of the maximum-entropy method of spectrum estimation and the realization that radar clutter may be modeled as an autoregressive process of relatively low order (Gibson, Haykin and Kesler, 1979; Bowyer, Rajasekaran and Gebhart, 1979). Inspired by the results of this model-dependent work on radar clutter and the earlier work on the innovations process by Kailath, a new algorithm (based on the likelihood ratio test) was derived by Metford and others to detect a discret-time stochastic process in additive white Gaussian noise (Metford, Haykin and Taylor, 1982). In 1984, Metford and Haykin extended the use of this algorithm to deal with the radar detection problem in which the clutter has unknown statistics (Metford and Haykin, 1984). The algorithm was given the name *innovations-based detection algorithm*. Using data collected from a *coherent*[19] airport surveillance radar operating in an air traffic control environment, they confirmed the superior performance of this new algorithm compared to other moving detection algorithms derived from classical detection theory.

In 1982, Basseville and Benveniste (independently of Metford and Haykin) also derived the same detection algorithm using an autoregressive modeling ap-

[19]In a coherent radar, information on both the amplitude and phase of the radar returns is available. On the other hand, in a noncoherent radar, only amplitude information is available.

proach to study the problem of sequential detection of abrupt changes in the spectrum of a discrete-time signal (Basseville and Benveniste, 1983).

In 1983, Kay considered the problem of detecting a known signal in colored Gaussian noise of unknown convariance (Kay, 1983). The result obtained by Kay is implicitly contained in the innovations-based detection algorithm.

(6) Adaptive Beamforming. The development of adaptive beamforming technology may be traced back to the invention of the *intermediate frequency (IF) sidelobe canceller* by Howells in the late 1950s. In a paper published in the 1976 Special Issue of the IEEE Transactions on Antennas and Propagation, Howells describes his personal observations on early work on adaptive antennas at the General Electric and Syracuse University Research Corporation (Howells, 1976). According to this historic report, Howells had developed by mid-1957 a side-lobe canceller capable of automatically nulling out the effect of one jammer. The sidelobe canceller uses a *primary* (high-gain) antenna and an *auxiliary omni-directional* (low-gain) antenna to form a two-element array with one degree of freedom that makes it possible to steer a deep null anywhere in the sidelobe region of the combined antenna pattern. In particular, a null is placed in the direction of the jammer, with only a minor perturbation of the main lobe. Subsequently, Howells patented the sidelobe canceller (Howells, 1965).

The second major contribution to adaptive array antennas was made by Applebaum in 1966. In a classic report, he derived the *control law* governing the operation of an adaptive array antenna, with a control loop for each element of the array (Applebaum, 1966). The algorithm derived by Applebaum was based on maximizing the signal-to-noise ratio (SNR) at the array antenna output for any type of noise environment. Applebaum's theory included the sidelobe canceller as a special case. His 1966 classic report was reprinted in the 1976 Special Issue of IEEE Transactions on Antennas and Propagation.

Another algorithm for the weight adjustment in adaptive array antennas was advanced independently in 1967 by Widrow and his co-workers at Stanford University. They based their theory on the simple and yet effective LMS algorithm (Widrow and others, 1967). The LMS algorithm itself had been invented by Widrow and Hoff in 1960. The 1967 paper by Widrow and others was not only the first publication in the open literature on adaptive array antenna systems, but also it is considered to be another classic of that era.

It is noteworthy that the maximum SNR algorithm (used by Applebaum) and the LMS algorithm (used by Widrow and his co-workers) for adaptive array antennas are rather similar. Both algorithms derive the control law for adaptive adjustment of the weights in the array antenna by sensing the correlation between element signals. Indeed, they both converge toward the optimum Wiener solution for stationary inputs (Gabriel, 1976).

A different method for solving the adaptive beamforming problem was proposed by Capon (1969). Capon realized that the poor performance of the delay-and-sum beamformer is due to the fact that its response along a direction of interest depends not only on the power of the incoming target signal but also undesirable contributions received from other sources of interference. To overcome this limitation of the delay-and-sum beamformer, Capon proposed a new beamformer in which the weight vector $\mathbf{w}(n)$ is chosen so as to *minimize the variance* (i.e., average power) of the beamformer output, subject to the constraint $\mathbf{w}^H(n) \, \mathbf{s}(\phi) = 1$ for all n, where $\mathbf{s}(\phi)$ is a prescribed *steering vector*. This constrained minimization yields an adaptive beamformer with *minimum-variance distortionless response* (MVDR).

A systolic array implementation of the adaptive MVDR beamformer has been proposed by Schreiber and Kuekes (1985). McWhirter has also proposed the use of systolic arrays for the design of adaptive beamformers (McWhirter, 1983; Ward and others, 1984). The structures proposed by Schreiber and Kuekes, and McWhirter are based on modifications of original ideas due to Kung and Gentleman for the use of systolic arrays to solve the linear least-squares problem (Kung and Leiserson, 1980; Gentleman and Kung, 1981).

A key assumption that is made in much of the literature on adaptive beamformers is that the interfering signals are *noncoherent* with respect to the target; that is, the phase differences between them and the target signal are *random*. In general, a pair of signals are said to be *coherent* if one of them is a scaled and delayed version of the other. The coherent interference problem arises, for example, when a radar target lies in close proximity to a reflecting land or sea surface, thereby giving rise to *multipath* propagation. In such an environment, the received signal consists of two components: one received *directly* from the *target* and the other from the *image* of the target produced by reflections from the land or sea surface. Another way in which coherent interference arises is when *smart jammers* are used in a hostile environment to induce such a phenomenon to occur deliberately. In any event, conventional forms of adaptive beamformers break down in the presence of coherent interfering signals. The earliest significant references on the effects of coherent interferences are those of White (1979), Gabriel (1980), and Widrow and others (1982). Shan and Kailath (1985) describe an adaptive beamformer that incorporates *spatial smoothing* as a means of overcoming the coherent interference problem. It appears that the use of spatial smoothing (in the context of direction finding) had been discussed earlier by Evans and others (1981, 1982). A form of spatial filtering, in conjunction with the modified forward–backward linear prediction method, has also been used by Haykin and others to resolve the low-angle tracking radar problem in the presence of coherent multipath propagation (Haykin, Greenlay, and Litva, 1985).

For review material on adaptive beamformers, see Gabriel (1976), Monzingo and Miller (1980), Hudson (1981), Owsley (1984), Wax (1985), Widrow and Stearns (1985). See also the book by Compton (to be published) and the special issue of IEEE Transactions on Antennas and Propagation (1986).

2

Stationary Discrete-time Stochastic Processes

2.1 INTRODUCTION

In this chapter we deal with two basic issues pertaining to the statistical characterization of a *stationary discrete-time stochastic process*. In the first part of the chapter (consisting of Sections 2.2 to 2.6), we discuss the various properties of the *correlation matrix* and *power spectral density* of such a process. In the second part of the chapter (consisting of Sections 2.7 to 2.10), we discuss the modeling of a stationary discrete-time stochastic process as the output of a linear discrete-time filter driven by *white noise* as input.

The term *stochastic process* or *random process* is used to describe the time evolution of a statistical phenomenon according to probabilistic laws. The time evolution of the phenomenon means that the stochastic process is a function of time, defined on some observation interval. The statistical nature of the phenomenon means that, before conducting an experiment, it is not possible to define exactly the way it evolves in time. Examples of a stochastic process include speech signals, television signals, radar signals, digital computer data, the output of a communication channel, seismological data, and noise.

The form of a stochastic process that is of interest to us is one that is defined at *discrete and uniformly spaced instants of time*. Such a restriction may arise naturally in practice, as in the case of radar signals or digital computer data. Alternatively, the stochastic process may be defined originally for a continuous range of real values of time; however, before processing, it is *sampled uniformly* in time, with the sampling rate chosen equal to or greater than twice the highest frequency component of the process (Haykin, 1983).

A stochastic process is *not* just a single function of time; rather, it represents, in theory, an infinite number of *different* realizations of the process. One particular realization of a discrete-time stochastic process is called a *discrete-time series* or simply *time series*. For convenience of notation, we will *normalize time with respect to the sampling period*. For example, the sequence $u(n)$, $u(n - 1)$, . . . , $u(n - M + 1)$ represents a time series that consists of the *present* observation $u(n)$ made at time n and $M - 1$ *past* observations of the process made at times $n - 1, . . . , n - M + 1$.

We say that a stochastic process is *strictly stationary* if its statistical properties are *invariant* to a shift of time. Specifically, for a discrete-time stochastic process represented by the time series $u(n)$, $u(n - 1)$, . . . , $u(n - M + 1)$ to be strictly stationary, the *joint probability density function* of these observations made at times $n, n - 1, . . . , n - M + 1$ must remain the same no matter what values we assign to n for fixed M.

For a detailed treatment of discrete-time stochastic processes, the reader is referred to Box and Jenkins (1976) and Priestley (1981).

2.2 PARTIAL CHARACTERIZATION OF A DISCRETE-TIME STOCHASTIC PROCESS

In practice, we usually find that it is not possible to determine (by means of suitable measurements) the joint probability density function for an arbitrary set of observations made on a stochastic process. Accordingly, we must content ourselves with a *partial* characterization of the process by specifying its first and second moments.

Consider a discrete-time stochastic process represented by the time series $u(n)$, $u(n - 1)$, . . . , $u(n - M + 1)$, which may be complex valued. We define the *mean-value function* of the process as

$$\mu(n) = E[u(n)] \tag{2.1}$$

where E denotes the *expectation operator*. We define the *autocorrelation function* of the process as

$$r(n, n - k) = E[u(n)u^*(n - k)], \qquad k = 0, \pm 1, \pm 2, . . . , \tag{2.2}$$

where the asterisk denotes *complex conjugation*. We define the *autocovariance function* of the process as

$$c(n, n - k) = E[(u(n) - \mu(n))(u(n - k) - \mu(n - k))^*], \qquad (2.3)$$
$$k = 0, \pm 1, \pm 2, \ldots,$$

From Eqs. (2.1) to (2.3), we see that the mean-value, autocorrelation, and autocovariance functions of the process are related by

$$c(n, n - k) = r(n, n - k) - \mu(n) \, \mu^*(n - k) \qquad (2.4)$$

For a partial characterization of the process, we therefore need to specify (1) the mean-value function $\mu(n)$, and (2) the autocorrelation function $r(n, n - k)$ or the autocovariance function $c(n, n - k)$ for various values of n and k that are of interest.

This form of partial characterization offers two important advantages:

1. It lends itself to practical measurements.
2. It is well suited to *linear* operations on stochastic processes.

For a discrete-time stochastic process that is strictly stationary, all three quantities defined in Eqs. (2.1) to (2.3) assume simpler forms. In particular, we find that the mean-value function of the process is a constant μ (say), so we may write

$$\mu(n) = \mu, \qquad \text{for all } n \qquad (2.5)$$

We also find that both the autocorrelation and autocovariance functions depend only on the *difference* between the observation times n and $n - k$, that is, k, as shown by

$$r(n, n - k) = r(k) \qquad (2.6)$$

and

$$c(n, n - k) = c(k) \qquad (2.7)$$

Note that when $k = 0$, corresponding to a time difference or *lag* of zero, $r(0)$ equals the *mean-square value* of $u(n)$:

$$r(0) = E[|u(n)|^2] \qquad (2.8)$$

and $c(0)$ equals the *variance* of $u(n)$:

$$c(0) = \sigma_u^2 \qquad (2.9)$$

The conditions of Eqs. (2.5) to (2.7) are *not* sufficient to guarantee that the discrete-time stochastic process is strictly stationary. However, a discrete-time stochastic process that is not strictly stationary, but for which these conditions hold, is said to be *wide-sense stationary*, *weakly stationary*, or *stationary to the second order*. A strictly stationary process $\{u(n)\}$ is stationary in the wide sense if and only if (Doob, 1953)

$$E[|u(n)|^2] < \infty, \text{ for all } n$$

This condition is ordinarily satisfied by stochastic processes encountered in the physical sciences and engineering. It is also of interest to note that if a *Gaussian* process is wide-sense stationary, the process is strictly stationary.

When the mean μ is zero, the autocorrelation and autocovariance functions of a weakly stationary process assume the same value. For convenience of analysis, from here on we will assume that the process has zero mean.

Correlation Matrix

Let the M-by-1 *observation vector* $\mathbf{u}(n)$ represent the elements of the time series $u(n), u(n-1), \ldots, u(n-M+1)$. To show the composition of the vector $\mathbf{u}(n)$ explicitly, we write

$$\mathbf{u}^T(n) = [u(n), u(n-1), \ldots, u(n-M+1)] \tag{2.10}$$

where the superscript T denotes *transposition*. We define the *correlation matrix* of a stationary discrete-time stochastic process represented by this time series as *the expectation of the outer product of the observation vector $\mathbf{u}(n)$ with itself*. Let $\mathbf{R}$ denote the M-by-M correlation matrix defined in this way. We thus write

$$\mathbf{R} = E\left[\mathbf{u}(n)\,\mathbf{u}^H(n)\right] \tag{2.11}$$

where the superscript H denotes *Hermitian transposition* (i.e., the operation of transposition combined with complex conjugation). By substituting Eq. (2.10) in (2.11) and using the condition of wide-sense stationarity, we may express the correlation matrix $\mathbf{R}$ in the expanded form:

$$\mathbf{R} = \begin{bmatrix} r(0) & r(1) & \cdots & r(M-1) \\ r(-1) & r(0) & \cdots & r(M-2) \\ \vdots & \vdots & \ddots & \vdots \\ r(-M+1) & r(-M+2) & \cdots & r(0) \end{bmatrix} \tag{2.12}$$

The element $r(0)$ on the main diagonal is always real-valued. For complex-valued data, the remaining elements of $\mathbf{R}$ assume complex values.

Ergodicity

The correlation matrix $\mathbf{R}$, as defined in Eq. (2.11), represents the result of *ensemble averaging*. If the stochastic process is *ergodic*, we may use *time averages* as approximations to the corresponding ensemble averages (Papoulis, 1984). The essence of the ergodicity condition is that each realization of the process in the form of a time series must eventually take on nearly all the modes of behavior of each other time series, provided that all such realizations are sufficiently long in duration. Thus, assuming that the discrete-time stochastic process of interest is *ergodic in the autocorrelation function*, we may write

$$\mathbf{R} = \lim_{N \to \infty} \frac{1}{N} \sum_{n=1}^{N} \mathbf{u}(n)\mathbf{u}^H(n) \tag{2.13}$$

Note that for the stochastic process to be ergodic it has to be stationary.

2.3 PROPERTIES OF THE CORRELATION MATRIX

The correlation matrix $\mathbf{R}$ plays a key role in the statistical analysis and design of discrete-time filters. It is therefore important that we understand its various properties and their implications. In particular, using the definition of Eq. (2.11), we find that the correlation matrix of a stationary discrete-time stochastic process has the following properties:

Property 1. *The correlation matrix of a stationary discrete-time stochastic process is Hermitian.*

We say that a *complex-valued* matrix is *Hermitian* if it is equal to its *conjugate transpose*. We may thus express the Hermitian property of the correlation matrix $\mathbf{R}$ by writing

$$\mathbf{R}^H = \mathbf{R} \tag{2.14}$$

This property follows directly from the definition of Eq. (2.11).

Another way of stating the Hermitian property of the correlation matrix $\mathbf{R}$ is to write

$$r(-k) = r^*(k) \tag{2.15}$$

where $r(k)$ is the correlation function of the stochastic process for a lag of k. Accordingly, for a weakly stationary process we only need M values of the autocorrelation function $r(k)$ for $k = 0, 1, \ldots, M - 1$ in order to completely define the correlation matrix $\mathbf{R}$.

Note that for the special case of *real-valued data*, the autocorrelation function $r(k)$ is real for all k, and the correlation matrix $\mathbf{R}$ is *symmetric*.

Property 2. *The correlation matrix of a stationary discrete-time stochastic process is Toeplitz.*

We say that a square matrix is *Toeplitz* if all the elements on its main diagonal are equal, and if the elements on any other diagonal parallel to the main diagonal are also equal.   From the expanded form of the correlation matrix **R** given in Eq. (2.12), we see that all the elements on the main diagonal are equal to $r(0)$, all the elements on the first diagonal above the main diagonal are equal to $r(1)$, all the elements along the first diagonal below the main diagonal are equal to $r(-1)$, and so on for the other diagonals.   We conclude therefore that the correlation matrix **R** is Toeplitz.

It is important to recognize, however, that the Toeplitz property of the correlation matrix **R** is a direct consequence of the assumption that the discrete-time stochastic process represented by the observation vector $\mathbf{u}(n)$ is weakly stationary. Indeed, we may state that if the discrete-time stochastic process is weakly stationary, then its correlation matrix **R** must be Toeplitz, and, conversely, if the correlation matrix **R** is Toeplitz, then the discrete-time stochastic process must be weakly stationary.

Property 3.   *The correlation matrix of a discrete-time stochastic process is always nonnegative definite and almost always positive definite.*

Let **x** be an arbitrary (nonzero) M-by-1 complex-valued vector.   Define the scalar random variable y as the *inner product* of **x** and the observation vector $\mathbf{u}(n)$, as shown by

$$y = \mathbf{x}^H \mathbf{u}(n)$$

Taking the Hermitian transpose of both sides and recognizing that y is a scalar, we get

$$y^* = \mathbf{u}^H(n)\mathbf{x}$$

where the asterisk denotes *complex conjugation*.   The mean-square value of the random variable y equals

$$
\begin{aligned}
E\,[|y|^2] &= E\,[yy^*] \\
&= E\,[\mathbf{x}^H \mathbf{u}(n)\mathbf{u}^H(n)\mathbf{x}] \\
&= \mathbf{x}^H E\,[\mathbf{u}(n)\mathbf{u}^H(n)]\mathbf{x} \\
&= \mathbf{x}^H \mathbf{R}\mathbf{x}
\end{aligned}
$$

where **R** is the correlation matrix as defined in Eq. (2.11).   The expression $\mathbf{x}^H \mathbf{R}\mathbf{x}$ is called a *Hermitian form*.   Since

$$E[|y|^2] \geq 0$$

it follows that

$$\mathbf{x}^H \mathbf{R}\mathbf{x} \geq 0 \tag{2.16}$$

A Hermitian form that satisfies this condition is said to be *nonnegative definite* or *positive semidefinite*.   Accordingly, we may state that the correlation matrix of a weakly stationary process is always nonnegative definite.

If the Hermitian form $\mathbf{x}^H \mathbf{R} \mathbf{x}$ satisfies the condition

$$\mathbf{x}^H \mathbf{R} \mathbf{x} > 0$$

for every nonzero $\mathbf{x}$, we say that the correlation matrix $\mathbf{R}$ is *positive definite*. This condition is satisfied for a weakly stationary process unless there are linear dependencies between the random variables that constitute the M elements of the observation vector $\mathbf{u}(n)$. Such a situation arises essentially only when the process $\{u(n)\}$ consists of the sum of K sinusoids with $K \le M$. In practice, we find that this idealized situation is so rare in occurrence that the correlation matrix $\mathbf{R}$ is almost always positive definite.

The positive definiteness of a correlation matrix implies that its determinant and all principal minors are greater than zero. For example, for $M = 2$, we must have

$$\begin{vmatrix} r(0) & r(1) \\ r^*(1) & r(0) \end{vmatrix} > 0$$

Similarly, for $M = 3$, we must have

$$\begin{vmatrix} r(0) & r(1) \\ r^*(1) & r(0) \end{vmatrix} > 0$$

$$\begin{vmatrix} r(0) & r(2) \\ r^*(2) & r(0) \end{vmatrix} > 0$$

$$\begin{vmatrix} r(0) & r(1) & r(2) \\ r^*(1) & r(0) & r(1) \\ r^*(2) & r^*(1) & r(0) \end{vmatrix} > 0$$

and so on for higher values of M. These conditions, in turn, imply that the correlation matrix is *nonsingular*. We say that a matrix is nonsingular if its inverse exists; otherwise, it is *singular*. Accordingly, we may state that a correlation matrix is almost always nonsingular.

Property 4. *When the elements that constitute the observation vector of a stationary discrete-time stochastic process are rearranged backward, the effect is equivalent to the transposition of the correlation matrix of the process.*

Let $\mathbf{u}^B(n)$ denote the M-by-1 vector obtained by rearranging the elements that constitute the observation vector $\mathbf{u}(n)$ *backward*. We illustrate this operation by writing

$$\mathbf{u}^{BT}(n) = [u(n - M + 1), u(n - M + 2), \ldots, (u(n)] \tag{2.17}$$

where the superscript B denotes the backward rearrangement of a vector. The correlation matrix of the vector $\mathbf{u}^B(n)$ equals, by definition,

$$E\left[\mathbf{u}^B(n)\,\mathbf{u}^{BH}(n)\right] = \begin{bmatrix} r(0) & r(-1) & \cdots & r(-M+1) \\ r(1) & r(0) & \cdots & r(-M+2) \\ \cdot & \cdot & \cdot & \cdot \\ \cdot & \cdot & \cdot & \cdot \\ \cdot & \cdot & \cdot & \cdot \\ r(M-1) & r(M-2) & \cdots & r(0) \end{bmatrix} \tag{2.18}$$

Hence, comparing the expanded correlation matrix of Eq. (2.18) with that of Eq. (2.12), we see that

$$E\left[\mathbf{u}^B(n)\,\mathbf{u}^{BH}(n)\right] = \mathbf{R}^T \tag{2.19}$$

which is the desired result.

Property 5.   *The correlation matrices $\mathbf{R}_M$ and $\mathbf{R}_{M+1}$ of a stationary discrete-time stochastic process, pertaining to M and $M + 1$ observations of the process, respectively, are related by*

$$\mathbf{R}_{M+1} = \left[\begin{array}{c|c} r(0) & \mathbf{r}^H \\ \hline \mathbf{r} & \mathbf{R}_M \end{array}\right] \tag{2.20}$$

or equivalently

$$\mathbf{R}_{M+1} = \left[\begin{array}{c|c} \mathbf{R}_M & \mathbf{r}^{B*} \\ \hline \mathbf{r}^{BT} & r(0) \end{array}\right] \tag{2.21}$$

where $r(0)$ is the autocorrelation of the process for a lag of zero, and

$$\mathbf{r}^H = [r(1), r(2), \ldots , r(M)] \tag{2.22}$$

and

$$\mathbf{r}^{BT} = [r(-M), r(-M+1), \ldots , r(-1)] \tag{2.23}$$

Note that in describing Property 5 we have added a subscript (M or $M + 1$) to the symbol for the correlation matrix in order to display dependence on the number of observations used to define this matrix. We will follow such a practice (in the context of the correlation matrix and other vector quantities) *only* when the issue at hand involves dependence on the number of observations or dimension of the matrix.

To prove the relation of Eq. (2.20), we express the correlation matrix $\mathbf{R}_{M+1}$

in its expanded form, partitioned as follows:

$$
\mathbf{R}_{M+1} =
\begin{bmatrix}
r(0) & r(1) & r(2) & \cdots & r(M) \\
\hline
r(-1) & r(0) & r(1) & \cdots & r(M-1) \\
r(-2) & r(-1) & r(0) & \cdots & r(M-2) \\
\vdots & \vdots & \vdots & \ddots & \vdots \\
r(-M) & r(-M+1) & r(-M+2) & \cdots & r(0)
\end{bmatrix}
\tag{2.24}
$$

Using Eqs. (2.12), (2.15), (2.22) in (2.24), we get the result given in Eq. (2.20). Note that according to this relation the observation vector $\mathbf{u}_{M+1}(n)$ is *partitioned* in the form

$$
\mathbf{u}_{M+1}(n) =
\begin{bmatrix}
u(n) \\
\hline
u(n-1) \\
u(n-2) \\
\vdots \\
u(n-M)
\end{bmatrix}
\tag{2.25}
$$

$$
=
\begin{bmatrix}
u(n) \\
\hline
\mathbf{u}_M(n-1)
\end{bmatrix}
$$

where the subscript $M + 1$ is intended to denote the fact that the vector $\mathbf{u}_{M+1}(n)$ has $M + 1$ elements, and likewise for $\mathbf{u}_M(n)$.

To prove the relation of Eq. (2.21), we express the correlation matrix $\mathbf{R}_{M+1}$ in its expanded form, partitioned as follows:

$$
\mathbf{R}_{M+1} =
\begin{bmatrix}
r(0) & r(1) & \cdots & r(M-1) & r(M) \\
r(-1) & r(0) & \cdots & r(M-2) & r(M-1) \\
\vdots & \vdots & \ddots & \vdots & \vdots \\
r(M+1) & r(-M+2) & \cdots & r(0) & r(1) \\
\hline
r(-M) & r(-M+1) & \cdots & r(-1) & r(0)
\end{bmatrix}
\tag{2.26}
$$

Here again, using Eqs. (2.12), (2.15), and (2.23) in (2.26), we get the result given in Eq. (2.21). Note that according to this second relation the observation vector

$\mathbf{u}_{M+1}(n)$ is partitioned in the alternative form

$$\mathbf{u}_{M+1}(n) = \begin{bmatrix} u(n) \\ u(n-1) \\ \cdot \\ \cdot \\ \cdot \\ u(n-M+1) \\ \hline u(n-M) \end{bmatrix} \tag{2.27}$$

$$= \begin{bmatrix} \mathbf{u}_M(n) \\ \hline u(n-M) \end{bmatrix}$$

2.4 THE EIGENVALUE PROBLEM

Let $\mathbf{R}$ denote the M-by-M correlation matrix of a stationary discrete-time stochastic process represented by the M-by-1 observation vector $\mathbf{u}(n)$. In general, this matrix may contain complex elements. We wish to find an M-by-1 vector $\mathbf{q}$ that satisfies the condition

$$\mathbf{R}\mathbf{q} = \lambda\mathbf{q} \tag{2.28}$$

for some constant λ. This condition states that the vector $\mathbf{q}$ is linearly transformed to the vector $\lambda\mathbf{q}$ by the transformation $\mathbf{R}$. Since λ is a constant, the vector $\mathbf{q}$ therefore has special significance in that it is left *invariant in direction* (in the M-dimensional space) by a linear transformation. For a typical M-by-M matrix $\mathbf{R}$, there will be M such vectors. To show this, we first rewrite Eq. (2.28) in the form

$$(\mathbf{R} - \lambda\mathbf{I})\mathbf{q} = \mathbf{0} \tag{2.29}$$

where $\mathbf{I}$ is the M-by-M identity matrix, and $\mathbf{0}$ is the M-by-1 null vector. The matrix $\mathbf{R} - \lambda\mathbf{I}$ has to be singular. Hence, Eq. (2.29) has a nonzero solution in the vector $\mathbf{q}$ if and only if the determinant of the matrix $(\mathbf{R} - \lambda\mathbf{I})$ equals zero; that is,

$$\det(\mathbf{R} - \lambda\mathbf{I}) = 0 \tag{2.30}$$

This determinant, when expanded, is clearly a polynomial in λ of degree M. We thus find that, in general, Eq. (2.30) has M distinct roots. Correspondingly, Eq. (2.28) has M solutions in the vector $\mathbf{q}$.

Equation (2.30) is called the *characteristic equation* of the matrix $\mathbf{R}$. Let λ_1, $\lambda_2, \ldots, \lambda_M$ denote the M roots of this equation. These roots are called the *eigenvalues* of the matrix $\mathbf{R}$. Although the M-by-M matrix $\mathbf{R}$ has M eigenvalues, they need not be distinct. For example, the M-by-M diagonal matrix

$$\mathbf{R} = \text{diag} (\sigma^2, \sigma^2, \ldots, \sigma^2)$$

has a single eigenvalue equal to σ^2, but with multiplicity M.

Let λ_i denote an eigenvalue of the matrix $\mathbf{R}$. Also, let $\mathbf{q}_i$ be a nonzero vector such that

$$\mathbf{R}\mathbf{q}_i = \lambda_i\mathbf{q}_i$$

The vector $\mathbf{q}_i$ is called the *eigenvector* associated with λ_i. An eigenvector can correspond to only one eigenvalue. However, an eigenvalue may have many eigenvectors. For example, if $\mathbf{q}_i$ is an eigenvector associated with eigenvalue λ_i, then so is $a\mathbf{q}_i$ for any scalar $a \neq 0$.

There is no best procedure for computing the eigenvalues of a matrix.[1] But there are certainly some bad procedures that should never be used. In particular, except for the simple case of a 2-by-2 matrix, use of the characteristic equation for eigenvalue computation is usually avoided.

2.5 PROPERTIES OF EIGENVALUES AND EIGENVECTORS

In this section we discuss the various properties of the eigenvalues and eigenvectors of the correlation matrix $\mathbf{R}$ of a stationary discrete-time stochastic process. Some of the properties derived here are direct consequences of the Hermitian property and the nonnegative definiteness of the correlation matrix $\mathbf{R}$, which were established in Section 2.3.

Property 1. *If $\lambda_1, \lambda_2, \ldots, \lambda_M$ denote the eigenvalues of the correlation matrix $\mathbf{R}$, then the eigenvalues of the matrix $\mathbf{R}^k$ equal $\lambda_1^k, \lambda_2^k, \ldots, \lambda_M^k$.*

Repeated premultiplication of both sides of Eq. (2.28) by the matrix $\mathbf{R}$ yields

$$\mathbf{R}^k\mathbf{q} = \lambda^k\mathbf{q}$$

This shows that (1) if λ is an eigenvalue of $\mathbf{R}$, then λ^k is an eigenvalue of $\mathbf{R}^k$, which is the desired result, and (2) every eigenvector of $\mathbf{R}$ is also an eigenvector of $\mathbf{R}^k$. Note that as a special case of this result, if λ is an eigenvalue of $\mathbf{R}$, then λ^{-1} is the eigenvalue of the inverse matrix $\mathbf{R}^{-1}$.

Property 2. *Let $\mathbf{q}_1, \mathbf{q}_2, \ldots, \mathbf{q}_M$ be the eigenvectors corresponding to the distinct eigenvalues $\lambda_1, \lambda_2, \ldots, \lambda_M$ of the M-by-M correlation matrix $\mathbf{R}$, respectively. Then the eigenvectors $\mathbf{q}_1, \mathbf{q}_2, \ldots, \mathbf{q}_M$ are linearly independent.*

---

[1]For a discussion of numerical methods for eigenvalue computation, see Stewart (1973).

We say that the eigenvectors $\mathbf{q}_1$, $\mathbf{q}_2$, . . . , $\mathbf{q}_M$ are *linearly dependent* if there are scalars v_1, v_2, . . . , v_M, not all zero, such that

$$\sum_{i=1}^{M} v_i \mathbf{q}_i = \mathbf{0} \tag{2.31}$$

If no such scalars exist, we say that the eigenvectors are *linearly independent*.

We will prove the validity of Property 2 by contradiction. Suppose that Eq. (2.31) holds for certain scalars v_i. Repeated multiplication of Eq. (2.31) by matrix **R** and the use of Eq. (2.28) yield the following set of M equations:

$$\sum_{i=1}^{M} v_i \lambda_i^{k-1} \mathbf{q}_i = \mathbf{0}, \qquad k = 1, 2, \ldots, M \tag{2.32}$$

This set of equations may be written in the form of a single matrix equation:

$$[v_1 \mathbf{q}_1, v_2 \mathbf{q}_2, \ldots, v_M \mathbf{q}_M]\mathbf{S} = \mathbf{0} \tag{2.33}$$

where

$$\mathbf{S} = \begin{bmatrix} 1 & \lambda_1 & \lambda_1^2 & \cdots & \lambda_1^{M-1} \\ 1 & \lambda_2 & \lambda_2^2 & \cdots & \lambda_2^{M-1} \\ \cdot & \cdot & \cdot & \cdot & \cdot \\ \cdot & \cdot & \cdot & \cdot & \cdot \\ \cdot & \cdot & \cdot & \cdot & \cdot \\ 1 & \lambda_M & \lambda_M^2 & \cdots & \lambda_M^{M-1} \end{bmatrix} \tag{2.34}$$

The matrix **S** is called a *Vandermonde matrix* (Strang, 1980). When the λ_i are distinct, the Vandermonde matrix **S** is nonsingular. Therefore, we may postmultiply Eq. (2.33) by the inverse matrix $\mathbf{S}^{-1}$, obtaining

$$[v_1 \mathbf{q}_1, v_2 \mathbf{q}_2, \ldots, v_M \mathbf{q}_M] = \mathbf{0}$$

Hence, each column $v_i \mathbf{q}_i = \mathbf{0}$. Since the eigenvectors $\mathbf{q}_i$ are not zero, this condition can be satisfied if and only if the v_i are all zero. This shows that the eigenvectors $\mathbf{q}_1$, $\mathbf{q}_2$, . . . , $\mathbf{q}_M$ cannot be linearly dependent if the corresponding eigenvalues λ_1, λ_2, . . . , λ_M are distinct. In other words, they are linearly independent.

We may put this property to an important use by having the linearly independent eigenvectors $\mathbf{q}_1$, $\mathbf{q}_2$, . . . , $\mathbf{q}_M$ serve as a *basis* for the representation of an arbitrary vector **w** with the same dimension as the eigenvectors themselves. In particular, we may express the arbitrary vector **w** as a linear combination of the eigenvectors $\mathbf{q}_1$, $\mathbf{q}_2$, . . . , $\mathbf{q}_M$ as follows:

$$\mathbf{w} = \sum_{i=1}^{M} v_i \mathbf{q}_i \tag{2.35}$$

where v_1, v_2, . . . , v_M are constants. Suppose now we apply a linear transformation to the vector **w** by premultiplying it by the matrix **R**, obtaining

$$\mathbf{Rw} = \sum_{i=1}^{M} v_i \mathbf{Rq}_i$$

By definition, we have $\mathbf{Rq}_i = \lambda_i \mathbf{q}_i$. Therefore, we may express the result of this linear transformation in the equivalent form

$$\mathbf{Rw} = \sum_{i=1}^{M} v_i \lambda_i \mathbf{q}_i$$

We thus see that when a linear transformation is applied to an arbitrary vector the eigenvectors remain independent of each other, and the effect of the transformation is simply to multiply each eigenvector by its respective eigenvalue.

Property 3.   *Let $\lambda_1, \lambda_2, \ldots, \lambda_M$ be the eigenvalues of the M-by-M correlation matrix* **R**.   *Then all these eigenvalues are real and nonnegative.*

To prove this property, we first use Eq. (2.28) to express the condition on the ith eigenvalue λ_i as

$$\mathbf{Rq}_i = \lambda_i \mathbf{q}_i, \qquad i = 1, 2, \ldots, M \tag{2.36}$$

Premultiplying both sides of this equation by $\mathbf{q}_i^H$, the Hermitian transpose of eigenvector $\mathbf{q}_i$, we get

$$\mathbf{q}_i^H \mathbf{Rq}_i = \lambda_i \mathbf{q}_i^H \mathbf{q}_i, \qquad i = 1, 2, \ldots, M \tag{2.37}$$

The inner product $\mathbf{q}_i^H \mathbf{q}_i$ is a positive scalar, representing the squared Euclidean length of the eigenvector $\mathbf{q}_i$, that is, $\mathbf{q}_i^H \mathbf{q}_i > 0$. We may therefore divide both sides of Eq. (2.37) by $\mathbf{q}_i^H \mathbf{q}_i$ and so express the ith eigenvalue λ_i as the ratio

$$\lambda_i = \frac{\mathbf{q}_i^H \mathbf{Rq}_i}{\mathbf{q}_i^H \mathbf{q}_i}, \qquad i = 1, 2, \ldots, M \tag{2.38}$$

Since the correlation matrix **R** is always nonnegative definite, the Hermitian form $\mathbf{q}_i^H \mathbf{Rq}_i$ in the numerator of this ratio is always nonegative; that is, $\mathbf{q}_i^H \mathbf{Rq}_i \geq 0$. Therefore, it follows from Eq. (2.38) that $\lambda_i \geq 0$ for all i. That is, all the eigenvalues of the correlation matrix **R** are always real and nonnegative.

When, however, the correlation matrix **R** is positive definite, which is almost always the case, we have $\mathbf{q}_i^H \mathbf{Rq}_i > 0$ and, correspondingly, $\lambda_i > 0$ for all i. That is, the eigenvalues of the correlation matrix **R** are almost always real and positive.

The ratio of the Hermitian form $\mathbf{q}_i^H \mathbf{Rq}_i$ to the inner product $\mathbf{q}_i^H \mathbf{q}_i$ on the right side of Eq. (2.38) is called the *Rayleigh quotient* of the vector $\mathbf{q}_i$. We may thus state that an eigenvalue of the correlation matrix equals the Rayleigh quotient of the corresponding eigenvector.

Property 4.   *Let* $\mathbf{q}_1$, $\mathbf{q}_2$, . . . , $\mathbf{q}_M$ *be the eigenvectors corresponding to the distinct eigenvalues* $\lambda_1, \lambda_2, \ldots, \lambda_M$ *of the M-by-M correlation matrix* $\mathbf{R}$, *respectively. Then the eigenvectors* $\mathbf{q}_1$, $\mathbf{q}_2$, . . . , $\mathbf{q}_M$ *are orthogonal to each other.*

Let $\mathbf{q}_i$ and $\mathbf{q}_j$ denote any two eigenvectors of the correlation matrix $\mathbf{R}$. We say that these two eigenvectors are *orthogonal* to each other if

$$\mathbf{q}_i^H \mathbf{q}_j = 0, \qquad i \neq j \tag{2.39}$$

Using Eq. (2.28), we may express the conditions on the eigenvectors $\mathbf{q}_i$ and $\mathbf{q}_j$ as follows, respectively,

$$\mathbf{R}\mathbf{q}_i = \lambda_i \mathbf{q}_i \tag{2.40}$$

and

$$\mathbf{R}\mathbf{q}_j = \lambda_j \mathbf{q}_j \tag{2.41}$$

Premultiplying both sides of Eq. (2.40) by the Hermitian-transposed vector $\mathbf{q}_j^H$, we get

$$\mathbf{q}_j^H \mathbf{R}\mathbf{q}_i = \lambda_i \mathbf{q}_j^H \mathbf{q}_i \tag{2.42}$$

Since the correlation matrix $\mathbf{R}$ is Hermitian, we have $\mathbf{R}^H = \mathbf{R}$. Also, from Property 3 we know that the eigenvalue λ_j is real for all j. Hence, taking the Hermitian transpose of both sides of Eq. (2.41) and using these two properties, we get

$$\mathbf{q}_j^H \mathbf{R} = \lambda_j \mathbf{q}_j^H \tag{2.43}$$

Postmultiplying both sides of Eq. (2.43) by the vector $\mathbf{q}_i$,

$$\mathbf{q}_j^H \mathbf{R}\mathbf{q}_i = \lambda_j \mathbf{q}_j^H \mathbf{q}_i \tag{2.44}$$

Subtracting Eq. (2.44) from (2.42);

$$(\lambda_i - \lambda_j)\mathbf{q}_j^H \mathbf{q}_i = 0 \tag{2.45}$$

Since the eigenvalues of the correlation matrix $\mathbf{R}$ are assumed to be distinct, we have $\lambda_i \neq \lambda_j$. Accordingly, the condition of Eq. (2.45) holds if and only if

$$\mathbf{q}_j^H \mathbf{q}_i = 0, \qquad i \neq j \tag{2.46}$$

which is the desired result. That is, the eigenvectors $\mathbf{q}_i$ and $\mathbf{q}_j$ are *orthogonal* to each other for $i \neq j$.

Property 5.   *Let* $\mathbf{q}_1$, $\mathbf{q}_2$, . . . , $\mathbf{q}_M$ *be the eigenvectors corresponding to the distinct eigenvalues* $\lambda_1, \lambda_2, \ldots, \lambda_M$ *of the M-by-M correlation matrix* $\mathbf{R}$, *respectively. Define the M-by-M matrix*

$$\mathbf{Q} = [\mathbf{q}_1, \mathbf{q}_2, \ldots, \mathbf{q}_M]$$

where

$$\mathbf{q}_i^H \mathbf{q}_j = \begin{cases} 1, & i = j \\ 0, & i \neq j \end{cases}$$

Define the M-by-M diagonal matrix

$$\mathbf{\Lambda} = \text{diag}(\lambda_1, \lambda_2, \ldots, \lambda_M)$$

Then the original matrix $\mathbf{R}$ *may be diagonalized as follows:*

$$\mathbf{Q}^H \mathbf{R} \mathbf{Q} = \mathbf{\Lambda}$$

The condition that $\mathbf{q}_i^H \mathbf{q}_i = 1$ for $i = 1, 2, \ldots, M$, requires that each eigenvector be *normalized* to have a *length of 1*. The *squared length* or *squared norm* of a vector $\mathbf{q}_i$ is defined as the inner product $\mathbf{q}_i^H \mathbf{q}_i$. The orthogonality condition that $\mathbf{q}_i^H \mathbf{q}_j = 0$, for $i \neq j$, follows from Property 4. When both of these conditions are simultaneously satisfied, that is,

$$\mathbf{q}_i^H \mathbf{q}_j = \begin{cases} 1, & i = j \\ 0, & i \neq j \end{cases} \tag{2.47}$$

we say the eigenvectors $\mathbf{q}_1, \mathbf{q}_2, \ldots, \mathbf{q}_M$ form an *orthonormal* set. By definition, the eigenvectors $\mathbf{q}_1, \mathbf{q}_2, \ldots, \mathbf{q}_M$ satisfy the equations [see Eq. (2.28)]

$$\mathbf{R} \mathbf{q}_i = \lambda_i \mathbf{q}_i, \qquad i = 1, 2, \ldots, M \tag{2.48}$$

The *M*-by-*M* matrix $\mathbf{Q}$ has as its columns the orthonormal set of eigenvectors $\mathbf{q}_1$, $\mathbf{q}_2, \ldots, \mathbf{q}_M$:

$$\mathbf{Q} = [\mathbf{q}_1, \mathbf{q}_2, \ldots, \mathbf{q}_M] \tag{2.49}$$

The *M*-by-*M* diagonal matrix $\mathbf{\Lambda}$ has the eigenvalues $\lambda_1, \lambda_2, \ldots, \lambda_M$ for the elements of its main diagonal:

$$\mathbf{\Lambda} = \text{diag}(\lambda_1, \lambda_2, \ldots, \lambda_M) \tag{2.50}$$

Accordingly, we may rewrite the set of M equations (2.48) as a single matrix equation:

$$\mathbf{R} \mathbf{Q} = \mathbf{Q} \mathbf{\Lambda} \tag{2.51}$$

Owing to the orthonormal nature of the eigenvectors, as defined in Eq. (2.47), we find that

$$\mathbf{Q}^H \mathbf{Q} = \mathbf{I}$$

Equivalently, we may write

$$\mathbf{Q}^{-1} = \mathbf{Q}^H \tag{2.52}$$

That is, the matrix $\mathbf{Q}$ is nonsingular with an inverse $\mathbf{Q}^{-1}$ equal to the Hermitian transpose of $\mathbf{Q}$. A matrix that has this property is called a *unitary matrix*.

Thus, premultiplying both sides of Eq. (2.51) by the Hermitian-transposed matrix $\mathbf{Q}^H$ and using the property of Eq. (2.52), we get the desired result:

$$\mathbf{Q}^H \mathbf{R} \mathbf{Q} = \mathbf{\Lambda} \tag{2.53}$$

This transformation is called the *unitary similarity transformation*.[2]

We have thus proved an important result: the correlation matrix $\mathbf{R}$ may be *diagonalized* by a unitary similarity transformation. Furthermore, the matrix $\mathbf{Q}$ that is used to diagonalize $\mathbf{R}$ has as its columns an orthonormal set of eigenvectors for $\mathbf{R}$. The resulting diagonal matrix $\mathbf{\Lambda}$ has as its diagonal elements the eigenvalues of $\mathbf{R}$.

By postmultiplying both sides of Eq. (2.51) by the inverse matrix $\mathbf{Q}^{-1}$ and then using the property of Eq. (2.52), we may also write

$$\mathbf{R} = \mathbf{Q}\mathbf{\Lambda}\mathbf{Q}^H$$

$$= \sum_{i=1}^{M} \lambda_i \mathbf{q}_i \mathbf{q}_i^H \tag{2.54}$$

where M is the dimension of matrix $\mathbf{R}$. This decomposition is known as the *spectral theorem*.

Property 6.   *Let $\lambda_1, \lambda_2, \ldots, \lambda_M$ be the eigenvalues of the M-by-M correlation matrix $\mathbf{R}$. Then the sum of these eigenvalues equals the trace of matrix $\mathbf{R}$.*

The *trace* of a square matrix is defined as the sum of the diagonal elements of the matrix. Taking the trace of both sides of Eq. (2.53), we may write

$$\text{tr}[\mathbf{Q}^H \mathbf{R} \mathbf{Q}] = \text{tr}[\mathbf{\Lambda}] \tag{2.55}$$

The diagonal matrix $\mathbf{\Lambda}$ has as its diagonal elements the eigenvalues of $\mathbf{R}$. Hence, we have

$$\text{tr}[\mathbf{\Lambda}] = \sum_{i=1}^{M} \lambda_i \tag{2.56}$$

---

[2]The unitary similarity transformation is a special case of the *singular value decomposition*. Let $\mathbf{A}$ be an L-by-M matrix of rank W. Then there are unitary matrices $\mathbf{X}$ and $\mathbf{Y}$ such that we may write

$$\mathbf{Y}^H \mathbf{A} \mathbf{X} = \begin{bmatrix} \mathbf{\Sigma} & \mathbf{0} \\ \mathbf{0} & \mathbf{0} \end{bmatrix}$$

where $\mathbf{\Sigma}$ is the diagonal matrix

$$\mathbf{\Sigma} = \text{diag}(\sigma_1, \sigma_2, \ldots, \sigma_W)$$

This transformation is called the *singular value decomposition*, and the σ's are called *singular values* of the matrix $\mathbf{A}$. If $\mathbf{A}$ is Hermitian, then the singular values of $\mathbf{A}$ are just the absolute values of the eigenvalues of $\mathbf{A}$. For a discussion of the singular value decomposition, see Section 7.12.

Since the correlation matrix $\mathbf{R}$ is Hermitian, and its eigenvalues are real and nonnegative, it follows that the singular values of a correlation matrix are the same as the eigenvalues of the matrix.

Using a rule in matrix algebra, we may write[3]

$$\text{tr}[\mathbf{Q}^H \mathbf{R} \mathbf{Q}] = \text{tr}[\mathbf{R} \mathbf{Q} \mathbf{Q}^H]$$

However, $\mathbf{Q}\mathbf{Q}^H$ equals the identity matrix $\mathbf{I}$ [this follows from Eq. (2.52)]. Hence, we have

$$\text{tr}[\mathbf{Q}^H \mathbf{R} \mathbf{Q}] = \text{tr}[\mathbf{R}]$$

Accordingly, we may rewrite Eq. (2.55) as

$$\text{tr}[\mathbf{R}] = \sum_{i=1}^{M} \lambda_i \qquad (2.57)$$

We have thus shown that the trace of the correlation matrix $\mathbf{R}$ equals the sum of its eigenvalues. Although in proving this result we used a property that requires the matrix $\mathbf{R}$ to be Hermitian with distinct eigenvalues, nevertheless, the result applies to any square matrix.

Property 7.   *The largest eigenvalue of the correlation matrix $\mathbf{R}$ is bounded by*

$$\lambda_{\max} \leq \sum_{k=0}^{M-1} |r(k)| \qquad (2.58)$$

and its smallest eigenvalue is bounded by

$$\lambda_{\min} \geq r(0) - \sum_{k=1}^{M-1} |r(k)| \qquad (2.59)$$

These two results follow directly from *Gershgorin's theorem*, which may be stated as follows (Ralston, 1965):

Let $\mathbf{A}$ be any M-by-M complex-valued matrix with element a_{ij}, where $i, j = 1, 2, \ldots, M$, and let $\lambda_1, \lambda_2, \ldots, \lambda_M$ be its eigenvalues. Denote by D_i the ith *Gershgorin disk* with center a_{ii} and radius

$$r_i = \sum_{\substack{j=1 \\ j \neq i}}^{M} |a_{ij}|, \qquad i = 1, 2, \ldots, M$$

Denote by Ω the union of these M disks. Then all the eigenvalues of matrix $\mathbf{A}$ lie inside Ω.

Applying this theorem to the special case of the correlation matrix $\mathbf{R}$, whose eigenvalues are real and nonnegative (by Property 3), we find that the eigenvalues of the correlation matrix $\mathbf{R}$ lie on a strip of the real axis, as shown in Fig. 2.1. From this diagram, we immediately deduce the results of Eqs. (2.58) and (2.59).

---

[3]This result follows from the following rule in matrix algebra. Let $\mathbf{A}$ be an M-by-N matrix and $\mathbf{B}$ be an N-by-M matrix. The trace of the matrix product $\mathbf{AB}$ equals the trace of $\mathbf{BA}$.

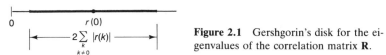

Figure 2.1   Gershgorin's disk for the eigenvalues of the correlation matrix **R**.

The lower bound of Eq. (2.59) is of limited practical value since the eigenvalues of a correlation matrix are always nonnegative and there is no guarantee that

$$r(0) > \sum_{k=1}^{M-1} |r(k)|$$

Property 8.   *The correlation matrix* **R** *is ill conditioned if the ratio of the largest eigenvalue to the smallest eigenvalue of* **R** *is large.*

To describe the ill condition or bad behavior of a matrix **A** quantitatively, we define the *condition number* of matrix **A** as follows (Strang, 1980):

$$\chi(\mathbf{A}) = \|\mathbf{A}\| \, \|\mathbf{A}^{-1}\| \tag{2.60}$$

where $\|\mathbf{A}\|$ is a *norm* of matrix **A**, and $\|\mathbf{A}^{-1}\|$ is the corresponding norm of the inverse matrix $\mathbf{A}^{-1}$. The norm of a matrix is a number assigned to the matrix that is in some sense a measure of the magnitude of the matrix. We find it natural to require that the norm of a matrix satisfy the following conditons:

1. $\|\mathbf{A}\| \geq 0$, $\|\mathbf{A}\| = 0$ if and only if $\mathbf{A} = \mathbf{0}$.
2. $\|\mathbf{cA}\| = |c| \, \|\mathbf{A}\|$, where c is any real number and $|c|$ is its magnitude.
3. $\|\mathbf{A} + \mathbf{B}\| \leq \|\mathbf{A}\| + \|\mathbf{B}\|$.
4. $\|\mathbf{AB}\| \leq \|\mathbf{A}\| \, \|\mathbf{B}\|$.

Condition 3 is the *triangle inequality*, and condition 4 is the *Schwartz inequality*. There are several ways of defining the norm $\|\mathbf{A}\|$, which satisfy the preceding conditions (Ralston, 1965). For our present discussion, however, we find it convenient to use the *spectral norm* defined as the square root of the largest eigenvalue of the matrix product $\mathbf{A}^H\mathbf{A}$, where $\mathbf{A}^H$ is the Hermitian transpose of **A**; that is,

$$\|\mathbf{A}\| = (\text{largest eigenvalue of } \mathbf{A}^H\mathbf{A})^{1/2} \tag{2.61}$$

Since for any matrix **A** the product $\mathbf{A}^H\mathbf{A}$ is always Hermitian and nonnegative definite, it follows that the eigenvalues of $\mathbf{A}^H\mathbf{A}$ are all real and nonnegative, as required. Moreover, from Eq. (2.38) we note that an eigenvalue of $\mathbf{A}^H\mathbf{A}$ equals the Rayleigh coefficient of the corresponding eigenvector. Squaring both sides of Eq. (2.61) and using this property, we may therefore write

$$\|\mathbf{A}\|^2 = \max \frac{\mathbf{x}^H\mathbf{A}^H\mathbf{A}\mathbf{x}}{\mathbf{x}^H\mathbf{x}}$$

$$= \max \frac{\|\mathbf{A}\mathbf{x}\|^2}{\|\mathbf{x}\|^2}$$

where $\|\mathbf{x}\|^2$ is the inner product of vector $\mathbf{x}$ with itself, and likewise for $\|\mathbf{Ax}\|^2$. We refer to $\|\mathbf{x}\|$ as the *Euclidean norm* or *length* of vector $\mathbf{x}$. We may thus express the norm of matrix $\mathbf{A}$ in the equivalent form

$$\|\mathbf{A}\| = \max \frac{\|\mathbf{Ax}\|}{\|\mathbf{x}\|} \tag{2.62}$$

According to this relation, the norm of $\mathbf{A}$ measures the largest amount by which any vector (eigenvector or not) is amplified by matrix multiplication, and the vector that is amplified the most is the eigenvector that corresponds to the largest eigenvalue of $\mathbf{A}^H\mathbf{A}$ (Strang, 1980).

Consider now the application of the definition in Eq. (2.61) to the correlation matrix $\mathbf{R}$. Since $\mathbf{R}$ is Hermitian, we have $\mathbf{R}^H = \mathbf{R}$. Hence, from Property 1 we deduce that if $\lambda_{\max}$ is the largest eigenvalue of $\mathbf{R}$, the largest eigenvalue of $\mathbf{R}^H\mathbf{R}$ equals $\lambda_{\max}^2$. Accordingly, the spectral norm of the correlation matrix $\mathbf{R}$ is

$$\|\mathbf{R}\|_S = \lambda_{\max}$$

Similarly, we may show that the spectral norm of $\mathbf{R}^{-1}$, the inverse of the correlation matrix, is

$$\|\mathbf{R}^{-1}\|_S = \frac{1}{\lambda_{\min}}$$

where $\lambda_{\min}$ is the smallest eigenvalue of $\mathbf{R}$. Thus, by adopting the spectral norm as the basis of the condition number, we have shown that the condition number of the correlation matrix $\mathbf{R}$ equals

$$\chi(\mathbf{R}) = \frac{\lambda_{\max}}{\lambda_{\min}} \tag{2.63}$$

This ratio is commonly referred to as the *eigenvalue spread* or the *eigenvalue ratio* of the correlation matrix.

Suppose that the correlation matrix $\mathbf{R}$ is *normalized* so that the magnitude of the largest element, $r(0)$, equals 1. Then, if the condition number or eigenvalue spread of the correlation matrix $\mathbf{R}$ is large, we find that the inverse matrix $\mathbf{R}^{-1}$ contains some very large elements. This behavior may cause trouble in solving a system of equations involving $\mathbf{R}^{-1}$. In such a case, we say that the correlation matrix $\mathbf{R}$ is *ill conditioned*.

2.6 POWER SPECTRAL DENSITY

The correlation sequence $\{r(k)\}$, $k = 0, \pm 1, \ldots, \pm(M - 1)$, or the M-by-M correlation matrix $\mathbf{R}$ represents the *time-domain description* of the second-order statistics of a stationary discrete-time stochastic process. By applying the *discrete-*

time Fourier transform to the correlation sequence $\{r(k)\}$, we may write

$$S(\omega) = \sum_{k=-(M-1)}^{M-1} r(k)e^{-j\omega k}, \qquad -\pi \leq \omega \leq \pi \qquad (2.64)$$

where ω is the *angular frequency*. We call the result of this Fourier transformation the *power spectral density* or *power spectrum* of the process; we use both terms interchangeably. Thus, the power spectral density $S(\omega)$ represents the *frequency-domain description* of the second-order statistics of the process.

We may recover the correlation sequence $\{r(k)\}$ by applying the *inverse discrete-time Fourier transform* to the power spectral density $S(\omega)$, as shown by

$$r(k) = \frac{1}{2\pi} \int_{-\pi}^{\pi} S(\omega)e^{j\omega k}\, d\omega, \qquad k = 0, \pm 1, \ldots \pm(M-1) \qquad (2.65)$$

Equations (2.64) and (2.65) constitute the discrete-time version of the *Wiener–Khintchine relations* (Priestley, 1981).

As a consequence of the Wiener-Khintchine relations, we may make the following statement concerning the transmission of a stationary discrete-time stochastic process through a discrete-time filter that is both *linear* and *time invariant*. Let the filter be characterized by the discrete transfer function $H(z)$, defined as the ratio of the *z-transform* of the filter output to the *z*-transform of the filter input. Suppose that we feed the filter with a stationary discrete-time stochastic process of power spectral density $S(\omega)$, as in Fig. 2.2. Let $S_o(\omega)$ denote the power spectral density of the filter output. We may then write (Priestley, 1981)

$$S_o(\omega) = |H(e^{j\omega})|^2 S(\omega) \qquad (2.66)$$

where $H(e^{j\omega})$ is the *frequency response* of the filter. The frequency response $H(e^{j\omega})$ equals the discrete transfer function $H(z)$ evaluated on the unit circle in the *z*-plane. The important feature of this result is that the value of the output spectral density at angular frequency ω depends purely on the squared *amplitude response* of the filter and the input power spectral density at the same angular frequency ω.

Properties of the Power Spectral Density

Property 1.   *The power spectral density of a stationary discrete-time stochastic process is periodic.*

From Eq. (2.64), we see that for any integer k (positive or negative)

$$S(\omega + 2k\pi) = S(\omega) \qquad (2.67)$$

Stationary process of power spectrum $S(\omega)$ → Discrete-time linear filter → Stationary process of power spectrum $S_o(\omega)$

Figure 2.2 Transmission of stationary process through discrete-time linear filter.

This shows that the power spectral density $S(\omega)$ is a periodic function of the angular frequency ω, with the period equal to 2π.

Property 2. *The power spectral density of a stationary discrete-time stochastic process is real.*

To prove this property, we rewrite Eq. (2.64) as

$$S(\omega) = r(0) + \sum_{k=1}^{M-1} r(k)e^{-j\omega k} + \sum_{k=-(M-1)}^{-1} r(k)e^{-j\omega k}$$

Replacing k with $-k$ in the third term on the right side of this equation, and recognizing that $r(-k) = r^*(k)$, we get

$$S(\omega) = r(0) + \sum_{k=1}^{M-1} [r(k)e^{-j\omega k} + r^*(k)e^{j\omega k}]$$

$$= r(0) + 2 \sum_{k=1}^{M-1} \text{Re}[r(k)e^{-j\omega k}] \tag{2.68}$$

where Re denotes the *real part operator*. Equation (2.68) shows that the power spectral density $S(\omega)$ is a real-valued function of ω.

Property 3. *The power spectral density of a stationary discrete-time stochastic process, in general, is not even.*

For a complex-valued stochastic process, we find that $S(-\omega) \neq S(\omega)$, indicating that $S(\omega)$ is not an even function of ω. If, however, the process is real valued, then $r(-k) = r(k)$, in which case we find that $S(-\omega) = S(\omega)$, and $S(\omega)$ is an even function of ω.

Property 4. *The mean-square value of a stationary discrete-time stochastic process equals, except for the scaling factor $1/2\pi$, the area under the power spectral density curve for $-\pi < \omega < \pi$.*

This property follows directly from Eq. (2.65), evaluated for $k = 0$. For this condition, we may thus write

$$r(0) = \frac{1}{2\pi} \int_{-\pi}^{\pi} S(\omega)\, d\omega \tag{2.69}$$

Since $r(0)$ equals the mean-square value of the process, we see that Eq. (2.69) is a mathematical description of Property 4.

Property 5. *The power spectral density of a stationary discrete-time stochastic process is nonnegative.*

To prove this property, suppose that the amplitude response of the filter in Fig. (2.2) is defined by

$$|H(e^{j\omega})| = \begin{cases} 1, & |\omega - \omega_c| \leq \Delta\omega \\ 0, & \text{remainder of the interval } -\pi \leq \omega \leq \pi \end{cases}$$

We assume that $\Delta\omega \ll \omega_c$. Then, using Eq. (2.66), we get

$$S_o(\omega) = \begin{cases} S(\omega_c), & |\omega - \omega_c| \leq \Delta\omega \\ 0, & \text{remainder of the interval } -\pi \leq \omega \leq \pi \end{cases}$$

Next, using Property 4, we may express the mean-square value of the filter output as

$$\text{Mean-square value of filter output} = \frac{1}{2\pi} \int_{-\pi}^{\pi} S_o(\omega) \, d\omega$$

$$= \frac{2}{\pi} S(\omega_c) \, \Delta\omega$$

Since the mean-square value of a random variable is always nonnegative, it follows that $S(\omega_c) \geq 0$ for all ω_c, which is the desired result.

Property 6. *The eigenvalues of the correlation matrix of a discrete-time stochastic process are bounded by the minimum and maximum values of the power spectral density of the process.*

Let λ_i and $\mathbf{q}_i$, $i = 1, 2, \ldots, M$, denote the eigenvalues of the M-by-M correlation matrix $\mathbf{R}$ of a discrete-time stochastic process $\{u(n)\}$ and their associated eigenvectors, respectively. From Eq. (2.38), we have

$$\lambda_i = \frac{\mathbf{q}_i^H \mathbf{R} \mathbf{q}_i}{\mathbf{q}_i^H \mathbf{q}_i}, \qquad i = 1, 2, \ldots, M$$

The Hermitian form in the numerator may be expressed in its expanded form as follows

$$\mathbf{q}_i^H \mathbf{R} \mathbf{q}_i = \sum_{k=1}^{M} \sum_{l=1}^{M} q_{ik}^* \, r(l - k) q_{il}$$

where q_{ik}^* is the kth element of the row vector $\mathbf{q}_i^H$, $r(l - k)$ is the k,lth element of the matrix $\mathbf{R}$, and q_{il} is the lth element of the column vector $\mathbf{q}_i$. Using the Wiener–Khintchine relation of Eq. (2.65), we may write

$$r(l - k) = \frac{1}{2\pi} \int_{-\pi}^{\pi} S(\omega) e^{j\omega(l-k)} \, d\omega$$

where $S(\omega)$ is the power spectral density of the process $\{u(n)\}$.   Hence,

$$\mathbf{q}_i^H \mathbf{R} \mathbf{q}_i = \frac{1}{2\pi} \sum_{k=1}^{M} \sum_{l=1}^{M} q_{ik}^* \, q_{il} \int_{-\pi}^{\pi} S(\omega) e^{j\omega(l-k)} \, d\omega$$

$$= \frac{1}{2\pi} \int_{-\pi}^{\pi} d\omega \, S(\omega) \sum_{k=1}^{M} q_{ik}^* e^{-j\omega k} \sum_{l=1}^{M} q_{il} e^{j\omega l}$$

Let the discrete-time Fourier transform of $q_{i1}^*, q_{i2}^*, \ldots, q_{iM}^*$ be denoted by

$$Q_i'(e^{j\omega}) = \sum_{k=1}^{M} q_{ik}^* e^{-j\omega k}$$

Therefore,

$$\mathbf{q}_i^H \mathbf{R} \mathbf{q}_i = \frac{1}{2\pi} \int_{-\pi}^{\pi} |Q_i'(e^{j\omega})|^2 \, S(\omega) d\omega$$

Similarly, we may show that

$$\mathbf{q}_i^H \mathbf{q}_i = \frac{1}{2\pi} \int_{-\pi}^{\pi} |Q_i'(e^{j\omega})|^2 \, d\omega$$

Accordingly, we may redefine the eigenvalue λ_i of the correlation matrix $\mathbf{R}$ in terms of the associated power spectral density as

$$\lambda_i = \frac{\displaystyle\int_{-\pi}^{\pi} |Q_i'(e^{j\omega})|^2 \, S(\omega) d\omega}{\displaystyle\int_{-\pi}^{\pi} |Q_i'(e^{j\omega})|^2 \, d\omega}$$

Let $S_{\min}$ and $S_{\max}$ denote the absolute minimum and maximum values of the power spectral density $S(\omega)$, respectively.   Then it follows that

$$\int_{-\pi}^{\pi} |Q_i'(e^{j\omega})|^2 S(\omega) d\omega \geq S_{\min} \int_{-\pi}^{\pi} |Q_i'(e^{j\omega})|^2 d\omega$$

and

$$\int_{-\pi}^{\pi} |Q_i'(e^{j\omega})|^2 S(\omega) d\omega \leq S_{\max} \int_{-\pi}^{\pi} |Q_i'(e^{j\omega})|^2 \, d\omega$$

Hence, we deduce that the eigenvalues λ_i are bounded by the maximum and minimum values of the associated power spectral density as follows:

$$S_{\min} \leq \lambda_i \leq S_{\max}, \qquad i = 1, 2, \ldots, M \tag{2.70}$$

Correspondingly, the eigenvalue spread $\chi(\mathbf{R})$ is bounded as

$$\chi(\mathbf{R}) = \frac{\lambda_{\max}}{\lambda_{\min}} \leq \frac{S_{\max}}{S_{\min}} \tag{2.71}$$

2.7 *AUTOREGRESSIVE MODELS*

The term *model* is used for any hypothesis that may be applied to explain or describe the hidden laws that are supposed to govern or constrain the generation of some data of interest.   The representation of a stochastic process by a model dates back to an idea that was originated by Yule (1927).   The idea is that a time series consisting of highly correlated observations may be generated by applying a series of statistically independent "shocks" to a linear filter, as in Fig. 2.3.   The shocks are random variables drawn from a fixed distribution that is usually assumed to be *Gaussian* with zero mean and constant variance.   Such a series of random variables constitutes a purely random process that is commonly referred to as *white noise*. The power spectral density of a white-noise process is *flat* (i.e., it has the same value at all frequencies).   The term "white noise" thus arises from analogy with "white light" in which all frequencies (i.e., colors) are present in equal amounts.

The stochastic model that is of special interest to us in this book is the *autoregressive (AR) model*.   The AR model has become widely used as a tool for fitting practical data in many different fields, notably geophysics, speech processing, and radar.   There are two basic reasons for the wide application of AR models:

1. The AR model is related to a fundamental theorem in the decomposition of time series, which was proposed by Wold (1938).
2. The parameters of an AR model can be computed efficiently by solving a system of *linear* equations known as the *Yule–Walker equations*.

Both of these issues are considered later.   We begin the discussion by defining an AR process.

We say that the time series $u(n)$, $u(n-1)$, . . . , $u(n-M)$ represents the realization of an autoregressive process of *order M* if it satisfies the difference equation

$$u(n) + a_1^* u(n-1) + \cdots + a_M^* u(n-M) = v(n) \qquad (2.72)$$

where a_1, a_2, . . . , a_M are constants called the *AR parameters*, and $\{v(n)\}$ is a white-noise process.   Except for the present value $u(n)$, each term on the left side of Eq. (2.72) represents an inner product.

We assume that the white noise $\{v(n)\}$ has zero mean and variance σ_v^2.   That is,

$$E[v(n)] = 0, \qquad \text{for all } n$$

Figure 2.3   Stochastic model.

and

$$E[v(n)v^*(k)] = \begin{cases} \sigma_v^2, & k = n \\ 0, & \text{otherwise} \end{cases} \tag{2.73}$$

To explain the reason for the name "autoregressive," we rewrite Eq. (2.72) in the form

$$u(n) = w_1^* u(n-1) + w_2^* u(n-2) + \cdots + w_M^* u(n-M) + v(n) \tag{2.74}$$

where $w_k = -a_k$. We thus see that the present value of the process, that is, $u(n)$, equals a *finite linear combination of past values* of the process, $u(n-1)$, . . . , $u(n-M)$, plus an *error term* $v(n)$. We now see the reason for the name "autoregressive." Specifically, a linear model

$$y = \sum_{k=1}^{M} w_k^* x_k + v$$

relating a *dependent* variable y to a set of *independent* variables $x_1, x_2, \ldots, x_M$ plus an error term v is often referred to as a *regression model*, and y is said to be "regressed" on $x_1, x_2, \ldots, x_M$. In Eq. (2.74), the variable $u(n)$ is *regressed* on previous values of *itself*; hence, the name "autoregressive."

The left side of Eq. (2.72) represents the *convolution* of the input sequence $\{u(n)\}$ and the sequence of parameters $\{a_n^*\}$. To highlight this point, we rewrite Eq. (2.72) in the form of a convolution sum:

$$\sum_{k=0}^{M} a_k^* u(n-k) = v(n) \tag{2.75}$$

where $a_0 = 1$. By taking the *z-transform*[4] of both sides of Eq. (2.75), we transform the convolution sum on the left side of the equation into a multiplication of the z-transforms of the two sequences $\{u(n)\}$ and $\{a_n^*\}$. Let $H_A(z)$ denote the z-transform of the sequence $\{a_n^*\}$:

$$H_A(z) = \sum_{n=0}^{M} a_n^* z^{-n} \tag{2.76}$$

Let $U(z)$ denote the z-transform of the input sequence $\{u(n)\}$:

$$U(z) = \sum_{n=0}^{\infty} u(n) \, z^{-n} \tag{2.77}$$

where z is a *complex variable*. We may thus transform the difference equation (2.72) into the equivalent form

$$H_A(z)U(z) = V(z) \tag{2.78}$$

---

[4]See Appendix C for a summary of the properties of the z-transform.

where

$$V(z) = \sum_{n=0}^{\infty} v(n)z^{-n} \qquad (2.79)$$

The z transform of Eq. (2.78) offers two interpretations, depending on whether the AR process $\{u(n)\}$ is viewed as the input or output of interest:

1. Given the AR process $\{u(n)\}$, we may use the filter shown in Fig. 2.4(a) to produce the white noise process $\{v(n)\}$ as output. The parameters of this filter bear a one-to-one correspondence with those of the AR process $\{u(n)\}$. Accordingly, this filter represents a *process analyzer* with discrete transfer function $H_A(z) = V(z)/U(z)$. The impulse response of the AR process analyzer, that is, the inverse z-transform of $H_A(z)$ has *finite duration*.

2. With the white noise $\{v(n)\}$ acting as input, we may use the filter shown in Fig. 2.4(b) to produce the AR process $\{u(n)\}$ as output. Accordingly, this second filter represents a *process generator*, whose transfer function equals

$$H_G(z) = \frac{U(z)}{V(z)}$$

$$= \frac{1}{H_A(z)}$$

$$= \frac{1}{\displaystyle\sum_{n=0}^{M} a_n^* z^{-n}} \qquad (2.80)$$

The impulse response of the AR process generator, that is, the inverse z-transform of $H_G(z)$ has *infinite duration*.

The filters shown in Fig. 2.4 consist of an interconnection of functional blocks that are used to represent three basic operations: (1) *addition*, (2) *multiplication* by a scalar, and (3) *storage*. The storage is represented by blocks labeled z^{-1}, the *unit-sample delay*. By applying z^{-1} to $u(n)$, we get $u(n - 1)$; similarly, by applying z^{-1} to $u(n - 1)$, we get $u(n - 2)$, and so on. Thus, by using a cascade of M unit-sample delays, we provide storage for the past values $u(n - 1)$, $u(n - 2), \ldots, u(n - M)$.

The AR process analyzer of Fig. 2.4(a) is an *all-zero filter*. It is so called because its transfer function $H_A(z)$ is completely defined by specifying the locations of its *zeros*. This filter is inherently stable.

The AR process generator of Fig. 2.4(b) is an *all-pole filter*. It is so called because its transfer function $H_G(z)$ is completely defined by specifying the locations of its *poles*, as shown by

$$H_G(z) = \frac{1}{(1 - p_1 z^{-1})(1 - p_2 z^{-1}) \cdots (1 - p_M z^{-1})} \qquad (2.81)$$

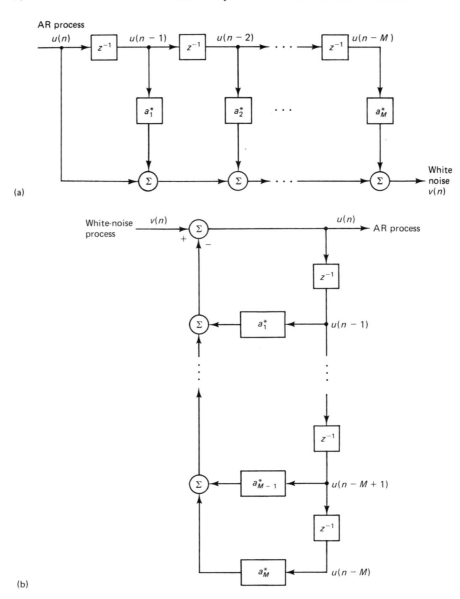

Figure 2.4   (a) AR process analyzer; (b) AR process generator.

The parameters $p_1, p_2, \ldots, p_M$ are *poles* of $H_G(z)$; they are defined by the roots of the *characteristic equation*

$$1 + a_1^* z^{-1} + a_2^* z^{-2} + \cdots + a_M^* z^{-M} = 0 \qquad (2.82)$$

We say that *a filter is stable if and only if its output is bounded for every bounded input.* In the case of a *causal* discrete-time linear filter, causal in that its impulse response is zero for negative time, this definition of stability leads to the following frequently used criterion (Tretter, 1976):

A causal filter with discrete transfer function H(z) is stable if and only if all the poles of H(z) lie inside the unit circle in the z-plane.

Accordingly, for the all-pole AR process generator of Fig. 2.4(b) to be stable, the roots of the characteristic equation (2.82) must all lie inside the unit circle in the z-plane. This is also a necessary and sufficient condition for wide-sense stationarity of the AR process produced by the model of Fig. 2.4(b). We will have more to say on the issue of stationarity later in the section.

The Wold Decomposition

Wold (1938) proved a fundamental theorem which states that any stationary discrete-time stochastic process may be decomposed into the sum of a *general linear process* and a *deterministic process*, with these two processes being uncorrelated with each other. More precisely, Wold proved the following result:

Any stationary discrete-time stochastic process $\{x(n)\}$ may be expressed in the form

$$x(n) = u(n) + s(n) \tag{2.83}$$

where

1. *$\{u(n)\}$ and $\{s(n)\}$ are uncorrelated processes,*
2. *$\{u(n)\}$ is a general linear process represented by*

$$u(n) = \sum_{k=0}^{\infty} b_k^* v(n - k) \tag{2.84}$$

with $b_0 = 1$, and

$$\sum_{k=0}^{\infty} |b_k|^2 < \infty,$$

and where $\{v(n)\}$ is a white-noise process uncorrelated with s(n); that is,

$$E[v(n)s^*(k)] = 0, \qquad \text{for all } n, k$$

3. *$\{s(n)\}$ is a deterministic process; that is, the process can be predicted from its own past with zero prediction variance.*

This result is known as *Wold's theorem*. A proof of this theorem is given in Priestley (1981).

According to Eq. (2.84), the general linear process $\{u(n)\}$ may be generated by feeding an *all-zero filter* with the white-noise process $\{v(n)\}$ as in Fig. 2.5(a). The zeros of the transfer function of this filter equal the roots of the equation:

$$\sum_{n=0}^{\infty} b_n^* z^{-n} = 0 \tag{2.85}$$

A solution of particular interest is an all-zero filter that is *minimum phase*. We say that this filter is minimum phase if these zeros lie inside the unit circle (Oppenheim and Schafer, 1975). In such a case we may replace the all-zero filter with an *equivalent* all-pole filter that has the same impulse response $\{h_n\} = \{b_n^*\}$, as in Fig. 2.5(b). This means that, except for a deterministic component, a stationary discrete-time stochastic process may also be represented as an AR process of sufficiently high order.

Solution of the Difference Equation Describing the Autoregressive Process

Equation (2.72) represents a *linear, constant coefficient, difference equation of order M*, in which $v(n)$ plays the role of *input* or *driving function* and $u(n)$ that of *output* or *solution*. By using the *classical method*[5] for solving such an equation, we may formally express the solution $u(n)$ as the sum of a *complementary function*, $u_c(n)$, and a *particular solution*, $u_p(n)$, as follows:

$$u(n) = u_c(n) + u_p(n) \tag{2.86}$$

The evaluation of the solution $u(n)$ may thus proceed in two stages:

1. The complementary function $u_c(n)$ is the solution of the *homogeneous equation*:

$$u(n) + a_1^* u(n-1) + a_2^* u(n-2) + \cdots + a_M^* u(n-M) = 0$$

In general, the complementary function $u_c(n)$ will therefore be of the form

$$u_c(n) = B_1 p_1^n + B_2 p_2^n + \cdots + B_M p_M^n \tag{2.87}$$

where $B_1, B_2, \ldots, B_M$ are arbitrary constants, and $p_1, p_2, \ldots, p_M$ are roots of the characteristic equation (2.82).

2. The particular solution $u_p(n)$ is defined by

$$u_p(n) = H_G(D)[v(n)] \tag{2.88}$$

where D is the *unit-delay operator*, and the operator $H_G(D)$ is obtained by substituting D for z^{-1} in the discrete-transfer function of Eq. (2.81). The unit-delay operator D has the property

$$D^k[u(n)] = u(n-k), \quad k = 0, 1, 2 \ldots \tag{2.89}$$

---

[5]We may also use the z-transform method to solve the difference equation (2.72). However, for the discussion here we find it more informative to use the classical method.

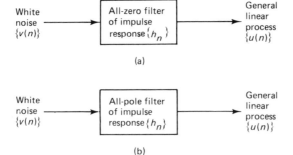

Figure 2.5 (a) Model, based on all-zero filter, for generating linear process $\{u(n)\}$; (b) model, based on all-pole filter, for generating general linear process $\{u(n)\}$. Both filters have the same impulse response.

The constants $B_1, B_2, \ldots, B_M$ are determined by the choice of *initial conditions* that equal M in number. It is customary to set

$$u(0) = 0$$

$$u(-1) = 0$$

$$\cdot$$
$$\cdot \qquad\qquad\qquad\qquad (2.90)$$
$$\cdot$$

$$u(-M + 1) = 0$$

This is equivalent to setting the output of the model in Fig. 2.4(b) as well as the succeeding $(M - 1)$ tap inputs equal to zero at time $n = 0$. Thus, by substituting these initial conditions into Eqs. (2.86)–(2.88), we obtain a set of M simultaneous equations that can be solved for the constants $B_1, B_2, \ldots, B_M$.

Property of Asymptotic Stationarity

The result of imposing the initial conditions of Eq. (2.90) on the solution $u(n)$ is to make the discrete-time stochastic process represented by this solution nonstationary. On reflection, it is clear that this must be so, since we have given a "special status" to the time point $n = 0$, and the property of *invariance under a shift of time origin* cannot hold, even for second-order moments. If, however, the solution $u(n)$ is able to "forget" its initial conditions, the resulting process is asymptotically stationary in the sense that it settles down to a stationary behavior as n approaches infinity (Priestley, 1981). This requirement may be achieved by choosing the parameters of the AR model in Fig. 2.4(b) such that the complementary function $u_c(n)$ decays to zero as n approaches infinity. From Eq. (2.87) we see that, for arbitrary constants in the equation, this requirement can be met if and only if

$$|p_k| < 1, \qquad \text{for all } k$$

Hence, *for asymptotic stationarity of the discrete-time stochastic process represented by the solution u(n), we require that all the poles of the filter in the AR model lie inside the unit circle in the z-plane.*

Correlation Function of the AR Process

Assuming that the condition for asymptotic stationarity is satisfied, we may derive an important recursive relation for the autocorrelation function of the resulting AR process $\{u(n)\}$ as follows. We first multiply both sides of Eq. (2.72) by $u^*(n - l)$ and then apply the expectation operator, thereby obtaining

$$E\left[\sum_{k=0}^{M} a_k^* u(n - k)u^*(n - l) \right] = E\left[v(n)u^*(n - l) \right] \qquad (2.91)$$

Next, we simplify the left side of Eq. (2.91) by interchanging the expectation and summation, and recognizing that the expectation $E[u(n - k)u^*(n - l)]$ equals the autocorrelation function of the AR process for a lag of $l - k$. We simplify the right side by observing that the expectation $E[v(n)u^*(n - l)]$ is zero for $l > 0$, since $u(n - l)$ only involves samples of the white-noise process at the filter input in Fig. 2.4(b) up to time $n - l$, which are uncorrelated with the white-noise sample $v(n)$. Accordingly, we simplify Eq. (2.91) as follows:

$$\sum_{k=0}^{M} a_k^* r(l - k) = 0, \qquad l > 0 \qquad (2.92)$$

where $a_0 = 1$. We thus see that the autocorrelation function of the AR process satisfies the difference equation

$$r(l) = w_1^* r(l - 1) + w_2^* r(l - 2) + \cdots + w_M^* r(l - M), \qquad l > 0 \qquad (2.93)$$

where $w_k = -a_k$, $k = 1, 2, \cdots, M$. Note that Eq. (2.93) is analogous to the difference equation satisfied by the AR process $\{u(n)\}$ itself.

We may express the general solution of Eq. (2.93) as follows:

$$r(m) = \sum_{k=1}^{M} C_k p_k^m \qquad (2.94)$$

where $C_1, C_2, \ldots, C_M$ are constants, and $p_1, p_2, \ldots, p_M$ are roots of the characteristic equation (2.82). Note that when the AR model of Fig. 2.4(b) satisfies the condition for asymptotic stationarity, $|p_k| < 1$ for all k, in which case the autocorrelation function $r(m)$ approaches zero as the lag m approaches infinity.

The exact form of the contribution made by a pole p_k in Eq. (2.94) depends on whether the pole is real or complex. When p_k is real, the corresponding contribution decays geometrically to zero as the lag m increases. We refer to such a contribution as a *damped exponential*. On the other hand, complex poles occur in conjugate pairs, and the contribution of a complex-conjugate pair of poles is in the form of a *damped sine wave*. We thus find that, in general, the autocorrelation function of an asymptotically stationary AR process consists of a mixture of damped exponentials and damped sine waves.

Yule–Walker Equations

Writing Eq. (2.93) for $l = 1, 2, \ldots, M$, we get a set of M simultaneous equations with the values $r(0)$, $r(1)$, $\ldots$, $r(M)$ of the autocorrelation function of the AR process as the known quantities and the AR parameters $a_1, a_2, \ldots, a_M$ as the unknowns. This set of equations may be expressed in the expanded matrix form

$$
\begin{bmatrix}
r(0) & r(1) & \cdots & r(M-1) \\
r(-1) & r(0) & \cdots & r(M-2) \\
\vdots & \vdots & \ddots & \vdots \\
r(-M+1) & r(-M+2) & \cdots & r(0)
\end{bmatrix}
\begin{bmatrix}
w_1 \\
w_2 \\
\vdots \\
w_M
\end{bmatrix}
$$

$$
=
\begin{bmatrix}
r(-1) \\
r(-2) \\
\vdots \\
r(-M)
\end{bmatrix}
\tag{2.95}
$$

where we have $w_k = -a_k$ and $r(-k) = r^*(k)$. The set of equations (2.95) is called the *Yule–Walker equations* (Yule, 1927; Walker, 1931).

We may express the Yule–Walker equations in the compact matrix form

$$\mathbf{Rw} = \mathbf{r} \tag{2.96}$$

and its solution as

$$\mathbf{w} = \mathbf{R}^{-1}\mathbf{r} \tag{2.97}$$

The vector $\mathbf{w}$ is defined by

$$\mathbf{w}^T = [w_1, w_2, \ldots, w_M]$$

The correlation matrix $\mathbf{R}$ is defined by Eq. (2.12), and the vector $\mathbf{r}$ is defined by Eq. (2.22).

Variance of the White Noise

For $l = 0$, we find that the expectation on the right side of Eq. (2.91) assumes the special form

$$
\begin{aligned}
E[v(n)u^*(n)] &= E[v(n)v^*(n)] \\
&= \sigma_v^2
\end{aligned}
\tag{2.98}
$$

where σ_v^2 is the variance of the zero-mean white noise $\{v(n)\}$. Accordingly, setting $l = 0$ in Eq. (2.91) and complex–conjugating both sides, we get the following formula for the variance of the white-noise process:

$$\sigma_v^2 = \sum_{k=0}^{M} a_k r(k) \tag{2.99}$$

where $a_0 = 1$.

Power Spectral Density of the AR Process

Earlier we indicated that, when a weakly stationary process is applied to a linear filter, the output power spectral density equals the squared amplitude response of the filter multiplied by the input power spectral density. In the model of Fig. 2.4(b), the power spectral density of the white-noise process applied to the filter input is a constant equal to σ_v^2, where σ_v^2 is the variance of this process. Hence, the power spectral density of the asymptotically stationary AR process produced at the filter output is defined by

$$
\begin{aligned}
S(\omega) &= \sigma_v^2 \, |H(e^{j\omega})|^2 \\[2mm]
&= \frac{\sigma_v^2}{\left|1 + \displaystyle\sum_{m=1}^{M} a_m^* e^{-jm\omega}\right|^2} \qquad -\pi \le \omega \le \pi
\end{aligned}
\tag{2.100}
$$

2.8 COMPUTER EXPERIMENT: AUTOREGRESSIVE PROCESS OF ORDER 2

To illustrate the theory developed above for the modeling of an AR process, we consider the example of a second-order AR process that is real valued.[6] Figure 2.6 shows the block diagram of the model used to generate this process. Its time-domain description is governed by the second-order difference equation

$$u(n) + a_1 u(n-1) + a_2 u(n-2) = v(n) \tag{2.101}$$

where $v(n)$ is drawn from a white-noise process of zero mean and variance σ_v^2. Figure 2.7(a) shows one realization of this white-noise process. The variance σ_v^2 is chosen to make the variance of $u(n)$ equal unity.

Conditions for Asymptotic Stationarity

The second-order AR process $\{u(n)\}$ is described by the characteristic equation

$$1 + a_1 z^{-1} + a_2 z^{-2} = 0 \tag{2.102}$$

---

[6]In this example, we follow the approach described by Box and Jenkins (1976).

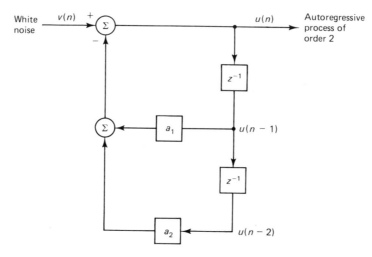

Figure 2.6   Model of (real valued) autoregressive process of order 2.

Let p_1 and p_2 denote the two roots of this equation:

$$p_1, p_2 = \frac{1}{2} \left(-a_1 \pm \sqrt{a_1^2 - 4a_2} \right) \tag{2.103}$$

To ensure the asymptotic stationarity of the AR process $\{u(n)\}$, we require that these two roots lie inside the unit circle in the z-plane.   That is, both p_1 and p_2 must have a magnitude less than 1.   This, in turn, requires that the AR parameters a_1 and a_2 lie in the triangular region defined by

$$-1 \le a_2 + a_1$$

$$-1 \le a_2 - a_1 \tag{2.104}$$

$$-1 \le a_2 \le 1$$

as shown in Fig. 2.8.

Autocorrelation Function

The autocorrelation function $r(m)$ of an asymptotically stationary AR process for lag m satisfies the difference equation (2.92).   Hence, using this equation, we obtain the following second-order difference equation for the autocorrelation function of a second-order AR process:

$$r(m) + a_1 r(m-1) + a_2 r(m-2) = 0, \quad m > 0 \tag{2.105}$$

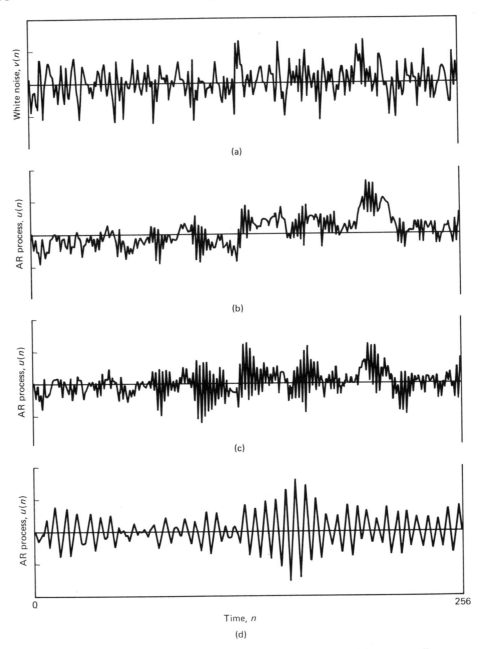

Figure 2.7 (a) one realization of white-noise input; (b), (c), (d) corresponding outputs of AR model of order 2 for parameters of Eqs. (2.108), (2.109), and (2.110), respectively.

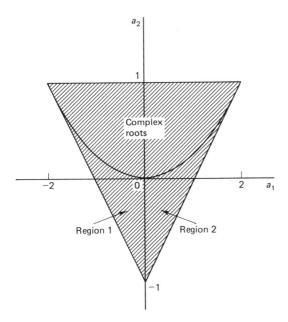

Figure 2.8 Permissible region for the AR parameters a_1 and a_2.

For the initial values, we have (as will be explained later)

$$r(0) = \sigma_u^2$$
$$r(1) = \frac{-a_1}{1 + a_2}\, \sigma_u^2 \qquad (2.106)$$

Thus, solving Eq. (2.105) for $r(m)$, we get (for $m > 0$)

$$r(m) = \sigma_u^2 \left[\frac{p_1(p_2^2 - 1)}{(p_2 - p_1)\,(p_1 p_2 + 1)}\, p_1^m - \frac{p_2(p_1^2 - 1)}{(p_2 - p_1)(p_1 p_2 + 1)}\, p_2^m \right] \qquad (2.107)$$

where p_1 and p_2 are defined by Eq. (2.103).

There are two specific cases to be considered, depending on whether the roots p_1 and p_2 are real or complex valued, as described next.

Case 1.   *Real Roots*

This case occurs when

$$a_1^2 - 4a_2 > 0$$

which corresponds to regions 1 and 2 below the parabolic boundary in Fig. 2.8. In region 1, the autocorrelation function remains positive as it damps out, corresponding to a positive dominant root. This situation is illustrated in Fig. 2.9(a) for the AR parameters

$$a_1 = -0.10$$
$$a_2 = -0.8 \qquad (2.108)$$

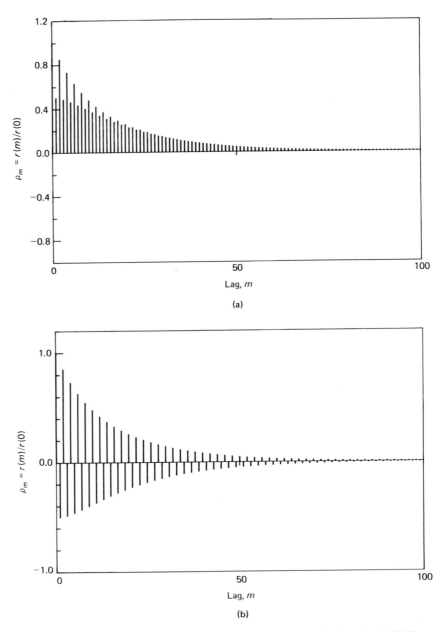

Figure 2.9 Plots of normalized autocorrelation function of real valued AR(2) process with eigenvalue spread $\chi(\mathbf{R}) = 3$: (a) $r(1) > 0$; (b) $r(1) < 0$; (c) conjugate roots.

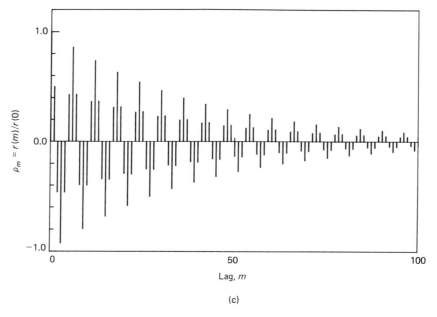

$$\rho_m = r(m)/r(0)$$

Lag, m

(c)

Figure 2.9 (*cont.*)

In Fig. 2.7(b), we show the time variation of the output of the model in Fig. 2.6 [with a_1 and a_2 assigned the values given in Eq. (2.108)]. This output is produced by the white-noise input shown in Fig. 2.7(a).

In region 2 of Fig. 2.8, the autocorrelation function alternates in sign as it damps out, corresponding to a negative dominant root. This situation is illustrated in Fig. 2.9(b) for the AR parameters

$$a_1 = 0.1$$
$$a_2 = -0.8$$

(2.109)

In Fig. 2.7(c) we show the time variation of the output of the model in Fig. 2.6 [with a_1 and a_2 assigned the values given in Eq. (2.109)]. This output is also produced by the white-noise input shown in Fig. 2.7(a).

Case 2. *Complex-conjugate Roots*
This occurs when

$$a_1^2 - 4a_2 < 0$$

which corresponds to the shaded region shown in Fig. 2.8 above the parabolic boundary. In this case, the autocorrelation function displays a pseudoperiodic behavior, as illustrated in Fig. 2.9(c) for the AR parameters

$$a_1 = -0.975$$
$$a_2 = 0.95$$

(2.110)

In Fig. 2.7(d) we show the time variation of the output of the model in Fig. 2.6 [with a_1 and a_2 assigned the values given in Eq. (2.110)], which is produced by the white-noise input shown in Fig. 2.7(a).

Yule–Walker Equations

Substituting the value $M = 2$ for the AR model order in Eq. (2.95), we get the following Yule–Walker equations for the second-order AR process:

$$\begin{bmatrix} r(0) & r(1) \\ r(1) & r(0) \end{bmatrix} \begin{bmatrix} w_1 \\ w_2 \end{bmatrix} = \begin{bmatrix} r(1) \\ r(2) \end{bmatrix} \tag{2.111}$$

where we have used the fact that $r(-1) = r(1)$ for a real-valued process. Solving Eq. (2.111) for w_1 and w_2, we get

$$w_1 = -a_1 = \frac{r(1)[r(0) - r(2)]}{r^2(0) - r^2(1)}$$

$$w_2 = -a_2 = \frac{r(0)r(2) - r^2(1)}{r^2(0) - r^2(1)} \tag{2.112}$$

We may also use Eq. (2.111) to express $r(1)$ and $r(2)$ in terms of the AR parameters a_1 and a_2 as follows:

$$r(1) = \frac{-a_1}{1 + a_2} \sigma_u^2$$

$$r(2) = \left(-a_2 + \frac{a_1^2}{1 + a_2} \right) \sigma_u^2 \tag{2.113}$$

where $\sigma_u^2 = r(0)$. This solution explains the initial values for $r(0)$ and $r(1)$ that were quoted in Eq. (2.106).

The conditions for asymptotic stationarity of the second-order AR process are given in terms of the AR parameters a_1 and a_2 in Eq. (2.104). Using the expressions for $r(1)$ and $r(2)$ in terms of a_1 and a_2, given in Eq. (2.113), we may reformulate the conditions for asymptotic stationarity as follows:

$$-1 < \rho_1 < 1$$

$$-1 < \rho_2 < 1 \tag{2.114}$$

$$\rho_1^2 < \frac{1}{2} (1 + \rho_2)$$

where ρ_1 and ρ_2 are the normalized *correlation coefficients* defined by

$$\rho_1 = \frac{r(1)}{r(0)} \tag{2.115}$$

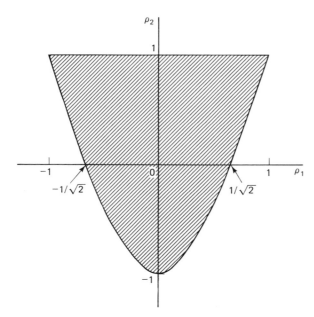

Figure 2.10 Permissible region for parameters of second-order AR process in terms of normalized correlation coefficients ρ_1 and ρ_2.

and

$$\rho_2 = \frac{r(2)}{r(0)}$$

Figure 2.10 shows the admissible region for ρ_1 and ρ_2.

Variance of the White-noise Process

Putting $M = 2$ in Eq. (2.99), we may express the variance of the white-noise process $\{v(n)\}$ as

$$\sigma_v^2 = r(0) + a_1 r(1) + a_2 r(2) \qquad (2.116)$$

Next, substituting Eq. (2.113) in (2.116), and solving for $\sigma_u^2 = r(0)$, we get

$$\sigma_u^2 = \left(\frac{1 + a_2}{1 - a_2}\right) \frac{\sigma_v^2}{[(1 + a_2)^2 - a_1^2]} \qquad (2.117)$$

For the three sets of AR parameters considered previously, we thus find that the variance of the white noise $\{v(n)\}$ has the values given in Table 2.1, assuming that $\sigma_u^2 = 1$.

TABLE 2.1

AR parameters variance

a_1	a_2	σ_v^2
-0.10	-0.8	0.27
0.1	-0.8	0.27
-0.975	0.95	0.0731

Power Spectral Density

Putting $M = 2$ in Eq. (2.100), we find that the power spectral density of the second-order AR process $\{u(n)\}$ equals

$$S(\omega) = \frac{\sigma_v^2}{|1 + a_1 e^{-j\omega} + a_2 e^{-j2\omega}|^2}, \qquad -\pi \le \omega \le \pi \qquad (2.118)$$

This is shown plotted in Fig. 2.11 for the three sets of AR parameter values listed in Table 2.1.

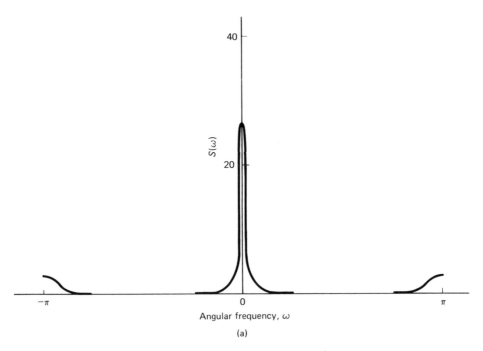

(a)

Figure 2.11 Power spectra of real valued AR(2) process with eigenvalue spread $\chi(\mathbf{R}) = 3$: (a) $r(1) > 0$; (b) $r(1) < 0$; (c) conjugate roots.

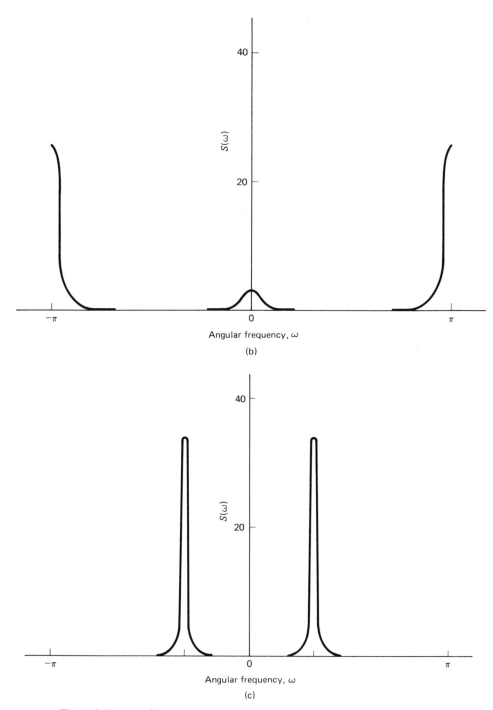

Figure 2.11   (*cont.*)

2.9 *OTHER STATIONARY STOCHASTIC MODELS*

The autoregressive model of Fig. 2.4(b), involving the use of an all-pole filter, constitutes one method of representing a stationary discrete-time stochastic process. Two other stationary stochastic models are shown in Figs. 2.12 and 2.13.

In Figure 2.12, the model consists of an *all-zero filter* driven by white noise. The result is a *moving-average (MA) process*, described by

$$u(n) = v(n) + b_1^*v(n - 1) + \cdots + b_K^*v(n - K) \qquad (2.119)$$

where $b_1, \ldots, b_K$ are constants called the *MA parameters*, and $\{v(n)\}$ is a white-noise process of zero mean and variance σ_v^2. Except for $v(n)$, each term on the right side of Eq. (2.119) represents an inner product. The *order* of the MA process equals K. The term moving average is a rather quaint one; nevertheless, its use is firmly established in the literature. Its usage arose in the following way: if we are given a complete realization of $\{v(n)\}$, we may compute $u(n)$ by constructing a *weighted average* of $v(n)$, $v(n - 1)$, ..., $v(n - K)$. From a computational viewpoint, the MA model differs from the AR model in that the determination of the MA parameters in the model of Fig. 2.12 requires solving a set of K *nonlinear* equations (Box and Jenkins, 1976; Priestley, 1981; Ulrych and Ooe, 1983), whereas the determination of the AR parameters requires the solution of *linear* equations. Note that in the Wold decomposition described by Eq. (2.83) the general linear process $\{u(n)\}$ is represented by an MA model.

In the model of Fig. 2.13, we use a filter with both poles and zeros. Accordingly, in this case, we obtain a mixed *autoregressive-moving average (ARMA) process*, described by

$$u(n) + a_1^*u(n - 1) + \cdots + a_M^*u(n - M)$$
$$= v(n) + b_1^*v(n - 1) + \cdots + b_K^*v(n - K) \qquad (2.120)$$

where $a_1, \ldots, a_M$ and $b_1, \ldots, b_K$ are called the *ARMA parameters*. Except for $u(n)$ on the left side and $v(n)$ on the right side of Eq. (2.120), all the other terms represent inner products. The *order* of the ARMA process equals (M, K). The problems that arise in the determination of the MA parameters are also encountered

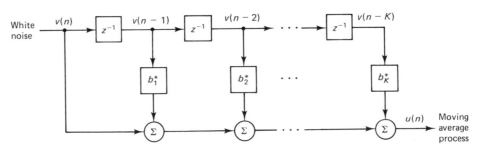

Figure 2.12   Moving-average model of order K.

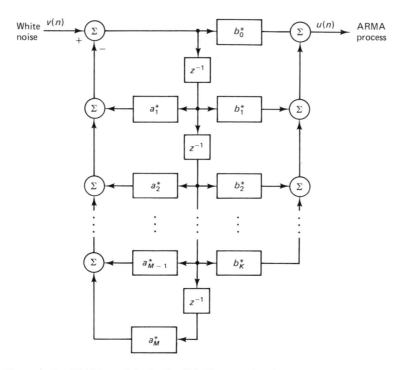

Figure 2.13   ARMA model of order (M, K), assuming that $M > K$.

in the determination of the ARMA parameters (Box and Jenkins, 1976; Priestley, 1981; Ulrych and Ooe, 1983).

2.10 SELECTING THE ORDER OF THE MODEL

The representation of a stochastic process by a linear model may be used for synthesis or analysis.   In *synthesis*, we generate a desired time series by assigning a prescribed set of values to the parameters of the model and feeding it with white noise of zero mean and prescribed variance.   In *analysis*, on the other hand, we estimate the parameters of the model by processing a given time series of finite length.   Insofar as the estimation is statistical, we need an appropriate measure of the fit between the model and the observed data.   This implies that, unless we have some prior information, the estimation procedure should include a criterion for selecting the *model order* (i.e., the number of independently adjusted parameters in the model).   In the case of an AR process defined by Eq. (2.72), the model order equals M.   In the case of an MA process defined by Eq. (2.119), the model order equals K.   In the case of an ARMA process defined by Eq. (2.120), the model order equals (M, K).   Various criteria for model-order selection are described in the literature (Priestley, 1981; Haykin, 1983).   In this section we describe two important criteria for selecting the order of the model, one of which was pioneered by Akaike (1973, 1974) and the other by Rissanen (1978) and Schwartz (1978). Akaike's criterion, being the older of the two, is more widely known. Both criteria, however, result from the use of information-theoretic arguments.

An Information-theoretic Criterion

Let $u_i = u(i)$, $i = 1, 2, \ldots, N$, denote the data obtained by N independent observations of a stationary discrete-time stochastic process, and $g(u_i)$ denote the probability density function of u_i. Let $f_U(u_i|\hat{\boldsymbol{\theta}}_m)$ denote the conditional probability density function of u_i, given $\hat{\boldsymbol{\theta}}_m$, where $\hat{\boldsymbol{\theta}}_m$ is the *estimated* vector of parameters that model the process. Let m be the order of the model so that we may write

$$\hat{\boldsymbol{\theta}}_m = \begin{bmatrix} \hat{\theta}_{1m} \\ \hat{\theta}_{2m} \\ \cdot \\ \cdot \\ \cdot \\ \hat{\theta}_{mm} \end{bmatrix} \tag{2.121}$$

We thus have several models that compete with each other to represent the process of interest. *An information-theoretic criterion* (AIC) proposed by Akaike selects the model for which the quantity

$$\text{AIC}(m) = -2L(\hat{\boldsymbol{\theta}}_m) + 2m \tag{2.122}$$

is a minimum. The function $L(\hat{\boldsymbol{\theta}}_m)$ is defined by

$$L(\hat{\boldsymbol{\theta}}_m) = \max \sum_{i=1}^{N} \ln f_U(u_i|\hat{\boldsymbol{\theta}}_m) \tag{2.123}$$

where ln denotes the natural logarithm. The criterion of Eq. (2.123) is derived by minimizing the *Kullback–Leibler mean information*,[7] which is used to provide a measure of the separation or distance between the "unknown" true probability

---

[7] In Akaike (1973, 1974, 1977) and Ulrych and Ooe (1983), the criterion of Eq. (2.122) is derived from the principle of minimizing the expectation $E[I(g;f(.|\boldsymbol{\theta})]$, where

$$I(g;f(.|\hat{\boldsymbol{\theta}}_m)) = \int_{-\infty}^{\infty} g(u) \ln g(u) \, du - \int_{-\infty}^{\infty} g(u) \ln f_U(u|\hat{\boldsymbol{\theta}}_m) \, du$$

We refer to $I(g;f(.|\boldsymbol{\theta}))$ as the *Kullback–Leibler mean information* for discrimination between $g(u)$ and $f_U(u|\hat{\boldsymbol{\theta}}_m)$ (Kullback and Leibler, 1951). The idea is to minimize the information added to the time series by modeling it as an AR, MA, or ARMA process of finite order, since any information added is virtually false information in a real-world situation. Since $g(u)$ is fixed and unknown, the problem reduces to one of maximizing the second term that makes up $I(g;f(.|\boldsymbol{\theta}))$.

density function $g(u)$ and the conditional probability density function $f_U(u|\hat{\boldsymbol{\theta}}_m)$ given by the model in the light of the observed data.

The function $L(\hat{\boldsymbol{\theta}}_m)$, constituting the first term on the right side of Eq. (2.122), except for a scalar, is recognized as the *log-likelihood* of the *maximum-likelihood estimates*[8] of the parameters in the model. The second term, $2m$, represents a *model complexity penalty* that makes AIC(m) an estimate of the Kullback–Leibler mean information.

The first term of Eq. (2.122) tends to decrease rapidly with model order m. On the other hand, the second term increases linearly with m. The result is that if we plot AIC(m) versus model order m the graph will, in general, show a definite minimum value, and the *optimum order* of the model is determined by that value of m at which AIC(m) attains its minimum value. The minimum value of AIC is called MAIC (minimum AIC).

Minimum Description Length Criterion

Rissanen (1978) and Schwartz (1978) question the theoretical validity of the structure-dependent term, $2m$, in Akaike's information-theoretic criterion of Eq. (2.122). They use entirely different approaches to solve the statistical model identification problem.

Rissanen uses information-theoretic ideas, starting with the notion that the number of digits (i.e., *length*) required to *encode* a set of observations $u(1)$, $u(2)$, ..., $u(N)$ depends on the model that is assumed to have generated the observed data. Accordingly, Rissanen selects the model that minimizes the code length of the observed data.[9]

Schwartz, on the other hand, uses a *Bayesian* approach. In particular, he studies the asymptotic behavior of Bayes estimators under a special class of *priors*. These priors put positive probability on the subspaces that correspond to the competing models. The decision is made by selecting the model that yields the *maximum a posteriori probability*.

It turns out that, in the large-sample limit, the two approaches taken by Rissanen and Schwartz yield essentially the same criterion, defined by (Wax and Kailath, 1985):

$$\text{MDL}(m) = -L(\hat{\boldsymbol{\theta}}_m) + \tfrac{1}{2}m \ln N \qquad (2.124)$$

where m is the number of independently adjusted parameters in the model, and N is the number of observations. As with Akaike's information-theoretic criterion, the function $L(\hat{\boldsymbol{\theta}}_m)$ is the log-likelihood of the maximum-likelihood estimates of the parameters in the model. In Eq. (2.124), we have chosen to refer to the

---

[8]For a discussion of the method of maximium likelihood, see Appendix A.

[9]The *minimum description length* of individual recursively definable objects has been studied by Kolmogorov (1968) and others.

criterion as MDL (for *minimum description length*) in recognition of the fact that Rissanen's derivation is more general; Schwartz's derivation is restricted to the case that the observations are independent and come from an exponential distribution.

Discussion

Basically, the minimum description length criterion of Eq. (2.124) differs from Akaike's information-theoretic criterion in Eq. (2.122) only in that the number of independently adjusted parameters in the model is multiplied by $\frac{1}{2} \ln N$, where N is the number of observations. Accordingly, we find that in general the minimum description length criterion leans more than Akaike's criterion toward lower-order models.

It has been shown by Rissanen (1978) and Wax and Kailath (1985) that the minimum description length criterion is a *consistent* model-order estimator, whereas Akaike's information-theoretic criterion is not. In this context we say that the criterion is consistent if it converges to the true model order as the sample size (i.e., the number of observations N) increases. For small sample sizes, however, the consistency of these two criteria is not as well defined; nevertheless, the results of computer simulation experiments appear to show that Akaike's information criterion may yield more accurate model-order estimates than the minimum description length criterion (Thomas, 1984).

2.11 THE INNOVATIONS PROCESS AND INNOVATIONS REPRESENTATION

We define the *innovations process* of a given discrete-time stochastic process, say, $\{u(n)\}$, as a *white noise* process, say, $\{v(n)\}$, that is related to $\{u(n)\}$ by a *causal and causally invertible transformation* (Kailath, 1968, 1970). The representation of the original process $\{u(n)\}$ by such a transformation is called the *innovations representation* or *canonical representation* of $\{u(n)\}$. The requirement that the transformation be causally invertible ensures that the given process $\{u(n)\}$ and its innovations process $\{v(n)\}$ are *probabilistically equivalent*. In particular, $\{v(n)\}$ and $\{u(n)\}$ contain the same "statistical information" since we can go back and forth in real time from one process to the other. Of course, $\{v(n)\}$ will generally be a simpler statistical process than $\{u(n)\}$. Moreover, since the values of $\{v(n)\}$ at different instants of time are *uncorrelated* with each other (because of the white noise assumption), we find that each observation $v(n)$ brings "new information," unlike the observation $u(n)$ that is, in general, *correlated* with past values of the original process; hence, the description of $\{v(n)\}$ as the innovations process.

In this section, we consider the innovations representation for the special case of a discrete-time stationary stochastic process $\{u(n)\}$ whose power spectral density

$S(\omega)$ is a *rational function* of $\exp(j\omega)$. This is the discrete-time version of the class of stochastic processes treated by Bode and Shannon (1950). The innovations representation of such a class of stochastic processes admits of a simple solution. Consider then a process $\{u(n)\}$ whose power spectral density can be expressed in the form of a product as follows:

$$S(\omega) = L(z)L(z^{-1})\Big|_{z=\exp(j\omega)} \qquad (2.125)$$

where $L(z)$ is a rational function in z^{-1} with all its poles and zeros confined to the interior of the unit circle in the z-plane. That is, $L(z)$ is minimum phase. Likewise, the reciprocal of $L(z)$, namely,

$$\Gamma(z) = \frac{1}{L(z)} \qquad (2.126)$$

is also minimum phase.

Suppose that the process $\{u(n)\}$ is applied to the input of a linear, causal, discrete-time filter with the transfer function $\Gamma(z)$, as in Fig. 2.14(a). Let $\{v(n)\}$ denote the resulting process developed at the output of the filter. The power spectral density of the process $\{v(n)\}$ equals

$$S_v(\omega) = |\Gamma(e^{j\omega})|^2 S(\omega)$$

$$= 1$$

This shows that the process $\{v(n)\}$ is white with unit variance. We therefore refer to the filter in Fig. 2.14(a) as a *whitening filter*. The process $\{v(n)\}$ represents the *innovations* of the original process $\{u(n)\}$. Let $\{\gamma_n\}$ denote the impulse response of the whitening filter. We may thus express the innovations process as the convolution of $\{\gamma_n\}$ and $\{u(n)\}$, as shown by

$$v(n) = \sum_{k=0}^{\infty} \gamma_k u(n - k) \qquad (2.127)$$

Given the innovations process $\{v(n)\}$, we may recover the original stochastic

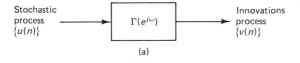

(a)

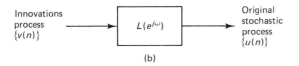

(b)

Figure 2.14 (a) Whitening filter; (b) inverse filter.

process $\{u(n)\}$ by applying $\{v(n)\}$ to a linear discrete-time filter of transfer function $L(z)$, as in Fig. 2.14(b). Let $\{l_n\}$ denote the impulse response of this filter. We may then express the original process as the convolution of $\{l_n\}$ and $\{v(n)\}$, as shown by

$$u(n) = \sum_{k=0}^{\infty} l_k v(n - k) \tag{2.128}$$

The filter in Fig. 2.14(b) is called the *inverse filter*.

The preceding discussion shows that a stochastic process $\{u(n)\}$ with rational power spectral density and its innovations process $\{v(n)\}$ are indeed probabilistically equivalent in the sense that either one of them may be expressed as the output of a linear, causal, discrete-time filter with the other process used as input.

Note that the AR process analyzer and generator shown in Fig. 2.4, represent special cases of the whitening and inverse filters, respectively.

For a real-valued process $\{u(n)\}$, the functions $L(z)$ and $L(z^{-1})$ must satisfy two conditions:

1. If z_i denotes a pole of $L(z)$, then $1/z_i$ is a zero of $L(z^{-1})$, and vice versa. This condition follows from the requirement that the power spectral density is an even function of the angular frequency, ω; that is, $S(-\omega) = S(\omega)$.
2. The poles and zeros of $L(z)$ or $L(z^{-1})$ are real or else they occur in complex-conjugate pairs. This second condition follows from the fact that the power spectral density is a real-valued function of ω.

The following example illustrates the meaning of these two conditions.

Example 1

Let

$$S(\omega) = \frac{8.5 - 4 \cos \omega}{5 - 4 \cos \omega}$$

denote the power spectral density of a discrete-time stochastic process $\{u(n)\}$. Since

$$\cos \omega = \tfrac{1}{2} (e^{j\omega} + e^{-j\omega})$$

we may rewrite the expression for the power spectral density, in terms of the variable z, as

$$L(z)L(z^{-1}) = \frac{8.5 - 2(z + z^{-1})}{5 - 2(z + z^{-1})}$$

$$= \frac{(z - 0.25)(z - 4)}{(z - 0.5)(z - 2)}$$

Assigning the poles and zeros of $L(z)$ $L(z^{-1})$ that lie inside the unit circle to $L(z)$

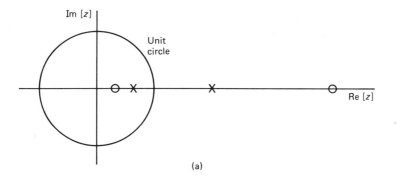

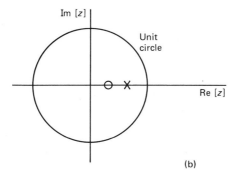

Figure 2.15   (a) Pole-zero pattern of $L(z)L(z^{-1})$; (b) pole-zero pattern of $L(z)$. The symbol $\times$ denotes a pole, and the symbol $\bigcirc$ denotes a zero.

and those that lie outside the unit circle to $L(z^{-1})$, we get (see Fig. 2.15)

$$L(z) = \frac{z - 0.25}{z - 0.5}$$

$$= \frac{1 - 0.25z^{-1}}{1 - 0.5z^{-1}}$$

and

$$L(z^{-1}) = \frac{z - 4}{z - 2}$$

The transfer function $L(z)$ defines the inverse filter.   The whitening filter has the transfer function

$$\Gamma(z) = \frac{1}{L(z)}$$

$$= \frac{1 - 0.5z^{-1}}{1 - 025z^{-1}}$$

The signal-flow graphs of the whitening filter and inverse filter are shown in Fig. 2.16.
Note that, in this example, the original process $\{u(n)\}$ represented by the power
spectral density is an ARMA process of order (1, 1).

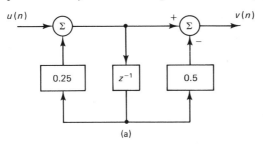

(a)

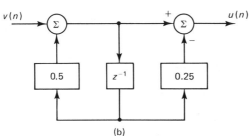

(b)

Figure 2.16 (a) Whitening filter; (b) inverse filter.

PROBLEMS

1. The sequences $\{y(n)\}$ and $\{u(n)\}$ are related by the difference equation

$$y(n) = u(n + a) - u(n - a)$$

where a is a constant. Evaluate the autocorrelation function of $\{y(n)\}$ in terms of that
of $\{u(n)\}$.

2. Show that, if a matrix $\mathbf{X}$ is positive definite, then the matrix $\mathbf{I}\mathrm{tr}[\mathbf{X}] - \mathbf{X}$ is positive
definite too, where tr $[\mathbf{X}]$ is the trace of $\mathbf{X}$ and $\mathbf{I}$ is the identity maxtrix of compatible
dimensions.

3. (a) Equation (2.20) relates the $(M + 1)$-by-$(M + 1)$ correlation matrix $\mathbf{R}_{M+1}$, pertaining
      to the observation vector $\mathbf{u}_{M+1}(n)$ taken from a stationary stochastic process, to the
      M-by-M correlation matrix $\mathbf{R}_M$ of the observation vector $\mathbf{u}_M(n)$ taken from the same
      process. Evaluate the inverse of the correlation matrix $\mathbf{R}_{M+1}$ in terms of the inverse
      of the correlation matrix $\mathbf{R}_M$.
   (b) Repeat your evaluation using Eq. (2.21).

4. Let $\mathbf{u}(n)$ denote an M-by-1 vector that is nonzero. Show that the M-by-M matrix $\mathbf{A}(n)$
   $= \mathbf{u}(n)\mathbf{u}^H(n)$ has rank 1. The rank of a matrix is defined as the number of linearly
   independent rows in the matrix.

5. Show that the eigenvalues of a triangular matrix equal the diagonal elements of the
   matrix.

6. The matrix $\mathbf{R}$ is said to be *normal* if $\mathbf{R}^H\mathbf{R} = \mathbf{R}\mathbf{R}^H$. Show that a triangular, normal matrix is diagonal.

7. Consider the Hermitian form $\mathbf{w}^H\mathbf{R}\mathbf{w}$, where $\mathbf{R}$ is an M-by-M correlation matrix and $\mathbf{w}$ is an M-by-1 vector. Let

$$\mathbf{R} = \mathbf{R}_r + j\mathbf{R}_i$$

where $\mathbf{R}_r$ and $\mathbf{R}_i$ are the real and imaginary parts of $\mathbf{R}$, respectively. Similarly, let $\mathbf{w}$ be expressed in terms of its real and imaginary parts as

$$\mathbf{w} = \mathbf{w}_r + j\mathbf{w}_i$$

Hence, show that the given Hermitian form may be expressed as

$$\mathbf{w}^H\mathbf{R}\mathbf{w} = \mathbf{w}_r^T\mathbf{R}_r\mathbf{w}_r + \mathbf{w}_i^T\mathbf{R}_r\mathbf{w}_i - 2\mathbf{w}_r^T\mathbf{R}_i\mathbf{w}_i$$

8. The discrete-time stochastic process $\{u(n)\}$ consists of K (uncorrelated) complex sinusoids plus additive white noise of zero mean and variance σ^2. The M-by-M correlation matrix of the process is denoted by $\mathbf{R}$. Show that the correlation matrix $\mathbf{R}$ may be expressed in the form

$$\mathbf{R} = \mathbf{S}\mathbf{D}\mathbf{S}^H + \sigma^2\mathbf{I}$$

where $\mathbf{S}$ is the M-by-K signal matrix determined by the angular frequencies of the sinusoids, $\mathbf{D}$ is a K-by-K diagonal matrix determined by their average power, and $\mathbf{I}$ is the M-by-M identity matrix.

9. For the discrete-time stochastic process $\{u(n)\}$ that was considered in Problem 8, specify the eigenvalues of the correlation matrix $\mathbf{R}$ of the process for the following three distinct conditions:

(a) The process $\{u(n)\}$ consists of K (uncorrelated) complex sinusoids of angular frequencies $\omega_1, \omega_2, \ldots, \omega_K$ and average powers $P_1, P_2, \ldots, P_K$, respectively.

(b) The process $\{u(n)\}$ consists of white noise of zero mean and variance σ^2.

(c) The process $\{u(n)\}$ consists of the uncorreleted sinusoids plus white noise.

How would you exploit these results to determine the number of complex sinusoids contained in the noisy process $\{u(n)\}$?

10. Let $\lambda_1, \lambda_2, \ldots, \lambda_M$ denote the eigenvalues of the correlation matrix of an observation vector $\mathbf{u}(n)$ taken from a stationary process of zero mean and variance σ_u^2. Show that

$$\sum_{i=1}^{M} \lambda_i = M\sigma_u^2$$

11. An M-by-M correlation matrix $\mathbf{R}$ is represented in terms of its eigenvalues $\lambda_1, \lambda_2, \ldots, \lambda_M$ and their associated eigenvectors $\mathbf{q}_1, \mathbf{q}_2, \ldots, \mathbf{q}_M$ as follows

$$\mathbf{R} = \sum_{i=1}^{M} \lambda_i \, \mathbf{q}_i\mathbf{q}_i^H$$

Show that the corresponding representation for the *square root* of matrix $\mathbf{R}$ is

$$\mathbf{R}^{1/2} = \sum_{i=1}^{M} \lambda_i^{1/2}\mathbf{q}_i\mathbf{q}_i^H$$

By definition, we have $\mathbf{R} = \mathbf{R}^{1/2}\,\mathbf{R}^{1/2}$

Using this result, describe a procedure for computing the square root of a square matrix.

12. Consider a stationary process $\{u(n)\}$ whose M-by-M correlation matrix equals $\mathbf{R}$. Show that the determinant of the correlation matrix $\mathbf{R}$ equals

$$\det(\mathbf{R}) = \prod_{i=1}^{M} \lambda_i$$

where $\lambda_1, \lambda_2, \ldots, \lambda_M$ are the eigenvalues of $\mathbf{R}$.

13. (a) Show that the product of two unitary matrices is also a unitary matrix
    (b) Show that the inverse of a unitary matrix is also a unitary matrix.

14. Show that the condition number of matrix $\mathbf{A}$ is unchanged when this matrix is multiplied by a unitary matrix of compatible dimensions.

15. Consider an L-by-M matrix $\mathbf{A}$. Show that the M-by-M matrix $\mathbf{A}^H\mathbf{A}$ and the L-by-L matrix $\mathbf{A}\mathbf{A}^H$ have the same nonzero eigenvalues.

16. A stochastic process $\{v(n)\}$ with a wide–band power spectrum is applied to a discrete-time linear filter whose amplitude response $|H(e^{j\omega})|$ is nonuniform. The maximum and minimum values of this response are denoted by $H_{\max}$ and $H_{\min}$, respectively. Let $\chi(\mathbf{R})$ denote the eigenvalue spread of the correlation matrix $\mathbf{R}$ of the stochastic process $\{u(n)\}$ produced at the output of the filter. Show that

$$\chi(\mathbf{R}) \simeq \left(\frac{H_{\max}}{H_{\min}}\right)^2$$

17. *Szegö's theorem* states that, if $g(.)$ is a continuous function, then

$$\lim_{m \to \infty} \frac{g(\lambda_1) + g(\lambda_2) + \cdots + g(\lambda_M)}{M} = \frac{1}{2\pi} \int_{-\pi}^{\pi} g[S(\omega)]\, d\omega$$

where $S(\omega)$ is the power spectral density of a stationary discrete-time stochastic process $\{u(n)\}$, and $\lambda_1, \lambda_2, \ldots, \lambda_M$ are the eigenvalues of the associated correlation matrix $\mathbf{R}$. It is assumed that the process $\{u(n)\}$ is limited to the interval $-\pi \leq \omega \leq \pi$. Using this theorem, show that

$$\lim_{m \to \infty} [\det(\mathbf{R})]^{1/M} = \exp\left(\frac{1}{2\pi} \int_{-\pi}^{\pi} \ln [S(\omega)]\, d\omega\right)$$

18. Consider a linear system of equations described by

$$\mathbf{R}\mathbf{w}_o = \mathbf{p}$$

where $\mathbf{R}$ is an M-by-M matrix, and $\mathbf{w}_o$ and $\mathbf{p}$ are M-by-1 vectors. The vector $\mathbf{w}_o$ represents the set of unknown parameters. Due to a combination of factors (e.g., measurement inaccuracies, computational errors), the matrix $\mathbf{R}$ is perturbed by a small amount $\delta\mathbf{R}$, producing a corresponding change $\delta\mathbf{w}$ in the vector of unknowns. Show that

$$\frac{\|\delta\mathbf{w}\|}{\|\mathbf{w}_o\|} \leq \chi(\mathbf{R})\frac{\|\delta\mathbf{R}\|}{\|\mathbf{R}\|}$$

where $\chi(\mathbf{R})$ is the condition number of $\mathbf{R}$, and $\|.\|$ denotes the norm of the quantity enclosed within. Comment on the practical significance of this result.

Hint: Use the inequality

$$\|\mathbf{Ax}\| \le \|\mathbf{A}\| \, \|\mathbf{x}\|$$

19. Consider a stationary Gaussian autoregressive (AR) process $\{u(n)\}$ of order M that is described by the real valued difference equation

$$u(n) + a_1 u(n-1) + \cdots + a_M u(n-M) = v(n)$$

where $\{v(n)\}$ is an independent, zero-mean, Gaussian noise process of variance σ_v^2, and $a_1, a_2, \ldots, a_M$ are the AR parameters. The observation vector $\mathbf{u}$, consisting of $u(1)$, $u(2), \ldots, u(N)$ as elements, is taken from the process $\{u(n)\}$. The joint probability density function of $\mathbf{u}$, given the AR parameter vector $\mathbf{a}$ (consisting of $a_0, a_1, \ldots, a_M$, with $a_0 = 1$) and white-noise variance σ_v^2, equals

$$f_{\mathbf{U}}(\mathbf{u}|\mathbf{a},\sigma_v^2) = \frac{1}{(2\pi)^{N/2}[\det (\mathbf{R})]^{1/2}} \exp\left(-\frac{1}{2}\mathbf{u}^T\mathbf{R}^{-1}\mathbf{u}\right)$$

where the correlation matrix $\mathbf{R} = E[\mathbf{uu}^T]$ and $\det (\mathbf{R})$ is its determinant.

By viewing $f(\mathbf{u}|\mathbf{a}, \sigma_v^2)$ as the likelihood function $l(\mathbf{a}, \sigma_v^2)$, the maximum-likelihood estimates of $\mathbf{a}$ and σ_v^2 are found by maximizing the likelihood function, or equivalently the log-likelihood function, with respect to $\mathbf{a}$ and σ_v^2. See Appendix A for details on the principle of maximum likelihood.

(a) Show that the autocorrelation function $r(k) = E[u(n)u(n-k)]$ may be expressed as

$$r(k) = \sigma_v^2 r_f(k)$$

where $r_f(k)$ is defined as the inverse z-transform of $H(z)H(1/z)$, where

$$H(z) = \frac{1}{1 + \displaystyle\sum_{i=1}^{M} a_i z^{-i}}$$

(b) Hence show that

$$f_{\mathbf{U}}(\mathbf{u}, \sigma_v^2) = \frac{1}{(2\pi)^{N/2}\sigma_v^N[\det (\mathbf{R}_f)]^{1/2}} \exp\left(-\frac{1}{2\sigma_v^2}\mathbf{u}^T\mathbf{R}_f^{-1}\mathbf{u}\right)$$

where $\mathbf{R}_f$ is an N-by-N correlation matrix whose jith element equals $r_f(i-j)$, with $i, j = 1, 2, \ldots, N$.

(c) By maximizing the log-likelihood function with respect to σ_v^2, show that the maximum-likelihood estimate of σ_v^2 equals

$$\sigma_{v,ml}^2 = \frac{1}{N}\mathbf{u}^T\mathbf{R}_f^{-1}\mathbf{u}$$

(d) For large data length N, assume that the term $[\det (\mathbf{R}_f^{-1})]^{1/N}$ is negligible; that is, it does not depend on $\mathbf{a}$. Using this approximation, show that the maximum-likelihood estimate of $\mathbf{a}$ is defined by

$$\mathbf{R}_f \mathbf{a}_{ml} = \begin{bmatrix} \sigma_{v,ml}^2 \\ 0 \end{bmatrix}$$

Hints: Evaluate the likelihood function $l(\mathbf{a})$ by substituting the maximum-likelihood estimate for σ_v^2 in $l(\mathbf{a}, \sigma_v^2)$. Hence, maximize this likelihood function, subject to the constraint that the first element a_0 in the vector $\mathbf{a}$ equals 1. You may use the *method of Lagrange multipliers* for the constrained optimization. See Appendix D for details on the method of Lagrange multipliers.

20. A first-order autoregressive (AR) process $\{u(n)\}$ that is real-valued satisfies the real-valued difference equation

$$u(n) + a_1 u(n - 1) = v(n)$$

where a_1 is a constant, and $\{v(n)\}$ is a white-noise process of variance σ_v^2. Such a process is also referred to as a *first-order Markov process*.

   (a) Show that if $\{v(n)\}$ has a nonzero mean the AR process $\{u(n)\}$ is nonstationary.

   (b) For the case when $\{v(n)\}$ has zero mean, and the constant a_1 satisfies the condition $|a_1| < 1$, show that the variance of $\{u(n)\}$ equals

$$\mathrm{Var}[u(n)] = \frac{\sigma_v^2}{1 - a_1^2}$$

   (c) For the conditions specified in part (b), find the autocorrelation function of the AR process $\{u(n)\}$. Sketch this autocorrelation function for the two cases $0 < a_1 < 1$ and $-1 < a_1 < 0$.

21. Consider an autoregressive process $\{u(n)\}$ of order 2, described by the difference equation

$$u(n) = u(n - 1) - 0.5u(n - 2) + v(n)$$

where $\{v(n)\}$ is a white-noise process of zero mean and variance 0.5.

   (a) Write the Yule–Walker equations for the process.

   (b) Solve these two equations for the autocorrelation function values $r(1)$ and $r(2)$.

   (c) Find the variance of $\{u(n)\}$.

22. A discrete-time stochastic process $\{x(n)\}$ that is real valued consists of an AR process $\{u(n)\}$ and additive white noise process $\{v_2(n)\}$. The AR component is described by the difference equation

$$u(n) + \sum_{k=1}^{M} a_k u(n - k) = v_1(n)$$

where $\{a_k\}$ are the set of AR parameters and $\{v_1(n)\}$ is a white noise process that is independent of $\{v_2(n)\}$. Show that $\{x(n)\}$ is an ARMA process described by

$$x(n) = -\sum_{k=1}^{M} a_k x(n - k) + \sum_{k=1}^{M} b_k e(n - k) + e(n)$$

where $\{e(n)\}$ is a white noise process. How are the MA parameters $\{b_k\}$ defined? How is the variance of $e(n)$ defined?

23. The power spectral density of stationary discrete-time stochastic process $\{u(n)\}$ is defined by

$$S(\omega) = \frac{25 - 24 \cos \omega}{17 - 8 \cos \omega}$$

   (a) Find the whitening filter and inverse filter for this process. Construct the signal-flow graphs of both filters.

(b) The whitening filter is to be approximated by an all-zero filter of order 3. Compute the coefficients of this approximation, and construct its signal-flow graph. Based on the approximation, find the corresponding (all-pole) inverse filter, and construct its signal-flow graph. Using the all-pole approximation for the inverse filter, determine the autoregressive process produced at its output in response to the (white) innovations process applied to the input.

(c) As a second approximation, determine the all-zero filter of order 3 for approximating the inverse filter computed in part (a). Based on this second approximation, find the corresponding (all-pole) whitening filter. Construct the signal-flow graphs of this new set of whitening and inverse filters. Determine the moving average process produced at the output of the inverse filter in response to the white (innovations) process applied to its input.

(d) Plot the power spectral density of the approximating autoregressive process in part (b) and that of the approximating moving average process in part (c), and compare them to that of the original process $\{u(n)\}$. Explain why the MA power spectrum approximates the given $S(\omega)$ more accurately than the AR power spectrum.

Computer-oriented Problems

24. Consider the second-order autoregressive (AR) process $\{u(n)\}$ described by

$$u(n) + a_1 u(n - 1) + a_2 u(n - 2) = v(n)$$

where a_1 and a_2 are real-valued constant coefficients, and $\{v(n)\}$ is a white-noise process of zero mean. The AR process $\{u(n)\}$ is normalized to have unit variance. In this problem the AR coefficients are varied in such a way that the characteristic equation of the process has complex conjugate roots. In particular, we wish to investigate the use of three sets of AR coefficients:

(i) $a_1 = -0.195$, $a_2 = 0.95$

(ii) $a_1 = -1.5955$, $a_2 = 0.95$

(iii) $a_1 = -1.9114$, $a_2 = 0.95$

For each set of parameters, perform the following computations:
(a) The eigenvalues of the correlation matrix of the AR process $\{u(n)\}$.
(b) The variance of the noise process $\{v(n)\}$.
(c) Using a random-noise generator for $\{v(n)\}$, plot the corresponding time variation of the AR process $\{u(n)\}$.
(d) Plot the autocorrelation function of the AR process $\{u(n)\}$ for varying lags.
(e) Plot the power spectrum of the AR process for varying frequency.

25. Repeat Problem 24 for the case when the coefficients are assigned each of the following three sets of values, in turn:

(i) $a_1 = -0.020$, $a_2 = -0.8$

(ii) $a_1 = -0.1636$, $a_2 = -0.8$

(iii) $a_1 = -0.1960$, $a_2 = -0.8$

In this case, the characteristic equation of the AR process has unequal roots.

3

Wiener Filter Theory

3.1 INTRODUCTION

As mentioned in Chapter 1, the *linear filtering problem*, which goes back to Gauss, was reintroduced and reformulated independently by Kolmogorov (1941) and Wiener (1942). Kolmogorov studied the discrete-time problem and solved it by using the *Wold decomposition* (Wold, 1938). Wiener, on the other hand, studied the continuous-time problem and derived the famous *Wiener–Hopf integral equation*, which requires a knowledge of correlation functions of the signal processes of interest. The solution of this equation is rather difficult for all but the simplest problems. The equivalent of this integral equation in the discrete-time case is known as the *normal equation*. The continuous-time solution to the Wiener–Hopf integral equation and the discrete-time solution to the normal equation are known collectively as *Wiener filters*.

In this chapter, we study the discrete-time formulation of Wiener filters for the general case of *complex-valued* time series, with the filter specified in terms of its impulse response. The reason for using complex-valued time series is that in many practical situations (e.g., communications, radar, sonar) the *baseband* signal of interest appears in complex form; the term *baseband* is used to designate the band of frequencies representing the original signal as delivered by a source of information.   The case of real-valued time series may of course be considered as a special case of this theory.   We limit the discussion, however, to linear filters with (1) an impulse response of *finite* duration, and (2) a single input and single output.

3.2 STATEMENT OF THE OPTIMUM FILTERING PROBLEM

Consider the linear transversal filter shown in block diagram form in Fig. 3.1.   The filter involves a combination of three basic operations: *storage*, *multiplication*, and *addition*, as described next:

1. The storage is represented by a cascade of $M - 1$ *one-sample delays*, with the block for each such unit labeled as z^{-1}.   We refer to the various points at which the one-sample delays are accessed as *tap points*.   The tap inputs are denoted by $u(n), u(n - 1), \ldots, u(n - M + 1)$.   Thus, with $u(n)$ viewed as the *current* value of the filter input, the remaining $M - 1$ tap inputs, $u(n - 1), \ldots, u(n - M + 1)$, represent *past* values of the input.

2. The *inner products* of tap inputs $u(n), u(n - 1), \ldots, u(n - M + 1)$ and *tap weights* $w_1, w_2, \ldots, w_M$ respectively, are formed by using a corresponding set of multipliers.   In particular, the multiplication involved in forming the inner product of $u(n)$ and w_1 is represented by a block labeled w_1^*, and so on for the other inner products.

3. The function of the adders is to sum the multiplier outputs to produce an overall output.

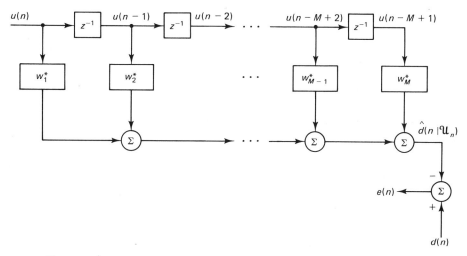

Figure 3.1   Transversal filter.

We refer to the number of delay elements in the filter as the *order* of the transversal filter. In the case of Fig. 3.1, the order equals $M - 1$. Note also that the impulse response of the transversal filter equals $\{h_k\} = \{w_k^*\}$, $k = 1, 2, \ldots, M$.

We assume that the discrete-time stochastic process represented by the tap inputs $u(n)$, $u(n - 1)$, $\ldots$, $u(n - M + 1)$ is weakly stationary. Also, without loss of generality, we assume that the mean value of the process is zero.

Denote the signal produced at the filter output at time n by $\hat{d}(n|\mathcal{U}_n)$. The intention is to use this filter output as an "estimate" of some *desired response* $d(n)$; hence, the use of a hat (ˆ) in the symbol representing the filter output. Also, the estimate assumes knowledge of the tap inputs $u(n)$, $u(n - 1)$, $\ldots$, $u(n - M + 1)$. The space spanned by these tap inputs is denoted by $\mathcal{U}_n$; hence, the inclusion of $\mathcal{U}_n$ in the symbol for the filter output. The filter output is related to the filter input and impulse response of the filter by the *convolution sum*

$$\hat{d}(n|\mathcal{U}_n) = \sum_{k=1}^{M} w_k^* u(n - k + 1) \tag{3.1}$$

The estimation problem is solved by designing the filter in Fig. 3.1 in such a way that the difference between $d(n)$, the sample value of the desired response at time n, and the actual filter output is made as "small" as possible in a statistical sense. The difference,

$$e(n) = d(n) - \hat{d}(n|\mathcal{U}_n) \tag{3.2}$$

is called the *estimation error*. In the Wiener theory, the *minimum mean-square error criterion* is used to optimize the filter. Specifically, the tap weights w_1, w_2, $\ldots$, w_M are chosen so as to minimize an *index of performance*, $J(\mathbf{w})$, defined as the mean-square value of the estimation error, or simply the *mean-squared error*:

$$J(\mathbf{w}) = E[e(n)e^*(n)] \tag{3.3}$$

where the asterisk denotes complex conjugation and $\mathbf{w}$ is the tap-weight vector represented by w_1, w_2, $\ldots$, w_M. The mean-squared error is a real and positive scalar. By minimizing $J(\mathbf{w})$, we obtain the best or *optimum* linear filter in the *minimum mean-square sense*.

3.3 ERROR-PERFORMANCE SURFACE

In developing the discrete-time version of the Wiener filter theory, it is convenient to use matrix notation. The composition of the *M-by-1 tap-weight vector* of the filter is shown by

$$\mathbf{w}^T = [w_1, w_2, \ldots, w_M] \tag{3.4}$$

and that of the *M*-by-1 *input vector* $\mathbf{u}(n)$ at time n is shown by

$$\mathbf{u}^T(n) = [u(n), u(n - 1), \ldots, u(n - M + 1)] \tag{3.5}$$

where the superscript T signifies *transposition*. Then, we may rewrite Eq. (3.1) in the form of an inner product of vectors as

$$\hat{d}(n|\mathcal{U}_n) = \mathbf{w}^H\mathbf{u}(n) \tag{3.6}$$

where the superscript H signifies *Hermitian transposition*. Equivalently, by taking the Hermitian transpose of both sides of Eq. (3.6), we may write

$$\hat{d}^*(n|\mathcal{U}_n) = \mathbf{u}^H(n)\mathbf{w} \tag{3.7}$$

Accordingly, we may express the estimation error as

$$e(n) = d(n) - \mathbf{w}^H\mathbf{u}(n) \tag{3.8}$$

or, in complex-conjugated form, as

$$e^*(n) = d^*(n) - \mathbf{u}^H(n)\mathbf{w} \tag{3.9}$$

Substituting Eqs. (3.8) and (3.9) in (3.3) for the mean-squared error, we get

$$J(\mathbf{w}) = E[(d(n) - \mathbf{w}^H\mathbf{u}(n))(d^*(n) - \mathbf{u}^H(n)\mathbf{w})] \tag{3.10}$$

Expanding the right side of Eq. (3.10), and recognizing that the tap-weight vector $\mathbf{w}$ is constant, we get

$$J(\mathbf{w}) = E[d(n)d^*(n)] - \mathbf{w}^H E[\mathbf{u}(n)d^*(n)] - E[d(n)\mathbf{u}^H(n)]\mathbf{w}$$
$$+ \mathbf{w}^H E[\mathbf{u}(n)\mathbf{u}^H(n)]\mathbf{w} \tag{3.11}$$

We assume that the tap-input vector $\mathbf{u}(n)$ and the desired response $d(n)$ are *jointly stationary*, so the four expectations in the right side of Eq. (3.11) may be interpreted as follows:

1. The expectation $E[d(n)d^*(n)]$ is equal to the variance of the desired response $d(n)$, assuming that $d(n)$ has zero mean. Denoting this variance by σ_d^2, we may write

$$\sigma_d^2 = E[d(n)d^*(n)] \tag{3.12}$$

2. The expectation $E[\mathbf{u}(n)d^*(n)]$ is equal to the M-by-1 *cross-correlation vector* between the tap-input vector $\mathbf{u}(n)$ and the desired response $d(n)$. Denoting this cross-correlation vector by $\mathbf{p}$, we may write

$$\mathbf{p} = E[\mathbf{u}(n)d^*(n)] \tag{3.13}$$

In expanded form, we have

$$\mathbf{p}^T = [p(0), p(-1), \ldots, p(1 - M)]$$

where

$$p(1 - k) = E[u(n - k + 1)d^*(n)], \qquad k = 1, 2, \ldots, M \tag{3.14}$$

3. The expectation $E[d(n)\mathbf{u}^H(n)]$ is equal to the Hermitian transpose of the cross-

correlation vector $\mathbf{p}$, as shown by

$$\mathbf{p}^H = E[d(n)\mathbf{u}^H(n)] \tag{3.15}$$

4. The expectation $E[\mathbf{u}(n)\mathbf{u}^H(n)]$ is equal to the M-by-M *correlation matrix* of the tap-input vector $\mathbf{u}(n)$. Denoting this correlation matrix by $\mathbf{R}$, we may write

$$\mathbf{R} = E[\mathbf{u}(n)\mathbf{u}^H(n)] \tag{3.16}$$

In expanded form, we have

$$\mathbf{R} = \begin{bmatrix} r(0) & r(1) & \cdots & r(M-1) \\ r(-1) & r(0) & \cdots & r(M-2) \\ \cdot & \cdot & \cdot & \cdot \\ \cdot & \cdot & \cdot & \cdot \\ \cdot & \cdot & \cdot & \cdot \\ r(-M+1) & r(-M+2) & \cdots & r(0) \end{bmatrix} \tag{3.17}$$

where the k, ith element with $k, i = 1, 2, \ldots, M$, is defined by

$$r(i - k) = E[u(n - k + 1)u^*(n - i + 1)] \tag{3.18}$$

The index k refers to the row number and the index i refers to the column number in the correlation matrix. Note that

$$r^*(i - k) = r(k - i) \tag{3.19}$$

Accordingly, we find that in the case of a stationary process the M-by-M correlation matrix $\mathbf{R}$ is uniquely defined by specifying the sequence of auto-correlation values of the input: $r(0), r(1), \ldots, r(M-1)$, which corresponds to lags of $0, 1, \ldots, M-1$, respectively.

Substituting the definitions of Eqs. (3.12), (3.13), (3.15), (3.16) in (3.11), we may rewrite the expression for the mean-squared error as

$$J(\mathbf{w}) = \sigma_d^2 - \mathbf{p}^H\mathbf{w} - \mathbf{w}^H\mathbf{p} + \mathbf{w}^H\mathbf{R}\mathbf{w} \tag{3.20}$$

Equation (3.20) states that, for the case when the tap-input vector $\mathbf{u}(n)$ and the desired response $d(n)$ are jointly stationary, the mean-squared error $J(\mathbf{w})$ is precisely a second-order function of the tap-weight vector $\mathbf{w}$. Accordingly, we may visualize the dependence of the mean-squared error $J(\mathbf{w})$ on the elements of the vector $\mathbf{w}$, that is, the tap weights $w_1, w_2, \ldots, w_M$, as a bowl-shaped surface with a unique minimum. We refer to this surface as the *error-performance surface* of the transversal filter.

The requirement is to design the filter so that it operates at the *bottom* or *minimum point* of the error-performance surface. At this point, the mean-squared error $J(\mathbf{w})$ attains its *minimum value*, denoted by $J_{\min}$ and, correspondingly, the tap-weight vector **w** attains its *optimum value*, dentoed by $\mathbf{w}_o$. The resultant transversal filter is said to be *optimum in the minimum mean-squared sense*.

To determine the optimum tap-weight vector $\mathbf{w}_o$, we first differentiate the mean-squared error $J(\mathbf{w})$ in Eq. (3.20) with respect to the tap-weight vector **w** and set the result equal to zero. The solution of this equation yields $\mathbf{w}_o$. To proceed with this evaluation, however, we need to know how to differentiate a scalar-valued function with respect to a complex-valued vector. This issue is discussed in the next section.

3.4 *DIFFERENTIATION WITH RESPECT TO A VECTOR*

Let g be a scalar-valued function of an M-by-1 vector **w**. Let the element w_k of the vector **w** be written as

$$w_k = a_k + jb_k \tag{3.21}$$

where a_k is the real part of w_k, and b_k is the imaginary part. Then g may be considered to be a function of the $2M$ variables a_k, b_k with $k = 1, 2, \ldots, M$. We assume that g is a differentiable function of these $2M$ variables. Then we define the derivative of g with respect to the vector **w** as the M-by-1 vector

$$\frac{dg}{d\mathbf{w}} = \begin{bmatrix} \dfrac{\partial g}{\partial a_1} + j\,\dfrac{\partial g}{\partial b_1} \\[2mm] \dfrac{\partial g}{\partial a_2} + j\,\dfrac{\partial g}{\partial b_2} \\[2mm] \cdot \\ \cdot \\ \cdot \\ \dfrac{\partial g}{\partial a_M} + j\,\dfrac{\partial g}{\partial b_M} \end{bmatrix} \tag{3.22}$$

We illustrate the application of the formula (3.22) by considering three examples that are of interest to us in the context of minimizing the mean-squared error $J(\mathbf{w})$.

Example 1

Let the scalar

$$g = \mathbf{c}^H\mathbf{w}$$

where both **c** and **w** are M-by-1 vectors. We write g in the expanded form

$$\begin{aligned} g &= \sum_{k=1}^{M} c_k^* w_k \\[2mm] &= \sum_{k=1}^{M} c_k^*(a_k + jb_k) \end{aligned} \tag{3.23}$$

Hence,

$$\frac{\partial g}{\partial a_k} = c_k^*, \quad k = 1, 2, \ldots, M \tag{3.24}$$

and

$$\frac{\partial g}{\partial b_k} = jc_k^*, \quad k = 1, 2, \ldots, M \tag{3.25}$$

Substituting Eqs. (3.24) and (3.25) into (3.22) and simplifying, we get the final result:

$$\frac{d}{d\mathbf{w}} (\mathbf{c}^H \mathbf{w}) = \mathbf{0} \tag{3.26}$$

where $\mathbf{0}$ is the M-by-1 null vector.

Example 2

Let the scalar

$$g = \mathbf{w}^H \mathbf{c}$$

where, as before, both $\mathbf{w}$ and $\mathbf{c}$ are M-by-1 vectors. In this case, we write

$$\begin{aligned} g &= \sum_{k=1}^{M} c_k w_k^* \\ &= \sum_{k=1}^{M} c_k (a_k - jb_k) \end{aligned} \tag{3.27}$$

Hence,

$$\frac{\partial g}{\partial a_k} = c_k \quad k = 1, 2, \ldots, M \tag{3.28}$$

and

$$\frac{\partial g}{\partial b_k} = -jc_k, \quad k = 1, 2, \ldots, M \tag{3.29}$$

Substituting Eqs. (3.28) and (3.29) into (3.22) and simplifying, we get the final result:

$$\frac{d}{d\mathbf{w}} (\mathbf{w}^H \mathbf{c}) = 2\mathbf{c} \tag{3.30}$$

Example 3

Let the scalar

$$g = \mathbf{w}^H \mathbf{Q} \mathbf{w}$$

where $\mathbf{w}$ is an M-by-1 vector and $\mathbf{Q}$ is an M-by-M matrix. Let

$$\mathbf{c}_1 = \mathbf{Q}^H \mathbf{w}$$

or, equivalently,

$$\mathbf{c}_1^H = \mathbf{w}^H \mathbf{Q}$$

Then we may put

$$g = \mathbf{c}_1^H \mathbf{w}$$

Treating $\mathbf{c}_1$ as a constant vector and differentiating g with respect to $\mathbf{w}$, we get (using the result of Example 1)

$$\frac{d}{d\mathbf{w}} (\mathbf{c}_1^H \mathbf{w}) = \mathbf{0}, \qquad \mathbf{c}_1 = \text{constant} \tag{3.31}$$

Next we put

$$\mathbf{c}_2 = \mathbf{Q}\mathbf{w}$$

and so write

$$g = \mathbf{w}^H \mathbf{c}_2$$

Then, treating $\mathbf{c}_2$ as a constant vector and differentiating g with respect to $\mathbf{w}$, we get (using the result of Example 2)

$$\frac{d}{d\mathbf{w}} (\mathbf{w}^H \mathbf{c}_2) = 2\mathbf{c}_2, \qquad \mathbf{c}_2 = \text{constant} \tag{3.32}$$

Summing the two contributions defined by Eqs. (3.31) and (3.32), we get the final result:

$$\frac{d}{d\mathbf{w}} (\mathbf{w}^H \mathbf{Q} \mathbf{w}) = 2\mathbf{Q}\mathbf{w} \tag{3.33}$$

3.5 THE NORMAL EQUATION

Having equipped ourselves with the rule for differentiating a scalar-valued function with respect to a vector, let us now return to Eq. (3.20), which defines the mean-squared error $J(\mathbf{w})$ as a function of the tap-weight vector $\mathbf{w}$.

We recognize that the variance σ_d^2 is a constant and, therefore, its derivative with respect to the tap-weight vector $\mathbf{w}$ is zero. Using the results of Eqs. (3.26), (3.30), and (3.33), we find that the derivatives of the remaining three terms on the right side of Eq. (3.20), with respect to $\mathbf{w}$, have the following values:

$$\frac{d}{d\mathbf{w}} (\mathbf{p}^H \mathbf{w}) = \mathbf{0}$$

$$\frac{d}{d\mathbf{w}} (\mathbf{w}^H \mathbf{p}) = 2\mathbf{p}$$

and

$$\frac{d}{d\mathbf{w}} (\mathbf{w}^H \mathbf{R} \mathbf{w}) = 2\mathbf{R}\mathbf{w}$$

Hence, we get the following expression for the *gradient vector*, ∇, defined as the derivative of the mean-squared error $J(\mathbf{w})$ with respect to the tap-weight vector $\mathbf{w}$:

$$\nabla = \frac{dJ(\mathbf{w})}{d\mathbf{w}}$$

$$= -2\mathbf{p} + 2\mathbf{R}\mathbf{w} \tag{3.34}$$

Let $\mathbf{w}_o$ denote the value of the *optimum tap-weight vector* for which the gradient vector ∇ equals a null vector. Hence, from Eq. (3.34) we get the result

$$\mathbf{R}\mathbf{w}_o = \mathbf{p} \tag{3.35}$$

Equation (3.35) is the discrete form of the Wiener–Hopf equation. It is also called the *normal equation*; the reason for this second name will be explained in the next section. Its solution yields the optimum tap-weight vector $\mathbf{w}_o$. To determine this solution, we premultiply both sides of Eq. (3.35) by $\mathbf{R}^{-1}$, the *inverse* of the correlation matrix, obtaining

$$\mathbf{w}_o = \mathbf{R}^{-1}\mathbf{p} \tag{3.36}$$

The computation of the optimum tap-weight vector $\mathbf{w}_o$ requires knowledge of two quantities: (1) the correlation matrix $\mathbf{R}$ of the tap-input vector $\mathbf{u}(n)$, and (2) the cross-correlation vector $\mathbf{p}$ between the the tap-input $\mathbf{u}(n)$ and the desired response $d(n)$.

It is instructive to express the normal equation in its expanded form. In particular, by substituting Eqs. (3.4), (3.14), and (3.17) in (3.35), we get

$$\begin{bmatrix} r(0) & r(1) & \cdots & r(M-1) \\ r(-1) & r(0) & \cdots & r(M-2) \\ \cdot & & & \cdot \\ \cdot & \cdot & \cdot & \cdot \\ \cdot & \cdot & \cdot & \cdot \\ r(-M+1) & r(-M+2) & \cdots & r(0) \end{bmatrix} \begin{bmatrix} w_{o1} \\ w_{o2} \\ \cdot \\ \cdot \\ \cdot \\ w_{oM} \end{bmatrix}$$

$$= \begin{bmatrix} p(0) \\ p(-1) \\ \cdot \\ \cdot \\ \cdot \\ p(1-M) \end{bmatrix} \tag{3.37}$$

We may rewrite this matrix relation in the equivalent form:

$$\sum_{i=1}^{M} w_{oi}\, r(i-k) = p(1-k), \qquad k = 1, 2, \ldots, M \tag{3.38}$$

3.6 *PRINCIPLE OF ORTHOGONALITY*

The normal equation (3.35) defines the tap-weight vector $\mathbf{w}_o$ of a transversal filter that is optimum in the minimum-mean-square sense. We may rewrite this equation by using the definitions given in Eqs. (3.13) and (3.16) for the cross-correlation vector $\mathbf{p}$ and correlation matrix $\mathbf{R}$, as follows:

$$E[\mathbf{u}(n)\mathbf{u}^H(n)]\mathbf{w}_o = E[\mathbf{u}(n)d^*(n)] \tag{3.39}$$

We note that, for a given set of parameters, the optimum tap-weight vector $\mathbf{w}_o$ is a constant. Hence, we may move it inside the expectation operator on the left side of Eq. (3.39) without affecting the result. Doing this, and then combining the resultant term with that on the right side of Eq. (3.39), we may rewrite this equation in the form

$$E[\mathbf{u}(n)(d^*(n) - \mathbf{u}^H(n)\,\mathbf{w}_o)] = \mathbf{0} \tag{3.40}$$

where $\mathbf{0}$ is the M-by-1 null vector. We now recognize that [see Eq. (3.9)]

$$e_o^*(n) = d^*(n) - \mathbf{u}^H(n)\mathbf{w}_o$$

where $e_o(n)$ is the estimation error resulting from use of the optimum filter. Accordingly, we may simplify Eq. (3.40) as

$$E[\mathbf{u}(n)e_o^*(n)] = \mathbf{0} \tag{3.41}$$

Equation (3.41) states that, when the transversal filter operates in its optimum condition, each element of the input vector $\mathbf{u}(n)$ and the estimation errors $e_o(n)$ are *orthogonal*. This result is known as the *principle of orthogonality*. Hence, we conclude that the two criteria "minimum mean-squared error" and "orthogonality between input and error" yield identical results.

Another useful property of the optimum filter is obtained by premultiplying both sides of Eq. (3.41) by the Hermitian transpose of the optimum tap-weight vector, obtaining

$$\mathbf{w}_o^H \, E[\mathbf{u}(n)e_o^*(n)] = 0 \tag{3.42}$$

Here, again, we may move $\mathbf{w}_o^H$ inside the expectation operator, and so write

$$E[\mathbf{w}_o^H \, \mathbf{u}(n)e_o^*(n)] = 0 \tag{3.43}$$

We note that the output of the optimum filter, in response to the input vector $\mathbf{u}(n)$, equals

$$\hat{d}(n|\mathcal{U}_n) = \mathbf{w}_o^H\mathbf{u}(n) \tag{3.44}$$

Hence, the inner product $\mathbf{w}_o^H\mathbf{u}(n)$ in Eq. (3.43) equals the optimum filter output. We may thus rewrite this equation as

$$E[\hat{d}(n|\mathcal{U}_n)e_o^*(n)] = 0 \tag{3.45}$$

which shows that, when the filter operates in its optimum condition, the estimate $\hat{d}(n|\mathcal{U}_n)$ at the filter output and the estimation error $e_o(n)$ are also orthogonal.

Equation (3.45) offers an interesting geometric interpretation of the conditions that exist at the output of the optimum filter, as illustrated in Fig. 3.2. In this figure, the desired response, the filter output, and the corresponding estimation error are represented by vectors labeled **d**, **d̂** and **e**$_o$ respectively. We see that for the optimum filter the vector representing the estimation error is *normal* to the vector representing the filter output. It is for this reason that Eq. (3.35), defining the optimum filter, is called the *normal equation*.

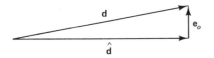

Figure 3.2 Relationship between the desired response $d(n)$, the estimate $\hat{d}(n|\mathcal{U}_n)$ at the filter output, and the error signal $e_o(n)$.

3.7 MINIMUM MEAN-SQUARED ERROR

Equation (3.20) defines the mean-squared error for any tap-weight vector **w**. When the tap-weight vector equals its optimum value $\mathbf{w}_o$, the mean-squared error attains its minimum value, $J_{\min}$, defined by

$$J_{\min} = \sigma_d^2 - \mathbf{p}^H\mathbf{w}_o - \mathbf{w}_o^H\mathbf{p} + \mathbf{w}_o^H\mathbf{R}\mathbf{w}_o$$

Since we have $\mathbf{R}\mathbf{w}_o = \mathbf{p}$, this expression simplifies as

$$\begin{align} J_{\min} &= \sigma_d^2 - \mathbf{p}^H\mathbf{w}_o \\ &= \sigma_d^2 - \mathbf{p}^H\mathbf{R}^{-1}\mathbf{p} \end{align} \tag{3.46}$$

Equivalently, we may write

$$J_{\min} = \sigma_d^2 - \sum_{k=1}^{M} w_{ok}p^*(1-k) \tag{3.47}$$

Another useful expression for the minimum mean-squared error, $J_{\min}$, may be derived by recognizing that

$$d(n) = \hat{d}(n|\mathcal{U}_n) + e_o(n) \tag{3.48}$$

For the optimum filter, the filter output $\hat{d}(n|\mathcal{U}_n)$ and estimated error $e_0(n)$ are orthogonal as in Eq. (3.45). Therefore, evaluating the mean-square values of both sides of Eq. (3.48) and using this orthogonality condition, we get

$$\sigma_d^2 = \sigma_{\hat{d}}^2 + J_{\min} \tag{3.49}$$

where σ_d^2 is the variance of the desired response $d(n)$, and $\sigma_{\hat{d}}^2$ is the variance of the estimate $\hat{d}(n|\mathcal{U}_n)$. This assumes that the desired response $d(n)$ and the tap-input vector $\mathbf{u}(n)$ have zero mean. Solving Eq. (3.49) for the minimum mean-squared error, we get

$$J_{\min} = \sigma_d^2 - \sigma_{\hat{d}}^2 \quad \geqslant 0 \tag{3.50}$$

This relation shows that for the optimum filter the minimum mean-squared error equals the difference between the variance of the desired response and the variance of the estimate it produces at the filter output.

It is convenient to normalize the expression in Eq. (3.50) in such a way that the minimum value of the mean-squared error always lies between zero and one. We may do this by dividing both sides of Eq. (3.50) by σ_d^2, obtaining

$$\frac{J_{\min}}{\sigma_d^2} = 1 - \frac{\sigma_{\hat{d}}^2}{\sigma_d^2} \qquad \sigma_d^2 \geqslant \sigma_{\hat{d}}^2 \tag{3.51}$$

Clearly, this is possible because σ_d^2 is never zero, except in the trivial case of a desired response $d(n)$ that is zero for all n. Let

$$\varepsilon = \frac{J_{\min}}{\sigma_d^2} \tag{3.52}$$

The quantity ε is called the *normalized mean-squared error*, in terms of which we may rewrite Eq. (3.51) in the form

$$\varepsilon = 1 - \frac{\sigma_{\hat{d}}^2}{\sigma_d^2} \tag{3.53}$$

We note that (1) the ratio ε can never be negative, and (2) the ratio $\sigma_d^2/\sigma_{\hat{d}}^2$ is always positive. We therefore have

$$0 \leq \varepsilon \leq 1 \tag{3.54}$$

If ε is zero, the optimum filter operates perfectly in the sense that there is complete agreement between the estimate $\hat{d}(n|\mathcal{U}_n)$ at the filter output and the desired response $d(n)$. On the other hand, if ε is unity, there is no agreement whatsoever between these two quantities; this corresponds to the worst possible situation.

For a given desired response $d(n)$, the filter performance always improves (i.e., ε decreases) as the number of taps, M, in the optimum transversal filter increases.

3.8 NUMERICAL EXAMPLE

To illustrate the filtering theory deveoped in the preceding, we consider the example depicted in Fig. 3.3. The desired response $\{d(n)\}$ is modeled as an AR process of

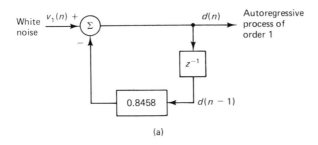

(a)

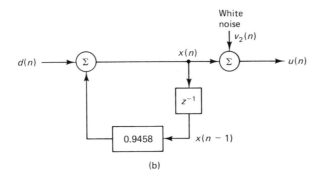

(b)

Figure 3.3 (a) Autoregressive model of desired response $\{d(n)\}$; (b) model of noisy communication channel.

order 1; that is, it may be produced by applying a white-noise process $\{v_1(n)\}$ of zero mean and variance $\sigma_1^2 = 0.27$ to the input of an all-pole filter of order 1, whose transfer function equals [see Fig. 3.3(a)]

$$H_1(z) = \frac{1}{1 + 0.8458z^{-1}}$$

The process $\{d(n)\}$ is applied to a communication channel modeled by the all-pole transfer function

$$H_2(z) = \frac{1}{1 - 0.9458z^{-1}}$$

The channel output $\{x(n)\}$ is corrupted by an additive white-noise process $\{v_2(n)\}$ of zero mean and variance $\sigma_2^2 = 0.1$, so a sample of the received signal $\{u(n)\}$ equals [see Fig. 3.3(b)]

$$u(n) = x(n) + v_2(n) \tag{3.55}$$

The white-noise processes $\{v_1(n)\}$ and $\{v_2(n)\}$ are uncorrelated. It is also assumed that $d(n)$ and $u(n)$, and therefore $v_1(n)$ and $v_2(n)$, are all real valued.

The requirement is to specify a Wiener filter consisting of a transversal filter with two taps, which operates on the received signal $\{u(n)\}$ so as to produce an estimate of the desired response that is optimum in the mean-square sense.

Statistical Characterization of the Desired Response {d(n)} and the Received Signal {u(n)}

We begin the analysis by considering the difference equations that characterize the various processes described by the models of Fig. 3.3. First, the generation of the desired response $\{d(n)\}$ is governed by the first-order difference equation

$$d(n) + a_1 d(n - 1) = v_1(n) \qquad (3.56)$$

where $a = 0.8458$. The variance of the process $\{d(n)\}$ equals (see Problem 2.20)

$$\sigma_d^2 = \frac{\sigma_1^2}{1 - a_1^2}$$

$$= \frac{0.27}{1 - (0.8458)^2} \qquad (3.57)$$

$$= 0.9486$$

The process $\{d(n)\}$ acts as input to the channel. Hence, from Fig. 3.3(b), we find that the channel output $\{x(n)\}$ is related to the channel input $\{d(n)\}$ by the first-order difference equation

$$x(n) + b_1 x(n - 1) = d(n) \qquad (3.58)$$

where $b_1 = -0.9458$. We also observe from the two parts of Fig. 3.3 that the channel output $\{x(n)\}$ may be generated by applying the white-noise process $\{v_1(n)\}$ to a second-order all-pole filter whose transfer function equals

$$H(z) = H_1(z)H_2(z)$$

$$= \frac{1}{(1 + 0.8458z^{-1})(1 - 0.9458z^{-1})} \qquad (3.59)$$

Accordingly, $\{x(n)\}$ is a second-order AR process described by the difference equation

$$x(n) + a_1 x(n - 1) + a_2 x(n - 2) = v(n) \qquad (3.60)$$

where $a_1 = -0.1$ and $a_2 = -0.8$. Note that both AR processes $\{d(n)\}$ and $\{x(n)\}$ are stationary.

To characterize the Wiener filter, we need to solve the normal equation (3.35). This equation requires knowledge of two quantities: (1) the correlation matrix **R** pertaining to the received signal $\{u(n)\}$, and (2) the cross-correlation vector **p** between $\{u(n)\}$ and the desired response $\{d(n)\}$. In our example, **R** is a 2-by-2 matrix and **p** is a 2-by-1 vector, since the transversal filter used to implement the Wiener filter is assumed to have two taps.

The received signal $\{u(n)\}$ consists of the channel output $\{x(n)\}$ plus the additive white noise $\{v_2(n)\}$. Since the processes $\{x(n)\}$ and $\{v_2(n)\}$ are uncorrelated, it follows that the correlation matrix $\mathbf{R}$ equals the correlation matrix of $\{x(n)\}$ plus the correlation matrix of $\{v_2(n)\}$. That is,

$$\mathbf{R} = \mathbf{R}_x + \mathbf{R}_2 \tag{3.61}$$

For the correlation matrix $\mathbf{R}_x$, we write [since the process $\{x(n)\}$ is real valued]

$$\mathbf{R}_x = \begin{bmatrix} r_x(0) & r_x(1) \\ r_x(1) & r_x(0) \end{bmatrix}$$

where $r_x(0)$ and $r_x(1)$ are the autocorrelation functions of the received signal $\{x(n)\}$ for lags of 0 and 1, respectively. From Section 2.8, we have

$$r_x(0) = \sigma_x^2$$

$$= \left(\frac{1 + a_2}{1 - a_2}\right) \frac{\sigma_1^2}{[(1 + a_2)^2 - a_1^2]}$$

$$= \left(\frac{1 - 0.8}{1 + 0.8}\right) \frac{0.27}{[(1 - 0.8)^2 - (0.1)^2]}$$

$$= 1$$

$$r_x(1) = \frac{-a_1}{1 + a_2}$$

$$= \frac{0.1}{1 - 0.8}$$

$$= 0.5$$

Hence,

$$\mathbf{R}_x = \begin{bmatrix} r_x(0) & r_x(1) \\ r_x(1) & r_x(0) \end{bmatrix}$$

$$= \begin{bmatrix} 1 & 0.5 \\ 0.5 & 1 \end{bmatrix} \tag{3.62}$$

Next we observe that since $\{v_2(n)\}$ is a white-noise process of zero mean and variance $\sigma_2^2 = 0.1$, the 2-by-2 correlation matrix $\mathbf{R}_2$ of this process equals

$$\mathbf{R}_2 = \begin{bmatrix} 0.1 & 0 \\ 0 & 0.1 \end{bmatrix} \tag{3.63}$$

Thus, substituting Eqs. (3.62) and (3.63) in (3.61), we find that the 2-by-2 correlation matrix of the received signal $\{x(n)\}$ equals

$$\mathbf{R} = \begin{bmatrix} 1.1 & 0.5 \\ 0.5 & 1.1 \end{bmatrix} \tag{3.64}$$

For the 2-by-1 cross-correlation vector $\mathbf{p}$, we write

$$\mathbf{p} = \begin{bmatrix} p(0) \\ p(1) \end{bmatrix}$$

where $p(0)$ and $p(1)$ are the cross-correlation functions between $\{d(n)\}$ and $\{u(n)\}$ for lags of 0 and 1, respectively. Since these two processes are real valued, we have $p(k - 1) = p(1 - k)$, where

$$p(k - 1) = E[d(n)u(n - k + 1)], \qquad k = 1, 2 \tag{3.65}$$

Substituting Eqs. (3.55) and (3.58) into (3.65), and recognizing that the channel output $\{x(n)\}$ is uncorrelated with the white-noise process $\{v_2(n)\}$, we get

$$p(k - 1) = r_x(k - 1) + b_1 r_x(k - 2), \qquad k = 1, 2$$

Putting $b_1 = -0.9458$ and using the element values for the correlation matrix $\mathbf{R}_x$ given in Eq. (3.62), we obtain

$$p(0) = r_x(0) + b_1 r_x(-1)$$
$$= 1 - 0.9458 \times 0.5$$
$$= 0.5272$$
$$p(1) = r_x(1) + b_1 r_x(0)$$
$$= 0.5 - 0.9458 \times 1$$
$$= -0.4458$$

Hence,

$$\mathbf{p} = \begin{bmatrix} 0.5272 \\ -0.4458 \end{bmatrix} \tag{3.66}$$

Error-performance Surface

The dependence of the mean-squared error on the 2-by-1 tap-weight vector $\mathbf{w}$ is defined by Eq. (3.20). Hence, substituting Eqs. (3.57), (3.64), and (3.66) into (3.20), we get

$$J(w_1, w_2) = 0.9486 - 2[0.5272, -0.4458] \begin{bmatrix} w_1 \\ w_2 \end{bmatrix}$$

$$+ [w_1, w_2] \begin{bmatrix} 1.1 & 0.5 \\ 0.5 & 1.1 \end{bmatrix} \begin{bmatrix} w_1 \\ w_2 \end{bmatrix}$$

$$= 0.9486 - 1.0544w_1 + 0.8916w_2 + w_1 w_2 + 1.1(w_1^2 + w_2^2)$$

Using a three-dimensional computer plot, the mean-squared error $J(w_1, w_2)$ is plotted versus the tap weights w_1 and w_2.   The result is shown in Fig. 3.4.   In this figure, the error performance is viewed obliquely at an angle of $25°$.

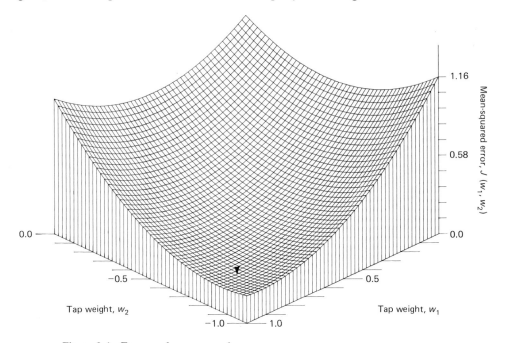

Figure 3.4   Error-performance surface.

The Wiener Filter

The 2-by-1 optimum tap-weight vector $\mathbf{w}_o$ of the Weiner filter is defined by Eq. (3.36).   In particular, it consists of the inverse matrix $\mathbf{R}^{-1}$ multipled by the cross-correlation vector $\mathbf{p}$.   Inverting the correlation matrix $\mathbf{R}$ of Eq. (3.64), we get

$$\mathbf{R}^{-1} = \begin{bmatrix} r(0) & r(1) \\ r(1) & r(0) \end{bmatrix}^{-1}$$

$$= \frac{1}{r^2(0) - r^2(1)} \begin{bmatrix} r(0) & -r(1) \\ -r(1) & r(0) \end{bmatrix} \qquad (3.67)$$

$$= \begin{bmatrix} 1.1456 & -0.5208 \\ -0.5208 & 1.1456 \end{bmatrix}$$

Hence, substituting Eqs. (3.66) and (3.67) into (3.36), we get the desired result:

$$\mathbf{w}_o = \begin{bmatrix} 1.1456 & -0.5208 \\ -0.5208 & 1.1456 \end{bmatrix} \begin{bmatrix} 0.5272 \\ -0.4458 \end{bmatrix}$$

$$= \begin{bmatrix} 0.8360 \\ -0.7853 \end{bmatrix}$$

Minimum Mean-squared Error

There only remains for us to evaluate the minimum value of the mean-squared error, $J_{\min}$, which results from the use of the optimum tap-weight vector $\mathbf{w}_o$. This minimum value is defined by Eq. (3.46). Hence, substituting Eqs. (3.57), (3.66), and (3.68) into the first line of (3.46), we get

$$J_{\min} = 0.9486 - [0.5272, -0.4458] \begin{bmatrix} 0.8360 \\ -0.7853 \end{bmatrix} \qquad (3.69)$$

$$= 0.1579$$

The point represented jointly by the optimum tap-weight vector $\mathbf{w}_o$ of Eq. (3.68) and the minimum mean-squared error of Eq. (3.69) defines the bottom of the error-performance surface. The location of this minimum point is indicated by an arrow in the three-dimensional computer plot of Fig. 3.4.

3.9 CANONICAL FORM OF ERROR-PERFORMANCE SURFACE

Subtracting the first line of Eq. (3.46) from (3.20) and then using the optimum solution of Eq. (3.35) to eliminate the cross-correlation vector $\mathbf{p}$ from the result of this subtraction, we find that the mean-squared error may be expressed in the quadratic form (see Problem 2)

$$J(\mathbf{w}) = J_{\min} + (\mathbf{w} - \mathbf{w}_o)^H \mathbf{R}(\mathbf{w} - \mathbf{w}_o) \qquad (3.70)$$

This equation shows explicitly the unique optimality of the minimizing tap-weight vector $\mathbf{w}_o$.

Although the quadratic form on the right side of Eq. (3.70) is quite informative, nevertheless, it is desirable to change the basis on which it is defined so that the representation of the error-performance surface is considerably simplified. To do this, we recall from Chapter 2 that the correlation matrix $\mathbf{R}$ of the tap-input vector may be expressed in terms of eigenvalues and eigenvectors as follows:

$$\mathbf{R} = \mathbf{Q}\mathbf{\Lambda}\mathbf{Q}^H \qquad (3.71)$$

where $\mathbf{\Lambda}$ is a diagonal matrix consisting of the eigenvalues $\lambda_1, \lambda_2, \ldots, \lambda_M$ of the correlation matrix, and the matrix $\mathbf{Q}$ has for its columns the eigenvectors $\mathbf{q}_1, \mathbf{q}_2, \ldots, \mathbf{q}_M$ associated with these eigenvalues, respectively. Hence, substituting Eq. (3.71) into (3.70), we get

$$J(\mathbf{w}) = J_{\min} + (\mathbf{w} - \mathbf{w}_o)^H \mathbf{Q}\mathbf{\Lambda}\mathbf{Q}^H(\mathbf{w} - \mathbf{w}_o) \qquad (3.72)$$

Define a *transformed* version of the difference between the tap-weight vector $\mathbf{w}$ and the optimum solution $\mathbf{w}_o$ as

$$\boldsymbol{v} = \mathbf{Q}^H(\mathbf{w} - \mathbf{w}_o) \qquad (3.73)$$

Then we may put the quadratic form of Eq. (3.72) into its *canonical form* defined by

$$J(\boldsymbol{v}) = J_{\min} + \boldsymbol{v}^H \boldsymbol{\Lambda} \boldsymbol{v} \tag{3.74}$$

This new formulation of the mean-squared error contains no cross-product terms, as shown by

$$J(\boldsymbol{v}) = J_{\min} + \sum_{k=1}^{M} \lambda_k v_k v_k^*$$

$$= J_{\min} + \sum_{k=1}^{M} \lambda_k |v_k|^2 \tag{3.75}$$

where v_k is the kth component of the vector $\boldsymbol{v}$. The feature that makes the canonical form of Eq. (3.75) a rather useful representation of the error-performance surface is the fact that the components of the transformed coefficient vector $\boldsymbol{v}$ consititute the *principal axes* of the error-performance surface. The practical significance of this result will become apparent in later chapters.

PROBLEMS

1. Show that the normal equation (3.35), defining the tap-weight vector $\mathbf{w}_o$ of the Wiener filter, and Eq. (3.46), defining the minimum mean-squared error $J_{\min}$, may be combined into a single matrix relation

    $$\mathbf{A} \begin{bmatrix} 1 \\ -\mathbf{w}_o \end{bmatrix} = \begin{bmatrix} J_{\min} \\ \mathbf{0} \end{bmatrix}$$

    The matrix $\mathbf{A}$ is the correlation matrix of the augmented vector

    $$\begin{bmatrix} d(n) \\ \mathbf{u}(n) \end{bmatrix}$$

    where $d(n)$ is the desired response and $\mathbf{u}(n)$ is the tap-input vector of the Wiener filter.

2. Equation (3.20) defines the mean-squared error $J(\mathbf{w})$ as a function of the tap-weight vector $\mathbf{w}$. Show that this expression may be reformulated as follows:

    $$J(\mathbf{w}) = J_{\min} + (\mathbf{w} - \mathbf{w}_o)^H \mathbf{R}(\mathbf{w} - \mathbf{w}_o)$$

    where $J_{\min}$ is the minimum mean-squared error, $\mathbf{w}_o$ is the optimum tap-weight vector, and $\mathbf{R}$ is the correlation matrix of the tap-input vector.

3. The minimum mean-squared error $J_{\min}$ is defined by [see the second line of Eq. (3.46)]

    $$J_{\min} = \sigma_d^2 - \mathbf{p}^H \mathbf{R}^{-1} \mathbf{p}$$

    where σ_d^2 is the variance of the desired response $d(n)$, $\mathbf{R}$ is the correlation matrix of the tap-input vector $\mathbf{u}(n)$, and $\mathbf{p}$ is the cross-correlation vector between $\mathbf{u}(n)$ and $d(n)$. By applying the unitary similarity transformation to the inverse of the correlation matrix, that is, $\mathbf{R}^{-1}$, show that

$$J_{\min} = \sigma_d^2 - \sum_{k=1}^{M} \frac{|\mathbf{q}_k^H \mathbf{p}|^2}{\lambda_k}$$

where λ_k is the kth eigenvalue of the correlation matrix $\mathbf{R}$, and $\mathbf{q}_k$ is the corresponding eigenvector. Note that $\mathbf{q}_k^H \mathbf{p}$ is a scalar.

4. An array processor consists of a primary sensor and an auxiliary sensor interconnected with each other. The output of the auxiliary sensor is weighted by w and then subtracted from the output of the primary sensor. Show that the mean-square value of the output of the array processor is minimized when the weight w attains the optimum value

$$w_o = \frac{E[u(1, n)u^*(2,n)]}{E[|u(2,n)|^2]}$$

where $u(1,n)$ and $u(2,n)$ are the primary- and auxiliary-sensor outputs at time n, respectively.

5. A linear array consists of M uniformly spaced sensors. The individual sensor outputs are weighted and then summed, producing the output

$$e(n) = \sum_{k=1}^{M} w_k^* u(k,n)$$

where $u(k,n)$ is the output of sensor k at time n, and w_k is the associated weight. The weights are chosen to minimize the mean-square value of $e(n)$, subject to the constraint

$$\mathbf{w}^H \mathbf{s} = 1$$

where $\mathbf{s}$ is a prescribed steering vector [see Section 1.4(h)].

By using the method of Lagrange multipliers, show that the optimum value of the vector $\mathbf{w}$ is defined by the *minimum-variance distortionless response (MVDR)* formula

$$\mathbf{w}_o = \frac{\mathbf{R}^{-1}\mathbf{s}}{\mathbf{s}^H \mathbf{R}^{-1}\mathbf{s}}$$

where $\mathbf{R}$ is the spatial correlation matrix of the linear array. For details on the method of Lagrange multipliers, see Appendix D.

6. Consider a discrete-time stochastic process $\{u(n)\}$ that consists of K (uncorrelated) complex sinusoids plus additive white noise of zero mean and variance σ^2. That is,

$$u(n) = \sum_{k=1}^{K} A_k e^{j\omega_k n} + v(n)$$

where the terms $A_k \exp(j\omega_k n)$ and the $v(n)$ refer to the kth sinusoid and noise, respectively. The process $\{u(n)\}$ is applied to a transversal filter with M taps, producing the output

$$e(n) = \mathbf{w}^H \mathbf{u}(n)$$

Assume that $M > K$. The requirement is to choose the tap-weight vector $\mathbf{w}$ so as to minimize the mean-square value of $e(n)$, subject to the multiple signal-protection constraint

$$\mathbf{S}^H \mathbf{w} = \mathbf{D}^{1/2}\mathbf{1}$$

where **S** is the M-by-K signal matrix whose kth column has 1, $\exp(-j\omega_k)$, ..., $\exp[-j\omega_k(M-1)]$ for its elements, **D** is the K-by-K diagonal matrix whose nonzero elements equal the average powers of the individual sinusoids, and the K-by-1 vector **1** has 1's for all its K elements. Using the method of Langrange multipliers, show that the value of the optimum weight vector that results from this constrained optimization equals

$$\mathbf{w}_o = \mathbf{R}^{-1}\mathbf{S}(\mathbf{S}^H\mathbf{R}^{-1}\mathbf{S})^{-1}\mathbf{D}^{1/2}\mathbf{1}$$

where **R** is the correlation matrix of the M-by-1 tap-input vector $\mathbf{u}(n)$. This formula represents a temporal generalization of the MVDR formula.

Hints: Refer to the solution of Problem 8 in Chapter 2 for the structure of the correlation matrix **R**. For details on the method of Lagrange multipliers, see Appendix D.

7. Consider the problem of detecting a known signal in the presence of additive noise. The noise is assumed to be Gaussian, to be independent of the signal, and to have zero mean and a positive definite correlation matrix $\mathbf{R}_N$. The aim of the problem is to show that under these conditions the three criteria minimum mean-squared error, maximum signal-to-noise ratio, and the likelihood ratio test yield identical designs for the transversal filter.

   Let $\{u(n)\}$, $n = 1, 2, \ldots, M$, denote a set of M real-valued data samples. Let $\{v(n)\}$, $n = 1, 2, \ldots, M$, denote a set of samples taken from a Gaussian noise process of zero mean. Finally, let $\{s(n)\}$, $n = 1, 2, \ldots, M$, denote samples of the signal. The detection problem is to determine whether the input consists of signal plus noise or noise alone. That is, the two hypotheses to be tested for are

   $$\text{hypothesis } H_2: \quad u(n) = s(n) + v(n), \qquad n = 1, 2, \ldots, M$$

   $$\text{hypothesis } H_1: \quad u(n) = v(n), \qquad n = 1, 2, \ldots, M$$

   (a) The *Wiener filter* minimizes the mean-squared error. Show that this criterion yields an optimum tap-weight vector for estimating s_k, the kth component of signal vector **s**, that equals

   $$\mathbf{w}_o = \frac{s_k}{1 + \mathbf{s}^T\mathbf{R}_N^{-1}\mathbf{s}} \mathbf{R}_N^{-1}\mathbf{s}$$

   Hint: To evaluate the inverse of the correlation matrix of $\{u(n)\}$ under hypothesis H_2, you may use the matrix inversion lemma. Let

   $$\mathbf{A} = \mathbf{B}^{-1} + \mathbf{C}\mathbf{D}^{-1}\mathbf{C}^T$$

   where **A**, **B** and **D** are positive-definite matrices. Then

   $$\mathbf{A}^{-1} = \mathbf{B} - \mathbf{B}\mathbf{C}(\mathbf{D} + \mathbf{C}^T\mathbf{B}\mathbf{C})^{-1}\mathbf{C}^T\mathbf{B}$$

   (b) The *maximum signal-to-noise ratio filter* maximizes the ratio

   $$\rho = \frac{\text{average power of filter output due to signal}}{\text{average power of filter output due to noise}}$$

   $$= \frac{E[(\mathbf{w}^T\mathbf{s})^2]}{E[(\mathbf{w}^T\mathbf{v})^2]}$$

Show that the tap-weight vector for which the output signal-to-noise ratio ρ is at maximum equals

$$\mathbf{w}_{SN} = \mathbf{R}_N^{-1}\mathbf{s}$$

Hint: Since $\mathbf{R}_N$ is positive definite, you may use $\mathbf{R}_N = \mathbf{R}_N^{1/2}\mathbf{R}_N^{1/2}$.

(c) The *likelihood ratio processor* computes the log-likelihood ratio and compares it to a threshold. If the threshold is exceeded, it decides in favor of hypothesis H_2; otherwise, it decides in favor of hypothesis H_1. The likelihood ratio is defined by

$$\Lambda = \frac{\mathbf{f}_U(\mathbf{u}|H_2)}{\mathbf{f}_U(\mathbf{u}|H_1)}$$

where $\mathbf{f}_U(\mathbf{u}|H_i)$ is the conditional joint probability density function of the observation vector $\mathbf{u}$, given that hypothesis H_i is true, where $i = 1, 2$. Show that the likelihood ratio test is equivalent to the test

$$\mathbf{w}_{ml}^T\mathbf{u} \underset{H_1}{\overset{H_2}{\gtrless}} \lambda$$

where λ is the threshold and

$$\mathbf{w}_{ml} = \mathbf{R}_N^{-1}\mathbf{s}$$

Hint: The joint probability density function of the M-by-1 Gaussian noise vector $\mathbf{v}$ (with zero mean and correlation matrix $\mathbf{R}_N$) equals

$$f_{\mathbf{v}}(\mathbf{v}) = \frac{1}{(2\pi)^{M/2}\,[\,\det(\mathbf{R}_N)]^{1/2}}\,\exp\left(-\frac{1}{2}\,\mathbf{v}^T\mathbf{R}_N^{-1}\mathbf{v}\right)$$

8. This problem is a continuation of the numerical example presented in Section 3.8. The aim of the problem is to explore the extent of the improvement that may result from using a more complex Wiener filter for the environment described in Section 3.8. To be specific, the new formulation of the Wiener filter has 3 taps.

(a) Find the 3-by-3 correlation matrix of the tap inputs of this filter and the 3-by-1 cross-correlation vector between the desired response and the tap inputs.

(b) Compute the 3-by-1 tap-weight vector of the Wiener filter, and also compute the new value for the minimum mean-squared error.

4

Linear Prediction

4.1 INTRODUCTION

One of the most celebrated problems in time series analysis is that of *predicting* a future value of a stationary discrete-time stochastic process, given a set of past sample values of the process. To be specific, consider the time series $u(n)$, $u(n - 1), \ldots, u(n - M)$, representing $(M + 1)$ samples of such a process up to and including time n. The operation of prediction may, for example, involve using the sample values $u(n - 1)$, $u(n - 2), \ldots, u(n - M)$ to make an estimate of $u(n)$. Let $\mathcal{U}_{n-1}$ denote the M-dimensional space spanned by $u(n - 1)$, $u(n - 2)$, $\ldots, u(n - M)$, and use $\hat{u}(n|\mathcal{U}_{n-1})$ to denote the *predicted value* of $u(n)$ given this set of samples.[1] In general, we may express this predicted value as some function

---

[1]If a space $\mathcal{U}_n$ consists of all linear combinations of random variables, $u_1, u_2, \ldots, u_n$, then these random variables *span* the space. In other words, every random variable in $\mathcal{U}_n$ can be expressed as some combination of the u's, as shown by

$$u = w_1^* u_1 + \cdots + w_n^* u_n$$

for some coefficients w_n.

122

θ of the given samples $u(n - 1), u(n - 2) \ldots , u(n - M)$ as follows:

$$\hat{u}(n|\mathcal{U}_{n-1}) = \theta(u(n - 1), \ldots , u(n - M)) \qquad (4.1)$$

We say that the prediction is *linear* when the function θ consists simply of a linear combination of the samples $u(n - 1), u(n - 2), \ldots , u(n - M)$, as shown by

$$\hat{u}(n|\mathcal{U}_{n-1}) = \sum_{k=1}^{M} w_{ok}^* \, u(n - k) \qquad (4.2)$$

where $w_{o1}, w_{o2}, \ldots , w_{oM}$ are constant coefficients. Our interest in this book is confined to linear prediction.

   In the operation described, the set of sample values $u(n - 1), u(n - 2)$, $\ldots , u(n - M)$ is used to make a prediction of the sample $u(n)$. This operation corresponds to one-step prediction into the future, measured with respect to time $n - 1$. Accordingly, we refer to this form of prediction as *one-step prediction in the forward direction* or simply *forward prediction*. In another form of prediction, we use the sample values $u(n), u(n - 1), \ldots , u(n - M + 1)$ to make a prediction of the sample $u(n - M)$. We refer to this second form of prediction as *backward prediction*.[2]

   In this chapter, we study forward linear prediction (FLP) as well as backward linear prediction (BLP). In particular, we use the Wiener filter theory to optimize the design of a forward or backward *predictor* in the minimum mean-square sense. The various properties of both predictors are discussed in detail. To apply the Wiener filter theory, we assume that the discrete-time stochastic process of interest is weakly stationary. For convenience of analysis, we also assume that the process has zero mean. As explained in Chapter 2, the correlation matrix of such a process has a Toeplitz structure. We will put this Toeplitz structure to good use in developing algorithms that are computationally efficient.

4.2 FORWARD LINEAR PREDICTION

Figure 4.1(a) shows a *forward predictor* that consists of a linear transversal filter with M tap weights $w_{o1}, w_{o2}, \ldots , w_{oM}$ and tap inputs $u(n - 1), u(n - 2), \ldots ,$ $u(n - M)$, respectively. We assume that these tap inputs are drawn from a stationary stochastic process of zero mean. We further assume that the tap weights are optimized in the mean-square sense in accordance with the Wiener filter theory. The predicted value equals $\hat{u}(n|\mathcal{U}_{n-1})$ as in Eq. (4.2). The desired response $d(n)$ equals $u(n)$, representing the actual sample value of the input process at time n.

---

[2]The term "backward prediction" is somewhat of a misnomer. A more appropriate description for this operation is "hindsight." Correspondingly, the use of "forward" in the associated operation of forward prediction is superfluous. Nevertheless, the terms "forward prediction" and "backward prediction" have become deeply imbedded in the literature on linear prediction.

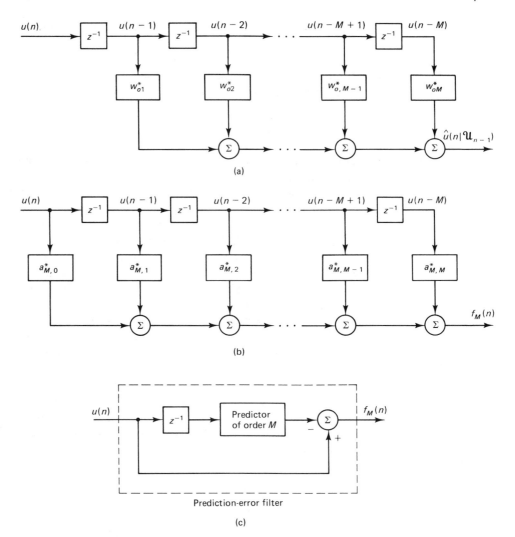

Figure 4.1 (a) One-step predictor; (b) prediction-error filter; (c) relationship between the predictor and the prediction-error filter.

We may thus write.

$$d(n) = u(n) \tag{4.3}$$

The *forward prediction error* equals the difference between the actual sample value $u(n)$ at time n and its predicted value $\hat{u}(n|\mathcal{U}_{n-1})$. We denote the forward prediction error by $f_M(n)$ and thus write

$$f_M(n) = u(n) - \hat{u}(n|\mathcal{U}_{n-1}) \tag{4.4}$$

The subscript M in the symbol for the forward prediction error signifies *order* of the predictor, defined as the number of unit-delay elements needed to store the given set of samples used to make the prediction. The reason for using the subscript will become apparent later in the chapter.

Let P_M denote the minimum mean-squared prediction error:

$$P_M = E[|f_M(n)|^2] \tag{4.5}$$

With the tap inputs assumed to have zero mean, the forward prediction error $f_M(n)$ will likewise have zero mean. Under this condition, P_M will also equal the variance of the forward prediction error. Yet another interpretation for P_M is that it may be viewed as the ensemble-averaged *forward prediction error power*, assuming that $f_M(n)$ is developed across a 1-Ω load. We will use the latter description to refer to P_M.

Let $\mathbf{w}_o$ denote the M-by-1 optimum tap-weight vector of the forward predictor in Fig. 4.1(a). We write it in expanded form as

$$\mathbf{w}_o^T = [w_{o1}, w_{o2}, \ldots, w_{oM}] \tag{4.6}$$

To solve the normal equation for the vector $\mathbf{w}_o$, we require knowledge of two quantities: (1) the M-by-M correlation matrix of the tap inputs $u(n - 1)$, $u(n - 2), \ldots, u(n - M)$, and (2) the M-by-1 cross-correlation vector between these tap inputs and the desired response $u(n)$. To evaluate P_M, we require a third quantity, the variance of $u(n)$. We now consider these three quantities, one by one:

1. The tap inputs $u(n - 1)$, $u(n - 2), \ldots, u(n - M)$ define the M-by-1 tap-input vector, $\mathbf{u}(n - 1)$, as shown by

$$\mathbf{u}^T(n - 1) = [u(n - 1), u(n - 2), \ldots, u(n - M)] \tag{4.7}$$

Hence, the correlation matrix of the tap inputs equals

$$\mathbf{R} = E[\mathbf{u}(n - 1)\mathbf{u}^H(n - 1)]$$

$$= \begin{bmatrix} r(0) & r(1) & \cdots & r(M - 1) \\ r(-1) & r(0) & \cdots & r(M - 2) \\ \vdots & \vdots & \ddots & \vdots \\ r(-M + 1) & r(-M + 2) & \cdots & r(0) \end{bmatrix} \tag{4.8}$$

where $r(k)$ is the autocorrelation function of the input process for lag k, $k = 0$, $1, \ldots, M - 1$. Note that the symbol used for the correlation matrix of the tap inputs in Fig. 4.1(a) is the same as that of the correlation matrix of the tap inputs in the transversal filter of Fig. 3.1. We are justified to do this since the input process in both cases is assumed to be weakly stationary, so the correlation matrix of the process is invariant to time shift.

2. The cross-correlation vector between the tap inputs $u(n - 1)$, ..., $u(n - M)$ and the desired response $u(n)$ equals

$$E[\mathbf{u}(n - 1)u^*(n)] = \begin{bmatrix} E[u(n - 1)u^*(n)] \\ E[u(n - 2)u^*(n)] \\ \vdots \\ E[u(n - M)u^*(n)] \end{bmatrix}$$

(4.9)

$$= \begin{bmatrix} r(-1) \\ r(-2) \\ \vdots \\ r(-M) \end{bmatrix}$$

$$= \mathbf{r}$$

3. The variance of $u(n)$ equals $r(0)$, since $u(n)$ has zero mean.

In Table 4.1, we summarize the various quantities pertaining to the Wiener filter of Fig. 3.1 and the corresponding quantities pertaining to the forward predictor of Fig. 4.1(a). The last column of this table pertains to the backward predictor, on which more will be said later.

Thus, using the correspondences of this table, we may adapt the normal equation (3.35) to solve the forward linear prediction (FLP) problem for

TABLE 4.1

Quantity	Wiener filter of Fig. 3.1	Forward predictor of Fig. 4.1(a)	Backward predictor of Fig. 4.2(a)
Tap-input vector	$\mathbf{u}(n)$	$\mathbf{u}(n - 1)$	$\mathbf{u}(n)$
Desired response	$d(n)$	$u(n)$	$u(n - M)$
Tap-weight vector	$\mathbf{w}_o$	$\mathbf{w}_o$	$\mathbf{g}$
Estimation error	$e(n)$	$f_M(n)$	$b_M(n)$
Correlation matrix of tap inputs	$\mathbf{R}$	$\mathbf{R}$	$\mathbf{R}$
Cross-correlation vector between tap inputs and desired response	$\mathbf{p}$	$\mathbf{r}$	$\mathbf{r}^{B*}$
Minimum mean-squared error	$J_{\min}$	P_M	P_M

stationary inputs and so write

$$\mathbf{R}\mathbf{w}_0 = \mathbf{r} \tag{4.10}$$

Similarly, the use of the first line of Eq. (3.46) yields the following value for the forward prediction–error power:

$$P_M = r(0) - \mathbf{r}^H\mathbf{w}_o \tag{4.11}$$

From Eqs. (4.10) and (4.11), we see that the M-by-1 tap-weight vector of the forward predictor and the forward prediction–error power are determined solely by the set of $(M + 1)$ autocorrelation function values of the input process for lags $0, 1, \ldots, M$.

Discussion

In comparing the normal equation (4.10) for forward predictor with the Yule–Walker equations (2.96) for autoregressive (AR) process, we see that these two systems of simultaneous equations are of exactly the same mathematical form. Furthermore, Eq. (4.11) defining the average power (i.e., variance) of the forward prediction error is also of the same mathematical form as Eq. (2.99) defining the variance of the white-noise process used to excite the autoregressive model.

For the case of an AR process for which we know the model order M, we may thus state that when a forward predictor is optimized in the mean-square sense, its tap weights take on the same values as the corresponding parameters of the process.   This relationship should not be surprising since the equation defining the forward prediction error and the difference equation defining the autoregressive model have the same mathematical form.   When the process is not autoregressive, however, the use of a predictor provides an approximation to the process.

4.3 FORWARD PREDICTION-ERROR FILTER

The forward predictor of Fig. 4.1(a) consists of M unit-delay elements and M tap weights $w_{o1}, w_{o2}, \ldots, w_{oM}$ that are fed with the respective samples $u(n - 1)$, $u(n - 2), \ldots, u(n - M)$ as inputs.   The resultant output is the predicted value of $u(n)$, which is defined by Eq. (4.2).   Hence, substituting Eq. (4.2) in (4.4), we may express the forward prediction error as

$$f_M(n) = u(n) - \sum_{k=1}^{M} w_{ok}^* u(n - k) \tag{4.12}$$

Let $a_{M,k}$, $k = 0, 1, \ldots, M$, denote the tap weights of a new transversal filter, which are related to the tap weights of the forward predictor as follows:

$$a_{M,k} = \begin{cases} 1, & k = 0 \\ -w_{ok}, & k = 1, 2, \ldots, M \end{cases} \tag{4.13}$$

Then we may combine the two terms on the right side of Eq. (4.12) into a single summation as follows:

$$f_M(n) = \sum_{k=0}^{M} a^*_{M,k} u(n - k) \tag{4.14}$$

This input–output relation is represented by the transversal filter shown in Fig. 4.1(b). A filter that operates on the set of samples $u(n)$, $u(n - 1), \ldots, u(n - M)$ to produce the forward prediction error $f_M(n)$ at its output is called a forward *prediction-error filter* (PEF).

The relationship between the forward prediction-error filter and the forward predictor is illustrated in block diagram form in Fig. 4.1(c). Note that the length of the prediction-error filter exceeds the length of the one-step prediction filter by 1. However, both filters have the same order, M, as they both involve the same number of delay elements for storage of past data.

Augmented Normal Equation for Forward Prediction

The normal equation (4.10) defines the tap-weight vector of the forward predictor, while Eq. (4.11) defines the resulting forward prediction-error power P_M. We may combine these two equations into a single matrix relation as follows:

$$\begin{bmatrix} r(0) & \mathbf{r}^H \\ \mathbf{r} & \mathbf{R} \end{bmatrix} \begin{bmatrix} 1 \\ -\mathbf{w}_o \end{bmatrix} = \begin{bmatrix} P_M \\ \mathbf{0} \end{bmatrix} \tag{4.15}$$

where $\mathbf{0}$ is the M-by-1 null vector. We observe that the $(M + 1)$-by-$(M + 1)$ matrix on the left side of Eq. (4.15) equals the correlation matrix of the tap inputs $u(n)$, $u(n - 1), \ldots, u(n - M)$ in the prediction-error filter of Fig 4.1(b); that is,

$$\begin{bmatrix} r(0) & \mathbf{r}^H \\ \hline \mathbf{r} & \mathbf{R} \end{bmatrix} = \begin{bmatrix} r(0) & r(1) & r(2) & \cdots & r(M) \\ \hline r(-1) & r(0) & r(1) & \cdots & r(M-1) \\ r(-2) & r(-1) & r(0) & \cdots & r(M-2) \\ \vdots & \vdots & \vdots & \vdots & \vdots \\ r(-M) & r(-M+1) & r(-M+2) & \cdots & r(0) \end{bmatrix} \tag{4.16}$$

The validity of this relation was discussed in Section 2.2 [see Eq. (2.24)]. We also observe from Eq. (4.13) that the $(M + 1)$-by-1 vector on the left side of Eq. (4.15) equals the *forward prediction-error filter vector*:

$$\begin{bmatrix} 1 \\ -\mathbf{w}_o \end{bmatrix} = \begin{bmatrix} a_{M,0} \\ a_{M,1} \\ \cdot \\ \cdot \\ \cdot \\ a_{M,M} \end{bmatrix} \qquad (4.17)$$

We may thus use Eqs. (4.16) and (4.17) to express the matrix relation of Eq. (4.15) in the expanded form:

$$\begin{bmatrix} r(0) & r(1) & \cdots & r(M) \\ r(-1) & r(0) & \cdots & r(M-1) \\ \cdot & \cdot & \cdot & \cdot \\ \cdot & \cdot & \cdot & \cdot \\ \cdot & \cdot & \cdot & \cdot \\ r(-M) & r(-M+1) & \cdots & r(0) \end{bmatrix} \begin{bmatrix} a_{M,0} \\ a_{M,1} \\ \cdot \\ \cdot \\ \cdot \\ a_{M,M} \end{bmatrix} = \begin{bmatrix} P_M \\ 0 \\ \cdot \\ \cdot \\ \cdot \\ 0 \end{bmatrix} \qquad (4.18)$$

We may also express the matrix relation of Eq. (4.18) as a system of $(M + 1)$ simultaneous equations as follows:

$$\sum_{l=0}^{M} a_{M,l} r(l - i) = \begin{cases} P_M, & i = 0 \\ 0, & i = 1, 2, \ldots, M \end{cases} \qquad (4.19)$$

We refer to Eq. (4.18) or (4.19) as the *augmented normal equation* of a forward prediction-error filter of order M.

Example 1

For the case of a prediction-error filter of order $M = 1$, Eq. (4.18) yields a pair of simultaneous equations described by

$$\begin{bmatrix} r(0) & r(1) \\ r(-1) & r(0) \end{bmatrix} \begin{bmatrix} a_{1,0} \\ a_{1,1} \end{bmatrix} = \begin{bmatrix} P_1 \\ 0 \end{bmatrix}$$

Solving for $a_{1,0}$ and $a_{1,1}$, we get

$$a_{1,0} = \frac{P_1}{\Delta_r} r(0)$$

$$a_{1,1} = -\frac{P_1}{\Delta_r} r(-1)$$

where Δ_r is the determinant of the correlation matrix; thus

$$\Delta_r = \begin{vmatrix} r(0) & r(1) \\ r(-1) & r(0) \end{vmatrix}$$

$$= r^2(0) - r(1)r(-1)$$

But $a_{1,0}$ equals 1. Hence,

$$P_1 = \frac{\Delta_r}{r(0)}$$

$$a_{1,1} = -\frac{r(-1)}{r(0)}$$

Consider next the case of a prediction-error filter of order $M = 2$. Equation (4.18) yields a system of three simultaneous equations, as shown by

$$\begin{bmatrix} r(0) & r(1) & r(2) \\ r(-1) & r(0) & r(1) \\ r(-2) & r(-1) & r(0) \end{bmatrix} \begin{bmatrix} a_{2,0} \\ a_{2,1} \\ a_{2,2} \end{bmatrix} = \begin{bmatrix} P_2 \\ 0 \\ 0 \end{bmatrix}$$

Solving for $a_{2,0}$, $a_{2,1}$ and $a_{2,2}$, we get

$$a_{2,0} = \frac{P_2}{\Delta_r} [r^2(0) - r(1)r(-1)]$$

$$a_{2,1} = -\frac{P_2}{\Delta_r} [r(-1)r(0) - r(1) \ r(-2)]$$

$$a_{2,2} = \frac{P_2}{\Delta_r} [r^2(-1) - r(0)r(-2)]$$

where Δ_r is the determinant of the correlation matrix:

$$\Delta_r = \begin{vmatrix} r(0) & r(1) & r(2) \\ r(-1) & r(0) & r(1) \\ r(-2) & r(-1) & r(0) \end{vmatrix}$$

The coefficient $a_{2,0}$ equals 1. Accordingly, we may express the prediction-error power P_2 as

$$P_2 = \frac{\Delta_r}{r^2(0) - r(1)r(-1)}$$

and the prediction-error filter coefficients $a_{2,1}$ and $a_{2,2}$ as

$$a_{2,1} = -\frac{r(-1)(r(0) - r(1)r(-2)}{r^2(0) - r(1)r(-1)}$$

$$a_{2,2} = \frac{r^2(-1) - r(0)r(-2)}{r^2(0) - r(1)r(-1)}$$

4.4 BACKWARD LINEAR PREDICTION

The form of linear prediction considered in Sections 4.2 and 4.3 is said to be in the *forward* direction. That is, given the time series $u(n)$, $u(n - 1)$, . . . , $u(n - M)$, we use the subset of M samples $u(n - 1)$, $u(n - 2)$, . . . , $u(n - M)$ to make a prediction of the sample $u(n)$. This operation corresponds to *one-step prediction into the future*, measured with respect to time $n - 1$. Naturally, we may also operate on this time series in the *backward* direction. That is, we may use the subset of M samples $u(n)$, $u(n - 1)$, . . . , $u(n - M + 1)$ to make a prediction of the sample $u(n - M)$. This second operation corresponds to *backward prediction by one step*, measured with respect to time $n - M + 1$.

Let $\mathcal{U}_n$ denote the M-dimensional space spanned by $u(n)$, $u(n - 1)$, . . . , $u(n - M + 1)$ that are used in making the backward prediction. Then, using this set of samples as tap inputs, we make a linear prediction of the sample $u(n - M)$ as shown by

$$\hat{u}(n - M|\mathcal{U}_n) = \sum_{k=1}^{M} g_k^* u(n - k + 1) \qquad (4.20)$$

where $g_1, g_2, \ldots, g_M$ are the tap weights. Figure 4.2(a) shows a representation of the backward predictor as described by Eq. (4.20). We assume that these tap weights are optimized in the mean-square sense in accordance with the Wiener filter theory.

In the case of backward prediction, the desired response equals

$$d(n) = u(n - M) \qquad (4.21)$$

The *backward prediction error* equals the difference between the actual sample value $u(n - M)$ and its predicted value $\hat{u}(n - M|\mathcal{U}_n)$. We denote the backward prediction error by $b_M(n)$ and thus write

$$b_M(n) = u(n - M) - \hat{u}(n - M|\mathcal{U}_n) \qquad (4.22)$$

Here, again, the subscript M in the symbol for the backward prediction error $b_M(n)$ signifies the number of unit-delay elements needed to store the given set of samples used to make the prediction, that is, the order of the predictor.

Let P_M denote the minimum mean-squared prediction error

$$P_M = E[|b_M(n)|^2] \qquad (4.23)$$

We may also view P_M as the ensemble-averaged *backward prediction-error power*, assuming that $b_M(n)$ is developed across a 1-Ω load.

Let **g** denote the M-by-1 optimum tap-weight vector of the backward predictor in Fig. 4.2(a). We express it in the expanded form as

$$\mathbf{g}^T = [g_1, g_2, \ldots, g_M] \qquad (4.24)$$

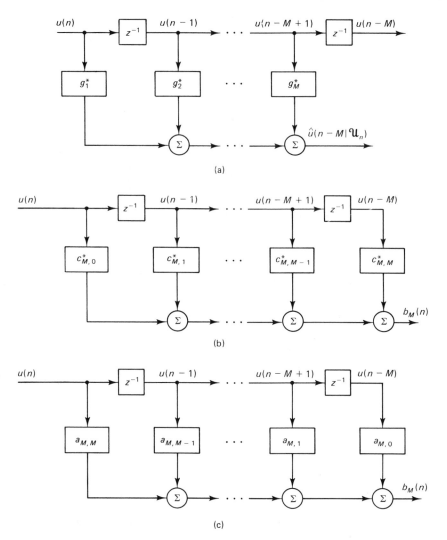

Figure 4.2   (a) Backward one-step predictor; (b) backward prediction-error filter; (c) backward prediction error filter defined in terms of the tap weights of the corresponding forward prediction–error filter.

To solve the normal equation for the vector **g**, we require knowledge of two quantities: (1) the M-by-M correlation matrix of the tap inputs $u(n)$, $u(n - 1)$, . . . , $u(n - M + 1)$; and (2) the M-by-1 cross-correlation vector between the desired response $u(n - M)$ and these tap inputs. To evaluate P_M, we need a third quantity, the variance of $u(n - M)$. We consider these three quantities in turn:

1. Let $\mathbf{u}(n)$ denote the M-by-1 tap-input vector in the backward predictor of Fig. 4.2(a). We write it in expanded form as

$$\mathbf{u}^T(n) = [u(n), u(n-1), \ldots, u(n-M+1)] \qquad (4.25)$$

The M-by-M correlation matrix of the tap inputs in Fig. 4.2(a) thus equals

$$\mathbf{R} = E[\mathbf{u}(n)\mathbf{u}^H(n)]$$

$$= \begin{bmatrix} r(0) & r(1) & \cdots & r(M-1) \\ r(-1) & r(0) & \cdots & r(M-2) \\ \cdot & \cdot & \cdot & \cdot \\ \cdot & \cdot & \cdot & \cdot \\ \cdot & \cdot & \cdot & \cdot \\ r(-M+1) & r(-M+2) & \cdots & r(0) \end{bmatrix} \qquad (4.26)$$

2. The M-by-1 cross-correlation vector between the tap inputs $u(n)$, $u(n-1)$, $\ldots$, $u(n-M+1)$ and the desired response $u(n-M)$ equals

$$E[\mathbf{u}(n)u^*(n-M)] = \begin{bmatrix} E[u(n)u^*(n-M)] \\ E[u(n-1)u^*(n-M)] \\ \cdot \\ \cdot \\ \cdot \\ E[u(n-M+1)u^*(n-M)] \end{bmatrix}$$

$$\qquad (4.27)$$

$$= \begin{bmatrix} r(M) \\ r(M-1) \\ \cdot \\ \cdot \\ \cdot \\ r(1) \end{bmatrix}$$

$$= \mathbf{r}^{B*}$$

where we have used the Hermitian property of the autocorrelation function. As usual, the superscript B denotes backward arrangement and the asterisk denotes complex conjugation.

3. The variance of the desired response $u(n-M)$ equals $r(0)$.

In the last column of Table 4.1, we summarize the various quantities pertaining to the backward predictor of Fig. 4.2(a).

Accordingly, using the correspondences of Table 4.1, we may adapt the normal equation (3.35) to solve the backward linear prediction (BLP) problem for stationary inputs, and so write

$$\mathbf{Rg} = \mathbf{r}^{B*} \tag{4.28}$$

Similarly, the use of Eq. (3.46) yields the following value for the backward prediction-error power:

$$P_M = r(0) - \mathbf{r}^{BT}\mathbf{g} \tag{4.29}$$

Here again we see that the M-by-1 tap-weight vector $\mathbf{g}$ of a backward predictor and the backward prediction-error power P_M are uniquely defined by knowledge of the set of autocorrelation function values of the process for lags 0, 1, . . . , M.

Relations between Backward and Forward Predictors

In comparing the normal equations (4.10) and (4.28), pertaining to forward prediction and backward prediction, respectively, we see that the vector on the right side of Eq. (4.28) differs from that of Eq. (4.10) in two respects: (1) its elements are arranged backward, and (2) they are complex conjugated. To correct for the first difference, we reverse the order in which the elements of the vector on the right side of Eq. (4.28) are arranged. This operation has the effect of replacing the left side of Eq. (4.28) by $\mathbf{R}^T\mathbf{g}^B$, where $\mathbf{R}^T$ is the transpose of the correlation matrix $\mathbf{R}$ and $\mathbf{g}^B$ is the backward version of the tap-weight vector $\mathbf{g}$ (see Problem 3). We may thus write

$$\mathbf{R}^T\mathbf{g}^B = \mathbf{r}^* \tag{4.30}$$

To correct for the remaining difference, we complex-conjugate both sides of Eq. (4.30), obtaining

$$\mathbf{R}^H\mathbf{g}^{B*} = \mathbf{r}$$

Since the correlation matrix $\mathbf{R}$ is Hermitian, that is, $\mathbf{R}^H = \mathbf{R}$, we may thus reformulate the normal equation for backward prediction as

$$\mathbf{Rg}^{B*} = \mathbf{r} \tag{4.31}$$

Now we may compare Eq. (4.31) with (4.10) and thus deduce the following fundamental relationship between the tap-weight vectors of a backward predictor and the corresponding forward predictor:

$$\mathbf{g}^{B*} = \mathbf{w}_o \tag{4.32}$$

Equation (4.32) states that we may transform a backward predictor into a forward predictor by reversing the sequence in which its tap-weights are positioned and also complex-conjugating them.

Next we wish to show that the ensemble-averaged error powers for backward prediction and forward prediction have exactly the same value. To do this, we first observe that the product $\mathbf{r}^{BT}\mathbf{g}$ equals $\mathbf{r}^T\mathbf{g}^B$, and so we may rewrite Eq. (4.29) as

$$P_M = r(0) - \mathbf{r}^T\mathbf{g}^B \tag{4.33}$$

Taking the complex conjugate of both sides of Eq. (4.33), and recognizing that both P_M and $r(0)$ are unaffected by this operation, since they are real-valued scalars, we get

$$P_M = r(0) - \mathbf{r}^H\mathbf{g}^{B*} \tag{4.34}$$

Comparing this result with Eq. (4.11) and using the equivalence of Eq. (4.32), we find that the backward prediction-error power has exactly the same value as the forward prediction-error power. Indeed, it is in anticipation of this equality that we have used the same symbol P_M to denote both the forward prediction-error power and the backward prediction-error power.

4.5 BACKWARD PREDICTION-ERROR FILTER

The backward prediction error $b_M(n)$ equals the difference between the desired response $u(n - M)$ at time $n - M$ and the linear prediction of it, given the samples $u(n), u(n - 1), \ldots, u(n - M + 1)$. This prediction is defined by Eq. (4.20). Therefore, substituting Eq. (4.20) in (4.22), we get

$$b_M(n) = u(n - M) - \sum_{k=1}^{M} g_k^* u(n - k + 1) \tag{4.35}$$

Define the tap weights of the backward prediction-error filter in terms of the corresponding backward predictor as follows:

$$c_{M,k} = \begin{cases} -g_{k+1}, & k = 0, 1, \ldots, M - 1 \\ 1, & k = M \end{cases} \tag{4.36}$$

Hence, we may rewrite Eq. (4.35) as [see Fig. 4.2(b)]

$$b_M(n) = \sum_{k=0}^{M} c_{M,k}^* u(n - k) \tag{4.37}$$

Equation (4.32) defines the tap-weight vector of the backward predictor in terms of that of the forward predictor. We may express the scalar version of this relation as

$$g_{M-k+1}^* = w_{ok}, \quad k = 1, 2, \ldots, M$$

or, equivalently

$$g_k = w_{o,M-k+1}^*, \quad k = 1, 2, \ldots, M \tag{4.38}$$

Hence, substituting Eq. (4.38) in (4.36), we get

$$c_{M,k} = \begin{cases} -w^*_{o,M-k}, & k = 0, 1, \ldots, M-1 \\ 1, & k = M \end{cases} \tag{4.39}$$

Thus, using the relationship between the tap weights of the forward prediction-error filter and those of the forward predictor as given in Eq. (4.13), we may write

$$c_{M,k} = a^*_{M,M-k}, \qquad k = 0, 1, \ldots, M \tag{4.40}$$

Accordingly, we may express the input–output relation of the backward prediction-error filter in the equivalent form

$$b_M(n) = \sum_{k=0}^{M} a_{M,M-k} u(n-k) \tag{4.41}$$

The input–output relation of Eq. (4.41) is depicted in Fig. 4.2(c). Comparison of this representation for a backward prediction-error filter with that of Fig. 4.1(b) for the corresponding forward prediction-error filter reveals that these two forms of a prediction-error filter for *stationary inputs* are uniquely related to each other. In particular, *we may transform a forward prediction-error filter into the corresponding backward prediction-error filter by reversing the sequence in which the tap weights are positioned and complex-conjugating them.* Note that in both figures, the respective tap inputs have the same values.

Augmented Normal Equation for Backward Prediction

The normal equation for backward prediction is defined by Eq. (4.28), and the resultant backward prediction-error power is defined by Eq. (4.29). We may combine these two equations into a single relation as follows:

$$\begin{bmatrix} \mathbf{R} & \mathbf{r}^{B*} \\ \mathbf{r}^{BT} & r(0) \end{bmatrix} \begin{bmatrix} -\mathbf{g} \\ 1 \end{bmatrix} = \begin{bmatrix} \mathbf{0} \\ P_M \end{bmatrix} \tag{4.42}$$

where $\mathbf{0}$ is the M-by-1 null vector. The $(M+1)$-by-$(M+1)$ matrix on the left side of Eq. (4.42) equals the correlation matrix of the tap inputs in the backward prediction-error filter of Fig. 4.2(c):

$$\begin{bmatrix} \mathbf{R} & \mathbf{r}^{B*} \\ \hline \mathbf{r}^{BT} & r(0) \end{bmatrix} = \begin{bmatrix} r(0) & r(1) & \cdots & r(M-1) & r(M) \\ & & & & \\ & & & & \\ r(-M+1) & r(-M+2) & \cdots & r(0) & r(1) \\ \hline r(-M) & r(-M+1) & \cdots & r(-1) & r(0) \end{bmatrix} \tag{4.43}$$

Hence, substituting Eqs. (4.43), (4.36) and (4.40) in (4.42), we get the expanded matrix relation:

$$
\begin{bmatrix}
r(0) & r(1) & \cdots & r(M-1) & r(M) \\
\cdot & \cdot & \cdot & \cdot & \cdot \\
\cdot & \cdot & \cdot & \cdot & \cdot \\
\cdot & \cdot & \cdot & \cdot & \cdot \\
r(-M+1) & r(-M+2) & \cdots & r(0) & r(1) \\
r(-M) & r(-M+1) & \cdots & r(-1) & r(0)
\end{bmatrix}
\begin{bmatrix}
a_{M,M}^{*} \\
\cdot \\
\cdot \\
\cdot \\
a_{M,1}^{*} \\
a_{M,0}^{*}
\end{bmatrix}
$$

$$
=
\begin{bmatrix}
0 \\
\cdot \\
\cdot \\
\cdot \\
0 \\
P_{M}
\end{bmatrix}
\tag{4.44}
$$

We may also express the matrix relation of Eq. (4.44) as a system of $(M+1)$ simultaneous equations:

$$
\sum_{l=0}^{M} a_{M,M-l}^{*} r(l-i) = \begin{cases} 0, & i = 0, 1, \ldots, M-1 \\ P_{M}, & i = M \end{cases}
\tag{4.45}
$$

We refer to Eq. (4.44) or (4.45) as the *augmented normal equation* of a backward prediction-error filter of order M.

Note that in the matrix form of the augmented normal equation for backward prediction defined by Eq. (4.44) the correlation matrix of the tap inputs has exactly the same form as in the corresponding equation (4.18). This is merely a restatement of the fact that the tap inputs in the backward prediction-error filter of Fig. 4.2(c) are exactly the same as those in the forward prediction-error filter of Fig. 4.1(b).

4.6 THE LEVINSON–DURBIN RECURSION

We now describe a direct method for computing the prediction-error filter coefficients and prediction-error power by solving the augmented normal equation. The method is recursive in nature and makes particular use of the Toeplitz structure of the correlation matrix of the tap inputs of the filter. It is known as the *Levinson–Durbin recursion*, so named in recognition of its use first by Levinson (1947) and then its independent reformulation at a later date by Durbin (1960). Basically, the procedure utilizes the solution of the augmented normal equation for a prediction-error filter of order $m-1$ to compute the corresponding solution for a prediction-error filter of order m (i.e., one order higher). The order $m = 1, 2, \ldots, M$, where M is the *final* order of the filter. The important virtue of the

Levinson–Durbin recursive procedure is its computational efficiency, in that its use results in a big saving in the number of operations (multiplications or divisions) and storage locations compared to standard methods such as the Gauss elimination method (Makhoul, 1975). To derive the Levinson–Durbin recursive procedure, we will use the matrix formulation of both forward and backward predictions in an elegant way (Burg, 1968, 1975).

Let the $(m + 1)$-by-1 vector $\mathbf{a}_m$ denote the tap-weight vector of a forward prediction-error filter of order m. The $(m + 1)$-by-1 tap-weight vector of the corresponding backward prediction-error filter is obtained by backward rearrangement of the elements of vector $\mathbf{a}_m$ and their complex conjugation. We denote the combined effect of these two operations by $\mathbf{a}_m^{B*}$. Let the m-by-1 vectors $\mathbf{a}_{m-1}$ and $\mathbf{a}_{m-1}^{B*}$ denote the tap-weight vectors of the corresponding forward and backward prediction-error filters of order $m - 1$, respectively. The Levinson–Durbin recursion may be stated in one of two equivalent ways:

1. The tap-weight vector of a *forward* prediction-error filter may be order-updated as follows:

$$\mathbf{a}_m = \begin{bmatrix} \mathbf{a}_{m-1} \\ 0 \end{bmatrix} + \Gamma_m \begin{bmatrix} 0 \\ \mathbf{a}_{m-1}^{B*} \end{bmatrix} \tag{4.46}$$

where Γ_m is a constant. The scalar version of this order-update is

$$a_{m,\,k} = a_{m-1,\,k} + \Gamma_m a_{m-1,\,m-k}^{*}, \qquad k = 0, 1, \ldots, m \tag{4.47}$$

where $a_{m,\,k}$ is the kth tap weight of a forward prediction-error filter of order m, and likewise for $a_{m-1,\,k}$. The element $a_{m-1,\,m-k}^{*}$ is the kth tap weight of a backward prediction-error filter of order $m - 1$. In Eq. (4.47), note that $a_{m-1,\,0} = 1$ and $a_{m-1,\,m} = 0$.

2. The tap-weight vector of a *backward* prediction-error filter may be order-updated as follows:

$$\mathbf{a}_m^{B*} = \begin{bmatrix} 0 \\ \mathbf{a}_{m-1}^{B*} \end{bmatrix} + \Gamma_m^{*} \begin{bmatrix} \mathbf{a}_{m-1} \\ 0 \end{bmatrix} \tag{4.48}$$

The scalar version of this order-update is

$$a_{m,\,m-k}^{*} = a_{m-1,\,m-k}^{*} + \Gamma_m^{*} a_{m-1,\,k}, \qquad k = 0, 1, \ldots, m \tag{4.49}$$

where $a_{m,\,m-k}^{*}$ is the kth tap weight of the backward prediction-error filter of order m, and the other elements are as defined previously.

The Levinson–Durbin recursion is usually formulated in the context of forward prediction, in vector form as in Eq. (4.46) or scalar form as in Eq. (4.47). The formulation of the recursion in the context of backward prediction, in vector form as in Eq. (4.48) or scalar form as in Eq. (4.49), follows directly from that of Eq. (4.46) or (4.47), respectively, through a combination of backward rearrangement and complex conjugation (see Problem 4).

To establish the condition that the constant Γ_m has to satisfy in order to justify the validity of the Levinson–Durbin recursion, we proceed in four stages as follows:

1. We premultiply both sides of Eq. (4.46) by $\mathbf{R}_{m+1}$, the $(m + 1)$-by-$(m + 1)$ correlation matrix of the tap inputs $u(n)$, $u(n - 1)$, . . . , $u(n - m)$ in the forward prediction-error filter of order m. For the left side of Eq. (4.46), we thus get [see Eq. (4.18)]

$$\mathbf{R}_{m+1}\mathbf{a}_m = \begin{bmatrix} P_m \\ \mathbf{0}_m \end{bmatrix} \tag{4.50}$$

where P_m is the forward prediction-error power, and $\mathbf{0}_m$ is the m-by-1 null vector. The subscripts in the matrix $\mathbf{R}_{m+1}$ and the vector $\mathbf{0}_m$ refer to their dimensions, whereas the subscripts in the vector $\mathbf{a}_m$ and the scalar P_m refer to prediction order.

2. For the first term on the right side of Eq. (4.46), we use the following partitioned form of the correlation matrix $\mathbf{R}_{m+1}$ (see Section 2.3).

$$\mathbf{R}_{m+1} = \begin{bmatrix} \mathbf{R}_m & \mathbf{r}_m^{B*} \\ \mathbf{r}_m^{BT} & r(0) \end{bmatrix}$$

where $\mathbf{R}_m$ is the m-by-m correlation matrix of the tap inputs $u(n)$, $u(n - 1)$, . . . , $u(n - m + 1)$, and $\mathbf{r}_m^{B*}$ is the cross-correlation vector between these tap inputs and desired response $u(n - m)$. We may thus write

$$\mathbf{R}_{m+1}\begin{bmatrix} \mathbf{a}_{m-1} \\ 0 \end{bmatrix} = \begin{bmatrix} \mathbf{R}_m & \mathbf{r}_m^{B*} \\ \mathbf{r}_m^{BT} & r(0) \end{bmatrix}\begin{bmatrix} \mathbf{a}_{m-1} \\ 0 \end{bmatrix}$$
$$= \begin{bmatrix} \mathbf{R}_m\mathbf{a}_{m-1} \\ \mathbf{r}_m^{BT}\mathbf{a}_{m-1} \end{bmatrix} \tag{4.51}$$

The augmented normal equation for the forward prediction-error filter of order $m - 1$ is

$$\mathbf{R}_m\mathbf{a}_{m-1} = \begin{bmatrix} P_{m-1} \\ \mathbf{0}_{m-1} \end{bmatrix} \tag{4.52}$$

where P_{m-1} is the prediction-error power for this filter, and $\mathbf{0}_{m-1}$ is the

$(m - 1)$-by-1 null vector.   Define the scalar

$$\Delta_{m-1} = \mathbf{r}_m^{BT} \mathbf{a}_{m-1}$$

$$= [r(-m), r(1 - m), \ldots, r(-1)] \begin{bmatrix} a_{m-1,\,0} \\ a_{m-1,\,1} \\ \cdot \\ \cdot \\ \cdot \\ a_{m-1,\,m-1} \end{bmatrix} \tag{4.53}$$

$$= \sum_{k=0}^{m-1} a_{m-1,\,k} r(k - m)$$

Substituting Eqs. (4.52) and (4.53) in (4.51), we may therefore write

$$\mathbf{R}_{m+1} \begin{bmatrix} \mathbf{a}_{m-1} \\ 0 \end{bmatrix} = \begin{bmatrix} P_{m-1} \\ \mathbf{0}_{m-1} \\ \Delta_{m-1} \end{bmatrix} \tag{4.54}$$

3. For the second term on the right side of Eq. (4.46), we use the following partitioned form of the correlation matrix $\mathbf{R}_{m+1}$ (see Section 2.2):

$$\mathbf{R}_{m+1} = \begin{bmatrix} r(0) & \mathbf{r}_m^H \\ \mathbf{r}_m & \mathbf{R}_m \end{bmatrix}$$

where $\mathbf{R}_m$ is the m-by-m correlation matrix of the tap inputs $u(n - 1)$, $u(n - 2), \ldots, u(n - m)$, and $\mathbf{r}_m$ is the m-by-1 cross-correlation vector between these tap inputs and desired response $u(n)$.   We may thus write

$$\mathbf{R}_{m+1} \begin{bmatrix} 0 \\ \mathbf{a}_{m-1}^{B*} \end{bmatrix} = \begin{bmatrix} r(0) & \mathbf{r}_m^H \\ \mathbf{r}_m & \mathbf{R}_m \end{bmatrix} \begin{bmatrix} 0 \\ \mathbf{a}_{m-1}^{B*} \end{bmatrix} \tag{4.55}$$

$$= \begin{bmatrix} \mathbf{r}_m^H \mathbf{a}_{m-1}^{B*} \\ \mathbf{R}_m \mathbf{a}_{m-1}^{B*} \end{bmatrix}$$

The scalar $\mathbf{r}_m^H \mathbf{a}_{m-1}^{B*}$ equals

$$\mathbf{r}_m^H \mathbf{a}_{m-1}^{B*} = [r(1), r(2), \ldots, r(m)] \begin{bmatrix} a_{m-1,\,m-1}^* \\ a_{m-1,m-2}^* \\ \cdot \\ \cdot \\ \cdot \\ a_{m-1,\,0}^* \end{bmatrix} \tag{4.56}$$

$$= \sum_{l=1}^m r(l) a_{m-1,\,m-l}^*$$

$$= \Delta_{m-1}^*$$

Also, the augmented normal equation for the backward prediction-error filter of order $m - 1$ states that

$$\mathbf{R}_m \mathbf{a}_{m-1}^{B*} = \begin{bmatrix} \mathbf{0}_{m-1} \\ P_{m-1} \end{bmatrix} \tag{4.57}$$

Substituting Eqs. (4.56) and (4.57) in (4.55), we may therefore write

$$\mathbf{R}_{m+1} \begin{bmatrix} 0 \\ \mathbf{a}_{m-1}^{B*} \end{bmatrix} = \begin{bmatrix} \Delta_{m-1}^* \\ \mathbf{0}_{m-1} \\ P_{m-1} \end{bmatrix} \tag{4.58}$$

4. Summarizing the results obtained in stages 1, 2, and 3 and, in particular, using Eqs. (4.50), (4.54) and (4.58), we may now state that the premultiplication of both sides of Eq. (4.46) by the correlation matrix $\mathbf{R}_{m+1}$ yields

$$\begin{bmatrix} P_m \\ \mathbf{0}_m \end{bmatrix} = \begin{bmatrix} P_{m-1} \\ \mathbf{0}_{m-1} \\ \Delta_{m-1} \end{bmatrix} + \Gamma_m \begin{bmatrix} \Delta_{m-1}^* \\ \mathbf{0}_{m-1} \\ P_{m-1} \end{bmatrix} \tag{4.59}$$

We conclude therefore that, if the order-update recursion of Eq. (4.46) holds, the results described by Eq. (4.59) are direct consequences of this recursion. Conversely, we may state that, if the conditions described by Eq. (4.59) apply, the tap-weight vector of a forward prediction-error filter may be order-updated as in Eq. (4.46).

From Eq. (4.59), we may make two important deductions:

1. By considering the first elements of the vectors on the left and right sides of Eq. (4.59), we have

$$P_m = P_{m-1} + \Gamma_m \Delta_{m-1}^* \qquad (4.60)$$

2. By considering the last elements of the vectors on the left and right sides of Eq. (4.59), we have

$$0 = \Delta_{m-1} + \Gamma_m P_{m-1} \qquad (4.61)$$

From Eq. (4.61), we see that the constant Γ_m has the value

$$\Gamma_m = -\frac{\Delta_{m-1}}{P_{m-1}} \qquad (4.62)$$

where Δ_{m-1} is itself defined by Eq. (4.53). Furthermore, eliminating Δ_{m-1} between Eqs. (4.60) and (4.61), we get the following relation for the order update of the prediction-error power:

$$P_m = P_{m-1} (1 - |\Gamma_m|^2) \qquad (4.63)$$

If the order m of the prediction-error filter increases, the corresponding value of the prediction-error power P_m normally decreases or else remains the same. Of course, P_m can never be negative. Hence, we must always have

$$0 \leq P_m \leq P_{m-1}, \qquad m \geq 1 \qquad (4.64)$$

For the elementary case of a prediction-error filter of order zero, we have

$$P_0 = r(0)$$

where $r(0)$ is the autocorrelation function of the input process for lag zero.

Starting with $m = 0$, and increasing the filter order by 1 at a time, we find that through the repeated application of Eq. (4.63) the prediction-error power for a prediction-error filter of *final* order M equals

$$P_M = P_0 \prod_{m=1}^{M} (1 - |\Gamma_m|^2) \qquad (4.65)$$

Interpretations of the Parameters Γ_m and Δ_{m-1}

The parameters Γ_m, $1 \leq m \leq M$, resulting from the application of the Levinson–Durbin recursion to a prediction-error filter of final order M, are called *reflection coefficients*. The use of this term comes from the analogy of Eq. (4.63) with transmission line theory, where (in the latter context) Γ_m may be considered as the reflection coefficient at the boundary between two sections with different characteristic impedances. Note that the condition on the reflection coefficient corresponding to that of Eq. (4.64) is

$$|\Gamma_m| \le 1, \qquad \text{for all } m \qquad\qquad (4.66)$$

From Eq. (4.47), we see that for a prediction-error filter of order m, the reflection coefficient Γ_m equals the *last* tap-weight $a_{m,m}$ of the filter. That is,

$$\Gamma_m = a_{m,m}$$

As for the parameter Δ_{m-1}, it may be interpreted as a cross-correlation between the forward prediction error $f_{m-1}(n)$ and the delayed backward prediction error $b_{m-1}(n-1)$. Specifically, we may write (see Problem 5)

$$\Delta_{m-1} = E[b_{m-1}(n-1)f_{m-1}^*(n)] \qquad\qquad (4.67)$$

where $f_{m-1}(n)$ is produced at the output of a forward prediction-error filter of order $m-1$ in response to the tap inputs $u(n)$, $u(n-1)\ldots$, $u(n-m+1)$, and $b_{m-1}(n-1)$ is produced at the output of a backward prediction-error filter of order $m-1$ in response to the tap inputs $u(n-1)$, $u(n-2)$, $\ldots$, $u(n-m)$.

Note that

$$f_0(n) = b_0(n) = u(n)$$

where $u(n)$ is the prediction-error filter input at time n. Accordingly, from Eq. (4.67) we find that the cross-correlation parameter has the *zero-order value*

$$\Delta_0 = E[b_0(n-1)f_0^*(n)]$$

$$= E[u(n-1)u^*(n)]$$

$$= r(-1)$$

where $r(-1)$ is the autocorrelation function of the input for a lag of -1.

We may also use Eqs. (4.62) and (4.67) to develop a second interpretation for the parameter Γ_m. In particular, since P_{m-1} may be viewed as the mean-square value of the forward prediction error $f_{m-1}(n)$, we may write

$$\Gamma_m = -\frac{E[b_{m-1}(n-1)f_{m-1}^*(n)]}{E[|f_{m-1}(n)|^2]} \qquad\qquad (4.68)$$

The right side of Eq. (4.68), except for the minus sign, is referred to as a *partial correlation (PARCOR) coefficient*. This terminology is widely used in the statistical literature (Box and Jenkins, 1976). Hence, the reflection coefficient, as defined here, is the negative of the PARCOR coefficient.

Application of the Levinson–Durbin Recursion

There are two possible ways of applying the Levinson–Durbin recursion to compute the prediction-error filter coefficients $a_{M,k}$, $k = 0, 1, \ldots, M$, and the prediction-error power P_M for a final prediction order M:

1. We have explicit knowledge of the autocorrelation function of the input process; in particular, we have $r(0)$, $r(1)$, . . . , $r(M)$, denoting the values of the autocorrelation function for lags, 0, 1, . . . , M, respectively. For example, we may obtain *unbiased* estimates of these parameters by means of the *time-average formula*

$$r(k) = \frac{1}{N - k} \sum_{n=1}^{N-k} u(n)u^*(n - k), \qquad k = 0, 1, \ldots, M \qquad (4.69)$$

where N is the total length of the input time series, with $N \gg M$. There are, of course, other estimators that we may use. In any event, given $r(0)$, $r(1)$, . . . , $r(M)$, the computation proceeds by using Eq. (4.53) for Δ_{m-1} and Eq. (4.63) for P_m. The recursion is initiated with $m = 0$, for which we have $P_0 = r(0)$ and $\Delta_0 = r(-1)$. Note also that $a_{m,0}$ equals 1 for all m, and $a_{m,k}$ is zero for all $k > m$. The computation is terminated when $m = M$. The resulting estimates of the prediction-error filter coefficients and prediction error power obtained by using this procedure are known as the *Yule–Walker estimates*.

2. We have explicit knowledge of the reflection coefficients Γ_1, Γ_2, . . . , Γ_M and the autocorrelation function $r(0)$ for a lag of zero. Later in the chapter we describe a procedure for estimating these reflection coefficients directly from the given data. In this second application of the Levinson–Durbin recursion, we only need the pair of relations

$$a_{m,k} = a_{m-1,k} + \Gamma_m a^*_{m-1,m-k}, \qquad k = 0, 1, \ldots, m$$

$$P_m = P_{m-1}(1 - |\Gamma_m|^2)$$

Here, again, the recursion is initiated with $m = 0$ and stopped when the order m reaches the final value M.

Example 2

To illustrate the second method for the application of the Levinson–Durbin recursion, suppose we are given the reflection coefficients Γ_1, Γ_2, Γ_3 and the average power P_0. The problem we wish to solve is to use these parameters to determine the corresponding tap weights $a_{3,1}$, $a_{3,2}$, $a_{3,3}$ and the prediction-error power P_3 for a prediction-error filter of order 3.

The application of the Levinson–Durbin recursion, described by Eqs. (4.47) and (4.63), yields the following results for $m = 1, 2, 3$:

1. Prediction-error filter order $m = 1$:

$$a_{1,0} = 1$$

$$a_{1,1} = \Gamma_1$$

$$P_1 = P_0(1 - |\Gamma_1|^2)$$

2. Prediction-error filter order $m = 2$:

$$a_{2,0} = 1$$

$$a_{2,1} = \Gamma_1 + \Gamma_2 \Gamma_1^*$$

$$a_{2,2} = \Gamma_2$$

$$P_2 = P_1(1 - |\Gamma_2|^2)$$

where P_1 is as defined above.

3. Prediction-error filter order $m = 3$:

$$a_{3,0} = 1$$

$$a_{3,1} = a_{2,1} + \Gamma_3 \Gamma_2^*$$

$$a_{3,2} = \Gamma_2 + \Gamma_3 a_{2,1}^*$$

$$a_{3,3} = \Gamma_3$$

$$P_3 = P_2(1 - |\Gamma_3|^2)$$

where $a_{2,1}$ and P_2 are as defined above.

The interesting point to observe from this example is that the Levinson–Durbin recursion yields not only the values of the tap weights and prediction-error power for the prediction-error filter of final order M but also the corresponding values of these parameters for the prediction-error filters of intermediate orders $M - 1, \ldots, 1$.

Inverse Form of the Levinson–Durbin Recursion

In the normal application of the Levinson–Durbin recursion, as illustrated in Example 2, we are given the set of reflection coefficients $\Gamma_1, \Gamma_2, \ldots, \Gamma_M$ and the requirement is to compute the corresponding set of tap weights $a_{M,1}, a_{M,2}, \ldots, a_{M,M}$ for a prediction-error filter of final order M. Of course, the remaining coefficient of the filter, $a_{M,0} = 1$. Frequently, however, the need arises to solve the following *inverse* problem: given the set of tap weights $a_{M,1}, a_{M,2}, \ldots, a_{M,M}$, solve for the corresponding set of reflection coefficients $\Gamma_1, \Gamma_2, \ldots, \Gamma_M$. We may solve this problem by applying the inverse form of Levinson–Durbin recursion, which we refer to simply as the *inverse recursion*.

To derive the inverse recursion, we first combine Eqs. (4.47) and (4.49), representing the scalar versions of the Levinson–Durbin recursion for forward and backward prediction-error filters, respectively, in matrix form as follows:

$$\begin{bmatrix} a_{m,k} \\ a_{m,m-k}^* \end{bmatrix} = \begin{bmatrix} 1 & \Gamma_m \\ \Gamma_m^* & 1 \end{bmatrix} \begin{bmatrix} a_{m-1,k} \\ a_{m-1,m-k}^* \end{bmatrix}, \qquad k = 0, 1, \ldots, m \qquad (4.70)$$

where the order $m = 1, 2, \ldots, M$. Then, solving Eq. (4.70) for the tap weight

$a_{m-1,k}$, we get

$$a_{m-1,k} = \frac{a_{m,k} - a_{m,m}a^*_{m,m-k}}{1 - |a_{m,m}|^2}, \qquad k = 0, 1, \ldots, m \tag{4.71}$$

where we have used the fact that $\Gamma_m = a_{m,m}$. We may now describe the procedure. Starting with the set of tap weights $\{a_{M,k}\}$ for which the prediction-error filter order equals M, we use the inverse recursion, Eq. (4.71), with decreasing filter order $m = M, M - 1, \ldots, 2$ to compute the tap weights of the corresponding prediction-error filters of order $M - 1, M - 2, \ldots, 1$, respectively. Finally, knowing the tap weights of all the prediction-error filters of interest (whose order ranges all the way from M down to 1), we use the fact that

$$\Gamma_m = a_{m,m}, \qquad m = M, M - 1, \ldots, 1$$

to determine the desired set of reflection coefficients $\Gamma_M, \Gamma_{M-1}, \ldots, \Gamma_1$. Example 3 illustrates the application of the inverse recursion.

Example 3

Suppose we are given the tap weights $a_{3,1}, a_{3,2}, a_{3,3}$ of a prediction-error filter of order 3, and the requirement is to determine the corresponding reflection coefficients $\Gamma_1, \Gamma_2, \Gamma_3$. Application of the inverse recursion, described by Eq. (4.71), for filter order $m = 3,2$ yields the following sets of tap weights:

1. Prediction-error filter of order 2 [corresponding to $m = 3$ in Eq. (4.71)]:

$$a_{2,1} = \frac{a_{3,1} - a_{3,3}\,a^*_{3,2}}{1 - |a_{3,3}|^2}$$

$$a_{2,2} = \frac{a_{3,2} - a_{3,3}a^*_{3,1}}{1 - |a_{3,3}|^2}$$

2. Prediction-error filter of order 1 [correponding to $m = 2$ in Eq. (4.71)]:

$$a_{1,1} = \frac{a_{2,1} - a_{2,2}\,a^*_{2,1}}{1 - |a_{2,2}|^2}$$

where $a_{2,1}$ and $a_{2,2}$ are as defined above.

Thus, the desired reflection coefficients are given by

$$\Gamma_3 = a_{3,3}$$

$$\Gamma_2 = a_{2,2}$$

$$\Gamma_1 = a_{1,1}$$

where $a_{3,3}$ is given, and $a_{2,2}$ and $a_{1,1}$ are as defined above.

4.7 *RELATION AMONG THE AUTOCORRELATION FUNCTION AND THE REFLECTION COEFFICIENTS*

It is customary to represent the second-order statistics of a stationary time series in terms of its autocorrelation function or, equivalently, the power spectrum. The autocorrelation function and power spectrum form a discrete-time Fourier transform pair (see Chapter 2). Another way of describing the second-order statistics of a stationary time series is to use the set of numbers $r(0), \Gamma_1, \Gamma_2, \ldots, \Gamma_M$, where $r(0)$ is the value of the autocorrelation function of the process for a lag of zero, and $\Gamma_1, \Gamma_2, \ldots, \Gamma_M$ are the reflection coefficients for a prediction-error filter of final order M. This is a consequence of the fact that the set of numbers $r(0), \Gamma_1, \Gamma_2, \ldots, \Gamma_M$ uniquely determines the corresponding set of autocorrelation function values $r(0), r(1), \ldots, r(M)$.

To prove this relationship, we first eliminate Δ_{m-1} between Eqs. (4.53) and (4.61), obtaining

$$\sum_{k=0}^{m-1} a_{m-1,k} r(k - m) = -\Gamma_m P_{m-1} \tag{4.72}$$

Solving Eq. (4.72) for $r(m) = r^*(-m)$ and recognizing that $a_{m-1,0}$ equals 1, we get

$$r(m) = -\Gamma_m^* P_{m-1} - \sum_{k=1}^{m-1} a_{m-1,k}^* r(m - k) \tag{4.73}$$

This is the desired recursive relation. If we are given the set of numbers $r(0), \Gamma_1, \Gamma_2, \ldots, \Gamma_M$, then by using Eq. (4.73), together with the Levinson–Durbin recursive equations (4.47) and (4.63), we may recursively generate the corresponding set of numbers $r(0), r(1), \ldots, r(M)$.

For the case when $|\Gamma_m| \leq 1$, we find from Eq. (4.73) that the permissible region for $r(m)$, the value of the autocorrelation function of the input signal for a lag of m, is the interior (including circumference) of a circle of radius P_{m-1} and center at

$$-\sum_{k=1}^{m-1} a_{m-1,k}^* r(m - k)$$

This is illustrated in Fig. 4.3.

Suppose now that we are given the set of autocorrelation function values $r(1), \ldots, r(M)$. Then we may recursively generate the corresponding set of numbers $\Gamma_1, \Gamma_2, \ldots, \Gamma_M$ by using

$$\Gamma_m = -\frac{1}{P_{m-1}} \sum_{k=0}^{m-1} a_{m-1,k} r(k - m) \tag{4.74}$$

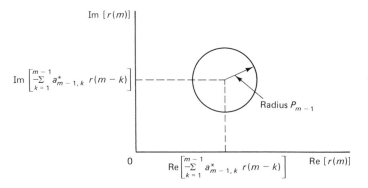

Figure 4.3 Permissible region for $r(m)$ for the case when $\Gamma_m \leq 1$.

which is obtained by solving Eq. (4.72) for Γ_m. In Eq. (4.74), it is assumed that P_{m-1} is nonzero. If P_{m-1} is zero, this would have been the result of $|\Gamma_{m-1}| = 1$, and the sequence of reflection coefficients $\Gamma_1, \Gamma_2, \ldots, \Gamma_{m-1}$ is terminated.

We conclude, therefore, that there is a one-to-one correspondence between the two sets of quantities $\{r(0), \Gamma_1, \Gamma_2, \ldots, \Gamma_M\}$ and $\{r(0), r(1), r(2), \ldots, r(M)\}$, in that if we are given the one we may uniquely determine the other in a recursive manner.

Example 4

Suppose we are given $r(0)$, and $\Gamma_1, \Gamma_2, \Gamma_3$ and the requirement is to compute $r(1)$, $r(2)$ and $r(3)$. We start with $m = 1$, for which Eq. (4.73) yields

$$r(1) = -P_0 \Gamma_1^*$$

where

$$P_0 = r(0)$$

For $m = 2$, the use of Eq. (4.73) yields

$$r(2) = -P_1 \Gamma_2^* - r(1)\Gamma_1^*$$

where

$$P_1 = P_0(1 - |\Gamma_1|^2)$$

Finally, for $m = 3$, the use of Eq. (4.73) yields

$$r(3) = -P_2 \Gamma_3^* - [a_{2,1}^* r(2) + \Gamma_2^* r(1)]$$

where

$$P_2 = P_1(1 - |\Gamma_2|^2)$$

$$a_{2,1} = \Gamma_1 + \Gamma_2 \Gamma_1^*$$

4.8 TRANSFER FUNCTION OF A FORWARD PREDICTION-ERROR FILTER

Let $H_{f,m}(z)$ denote the *transfer function* of a forward prediction-error filter of order m, and whose impulse response is defined by the sequence of numbers $a^*_{m,k}$, $k = 0, 1, \ldots, m$, as illustrated in Fig. 4.1(b) for $m = M$. By definition, the transfer function of a discrete-time filter equals the *z-transform* of its impulse response (Oppenheim and Schafer, 1975). We may therefore write

$$H_{f,m}(z) = \sum_{k=0}^{m} a^*_{m,k} z^{-k} \tag{4.75}$$

where z^{-1} is the unit-sample delay. Based on the Levinson–Durbin recursion, in particular Eq. (4.47), we may relate the coefficients of this filter of order m to those of a corresponding prediction-error filter of order $m - 1$ (i.e., one order smaller). Therefore, substituting Eq. (4.47) into (4.75), we get

$$
\begin{aligned}
H_{f,m}(z) &= \sum_{k=0}^{m} a^*_{m-1,k} z^{-k} + \Gamma^*_m \sum_{k=0}^{m} a_{m-1,m-k} z^{-k} \\
&= \sum_{k=0}^{m-1} a^*_{m-1,k} z^{-k} + \Gamma^*_m z^{-1} \sum_{k=0}^{m-1} a_{m-1,m-1-k} z^{-k}
\end{aligned}
\tag{4.76}
$$

where, in the second line, we have used the fact that $a_{m-1,m} = 0$. The sequence of numbers $a^*_{m-1,k}$, $k = 0, 1, \ldots, m - 1$, defines the impulse response of a forward prediction-error filter of order $m - 1$. Hence, we may write

$$H_{f,m-1}(z) = \sum_{k=0}^{m-1} a^*_{m-1,k} z^{-k} \tag{4.77}$$

The sequence of numbers $a_{m-1,m-1-k}$, $k = 0, 1, \ldots, m - 1$, defines the impulse response of a backward prediction-error filter of order $m - 1$; this is illustrated in Fig. 4.2(c) for the case of prediction order $m = M$. Hence, the second summation on the right side of Eq. (4.76) represents the transfer function of this backward prediction-error filter. Let $H_{b,m-1}(z)$ denote this transfer function, as shown by

$$H_{b,m-1}(z) = \sum_{k=0}^{m-1} a_{m-1,m-1-k} z^{-k} \tag{4.78}$$

Hence, substituting Eqs. (4.77) and (4.78) in (4.76), we may write

$$H_{f,m}(z) = H_{f,m-1}(z) + \Gamma^*_m z^{-1} H_{b,m-1}(z) \tag{4.79}$$

The order update recursion of Eq. (4.79) shows that, given the reflection coefficient Γ_m and the transfer functions of the forward and backward prediction-error filters of order $m - 1$, the transfer function of the corresponding forward prediction-error filter of order m is uniquely determined.

Minimum-Phase Property

On the unit circle in the z-plane (i.e., for $|z| = 1$), we find that

$$|H_{f,m-1}(z)| = |H_{b,m-1}(z)|, \qquad |z| = 1$$

This is readily proved by substituting $z = \exp(j\omega)$, $-\pi \leq \omega \leq \pi$, in Eqs. (4.77) and (4.78). Suppose that the reflection coefficient Γ_m satisfies the requirement $|\Gamma_m| < 1$ for all m. Then we find that on the unit circle in the z-plane the magnitude of the second term in the right side of Eq. (4.79) satisfies the conditions

$$|\Gamma_m^* z^{-1} H_{b,m-1}(z)| < |H_{b,m-1}(z)| = |H_{f,m-1}(z)|, \qquad |z| = 1 \qquad (4.80)$$

At this stage in our discussion, it is useful to recall *Rouché's theorem* from the theory of complex variables (Levinson and Redheffer, 1970). Rouché's theorem states that

If a function $F(z)$ is analytic upon a contour C in the z-plane and within the region enclosed by this contour, and if a second function $G(z)$, in addition to satisfying the same analyticity conditions, also fulfills the condition $|G(z)| < |F(z)|$ on the contour C, then the function $F(z) + G(z)$ has the same number of zeros within the region enclosed by the contour C as does the function $F(z)$.

Ordinarily, the enclosed contour C is traversed in the *counterclockwise* direction, and the region enclosed by the contour lies to the *left* of it, as illustrated in Fig. 4.4. We say that a function is analytic upon the contour C and within the region enclosed by it if the function has a continuous derivative everywhere upon the contour C and within the region enclosed by this contour. For this requirement to be satisfied, the function must have no poles upon the contour C or inside the region enclosed by the contour.

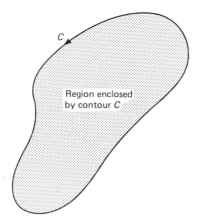

Figure 4.4 Contour C (traversed in the counterclockwise direction) and the region enclosed by it.

Let the contour C be the unit circle in the z plane, which is traversed in the *clockwise* direction, as in Fig. 4.5. According to the convention described, this assumption implies that the region enclosed by the contour C lies *outside* the unit circle in the z-plane.

Let the functions $F(z)$ and $G(z)$ be identified with the two terms in the right side of Eq. (4.79), as shown by

$$F(z) = H_{f,m-1}(z) \tag{4.81}$$

$$G(z) = \Gamma_m^* z^{-1} H_{b,m-1}(z) \tag{4.82}$$

We observe that:

1. The functions $F(z)$ and $G(z)$ have no poles inside the contour C defined in Fig. 4.5. Indeed, their derivatives are continuous throughout the region enclosed by this contour. Therefore, they are both analytic everywhere upon the unit circle and the region outside it.
2. In view of Eq. (4.80), we have $|G(z)| < |F(z)|$ on the unit circle.

Accordingly, the functions $F(z)$ and $G(z)$ defined by Eqs. (4.81) and (4.82), respectively, satisfy all the conditions required by Rouché's theorem with respect to the contour C defined as the unit circle in Fig. 4.5.

Suppose that $H_{f,m-1}(z)$ and therefore $F(z)$ are known to have no zeros outside the unit circle in the z-plane. Then, by applying Rouché's theorem, we find that $F(z) + G(z)$, or, equivalently, $H_{f,m}(z)$ also has no zeros on or outside the unit circle in the z plane.

In particular, for $m = 0$, the transfer function $H_{f,0}(z)$ is a constant equal to 1; therefore, it has no zeros at all. Using the result just derived, we may state that since $H_{f,0}(z)$ has no zeros outside the unit circle, then $H_{f,1}(z)$ will also have no zeros in this region of the z-plane, provided that $|\Gamma_1| < 1$. Indeed, we can

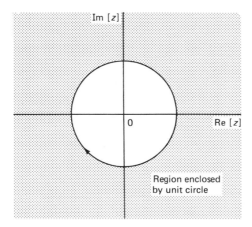

Figure 4.5 Unit circle (traversed in the clockwise direction) used as contour C.

easily prove this result by noting that

$$H_{f,1}(z) = a_{1,0}^* + a_{1,1}^* z^{-1}$$

$$= 1 + \Gamma_1^* z^{-1}$$

Hence, $H_{f,1}(z)$ has a single zero located at $z = -\Gamma_1^*$ and a pole at $z = 0$. With the reflection coefficient Γ_1 constrained by the condition $|\Gamma_1| < 1$, it follows that this zero must lie inside the unit circle. In other words, $H_{f,1}(z)$ has no zeros on or outside the unit circle. If $H_{f,1}(z)$ has no zeros on or outside the unit circle, then $H_{f,2}(z)$ will also have no zeros on or outside the unit circle provided that $|\Gamma_2| < 1$, and so on.

We may thus state that the transfer function $H_{f,m}(z)$ of a forward prediction-error filter of order m has no zeros on or outside the unit circle in the z-plane for all values of m, if and only if the reflection coefficients satisfy the condition $|\Gamma_m| < 1$ for all m. Such a filter is said to be *minimum phase* in the sense that, for a specified amplitude response, it has the minimum phase response possible for all values of z on the unit circle (Oppenheim and Schafer, 1975).

4.9 TRANSFER FUNCTION OF A BACKWARD PREDICTION-ERROR FILTER

Consider a backward prediction-error filter of order m whose impulse response is denoted by the sequence of numbers $a_{m,m-k}$, $k = 0, 1, \ldots, m$, as illustrated in Fig. 4.2(c) for $m = M$. Let $H_{b,m}(z)$ denote the transfer function of this filter:

$$H_{b,m}(z) = \sum_{k=0}^{m} a_{m,m-k} z^{-k} \tag{4.83}$$

Equation (4.49) describes the scalar version of the Levinson–Durbin recursion for a backward prediction-error filter of order m. Hence, substituting Eq. (4.49) in (4.83), we get

$$H_{b,m}(z) = z^{-1} H_{b,m-1}(z) + \Gamma_m H_{f,m-1}(z) \tag{4.84}$$

where $H_{b,m-1}$ and $H_{f,m-1}(z)$ are defined by Eqs. (4.78) and (4.77) for order $m-1$, respectively. The order-update recursion of Eq. (4.84) shows that, given the reflection coefficient Γ_m and the transfer functions of forward and backward prediction-error filters of order $m-1$, the transfer function of the corresponding backward prediction-error filter of order m is uniquely determined.

Relation between the Transfer Functions of Backward and Forward Prediction-Error Filters

We also find it useful to develop the relationship between the transfer functions of backward and forward prediction-error filters of the same order, such that if we are given one we may determine the other. To do this, we first evaluate

$H_{f,m}^{*}(z)$, the complex conjugate of the transfer function of a forward prediction-error filter of order m, and so write [see Eq. (4.75)]

$$H_{f,m}^{*}(z) = \sum_{k=0}^{m} a_{m,k}(z^{*})^{-k} \tag{4.85}$$

Replacing z by the reciprocal of its complex conjugate z^{*}, we may rewrite Eq. (4.85) as

$$H_{f,m}^{*}\left(\frac{1}{z^{*}}\right) = \sum_{k=0}^{m} a_{m,k} z^{k}$$

Next, replacing k by m-k, we get

$$H_{f,m}^{*}\left(\frac{1}{z^{*}}\right) = z^{m} \sum_{k=0}^{m} a_{m,m-k} z^{-k} \tag{4.86}$$

The summation on the right of Eq. (4.86) equals $H_{b,m}(z)$, the transfer function of a backward prediction-error filter of order m. We thus get the desired relation:

$$H_{b,m}(z) = z^{-m} H_{f,m}^{*}\left(\frac{1}{z^{*}}\right) \tag{4.87}$$

where $H_{f,m}^{*}(1/z^{*})$ is obtained by complex-conjugating $H_{f,m}(z)$, the transfer function of a forward prediction-error filter of order m, and replacing z by the reciprocal of z^{*}. Equation (4.87) states that multiplication of the new function obtained in this way by z^{-m} yields $H_{b,m}(z)$, the transfer function of the corresponding backward prediction-error filter.

Maximum-Phase Property

Let the transfer function $H_{f,m}(z)$ be expressed in its factored form as follows:

$$H_{f,m}(z) = \prod_{i=1}^{m} (1 - z_{i}z^{-1}) \tag{4.88}$$

where z_{i}, $i = 1, 2, \ldots, m$, denote the zeros of the forward prediction-error filter. Hence, substituting Eq. (4.88) into (4.87), we may express the transfer function of the corresponding backward prediction-error filter in the factored form

$$\begin{aligned} H_{b,m}(z) &= z^{-m} \prod_{i=1}^{m} (1 - z_{i}^{*}z) \\ &= \prod_{i=1}^{m}(z^{-1} - z_{i}^{*}) \end{aligned} \tag{4.89}$$

The zeros of this transfer function are located at $1/z_{i}^{*}$, $i = 1, 2, \ldots, m$. That is, the zeros of the backward and forward prediction-error filters are the *inverse* of each other with respect to the unit circle in the z plane. The geometric nature

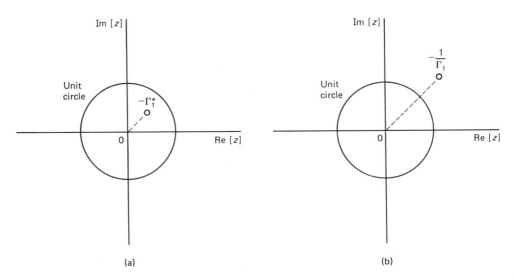

Figure 4.6 (a) Zero of forward prediction-error filter at $z = -\Gamma_1^*$; (b) corresponding zero of backward prediction-error filter at $z = -1/\Gamma_1$.

of this relationship is illustrated for $m = 1$ in Fig. 4.6. The forward prediction-error filter has a zero at $z = -\Gamma_1^*$, as in Fig. 4.6(a), and the backward prediction-error filter has a zero at $z = -1/\Gamma_1$, as in Fig. 4.6(b). In this figure, it is assumed that the reflection coefficient Γ_1 has a complex value.

Consequently, backward prediction-error filters have all their zeros located outside the unit circle in the z plane, in accordance with $|\Gamma_m| < 1$ for all m. Such filters are said to be *maximum phase* (Oppenheim and Schafer, 1975).

Relation between the Backward and Forward Prediction Errors

The transfer function $H_{b,m}(z)$ of a backward prediction-error filter of order m equals (by definition)

$$H_{b,m}(z) = \frac{B_m(z)}{U(z)} \tag{4.90}$$

where $B_m(z)$ is the z-transform of the backward prediction error sequence $\{b_m(n)\}$ at the output of the filter, and $U(z)$ is the z-transform of the signal $\{u(n)\}$ applied to the filter input; see Fig. 4.2(c) for $m = M$. Similarly, the transfer function $H_{f,m}(z)$ of the corresponding forward prediction-error filter equals

$$H_{f,m}(z) = \frac{F_m(z)}{U(z)} \tag{4.91}$$

where $F_m(z)$ is the z-transform of the forward prediction-error sequence $\{f_m(n)\}$ at the output of the filter, and $U(z)$ is, as before, the z-transform of the filter input;

see Fig. 4.2(b) for $m = M$. Solving Eqs. (4.90) and (4.91) for $B_m(z)$, we get

$$B_m(z) = \frac{H_{b,m}(z)}{H_{f,m}(z)} F_m(z) \tag{4.92}$$

To evaluate the ratio $H_{b,m}(z)/H_{f,m}(z)$, we divide Eq. (4.89) by (4.88), obtaining the result

$$\frac{H_{b,m}(z)}{H_{f,m}(z)} = \prod_{i=1}^{m} \left(\frac{z^{-1} - z_i^*}{1 - z_i z^{-1}} \right) \tag{4.93}$$

where z_i, $i = 1, 2, \ldots, m$, denote the zeros of the transfer function of the foward prediction-error filter. Owing to the minimum-phase property of this filter, we observe that $|z_i| < 1$ for all i. Hence, the poles and zeros of the transfer function in Eq. (4.93) are the inverse of each other with respect to the unit circle in the z-plane, with the poles located inside the unit circle and the zeros located outside. Figure 4.7 shows the geometry of this relationship for a typical pole-zero factor. A filter with such a transfer function is called an *all-pass filter*. It has the property that its amplitude response equals 1 for all values of z on the unit circle. In other words, it passes all frequencies with no amplitude distortion; hence, the name.

Accordingly, we may convert a forward prediction-error sequence $\{f_m(n)\}$ into the corresponding backward prediction-error sequence $\{b_m(n)\}$ by passing it through an all-pass filter of order m, as in Fig. 4.8. This filter is stable, since all its poles are located inside the unit circle in the z-plane. Note, however, the opposite conversion from $\{b_m(n)\}$ to $\{f_m(n)\}$ is not practically feasible, because the all-pass filter would then have its poles located outside the unit circle and, therefore, be unstable.

Note also that, since the amplitude response of the all-pass filter in Fig. 4.8 equals 1 for all frequencies, the forward prediction-error sequence $\{f_m(n)\}$ and the backward prediction-error sequence $\{b_m(n)\}$ have exactly the same power spectral density. This is intuitively satisfying since both prediction-error sequences have the same average power for stationary inputs.

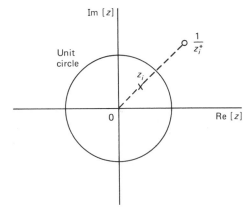

Figure 4.7 Geometry of typical pole–zero factor for all–pass filter.

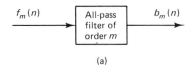

(a)

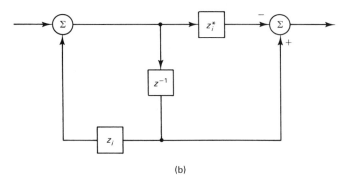

(b)

Figure 4.8 (a) Conversion of forward prediction-error sequence $\{f_m(n)\}$ into backward prediction-error sequence $\{b_m(n)\}$; (b) ith section of all-pole filter, $i = 1, 2, \ldots, m$.

4.10 THE SCHUR-COHN TEST

The test described in Section 4.8 for the minimum-phase condition of a forward prediction-error of order M is relatively simple to apply if we know the associated set of reflection coefficients $\Gamma_1, \Gamma_2, \ldots, \Gamma_M$. For the filter to be minimum phase [i.e., for all the zeros of the transfer function of the filter, $H_{f,m}(z)$, to lie inside the unit circle], we simply require that $|\Gamma_m| < 1$ for all m. Suppose, however, that instead of these reflection coefficients we are given the tap weights of the filter, $a_{M,1}, a_{M,2}, \ldots a_{M,M}$. In this case we first apply the inverse recursion [described by Eq. (4.71)] to compute the corresponding set of reflection coefficients $\Gamma_1, \Gamma_2, \ldots, \Gamma_M$. Then, as before, we check whether or not $|\Gamma_m| < 1$ for all m.

The method just described for determining whether or not $H_{f,m}(z)$ has zeros inside the unit circle, given the coefficients $a_{M,1}, a_{M,2}, \ldots, a_{M,M}$, is essentially the same as the *Schur-Cohn test*.[3]

To formulate the Schur-Cohn test, let

$$x(z) = a_{M,M}z^M + a_{M,M-1}z^{M-1} + \cdots + a_{M,0} \qquad (4.94)$$

which is a polynomial in z, with $x(0) = a_{M,0} = 1$. Define

$$\begin{aligned} x'(z) &= z^M x^*(1/z^*) \\ &= a_{M,M}^* + a_{M,M-1}^* z + \cdots + a_{M,0}^* z^M \end{aligned} \qquad (4.95)$$

which is the *reciprocal polynomial* associated with $x(z)$. The polynomial $x(z)$ is so called since its zeros are the reciprocals of the zeros of $x(z)$. For $z = 0$, we

---

[3]The classical Schur–Cohn test is discussed in Marden (1949) and Tretter (1976). The origin of the test can be traced back to Schur (1917) and Cohn (1922); hence, the name. The test is also referred to as the Lehmer-Schur method (Ralston, 1965); this is in recognition of the application of Schur's theorem by Lehmer (1961).

have $x'(0) = a_{M,M}^*$. Next, define the linear combination

$$T[x(z)] = a_{M,0}^* x(z) - a_{M,M} x'(z) \tag{4.96}$$

so that, in particular, the value

$$T[x(0)] = a_{M,0}^* x(0) - a_{M,M} x'(0)$$
$$= 1 - |a_{M,M}|^2 \tag{4.97}$$

is real. Note also that $T[x(z)]$ has no term in z^M. Repeat this operation as far as possible, so that if we define

$$T^i[x(z)] = T\{T^{i-1}[x(z)]\} \tag{4.98}$$

we generate a finite sequence of polynomials in z of *decreasing* order. Let it be assumed that:

1. The coefficient $a_{m,0}$ is nonzero.
2. The polynomial $x(z)$ has no zero on the unit circle.
3. The integer m is the smallest for which

$$T^m[x(z)] = 0, \text{ where } m \le M + 1.$$

Then, we may state the Schur-Cohn theorem as follows (Lehmer, 1961):

> *If for some i such that $1 \le i < m$, we have $T^i[x(0)] < 0$, then $x(z)$ has at least one zero inside the unit circle. If, on the other hand, $T^i[x(0)] > 0$ for $1 \le i < m$, and $T^{m-1}[x(z)]$ is a constant, then no zero of $x(z)$ lies inside the unit circle.*

To apply this theorem to determine whether or not the polynomial $x(z)$ of Eq. (4.94), with $a_{M,0} \ne 0$, has a zero inside the unit circle, we proceed as follows (Ralston, 1965):

1. Calculate $T[x(z)]$. Is $T[x(0)]$ negative? If so, there is a zero inside the unit circle; if not, proceed to step 2.
2. Calculate $T^i[x(z)]$, $i = 1, 2, \ldots$, until $T^i[x(0)] < 0$ for $i < m$, or $T^i[x(0)] > 0$ for $i < m$. If the former occurs, there is a zero inside the unit circle. If the latter occurs, and if $T^{m-1}[x(z)]$ is a constant, then there is no zero inside the unit circle.

Note that when the polynomial $x(z)$ has zeros inside the unit circle, this algorithm does not tell us how many; rather, it only confirms their existence.

The connection between the Schur-Cohn method and the inverse recursion is readily established by observing that (see Problem 6):

1. The polynomial $x(z)$ is related to the transfer function of a backward prediction-error filter of order M as follows

$$x(z) = z^M H_{b,M}(z) \tag{4.99}$$

Accordingly, if the Schur-Cohn test indicates that $x(z)$ has zero(s) inside the
unit circle, we may conclude that the transfer function $H_{b,M}(z)$ is *not* maximum
phase.

2. The reciprocal polynomial $x'(z)$ is related to the transfer function of the
   corresponding forward prediction-error filter of order M as follows

$$x'(z) = z^M H_{f,M}(z) \tag{4.100}$$

Accordingly, if the Schur-Cohn test indicates that the original polynomial
$x(z)$, with which $x'(z)$ is associated, has zero(s) inside the unit circle, we may
then conclude that the transfer function $H_{f,M}(z)$ is *not* minimum phase.

3. In general, we have

$$T^i[x(0)] = \prod_{j=0}^{i-1} (1 - |a_{M-j,M-j}|^2), \qquad 1 \le i \le M \tag{4.101}$$

and

$$H_{b,M-i}(z) = \frac{z^{i-M} T^i[x(z)]}{T^i[x(0)]} \tag{4.102}$$

where $H_{b,M-i}(z)$ is the transfer function of the backward prediction-error
filter of order $M - i$.

4.11 WHITENING PROPERTY OF PREDICTION-ERROR FILTERS

By definition, a *white-noise* process consists of a sequence of uncorrelated random
variables. Thus, assuming that such a process, denoted by $\{v(n)\}$, has zero mean
and variance σ_v^2, we may write (see Section 2.7)

$$E[v(k)v^*(n)] = \begin{cases} \sigma_v^2, & k = n \\ 0, & k \ne n \end{cases} \tag{4.103}$$

Accordingly, we say that white noise is purely unpredictable in the sense that the
value of the process at time n is uncorrelated with all past values of the process
up to and including time $n - 1$ (and, indeed, with all future values of the process,
too).

We may now state another important property of a prediction-error filter.
In theory, a prediction-error filter is capable of whitening a stationary discrete-
time stochastic process applied to its input, provided that the order of the filter is
high enough. Basically, prediction relies on the presence of correlation between

adjacent samples of the input process.   The implication of this is that, as we increase the order of the prediction-error filter, we successively reduce the correlation between adjacent samples of the input process, until ultimately we reach a point at which the filter has a high enough order to produce an output process that consists of a sequence of uncorrelated samples.   The whitening of the original process applied to the filter input will have thereby been accomplished.

Implications of the Whitening Property of a Prediction-error Filter and the Autoregressive Modeling of a Stationary Stochastic Process

The whitening property of a prediction-error filter, operating on a stationary discrete-time stochastic process, is intimately related to the *autoregressive modeling* of the process.   Indeed, we may view these two operations as *complementary*, as illustrated in Fig. 4.9.   Part (a) of the figure depicts a forward prediction-error filter of order M, whereas part (b) depicts the corresponding autoregressive model. We may make the following observations:

1. We may view the operation of prediction-error filtering applied to a stationary process $\{u(n)\}$ as one of *analysis*.   In particular, we may use such an operation to whiten the process $\{u(n)\}$ by choosing the prediction-error filter order M sufficiently large, in which case the prediction error process $\{f_M(n)\}$ at the filter output consists of uncorrelated samples.   When this unique condition has been established, the original stochastic process $\{u(n)\}$ is represented by the tap weights of the filter, $\{a_{M,k}\}$, and the prediction error power, P_M.

2. We may view the autoregressive (AR) modeling of the stationary process $\{u(n)\}$ as one of *synthesis*.   In particular, we may generate the AR process $\{u(n)\}$ by applying a white-noise process $\{v(n)\}$ of zero mean and variance σ_v^2 to the input of an *inverse* filter whose parameters are set equal to the AR parameters $\{w_{ok}\}$.

The two filter structures of Fig. 4.9 constitute a *matched pair*, with their parameters related as follows:

and
$$a_{M,k} = -w_{ok}, \qquad k = 1, 2, \ldots, M$$
$$P_M = \sigma_v^2$$

The prediction-error filter of Fig. 4.9(a) is an *all-zero filter with an impulse response of finite duration*.   On the other hand, the inverse filter in the AR model of Fig. 4.9(b) is an *all-pole filter with an impulse response of infinite duration*.   The prediction-error filter is minimum phase, with the zeros of its transfer function located at exactly the same positions (inside the units circle in the z-plane) as the poles of the transfer function of the inverse filter in part (b) of the figure.   This assures

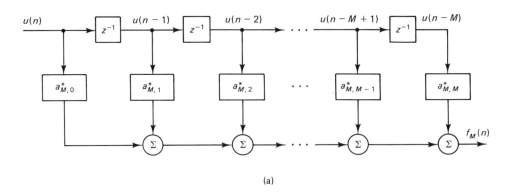

(a)

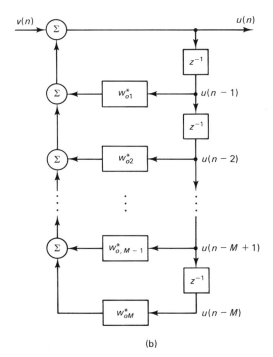

(b)

Figure 4.9 (a) Prediction-error (all-zero) filter; (b) autoregressive (all-pole) model
with $w_{ok} = -a_{M,k}$, $k = 1, 2, \ldots, M$; the input $\{v(n)\}$ is white noise.

the stability of the inverse filter or, equivalently, the asymptotic stationarity of the
AR process generated at the output of this filter.

The principles described provide the basis of *linear predictive coding (LPC)*
vocorders for the transmission and reception of digitized speech.   Note that the

(all-zero) prediction-error filter and the (all-pole) inverse filter in Fig. 4.9 for an AR process play the same roles as the whitening filter and inverse filter of Fig. 2.14 for a process with rational power spectrum, respectively.

4.12 EIGENVECTOR REPRESENTATIONS OF PREDICTION-ERROR FILTERS

The study of a prediction-error filter would be incomplete without considering its representation in terms of the eigenvalues (and associated eigenvectors) of the correlation matrix of the tap inputs in the filter. To develop such a representation, we first rewrite the augmented normal equation (4.18), pertaining to a forward prediction-error filter of order M, in the compact matrix form

$$\mathbf{R}_{M+1}\mathbf{a}_M = P_M\boldsymbol{\delta}_{M+1} \tag{4.104}$$

where $\mathbf{R}_{M+1}$ is the $(M + 1)$-by-$(M + 1)$ correlation matrix of the tap inputs $u(n)$, $u(n - 1), \ldots, u(n - M)$ in the filter of Fig. 4.1(b), $\mathbf{a}_M$ is the $(M + 1)$-by-1 tap-weight vector of the filter, and the scalar P_M is the prediction error power. The $(M + 1)$-by-1 vector $\boldsymbol{\delta}_{M+1}$ is called the *first unit vector*; it has unity for its first element and zero for all the others. We illustrate this by writing

$$\boldsymbol{\delta}_{M+1}^T = [1, 0, \ldots, 0] \tag{4.105}$$

Solving Eq. (4.104) for $\mathbf{a}_M$, we get

$$\mathbf{a}_M = P_M\mathbf{R}_{M+1}^{-1}\boldsymbol{\delta}_{M+1} \tag{4.106}$$

where $\mathbf{R}_{M+1}^{-1}$ is the inverse of the correlation matrix $\mathbf{R}_{M+1}$. Using the eigenvalue–eigenvector representation of a correlation matrix, which was discussed in Section 2.5, we may express the inverse matrix $\mathbf{R}_{M+1}^{-1}$ as follows:

$$\mathbf{R}_{M+1}^{-1} = \mathbf{Q}\boldsymbol{\Lambda}^{-1}\mathbf{Q}^H \tag{4.107}$$

where $\boldsymbol{\Lambda}$ is an $(M + 1)$-by-$(M + 1)$ diagonal matrix consisting of the eigenvalues of the correlation matrix $\mathbf{R}_{M+1}$, and $\mathbf{Q}$ is an $(M + 1)$-by-$(M + 1)$ matrix whose columns are the associated eigenvectors. That is,

$$\boldsymbol{\Lambda} = \text{diag}\,[\lambda_0, \lambda_1, \ldots, \lambda_M] \tag{4.108}$$

and

$$\mathbf{Q} = [\mathbf{q}_0, \mathbf{q}_1, \ldots, \mathbf{q}_M] \tag{4.109}$$

where $\lambda_0, \lambda_1, \ldots, \lambda_M$ are the real-valued eigenvalues of the correlation matrix $\mathbf{R}_{M+1}$, and $\mathbf{q}_0, \mathbf{q}_1, \ldots, \mathbf{q}_M$ are the respective eigenvectors. Thus, substituting

Eqs. (4.107), (4.108), and (4.109) in (4.106), we get

$$\mathbf{a}_M = P_M \mathbf{Q} \mathbf{\Lambda}^{-1} \mathbf{Q}^H \boldsymbol{\delta}_{M+1}$$

$$= P_M [\mathbf{q}_0, \mathbf{q}_1, \ldots, \mathbf{q}_M] \operatorname{diag}[\lambda_0^{-1}, \lambda_1^{-1}, \ldots, \lambda_M^{-1}] \begin{bmatrix} \mathbf{q}_0^H \\ \mathbf{q}_1^H \\ \cdot \\ \cdot \\ \cdot \\ \mathbf{q}_M^H \end{bmatrix} \begin{bmatrix} 1 \\ 0 \\ \cdot \\ \cdot \\ \cdot \\ 0 \end{bmatrix} \qquad (4.110)$$

$$= P_M \sum_{k=0}^{M} \left(\frac{q_{k0}^*}{\lambda_k} \right) \mathbf{q}_k$$

where q_{k0} is the first element of the kth eigenvector of the correlation matrix $\mathbf{R}_{M+1}$.

We note that the first element of the forward prediction-error filter vector $\mathbf{a}_M$ equals 1. Therefore, using this fact, we find from Eq. (4.110) that the prediction-error power equals

$$P_M = \frac{1}{\displaystyle\sum_{k=0}^{M} |q_{k0}|^2 \lambda_k^{-1}} \qquad (4.111)$$

Thus, we see from Eqs. (4.110) and (4.111) that the tap-weight vector of a forward prediction-error filter of order M and the resultant prediction-error power are uniquely defined by specifying the $(M + 1)$ eigenvalues and the corresponding $(M + 1)$ eigenvectors of the $(M + 1)$-by-$(M + 1)$ correlation matrix of the tap inputs of the filter.

From Eq. (4.110), we readily find that the tap-weight vector of the corresponding backward prediction-error filter may be represented in the eigenvalue–eigenvector form as

$$\mathbf{a}_M^{B*} = P_M \sum_{k=0}^{M} \left(\frac{q_{k0}}{\lambda_k} \right) \mathbf{q}_k^{B*} \qquad (4.112)$$

where $\mathbf{q}_k^{B*}$ is obtained by rearranging the elements of the eigenvector $\mathbf{q}_k$ in reverse order and complex-conjugating them.

4.13 LATTICE PREDICTORS

A forward or backward prediction-error filter may be realized using a tapped-delay-line structure, as shown in Figs. 4.1 and 4.2. There exists, however, another important device known as a *lattice predictor* that combines the forward and backward prediction-error filtering operations into a single structure. The lattice predictor consists of a number of stages connected in cascade, with each stage in the form of a lattice; hence, the name. The number of stages in a lattice predictor equals the prediction-error filter order. Thus, for a prediction-error filter of order m, there are m stages in the lattice realization of the filter.

Order-update Recursions for the Prediction Errors

The input–output relations that characterize a lattice predictor may be derived in various ways, depending on the particular form in which the Levinson–Durbin recursion is utilized. For the derivation presented here, we start with the matrix formulations of this recursion given by Eqs. (4.46) and (4.48) that pertain to the forward and backward operations of a prediction-error filter, respectively. For convenience of presentation, we reproduce these two relations here:

$$\mathbf{a}_m = \begin{bmatrix} \mathbf{a}_{m-1} \\ 0 \end{bmatrix} + \Gamma_m \begin{bmatrix} 0 \\ \mathbf{a}_{m-1}^{B*} \end{bmatrix} \tag{4.113}$$

$$\mathbf{a}_m^{B*} = \begin{bmatrix} 0 \\ \mathbf{a}_{m-1}^{B*} \end{bmatrix} + \Gamma_m^* \begin{bmatrix} \mathbf{a}_{m-1} \\ 0 \end{bmatrix} \tag{4.114}$$

The $(m + 1)$-by-1 vector $\mathbf{a}_m$ and the m-by-1 vector $\mathbf{a}_{m-1}$ refer to forward prediction-error filters of order m and $m - 1$, respectively. The $(m + 1)$-by-1 vector $\mathbf{a}_m^{B*}$ and the m-by-1 vector $\mathbf{a}_{m-1}^{B*}$ refer to the corresponding backward prediction-error filters of order m and $m - 1$, respectively. The scalar Γ_m is the associated reflection coefficient.

Consider first the forward prediction-error filter of order m, with its tap inputs denoted by $u(n)$, $u(n - 1)$, . . . , $u(n - m)$. We may partition $\mathbf{u}_{m+1}(n)$, the $(m + 1)$-by-1 tap-input vector of this filter, in the form

$$\mathbf{u}_{m+1}(n) = \begin{bmatrix} \mathbf{u}_m(n) \\ \hline u(n - m) \end{bmatrix} \tag{4.115}$$

or in the equivalent form

$$\mathbf{u}_{m+1}(n) = \begin{bmatrix} u(n) \\ \hline \mathbf{u}_m(n - 1) \end{bmatrix} \tag{4.116}$$

Form the inner product of the $(m + 1)$-by-1 vectors $\mathbf{a}_m$ and $\mathbf{u}_{m+1}(n)$. This is done by premultiplying $\mathbf{u}_{m+1}(n)$ by the Hermitian transpose of $\mathbf{a}_m$. Thus, using Eq. (4.113) for $\mathbf{a}_m$, we may treat the terms resulting from this multiplication, as follows:

1. For the left side of Eq. (4.113), premultiplication of $\mathbf{u}_{m+1}(n)$ by $\mathbf{a}_m^H$ yields

$$f_m(n) = \mathbf{a}_m^H \mathbf{u}_{m+1}(n) \tag{4.117}$$

where $f_m(n)$ is the forward prediction error produced at the output of the forward prediction-error filter of order m.

2. For the first term on the right side of Eq. (4.113), we use the partitioned form of $\mathbf{u}_{m+1}(n)$ given in Eq. (4.115). We may therefore write

$$[\mathbf{a}_{m-1}^H \mathbin{\vdots} 0] \, \mathbf{u}_{m+1}(n) = [\mathbf{a}_{m-1}^H \mathbin{\vdots} 0] \begin{bmatrix} \mathbf{u}_m(n) \\ \hline u(n-m) \end{bmatrix}$$

$$= \mathbf{a}_{m-1}^H \mathbf{u}_m(n) \qquad\qquad (4.118)$$

$$= f_{m-1}(n)$$

where $f_{m-1}(n)$ is the forward prediction error produced at the output of the forward prediction-error filter of order $m-1$.

3. For the second matrix term on the right side of Eq. (4.113), we use the partitioned form of $\mathbf{u}_{m+1}(n)$ given in Eq. (4.116). We may therefore write

$$[0 \mathbin{\vdots} \mathbf{a}_{m-1}^{BT}] \, \mathbf{u}_{m+1}(n) = [0 \mathbin{\vdots} \mathbf{a}_{m-1}^{BT}] \begin{bmatrix} u(n) \\ \hline \mathbf{u}_m(n-1) \end{bmatrix}$$

$$= \mathbf{a}_{m-1}^{BT} \mathbf{u}_m(n-1) \qquad\qquad (4.119)$$

$$= b_{m-1}(n-1)$$

where $b_{m-1}(n-1)$ is the *delayed* backward prediction error produced at the output of the backward prediction-error filter of order $m-1$.

Combining the results of the multiplication, described by Eqs. (4.117), (4.118) and (4.119), we may thus write

$$f_m(n) = f_{m-1}(n) + \Gamma_m^* b_{m-1}(n-1) \qquad\qquad (4.120)$$

Consider next the backward prediction-error filter of order m, with its tap inputs denoted by $u(n)$, $u(n-1)$, ..., $u(n-m)$. Here again we may express $\mathbf{u}_{m+1}(n)$, the $(m+1)$-by-1 tap-input vector of this filter, in the partitioned form of Eq. (4.115) or that of Eq. (4.116). In this case, the terms resulting from the formation of the inner product of the vectors $\mathbf{a}_m^{B*}$ and $\mathbf{u}_{m+1}(n)$ are treated as follows:

1. For the left side of Eq. (4.114), premultiplication of $\mathbf{u}_{m+1}(n)$ by the Hermitian transpose of $\mathbf{a}_m^{B*}$ yields

$$b_m(n) = \mathbf{a}_m^{BT} \, \mathbf{u}_{m+1}(n) \qquad\qquad (4.121)$$

where $b_m(n)$ is the backward prediction error produced at the output of the backward prediction-error filter of order m.

2. For the first term on the right side of Eq. (4.114), we use the partitioned

form of the tap-input vector $\mathbf{u}_{m+1}(n)$ given in Eq. (4.116). Thus, multiplying the Hermitian transpose of this term by $\mathbf{u}_{m+1}(n)$, we get

$$[0 \mathbin{\vdots} \mathbf{a}^{BT}_{m-1}] \, \mathbf{u}_{m+1}(n) = [0 \mathbin{\vdots} \mathbf{a}^{BT}_{m-1}] \begin{bmatrix} u(n) \\ \hline \\ \mathbf{u}_m(n-1) \end{bmatrix}$$

$$= \mathbf{a}^{BT}_{m-1} \, \mathbf{u}_m(n-1) \qquad\qquad (4.122)$$

$$= b_{m-1}(n-1)$$

3. For the second matrix term on the right side of Eq. (4.114), we use the partitioned form of the top-input vector $\mathbf{u}_{m+1}(n)$ given in Eq. (4.115). Thus, multiplying the Hermitian transpose of this term by $\mathbf{u}_{m+1}(n)$, we get

$$[\mathbf{a}^H_{m-1} \mathbin{\vdots} 0] \, \mathbf{u}_{m+1}(n) = [\mathbf{a}^H_{m-1} \mathbin{\vdots} 0] \begin{bmatrix} \mathbf{u}_m(n) \\ \hline \\ u(n-m) \end{bmatrix}$$

$$= \mathbf{a}^H_{m-1} \, \mathbf{u}_m(n) \qquad\qquad (4.123)$$

$$= f_{m-1}(n)$$

Combining the results of Eqs. (4.121), (4.112), and (4.123), we thus find that the inner product of $\mathbf{a}^{B*}_m$ and $\mathbf{u}_{m+1}(n)$ yields

$$b_m(n) = b_{m-1}(n-1) + \Gamma_m f_{m-1}(n) \qquad\qquad (4.124)$$

Equations (4.120) and (4.124) are the desired pair of *order-update recursions* that characterize stage m of the lattice predictor. They are reproduced here in matrix form as follows

$$\begin{bmatrix} f_m(n) \\ b_m(n) \end{bmatrix} = \begin{bmatrix} 1 & \Gamma^*_m \\ \Gamma_m & 1 \end{bmatrix} \begin{bmatrix} f_{m-1}(n) \\ b_{m-1}(n-1) \end{bmatrix}, \qquad m = 1, 2, \ldots, M \qquad (4.125)$$

We may view $b_{m-1}(n-1)$ as the result of applying the unit-delay operator z^{-1} to the backward prediction error $b_{m-1}(n)$; that is,

$$b_{m-1}(n-1) = z^{-1}[b_{m-1}(n)] \qquad\qquad (4.126)$$

Thus, using Eqs. (4.125) and (4.126), we may represent stage m of the lattice predictor by the signal-flow graph shown in Fig. 4.10(a). Except for the branch pertaining to the block labeled z^{-1}, this signal-flow graph has the appearance of a lattice; hence, the name lattice predictor. Note also that the parameterization of stage m of the lattice predictor is uniquely defined by the reflection coefficient Γ_m.

The first application of lattice filters in on-line adaptive signal processing was apparently made by Itakura and Saito (1971) in the field of speech analysis. Equiv-

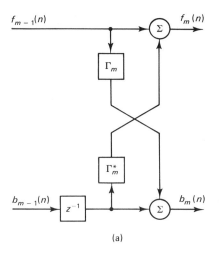

(a)

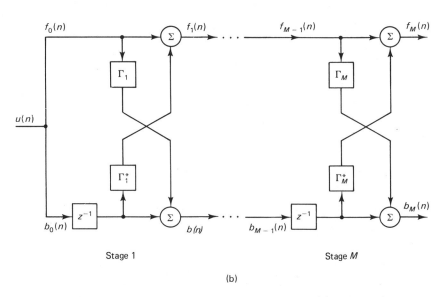

(b)

Figure 4.10 Signal-flow graph for stage m of a lattice predictor; (b) lattice equivalent model of prediction-error filter of order M.

alent lattice-filter models, however, were familiar in geophysical signal processing as "layered earth models" (Robinson, 1967; Burg, 1968). It is also of interest to note that such lattice filters have been well studied in network theory, especially in the cascade synthesis of multiport networks (Dewilde, 1969).

For the elementary case of $m = 0$, we simply have

$$f_0(n) = b_0(n) = u(n) \tag{4.127}$$

where $u(n)$ is the input signal at time n. Therefore, starting with $m = 0$, and progressively increasing the order of the filter by 1, we obtain the *lattice equivalent model* shown in Fig. 4.10(b) for a prediction-error filter of final order M. In this figure we merely require knowledge of the complete set of reflection coefficients $\Gamma_1, \Gamma_2, \ldots, \Gamma_M$, one for each stage of the filter.

The lattice filter represents a highly efficient structure for generating the sequence of forward prediction errors and the corresponding sequence of backward prediction errors simultaneously. Consider, for example, the requirement to generate the sequence of backward prediction errors $b_0(n), b_1(n), \ldots, b_M(n)$. The element $b_0(n)$ equals the input $u(n)$. To generate $b_1(n), b_2(n), \ldots, b_M(n)$ using a tapped-delay-line implementation, we need a total of M separate filters. Specifically, to generate $b_1(n)$, we need a backward prediction-error filter of order 1. To generate $b_2(n)$, we need a second backward prediction-error filter of order 2, and so on for the other elements of the sequence. On the other hand, we can generate the required sequence of backward prediction errors by using a single lattice filter that consists of M stages (i.e., of order M). In addition, we also generate the corresponding sequence of forward prediction errors by using the lattice filter.

The lattice filter is *modular* in structure; hence, if the requirement calls for increasing the order of the predictor, we simply add one or more stages (as desired) without affecting earlier computations.

4.14 CORRELATION PROPERTIES OF LATTICE FILTERS

The lattice filter of Fig. 4.10 exhibits some interesting correlation properties between the forward and backward prediction errors developed at the various stages of the filter (Griffiths, 1977, Makhoul, 1978). Basically, these properties are consequences of the principle of orthogonality described in Section 3.6.

Property 1. *The forward prediction error $f_m(n)$ and the input signal $u(n)$ are orthogonal:*

$$E[f_m(n)u^*(n - k)] = 0, \quad 1 \le k \le m \tag{4.128}$$

Similarly, the backward prediction error $b_m(n)$ and the input signal $u(n)$ are orthogonal:

$$E[b_m(n)u^*(n - k)] = 0, \quad 0 \le k \le m - 1 \tag{4.129}$$

Note the difference between the ranges of the index k in Eqs. (4.128) and (4.129).

Equations (4.128) and (4.129) are both restatements of the principle of orthogonality. By definition, the forward prediction error $f_m(n)$ equals the difference between $u(n)$ and the prediction of $u(n)$, given the tap inputs $u(n - 1), u(n - 2), \ldots, u(n - m)$. By the principle of orthogonality the error $f_m(n)$ is orthogonal to $u(n - k)$, $k = 1, 2, \ldots, m$. This proves Eq. (4.128).

The backward prediction error, by definition, equals the difference between $u(n - m)$ and the prediction of $u(n - m)$, given the tap inputs $u(n)$, $u(n - 1)$, ..., $u(n - m + 1)$. Here, again, by the principle of orthogonality the error $b_m(n)$ is orthogonal to $u(n - k)$, $k = 0, 1, \ldots, m - 1$. This proves Eq. (4.129).

Property 2. *The cross-correlation of the forward prediction error $f_m(n)$ and the input $u(n)$ equals the cross-correlation of the backward prediction error $b_m(n)$ and the time-shifted input $u(n - m)$, as shown by*

$$E[f_m(n)u^*(n)] = E[b_m(n)u^*(n - m)] = P_m \tag{4.130}$$

where P_m is the corresponding prediction-error power.

To prove the first part of this property, we note that $u(n)$ equals the forward prediction error $f_m(n)$ plus the prediction of $u(n)$, given the samples $u(n - 1)$, $u(n - 2)$, ..., $u(n - m)$. Since this prediction is orthogonal to the error $f_m(n)$ [which is a corollary to the principle of orthogonality], it follows that

$$E[f_m(n)u^*(n)] = E[f_m(n)f_m^*(n)]$$

$$= P_m$$

To prove the second part of the property, we note that $u(n - m)$ equals the backward prediction error $b_m(n)$ plus the prediction of $u(n - m)$, given the samples $u(n)$, $u(n - 1)$, ..., $u(n - m + 1)$. Since this prediction is orthogonal to the error $b_m(n)$, it follows that

$$E[b_m(n)u^*(n - m)] = E[b_m(n)b_m^*(n)]$$

$$= P_m$$

This completes the proof of Eq. (4.130).

Property 3. *The backward prediction errors are orthogonal to each other, as shown by*

$$E[b_m(n)b_i^*(n)] = \begin{cases} P_m, & m = i \\ 0, & m \neq i \end{cases} \tag{4.131}$$

The forward prediction errors do not, however, exhibit the same orthogonality property as the backward prediction errors; rather, they are correlated, as shown by

$$E[f_m(n)f_i^*(n)] = P_m, \qquad m \geq i \tag{4.132}$$

Without loss of generality, we may assume that $m \geq i$. To prove Eq. (4.131), we express the backward prediction error $b_i(n)$ in terms of the input $u(n)$ as the convolution sum

$$b_i(n) = \sum_{k=0}^{i} a_{i,i-k}u(n-k) \tag{4.133}$$

where $a_{i,i-k}^*$, $k = 0, 1, \ldots, i$, are the coefficients of a backward prediction-error filter of order i. Hence, we may write

$$E[b_m(n)b_i^*(n)] = E\left[b_m(n) \sum_{k=0}^{i} a_{i,i-k}^* u^*(n-k) \right] \qquad (4.134)$$

Now, by Property 1, we have

$$E[b_m(n)u^*(n-k)] = 0, \qquad 0 \le k \le m-1$$

For the case when $m > i$, and with $0 \le k \le i$, we therefore find that all the expectation terms inside the summation on the right side of Eq. (4.134) are zero. Correspondingly,

$$E[b_m(n)b_i^*(n)] = 0, \qquad m \ne i$$

When $m = i$, Eq. (4.134) reduces to

$$E[b_m(n)b_i^*(n)] = E[b_m(n)b_m^*(n)]$$

$$= P_m, \qquad m = i$$

This completes the proof of Eq. (4.131).

To prove Eq. (4.132), we express the forward prediction error $f_i(n)$ in terms of the input $u(n)$ as the convolution sum

$$f_i(n) = \sum_{k=0}^{i} a_{i,k}^* u(n-k) \qquad (4.135)$$

where $a_{i,k}$, $k = 0, 1, \ldots i$, are the coefficients of a forward prediction-error filter of order i. Hence,

$$E[f_m(n)f_i^*(n)] = E\left[f_m(n) \sum_{k=0}^{i} a_{i,k} u^*(n-k) \right] \qquad (4.136)$$

$$= E[f_m(n)u^*(n)] + \sum_{k=1}^{i} a_{i,k}E[f_m(n)u^*(n-k)]$$

where we have used the fact that $a_{i,0} = 1$. However, by Property 1, we have

$$E[f_m(n)u^*(n-k)] = 0, \qquad 1 \le k \le m$$

Also, by Property 2, we have

$$E[f_m(n)u^*(n)] = P_m$$

Therefore, Eq. (4.136) reduces to

$$E[f_m(n)f_i^*(n)] = P_m, \qquad m \ge i$$

This completes the proof of Eq. (4.132).

Property 4. *The time-shifted versions of the forward and backward prediction errors are orthogonal, as shown by, respectively,*

$$E[f_m(n)f_i^*(n-l)] = E[f_m(n+l)f_i^*(n)] = 0, \qquad 1 \le l \le m-i \qquad (4.137)$$

$$m > i$$

$$E[b_m(n)b_i^*(n-l)] = E[b_m(n+l)b_i^*(n)] = 0, \qquad 0 \le l \le m-i-1 \qquad (4.138)$$

$$m > i$$

where l is an integer lag.

To prove Eq. (4.137), we use Eq. (4.135) to write

$$
\begin{aligned}
E[f_m(n)f_i^*(n-l)] &= E\left[f_m(n) \sum_{k=0}^{i} a_{i,k} u^*(n-l-k) \right] \\
&= \sum_{k=0}^{i} a_{i,k} E[f_m(n)u^*(n-l-k)]
\end{aligned}
\qquad (4.139)
$$

By Property 1, we have

$$E[f_m(n)u^*(n-l-k)] = 0, \qquad 1 \le l+k \le m \qquad (4.140)$$

In the summation on the right side of Eq. (4.139), we have $0 \le k \le i$. For the orthogonality relationship of Eq. (4.140) to hold for all values of k inside this range, the lag l must correspondingly satisfy the condition $1 \le l \le m-i$. Thus, with the lag l bounded in this way, and with $m > i$, all the expectation terms inside the summation on the right side of Eq. (4.139) are zero. We therefore have

$$E[f_m(n)f_i^*(n-l)] = 0, \qquad 1 \le l \le m-i$$

$$m > i$$

By definition, we have

$$E[f_m(n)f_i^*(n-l)] = E[f_m(n+l)f_i^*(n)]$$

Therefore, if the expectation $E[f_m(n)f_i^*(n-l)]$ is zero, then so is the expectation $E[f_m(n+l)f_i^*(n)]$. This completes the proof of Eq. (4.137).

To prove Eq. (4.138), we use Eq. (4.133) to write

$$
\begin{aligned}
E[b_m(n)b_i^*(n-l)] &= E\left[b_m(n) \sum_{k=0}^{i} a_{i,i-k}^* u^*(n-l-k) \right] \\
&= \sum_{k=0}^{i} a_{i,i-k}^* E[b_m(n)u^*(n-l-k)]
\end{aligned}
\qquad (4.141)
$$

By Property 1, we have

$$E[b_m(n)u^*(n-l-k)] = 0, \qquad 0 \le l+k \le m-1 \qquad (4.142)$$

For the orthogonality relationship to hold for $0 \leq k \leq i$, the lag l must satisfy the condition $0 \leq l \leq m-i-1$. Then, with $m > i$, we find that all the expectations inside the summation on the right side of Eq. (4.141) are zero. We therefore have

$$E[b_m(n)b_i^*(n-l)] = 0, \qquad 0 \leq l \leq m-i-1$$
$$m > i$$

By definition, we have

$$E[b_m(n)b_i^*(n-l)] = E[b_m(n+l)b_i^*(n)]$$

Hence, if the expectation $E[b_m(n)b_i^*(n-l)]$ is zero, then so is the expectation $E[b_m(n+l)b_i^*(n)]$. This completes the proof of Eq. (4.138).

Property 5. *The time-shifted foward prediction errors $f_m(n+m)$ and $f_i(n+i)$ are orthogonal, as shown by*

$$E[f_m(n+m)f_i^*(n+i)] = \begin{cases} P_m, & m = i \\ 0, & m \neq i \end{cases} \qquad (4.143)$$

The corresponding time-shifted backward prediction errors $b_m(n+m)$ and $b_i(n+i)$, on the other hand, are correlated as shown by

$$E[b_m(n+m)b_i^*(n+i)] = P_m, \qquad m \geq i \qquad (4.144)$$

Equations (4.143) and (4.144) are the duals of Eqs. (4.131) and (4.132).

Without loss of generality, we may assume $m \geq i$. To prove Eq. (4.143), we first recognize that

$$E[f_m(n+m)f_i^*(n+i)] = E[f_m(n)f_i^*(n-m+i)]$$
$$= E[f_m(n)f_i^*(n-l)] \qquad (4.145)$$

where

$$l = m - i$$

Therefore, with $m > i$, we find from Property 4 that the expectation in Eq. (4.145) is zero. When, however, $m = i$, the lag l is zero, and this expectation equals P_m, the mean-square value of $f_m(n)$. This completes the proof of Eq. (4.143).

To prove Eq. (4.144), we recognize that

$$E[b_m(n+m)b_i^*(n+i)] = E[b_m(n)b_i^*(n-m+i)]$$
$$= E[b_m(n)b_i^*(n-l)] \qquad (4.146)$$

where

$$l = m - i$$

This value of l lies outside the range for which the expectation in Eq. (4.146) is zero [see Eq. (4.138)]. This means that $b_m(n+m)$ and $b_i(n+i)$ are correlated.

To determine this correlation, we use Eq. (4.133) to write

$$E[b_m(n+m)b_i^*(n+i)] = E\left[b_m(n+m)\sum_{k=0}^{i}a_{i,i-k}^*u^*(n+i-k)\right]$$

$$= E[b_m(n+m)u^*(n)] \qquad (4.147)$$

$$+ \sum_{k=0}^{i-1}a_{i,i-k}^* E[b_m(n+m)u^*(n+i-k)]$$

where we have used $a_{i,0} = 1$. By Property 1, we have

$$E[b_m(n+m)u^*(n+i-k)] = E[b_m(n)u^*(n+i-k-m)] \qquad (4.148)$$

$$= 0, \qquad 0 \le k+m-i \le m-1$$

The orthogonality relationship of Eq. (4.148) holds for $i-m \le k \le i-1$. The summation on the right side of Eq. (4.147) applies for $0 \le k \le i-1$. Hence, with $m \ge i$, all the expectation terms inside this summation are zero. Correspondingly, Eq. (4.147) reduces to

$$E[b_m(n+m)b_i^*(n+i)] = E[b_m(n+m)u^*(n)]$$

$$= E[b_m(n)u^*(n-m)]$$

$$= P_m, \qquad m \ge i$$

where we have made use of Property 2. This completes the proof of Eq. (4.144).

Property 6.   *The forward and backward prediction errors exhibit the following cross-correlation property:*

$$E[f_m(n)b_i^*n)] = \begin{cases} \Gamma_i^*P_m, & m \ge i \\ 0, & m < i \end{cases} \qquad (4.149)$$

To prove this property, we use Eq. (4.133) to write

$$E[f_m(n)b_i^*(n)] = E\left[f_m(n)\sum_{k=0}^{i}a_{i,i-k}^*u^*(n-k)\right] \qquad (4.150)$$

$$= a_{i,i}^*E[f_m(n)u^*(n)] + \sum_{k=1}^{i}a_{i,\,i-k}^*E[f_m(n)u^*(n-k)]$$

By Property 1, we have

$$E[f_m(n)u^*(n-k)] = 0, \qquad 1 \le k \le m$$

Assuming that $m \ge i$, we therefore find that all the expectation terms in the second term (summation) on the right side of Eq. (4.150) are zero. Hence,

$$\sum_{k=1}^{i}a_{i,i-k}^*E[f_m(n)u^*(n-k)] = 0$$

By Property 2, we have

$$E[f_m(n)u^*(n)] = P_m$$

Therefore, with $a_{i,i} = \Gamma_i$, we find that Eq. (4.150) reduces to

$$E[f_m(n)b_i^*(n)] = \Gamma_i^* P_m, \qquad m \geq i$$

For the case when $m < i$, we adapt Eq. (4.135) to write

$$E[f_m(n)b_i^*(n)] = E\left[\sum_{k=0}^{m} a_{m,k}^* u(n-k)b_i^*(n)\right]$$

$$= \sum_{k=0}^{m} a_{m,k}^* E[u(n-k)b_i^*(n)] \qquad (4.151)$$

By Property 1, we have

$$E[b_i(n)u^*(n-k)] = 0, \qquad 0 \leq k \leq i-1$$

Therefore, with $m < i$, we find that all the expectation terms inside the summation on the right side of Eq. (4.151) are zero. Thus,

$$E[f_m(n)b_i^*(n)] = 0, \qquad m < i$$

This completes the proof of Property 6.

4.15 GENERATION OF THE SEQUENCE OF BACKWARD PREDICTION ERRORS VIEWED AS A GRAM–SCHMIDT ORTHOGONALIZATION PROCESS

To develop further insight into the operation of a lattice filter, we now show that the sequence of backward prediction errors $b_0(n)$, $b_1(n)$, . . . , $b_m(n)$ produced by such a structure, consisting of m stages in cascade, may be viewed as a form of the *Gram–Schmidt orthogonalization procedure* applied to the corresponding sequence of input samples $u(n)$, $u(n-1)$, . . . , $u(n-m)$.

In the context of the issue of interest to us here, we may state the Gram–Schmidt orthogonalization process as follows (Strang, 1980):

Given $m + 1$ linearly independent (correlated) random variables u_0, u_1, . . . , u_m, it is always possible to construct an orthogonal system of $m + 1$ random variables v_0, v_1, . . . , v_m, each of which is a linear combination of u_0, u_1, . . . , u_m.

We illustrate this procedure for $m = 2$. It is a straightforward matter to extend the procedure to any value of m.

Example 5

Given the linearly independent (correlated) random variables u_0, u_1, u_2, we proceed in three steps as described next in order to determine the orthogonal system of random variables v_0, v_1, v_2:

1. Put

$$v_0 = u_0 \qquad\qquad (4.152)$$

2. Put

$$v_1 = \alpha_{1,1}u_0 + u_1 \qquad\qquad (4.153)$$

where $\alpha_{1,1}$ is a constant coefficient to be determined. The correlation of v_1 and v_0 equals

$$E[v_0 v_1^*] = E[u_0(\alpha_{1,1}^* u_0^* + u_1^*)]$$

$$= \alpha_{1,1}^* E[u_0 u_0^*] + E[u_0 u_1^*]$$

$$= \alpha_{1,1}^* r(0) + r(1)$$

The requirement is to choose the coefficient $\alpha_{1,1}$ so that v_1 and v_0 are orthogonal; that is,

$$E[v_0 v_1^*] = 0$$

Hence,

$$\alpha_{1,1}^* r(0) + r(1) = 0$$

Solving for $\alpha_{1,1}^*$, we get

$$\alpha_{1,1}^* = -\frac{r(1)}{r(0)}$$

3. Put

$$v_2 = \alpha_{2,2}u_0 + \alpha_{2,1}u_1 + u_2 \qquad\qquad (4.154)$$

where $\alpha_{2,2}$ and $\alpha_{2,1}$ are constant coefficients to be determined. The correlation of v_2 and v_0 equals

$$E[v_0 v_2^*] = E[u_0(\alpha_{2,2}^* u_0^* + \alpha_{2,1}^* u_1^* + u_2^*)]$$

$$= \alpha_{2,2}^* E[u_0 u_0^*] + \alpha_{2,1}^* E[u_0 u_1^*] + E[u_0 u_2^*]$$

$$= \alpha_{2,2}^* r(0) + \alpha_{2,1}^* r(1) + r(2)$$

The correlation of v_2 and v_1 equals

$$E[v_1 v_2^*] = E[(\alpha_{1,1}u_0 + u_1)(\alpha_{2,2}^* u_0^* + \alpha_{2,1}^* u_1^* + u_2^*)]$$

$$= \alpha_{1,1}E[v_0 v_2^*] + \alpha_{2,2}^* E[u_1 u_0^*] + \alpha_{2,1}^* E[u_1 u_1^*] + E[u_1 u_2^*]$$

$$= \alpha_{1,1}E[v_0 v_2^*] + \alpha_{2,2}^* r(-1) + \alpha_{2,1}^* r(0) + r(1)$$

The requirement is to choose the coefficients $\alpha_{2,2}$ and $\alpha_{2,1}$ so that v_2 is orthogonal to both v_0 and v_1; that is,

$$E[v_0 v_2^*] = 0$$

and

$$E[v_1 v_2^*] = 0$$

Hence,

$$\alpha_{2,2}^* r(0) + \alpha_{2,1}^* r(1) + r(2) = 0$$

$$\alpha_{2,2}^* r(-1) + \alpha_{2,1}^* r(0) + r(1) = 0$$

Solving this pair of simultaneous equations for $\alpha_{2,1}^*$ and $\alpha_{2,2}^*$, we get

$$\alpha_{2,1}^* = \frac{r(2)r(-1) - r(0)r(1)}{r^2(0) - r(1)r(-1)}$$

and

$$\alpha_{2,2}^* = \frac{r^2(1) - r(0)r(2)}{r^2(0) - r(1)r(-1)}$$

Rewriting Eqs. (4.152), (4.153), and (4.154) in matrix form, we have

$$\begin{bmatrix} v_0 \\ v_1 \\ v_2 \end{bmatrix} = \begin{bmatrix} 1 & 0 & 0 \\ \alpha_{1,1} & 1 & 0 \\ \alpha_{2,2} & \alpha_{2,1} & 1 \end{bmatrix} \begin{bmatrix} u_0 \\ u_1 \\ u_2 \end{bmatrix} \tag{4.155}$$

where the coefficients $\alpha_{1,1}$, and $\alpha_{2,1}$, $\alpha_{2,2}$ are as defined previously.

We now wish to relate the results of this example to the backward prediction errors produced by a lattice filter consisting of two stages, as in Fig. 4.11. These two stages are characterized by the reflection coefficients Γ_1 and Γ_2. In particular, we wish to express the backward prediction errors in terms of the input samples. At this input of the filter, we have

$$b_0 = u(n) \tag{4.156}$$

At the output of the first stage, we have

$$b_1(n) = a_{1,1}u(n) + u(n-1) \tag{4.157}$$

where

$$a_{1,1} = \Gamma_1$$

Finally, at the output of the second stage, we have

$$b_2(n) = a_{2,2}u(n) + a_{2,1}u(n-1) + u(n-2) \tag{4.158}$$

where

$$a_{2,1} = \Gamma_1 + \Gamma_2\Gamma_1^*$$

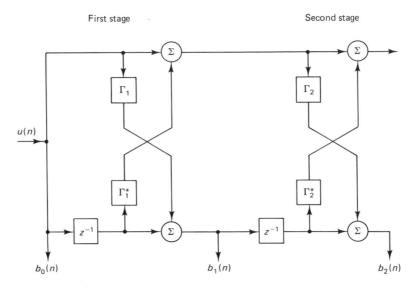

Figure 4.11 Two-stage lattice predictor with the backward prediction errors shown as the variables of interest.

and

$$a_{2,2} = \Gamma_2$$

Combining Eqs. (4.156), (4.157), and (1.458) into a single matrix relation, we have

$$\begin{bmatrix} b_0(n) \\ b_1(n) \\ b_2(n) \end{bmatrix} = \begin{bmatrix} 1 & 0 & 0 \\ a_{1,1} & 1 & 0 \\ a_{2,2} & a_{2,1} & 1 \end{bmatrix} \begin{bmatrix} u(n) \\ u(n-1) \\ u(n-2) \end{bmatrix} \qquad (4.159)$$

Comparing Eqs. (4.155) and (4.159), we may identify

$$\begin{bmatrix} u_0 \\ u_1 \\ u_2 \end{bmatrix} = \begin{bmatrix} u(n) \\ u(n-1) \\ u(n-2) \end{bmatrix}$$

$$\begin{bmatrix} v_0 \\ v_1 \\ v_2 \end{bmatrix} = \begin{bmatrix} b_0(n) \\ b_1(n) \\ b_2(n) \end{bmatrix}$$

and

$$\alpha_{1,1} = a_{1,1}$$

$$\alpha_{2,1} = a_{2,1}$$

$$\alpha_{2,2} = a_{2,2}$$

Discussion

We may now make the following observations and generalizations:

1. The generation of a sequence of backward prediction errors $b_0(n)$, $b_1(n)$, . . . , $b_M(n)$ by a lattice filter consisting of M stages is equivalent to a Gram–Schmidt orthogonalization process applied recursively to a corresponding sequence of input samples $u(n)$, $u(n - 1)$, . . . , $u(n - M)$. This, however, is merely a restatement of Property 3, Section 4.14, stating that the backward prediction errors are orthogonal to each other.

2. To transform the tap-input vector $\mathbf{u}(n)$ into a backward prediction-error vector $\mathbf{b}(n)$, we premultiply the input vector by a lower triangular matrix $\mathbf{L}$ with 1's along the main diagonal. The nonzero elements along each row of the matrix $\mathbf{L}$ are defined by the tap weights of the backward prediction-error filter whose order corresponds to the position of the pertinent row in the matrix. We may thus write, in general,

$$\mathbf{b}(n) = \mathbf{L}\mathbf{u}(n) \tag{4.160}$$

where

$$\mathbf{b}^T(n) = [b_0(n), b_1(n), \ldots, b_M(n)] \tag{4.161}$$

$$\mathbf{u}^T(n) = [u(n), u(n - 1), \ldots, u(n - M)] \tag{4.162}$$

and

$$\mathbf{L} = \begin{bmatrix} 1 & 0 & \cdots & 0 \\ a_{1,1} & 1 & \cdots & 0 \\ \cdot & \cdot & \cdot & \cdot \\ \cdot & \cdot & \cdot & \cdot \\ \cdot & \cdot & \cdot & \cdot \\ a_{M,M} & a_{M,M-1} & \cdots & 1 \end{bmatrix} \tag{4.163}$$

Note that the determinant of matrix $\mathbf{L}$ equals unity. The matrix $\mathbf{L}$ is therefore nonsingular.

3. There is a one-to-one correspondence between the input vector $\mathbf{u}(n)$ and the

backward prediction-error vector $\mathbf{b}(n)$ in that the one may be obtained from the other without any loss of information.

The transformation of the input vector $\mathbf{u}(n)$ into the backward prediction-error vector $\mathbf{b}(n)$, as in Eq. (4.160), has many implications. In particular, it provides the mathematical basis for *joint-process estimation*, an issue that is discussed in Section 4.18. The transformation is also intimately related to a special decomposition known as the *Cholesky decomposition* (Morf, 1974); this issue is discussed in Section 4.17 and Problem 13.

4.16 EQUIVALENCE OF THE INPUT VECTOR AND THE BACKWARD PREDICTION-ERROR VECTOR FROM AN INFORMATION-THEORETIC VIEWPOINT

In this section we wish to substantiate the statement that the input vector $\mathbf{u}(n)$, consisting of the elements $u(n), u(n-1), \ldots, u(n-M)$, and the corresponding backward prediction-error vector $\mathbf{b}(n)$, consisting of the elements $b_0(n), b_1(n), \ldots, b_M(n)$, are probabilistically equivalent in that they contain exactly the same amount of information. In particular, we will show that both of these random vectors have the same *entropy,* which provides a quantitative measure of the information content of a random vector (Shannon, 1948).

The entropy of the random input vector $\mathbf{u}(n)$ is defined by (Shannon, 1948)

$$H_{\mathbf{u}} = -\int_{-\infty}^{\infty} f_{\mathbf{U}}(\mathbf{u})\, \ln[f_{\mathbf{U}}(\mathbf{u})]d\mathbf{u} \qquad (4.164)$$

where $f_{\mathbf{U}}(\mathbf{u})$ is the joint probability density function of the input vector $\mathbf{u}(n)$. Note that the integral in Eq. (4.164) is a multiple integral, with the integration variables consisting of the individual elements of the input vector $\mathbf{u}(n)$.

The backward prediction-error vector $\mathbf{b}(n)$ is related to the input vector $\mathbf{u}(n)$ by Eq. (4.160). Since this transformation is one-to-one, it is *reversible.* Hence, we may write

$$\mathbf{u}(n) = \mathbf{L}^{-1}\,\mathbf{b}(n) \qquad (4.165)$$

where $\mathbf{L}^{-1}$ is the inverse of the transformation matrix $\mathbf{L}$. We may thus express the joint probability density function $f_{\mathbf{B}}(\mathbf{b})$ of the backward prediction-error vector in terms of the joint probability density function $f_{\mathbf{U}}(\mathbf{u})$ of the input vector as follows[4]

$$f_{\mathbf{B}}(\mathbf{b}) = |\det(\mathbf{L}^{-1})| f_{\mathbf{U}}(\mathbf{L}^{-1}\mathbf{b}) \qquad (4.166)$$

---

[4]Consider the *one-to-one* transformation of two vectors $\mathbf{x}$ and $\mathbf{y}$ that is described by

$$\mathbf{y} = \mathbf{G}\mathbf{x}$$

where $\mathbf{G}$ is the transformation matrix. Let $\mathbf{G}^{-1}$ denote the inverse of matrix $\mathbf{G}$. The transformation where the term within the absolute value signs is called the *Jacobian* of the transformation. We also note that

$$du = |\det(\mathbf{L}^{-1})|d\mathbf{b} \tag{4.167}$$

The entropy of the backward prediction-error vector $\mathbf{b}(n)$ is defined by

$$
\begin{aligned}
H_\mathbf{b} &= -\int_{-\infty}^{\infty} f_\mathbf{B}(\mathbf{b}) \ln[f_\mathbf{B}(\mathbf{b})] \, d\mathbf{b} \\
&= -\int_{-\infty}^{\infty} |\det(\mathbf{L}^{-1})| f_\mathbf{U}(\mathbf{L}^{-1}\mathbf{b}) \ln[|\det(\mathbf{L}^{-1})| f_\mathbf{U}(\mathbf{L}^{-1}\mathbf{b})] \, d\mathbf{b} \\
&= -\int_{-\infty}^{\infty} f_\mathbf{U}(\mathbf{u}) \ln[|\det(\mathbf{L}^{-1})| f_\mathbf{U}(\mathbf{u})] \, d\mathbf{u}
\end{aligned}
$$

where in the second line we have used Eq. (4.166) and in the third line we have used Eqs. (4.165) and (4.167). We now recognize that since the transformation matrix $\mathbf{L}$ has unity for all of its diagonal elements, as shown in Eq. (4.163), the Jacobian of the transformation equals unity. We may therefore simplify the entropy $H_\mathbf{b}$ as

$$
\begin{aligned}
H_\mathbf{b} &= -\int_{-\infty}^{\infty} f_\mathbf{U}(\mathbf{u}) \ln[f_\mathbf{U}(\mathbf{u})] \, d\mathbf{u} \\
&= H_\mathbf{u}
\end{aligned}
$$

We have thus shown that the backward prediction-error vector $\mathbf{b}(n)$ has the same entropy, and therefore, contains the same amount of information, as the input vector $\mathbf{u}(n)$.

Example 6

Consider the special case of a Gaussian-distributed, real-valued random vector $\mathbf{u}(n)$ of zero mean and correlation matrix $\mathbf{R}$. Hence, we may express the joint probability density function of $\mathbf{u}(n)$, assumed to consist of $M + 1$ elements, as

$$f_\mathbf{U}(\mathbf{u}) = [2\pi\det(\mathbf{R})]^{-(M+1)/2}\exp[-\tfrac{1}{2}\,\mathbf{u}^T\mathbf{R}^{-1}\mathbf{u}] \tag{4.168}$$

---

is reversible, so that the inverse transformation

$$\mathbf{x} = \mathbf{G}^{-1}\mathbf{y}$$

also exists for all $\mathbf{x}$ and $\mathbf{y}$. The *Jacobian* of the transformation, by definition, is the determinant

$$J = \det(\mathbf{G}^{-1})$$

Suppose we are given the joint probability density function $f_\mathbf{X}(\mathbf{x})$. To find the joint probability density function $f_\mathbf{Y}(\mathbf{y})$, we write (Wozencraft and Jacobs, 1965)

$$f_\mathbf{Y}(\mathbf{y}) = |J| f_\mathbf{X}(\mathbf{G}^{-1}\mathbf{y})$$

where $|J|$ is the absolute value of the Jacobian associated with the transformation matrix $\mathbf{G}$.

Substituting Eq. (4.168) into (4.164) and simplifying, we get the following result, except for an additive constant:

$$H_u = \tfrac{1}{2}\ln[\det(\mathbf{R})] \tag{4.169}$$

Since the random vectors $\mathbf{u}(n)$ and $\mathbf{b}(n)$ have the same entropy, it follows that the correlation matrix $\mathbf{R}$ has the same determinant as that of the correlation matrix of the prediction-error vector $\mathbf{b}$. The correlation matrix of $\mathbf{b}$ is a *diagonal* matrix as shown by

$$\begin{aligned}\mathbf{D} &= E[\mathbf{b}(n)\mathbf{b}^H(n)] \\ &= \mathrm{diag}(P_0, P_1, \ldots, P_M)\end{aligned} \tag{4.170}$$

where we have used the property

$$E[b_k(n)b_l^*(n)] = \begin{cases} P_k, & k=l \\ 0, & k \neq l \end{cases}$$

The correlation matrix $\mathbf{D}$ in Eq. (4.170) has a determinant equal to the product of the individual prediction-error powers, $P_0, P_1, \ldots, P_M$. Thus, for the case of a prediction-error filter of order M, we have

$$\det(\mathbf{R}) = \prod_{k=0}^{M} P_k \tag{4.171}$$

Correspondingly, we may express the entropy of a Gaussian-distributed random vector $\mathbf{u}(n)$ of zero mean in terms of the set of prediction-error powers $P_0, P_1, \ldots, P_M$ as follows:

$$H_u = \frac{1}{2} \sum_{k=0}^{M} \ln P_k \tag{4.172}$$

4.17 EQUIVALENCE OF THE POSITIVE DEFINITENESS OF THE CORRELATION MATRIX OF THE TAP INPUTS AND THE MINIMUM-PHASE PROPERTY OF THE PREDICTION-ERROR FILTER

In Section 4.8 we showed that the forward prediction-error filter of order M is minimum phase if the associated reflection coefficients $\Gamma_1, \Gamma_2, \ldots, \Gamma_M$ satisfy the condition that $|\Gamma_m| < 1$ for all m. In this section we show that this condition is satisfied if $\mathbf{R}$, the correlation matrix of the tap inputs in the filter, is positive definite.

We prove this property by using the transformation described by Eq. (4.160). In particular, substituting this transformation [from the tap-input vector $\mathbf{u}(n)$ into the vector of backward prediction errors $\mathbf{b}(n)$] into Eq. (4.170), we find that the correlation matrix of $\mathbf{b}(n)$ equals

$$\begin{aligned}\mathbf{D} &= E[\mathbf{L}\mathbf{u}(n)\mathbf{u}^H(n)\mathbf{L}^H] \\ &= \mathbf{L}E[\mathbf{u}(n)\mathbf{u}^H(n)]\mathbf{L}^H \\ &= \mathbf{L}\mathbf{R}\mathbf{L}^H\end{aligned} \tag{4.173}$$

where $\mathbf{R}$ is the correlation matrix of the tap-input vector $\mathbf{u}(n)$, and the triangular transformation matrix $\mathbf{L}$ is defined by Eq. (4.163). The matrix $\mathbf{L}$ is nonsingular. We now make use of the following theorem in matrix algebra (Strang, 1980):

Let $\mathbf{R}$ be an $(M + 1)$-by-$(M + 1)$ positive-definite matrix. Then $\mathbf{LRL}^H$ is a positive-definite matrix for any nonsingular $(M + 1)$-by-$(M + 1)$ matrix $\mathbf{L}$.

For the $(M + 1)$-by-$(M + 1)$ diagonal matrix $\mathbf{D}$ to be positive definite, we require that all of its diagonal elements be individually positive; that is,

$$P_m > 0, \qquad m = 0, 1, \ldots, M \tag{4.174}$$

We note that $P_0 = E[|b_0(n)|^2] = E[|u(n)|^2]$; hence, we always have $P_0 > 0$. Next, from Eq. (4.63), we deduce that

$$P_1 = P_0(1 - |\Gamma_1|^2)$$

Therefore, with $P_0 > 0$, it is necessary that $|\Gamma_1| < 1$ for $P_1 > 0$. Continuing in this fashion, we find that the condition of Eq. (4.174) is equivalent to

$$|\Gamma_m| < 1, \qquad m = 1, 2, \ldots, M \tag{4.175}$$

We may thus state that:

1. If the $(M + 1)$-by-$(M + 1)$ correlation matrix of the tap-input vector $\mathbf{u}(n)$ is positive definite, then all the reflection coefficients $\Gamma_1, \Gamma_2, \ldots, \Gamma_M$ have absolute values less than 1. Correspondingly, the forward prediction-error filter is minimum phase.
2. Conversely, if all the reflection coefficients $\Gamma_1, \Gamma_2, \ldots, \Gamma_M$ have absolute values less than 1, the correlation matrix of the corresponding tap-input vector $\mathbf{u}(n)$ is positive definite.

Cholesky Decomposition

It is of interest to note that when the correlation matrix $\mathbf{R}$ of the tap-input vector $\mathbf{u}(n)$ is positive definite, and consequently the correlation matrix $\mathbf{D}$ of $\mathbf{b}(n)$, the vector of backward prediction errors, is also positive definite, then Eq. (4.173) may be rewritten in the form

$$\mathbf{R}^{-1} = \mathbf{L}^H \mathbf{D}^{-1} \mathbf{L} \tag{4.176}$$
$$= (\mathbf{D}^{-1/2}\mathbf{L})^H(\mathbf{D}^{-1/2}\mathbf{L})$$

where $\mathbf{R}^{-1}$ is the inverse of the correlation matrix $\mathbf{R}$, and the inverse matrix $\mathbf{D}^{-1}$ is defined by

$$\mathbf{D}^{-1} = \mathrm{diag}(P_0^{-1}, P_1^{-1}, \ldots, P_M^{-1}) \tag{4.177}$$

The matrix $\mathbf{D}^{-1/2}$, the *square root* of $\mathbf{D}^{-1}$, is likewise a diagonal matrix defined by

$$\mathbf{D}^{-1/2} = \mathrm{diag}(P_0^{-1/2}, P_1^{-1/2}, \ldots, P_M^{-1/2}) \tag{4.178}$$

The transformation of Eq. (4.176) represents the *Cholesky decomposition* of the inverse matrix $\mathbf{R}^{-1}$ (Stewart, 1973).

4.18 JOINT–PROCESS ESTIMATION

In this section, we use the lattice predictor, as a subsystem, to solve a *joint-process estimation problem* that is optimal in the mean-square sense (Griffiths, 1977; Makhoul, 1978). In particular, we consider the minimum mean-square estimation of a process $\{d(n)\}$, termed the desired response, by using a set of *observables* derived from a related process $\{u(n)\}$. We assume that the processes $\{d(n)\}$ and $\{u(n)\}$ are jointly stationary. This estimation problem is similar to that considered in Chapter 3, with one basic difference. In Chapter 3 we used samples of the process $\{u(n)\}$ as the observables directly. Our approach here is different in that for the observables we use the set of backward prediction errors obtained by feeding the input of a multistage lattice predictor with samples of the process $\{u(n)\}$. The fact that the backward prediction errors are orthogonal to each other simplifies the solution to the problem significantly.

The structure of the *joint-process estimator* is shown in Fig. 4.12. This device performs two optimum estimations jointly (hence, the name):

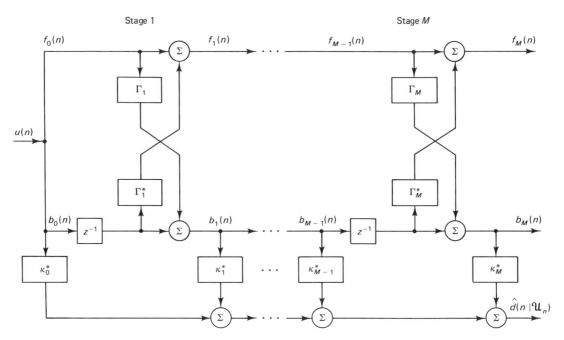

Figure 4.12 Lattice-based structure for joint-process estimation.

1. The *lattice predictor*, consisting of a cascade of M stages, characterized individually by the reflection coefficients $\Gamma_1, \Gamma_2, \ldots, \Gamma_M$, performs predictions (of varying orders) on the input. In particular, it transforms the sequence of (correlated) input samples $u(n), u(n-1), \ldots, u(n-M)$ into a corresponding sequence of (uncorrelated) backward prediction errors $b_0(n), b_1(n), \ldots, b_M(n)$.

2. The *multiple regression filter*, characterized by the set of weights $\kappa_0, \kappa_1, \ldots, \kappa_M$, operates on the sequence of backward prediction errors $b_0(n), b_1(n), \ldots, b_M(n)$ as inputs, respectively, to produce an *estimate* of the desired response $d(n)$. The resulting estimate is defined as the sum of the respective inner products of these two sets of quantities, as shown by

$$\hat{d}(n|\mathcal{U}_n) = \sum_{i=0}^{M} \kappa_i^* b_i(n) \tag{4.179}$$

where $\mathcal{U}_n$ is the space spanned by the inputs $u(n), u(n-1), \ldots, u(n-M)$. We may rewrite Eq. (4.179) in matrix form as follows

$$\hat{d}(n|\mathcal{U}_n) = \boldsymbol{\kappa}^H \mathbf{b}(n) \tag{4.180}$$

where $\boldsymbol{\kappa}$ is an $(M+1)$-by-1 vector defined by

$$\boldsymbol{\kappa}^T = [\kappa_0, \kappa_1, \ldots, \kappa_M] \tag{4.181}$$

We refer to $\kappa_0, \kappa_1, \ldots, \kappa_M$ as the *regression coefficients* of the estimator, and to $\boldsymbol{\kappa}$ as the *regression vector*.

Let $\mathbf{D}$ denote the $(M+1)$-by-$(M+1)$ correlation matrix of $\mathbf{b}(n)$, the $(M+1)$-by-1 vector of backward prediction errors, defined in Eq. (4.170). Let $\mathbf{s}$ denote the $(M+1)$-by-1 cross-correlation vector between the backward prediction errors and the desired response:

$$\mathbf{s} = E[\mathbf{b}(n)d^*(n)] \tag{4.182}$$

Therefore, applying the normal equation to our present situation, we find that the optimum tap-weight vector $\boldsymbol{\kappa}_o$ is defined by

$$\mathbf{D}\boldsymbol{\kappa}_o = \mathbf{s} \tag{4.183}$$

Solving for $\boldsymbol{\kappa}_o$, we get

$$\boldsymbol{\kappa}_o = \mathbf{D}^{-1}\mathbf{s} \tag{4.184}$$

where the inverse matrix $\mathbf{D}^{-1}$ is a diagonal matrix, defined in terms of various prediction error powers as in Eq. (4.177). Note that, unlike the ordinary transversal filter realization of the Wiener filter, the computation of the tap-weight vector $\boldsymbol{\kappa}_o$ in the joint-process estimator of Fig. 4.12 is relatively simple to accomplish.

4.19 *INVERSE FILTERING USING THE LATTICE STRUCTURE*

The multistage lattice predictor of Fig. 4.10(b) may be viewed as an *analyzer*. That is, it enables us to represent an autoregressive (AR) process $\{u(n)\}$ by a corresponding sequence of reflection coefficients $\{\Gamma_m\}$. By rewiring this multistage lattice predictor in the manner depicted in Fig. 4.13, we may use this new structure as a *synthesizer* or *inverse filter*. That is, given the sequence of reflection coefficients $\{\Gamma_m\}$, we may reproduce the original AR process by applying a white-noise process $\{v(n)\}$ to the input of the structure in Fig. 4.13. This lattice inverse filter differs from the inverse filter of Fig. 4.9(b) (based on a tapped-delay-line structure) in that it produces a truly stationary time series from the very first sample. On the other hand, the inverse filter of Fig. 4.9(b) produces transients due to nonstationary initial conditions, with the result that the time series produced at its output is only asymptotically stationary.

We illustrate the operation of the lattice inverse filter with an example.

Example 7

Consider the two-stage lattice inverse filter of Fig. 4.14. There are four possible paths that can contribute to the makeup of the sample $u(n)$ at the output, as illustrated in Fig. 4.15:

$$u(n) = v(n) - \Gamma_1^* \, u(n-1) - \Gamma_1\Gamma_2^* u(n-1) - \Gamma_2^* u(n-2)$$

$$= v(n) - (\Gamma_1^* + \Gamma_1\Gamma_2^*)u(n-1) - \Gamma_2^* u(n-2)$$

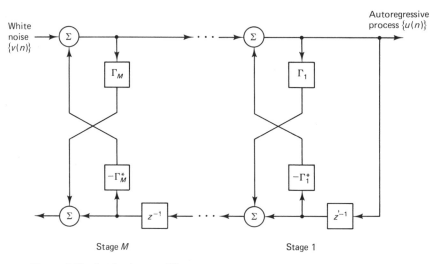

Figure 4.13   Lattice inverse filter.

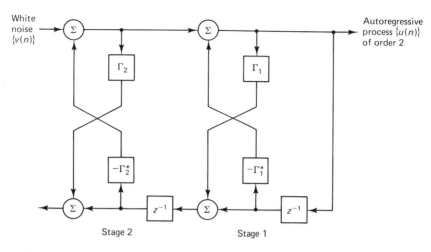

Figure 4.14   Lattice inverse filter of order 2.

From Example 2, we recall that

$$a_{2,1} = \Gamma_1 + \Gamma_1^*\Gamma_2$$

$$a_{2,2} = \Gamma_2$$

We may therefore express the mechanism governing the generation of process $\{u(n)\}$ as follows:

$$u(n) + a_{2,1}^*u(n-1) + a_{2,2}^*u(n-2) = v(n)$$

which is recognized as the difference equation of a second-order AR process.

4.20 THE BURG FORMULA

In previous sections of the chapter, we have shown that the reflection coefficients $\Gamma_1, \Gamma_2, \ldots, \Gamma_M$ that characterize a multistage lattice predictor of order M play a key role in different aspects of the linear prediction problem. We finish the chapter by describing a *block-processing* method for estimating these reflection coefficients directly from the observed data. The method relies on the decoupling property of a lattice predictor for *stationary* inputs. In particular, this property makes it possible to accomplish the *global* optimization of a multistage lattice predictor as a sequence of *local* optimization problems, one at each stage of the lattice predictor. Accordingly, it is a straightforward matter to increase the order of the predictor by simply adding one or more stages, as required, without affecting the earlier design computations. For example, suppose we have optimized the design of a lattice predictor consisting of M stages. To increase the order of the filter by 1,

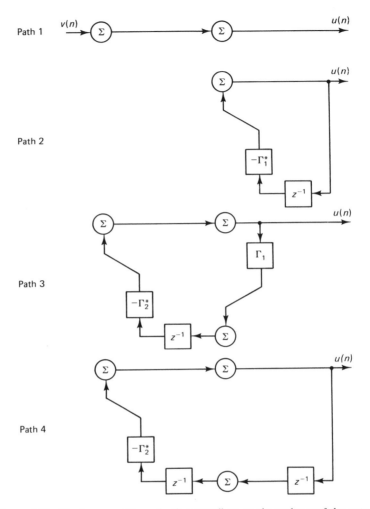

Figure 4.15 The four possible paths that contribute to the makeup of the output $u(n)$.

we simply add a new stage that is locally optimized, leaving the optimum design of the earlier stages unchanged.

Consider then stage m of the lattice predictor shown in Fig. 4.10(a), for which the input–output relations are given in matrix form in Eq. (4.125). These relations are reproduced here for convenience:

$$f_m(n) = f_{m-1}(n) + \Gamma_m^* b_{m-1}(n-1) \qquad (4.185)$$

$$b_m(n) = b_{m-1}(n-1) + \Gamma_m f_{m-1}(n-1) \qquad (4.186)$$

where $m = 1, 2, \ldots, M$, and M is the final order of the predictor. Several methods may be used to optimize the design of this stage, depending on the criterion of interest (Makhoul, 1977). In one particular method that yields a design with interesting properties (which conform to the lattice predictor theory), the reflection Γ_m is chosen so as to minimize the sum of the mean-square values of the forward and backward prediction errors. Let the index of performance J_m denote this sum at the output of stage m of the lattice predictor:

$$J_m = E[|f_m(n)|^2] + E[|b_m(n)|^2] \tag{4.187}$$

Substituting Eqs. (4.185) and (4.186) in (4.187), we get

$$
\begin{aligned}
J_m = {} & \{E[|f_{m-1}(n)|^2] + E[|b_{m-1}(n-1)|^2]\}[1 + |\Gamma_m|^2] \\
& + 2\Gamma_m E[f_{m-1}(n)b_{m-1}^*(n-1)] \\
& + 2\Gamma_m^* E[b_{m-1}(n-1)f_{m-1}^*(n)]
\end{aligned}
\tag{4.188}
$$

In general, the reflection coefficient Γ_m is a complex valued scalar. We may therefore treat the differentiation of J_m with respect to Γ_m as a special case of the rule described in Section 3.4. We may thus determine the complex-valued gradient of J_m with respect to Γ_m as follows:

$$
\begin{aligned}
\frac{\partial J_m}{\partial \Gamma_m} = {} & 2\Gamma_m\{E[|f_{m-1}(n)|^2] + E[|b_{m-1}(n-1)|^2]\} \\
& + 4E[b_{m-1}(n-1)f_{m-1}^*(n)]
\end{aligned}
\tag{4.189}
$$

Putting this gradient equal to zero, we find that the optimum value of the reflection coefficient, for which the index of performance J_m in Eq. (4.188) is minimum, equals

$$\Gamma_{m,o} = -\frac{2E[b_{m-1}(n-1)f_{m-1}^*(n)]}{E[|f_{m-1}(n)|^2 + |b_{m-1}(n-1)|^2]}, \qquad m = 1, 2, \ldots, M \tag{4.190}$$

Equation (4.190) for the reflection coefficient is known as the *Burg formula* (Burg, 1968).[5] It use offers two interesting properties (see Problem 16):

1. The reflection coefficient $\Gamma_{m,o}$ satisfies the condition

$$|\Gamma_{m,o}| \le 1, \qquad \text{for all } m \tag{4.191}$$

In other words, the Burg formula always yields a minimum–phase design for the lattice predictor.

2. The mean-square values of the forward and backward prediction errors at the output of stage m are related to those at the input as follows, respectively:

$$E[|f_m(n)|^2] = (1 - |\Gamma_{m,o}|^2)E[|f_{m-1}(n)|^2] \tag{4.192}$$

and

$$E[|b_m(n)|^2] = (1 - |\Gamma_{m,o}|^2)E[|b_{m-1}(n-1)|^2] \tag{4.193}$$

[5]The 1968 paper by Burg is reproduced in the book by Childers (1978).

Block Estimation

In the *block estimation* procedure, the available data are divided into individual blocks, each of length N. For each block of data, the Burg formula of Eq. (4.190) is used to estimate a value of the reflection coefficient for each stage of the lattice predictor. Typically, for a given stage in the lattice predictor, the reflection coefficient varies from one block of data to another.

By substituting time averages for the expectations in the numerator and denominator of Eq. (4.190), we may express the estimate of $\Gamma_{m,o}$ as

$$\hat{\Gamma}_m = -\frac{2 \sum_{n=m+1}^{N} b_{m-1}(n-1) f_{m-1}^*(n)}{\sum_{n=m+1}^{N} [|f_{m-1}(n)|^2 + |b_{m-1}(n-1)|^2]}, \qquad m = 1, 2, \ldots, M \qquad (4.194)$$

where N is the length of the block of input data, and $f_0(n) = b_0(n) = u(n)$.

With a lattice predictor of m stages, each of which contains a single unit-delay element, and with the input $u(n)$ zero for $n \leq 0$, we find that *all* the samples in the input data contribute to the outputs of stage m in the predictor for the first time at $n = m + 1$; hence, the use of this value for the lower limits of the summation terms in Eq. (4.194). Note also that the estimate $\hat{\Gamma}_m$ for the mth reflection coefficient is dependent on data length N. The choice of N is usually dictated by two conflicting factors. First, it should be large enough to smooth out the effects of noise in computing the time averages in the numerator and denominator of Eq. (4.194). Second, it should be small enough to ensure quasi-statistical stationarity of the input data during the computations.

The block-estimation approach usually requires a large amount of computation, as well as a large amount of storage. Furthermore, in this approach, we find that for any stage of the predictor the estimate of the reflection coefficient at time $n + 1$ does not depend in a simple way on its previous estimate at time n. This behavior is to be contrasted with the adaptive estimation procedure to be described in Chapter 9.

PROBLEMS

1. The augmented normal equation (4.18) of a forward prediction-error filter was derived by first optimizing the linear prediction filter in the mean-square sense and then combining the two resultants: the normal equation for the tap-weight vector and the minimum mean-squared prediction error. This problem addresses the issue of deriving Eq. (4.18) directly by proceeding as follows:
   (a) Formulate the expression for the mean-square value of the forward prediction error as a function of the tap-weight vector of the forward prediction-error filter.
   (b) Minimize this mean-squared prediction error, subject to the constraint that the leading element of the tap-weight vector of the forward prediction-error filter equals 1.

Hint: Use the method of Lagrange multipliers to solve the constrained optimization problem. For details of this method, see Appendix D. This hint also applies to part (b) of Problem 2.

2. The augmented normal equation (4.44) of a backward prediction-error filter was derived indirectly in Section 4.5. This problem addresses the issue of deriving Eq. (4.44) directly by proceeding as follows:
   (a) Formulate the expression for the mean-square value of the backward prediction error in terms of the tap-weight vector of the backward prediction-error filter.
   (b) Minimize this mean-squared prediction error, subject to the constraint that the last element of the tap-weight vector of the backward prediction-error filter equals 1.

3. (a) Equation (4.28) defines the normal equation for backward linear prediction. This equation is reproduced here for convenience:

$$\mathbf{Rg} = \mathbf{r}^{B*}$$

   where $\mathbf{g}$ is the tap-weight vector of the predictor, $\mathbf{R}$ is the correlation matrix of the tap inputs $u(n)$, $u(n-1)$, . . . , $u(n-M+1)$, and $\mathbf{r}^{B*}$ is the cross-correlation vector between these tap inputs and the desired response $u(n-M)$. Show that if the elements of the column vector $\mathbf{r}^{B*}$ are rearranged in reverse order the effect of this reversal is to modify the normal equation as

$$\mathbf{R}^T\mathbf{g}^B = \mathbf{r}^*$$

   (b) Show that the inner products $\mathbf{r}^{BT}\mathbf{g}$ and $\mathbf{r}^T\mathbf{g}^B$ are equal.

4. Equation (4.46) defines the Levinson–Durbin recursion for forward linear prediction. By rearranging the elements of the tap-weight vector $\mathbf{a}_m$ backward and then complex-conjugating them, reformulate the Levinson–Durbin recursion for backward linear prediction as in Eq. (4.48).

5. Starting with the definition of Eq. (4.53) for Δ_{m-1}, show that Δ_{m-1} equals the cross-correlation between the delayed backward prediction error $b_{m-1}(n-1)$ and the forward prediction error $f_{m-1}(n)$.

6. Develop in detail the relationship between the Schur–Cohn method and the inverse recursion as outlined by Eqs. (4.99) through (4.102).

7. Consider an autoregressive process $\{u(n)\}$ of order 2, described by the difference equation

$$u(n) = u(n-1) - 0.5u(n-2) + v(n)$$

   where $\{v(n)\}$ is a white-noise process of zero mean and variance 0.5.
   (a) Find the average power of $\{u(n)\}$.
   (b) Find the reflection coefficients Γ_1 and Γ_2.
   (c) Find the average prediction-error powers P_1 and P_2.

8. Using the one-to-one correspondence between the two sequences of numbers $\{r(0), \Gamma_1, \Gamma_2\}$ and $\{r(0), r(1), r(2)\}$, compute the autocorrelation function values $r(1)$ and $r(2)$ that correspond to the reflection coefficients Γ_1 and Γ_2 found in Problem 7 for the second-order autoregressive process $\{u(n)\}$.

9. In Section 4.8, we presented a proof of the minimum-phase property of a prediction-error filter by using Rouché's theorem. In this problem, we explore another proof of this property by contradiction. Consider Fig. P4.1, which shows the prediction-error

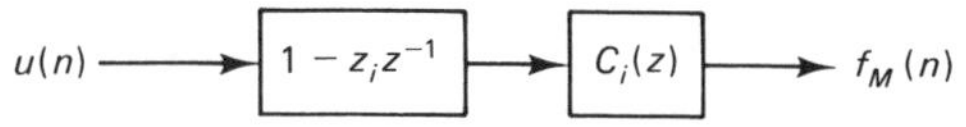

$$u(n) \longrightarrow \boxed{1 - z_i z^{-1}} \longrightarrow \boxed{C_i(z)} \longrightarrow f_M(n)$$

Figure P4.1

filter (of order M) represented as the cascade of two functional blocks, one with transfer function $C_i(z)$ and the other with its transfer function equal to the zero factor $(1 - z_i z^{-1})$. Let $S(\omega)$ denote the power spectral density of the process $\{u(n)\}$ applied to the input of the prediction-error filter.

(a) Show that the mean-square value of the forward prediction error $f_M(n)$ equals

$$\varepsilon = \int_{-\pi}^{\pi} S(\omega) |C_i(e^{j\omega})|^2 \left[1 - 2\rho_i \cos(\omega - \omega_i) + \rho_i^2\right] d\omega$$

where $z_i = \rho_i e J \omega_i$. Hence, evaluate the derivative $\partial \varepsilon / \partial \rho_i$.

(b) Suppose that $\rho_i > 1$ so that the complex zero lies outside the unit circle. Hence, show that under this condition $\partial \varepsilon / \partial \rho_i > 0$. Is such a condition possible and at the same time the filter operates at its optimum condition? What conclusion can you draw from your answers?

10. (a) Construct the two-stage lattice predictor for the second-order autoregressive process $\{u(n)\}$ considered in Problem 7.

(b) Given a white-noise process $\{v(n)\}$, construct the two-stage lattice synthesizer for generating the autoregressive process $\{u(n)\}$. Check your answer against the second-order difference equation for the process $\{u(n)\}$ that was considered in Problem 7.

11. An autoregressive process $\{u(n)\}$ of order L is applied to a prediction-error filter of order M. Discuss the nature of the process $\{f_M(n)\}$ produced at the output of the filter for the following three distinct cases:

(a) $L > M$

(b) $L = M$

(c) $L < M$

12. In a *normalized* lattice predictor, the forward and backward prediction errors at the various stages of the predictor are all normalized to have *unit variance*. Such an operation makes it possible to utilize the full dynamic range of multipliers used in the hardware implementation of a lattice predictor. For stage m of the normalized lattice predictor, the normalized forward and backward prediction errors are defined as follows, respectively:

$$\bar{f}_m(n) = \frac{f_m(n)}{P_m^{1/2}}$$

and

$$\bar{b}_m(n) = \frac{b_m(n)}{P_m^{1/2}}$$

where P_m is the average power (or variance) of the forward prediction error $f_m(n)$ or that of the backward prediction error $b_m(n)$. Show that the structure of stage m of the normalized lattice predictor is as shown in Fig. P4.2 for real-valued data.

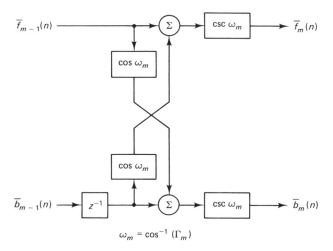

$$\omega_m = \cos^{-1}(\Gamma_m)$$

Figure P4.2

13. **(a)** Consider the matrix product **LR** that appears in the decomposition of Eq. (4.173), where the $(M + 1)$-by-$(M + 1)$ lower triangular matrix **L** is defined in Eq. (4.163) and **R** is the $(M + 1)$-by-$(M + 1)$ correlation matrix. Let **Y** denote this matrix product, and let y_{mk} denote the mkth element of **Y**. Hence, show that

$$y_{mm} = P_m, \qquad m = 0, 1, \ldots, M$$

where P_m is the prediction error power for order m.

(b) Show that the mth column of matrix **Y** is obtained by passing the autocorrelation sequence $\{r(0), r(1), \ldots, r(m)\}$ through a corresponding sequence of backward prediction-error filters represented by the transfer functions $H_{b,0}(z)$, $H_{b,1}(z)$, ..., $H_{b,m}(z)$.

(c) Suppose we apply the autocorrelation sequence $\{r(0), r(1), \ldots, r(m)\}$ to the input of a lattice predictor of order m. Show that the variables appearing at the various points on the lower line of the predictor at time m equal the elements of the mth column of matrix **Y**.

(d) For the situation described in part (c), show that the lower output of stage m in the predictor at time m equals P_m, and that the upper output of this same stage at time $m + 1$ equals Δ_m^*. How is the ratio of these two outputs related to the reflection coefficient of stage $m + 1$?

(e) Use the results of part (d) to develop a recursive procedure for computing the sequence of reflection coefficients from the autocorrelation sequence.

14. In Section 4.19, we considered the use of an inverse lattice filter as the generator of an autoregressive process. The lattice inverse filter may also be used to efficiently compute the autocorrelation sequence $r(1), r(2), \ldots, r(m)$ normalized with respect to $r(0)$. The procedure involves initializing the states (i.e., unit-delay elements) of the lattice inverse filter to $1, 0, \ldots, 0$ and then allowing the filter to operate with zero input. In effect, this procedure provides a lattice interpretation of Eq. (4.73) that relates the autocorrelation sequence $\{r(0), r(1), \ldots, r(M)\}$ and the augmented sequence of reflection

coefficients $\{r(0), \Gamma_1, \ldots, \Gamma_M\}$. Demonstrate the validity of this procedure for the following values of final order M:

(a) $M = 1$

(b) $M = 2$

(c) $M = 3$

15. Consider the problem of optimizing stage m of the lattice predictor. The index of performance to be used in the optimization is described by

$$J_m(\Gamma_m) = aE[|f_m(n)|^2] + (1 - a)E[|b_m(n)|^2]$$

where a is a constant that lies between zero and one; $f_m(n)$ and $b_m(n)$ denote the forward and backward prediction errors at the output of stage m, respectively.

(a) Show that the optimum value of the reflection coefficient Γ_m for which J_m is at minimum equals

$$\Gamma_{m,o}(a) = -\frac{E[b_{m-1}(n - 1)f^*_{m-1}(n)]}{(1 - a)E[|f_{m-1}(n)|^2] + aE[|b_{m-1}(n - 1)|^2]}$$

(b) Evaluate $\Gamma_{m,o}(a)$ for each of the following three special conditions:

$$\text{(i)} \quad a = 1$$

$$\text{(ii)} \quad a = 0$$

$$\text{(iii)} \quad a = \tfrac{1}{2}$$

Notes: When the parameter $a = 1$, the index of performance reduces to

$$J_m(\Gamma_m) = E[|f_m(n)|^2]$$

We refer to this criterion as the *forward method*.

When the parameter $a = 0$, the index of performance reduces to

$$J_m(\Gamma_m) = E[|b_m(n)|^2]$$

We refer to this criterion as the *backward method*.

When the parameter $a = \tfrac{1}{2}$, the formula for $\Gamma_{m,o}(a)$ reduces to the Burg formula.

16. Let $\Gamma_{m,o}(1)$ and $\Gamma_{m,o}(0)$ denote the optimum values of the reflection coefficient Γ_m for stage m of the lattice predictor using the forward method and backward method, respectively, as determined in Problem 15.

(a) Show that the optimum value of $\Gamma_{m,o}$ obtained from the Burg formula equals the harmonic mean of the two values $\Gamma_{m,o}(1)$ and $\Gamma_{m,o}(0)$, as shown by

$$\frac{2}{\Gamma_{m,o}} = \frac{1}{\Gamma_{m,o}(1)} + \frac{1}{\Gamma_{m,o}(0)}$$

(b) Using the result of part (a), show that

$$|\Gamma_{m,o}| \leq 1, \quad \text{for all } m$$

(c) For the case of a lattice predictor using the Burg formula, show that the mean-square values of the forward and backward prediction errors at the output of stage m are related to those at the input as follows, respectively:

$$E[|f_m(n)|^2] = (1 - |\Gamma_{m,o}|^2)E[|f_{m-1}(n)|^2]$$

and

$$E[|b_m(n)|^2] = (1 - |\Gamma_{m,\,o}|^2)E[|b_{m-1}(n-1)|^2]$$

17. Consider a linear array that consists of a total of N uniformly spaced antenna elements. A *subaperture* is formed by using a contiguous subset of $M + 1$ elements. Assume that $N > M + 1$ so that it becomes possible to set up several such subapertures.

   (a) Show that a prediction-error filter operating on the $M + 1$ element outputs of a subaperture is identical in configuration to a sidelobe canceller with one of the end elements of the subaperture providing the main beam and the remaining M elements of the subaperture treated as auxiliaries.

   (b) Set up a lattice-equivalent model for the prediction-error filter that operates on the $M + 1$ element outputs of a subaperture.

   (c) Suppose that a total of K snapshots of data are available for processing, with a *snapshot* defined as one simultaneous sampling of the aperture outputs at all elements of the array. Assume that the environment in which the array operates is stationary. Hence, modify the Burg formula for estimating the reflection coefficients of the lattice-equivalent model so that it includes not only spatial averaging but also temporal averaging.

In parts (a) and (b) of the problem, illustrate your answer for $M = 3$.

5

Adaptive Transversal Filters Using Gradient-Vector Estimation

5.1 INTRODUCTION

In this chapter, we develop the theory of a closed-loop realization of *adaptive filters* built around a transversal or tapped-delay-line filter that is characterized by an impulse response of finite duration, and which does not require complete *a priori* knowledge of the statistics of the signals to be filtered. As discussed in Chapter 1, two basic processes take place in an adaptive filter:

1. The *adaptive process*, which involves the automatic adjustment of the tap weights of the filter in accordance with some algorithm.

2. The *filtering process*, which involves (1) multiplying the tap inputs by the corresponding set of tap weights resulting from the adaptive process to produce an estimate of the desired response, and (2) generating an estimation error by comparing this estimate with the actual value of the desired responses. The estimation error is in turn used to actuate the adaptive process, thereby closing the feedback loop.

Various algorithms have been developed for implementing the adaptive process. In this chapter we study a widely used algorithm known as the *least-mean-square (LMS) algorithm* (Widrow and Hoff, 1960). A significant feature of the LMS algorithm is its simplicity; its does not require measurements of the pertinent correlation functions, nor does it require matrix inversion. Indeed, it is the simplicity of the LMS algorithm that has made it the standard against which other adaptive filtering algorithms are benchmarked.

We begin the study by describing an old optimization technique known as the *method of steepest descent*, which is a recursive method for finding the minimum point of the error-performance surface without knowledge of the error-performance surface itself. This method provides some heuristics for writing the expression for the LMS algorithm.

We then go on to analyze the important characteristics of the LMS algorithm when operating in a *stationary environment*. This analysis will not only help us develop a deep understanding of the LMS algorithm in its own right, but also it gives us invaluable insight into the operation of more complicated adaptive filtering algorithms.

5.2 STRUCTURE OF THE ADAPTIVE FILTER

The structure of the adaptive filter is shown in the block-diagram form in Fig. 5.1. It consists of two basic parts: (1) a transversal filter with adjustable tap weights whose values at time n are denoted $w_1(n), w_2(n), \ldots, w_M(n)$, and (2) a mechanism for adjusting these tap weights in an adaptive manner. During the filtering process,

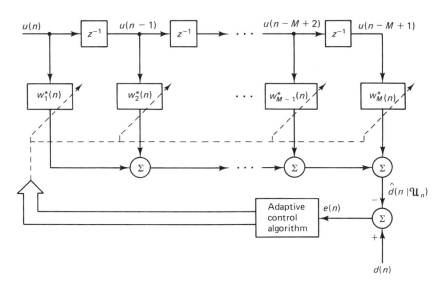

Figure 5.1   Structure of adaptive transversal filter.

an additional signal $d(n)$, called the *desired response*, is supplied along with the usual tap inputs. In effect, the desired response provides a frame of reference for adjusting the tap weights of the filter. The vector of tap inputs at time n is denoted by $\mathbf{u}(n)$, and the corresponding *estimate* of the desired response at the filter output is denoted $\hat{d}(n|\mathcal{U}_n)$, where $\mathcal{U}_n$ is the space spanned by the tap inputs $u(n)$, $u(n - 1), \ldots, u(n - M + 1)$. By comparing this estimate with the actual value of the desired response $d(n)$, we produce an *estimation error* denoted by $e(n)$. We may thus write

$$\begin{aligned} e(n) &= d(n) - \hat{d}(n|\mathcal{U}) \\ &= d(n) - \mathbf{w}^H(n)\mathbf{u}(n) \end{aligned} \tag{5.1}$$

where the term $\mathbf{w}^H(n)\mathbf{u}(n)$ is the inner product of the tap-weight vector $\mathbf{w}(n)$ and the tap input vector $\mathbf{u}(n)$. The expanded form of the tap-weight vector is described by

$$\mathbf{w}^T(n) = [w_1(n), w_2(n), \ldots, w_M(n)] \tag{5.2}$$

and that of the tap-input vector by

$$\mathbf{u}^T(n) = [u(n), u(n - 1), \ldots, u(n - M + 1)] \tag{5.3}$$

If the tap-input vector $\mathbf{u}(n)$ and the desired $d(n)$ are jointly stationary, then the mean-squared error $J(n)$ at time n is a quadratic function of the tap-weight vector, and so we may write [see Eq. (3.20)]

$$J(n) = \sigma_d^2 - \mathbf{w}^H(n)\mathbf{p} - \mathbf{p}^H\mathbf{w}(n) + \mathbf{w}^H(n)\mathbf{R}\mathbf{w}(n) \tag{5.4}$$

where σ_d^2 = variance of the desired response $d(n)$

$\mathbf{p}$ = cross-correlation vector between the tap-input vector $\mathbf{u}(n)$

and the desired response $d(n)$

$\mathbf{R}$ = correlation matrix of the tap-input vector $\mathbf{u}(n)$

Thus we may visualize the dependence of the mean-squared error $J(n)$ on the elements of the tap-weight vector $\mathbf{w}(n)$ as a *bowl-shaped surface* with a unique minimum. We refer to this surface as the *error-performance surface* of the adaptive filter. The adaptive process has the task of continually seeking the *bottom* or *minimum point* of this surface. At the minimum point of the error-performance surface, the tap-weight vector takes on the optimum value $\mathbf{w}_o$, which is defined by the normal equation [see Eq. (3.35)]

$$\mathbf{R}\mathbf{w}_o = \mathbf{p} \tag{5.5}$$

The minimum mean-squared error equals [see Eq. (3.46)]

$$J_{\min} = \sigma_d^2 - \mathbf{p}^H\mathbf{w}_o \tag{5.6}$$

For an adaptive transversal filter operating in a stationary environment, the error-performance surface has a constant shape as well as orientation. The problem is then simply one of designing the adaptive filter so that it operates at or near the minimum point of this surface. When, however, the adaptive filter operates in a nonstationary environment, the bottom of the error-performance surface continually moves, while the orientation and curvature of the surface may be changing too. Therefore, when the inputs are nonstationary, the adaptive filter has the task of not only seeking out the bottom of the error performance surface, but also continually *tracking* it.

5.3 *THE METHOD OF STEEPEST DESCENT*

The requirement that an adaptive transversal filter has to satisfy is to find a solution for its tap-weight vector that satisfies the normal equation (5.5). One way of doing this would be to solve this equation by some analytical means. Although, in general, this procedure is quite straightforward, nevertheless, it presents serious computational difficulties, especially when the filter contains a large number of tap weights and when the input data rate is high.

An alternative procedure is to use the *method of steepest descent*, which is one of the oldest methods of optimization (Murray, 1972). To find the minimum value of the mean-squared error, $J_{\min}$, by the method of steepest descent, we proceed as follows:

1. We begin with an initial value $\mathbf{w}(0)$ for the tap-weight vector, which is chosen arbitrarily. The value $\mathbf{w}(0)$ provides an initial guess as to where the minimum point of the error-performance surface may be located. Typically, $\mathbf{w}(0)$ is set equal to the null vector.

2. Using this initial or present guess, we compute the *gradient vector*, which is defined as the gradient of the mean-squared error $J(n)$, evaluated with respect to the tap-weight vector $\mathbf{w}(n)$ at time n (i.e., the nth iteration).

3. We compute the next guess at the tap-weight vector by making a change in the initial or present guess in a direction opposite to that of the gradient vector.

4. We go back to step 2 and repeat the process.

It is intuitively reasonable that successive corrections to the tap-weight vector in the direction of the negative of the gradient vector (i.e., in the direction of the steepest descent of the error-performance surface) should eventually lead to the minimum mean-squared error $J_{\min}$, at which point the tap-weight vector assumes its optimum value $\mathbf{w}_o$.

Let $\boldsymbol{\nabla}(n)$ denote the value of the *gradient vector* at time n. Let $\mathbf{w}(n)$ denote the value of the tap-weight vector at time n. According to the method of steepest descent, the updated value of the tap-weight vector at time $n + 1$ is computed by using the simple recursive relation

$$\mathbf{w}(n + 1) = \mathbf{w}(n) + \tfrac{1}{2}\mu[-\mathbf{\nabla}(n)] \tag{5.7}$$

where μ is a positive real-valued constant.

Differentiating the mean-squared error $J(n)$ of Eq. (5.4) with respect to the tap-weight vector $\mathbf{w}(n)$, we get the following value for the gradient vector [see also Eq. (3.34)]:

$$
\begin{aligned}
\mathbf{\nabla}(n) &= \frac{\partial J(n)}{\partial \mathbf{w}(n)} \\
&= -2\mathbf{p} + 2\mathbf{R}\mathbf{w}(n)
\end{aligned}
\tag{5.8}
$$

For the application of the steepest-descent algorithm, we assume that the correlation matrix $\mathbf{R}$ and the cross-correlation vector $\mathbf{p}$ are known so that we may compute the gradient vector $\mathbf{\nabla}(n)$ for a given value of the tap-weight vector $\mathbf{w}(n)$. Thus, substituting Eq. (5.8) in (5.7), we may compute the updated value of the tap-weight vector $\mathbf{w}(n + 1)$ by using the simple recursive relation

$$\mathbf{w}(n + 1) = \mathbf{w}(n) + \mu[\mathbf{p} - \mathbf{R}\mathbf{w}(n)], \qquad n = 0, 1, 2, \ldots, \tag{5.9}$$

We observe that the parameter μ controls the size of the incremental correction applied to the tap-weight vector as we proceed from one iteration cycle to the next. We therefore refer to μ as the *step-size parameter* or *weighting constant*. Equation (5.9) describes the mathematical formulation of the steepest-descent algorithm. For obvious reasons, this recursive equation is also referred to as the *deterministic gradient algorithm*.

We may view the steepest-descent algorithm of Eq. (5.9) as a *feedback model*, as illustrated by the *signal-flow graph* shown in Fig. 5.2. This model is multidimensional in the sense that the "signals" at the *nodes* of the graph consist of vectors and that the *transmittance* of each branch of the graph is a scalar or a square matrix. For each branch of the graph, the signal vector flowing out equals the signal vector flowing in multiplied by the transmittance matrix of the branch. For two branches

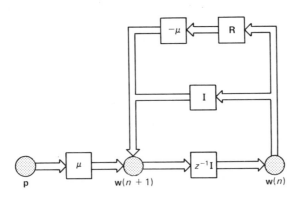

Figure 5.2 Signal-flow graph representation of the steepest-descent algorithm.

connected in parallel, the overall transmittance matrix equals the sum of the transmittance matrices of the individual branches. For two branches connected in cascade, the overall transmittance matrix equals the product of the individual transmittance matrices arranged in the same order as the pertinent branches. Finally, the symbol z^{-1} is the unit-delay operator, and $z^{-1}\mathbf{I}$ is the transmittance matrix of a unit-delay branch representing a delay of one iteration cycle.

5.4 STABILITY OF THE STEEPEST-DESCENT ALGORITHM

Since the steepest-descent algorithm involves the presence of *feedback*, as exemplified by the model of Fig. 5.2, the algorithm is subject to the possibility of its becoming *unstable*. From the feedback model of Fig. 5.2, we observe that the *stability performance* of the steepest-descent algorithm in determined by two factors: (1) the step-size parameter μ, and (2) the correlation matrix $\mathbf{R}$ of the tap-input vector $\mathbf{u}(n)$, as these two parameters completely control the transfer function of the *feedback loop*.

To determine *the condition for the stability* of the steepest-descent algorithm, we examine the *natural modes* of the algorithm (Widrow, 1970). In particular, we use the representation of the correlation matrix $\mathbf{R}$ in terms of its eigenvalues and eigenvectors to define a transformed version of the tap-weight vector. We begin the analysis by defining a *weight-error vector* at time n as

$$\mathbf{c}(n) = \mathbf{w}(n) - \mathbf{w}_o \qquad (5.10)$$

where $\mathbf{w}_o$ is the optimum value of the tap-weight vector, as defined by the normal equation (5.5). Then, eliminating the cross-correlation vector $\mathbf{p}$ between Eqs. (5.5) and (5.9), and rewriting the result in terms of the weight-error vector $\mathbf{c}(n)$, we get

$$\mathbf{c}(n + 1) = (\mathbf{I} - \mu\mathbf{R})\mathbf{c}(n) \qquad (5.11)$$

where $\mathbf{I}$ is the identity matrix. Equation (5.11) is represented by the feedback model shown in Fig. 5.3. This diagram further emphasizes the fact that the stability performance of the steepest-descent algorithm is controlled exclusively by μ and $\mathbf{R}$.

Using the unitary similarity transformation, we may express the correlation matrix $\mathbf{R}$ as follows (see Chapter 2):

$$\mathbf{R} = \mathbf{Q}\mathbf{\Lambda}\mathbf{Q}^H \qquad (5.12)$$

The matrix $\mathbf{Q}$ has as its columns an orthogonal set of *eigenvectors* associated with the eigenvalues of the matrix $\mathbf{R}$. The matrix $\mathbf{Q}$ is called the *unitary matrix* of the transformation. The matrix $\mathbf{\Lambda}$ is a *diagonal* matrix and has as its diagonal elements the eigenvalues of the correlation matrix $\mathbf{R}$. These eigenvalues, denoted by λ_1, $\lambda_2, \ldots, \lambda_M$, are all positive and real. Each eigenvalue is associated with a corresponding eigenvector or column of matrix $\mathbf{Q}$. Substituting Eq. (5.12) in

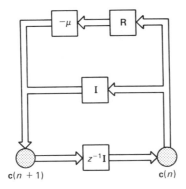

Figure 5.3 Signal-flow graph representation of the steepest-descent algorithm
based on the weight error vector.

(5.11), we get

$$c(n + 1) = (I - \mu Q \Lambda Q^H)c(n) \tag{5.13}$$

Premultiplying both sides of this equation by Q^H and using the property of the
unitary matrix Q that Q^H equals the inverse Q^{-1} (see Chapter 2), we get

$$Q^H c(n + 1) = (I - \mu \Lambda)Q^H c(n) \tag{5.14}$$

We now define a new set of coordinates as follows:

$$\begin{aligned}
v(n) &= Q^H c(n) \\
&= Q^H[w(n) - w_o]
\end{aligned} \tag{5.15}$$

Accordingly, we may rewrite Eq. (5.13) in the desired form:

$$v(n + 1) = (I - \mu \Lambda)v(n) \tag{5.16}$$

The initial value of $v(n)$ equals

$$v(0) = Q^H[w(0) - w_o] \tag{5.17}$$

Assuming that the initial tap-weight vector $w(0)$ is zero, Eq. (5.17) reduces to

$$v(0) = -Q^H w_o \tag{5.18}$$

For the kth natural mode of the steepest-descent algorithm, we thus have

$$v_k(n + 1) = (1 - \mu \lambda_k)v_k(n), \qquad k = 1, 2, \ldots, M \tag{5.19}$$

where λ_k is the kth eigenvalue of the correlation matrix R. This equation is
represented by the scalar-valued feedback model of Fig. 5.4, where z^{-1} is the unit-

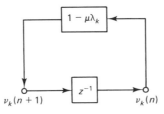

Figure 5.4 Signal-flow graph representation of the kth natural mode of the
steepest-descent algorithm.

delay operator. Clearly, the structure of this model is much simpler than that of the original matrix-valued feedback model of Fig. 5.2. These two models represent different and yet equivalent ways of viewing the steepest-descent algorithm.

Equation (5.19) is a homogeneous *difference equation of the first order*. Assuming that $v_k(n)$ has the initial value $v_k(0)$, we readily obtain the solution

$$v_k(n) = (1 - \mu\lambda_k)^n v_k(0), \qquad k = 1, 2, \ldots, M \qquad (5.20)$$

Since all eigenvalues of the correlation matrix $\mathbf{R}$ are positive and real, the response $v_k(n)$ will exhibit no oscillations. Furthermore, as illustrated in Fig. 5.5, the numbers generated by Eq. (5.20) represent a *geometric series* with a geometric ratio equal to $1 - \mu\lambda_k$. For *stability* or *convergence* of the steepest-descent algorithm, the magnitude of this geometric ratio must be less than 1 for all k. That is, provided we have

$$-1 < 1 - \mu\lambda_k < 1, \qquad \text{for all } k$$

then as the number of iterations, n, approaches infinity, all the natural modes of the steepest-descent algorithm die out, irrespective of the initial conditions. This is equivalent to saying that the tap-weight vector $\mathbf{w}(n)$ approaches the optimum solution $\mathbf{w}_o$ as n approaches infinity.

Since the eigenvalues of the correlation matrix $\mathbf{R}$ are all real and positive, it therefore follows that the necessary and sufficient condition for the convergence or stability of the steepest-descent algorithm is that the step-size parameter μ satisfy the following condition:

$$0 < \mu < \frac{2}{\lambda_{\max}} \qquad (5.21)$$

where $\lambda_{\max}$ is the largest eigenvalue of the correlation matrix $\mathbf{R}$.

Referring to Fig. 5.5, we see that an exponential envelope of *time constant* τ_k can be fitted to the geometric series by assuming the unit of time to be the duration of one iteration cycle and by choosing the time constant τ_k such that

$$1 - \mu\lambda_k = \exp\left(-\frac{1}{\tau_k}\right)$$

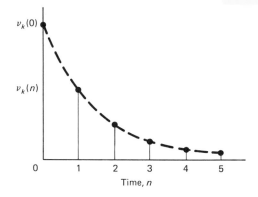

Figure 5.5 Variation of the kth natural mode of the steepest-descent algorithm with time, assuming that the magnitude of $1 - \mu\lambda_k$ is less than 1.

Hence, the kth time constant τ_k can be expressed in terms of the step-size parameter μ and the kth eigenvalue as follows:

$$\tau_k = \frac{-1}{\ln(1 - \mu\lambda_k)} \tag{5.22}$$

The time constant τ_k defines the time required for the amplitude of the kth natural mode $v_k(n)$ to decay to $1/e$ of its initial value $v_k(0)$, where e is the base of the natural logarithm. For the special case of slow adaptation, for which the step-size parameter μ is small, we may approximate the time constant τ_k as

$$\tau_k \simeq \frac{1}{\mu\lambda_k}, \qquad \mu \ll 1 \tag{5.23}$$

We may now formulate the transient behavior of the original tap-weight vector $\mathbf{w}(n)$. In particular, premultiplying both sides of Eq. (5.15) by $\mathbf{Q}$, using the fact that $\mathbf{Q}\mathbf{Q}^H = \mathbf{I}$, and solving for $\mathbf{w}(n)$, we get the desired result:

$$\mathbf{w}(n) = \mathbf{w}_o + \mathbf{Q}\mathbf{v}(n)$$

$$= \mathbf{w}_o + [\mathbf{q}_1, \mathbf{q}_2, \ldots, \mathbf{q}_M] \begin{bmatrix} v_1(n) \\ v_2(n) \\ \cdot \\ \cdot \\ \cdot \\ v_M(n) \end{bmatrix} \tag{5.24}$$

$$= \mathbf{w}_o + \sum_{k=1}^{M} \mathbf{q}_k v_k(n)$$

where $\mathbf{q}_1, \mathbf{q}_2, \ldots, \mathbf{q}_M$ are the eigenvectors associated with the eigenvalues $\lambda_1, \lambda_2, \ldots, \lambda_M$ of the correlation matrix $\mathbf{R}$, respectively, and the kth natural mode $v_k(n)$ is defined by Eq. (5.20). Thus, substituting Eq. (5.20) in (5.24), we find that the transient behavior of the ith tap weight is described by (Griffiths, 1975)

$$w_i(n) = w_{oi} + \sum_{k=1}^{N} q_{ki}v_k(0)(1 - \mu\lambda_k)^n, \qquad i = 1, 2, \ldots, M \tag{5.25}$$

where w_{oi} is the optimum value of the ith tap weight, and q_{ki} is the ith element of the kth eigenvector $\mathbf{q}_k$.

Equation (5.25) shows that each tap weight in the steepest-descent algorithm converges as the weighted sum of exponentials of the form $(1 - \mu\lambda_k)^n$. The time τ_k required for each term to reach $1/e$ of its initial value is given by Eq. (5.22). However, the *overall time constant*, τ_a, defined as the time required for the summation term in Eq. (5.25) to decay to $1/e$ of its initial value, cannot be expressed in a simple closed form similar to Eq. (5.22). Nevertheless, the *slowest rate of convergence* is attained when $q_{ki}v_k(0)$ is zero for all k except for that corresponding to the smallest eigenvalue $\lambda_{\min}$ of matrix $\mathbf{R}$, so the upper bound on τ_a is defined by $-1/\ln(1 - \mu\lambda_{\min})$. The *fastest rate of convergence* is attained when all the $q_{ik}v_k(0)$ are zero except for that corresponding to the largest eigenvalue $\lambda_{\max}$, and so the lower bound on τ_a is defined by $-1/\ln(1 - \mu\lambda_{\max})$. Accordingly, the overall time constant τ_a for any tap weight of the steepest-descent algorithm is bounded as follows (Griffiths, 1975):

$$\frac{-1}{\ln(1 - \mu\lambda_{\max})} \leq \tau_a \leq \frac{-1}{\ln(1 - \mu\lambda_{\min})} \tag{5.26}$$

We see therefore that, when the eigenvalues of the correlation matrix $\mathbf{R}$ are widely spread (i.e., the correlation matrix of the tap inputs is ill conditioned), the settling time of the steepest-descent algorithm is limited by the smallest eigenvalues or the slowest modes.

5.5 *THE MEAN-SQUARED ERROR*

We may develop further insight into the operation of the steepest-descent algorithm by examining the formula for the mean-squared error. At any time n, the value of the mean-squared error $J(n)$ is given by [see Eq. (3.75)]

$$J(n) = J_{\min} + \sum_{k=1}^{M} \lambda_k |v_k(n)|^2 \tag{5.27}$$

where $J_{\min}$ is the minimum mean-squared error. The transient behavior of the kth natural mode, $v_k(n)$, is defined by Eq. (5.20). Hence substituting Eq. (5.20) into (5.27), we get

$$J(n) = J_{\min} + \sum_{k=1}^{M} \lambda_k (1 - \mu\lambda_k)^{2n} |v_k(0)|^2 \tag{5.28}$$

where $v_k(0)$ is the initial value of $v_k(n)$. When the steepest-descent algorithm is convergent, that is, the step-size parameter μ is chosen within the bounds defined by Eq. (5.21), we see that, irrespective of the initial conditions,

$$\lim_{n\to\infty} J(n) = J_{\min} \tag{5.29}$$

The curve obtained by plotting the mean-squared error $J(n)$ versus the number of iterations, n is called a *learning curve*. From Eq. (5.28), we see that the learning curve of the steepest-descent algorithm consists of a sum of exponentials, each of which corresponds to a natural mode of the algorithm. In general, the number of natural modes equals the number of tap weights. In going from the initial value $J(0)$ to the final value $J_{\min}$, the exponential decay for the kth natural mode has a time constant equal to

$$\tau_{k,\text{mse}} \simeq \frac{-1}{2\ln(1 - \mu\lambda_k)} \tag{5.30}$$

For small values of the step-size parameter μ, we may approximate this time constant as

$$\tau_{k,\text{mse}} \simeq \frac{1}{2\mu\lambda_k} \tag{5.31}$$

5.6 EXAMPLE: EFFECTS OF EIGENVALUE SPREAD AND STEP-SIZE PARAMETER ON THE STEEPEST-DESCENT ALGORITHM

In this example, we examine the transient behavior of the steepest-descent algorithm applied to a predictor that operates on a real-valued autoregressive (AR) process. Figure 5.6 shows the structure of the predictor, assumed to contain two tap weights that are denoted by $w_1(n)$ and $w_2(n)$; the dependence of these tap weights on time n emphasizes the transient condition of the predictor. The AR process $\{u(n)\}$ is described by the second-order difference equation

$$u(n) + a_1 u(n - 1) + a_2 u(n - 2) = v(n) \tag{5.32}$$

where the sample $v(n)$ is drawn from a white-noise process of zero mean and variance σ_v^2. The AR parameters a_1 and a_2 are chosen so that the roots of the characteristic equation

$$1 + a_1 z^{-1} + a_2 z^{-2} = 0$$

are complex; that is, $a_1^2 < 4a_2$. The particular values assigned to a_1 and a_2 are determined by the desired eigenvalue spread $\chi(\mathbf{R})$. For specified values of a_1 and a_2, the variance σ_v^2 of the white-noise process is chosen to make the process $\{u(n)\}$ have variance $\sigma_u^2 = 1$.

The requirement is to evaluate the transient behavior of the steepest-descent algorithm for the following two conditions:

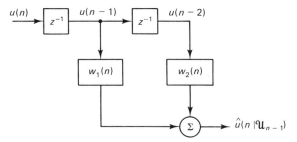

Figure 5.6 Two-tap predictor for real valued input.

1. Varying eigenvalue spread $\chi(\mathbf{R})$ and fixed step-size parameter μ.
2. Varying step-size parameter μ and fixed eigenvalue spread $\chi(\mathbf{R})$.

Characterization of the AR Process

Since the predictor of Fig. 5.6 has two tap weights and the AR process $\{u(n)\}$ is real valued, it follows that the correlation matrix $\mathbf{R}$ of the tap inputs is a 2-by-2 symmetric matrix:

$$\mathbf{R} = \begin{bmatrix} r(0) & r(1) \\ r(1) & r(0) \end{bmatrix}$$

where

$$r(0) = \sigma_u^2$$

$$r(1) = \frac{-a_1}{1 + a_2}\, \sigma_u^2$$

$$\sigma_u^2 = \left(\frac{1 + a_2}{1 - a_2}\right) \frac{\sigma_v^2}{(1 + a_2)^2 - a_1^2}$$

The two eigenvalues of $\mathbf{R}$ equal

$$\lambda_1 = \left(1 - \frac{a_1}{1 + a_2}\right)\sigma_u^2$$

$$\lambda_2 = \left(1 + \frac{a_1}{1 + a_2}\right)\sigma_u^2$$

Hence, the eigenvalue spread equals (assuming that a_1 is negative)

$$\chi(\mathbf{R}) = \frac{\lambda_1}{\lambda_2} = \frac{1 - a_1 + a_2}{1 + a_1 + a_2}$$

The eigenvectors $\mathbf{q}_1$ and $\mathbf{q}_2$ associated with the respective eigenvalues λ_1 and λ_2 equal

$$\mathbf{q_1} = \frac{1}{\sqrt{2}}\begin{bmatrix} 1 \\ 1 \end{bmatrix}$$

$$\mathbf{q}_2 = \frac{1}{\sqrt{2}}\begin{bmatrix} 1 \\ -1 \end{bmatrix}$$

both of which are normalized to unit length.

Experiment 1: Varying Eigenvalue Spread

In this experiment, the step-size parameter μ is fixed at 0.3, and the evaluations are made for the 4 sets of AR parameters given in Table 5.1.

For a given set of parameters, we use a two-dimensional plot of the transformed tap-weight error $v_1(n)$ versus $v_2(n)$ to display the transient behavior of the steepest-descent algorithm. In particular, the use of Eq. (5.20) yields

$$
\boldsymbol{v}(n) = \begin{bmatrix} v_1(n) \\ v_2(n) \end{bmatrix}
$$
$$
= \begin{bmatrix} (1 - \mu\lambda_1)^n v_1(0) \\ (1 - \mu\lambda_2)^n v_2(0) \end{bmatrix}, \qquad n = 1, 2, \ldots
\tag{5.33}
$$

To calculate the initial value $\boldsymbol{v}(0)$, we use Eq. (5.18), assuming that the initial value $\mathbf{w}(0)$ of the tap-weight vector $\mathbf{w}(n)$ is zero. This equation requires knowledge of the optimum tap-weight vector $\mathbf{w}_o$. Now when the two-tap predictor of Fig. 5.6 is optimized, with the second-order AR process of Eq. (5.32) supplying the tap inputs, we find that the optimum tap-weight vector equals

$$
\mathbf{w}_o = \begin{bmatrix} -a_1 \\ -a_2 \end{bmatrix}
$$

and the minimum mean-squared error equals

$$
J_{\min} = \sigma_v^2
$$

Accordingly, the use of Eq. (5.18) yields the initial value:

$$
\boldsymbol{v}(0) = \begin{bmatrix} v_1(0) \\ v_2(0) \end{bmatrix}
$$
$$
= \frac{-1}{\sqrt{2}} \begin{bmatrix} 1 & 1 \\ 1 & -1 \end{bmatrix} \begin{bmatrix} -a_1 \\ -a_2 \end{bmatrix}
\tag{5.34}
$$
$$
= \frac{1}{\sqrt{2}} \begin{bmatrix} a_1 + a_2 \\ a_1 - a_2 \end{bmatrix}
$$

Thus, for specified parameters, we use Eq. (5.34) to compute the initial value $\boldsymbol{v}(0)$, and then use Eq. (5.33) to compute $\boldsymbol{v}(1)$, $\boldsymbol{v}(2)$, ... By joining the points defined

TABLE 5.1

Case	AR parameters		Eigenvalues		Eigenvalue spread $\chi = \lambda_1/\lambda_2$	Minimum mean-squared Error $J_{\min} = \sigma_v^2$
	a_1	a_2	λ_1	λ_2		
1	-0.1950	0.95	1.1	0.9	1.22	0.0965
2	-0.9750	0.95	1.5	0.5	3	0.0731
3	-1.5955	0.95	1.818	0.182	10	0.0322
4	-1.9114	0.95	1.957	0.0198	100	0.0038

by these values of $\boldsymbol{v}(n)$ for varying time n, we obtain a *trajectory* that describes the transient behavior of the steepest-descent algorithm for the particular set of parameters.

It is informative to include in the two-dimentional plot of $v_1(n)$ versus $v_2(n)$ loci representing Eq. (5.27) for fixed values of n. For our example, Eq. (5.27) yields

$$J(n) - J_{\min} = \lambda_1 v_1^2(n) + \lambda_2 v_2^2(n) \tag{5.35}$$

When $\lambda_1 = \lambda_2$ and time n is fixed, Eq. (5.35) represents a circle with center at the origin and radius equal to the square root of $[J(n) - J_{\min}]/\lambda$, where λ is the common value of the two eigenvalues. When, on the other hand, $\lambda_1 \neq \lambda_2$, Eq. (5.35) represents (for fixed n) an ellipse with major axis equal to the square root of $[J(n) - J_{\min}]/\lambda_1$ and minor axis equal to the square root of $[J(n) - J_{\min}]/\lambda_2$.

Case 1: Eigenvalue Spread $\chi(R) = 1.22$ For the parameter values given for Case 1 in Table 5.1, the eigenvalue spread $\chi(\mathbf{R})$ equals 1.22; that is, the eigenvalues λ_1 and λ_2 are approximately equal. The use of these parameter values in Eqs. (5.33) and (5.34) yields the trajectory of $[v_1(n), v_2(n)]$ shown in Fig. 5.7(a), with n as running parameter, and their use in Eq. (5.35) yields the (approximately) circular loci shown for fixed values of $J(n)$, corresponding to $n = 0, 1, 2, 3, 4, 5$.

We may also display the transient behavior of the steepest-descent algorithm by plotting the tap weight $w_1(n)$ versus $w_2(n)$. In particular, for our example the use of Eq. (5.24) yields the tap-weight vector

$$\begin{aligned}
\mathbf{w}(n) &= \begin{bmatrix} w_1(n) \\ w_2(n) \end{bmatrix} \\
&= \begin{bmatrix} -a_1 + \left(v_1(n) + v_2(n)\right)/\sqrt{2} \\ -a_2 + \left(v_1(n) - v_2(n)\right)/\sqrt{2} \end{bmatrix}
\end{aligned} \tag{5.36}$$

The corresponding trajectory of $[w_1(n), w_2(n)]$, with n as a running parameter, obtained by using Eq. (5.36), is shown plotted in Fig. 5.8(a). Here again we have included the loci of $[w_1(n), w_2(n)]$ for fixed values of $J(n)$ corresponding to $n = 0, 1, 2, 3, 4, 5$. Note that these loci, unlike Fig. 5.7(a), are ellipsoidal.

Case 2: Eigenvalue Spread $\chi(R) = 3$ The use of the parameter values for Case 2 in Eqs. (5.33) and (5.34) yields the trajectory of $[v_1(n), v_2(n)]$ shown in Fig. 5.7(b), with n as running parameter, and their use in Eq. (5.35) yields the ellipsoidal loci shown for the fixed values of $J(n)$ for $n = 0, 1, 2, 3, 4, 5$. Note that for this set of parameter values the initial value $v_2(0)$ is approximately zero, so the initial value $\boldsymbol{v}(0)$ lies practically on the v_1-axis.

The corresponding trajectory of $[w_1(n), w_2(n)]$, with n as running parameter, is shown in Fig. 5.8(b).

Case 3: Eigenvalue Spread $\chi(R) = 10$ For this case, the application of Eqs. (5.33) and (5.34) yields the trajectory of $[v_1(n), v_2(n)]$ shown in Fig. 5.7(c), with n as running parameter, and the application of Eq. (5.35) yields the ellipsoidal loci included in this figure for fixed values of $J(n)$ for $n = 0, 1, 2, 3, 4, 5$. The corresponding trajectory of $[w_1(n), w_2(n)]$, with n as running parameter, is shown in Fig. 5.8(c).

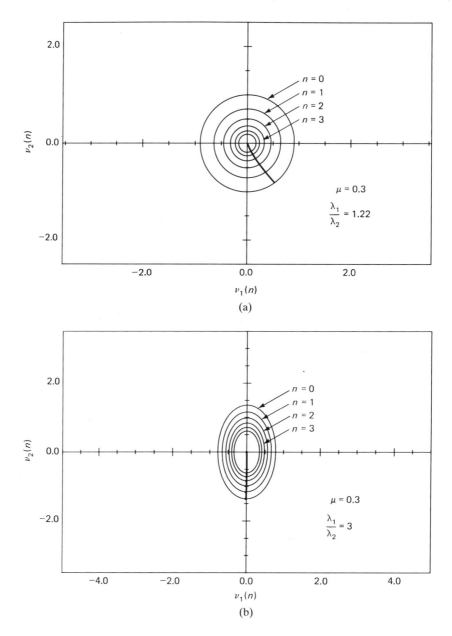

Figure 5.7 Loci of $v_1(n)$ versus $v_2(n)$ for the steepest descent alogrithm with step-size parameter $\mu = 0.3$ and varying eigenvalue spread: (a) $\chi(\mathbf{R}) = 1.22$; (b) $\chi(\mathbf{R}) = 3$; (c) $\chi(\mathbf{R}) = 10$; (d) $\chi(\mathbf{R}) = 100$.

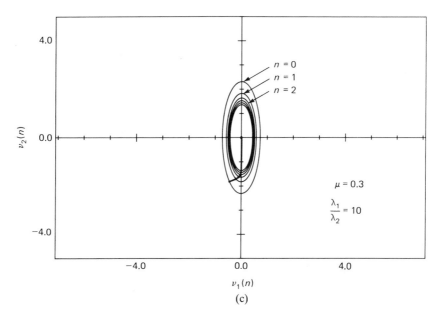

(c)

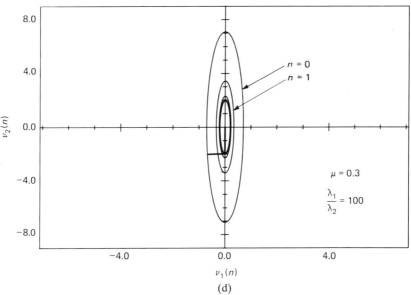

(d)

Figure 5.7   (*cont.*)

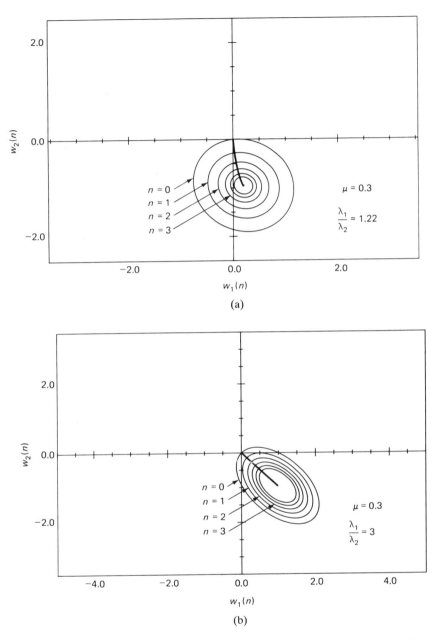

Figure 5.8 Loci of $w_1(n)$ versus $w_2(n)$ for the steepest descent algorithm with step–size parameter $\mu = 0.3$ and varying eigenvalue spread: (a) $\chi(\mathbf{R}) = 1.22$; (b) $\chi(\mathbf{R}) = 3$; (c) $\chi(\mathbf{R}) = 10$; (d) $\chi(\mathbf{R}) = 100$.

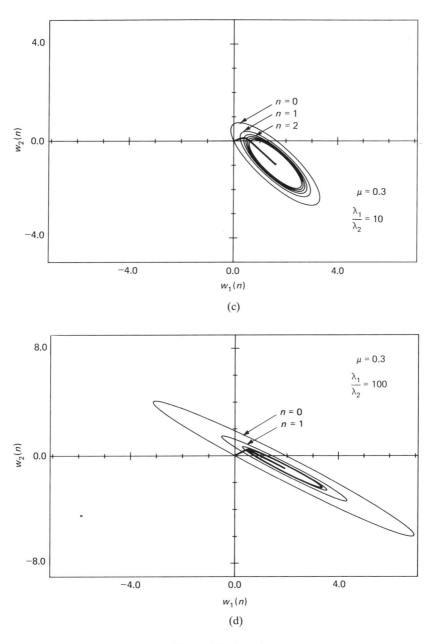

Figure 5.8 *(cont.)*

Case 4: Eigenvalue Spread $\chi(R) = 100$ For this case, application of the preceding equations yields the results shown in Fig. 5.7(d) for the trajectory of $[v_1(n), v_2(n)]$ and the ellipsoidal loci for fixed values of $J(n)$. The corresponding trajectory of $[w_1(n), w_2(n)]$ is shown in Fig. 5.8(d).

In Fig. 5.9 we have plotted the mean-squared error $J(n)$ versus n for the four eigenvalue spreads 1.22, 3, 10, and 100. We see that as the eigenvalue spread increases (and the input process becomes more correlated), the minimum mean-squared error J_{min} decreases. This makes intuitive sense: the predictor should do a better job tracking a highly correlated input process than a weakly correlated one.

Experiment 2: Varying Step-size Parameter

In this experiment the eigenvalue spread is fixed at $\chi(\mathbf{R}) = 10$, and the step-size parameter μ is varied. In particular, we examine the transient behavior of the steepest descent algorithm for $\mu = 0.3, 1.0$. The corresponding results in terms of the transformed tap-weight errors $v_1(n)$ and $v_2(n)$ are shown in parts (a) and (b) of Fig. 5.10, respectively. The results included in part (a) of this figure are the same as those in Fig. 5.7(c). Note also that in accordance with Eq. (5.21), the critical value of the step-size parameter equals $\mu_{max} = 2/\lambda_{max} = 1.1$, which is slightly in excess of the actual value $\mu = 1$ used in Fig. 5.10(b).

The results for $\mu = 0.3, 1.0$ in terms of the tap weights $w_1(n)$ and $w_2(n)$ are shown in parts (a) and (b) of Fig. 5.11, respectively. Here again, the results included in part (a) of the figure are the same as those in Fig. 5.8(c).

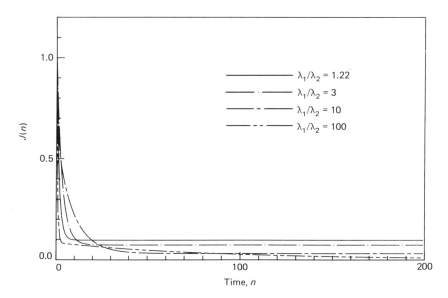

Figure 5.9 Learning curves of steepest descent algorithm with step–size parameter $\mu = 0.3$ and varying eigenvalue spread: (a) $\chi(\mathbf{R}) = 1.22$; (b) $\chi(\mathbf{R}) = 3$; (c) $\chi(\mathbf{R}) = 10$; (d) $\chi(\mathbf{R}) = 100$.

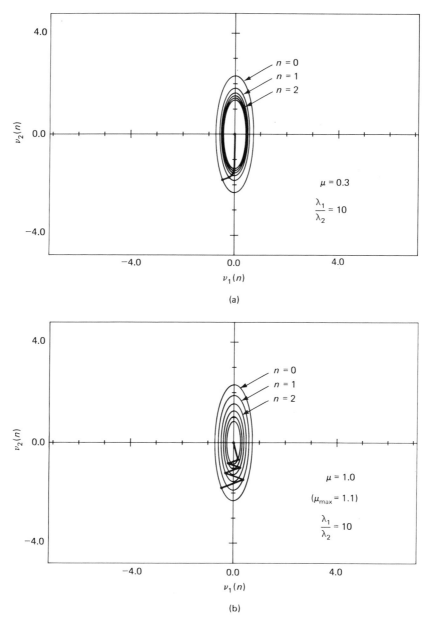

Figure 5.10 Loci of $v_1(n)$ versus $v_2(n)$ for the steepest descent algorithm with eigenvalue spread $\chi(\mathbf{R}) = 10$ and varying step–size parameters: (a) overdamped, $\mu = 0.3$; (b) underdamped, $\mu = 1.0$.

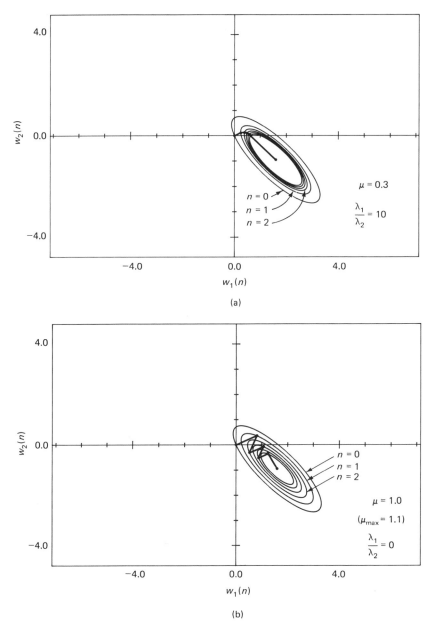

Figure 5.11 Loci of $w_1(n)$ versus $w_2(n)$ for the steepest descent algorithm with eigenvalue spread $\chi(\mathbf{R}) = 10$ and varying step–size parameters: (a) overdamped, $\mu = 0.3$; (b) underdamped, $\mu = 1.0$.

Discussion

Based on the results presented for Experiments 1 and 2, we may make the following observations:

1. The trajectory of $[v_1(n), v_2(n)]$, with the number of iterations n as running parameter, is normal to the locus of $[v_1(n), v_2(n)]$ for fixed $J(n)$. Likewise, the trajectory of $[w_1(n), w_2(n)]$, with n as running parameter, is normal to the locus of $[w_1(n), w_2(n)]$ for fixed $J(n)$.

2. When the eigenvalues λ_1 and λ_2 are equal, the trajectory of $[v_1(n), v_2(n)]$ or that of $[w_1(n), w_2(n)]$, with n as running parameter, is a straight line. This is illustrated in Fig. 5.7(a) or 5.8(a) for which the eigenvalues λ_1 and λ_2 are approximately equal.

3. When the conditions are right for the initial value $v(0)$ of the transformed tap-weight error vector $v(n)$ to lie on the v_1-axis or v_2-axis, the trajectory of $[v_1(n), v_2(n)]$, with n as running parameter, is a straight line. This is illustrated in Fig. 5.7(b), where $v_2(0)$ is approximately zero. Correspondingly, the trajectory of $[w_1(n), w_2(n)]$, with n as running parameter, is also a straight line, as illustrated in Fig. 5.8(b).

4. Except for two special cases, (1) equal eigenvalues, and (2) the right choice of initial conditions, the trajectory of $[v_1(n), v_2(n)]$, with n as running parameter, follows a curved path, as illustrated in Fig. 5.7(c). Correspondingly, the trajectory of $[w_1(n), w_2(n)]$, with n as running parameter, also follows a curved path, as illustrated in Fig. 5.8(c). When the eigenvalue spread is very high (i.e., the input data are very highly correlated), the trajectories of $[v_1(n), v_2(n)]$ and $[w_1(n), w_2(n)]$ develop distinct bends. This is illustrated in Figs. 5.7(d) and 5.8(d), respectively, for the case of $\chi(\mathbf{R}) = 100$.

5. The steepest-descent algorithm converges fastest when the eigenvalues λ_1 and λ_2 are equal or the starting point of the algorithm is chosen right, for which cases the trajectory formed by joining the points $v(0), v(1), v(2), \ldots$, is a straight line, the shortest possible path.

6. For fixed step-size parameter μ, as the eigenvalue spread $\chi(\mathbf{R})$ increases (i.e., the correlation matrix $\mathbf{R}$ of the tap inputs becomes more ill conditioned), the ellipsoidal loci of $[v_1(n), v_2(n)]$ for fixed values of $J(n)$, for $n = 0, 1, 2, \ldots$, become increasingly narrower (i.e., the minor axis becomes smaller) and more crowded.

7. When the step-size parameter μ is small, the transient behavior of the steepest-descent algorithm is *overdamped* in that the trajectory formed by joining the points $v(0), v(1), v(2), \ldots$, follows a continuous path. When, on the other hand, μ approaches the maximum allowable value, $\mu_{\max} = 2/\lambda_{\max}$, the transient behavior of the steepest-descent algorithm is *underdamped* in that this trajectory exhibits oscillations. These two different forms of transient behavior are illustrated in parts (a) and (b) of Fig. 5.10 in terms of $v_1(n)$ and

$v_2(n)$.   The corresponding results in terms of $w_1(n)$ and $w_2(n)$ are presented in parts (a) and (b) of Fig. 5.11.

The conclusion to be drawn from these observations is that the transient behavior of the steepest-descent algorithm is highly sensitive to variations in the step-size parameter μ and the eigenvalue spread of the correlation matrix of the tap inputs.

5.7 THE LEAST-MEAN-SQUARE ADAPTATION ALGORITHM

If it were possible to make exact measurements of the gradient vector at each iteration, and if the step-size parameter μ is suitably chosen, then the tap-weight vector computed by using the method of steepest descent would indeed converge to the optimum Wiener solution.   In reality, however, exact measurements of the gradient vector are not possible, and the gradient vector must be estimated from the available data.   In other words, the tap-weight vector is updated in accordance with an algorithm that adapts to the incoming data.   One such algorithm is the *least-mean-square (LMS) algorithm* (Widrow and Hoff, 1960; Widrow, 1970).   A significant feature of the LMS algorithm is its simplicity; it does not require measurements of the pertinent correlation functions, nor does it require matrix inversion.

To develop an estimate of the gradient vector $\nabla(n)$, the most obvious strategy is to substitute estimates of the correlation matrix $\mathbf{R}$ and the cross-correlation vector $\mathbf{p}$ in the formula of Eq. (5.8), which is reproduced here for convenience:

$$\nabla(n) = -2\mathbf{p} + 2\mathbf{R}\mathbf{w}(n) \tag{5.37}$$

The *simplest* choice of estimators for $\mathbf{R}$ and $\mathbf{p}$ is to use *instantaneous estimates* that are based on sample values of the tap-input vector and desired response, as defined by, respectively,

and
$$\hat{\mathbf{R}}(n) = \mathbf{u}(n)\mathbf{u}^H(n) \tag{5.38}$$

$$\hat{\mathbf{p}}(n) = \mathbf{u}(n)d^*(n) \tag{5.39}$$

Correspondingly, the instantaneous estimate of the gradient vector is as follows:

$$\hat{\nabla}(n) = -2\mathbf{u}(n)d^*(n) + 2\mathbf{u}(n)\mathbf{u}^H(n)\hat{\mathbf{w}}(n) \tag{5.40}$$

This estimate is *unbiased* in that its expected value equals the true value of the gradient vector.   Note also that the estimate $\hat{\nabla}(n)$ equals the gradient of the instantaneous squared error $|e(n)|^2$ (see Problem 1).

Substituting the estimate of Eq. (5.40) for the gradient vector $\nabla(n)$ in the steepest-descent algorithm as described in Eq. (5.7), we get a new recursive relation for updating the tap-weight vector:

$$\hat{\mathbf{w}}(n+1) = \hat{\mathbf{w}}(n) + \mu\mathbf{u}(n)[d^*(n) - \mathbf{u}^H(n)\hat{\mathbf{w}}(n)] \tag{5.41}$$

Here we have used a hat over the symbol for the tap-weight vector to distinguish it from the value obtained by using the steepest descent algorithm. Equivalently, we may write the result in the form of a pair of relations as follows:

$$e(n) = d(n) - \hat{\mathbf{w}}^H(n)\mathbf{u}(n) \tag{5.42}$$

$$\hat{\mathbf{w}}(n+1) = \hat{\mathbf{w}}(n) + \mu\mathbf{u}(n)\dot{e}^*(n) \tag{5.43}$$

Equation (5.42) defines the estimation error $e(n)$, the computation of which is based on the *current estimate* of the tap-weight vector, $\hat{\mathbf{w}}(n)$. Note also that the second term, $\mu\mathbf{u}(n)e^*(n)$, on the right side of Eq. (5.43) represents the *correction* that is applied to the current estimate of the tap-weight vector, $\hat{\mathbf{w}}(n)$. The iterative procedure is started with the initial guess $\hat{\mathbf{w}}(0)$. A convenient choice for this initial guess is the null vector; we may thus set $\hat{\mathbf{w}}(0) = \mathbf{0}$.

The algorithm described by Eq. (5.41), or, equivalently, by Eqs. (5.42) and (5.43), is the complex form of the adaptive *least-mean-square* (LMS) *algorithm* (Widrow and others, 1975).[1] It is also known as the *stochastic gradient algorithm*. Note, however, that the allowed set of directions along which we "step" from one iteration cycle to the next is quite random and cannot therefore be thought of as being gradient directions. Nevertheless, tradition dominates, and the name "stochastic gradient algorithm" is sometimes used.

Figure 5.12 shows a signal-flow graph representation of the LMS algorithm in the form of a feedback model. This model bears a close resemblance to the feedback model of Fig. 5.2 describing the steepest-descent algorithm.

The instantaneous estimates of $\mathbf{R}$ and $\mathbf{p}$ given in Eqs. (5.38) and (5.39), respectively, have relatively large variances. At first sight, it may therefore seem that the LMS algorithm is incapable of good performance since the algorithm uses these instantaneous estimates. However, we must remember that the LMS algorithm is recursive in nature, with the result that the algorithm itself effectively averages these estimates, in some sense, during the course of adaptation.

The procedure described for the derivation of the LMS algorithm may be viewed as an approximation of the steepest-descent algorithm, using instantaneous estimates of (1) the correlation matrix $\mathbf{R}$ of the input vector $\mathbf{u}(n)$, and (2) the cross-correlation vector $\mathbf{p}$ between $\mathbf{u}(n)$ and the desired response $d(n)$. Sharpe and Nolte (1981) describe another procedure for deriving the LMS algorithm. They start with the solution to the normal equation and use a finite summation for the inverse of the correlation matrix. Next, they use the instantaneous estimates of Eqs. (5.38) and (5.39) for $\mathbf{R}$ and $\mathbf{p}$ in the resultant recursive equation, thereby obtaining the same formula for the LMS algorithm (see Problem 2).

---

[1]The complex form of the LMS algorithm, as originally proposed by Widrow and others (1975), differs slightly from that described in Eqs. (5.42) and (5.43). Widrow and other authors have based their derivation on the definition $\mathbf{R} = E[\mathbf{u}^*(n)\mathbf{u}^T(n)]$ for the correlation matrix. On the other hand, the LMS algorithm described by Eqs. (5.42) and (5.43) is based on the definition $\mathbf{R} = E[\mathbf{u}(n)\mathbf{u}^H(n)]$ for the correlation matrix. The adoption of the latter definition for the correlation matrix of complex–valued data is the natural extension of the definition for real–valued data.

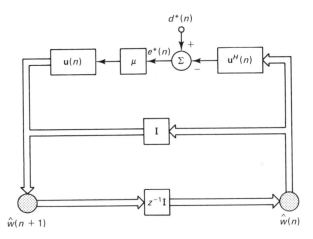

Figure 5.12 Signal-flow graph representation of the LMS algorithm.

5.8 *THE FUNDAMENTAL ASSUMPTION*

Although the initial tap-weight vector $\hat{\mathbf{w}}(0)$ is usually a known constant, the application of the LMS algorithm results in the propagation of randomness of the data processes into the estimate of the tap-weight vector. Accordingly, in analyzing the performance of the LMS algorithm, we have to treat the tap-weight vector estimate $\hat{\mathbf{w}}(n)$ at time $n > 0$ as a random vector. The problem is further complicated by the fact that, during adaptation, the estimation error $e(n)$, and therefore the tap-weight vector estimate $\hat{\mathbf{w}}(n)$, are nonstationary. Accordingly, quantities of interest, such as the first two moments of $\hat{\mathbf{w}}(n)$ and the mean-squared error $J(n)$ are all functions of the number of iterations, n.

To make the statistical analysis of the LMS algorithm mathematically tractable, we use the *fundamental assumption*, which consists of four parts as follows:

1. Each sample vector $\mathbf{u}(n)$ of the input process is statistically independent of all previous sample vectors $\mathbf{u}(k)$, $k = 0, 1, \ldots, n-1$, as shown by

$$E[\mathbf{u}(n)\mathbf{u}^H(k)] = \mathbf{0}, \qquad k = 0, 1, \ldots, n-1 \tag{5.44}$$

2. Each sample vector $\mathbf{u}(n)$ of the input process is statistically independent of all previous samples of the desired response $d(k)$, $k = 0, 1, \ldots, n-1$, as shown by

$$E[\mathbf{u}(n)d^*(k)] = \mathbf{0}, \qquad k = 0, \ldots, n-1 \tag{5.45}$$

3. The sample $d(n)$ of the desired response is dependent on the corresponding sample vector $\mathbf{u}(n)$ of the input process, but statistically independent of all previous samples of the desired response.

4. The tap-input vector $\mathbf{u}(n)$ and the desired response $d(n)$ consist of *mutually Gaussian-distributed* random variables for all n.

The statistical analysis of the LMS algorithm based on the fundamental assumption is called the *independence theory*.

From Eq. (5.41), we observe that the tap-weight vector $\hat{\mathbf{w}}(n+1)$ at time $n+1$ depends *only* on three inputs:

1. The previous sample vectors of the input process, $\mathbf{u}(n)$, $\mathbf{u}(n-1)$, . . . , $\mathbf{u}(0)$.
2. The previous samples of the desired response, $d(n)$, $d(n-1)$, . . . , $d(0)$.
3. The initial value of the tap-weight vector, $\hat{\mathbf{w}}(0)$.

Accordingly, in view of points 1 and 2 in the fundamental assumption, we find that the tap-weight vector $\hat{\mathbf{w}}(n+1)$ is independent of both $\mathbf{u}(n+1)$ and $d(n+1)$. This is a very useful observation and one that will be used repeatedly in the sequel. The significance of the other two points in the fundamental assumption will become apparent as we proceed with the analysis.

Clearly, there are many practical problems for which the input process and the desired response do not satisfy the fundamental assumption. Nevertheless, experience with the LMS algorithm has shown that the independence theory retains sufficient information about the structure of the adaptive process for the results of the theory to serve as reliable design guidelines, even for some problems having highly dependent data samples.

Another point that is noteworthy is the fact that, in the fundamental assumption, we have emphasized statistical independence of samples, rather than uncorrelatedness. It is indeed true that independent samples are equivalent to uncorrelated samples wherever the samples are Gaussian distributed. Nevertheless, the analysis may be extended to certain other distributions by making corrections for higher-order moments, provided the samples are independent (Senne, 1968).

In the next two sections, we use the fundamental assumption to derive the first two moments of the estimate of the tap-weight vector: (1) the average value and (2) the correlation matrix. We find it more convenient, however, to work with the weight-error vector rather than the tap-weight vector itself. To avoid confusion with the notation used in the study of the steepest-descent algorithm, we define the weight-error vector for the LMS algorithm as follows:

$$\boldsymbol{\varepsilon}(n) \ = \ \hat{\mathbf{w}}(n) \ - \ \mathbf{w}_o \qquad\qquad (5.46)$$

where, as before, $\mathbf{w}_o$ denotes the optimum Wiener solution for the tap-weight vector, and $\hat{\mathbf{w}}(n)$ is the estimate produced by the LMS algorithm at time n.

5.9 AVERAGE TAP-WEIGHT VECTOR

Subtracting the optimum tap-weight vector $\mathbf{w}_o$ from both sides of Eq. (5.41), and using the definition of Eq. (5.46) to eliminate $\hat{\mathbf{w}}(n)$ from the correction term on the right side of Eq. (5.41), we may rewrite the LMS algorithm in terms of the weight-error vector $\varepsilon(n)$ as follows:

$$\varepsilon(n+1) = [\mathbf{I} - \mu\mathbf{u}(n)\mathbf{u}^H(n)]\varepsilon(n) + \mu[\mathbf{u}(n)d^*(n) - \mathbf{u}(n)\mathbf{u}^H(n)\mathbf{w}_o] \qquad (5.47)$$

where $\mathbf{I}$ is the identity matrix. As a consequence of the fundamental assumption, we observe that the tap-weight vector $\hat{\mathbf{w}}(n)$ is independent of the input vector $\mathbf{u}(n)$, which was justified in Section 5.8. Correspondingly, the weight error vector $\varepsilon(n)$ is also independent of $\mathbf{u}(n)$. Hence, taking the mathematical expectation of both sides of Eq. (5.47), and using the independence of $\varepsilon(n)$ from $\mathbf{u}(n)$, we get

$$
\begin{aligned}
E[\varepsilon(n+1)] &= E[(\mathbf{I} - \mu\mathbf{u}(n)\mathbf{u}^H(n))\varepsilon(n)] \\
&\quad + \mu E[\mathbf{u}(n)d^*(n) - \mathbf{u}(n)\mathbf{u}^H(n)\mathbf{w}_o] \\
&= (\mathbf{I} - \mu E[\mathbf{u}(n)\mathbf{u}^H(n)])E[\varepsilon(n)] \qquad (5.48) \\
&\quad + \mu(E[\mathbf{u}(n)d^*(n)] - E[\mathbf{u}(n)\mathbf{u}^H(n)]\mathbf{w}_o) \\
&= (\mathbf{I} - \mu\mathbf{R})E[\varepsilon(n)] + \mu(\mathbf{p} - \mathbf{R}\mathbf{w}_o)
\end{aligned}
$$

where we have used the definitions of the correlation matrix $\mathbf{R}$ and the cross-correlation vector $\mathbf{p}$. That is,

$$\mathbf{R} = E[\mathbf{u}(n)\mathbf{u}^H(n)]$$

and

$$\mathbf{p} = E[\mathbf{u}(n)d^*(n)]$$

We recognize from the normal equation that

$$\mathbf{R}\mathbf{w}_o = \mathbf{p}$$

Accordingly, the second term on the right side of Eq. (5.48) equals zero. We may thus simplify this equation as follows:

$$E[\varepsilon(n+1)] = (\mathbf{I} - \mu\mathbf{R})E[\varepsilon(n)] \qquad (5.49)$$

Comparing Eqs. (5.11) and (5.49), we observe that they are of exactly the same mathematical form. In particular, the average weight-error vector, $E[\varepsilon(n)]$, in Eq. (5.49) pertaining to the LMS algorithm has the same role as the weight-error vector $\mathbf{c}(n)$ in Eq. (5.11) describing the steepest-descent algorithm. From our study of the steepest-descent algorithm in Section 5.4, we recall that $\mathbf{c}(n)$ converges to zero as n approaches infinity, provided that Eq. (5.21) is satisfied. From the analogy between Eqs. (5.11) and (5.49), we therefore deduce that for the LMS algorithm the mean of $\varepsilon(n)$ converges to zero as n approaches infinity, provided that the following condition is satisfied (Widrow, 1970; Ungerboeck, 1972):

$$0 < \mu < \frac{2}{\lambda_{\max}} \qquad (5.50)$$

where $\lambda_{\max}$ is the largest eigenvalue of the correlation matrix $\mathbf{R}$. In other words, provided that the step-size parameter μ is set within the bounds defined by Eq. (5.50), then the mean of the tap-weight vector $\hat{\mathbf{w}}(n)$ computed by using the LMS algorithm converges to the optimum Weiner solution $\mathbf{w}_o$ as the number of iterations, n, approaches infinity.   Under this condition, we say that the LMS algorithm is *convergent in the mean.*

Also, as with the steepest-descent algorithm, we find that when the eigenvalues of the correlation matrix $\mathbf{R}$ are widely spread, the time taken by the average tap-weight vector to converge is primarily limited by the smallest eigenvalues.

5.10 *WEIGHT-ERROR CORRELATION MATRIX*

We next wish to develop a recursive equation for the time evolution of the correlation matrix of the weight-error vector $\boldsymbol{\varepsilon}(n)$.   Using $\mathbf{K}(n)$ to denote this correlation matrix at time n, we have, by definition,

$$\mathbf{K}(n) = E[\boldsymbol{\varepsilon}(n)\boldsymbol{\varepsilon}^H(n)] \tag{5.51}$$

The procedure we will use to derive the recursive equation is simply to write the definition of $\mathbf{K}(n + 1)$ by using Eq. (5.51), substitute Eq. (5.47) for $\boldsymbol{\varepsilon}(n + 1)$, and do the ensemble average using the fundamental assumption.

Taking the Hermitian transpose of both sides of Eq. (5.47), we get

$$\boldsymbol{\varepsilon}^H(n + 1) = \boldsymbol{\varepsilon}^H(n)[\mathbf{I} - \mu\mathbf{u}(n)\mathbf{u}^H(n)]$$
$$+ \mu[d(n)\mathbf{u}^H(n) - \mathbf{w}_o^H\mathbf{u}(n)\mathbf{u}^H(n)] \tag{5.52}$$

To evaluate the correlation matrix $\mathbf{K}(n + 1)$, we take the expectation of the outer product $\boldsymbol{\varepsilon}(n + 1)\boldsymbol{\varepsilon}^H(n + 1)$.   In doing this, various cross terms arise as a result of multiplication, and the computations naturally fall into four steps, as follows:

Step 1.   We consider first the term

$$E[(\mathbf{I} - \mu\mathbf{u}(n)\mathbf{u}^H(n))\boldsymbol{\varepsilon}(n)\boldsymbol{\varepsilon}^H(n) \,(\mathbf{I} - \mu\mathbf{u}(n)\mathbf{u}^H(n))]$$
$$= E[\boldsymbol{\varepsilon}(n)\boldsymbol{\varepsilon}^H(n)] - \mu E[\mathbf{u}(n)\mathbf{u}^H(n)\boldsymbol{\varepsilon}(n)\boldsymbol{\varepsilon}^H(n)] - \mu E[\boldsymbol{\varepsilon}(n)\boldsymbol{\varepsilon}^H(n)\mathbf{u}(n)\mathbf{u}^H(n)]$$
$$+ \mu^2 E[\mathbf{u}(n)\mathbf{u}^H(n)\boldsymbol{\varepsilon}(n)\boldsymbol{\varepsilon}^H(n)\mathbf{u}(n)\mathbf{u}^H(n)]$$
$$= \mathbf{K}(n) - \mu E[\mathbf{u}(n)\mathbf{u}^H(n)]E[\boldsymbol{\varepsilon}(n)\boldsymbol{\varepsilon}^H(n)] - \mu E[\boldsymbol{\varepsilon}(n)\boldsymbol{\varepsilon}^H(n)]E[\mathbf{u}(n)\mathbf{u}^H(n)]$$
$$+ \mu^2 E[\mathbf{u}(n)\mathbf{u}^H(n)\boldsymbol{\varepsilon}(n)\boldsymbol{\varepsilon}^H(n)\mathbf{u}(n)\mathbf{u}^H(n)]$$
$$= \mathbf{K}(n) - \mu\mathbf{R}\mathbf{K}(n) - \mu\mathbf{K}(n)\mathbf{R} + \mu^2 E[\mathbf{u}(n)\mathbf{u}^H(n)\boldsymbol{\varepsilon}(n)\boldsymbol{\varepsilon}^H(n)\mathbf{u}(n)\mathbf{u}^H(n)] \tag{5.53}$$

In this relation, we have used Eq. (5.51) and the definition

$$\mathbf{R} = E[\mathbf{u}(n)\mathbf{u}^H(n)]$$

where $\mathbf{R}$ denotes the correlation matrix of the tap-input vector $\mathbf{u}(n)$.   The last term on the right side of Eq. (5.53) involves fourth-order moments of sample vectors of the input process. These higher-order moments may be evaluated by using the formula for the fourth-order moment of a Gaussian process (Deutsch, 1965). In effect, we invoke part 4 of the fundamental assumption. Specifically, if x_1, x_2, x_3, and x_4 denote four different samples of a Gaussian process, we may write

$$E[x_1 x_2 x_3 x_4] = E[x_1 x_2]E[x_3 x_4] + E[x_1 x_3]E[x_2 x_4] + E[x_1 x_4]E[x_2 x_3] \qquad (5.54)$$

To use this formula, we express the M-by-M matrix representing the multiple product $\mathbf{u}(n)\mathbf{u}^H(n)\boldsymbol{\varepsilon}(n)\boldsymbol{\varepsilon}^H(n)\mathbf{u}(n)\mathbf{u}^H(n)$ as a multiple sum of the elements of the component vectors. Let the brace notation $\{x_{ij}(n)\}$ denote this matrix having elements $x_{ij}(n)$, with $i, j = 1, 2, \ldots, M$. We may then write

$$\{x_{ij}(n)\} = \mathbf{u}(n)\mathbf{u}^H(n)\boldsymbol{\varepsilon}(n)\boldsymbol{\varepsilon}^H(n)\mathbf{u}(n)\mathbf{u}^H(n)$$

$$= \left\{ \sum_{l=1}^{M} \sum_{p=1}^{M} \left(\mathbf{u}(n)\mathbf{u}^H(n)\right)_{il} \left(\boldsymbol{\varepsilon}(n)\boldsymbol{\varepsilon}^H(n)\right)_{lp} \left(\mathbf{u}(n)\mathbf{u}^H(n)\right)_{pj} \right\} \qquad (5.55)$$

where

$$\left(\mathbf{u}(n)\mathbf{u}^H(n)\right)_{il} = \text{element on } i\text{th row and } l\text{th column of } \mathbf{u}(n)\mathbf{u}^H(n)$$

$$= u(n - i + 1)u^*(n - l + 1) \qquad (5.56)$$

$$\left(\boldsymbol{\varepsilon}(n)\boldsymbol{\varepsilon}^H(n)\right)_{lp} = \text{element on } l\text{th row and } p\text{th column of } \boldsymbol{\varepsilon}(n)\boldsymbol{\varepsilon}^H(n)$$

$$= \varepsilon_l(n)\varepsilon_p^*(n) \qquad (5.57)$$

$$\left(\mathbf{u}(n)\mathbf{u}^H(n)\right)_{pj} = \text{element on } p\text{th row and } j\text{th column of } \mathbf{u}(n)\mathbf{u}^H(n)$$

$$= u(n - p + 1)u^*(n - j + 1) \qquad (5.58)$$

Thus, any term inside the double summation in Eq. (5.55) may be regrouped as the product of two distinct components. One component, involving tap inputs only, is written as

$$u(n - i + 1)u^*(n - l + 1)u(n - p + 1)u^*(n - j + 1)$$

The second component, involving tap-weight errors only, is written as

$$\varepsilon_l(n)\varepsilon_p^*(n)$$

As a consequence of the fundamental assumption, these two components may be treated independent of each other. This means that the expected value of any term inside the double summation in Eq. (5.55) may be expressed as the product of the expected values of these two components.

    To determine the expected value of the first component, we use the formula of Eq. (5.54) to write

$$E[u(n - i + 1)u^*(n - l + 1)u(n - p + 1)u^*(n - j + 1)]$$

$$= E[u(n - i + 1)u^*(n - l + 1)]E[u(n - p + 1)u^*(n - j + 1)]$$

$$+ E[u(n - i + 1)u(n - p + 1)]E[u^*(n - l + 1)u^*(n - j + 1)]$$

$$+ E[u(n - i + 1)u^*(n - j + 1)]E[u^*(n - l + 1)u(n - p + 1)] \quad (5.59)$$

$$= r(l - i)r(j - p)$$

$$+ E[u(n - i + 1)u(n - p + 1)]E[u^*(n - l + 1)u^*(n - j + 1)]$$

$$+ r(j - i)r(l - p)$$

where the autocorrelation functions $r(l - i)$, $r(j - p)$, $r(j - i)$, and $r(l - p)$ are defined for different lags in the usual way. On the other hand, the quantities $E[u(n - i + 1)u(n - p + 1)]$ and $E[u^*(n - l + 1)u^*(n - j + 1)]$ are so seldom used in practice that we do not adopt any special notations for them.

As for the second component, its expected value equals the lpth element of the weight-error correlation matrix $\mathbf{K}(n)$. Let

$$k_{lp}(n) = E[\varepsilon_l(n)\varepsilon_p^*(n)] \quad (5.60)$$

Next, we recognize that

$$E[\{x_{ij}(n)\}] = \{E[x_{ij}(n)]\} \quad (5.61)$$

where $x_{ij}(n)$ is represented by the double summation in Eq. (5.55). Moreover, we may interchange the operations of expectation and double summation. Accordingly, we may express the expectation of the matrix $\{x_{ij}(n)\}$ in the form

$$E[\{x_{ij}(n)\}] = \{r(j - i) \sum_{l=1}^{M} \sum_{p=1}^{M} r(l - p)k_{lp}(n)\}$$

$$+ \left\{ \sum_{l=1}^{M} \sum_{p=1}^{M} r(l - i)r(j - p)k_{lp}(n) \right\}$$

$$\quad (5.62)$$

$$+ \left\{ \sum_{l=1}^{M} \sum_{p=1}^{M} E[u(n - i + 1)u(n - p + 1)] \right.$$

$$\left. \cdot E[u^*(n - l + 1)u^*(n - j + 1)] \, k_{lp}(n) \right\}$$

Now we observe that the last expectation on the right side of Eq. (5.53) is multiplied by μ^2. Correspondingly, all three terms in Eq. (5.62) are multiplied by μ^2. Since the step-size parameter μ is usually small, particularly when the adaptation is slow, we would be justified to neglect all three terms (unless one of them is relatively large). The first term on the right side of Eq. (5.62) can be M times larger than the other two and, therefore, this is the only significant term that we need to keep. Note that M equals the number of taps. Accordingly, we may write

$$E[\mathbf{u}(n)\mathbf{u}^H(n)\boldsymbol{\varepsilon}(n)\boldsymbol{\varepsilon}^H(n)\mathbf{u}(n)\mathbf{u}^H(n)] \simeq \left\{ r(j-i) \sum_{l=1}^{M} \sum_{p=1}^{M} r(l-p)k_{lp}(n) \right\}$$

$$= \mathbf{R}\mathrm{tr}[\mathbf{R}\mathbf{K}(n)] \tag{5.63}$$

where tr[ ] is the *matrix trace operator*.

Step 2. We next consider the term

$$\mu E[(\mathbf{u}(n)d^*(n) - \mathbf{u}(n)\mathbf{u}^H(n)\mathbf{w}_o)\boldsymbol{\varepsilon}^H(n)(\mathbf{I} - \mu\mathbf{u}(n)\mathbf{u}^H(n))]$$

We may identify two contributions in this term: one linear in μ, and the other involving μ^2. The contribution linear in μ equals

$$\mu E[\mathbf{u}(n)(d^*(n) - \mathbf{u}^H(n)\mathbf{w}_o)\boldsymbol{\varepsilon}^H(n)] = \mu E[\mathbf{u}(n)d^*(n) - \mathbf{u}(n)\mathbf{u}^H(n)\mathbf{w}_o]E[\boldsymbol{\varepsilon}^H(n)] \tag{5.64}$$

$$= \mu(\mathbf{p} - \mathbf{R}\mathbf{w}_o)E[\boldsymbol{\varepsilon}^H(n)]$$

However, from the normal equation we have

$$\mathbf{R}\mathbf{w}_o = \mathbf{p}$$

Accordingly, the contribution defined by Eq. (5.64) is zero.

The contribution involving the factor μ^2 equals

$$-\mu^2 E[\mathbf{u}(n)(d^*(n) - \mathbf{u}^H(n)\mathbf{w}_o)\boldsymbol{\varepsilon}^H(n)\mathbf{u}(n)\mathbf{u}^H(n)]$$

$$= -\mu^2 E[\mathbf{u}(n)d^*(n)\boldsymbol{\varepsilon}^H(n)\mathbf{u}(n)\mathbf{u}^H(n)] \tag{5.65}$$

$$+ \mu^2 E[\mathbf{u}(n)\mathbf{u}^H(n)\mathbf{w}_o\boldsymbol{\varepsilon}^H(n)\mathbf{u}(n)\mathbf{u}^H(n)]$$

Following a similar procedure to that used in step 1, we find that the dominant contributions made by the two terms on the right side of Eq. (5.65) also cancel each other. Specifically, the first expectation on the right side of this equation may be approximated as follows:

$$E[\mathbf{u}(n)d^*(n)\boldsymbol{\varepsilon}^H(n)\mathbf{u}(n)\mathbf{u}^H(n)] \simeq \mathbf{R}E[\boldsymbol{\varepsilon}^H(n)]\mathbf{p} \tag{5.66}$$

For the second term on the right side of Eq. (5.65), we may use the following approximation:

$$E[\mathbf{u}(n)\mathbf{u}^H(n)\mathbf{w}_o\boldsymbol{\varepsilon}^H(n)\mathbf{u}(n)\mathbf{u}^H(n)] \simeq \mathbf{R}\mathrm{tr}[\mathbf{R}\mathbf{w}_o E[\boldsymbol{\varepsilon}^H(n)]$$

$$= \mathbf{R}E[\boldsymbol{\varepsilon}^H(n)]\mathbf{R}\mathbf{w}_o \tag{5.67}$$

where we have used the fact that

$$\mathrm{tr}[\mathbf{R}\mathbf{w}_o E[\boldsymbol{\varepsilon}^H(n)]] = E[\boldsymbol{\varepsilon}^H(n)]\mathbf{R}\mathbf{w}_o$$

Using the two approximations of Eqs. (5.66) and (5.67) in (5.65), taking out the common factor $\mathbf{R}\,E[\boldsymbol{\varepsilon}^H(n)]$, and again using the normal equation, we find that the final result is zero.

Step 3.   The third term to be considered is

$$\mu E[I - \mu\mathbf{u}(n)\mathbf{u}^H(n)]\boldsymbol{\varepsilon}(n)[d(n)\mathbf{u}^H(n) - \mathbf{w}_o^H\mathbf{u}(n)\mathbf{u}^H(n)]$$

This term is similar in form to that considered in step 2. We may therefore approximate this term by zero, too.

Step 4.   The final term to be considered is

$$\mu^2 E[(\mathbf{u}(n)d^*(n) - \mathbf{u}(n)\mathbf{u}^H(n)\mathbf{w}_o)(d(n)\mathbf{u}^H(n) - \mathbf{w}_o^H\mathbf{u}(n)\mathbf{u}^H(n))]$$

$$= \mu^2 E[\mathbf{u}(n)d^*(n)d(n)\mathbf{u}^H(n)] - \mu^2 E[\mathbf{u}(n)\mathbf{u}^H(n)\mathbf{w}_o d(n)\mathbf{u}^H(n)] \qquad (5.68)$$

$$- \mu^2 E[\mathbf{u}(n)d^*(n)\mathbf{w}_o^H\mathbf{u}(n)\mathbf{u}^H(n)] + \mu^2 E[\mathbf{u}(n)\mathbf{u}^H(n)\mathbf{w}_o\mathbf{w}_o^H\mathbf{u}(n)\mathbf{u}^H(n)]$$

Using a procedure similar to that detailed in steps 1 and 2, we may show that

$$E[\mathbf{u}(n)d^*(n)d(n)\mathbf{u}^H(n)] \simeq \mathbf{R}\sigma_d^2 \qquad (5.69)$$

$$E[\mathbf{u}(n)\mathbf{u}^H(n)\mathbf{w}_o d(n)\mathbf{u}^H(n)] \simeq \mathbf{R}\,\mathbf{p}^H\mathbf{w}_o \qquad (5.70)$$

$$E[\mathbf{u}(n)d^*(n)\mathbf{w}_o^H\mathbf{u}(n)\mathbf{u}^H(n)] \simeq \mathbf{R}\,\mathbf{w}_o^H\mathbf{p} \qquad (5.71)$$

$$E[\mathbf{u}(n)\mathbf{u}^H(n)\mathbf{w}_o\mathbf{w}_o^H\mathbf{u}(n)\mathbf{u}^H(n)] \simeq \mathbf{R}\,\mathrm{tr}[\mathbf{R}\,\mathbf{w}_o\mathbf{w}_o^H]$$

$$= \mathbf{R}\,\mathbf{w}_o^H\mathbf{R}\,\mathbf{w}_o \qquad (5.72)$$

Substituting these approximations in Eq. (5.68), we get

$$\mu^2 E[(\mathbf{u}(n)d^*(n) - \mathbf{u}(n)\mathbf{u}^H(n)\mathbf{w}_o)(d(n)\mathbf{u}^H(n) - \mathbf{w}_o^H\mathbf{u}(n)\mathbf{u}^H(n))]$$

$$= \mu^2 \mathbf{R}[\sigma_d^2 - \mathbf{p}^H\mathbf{w}_o - \mathbf{w}_o^H\mathbf{p} + \mathbf{w}_o^H\mathbf{R}\,\mathbf{w}_o] \qquad (5.73)$$

$$= \mu^2 \mathbf{R}J_{\min}$$

where we have used the formulas for the normal equation and the minimum mean-squared error $J_{\min}$.

We may now combine the results of steps 1 through 4, and thus describe the time evolution of the weight-error correlation matrix by the following difference equation (Mazo, 1979):

$$\mathbf{K}(n + 1) = \mathbf{K}(n) - \mu[\mathbf{R}\,\mathbf{K}(n) + \mathbf{K}(n)\mathbf{R}]$$

$$+ \mu^2 \mathbf{R}\,\mathrm{tr}[\mathbf{R}\,\mathbf{K}(n)] + \mu^2 J_{\min}\mathbf{R} \qquad (5.74)$$

The last term, $\mu^2 J_{\min}\mathbf{R}$, on the right side of Eq. (5.74) prevents $\mathbf{K}(n) = \mathbf{0}$ from being a solution to this equation. Accordingly, the correlation matrix $\mathbf{K}(n)$ is prevented from going to zero by this small forcing term. In particular, the weight-error vector $\boldsymbol{\varepsilon}(n)$ only approaches zero, but then executes small fluctuations about zero.

We now demonstrate that even though the recursive relation of Eq. (5.74) is an approximation, nevertheless, the positive-definite character of the correlation matrix $\mathbf{K}(n)$ is preserved in this equation (Mazo, 1979). We do this demonstration by induction on n.

We observe that the correlation matrix $\mathbf{R}$ of the tap-input vector is Hermitian and positive definite. Hence, it follows from Eq. (5.74) that if $\mathbf{K}(n)$ is Hermitian, then so will $\mathbf{K}(n + 1)$ be.

Next we rewrite the right side of Eq. (5.74) as follows:

$$(\mathbf{I} - \mu\mathbf{R})\mathbf{K}(n)(\mathbf{I} - \mu\mathbf{R}) + \mu^2(\mathbf{R}\,\mathrm{tr}[\mathbf{R}\,\mathbf{K}(n)] - \mathbf{R}\,\mathbf{K}(n)\mathbf{R}) + \mu^2 J_{\min}\mathbf{R} \qquad (5.75)$$

Since $\mathbf{R}$ is positive definite, it follows that the first term of this expression is also positive definite provided that $\mathbf{K}(n)$ is positive definite. Clearly, the last term of this expression is always positive definite. There now only remains for us to show that the second term is also positive definite. We rewrite this second term, except for μ^2, in the form

$$\sqrt{\mathbf{R}}\,\left(\mathrm{tr}[\sqrt{\mathbf{R}}\,\mathbf{K}(n)\,\sqrt{\mathbf{R}}]\mathbf{I} - \sqrt{\mathbf{R}}\,\mathbf{K}(n)\,\sqrt{\mathbf{R}}\right)\sqrt{\mathbf{R}} \qquad (5.76)$$

where $\sqrt{\mathbf{R}}$ is the *square root* of $\mathbf{R}$. In Eq. (5.76) we have used the property[2]

$$\mathrm{tr}[\mathbf{R}\,\mathbf{K}(n)] = \mathrm{tr}[\sqrt{\mathbf{R}}\,\sqrt{\mathbf{R}}\mathbf{K}(n)]$$
$$= \mathrm{tr}[\sqrt{\mathbf{R}}\mathbf{K}(n)\,\sqrt{\mathbf{R}}]$$

The matrix $\sqrt{\mathbf{R}}\mathbf{K}(n)\,\sqrt{\mathbf{R}}$ is positive definite, provided that $\mathbf{K}(n)$ is. Next, we use the following property (see Problem 2.2): *If a matrix $\mathbf{X}$ is positive definite, then $\mathrm{tr}[\mathbf{X}]\mathbf{I} - \mathbf{X}$ is positive definite too.* Hence, it follows that the expression in Eq. (5.76) or, equivalently, the second term of the expression in Eq. (5.75) is positive definite. We conclude therefore that the complete expression in Eq. (5.75) and, therefore, $\mathbf{K}(n + 1)$ is positive definite provided that $\mathbf{K}(n)$ is. The proof by induction is completed by noting that $\mathbf{K}(0)$ is positive definite, where

$$\begin{aligned}
\mathbf{K}(0) &= \boldsymbol{\varepsilon}(0)\,\boldsymbol{\varepsilon}^H(0) \\
&= [\mathbf{w}(0) - \mathbf{w}_o][\mathbf{w}^H(0) - \mathbf{w}_o^H]
\end{aligned} \qquad (5.77)$$

In summary, Eq. (5.74) provides us with a relation to recursively update the weight-error correlation matrix $\mathbf{K}(n)$, starting with $n = 0$, for which we have $\mathbf{K}(0)$. Furthermore, even though this relation is an approximation, nevertheless, after each iteration it does yield a positive-definite answer for the updated value of the weight-error correlation matrix.

---

[2]Let $\mathbf{X}$ denote an M-by-N matrix and $\mathbf{Y}$ denote an N-by-M matrix. The trace of matrix product $\mathbf{XY}$ equals the trace of matrix product $\mathbf{YX}$; that is,

$$\mathrm{tr}[\mathbf{XY}] = \mathrm{tr}[\mathbf{YX}]$$

5.11 AVERAGE MEAN-SQUARED ERROR

Ideally, the minimum mean-squared error $J_{\min}$ is realized when the coefficient vector $\mathbf{w}(n)$ of the transversal filter approaches the optimum value $\mathbf{w}_o$, defined by the normal equation. Indeed, as shown in Section 5.5, the steepest-descent algorithm does realize this idealized condition as the number of iterations, n, approaches infinity. The steepest-descent algorithm has the capability to do this, because it uses *exact* measurements of the gradient vector at each iteration of the algorithm. On the other hand, the LMS algorithm relies on a *noisy* estimate for the gradient vector, with the result that the tap-weight vector estimate $\hat{\mathbf{w}}(n)$ only approaches the optimum value $\mathbf{w}_o$ after a large number of iterations and then executes small fluctuations about $\mathbf{w}_o$. Consequently, use of the LMS algorithm, after a large number of iterations, results in a mean-squared error $J(\infty)$ that is greater than the minimum mean-squared error $J_{\min}$. The amount by which the actual value of $J(\infty)$ is greater than $J_{\min}$ is called the *excess mean-squared error*.

There is another basic difference between the steepest-descent algorithm and the LMS algorithm. In Section 5.5, we showed that the steepest-descent algorithm has a well-defined learning curve, obtained by plotting the mean-squared error versus the number of iterations. For this algorithm, the learning curve consists of a sum of decaying exponentials, the number of which equals (in general) the number of tap coefficients. On the other hand, in individual applications of the LMS algorithm, we find that the learning curve consists of noisy, decaying exponentials. The amplitude of the noise usually becomes smaller as the step-size parameter μ is reduced.

Imagine now an ensemble of adaptive transversal filters. Each filter is assumed to use the LMS algorithm with the same step-size parameter μ and the same initial tap-weight vector $\hat{\mathbf{w}}(0)$. Also, each adaptive filter has individual stationary ergodic inputs that are selected at random from the same statistical population. If, at each time n, we compute the ensemble average of the noisy learning curves for this ensemble of adaptive filters, we find that the resultant consists of a sum of decaying exponentials. In practice, we usually find that this smooth ensemble-averaged learning curve is closely realized by averaging out 50 to 200 independent trials of the LMS algorithm.

Thus, we may use an *average mean-squared* error, denoted by $E[J(n)]$, to describe the dynamic behavior of the LMS algorithm. The need for ensemble averaging arises because, as mentioned previously, the estimation error process $\{e(n)\}$ is nonstationary during the adaptation process, as the tap-weight vector estimate $\hat{\mathbf{w}}(n)$ adapts toward the optimum value $\mathbf{w}_o$.

In terms of the tap-weight error vector, $\boldsymbol{\varepsilon}(n)$, we may express the mean-squared error, $J(n)$, as follows:

$$
\begin{aligned}
J(n) &= J_{\min} + \big(\mathbf{w}(n) - \mathbf{w}_o\big)^H \mathbf{R}\big(\mathbf{w}(n) - \mathbf{w}_o\big) \\
&= J_{\min} + \boldsymbol{\varepsilon}^H(n)\, \mathbf{R}\, \boldsymbol{\varepsilon}(n)
\end{aligned}
\tag{5.78}
$$

This equation shows explicitly the unique optimality of the minimizing tap-weight vector $\mathbf{w}_o$. We define the *excess mean-squared error* as the difference between the mean-squared error, $J(n)$, produced by the adaptive algorithm at time n and the minimum value, $J_{\min}$, pertaining to the optimum Wiener solution. Denoting the excess mean-squared error by $J_{\text{ex}}(n)$, we have

$$J_{\text{ex}}(n) = J(n) - J_{\min} \tag{5.79}$$
$$= \boldsymbol{\varepsilon}^H(n)\mathbf{R}\boldsymbol{\varepsilon}(n)$$

In Eq. (5.79), the vector $\boldsymbol{\varepsilon}(n)$ is random, and, in fact, its value at time n depends on the entire history of the input data processes from the start of adaptation. The measure we use for the progress of the adaptive algorithm is $E[J(n)]$, the *ensemble average of the mean-squared error* $J(n)$ at time n over all samples of the input data processes. We refer to the curve obtained by plotting $E[J(n)]$ versus n as the *ensemble-averaged learning curve* of the LMS algorithm. The average value $E[J(n)]$ equals $J_{\min}$ plus $E[J_{\text{ex}}(n)]$. The information we require about $E[J_{\text{ex}}(n)]$ is contained in the weight-error correlation matrix $\mathbf{K}(n)$, as shown by

$$E[J_{\text{ex}}(n)] = E[\boldsymbol{\varepsilon}^H(n)\mathbf{R}\boldsymbol{\varepsilon}(n)] \tag{5.80}$$
$$= \text{tr}[\mathbf{R}\mathbf{K}(n)]$$

To compute $\mathbf{K}(n)$, we may use the recursive relation of Eq. (5.74). However, when the mean-squared error is of primary interest, another form of this equation obtained by a simple rotation of coordinates is more useful. The particular rotation of coordinates we have in mind is described by the unitary similarity transformation of Eq. (5.12), reproduced here in the form

$$\mathbf{Q}^H\mathbf{R}\mathbf{Q} = \boldsymbol{\Lambda} \tag{5.81}$$

where $\boldsymbol{\Lambda}$ is a diagonal matrix consisting of the eigenvalues of the correlation matrix $\mathbf{R}$, and $\mathbf{Q}$ is the unitary matrix consisting of the eigenvectors associated with these eigenvalues. Note that the matrix $\boldsymbol{\Lambda}$ is real valued. Furthermore, let

$$\mathbf{Q}^H\mathbf{K}(n)\mathbf{Q} = \mathbf{X}(n) \tag{5.82}$$

In general, $\mathbf{X}(n)$ is not a diagonal matrix. Using Eqs. (5.81) and (5.82), we have

$$\text{tr}[\mathbf{R}\mathbf{K}(n)] = \text{tr}[\mathbf{Q}\boldsymbol{\Lambda}\mathbf{Q}^H\mathbf{Q}\mathbf{X}(n)\mathbf{Q}^H] \tag{5.83}$$
$$= \text{tr}[\mathbf{Q}\boldsymbol{\Lambda}\mathbf{X}(n)\mathbf{Q}^H]$$

where we have used the property $\mathbf{Q}^H\mathbf{Q} = \mathbf{I}$. Next we use the property[3]

$$\text{tr}[\mathbf{R}\mathbf{K}(n)] = \text{tr}[\mathbf{Q}^H\mathbf{Q}\boldsymbol{\Lambda}\mathbf{X}(n)] \tag{5.84}$$
$$= \text{tr}[\boldsymbol{\Lambda}\mathbf{X}(n)]$$

---

[3] In the first line of Eq. (5.84), we have used the property that the trace of the product of two matrices is the same whether one matrix is premultiplied or postmultiplied by the other matrix. See the footnote following Eq. (5.76).

where we have again used $\mathbf{Q}^H\mathbf{Q} = \mathbf{I}$.   Accordingly, we have

$$E[J_{\text{ex}}(n)] = \text{tr}[\mathbf{\Lambda}\mathbf{X}(n)] \tag{5.85}$$

Since $\mathbf{\Lambda}$ is a diagonal matrix, we may also write

$$E[J_{\text{ex}}(n)] = \sum_{i=1}^{M} \lambda_i x_i(n) \tag{5.86}$$

where the $x_i(n)$, $i = 1, 2, \ldots, M$, are the diagonal elements of the matrix $\mathbf{X}(n)$, and λ_i are the eigenvalues of the correlation matrix $\mathbf{R}$.

Next, using the transformations described by Eqs. (5.81) and (5.82), we may rewrite the recursive equation (5.74) in terms of $\mathbf{X}(n)$ and $\mathbf{\Lambda}$ as follows:

$$\mathbf{X}(n + 1) = \mathbf{X}(n) - \mu[\mathbf{\Lambda}\mathbf{X}(n) + \mathbf{X}(n)\mathbf{\Lambda}]$$
$$+ \mu^2\mathbf{\Lambda} \, \text{tr}[\mathbf{\Lambda}\mathbf{X}(n)] + \mu^2 J_{\min} \mathbf{\Lambda} \tag{5.87}$$

We observe from Eq. (5.86) that $E[J_{\text{ex}}(n)]$ depends only on the $x_i(n)$.   This suggests that we look at the diagonal terms of the recursive equation (5.87).   Because of the form of this equation, the x_i decouple from the off-diagonal terms, and so we have

$$x_i(n + 1) = x_i(n) - 2\mu\lambda_i x_i(n) + \mu^2\lambda_i \sum_{j=1}^{M} \lambda_j x_j(n)$$
$$+ \mu^2 J_{\min}\lambda_i, \qquad i = 1, 2, \ldots, M \tag{5.88}$$

Define the M-by-1 vectors $\mathbf{x}(n)$ and $\boldsymbol{\lambda}$ as follows, respectively:

$$\mathbf{x}^T(n) = [x_1(n), x_2(n), \ldots, x_M(n)] \tag{5.89}$$

and

$$\boldsymbol{\lambda}^T = [\lambda_1, \lambda_2, \ldots, \lambda_M] \tag{5.90}$$

Then we may rewrite Eq. (5.88) in matrix form as

$$\mathbf{x}(n + 1) = \mathbf{B}\mathbf{x}(n) + \mu^2 J_{\min}\boldsymbol{\lambda} \tag{5.91}$$

where $\mathbf{B}$ is an M-by-M matrix with elements

$$b_{ij} = \begin{cases} (1 - \mu\lambda_i)^2, & i = j \\ \mu^2\lambda_i\lambda_j, & i \neq j \end{cases} \tag{5.92}$$

From Eq. (5.92), we see that the matrix $\mathbf{B}$ is symmetric, and its elements, b_{ij}, are all positive and real valued.   The matrix $\mathbf{B}$, however, is not necessarily positive definite.

Equation (5.91) is a difference equation of order 1 in matrix form.   Therefore, assuming an initial value $\mathbf{x}(0)$, the solution to this equation is

$$\mathbf{x}(n) = \mathbf{B}^n\mathbf{x}(0) + \mu^2 J_{\min} \sum_{i=0}^{n-1} \mathbf{B}^i\boldsymbol{\lambda} \tag{5.93}$$

By analogy with the formula for the sum of a geometric series, we may express the finite sum $\sum_{i=0}^{n-1} \mathbf{B}^i$ as follows:

$$\sum_{i=0}^{n-1} \mathbf{B}^i = (\mathbf{I} - \mathbf{B}^n)(\mathbf{I} - \mathbf{B})^{-1} \tag{5.94}$$

where $\mathbf{I}$ is the identity matrix. Substituting Eq. (5.94) in (5.93), we thus get

$$\mathbf{x}(n) = \mathbf{B}^n[\mathbf{x}(0) - \mu^2 J_{\min}(\mathbf{I} - \mathbf{B})^{-1}\boldsymbol{\lambda}] + \mu^2 J_{\min}(\mathbf{I} - \mathbf{B})^{-1}\boldsymbol{\lambda} \tag{5.95}$$

The first term on the right side of Eq. (5.95) is the *transient* component of the vector $\mathbf{x}(n)$, and the second term is the *steady-state* component. Since the matrix $\mathbf{B}$ is symmetric, we may apply to it a unitary similarity transformation similar to that described by Eq. (5.81). We may thus write

$$\mathbf{G}^H \mathbf{B} \mathbf{G} = \mathbf{C} \tag{5.96}$$

The matrix $\mathbf{C}$ is a diagonal matrix with elements $c_i = 1, 2, \ldots, M$, which are the eigenvalues of $\mathbf{B}$. The matrix $\mathbf{G}$ is the unitary matrix whose ith column is the eigenvector $\mathbf{g}_i$ of $\mathbf{B}$, associated with eigenvalue c_i. Because of the property

$$\mathbf{G}\mathbf{G}^H = \mathbf{I} \tag{5.97}$$

we find that

$$\mathbf{B}^n = \mathbf{G}\mathbf{C}^n\mathbf{G}^H \tag{5.98}$$

Hence, we may rewrite Eq. (5.95) in the form

$$\mathbf{x}(n) = \mathbf{G}\mathbf{C}^n\mathbf{G}^H[\mathbf{x}(0) - \mu^2 J_{\min}(\mathbf{I} - \mathbf{B})^{-1}\boldsymbol{\lambda}] + \mu^2 J_{\min}(\mathbf{I} - \mathbf{B})^{-1}\boldsymbol{\lambda} \tag{5.99}$$

Since $\mathbf{C}$ is a diagonal matrix, we have

$$\mathbf{C}^n = \operatorname{diag}(c_1^n, c_2^n, \ldots, c_M^n) \tag{5.100}$$

It follows therefore that the solution of Eq. (5.99), and therefore that of Eq. (5.95), is stable if and only if the eigenvalues of matrix $\mathbf{B}$ all have a magnitude less than 1, as shown by

$$-1 < c_i < 1, \qquad \text{for all } i \tag{5.101}$$

When this condition is satisfied, the transient component in Eq. (5.99) or (5.95) decays to zero as the number of iterations, n, approaches infinity, leaving the steady-state component as the only component. We may then write

$$\mathbf{x}(\infty) = \mu^2 J_{\min}(\mathbf{I} - \mathbf{B})^{-1}\boldsymbol{\lambda} \tag{5.102}$$

Substituting Eq. (5.102) in (5.99), we may rewrite the solution as

$$\mathbf{x}(n) = \mathbf{G}\mathbf{C}^n\mathbf{G}^H[\mathbf{x}(0) - \mathbf{x}(\infty)] + \mathbf{x}(\infty) \tag{5.103}$$

In view of the diagonal nature of matrix $\mathbf{C}^n$, and since the unitary matrix $\mathbf{G}$ consists of the eigenvectors of $\mathbf{B}$ as its columns, we may express the matrix product $\mathbf{G}\mathbf{C}^n\mathbf{G}^H$

as follows:

$$\mathbf{GC}^n\mathbf{G}^H = [\mathbf{g}_1, \mathbf{g}_2, \ldots, \mathbf{g}_M] \begin{bmatrix} c_1^n & 0 & \cdot & \cdot & \cdot & 0 \\ 0 & c_2^n & \cdot & \cdot & \cdot & 0 \\ \cdot & & \cdot & & & \cdot \\ \cdot & & & \cdot & & \cdot \\ \cdot & & & & \cdot & \cdot \\ 0 & \cdot & \cdot & \cdot & \cdot & c_M^n \end{bmatrix} \begin{bmatrix} \mathbf{g}_1^H \\ \mathbf{g}_2^H \\ \cdot \\ \cdot \\ \cdot \\ \mathbf{g}_M^H \end{bmatrix}$$

(5.104)

$$= \sum_{i=1}^{M} c_i^n \mathbf{g}_i \mathbf{g}_i^H$$

Accordingly, we may rewrite Eq. (5.103) in the equivalent form

$$\mathbf{x}(n) = \sum_{i=1}^{M} c_i^n \mathbf{g}_i \mathbf{g}_i^H [\mathbf{x}(0) - \mathbf{x}(\infty)] + \mathbf{x}(\infty) \qquad (5.105)$$

The average value of the excess mean-squared error equals [see Eq. (5.86)]

$$E[J_{\mathrm{ex}}(n)] = \boldsymbol{\lambda}^T \mathbf{x}(n)$$

$$= \sum_{i=1}^{n} c_i^n \boldsymbol{\lambda}^T \mathbf{g}_i \mathbf{g}_i^H [\mathbf{x}(0) - \mathbf{x}(\infty)] + \boldsymbol{\lambda}^T \mathbf{x}(\infty) \qquad (5.106)$$

$$= \sum_{i=1}^{n} c_i^n \boldsymbol{\lambda}^T \mathbf{g}_i \mathbf{g}_i^H [\mathbf{x}(0) - \mathbf{x}(\infty)] + E[J_{\mathrm{ex}}(\infty)]$$

where

$$E[J_{\mathrm{ex}}(\infty)] = \boldsymbol{\lambda}^T \mathbf{x}(\infty)$$

$$= \sum_{j=1}^{M} \lambda_j x_j(\infty) \qquad (5.107)$$

In Eq. (5.106), the first term on the right side describes the transient behavior of the average excess mean-squared error, whereas the second term represents its value after adaptation (i.e., its steady-state value).

We now wish to derive an expression for the steady-state component $E[J_{\mathrm{ex}}(\infty)]$ in terms of the eigenvalues of the original correlation matrix $\mathbf{R}$. We may do this by premultiplying $\mathbf{x}(\infty)$ in Eq. (5.102) by $\boldsymbol{\lambda}^T$ and manipulating the resultant vector product. However, a simpler approach is to go back to Eq. (5.88). Then, putting

$n = \infty$, and summing both sides with respect to i, we get the desired formula[4]:

$$E[J_{ex}(\infty)] = \frac{\mu J_{min} \sum_{i=1}^{M} \lambda_i}{2 - \mu \sum_{i=1}^{M} \lambda_i} \tag{5.108}$$

Note that we may also rewrite this expression in the form

$$E[J_{ex}(\infty)] = \frac{\mu J_{min}}{2 - \mu \sum_{i=1}^{M} \lambda_i} \boldsymbol{\lambda}^T \mathbf{1} \tag{5.109}$$

where $\mathbf{1}$ is an M by an *unit vector* (with all its elements equal to 1). Hence, using Eq. (5.109) and the first line of Eq. (5.107), we deduce that the steady-state vector $\mathbf{x}(\infty)$ equals

$$\mathbf{x}(\infty) = \frac{\mu J_{min}}{2 - \mu \sum_{i=1}^{M} \lambda_i} \mathbf{1} \tag{5.110}$$

Finally, substituting Eq. (5.109) in (5.106), and using the first line of Eq. (5.79), we may express the time evolution of the average mean-squared error by the equation (Ungerboeck, 1972)

$$E[J(n)] = \sum_{i=1}^{M} \gamma_i c_i^n + \frac{2J_{min}}{2 - \mu \sum_{i=1}^{M} \lambda_i} \tag{5.111}$$

where

$$\gamma_i = \boldsymbol{\lambda}^T \mathbf{g}_i \mathbf{g}_i^H \left[\mathbf{x}(0) - \frac{\mu J_{min}}{2 - \mu \sum_{i=1}^{M} \lambda_i} \mathbf{1} \right], \qquad i = 1, 2, \ldots, M \tag{5.112}$$

Note that the ith element of the initial value $\mathbf{x}(0)$ is related to the initial value of the weight-error vector $\boldsymbol{\varepsilon}(0) = \mathbf{w}(0) - \mathbf{w}_o$ as follows [see Eq. (5.82)]:

$$x_i(0) = E[\mathbf{q}_i^H \boldsymbol{\varepsilon}(0) \boldsymbol{\varepsilon}^H(0) \mathbf{q}_i], \qquad i = 1, 2, \ldots, M \tag{5.113}$$

where $\mathbf{q}_i$ is the eigenvector of the correlation matrix $\mathbf{R}$ associated with eigenvalue λ_i.

5.12 PROPERTIES OF THE TRANSIENT BEHAVIOR OF THE AVERAGE MEAN-SQUARED ERROR

We may now summarize the properties of the transient behavior of the average mean-squared error, $E[J(n)]$. These properties provide us with a deeper under-

---

[4]Note that when n approaches infinity, $x_i(n+1)$ and $x_i(n)$ become equal in the limit; we may therefore cancel them out.

standing of the operation of the LMS algorithm in a stationary environment (Ungerboeck, 1972). They all follow from Eq. (5.111).

Property 1.   *The transient component of the mean-squared error, $E[J(n)]$, does not exhibit oscillations.*

The transient component of $E[J(n)]$ equals

$$\sum_{i=1}^{M} \gamma_i c_i^n,$$

where the γ_i are constant coefficients and the c_i, $i = 1, 2, \ldots, M$ are the eigenvalues of matrix **B**. These eigenvalues are all real numbers, since **B** is a symmetric matrix [see Eq. (5.92)]. Hence, the ensemble-averaged learning curve, that is, a plot of the average mean-squared error, $E[J(n)]$, versus the number of iterations, n, consists only of exponentials.

Note, however, that the learning curve represented by a plot of the mean-squared error $J(n)$, without averaging, versus n consists of *noisy* exponentials. The amplitude of the noise becomes smaller as the step-size parameter μ is reduced.

Property 2.   Convergence of the Average Mean-Squared Error.   *The average mean-squared error, $E[J(n)]$, converges to a steady state value equal to $J_{min} + J_{ex}(\infty)$ if and only if the step-size parameter μ satisfies the condition*

$$0 < \mu < \frac{2}{\sum_{i=1}^{M} \lambda_i} \tag{5.114}$$

*where the λ_i, $i = 1, 2, \ldots, M$, are the eigenvalues of the correlation matrix **R**, and M is the number of taps. When this condition is satisfied, we say that the LMS algorithm is convergent in the mean square.*

The transient component of the average mean-squared error, $E[J(n)]$, decays to zero if and only if the eigenvalues of matrix **B** satisfy the condition

$$-1 < c_i < 1, \qquad \text{for all } i \tag{5.115}$$

We now have to show that the two convergence conditions of Eqs. (5.114) and (5.115) are equivalent to each other.

Let **g** be an eigenvector of matrix **B**, associated with eigenvalue c. Then, by definition we have

$$\mathbf{Bg} = c\mathbf{g} \tag{5.116}$$

In expanded form, this equation reads

$$
\begin{bmatrix}
(1 - \mu\lambda_1)^2 & \mu^2\lambda_1\lambda_2 & \cdots & \mu\lambda_1\lambda_M \\
\mu^2\lambda_2\lambda_1 & (1 - \mu\lambda_2)^2 & \cdots & \mu^2\lambda_2\lambda_M \\
\vdots & \vdots & \ddots & \vdots \\
\mu^2\lambda_M\lambda_1 & \mu^2\lambda_M\lambda_2 & \cdots & (1 - \mu\lambda_M)^2
\end{bmatrix}
\begin{bmatrix}
g_1 \\
g_2 \\
\vdots \\
g_M
\end{bmatrix}
$$

$$
= c
\begin{bmatrix}
g_1 \\
g_2 \\
\vdots \\
g_M
\end{bmatrix}
\tag{5.117}
$$

or, equivalently,

$$
g_i - 2\mu\lambda_i g_i + \mu^2 \left(\sum_{j=1}^{M} \lambda_j g_j \right) \lambda_i = c g_i, \qquad i = 1, 2, \ldots, M \tag{5.118}
$$

with the g_i denoting the elements of $\mathbf{g}$. Solving Eq. (5.118) for g_i, we thus get

$$
g_i = -\mu^2 \left(\sum_{j=1}^{M} \lambda_j g_j \right) \frac{\lambda_i}{1 - c - 2\mu\lambda_i}, \qquad i = 1, 2, \ldots, M \tag{5.119}
$$

Based on the result of Eq. (5.119), there are two separate cases to be considered (Mazo, 1979):

Case 1. Whenever the correlation matrix $\mathbf{R}$ of the tap-input vector has an eigenvalue $\lambda_i = 0$, we find from Eq. (5.118) that matrix $\mathbf{B}$ has an eigenvalue $c = 1$ for all μ. Correspondingly, the eigenvector $\mathbf{g}$ has element $g_i = 1$ and all its remaining $M - 1$ elements equal to zero. Since such eigenvalues of the matrix $\mathbf{B}$ do not change with the step-size parameter μ, they are of no interest here.

Case 2. Suppose we identify the nonzero eigenvalues of the correlation matrix $\mathbf{R}$ as $\overline{\lambda}_i$, $i = 1, 2, \ldots, \overline{M}$, where $\overline{M} < M$. Then, we are concerned with

$$
b_{ij} = \begin{cases} (1 - \mu\overline{\lambda}_i)^2, & i = j \\ \mu^2\overline{\lambda}_i\overline{\lambda}_j, & i \neq j \end{cases} \tag{5.120}
$$

in a space of reduced dimension $\overline{M}$. For a step-size parameter μ that is small enough, the eigenvalues of matrix $\mathbf{B}$ are approximately $1 - 2\mu\overline{\lambda}_i$. Since $\mu > 0$, these eigenvalues of $\mathbf{B}$ are necessarily less than 1. Suppose we now increase μ until possibly one of the eigenvalues of $\mathbf{B}$ equals ± 1, representing the convergence boundaries [see Eq. (5.115)]. We wish to find the critical value of μ for which this possibility occurs. We observe that the elements b_{ij} in Eq. (5.120) are all strictly positive (except, at the very most, $\overline{M}$ values of μ when $\mu\lambda_i = 0$ for $i = 1$, $2, \ldots, \overline{M}$). The matrix **B** is thus said to be *positive* since all its elements are positive. Hence, we may use the *Perron theorem*,[5] which applies to a positive square matrix such as matrix **B**. The Perron theorem states that (Bellman, 1960):

*If **B** is a positive matrix, there is a unique eigenvalue of **B**, which has the largest magnitude. This eigenvalue is positive and simple (i.e., of multiplicity 1), and its associated eigenvector consists entirely of positive elements.*

Accordingly, the magnitude of the largest eigenvalue of matrix **B** may be taken to be associated with a positive eigenvalue. Thus, in Eq. (5.119) [reinterpreted to match the notation of Eq. (5.120)], we may set the eigenvalue $c = 1$, multiply both sides of the equation by λ_i, and then sum on i. The results of all these operations yields the critical value of the step-size parameter as

$$
\begin{aligned}
\mu_{\text{crit}} &= \frac{2}{\sum_{i=1}^{\overline{M}} \overline{\lambda}_i} \\
&= \frac{2}{\sum_{i=1}^{M} \lambda_i}
\end{aligned}
\tag{5.121}
$$

Accordingly, the LMS algorithm is stable in the sense that the average mean-squared error converges to a steady-state value if and only if the step-size parameter μ satisfies the condition of Eq. (5.114). This completes the proof of Property 2.

The stability condition of Eq. (5.114) has an important practical interpretation. In particular, we may use the following relation (see Problem 10, Chapter 2):

$$
\sum_{i=1}^{M} \lambda_i = \text{sum of the mean-square values of the } M \text{ tap inputs}
\tag{5.122}
$$

Using the term *total input power* to refer to the sum of the mean-square values of tap inputs $u(n), u(n-1), \ldots, u(n-M-1)$, we may restate the stability condition of Eq. (5.114) as follows:

$$
0 < \mu < \frac{2}{\text{total input power}}
\tag{5.123}
$$

Note that for a stationary input the total input power equals $Mr(0)$, where M is the number of taps and $r(0)$ is the autocorrelation function of the tap inputs for zero lag.

The stability condition of Eq. (5.114) imposes a smaller upper bound on the step-size parameter μ than that of Eq. (5.50), which guarantees the convergence

---

[5]The Perron theorem is also known as the *Perron–Frobenius theorem*.

of the LMS algorithm in the mean.   This follows from the fact that

$$\lambda_{\max} \leq \sum_{i=1}^{M} \lambda_i$$

We conclude, therefore, that the condition of Eq. (5.114) is the necessary and sufficient condition for the *overall stability* of the LMS algorithm.

Property 3.   *A small eigenvalue of the correlation matrix* **R** *of the tap-input vector leads to a slowly converging term in the transient component of the average mean-squared error,* $E[J(n)]$. *But the slower the term converges relative to the terms due to the other eigenvalues of* **R**, *the smaller the probability that this term contributes significantly to the average mean-squared error.*

To prove this property, we observe from case 1 considered previously that, if $\lambda_i \simeq 0$, then for matrix **B** we have an eigenvalue $c_i \simeq 1$ and an associated eigenvector $\mathbf{g}_i$ that has zeros for all its elements except for the ith element, which equals 1.   This means that $\boldsymbol{\lambda}^T \mathbf{g}_i \simeq 0$.   Now the corresponding term in the transient component of $E[J(n)]$ equals $\gamma_i c_i^n$.   This term decays slowly since $c_i \simeq 1$, but its amplitude $\gamma_i \simeq 0$ since γ_i is proportional to $\boldsymbol{\lambda}^T \mathbf{g}_i$ [see Eq. (5.112)].   This completes the proof of Property 3.

In general, it is true that a large spread of the eigenvalues of correlation matrix **R** of the tap-input vector results in a slow convergence.   However, because the slower-converging terms in the transient component of Eq. (5.111) are usually given smaller weights, their contribution to the transient component is correspondingly smaller.   We conclude therefore that a spread of the eigenvalues of the correlation matrix **R** tends to affect the convergence of the average mean-squared error, $E[J(n)]$, to a lesser extent than the convergence of the average tap-weight vector, $E[\hat{\mathbf{w}}(n)]$.

Property 4: Misadjustment.   *The misadjustment,* $\mathcal{M}$, *defined as the dimensionless ratio of the steady-state value of the average excess mean-squared error to the minimum mean-squared error, equals*

$$\mathcal{M} = \frac{\mu \sum_{i=1}^{M} \lambda_i}{2 - \mu \sum_{i=1}^{M} \lambda_i} \tag{5.124}$$

To prove this property, we may use Eq. (5.108), which defines the steady-state value of the average excess mean-squared error.   This equation is reproduced here for convenience:

$$E[J_{\text{ex}}(\infty)] = \frac{\mu J_{\min} \sum_{i=1}^{M} \lambda_i}{2 - \mu \sum_{i=1}^{M} \lambda_i}$$

The misadustment is defined by (Widrow, 1970)

$$\mathcal{M} = \frac{E[J_{ex}(\infty)]}{J_{min}}$$

Therefore, dividing the expression for $E[J_{ex}(\infty)]$ by J_{min}, we get the result of Eq. (5.124).

When the step-size parameter μ is small enough to make

$$\mu \sum_{i=1}^{M} \lambda_i << 2$$

we find from Eq. (5.124) that the misadjustment $\mathcal{M}$ varies linearly with μ.

The misadjustment $\mathcal{M}$ provides a useful measure of the cost of adaptivity. Thus, for example, a misadjustment of 10 percent means that the adaptive algorithm produces an average mean-squared error (after adaptation) that is 10 percent greater than the minimum mean-squared error J_{min}. Also, it is of interest to note that, when the eigenvalues of the correlation matrix **R** are widely spread, Eq. (5.124) shows that the misadjustment is primarily limited by the larger eigenvalues.

5.13 SUMMARY OF THE LMS ALGORITHM

Parameters: M = number of taps
μ = step-size parameter

$$0 < \mu < \frac{2}{\text{total input power}}$$

Total input power = $Mr(0)$

Initial conditions: $\hat{\mathbf{w}}(0) = \mathbf{0}$

Data:
(a) Given: $\mathbf{u}(n)$ = M-by-1 tap-input vector at time n
$d(n)$ = desired response at time n

(b) To be computed: $\hat{\mathbf{w}}(n+1)$ = estimate of tap-weight vector at time $n+1$

Computation $(n = 0, 1, 2, \ldots)$:
$$e(n) = d(n) - \hat{\mathbf{w}}^H(n)\mathbf{u}(n)$$
$$\hat{\mathbf{w}}(n+1) = \hat{\mathbf{w}}(n) + \mu\mathbf{u}(n)e^*(n)$$

Note:With the initial value $\hat{\mathbf{w}}(0)$ of the tap-weight vector set equal to zero and with all the tap inputs of the transversal filter set equal to zero initially, we find from the second relation that $\hat{\mathbf{w}}(1)$ is also zero. Accordingly, after the first iteration of the LMS algorithm, we always find that $e(1) = d(1)$.

5.14 DISCUSSION

The feature that distinguishes the LMS algorithm from other adaptive algorithms is the simplicity of its implementation, be that in software or hardware form. The simplicity of the LMS algorithm is exemplified by the pair of equations that are involved in its computation (see the previous discussion).

The three principal factors affect the response of the LMS algorithm: the step-size parameter μ, the number of taps M, and the eigenvalues of the correlation matrix $\mathbf{R}$ of the tap-input vector. In the light of the analysis presented in Sections 5.9 through 5.12 using the independence theory, their individual effects may be summarized as follows:

1. When a small value is assigned to μ, the adaptation is slow, which is equivalent to the LMS algorithm having a long "memory." Correspondingly, the excess mean-squared error after adaptation is small, on the average, because of the large amount of data used by the algorithm to estimate the gradient vector. On the other hand, when μ is large, the adaptation is relatively fast, but at the expense of an increase in the average excess mean-squared error after adaptation. In this case, less data enter the estimation; hence, a degraded estimation error performance. Thus, the parameter μ may also be viewed as the *memory* of the LMS algorithm in the sense that it determines the weighting applied to the tap inputs.

2. The convergence properties of the average mean-squared error $E[J(n)]$ depend, unlike the average tap-weight vector, on the number of taps, M. The necessary and sufficient condition for $E[J(n)]$ to be convergent is

$$0 < \mu < \frac{2}{\sum_{i=1}^{M} \lambda_i}$$

where the λ_i, $i = 1, 2, \ldots, M$, are the eigenvalues of the correlation matrix $\mathbf{R}$ of the tap inputs. This stability condition may also be stated in the equivalent form

$$0 < \mu < \frac{2}{\text{total input power}}$$

where the total input power refers to the sum of the mean-square values of the individual tap inputs $u(n), u(n-1), \ldots, u(n-M+1)$. When this condition is satisfied, we say that the LMS algorithm is convergent in the mean square. On the other hand, the necessary and sufficient condition for the average tap-weight vector $E[\hat{\mathbf{w}}(n)]$ to be convergent is

$$0 < \mu < \frac{2}{\lambda_{\max}}$$

where $\lambda_{\max}$ is the largest eigenvalue of $\mathbf{R}$. When this second condition is satisfied, we say that the LMS algorithm is convergent in the mean.   Since

$$\lambda_{\max} \le \sum_{i=1}^{M} \lambda_i$$

we see that, by choosing the step-size parameter μ to satisfy the convergence condition for $E[J(n)]$, we automatically satisfy the convergence condition for $E[\hat{\mathbf{w}}(n)]$.

3. When the eigenvalues of the correlation matrix $\mathbf{R}$ are widely spread, the excess mean-squared error produced by the LMS algorithm is primarily determined by the largest eigenvalues, and the time taken by $E[\hat{\mathbf{w}}(n)]$ to converge is limited by the smallest eigenvalues.   However, the speed of convergence of $E[J(n)]$ is affected by a spread of the eigenvalues of $\mathbf{R}$ to a lesser extent than the convergence of $E[\hat{\mathbf{w}}(n)]$.   In any event, when the eigenvalue spread is large (i.e., the correlation matrix of the tap inputs is ill conditioned), the LMS algorithm slows down in that it requires a large number of iterations for it to converge, the very condition for which effective operation of the algorithm is required.

The mathematical details of the results summarized here were derived using the independence theory.   This theory invokes the assumption that the sequence of random vectors that direct the "hunting" of the tap-weight vector toward the optimum Wiener solution are statistically independent.   Even though in reality this assumption is often far from true, nevertheless, the results predicted by the independence theory are usually found to be in excellent agreement with experiments and computer simulations.

A basic limitation of the independence theory, however, is the fact that it ignores the statistical dependence between the "gradient" directions as the algorithm proceeds from one iteration to the next.   This statistical dependence results from the *shifting property* of the input data.   At time n, the gradient vector is proportional to the corresponding sample value of the tap-input vector, shown by

$$\mathbf{u}^T(n) = [u(n), u(n-1), \ldots, u(n-M+1)]$$

At time $n+1$, it is proportional to the updated sample value of the tap-input vector, shown by

$$\mathbf{u}^T(n+1) = [u(n+1), u(n), \ldots, u(n-M+2)]$$

Thus, with the arrival of the new sample $u(n+1)$, the oldest sample $u(n-M+1)$ is discarded from $\mathbf{u}(n)$, and the remaining samples $u(n), u(n-1), \ldots, u(n-M+2)$ are shifted back in time by one time unit.   We see therefore that the tap-input vectors, and correspondingly the gradient directions, are indeed statistically dependent.

In recent years a few attempts have been made to analyze the response of the LMS algorithm, taking into account the statistical dependence of the tap-input vectors. Two significant contributions in this regard are the papers by Mazo (1979) and Jones, Cavin, and Reed (1982).

Mazo considers a binary baseband adaptive equilization problem wherein the communication channel is represented by a finite-impulse response model. He develops an exact theory for computing the average weight-error vector and average mean-squared error produced by the LMS algorithm after adaptation (i.e., when steady-state conditions are established). By using a perturbation analysis in the step-size parameter μ, Mazo shows that when this parameter has a small value the difference between the results of the independence theory and the exact theory (which takes into account the statistical dependence of the gradient directions) is often small.

Jones, Cavin, and Reed also present a detailed mathematical analysis of a class of gradient-based adaptive filtering algorithms (which includes the LMS algorithm as a special case) with dependent input data. A particular structural representation for this statistical dependence is assumed, which appears to arise in a significantly wide class of physical problems to justify the detailed treatment. The procedure they used is different from Mazo's in that they first imbed the adaptive algorithm in a higher-order stochastic system from which the relevant moments can, in principle, be computed recursively. For sufficiently small values of the step-size parameter μ, the new results derived by Jones, Cavin, and Reed for the average weight-error vector and the average mean-squared error under steady-state conditions are likely to differ little from those obtained by using the independence theory.

Thus, while the ultimate justification of the independence theory must remain empirical, nevertheless, the exact theory presented by Mazo and that presented by Jones, Cavin, and Reed make the success of the independence theory mathematically plausible.

5.15 COMPUTER EXPERIMENT ON ADAPTIVE PREDICTION

In this experiment we use the LMS algorithm to adaptively estimate the tap weights of the linear predictor shown in Fig. 5.13. This predictor was studied in Section 5.6, using the steepest descent algorithm. As before, we assume that (1) the predictor has two tap weights, and (2) the tap inputs $u(n-1)$ and $u(n-2)$ are drawn from the *real-valued* AR process $\{u(n)\}$ described by the second-order difference equation (5.32).

The purpose of the experiment is to study the sensitivity of the LMS algorithm (in the context of this predictor) to variations in (1) the step-size parameter μ, and (2) the eigenvalue spread $\chi(\mathbf{R})$ of the correlation matrix of the tap inputs $u(n-1)$ and $u(n-2)$.

The tap-input vector (used in the filtering process) equals

$$\mathbf{u}(n-1) = \begin{bmatrix} u(n-1) \\ u(n-2) \end{bmatrix}$$

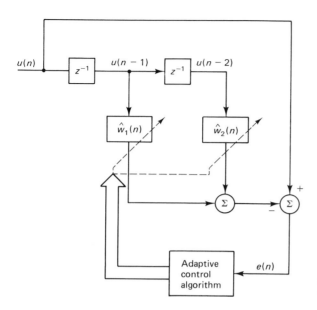

Figure 5.13 Adaptive two-tap predictor with real valued input.

The desired response equals $u(n)$. Denote the estimate of the two-tap weight vector at time n by

$$\hat{\mathbf{w}}(n) = \begin{bmatrix} \hat{w}_1(n) \\ \hat{w}_2(n) \end{bmatrix}$$

We may therefore express the LMS algorithm for the operation of this predictor (with *real* inputs) as follows:

$$e(n) = u(n) - \hat{\mathbf{w}}^T(n)\mathbf{u}(n-1) \tag{5.125}$$

$$\hat{\mathbf{w}}(n+1) = \hat{\mathbf{w}}(n) + \mu\mathbf{u}(n-1)e(n) \tag{5.126}$$

The algorithm starts with the initial condition $\hat{\mathbf{w}}(0) = \mathbf{0}$.

As mentioned previously, the AR process $\{u(n)\}$ is described by the second-order difference equation (5.32) reproduced here for convenience:

$$u(n) + a_1u(n-1) + a_2u(n-2) = v(n)$$

The zero-mean white-noise process $\{v(n)\}$ has its variance σ_v^2 adjusted to make the AR process $\{u(n)\}$ have unit variance. Also, we have $a_1^2 < 4a_2$, which makes the roots of the characteristic equation of the AR process $\{u(n)\}$ assume complex conjugate values. Within the bounds defined by this equation, the values assigned to the AR parameters a_1 and a_2 are determined by the desired eigenvalue spread $\chi(\mathbf{R})$ of the correlation matrix of the tap inputs $u(n-1)$ and $u(n-1)$.

For one trial of the experiment, 512 samples of white-noise process $\{v(n)\}$ are obtained from a *random-number generator* of zero mean and adjustable variance. For each value of n in the range $1 \leq 512$, the LMS algorithm of Eqs. (5.125) and (5.126) is applied and the squared magnitude of the estimation error $e(n)$ is computed.   The experiment is repeated 200 times, each time using an *independent* realization of the process $\{v(n)\}$.   The average squared error is then determined by computing the ensemble average of $e^2(n)$ over these 200 independent trials of the experiment.   The result is an *approximation* to the ensemble-averaged learning curve of the adaptive predictor for the preassigned values of the step-size parameter μ and the eigenvalue spread $\chi(\mathbf{R})$.

The results of these computations are shown in parts (a), (b) and (c) of Figure 5.14, corresponding to the eigenvalue spread $\chi(\mathbf{R}) = 3, 10, 100$, respectively.   In each part of the figure, two values of the step-size parameter are used: $\mu = 0.005$, 0.05

In Table 5.2 we have also summarized the following experimental results (for varying step-size parameter μ and eigenvalue spread $\chi(\mathbf{R})$):

1. The tap weights $\hat{w}_1(\infty)$ and $\hat{w}_2(\infty)$: These are obtained by averaging the steady-state values of the estimates $\hat{w}_1(n)$ and $\hat{w}_2(n)$, respectively, over 200 independent trials of the experiment.

2. The average mean-squared error $J(\infty)$: This is obtained by first time–averaging 200 samples of $e^2(n)$, starting from the steady-state tap values, and then ensemble averaging over 200 independent trials of the experiment.

3. Misadjustment, $\mathcal{M}$: This is obtained by computing $[J(\infty) - J_{\min}]/J_{\min}$, where $J_{\min}$ is the minimum mean-square error as calculated from theory.

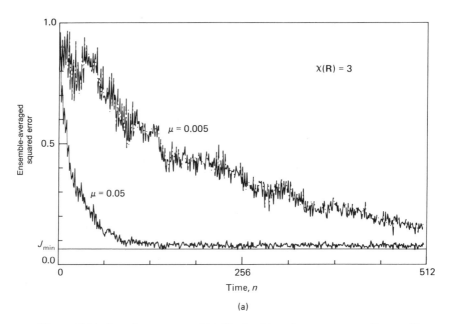

(a)

Figure 5.14 Learning curves of LMS algorithm for two-tap predictor with $\mu = 0.005$, 0.05 and varying eigenvalue spread: (a) $\chi(\mathbf{R}) = 3$; (b) $\chi(\mathbf{R}) = 10$; (c) $\chi(\mathbf{R}) = 100$.

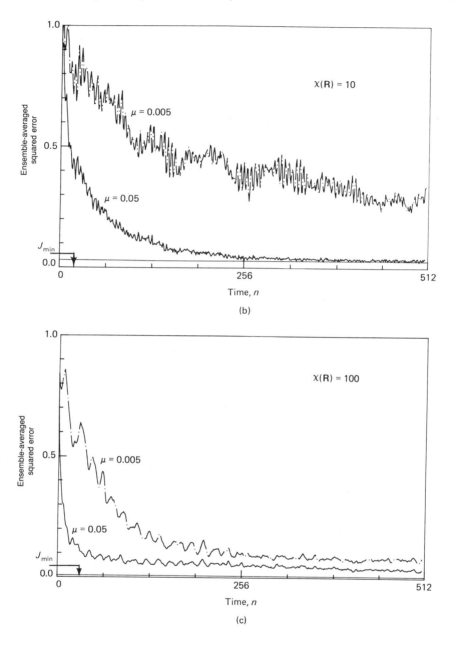

(b)

(c)

In Table 5.2, we have also included the corresponding theoretical results.

Based on the results of Fig. 5.14 and Table 5.2, we see that the LMS algorithm is highly sensitive to variations in the step-size parameter μ and the eigenvalue spread $\chi(\mathbf{R})$. In particular, we may make the following observations:

1. For fixed eigenvalue spread, increasing the step-size parameter μ reduces the number of iterations required for the LMS algorithm to reach steady-state conditions. An increase in μ, however, causes a corresponding increase in the misadjustment $\mathcal{M}$.

2. For fixed step-size parameter μ, increasing the eigenvalue spread $\chi(\mathbf{R})$ increases the number of iterations required to reach steady-state conditions.

3. There is close agreement between theory and experiment insofar as steady-state parameters (i.e., average tap-weight vector, excess mean-squared error, and misadjustment) are concerned.

TABLE 5.2

| | Tap weights | | | Theory | | | | | Measured | | | | | |
| | | | | $\mu = 0.05$ | | $\mu = 0.005$ | | | $\mu = 0.05$ | | | | $\mu = 0.005$ | | |
$\chi(\mathbf{R})$	w_1	w_2	$J_{\min}$	$J_{\text{ex}}(\infty)$	$\mathcal{M}$	$j_{\text{ex}}(\infty)$	$\mathcal{M}$	$\hat{w}_1$	$\hat{w}_2$	$J_{\text{ex}}(\infty)$	$\mathcal{M}$	$\hat{w}_1$	$\hat{w}_2$	$J_{\text{ex}}(\infty)$	$\mathcal{M}$
3	0.9750	-0.95	0.0731	0.0039	5.26%	0.0004	0.67%	0.9679	-0.9207	0.0043	5.88%	0.9762	-0.9471	0.0006	0.82%
10	1.5955	-0.95	0.0322	0.0017	5.26%	0.0002	0.67%	1.5876	-0.9395	0.0019	5.9%	1.5949	-0.9495	0.0003	0.93%
100	1.9114	-0.95	0.0038	0.0002	5.26%	0.25×10^{-4}	0.67%	1.9082	-0.9499	0.0003	7.89%	1.9111	-0.9502	0.0001	2.63%

5.16 COMPUTER EXPERIMENT ON ADAPTIVE EQUALIZATION

In this second computer experiment we study the use of the LMS algorithm for *adaptive equalization* of a linear dispersive channel that produces (unknown) distortion. Here again we assume that the data are all *real valued*. Figure 5.15 shows the block diagram of the system used to carry out the study. Random-number generator 1 provides the test signal, $\{a_n\}$, used for probing the channel, whereas random-number generator 2 serves as the source of additive white noise $\{v(n)\}$ that corrupts the channel output. These two random-number generators are independent of each other. The adaptive equalizer has the task of correcting for the distortion produced by the channel in the presence of the additive white noise. Random-number generator 1, after suitable delay, also supplies the desired response applied to the adaptive equalizer.[6]

The random sequence $\{a(n)\}$ applied to the channel input is in *polar* form, with $a(n) = \pm 1$, so the sequence $\{a(n)\}$ has zero mean. The impulse response of

---

[6]In an operational system, the receiver is physically separate from the transmitter. As discussed in Chapter 1, one method to produce the test signal for probing the channel is to use a maximal-length sequence generator. At the receiver, an identical maximal-length sequence generator is used, which is synchronized to that at the transmitter. This second generator supplies the desired response to the adaptive equalizer.

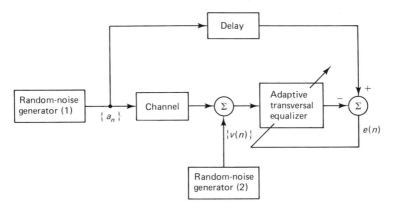

Figure 5.15 Block diagram of adaptive equalizer experiment.

the channel is described by the raised cosine:[7]

$$h_n = \begin{cases} \dfrac{1}{2}\left[1 + \cos\left(\dfrac{2\pi}{W}(n - 2)\right)\right], & n = 1, 2, 3 \\ 0, & \text{otherwise} \end{cases} \qquad (5.127)$$

where the parameter W controls the amount of amplitude distortion produced by the channel, with the distortion increasing with W.

Equivalently, the parameter W controls the eigenvalue spread $\chi(\mathbf{R})$ of the correlation matrix of the tap inputs in the equalizer, with the eigenvalue spread increasing with W. The sequence $\{v(n)\}$, produced by the second random generator, has zero mean and variance $\sigma_v^2 = 0.001$.

The equalizer has $M = 11$ taps. Since the channel has an impulse response $\{h_n\}$ that is symmetric about time $n = 2$, as depicted in Fig. 5.16(a), it follows that the optimum tap weights $\{w_{on}\}$ of the equalizer are likewise symmetric about time $n = 5$, as depicted in Fig. 5.16(b). Accordingly, the channel input $\{a(n)\}$ is delayed by $2 + 5 = 7$ samples to provide the desired response for the equalizer.

The experiment is in two parts that are intended to evaluate the response of the adaptive equalizer using the LMS algorithm to changes in the eigenvalue spread $\chi(\mathbf{R})$ and step-size parameter μ. Before proceeding to describe the results of the experiment, however, we first compute the eigenvalues of the correlation matrix $\mathbf{R}$ of the 11 tap inputs in the equalizer.

Correlation Matrix of the Equalizer Input

The first tap input of the equalizer at time n equals

$$u(n) = \sum_{k=1}^{3} h_k a(n - k) + v(n)$$

[7]The parameters specified in this experiment closely follow the paper by Satorius and Alexander (1979).

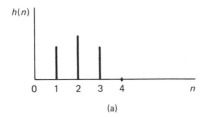

(a)

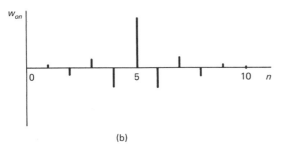

(b)

Figure 5.16 (a) Impulse response of channel; (b) impulse response of optimum transversal equalizer.

where all the parameters are real valued. Hence, the correlation matrix $\mathbf{R}$ of the 11 tap inputs of the equalizer, $u(n-1)$, $u(n-1)$, . . . , $u(n-10)$, is a symmetric 11-by-11 matrix. Also, since the impulse response $\{h_n\}$ has nonzero values only for $n = 1, 2, 3$, and the noise process $\{v(n)\}$ is white with zero mean and variance σ_v^2, the correlation matrix $\mathbf{R}$ is *quintdiagonal*. That is, the only nonzero elements of $\mathbf{R}$ are on the main diagonal and the four diagonals directly above and below it, two or either side, as shown by the special structure

$$
\mathbf{R} = \begin{bmatrix}
r(0) & r(1) & r(2) & 0 & \cdots & 0 \\
r(1) & r(0) & r(1) & r(2) & \cdots & 0 \\
r(2) & r(1) & r(0) & r(1) & \cdots & 0 \\
0 & r(2) & r(1) & r(0) & \cdots & 0 \\
\cdot & \cdot & \cdot & \cdot & \cdot & \cdot \\
\cdot & \cdot & \cdot & \cdot & \cdot & \cdot \\
\cdot & \cdot & \cdot & \cdot & \cdot & \cdot \\
0 & 0 & 0 & 0 & \cdots & r(0)
\end{bmatrix}
\tag{5.128}
$$

where $r(0) = h_1^2 + h_2^2 + h_3^2 + \sigma_v^2$

$r(1) = h_1 h_2 + h_2 h_3$

$$r(2) = h_1 h_3$$

The variance $\sigma_v^2 = 0.001$, and h_1, h_2, h_3 are determined by the value assigned to parameter W in Eq. (5.127).

In Table 5.3, we have listed (1) values of the autocorrelation function $r(l)$ for lag $l = 0, 1, 2$, and (2) the smallest eigenvalue, λ_{min}, the largest eigenvalue, λ_{max}, and the eigenvalue spread $\chi(\mathbf{R}) = \lambda_{max}/\lambda_{min}$. We thus see that the eigenvalue spread ranges from 6.0782 (for $W = 2.9$) to 46.8216 (for $W = 3.5$).

Experiment 1: Effect of Eigenvalue Spread

For the first part of the experiment, the step-size parameter was held fixed at $\mu = 0.075$. This is less than the critical value determined by Eq. (5.123) for the worst eigenvalue spread of 46.8216 (corresponding to $W = 3.5$).

$$\mu_{crit} = \frac{2}{\text{total input power}}$$

$$= \frac{2}{Mr(0)}$$

$$= 0.14$$

The choice of $\mu = 0.075$ therefore assures the convergence of the adaptive equalizer (in both the mean and mean square) for all the conditions listed in Table 5.3.

For each eigenvalue spread, an approximation to the ensemble-averaged learning curve of the adaptive equalizer is obtained by averaging the instantaneous squared error "$e^2(n)$, versus n" curve over 200 independent trials of the computer experiment. The results of this computation are shown in Fig. 5.17.

We thus see from Fig. 5.17 that increasing the eigenvalue spread $\chi(\mathbf{R})$ has the effect of slowing down the rate of convergence of the adaptive equalizer and also increasing the steady-state value of the average squared error. For example, when $\chi(\mathbf{R}) = 6.0782$, approximately 80 iterations are required for the adaptive equalizer to converge in the mean square, and the average-squared error (after 500 iterations) approximately equals 0.003. On the other hand, when $\chi(\mathbf{R}) = 46.8216$ (i.e., the equalizer input is ill conditioned), the equalizer requires approximately 200 iterations to converge in the mean square, and the resulting average squared error (after 500 iterations) approximately equals 0.03.

In Fig. 5.18, we have plotted the ensemble-averaged impulse response of the adaptive equalizer after 1000 iterations for each of the four eigenvalue spreads of

TABLE 5.3

W	$=2.9$,	3.1,	3.3,	3.5
$r(0)$	1.0963	1.1568	1.2264	1.3022
$r(1)$	0.4388	0.5596	0.6729	0.7774
$r(2)$	0.0481	0.0783	0.1132	0.1511
λ_{min}	0.3339	0.2136	0.1256	0.0656
λ_{max}	2.0295	2.3761	2.7263	3.0707
$\chi(\mathbf{R}) = \lambda_{max}/\lambda_{min}$	6.0782	11.1238	21.7132	46.8216

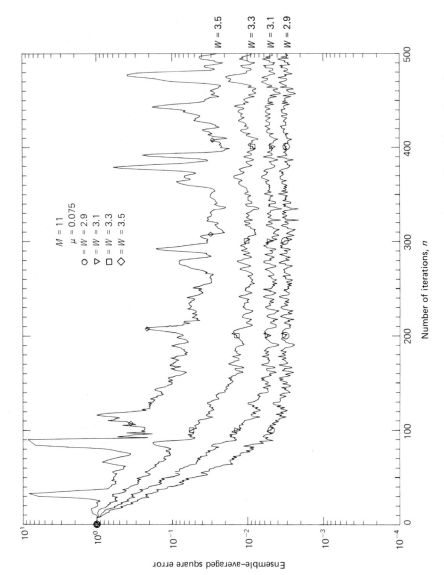

Figure 5.17 Learning curves of the LMS algorithm for adaptive equalizer with number of taps $M = 11$, step-size parameter $\mu = 0.075$, and varying eigenvalue spread $\chi(\mathbf{R})$.

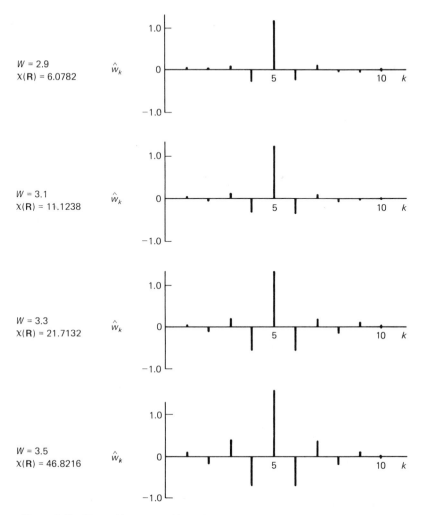

Figure 5.18 Ensemble-averaged impulse response of the adaptive equalizer (after 1000 iterations) for each of four different eigenvalue spreads.

interest. As before, the ensemble averaging was carried out over 200 independent trials of the experiment. We see that in each case the ensemble-averaged impulse response of the adaptive equalizer is very close to being symmetric with respect to the center tap, as expected. The variation in the impulse response from one eigenvalue spread to another merely reflects the effect of a corresponding change in the impulse response of the channel.

Experiment 2: Effect of Step-size Parameter

For the second part of the experiment, the parameter W in Eq. (5.127) was

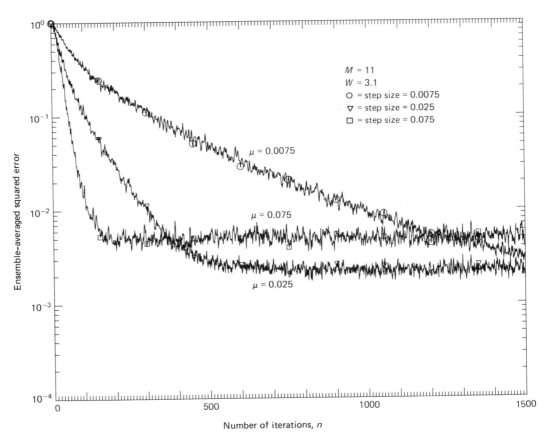

Figure 5.19 Learning curves of the LMS algorithm for adaptive equalizer with number of taps $M = 11$, fixed eigenvalue spread, and varying step–size parameter μ.

fixed at 3.1, yielding an eigenvalue spread of 11.1238 for the correlation matrix of the tap inputs in the equalizer. The step-size parameter μ was this time assigned one of three values: 0.075, 0.025, 0.0075.

Figure 5.19 shows the results of this computation. As before, each learning curve is the result of ensemble averaging the instantaneous squared error "$e^2(n)$ versus n" curve over 200 independent trials of the computer experiment.

The results confirm that the rate of convergence of the adaptive equalizer is highly dependent on the step-size parameter μ. For large step-size parameter ($\mu = 0.075$), the equalizer converged to steady-state conditions in approximately 120 iterations. On the other hand, when μ is small (equal to 0.0075), the rate of convergence slowed down by more than an order of magnitude. The results also show that the steady-state value of the average squared error (and hence the misadjustment) increases, with increasing μ.

5.17 OPERATION OF THE LMS ALGORITHM IN A NONSTATIONARY ENVIRONMENT

The analysis of the LMS algorithm presented previously has been limited in large measure to a stationary environment. We now briefly consider the ability of the LMS algorithm to operate in a nonstationary environment.

Nonstationarity of an environment may arise in practice in one of two ways:

1. The frame of reference provided by the desired response may be time varying. Such a situation arises, for example, in system identification when an adaptive transversal filter is used to model a time-varying system. In this case, the correlation matrix of the tap inputs of the adaptive transversal filter remains fixed (as in a stationary environment), whereas the cross-correlation vector between the tap inputs and the desired response assumes a time-varying form.

2. The stochastic process supplying the tap inputs of the adaptive filter is nonstationary. This situation arises, for example, when an adaptive transversal filter is used to equalize a time-varying channel. In this second case, both the correlation matrix of the tap inputs in the adaptive transversal filter and the cross-correlation vector of the tap inputs and the desired response assume time-varying forms.

In any event, when an adaptive filter operates in a nonstationary environment, the optimum tap-weight vector assumes a time-varying form in that its value changes from one iteration to the next. Then the LMS algorithm has the task of not only seeking the minimum point of the error-performance surface but also tracking the continually changing position of this minimum point.

Let $\mathbf{w}_o(n)$ denote the time-varying optimum tap-weight vector of a transversal filter that operates in a nonstationary environment, where n denotes the iteration number. Application of the LMS algorithm causes the tap-weight vector estimate $\hat{\mathbf{w}}(n)$ of the adaptive filter to attempt to best match the unknown value $\mathbf{w}_o(n)$. At the nth instant, the *tap-weight error vector* may be expressed as

$$\begin{aligned}
\boldsymbol{\varepsilon}(n) &= \hat{\mathbf{w}}(n) - \mathbf{w}_o(n) \\
&= \big(\hat{\mathbf{w}}(n) - E[\hat{\mathbf{w}}(n)]\big) + \big(E[\hat{\mathbf{w}}(n)] - \mathbf{w}_o(n)\big)
\end{aligned} \tag{5.129}$$

where the expectations are averaged over the ensemble. Two components of error are identified in Eq. (5.129):

1. Any difference between the individual tap-weight vectors of the adaptive filter and their ensemble mean, $E[\hat{\mathbf{w}}(n)]$, is due to errors in the estimate used for its gradient vector. This difference is called the *weight vector noise*. In Eq. (5.129), it is represented by the term

$$\boldsymbol{\varepsilon}_1(n) = \hat{\mathbf{w}}(n) - E[\hat{\mathbf{w}}(n)] \tag{5.130}$$

2. Any difference between $E[\hat{\mathbf{w}}(n)]$, the ensemble mean of the tap-weight vector, and $\mathbf{w}_o(n)$, the target value, is due to lag in the adaptive process. This difference is called the *weight vector lag*. In Eq. (5.129), it is represented by the term

$$\boldsymbol{\varepsilon}_2(n) = E[\hat{\mathbf{w}}(n)] - \mathbf{w}_o(n) \tag{5.131}$$

When the LMS algorithm is applied to a stationary environment, $\mathbf{w}_o(n)$ assumes a constant value that equals $E[\hat{\mathbf{w}}(n)]$, with the result that the weight vector lag is zero. It is therefore the presence of the weight vector lag that distinguishes the operation of the LMS algorithm in a nonstationary environment from that in a stationary one.

To evaluate the contribution made by the weight vector lag, we consider an example of time-varying system identification (Widrow and others, 1976). The unknown system and the LMS adaptive filter (used to model the unknown system) are depicted in Fig. 5.20. To be specific, we assume the following:

1. The unknown system consists of a transversal filter whose tap weights (M in number) undergo independent stationary *first-order Markov processes*.[8] That is, the elements of the tap-weight vector $\mathbf{w}_o(n)$ characterizing this filter originate from a corresponding set of independent white-noise excitations (each with zero mean and variance σ^2) applied to a bank of one-pole low-pass digital filters, each with a transfer function equal to $1/(1 - az^{-1})$. The parameter a controls the *time constant of nonstationarity*, which equals $1/(1 - a)$.

2. The tap-input vector $\mathbf{u}(n)$, applied to both the unknown system and the LMS adaptive filter, is stationary with zero mean and correlation matrix $\mathbf{R}$.

3. The output of the unknown system is corrupted by additive white noise $\{v(n)\}$ of zero mean and variance σ_v^2. The presence of this noise prevents a perfect match between the unknown system and the LMS adaptive filter.

The time-varying tap-weight vector of the unknown system, denoted by $\mathbf{w}_o(n)$, represents the "target" to be tracked by the LMS adaptive filter. Whenever $\hat{\mathbf{w}}(n)$, the tap-weight vector of the LMS adaptive filter, equals $\mathbf{w}_o(n)$, the minimum mean-squared error $J_{\min} = \sigma_v^2$.

The desired response $d(n)$ applied to the adaptive filter equals the overall output of the unknown system. Since this system is time-varying, the desired response is correspondingly nonstationary. Accordingly, with the correlation matrix of the tap inputs having the fixed value $\mathbf{R}$, we find that the LMS adaptive filter

---

[8]We say a random vector $\mathbf{w}_o(n)$ is *Markov of first order*, or simply *Markov*, if the conditional joint probability density function of $\mathbf{w}_o(n)$ conditioned on all its past (given) values is the same as using the value in the immediate past. That is, if for every n, the relation

$$f_{\mathbf{w}}(\mathbf{w}_o(n)|\mathbf{w}_o(n - 1), \ldots, \mathbf{w}_o(1)) = f_{\mathbf{w}}(\mathbf{w}_o(n)|\mathbf{w}_o(n-1))$$

holds, the model represented by the random vector $\mathbf{w}_o(n)$ is Markov.

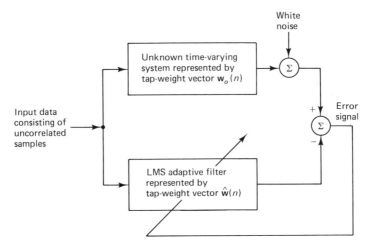

Figure 5.20   Modeling of an unknown system by an LMS adaptive filter

has a quadratic bowl-shaped error-performance surface whose position varies continually with time, while its minimum point, eigenvalues, and eigenvectors all remain fixed.

The average excess mean-squared error equals [see the first line of Eq. (5.80)]

$$E[J_{ex}(n)] = E[\boldsymbol{\epsilon}^H(n)\mathbf{R}\boldsymbol{\epsilon}(n)] \tag{5.132}$$

Therefore, substituting Eqs. (5.129) to (5.131) in (5.132) and expanding, we get

$$\begin{aligned} E[J_{ex}(n)] &= E[(\boldsymbol{\epsilon}_1^H(n) + \boldsymbol{\epsilon}_2^H(n))\mathbf{R}(\boldsymbol{\epsilon}_1(n) + \boldsymbol{\epsilon}_2(n))] \\ &= E[\boldsymbol{\epsilon}_1^H(n)\mathbf{R}\boldsymbol{\epsilon}_1(n)] + E[\boldsymbol{\epsilon}_2^H(n)\mathbf{R}\boldsymbol{\epsilon}_2(n)] \\ &\quad + 2E[\boldsymbol{\epsilon}_1^H(n)\mathbf{R}\boldsymbol{\epsilon}_2(n)] \end{aligned} \tag{5.133}$$

The last term on the right side of Eq. (5.133) equals zero. To show this, we recognize that $\mathbf{w}_o(n)$ is constant over the ensemble and thus write

$$\begin{aligned} E[\boldsymbol{\epsilon}_1^H(n)\mathbf{R}\boldsymbol{\epsilon}_2(n)] &= E[(\hat{\mathbf{w}}(n) - E[\hat{\mathbf{w}}(n)])^H \mathbf{R}(E[\hat{\mathbf{w}}(n)] - \mathbf{w}_o(n))] \\ &= E[\hat{\mathbf{w}}^H(n)]\mathbf{R}E[\hat{\mathbf{w}}(n)] - E[\hat{\mathbf{w}}^H(n)]\mathbf{R}E[\hat{\mathbf{w}}(n)] \\ &\quad - E[\hat{\mathbf{w}}^H(n)]\mathbf{R}\mathbf{w}_o(n) + E[\hat{\mathbf{w}}^H(n)]\mathbf{R}\mathbf{w}_o(n) \\ &= \mathbf{0} \end{aligned}$$

Hence, we may simplify Eq. (5.133) as follows:

$$E[J_{ex}(n)] = E[\boldsymbol{\epsilon}_1^H(n)\mathbf{R}\boldsymbol{\epsilon}_1(n)] + E[\boldsymbol{\epsilon}_2^H(n)\mathbf{R}\boldsymbol{\epsilon}_2(n)] \tag{5.134}$$

To a first order of approximation, we may evaluate the average excess mean-squared error in Eq. (5.134) by considering the contributions made by the weight vector noise $\boldsymbol{\epsilon}_1(n)$ and the weight vector lag $\boldsymbol{\epsilon}_2(n)$, one at a time.

Thus, to evaluate the contribution due to the weight vector noise $\varepsilon_1(n)$, represented by the first term on the right side of Eq. (5.134), we may set $\varepsilon_2(n)$ equal to zero. This is equivalent to putting $E[\hat{\mathbf{w}}(n)] = \mathbf{w}_o$, in which case we find that the expectation of $\varepsilon_1^H(n)\mathbf{R}\varepsilon_1(n)$ equals the ensemble-averaged value of the excess mean-squared error [see Eq. (5.80)]. The important thing to note about this contribution is that it is proportional to the step-size parameter μ for the case of slow adaptation (for which μ is small).

Next, to evaluate the contribution due to the weight vector lag $\varepsilon_2(n)$, represented by the second term on the right side of Eq. (5.134), we may set that $\varepsilon_1(n)$ equal to zero. This is equivalent to putting $E[\hat{\mathbf{w}}(n)] = \hat{\mathbf{w}}(n)$.

In cases of interest, the step-size parameter μ is small and the time constant of nonstationarity is large [i.e., the parameter a in the first-order Markov model for the target weight vector $\mathbf{w}_o(n)$ is close to 1]. Under these conditions, we find that the expectation of $\varepsilon_2^H(n)\mathbf{R}\varepsilon_2(n)$ is inversely proportional to the step-size parameter μ (Widrow and others, 1976). See also Problem 9.

From Eq. (5.134), the average excess mean-squared error is the sum of components due to weight vector noise $\varepsilon_1(n)$ and weight vector lag $\varepsilon_2(n)$. It follows therefore that the total misadjustment of the LMS adaptive filter is likewise the sum of two misadjustment components, one directly proportional to the step-size parameter μ and the other inversely proportional to μ. This suggests that the optimum choice of μ (which results in the minimum value of total misadjustment) occurs when these two contributions to misadjustment are equal. That is, the rate of adaptation is optimized when the loss of performance due to gradient vector lag equals that due to weight vector noise. Figure 5.21 illustrates the trade-offs involved in adjusting the step-size parameter μ for minimization of the total misadjustment due to the combined effects of weight vector noise and weight vector

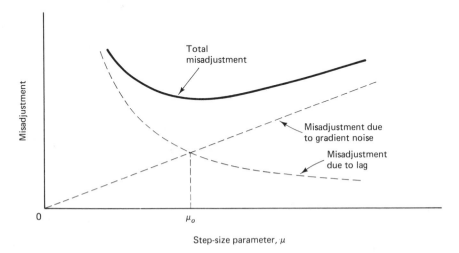

Figure 5.21 Total misadjustment versus step–size parameter, μ.

lag. The optimum value of the step-size parameter, denoted by μ_o, is shown in the figure.

5.18 DIGITAL IMPLEMENTATION OF THE LMS ALGORITHM

A study of the LMS algorithm would be incomplete without some discussion of the effects of *roundoff errors* that arise when the algorithm is implemented *digitally*. It is rather ironic that despite the practical importance of the subject, not that many papers have been written on it. This may not be surprising because mathematical analysis of the problem is indeed difficult. Nevertheless, some significant results have been reported in the literature (Gitlin, Mazo and Taylor, 1973; Gitlin and Weinstein, 1979; Caraiscos and Liu, 1984).

In a digital implementation of the LMS algorithm, the adjustable tap weights as well as the signal levels in the algorithm are *quantized* to within a least significant digit. By so doing, we introduce a new source of error, namely, *quantization*. In general, the presence of quantization precludes the tap weights reaching their optimum Wiener setting. Also, under certain conditions, it can cause a large increase in the total output mean squared error, compared to the pure analog (infinite precision) form of the algorithm.

In this section, we present a brief discussion of the effects of quantization errors using fixed-point arithmetic.[9] Throughout the section, we assume that the data and tap weights are *real valued*.

Mathematical Model of the LMS Algorithm Using Fixed-point Arithmetic

We assume that the input data samples are properly scaled, so that their values lie between -1 and $+1$. Each data sample is represented by B_d bits plus sign, yielding a quantizing interval defined by

$$\delta q = 2^{-B_d} \tag{5.135}$$

Assuming that a quantization error is *uniformly distributed* within a quantizing interval, we find that the quantization error resulting from the digital representation of data samples has the variance

$$\sigma_d^2 = \frac{\delta q^2}{12}$$
$$= \frac{2^{-2B_d}}{12} \tag{5.136}$$

---

[9]The presentation is largely based on the paper by Caraiscos and Liu (1984). In this paper, both the fixed-point and floating-point implementations of the LMS algorithm are considered.

Similarly, each tap weight is represented by B_w bits plus sign. The associated quantizing error will therefore have the variance

$$\sigma_w^2 = \frac{2^{-2B_w}}{12} \qquad (5.137)$$

We assume that during the course of computation no *overflow* occurs. Hence, additions do not introduce any error, while each multiplication introduces an error after the product is quantized.

As a matter of convention, we use unprimed and primed symbols to represent quantities of infinite and finite precision, respectively. Figure 5.22 shows a block diagram of the LMS algorithm using fixed-point arithmetic. The scaling factor a has been introduced to prevent overflow of the elements of the tap-weight vector $\mathbf{w}'(n)$ and the estimate $\hat{y}'(n)$. In particular, for the desired response and its estimate, we write

$$y'(n) = ad'(n) \qquad (5.138)$$

and

$$\hat{d}'(n) = \frac{1}{a} \hat{y}'(n) \qquad (5.139)$$

Multiplication by the factor a is implemented by a right shift by one or more bits, while multiplication by $1/a$ is implemented by a left shift by a similar number of bits. Usually, the required value of a is not expected to be small.

We use the symbol $fx[\ ]$ to denote the fixed-point arithmetic representation

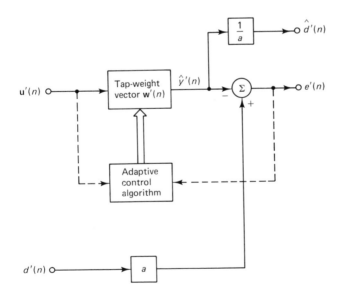

Figure 5.22 Fixed-point arithmetic model of the LMS algorithm for real valued data.

of the quantity contained within.   For the input date we thus write

$$\mathbf{u}'(n) = fx[\mathbf{u}(n)] \qquad (5.140)$$
$$= \mathbf{u}(n) + \delta\mathbf{u}(n)$$

and

$$y'(n) = fx[y(n)] \qquad (5.141)$$
$$= y(n) + \delta y(n)$$

where $\delta\mathbf{u}(n)$ and $\delta y(n)$ are quantization errors.   Similarly, for the tap-weight vector, we write

$$\mathbf{w}'(n) = fx[\mathbf{w}(n)] \qquad (5.142)$$
$$= \mathbf{w}(n) + \delta\mathbf{w}(n)$$

where $\delta\mathbf{w}(n)$ is the quantization error vector.

We may now express the digital (finite precision) form of the LMS algorithm for the case of *real data* as follows

$$e'(n) = y'(n) - fx[\mathbf{u}'^T(n)\,\mathbf{w}'(n)] \qquad (5.143)$$

$$\mathbf{w}'(n+1) = \mathbf{w}'(n) + fx[\mu\mathbf{u}'(n)e'(n)] \qquad (5.144)$$

where $e'(n)$ is the estimation error.   Using Eqs. (5.140) and (5.142), and ignoring all error terms higher than first order, we may express the computed output as

$$\hat{y}'(n) = fx[\mathbf{u}'^T(n)\,\mathbf{w}'(n)]$$
$$= \mathbf{u}^T(n)\,\mathbf{w}(n) + \delta\mathbf{u}^T(n)\,\mathbf{w}(n) + \mathbf{u}^T(n)\,\delta\mathbf{w}(n) + \beta(n) \qquad (5.145)$$

where $\beta(n)$ is the quantization error resulting from computation of the inner product $\mathbf{u}'^T(n)\,\mathbf{w}'(n)$.   The variance of $\beta(n)$ equals $c\sigma_d^2$, where the constant c depends on how this inner product is computed.   If all M scalar products contained in this inner product are computed without quantization, summed, and the final result is then quantized in B_d bits plus sign, the variance of $\beta(n)$ equals σ_d^2 and $c = 1$.   If, on the other hand, the scalar products are quantized individually and then summed, the variance of $\beta(n)$ equals $M\sigma_d^2$ and $c = M$, where M is the number of taps.

Substituting Eqs. (5.141) and (5.145) into (5.143), we get the following result for the estimation error

$$e'(n) = y(n) - \mathbf{u}^T(n)\mathbf{w}(n) - \delta\mathbf{u}^T(n)\mathbf{w}(n) - \mathbf{u}^T(n)\delta\mathbf{w}(n) + \delta y(n) - \beta(n) \qquad (5.146)$$

The *total output error* equals

$$d(n) - \hat{d}'(n) = \frac{1}{a}\,e(n) - \frac{1}{a}\,[\delta\mathbf{u}^T(n)\mathbf{w}(n) + \mathbf{u}^T(n)\delta\mathbf{w}(n) + \beta(n)] \qquad (5.147)$$

where

$$e(n) = y(n) - \mathbf{u}^T(n)\,\mathbf{w}(n) \qquad (5.148)$$

The first term, $e(n)/a$, on the right side of Eq. (5.147) is the estimation error produced by the pure analog (infinite precision) form of the LMS algorithm. The second term, $(1/a)[\delta\mathbf{u}^T(n)\mathbf{w}(n) + \mathbf{u}^T(n)\delta\mathbf{w}(n) + \beta(n)]$ is the accumulated error produced by the use of finite precision arithmetic.

The elements of the quantization error vector $\delta\mathbf{u}(n)$ and the quantization error $\delta y(n)$ are modeled as white noise sequences that are independent of the signals and independent of each other (Oppenheim and Schaefer, 1975). They have zero mean and variance σ_d^2. Likewise, the quantization error $\beta(n)$ is modeled as a white noise sequence that is independent of the signals and independent of the other quantization errors. It has zero mean and variance $c\sigma_d^2$, where c is defined previously.

By invoking the independence theory (see Section 5.8) and the modeling assumptions made above, Caraiscos and Liu show that the terms $\delta\mathbf{u}^T(n)\mathbf{w}(n)$, $\mathbf{u}^T(n)\delta\mathbf{w}(n)$, and $\beta(n)$ are uncorrelated with each other, and that their sum is uncorrelated with $e(n)$ (Caraiscos and Liu, 1984). Accordingly, the two terms on the right side of Eq. (5.147) are uncorrelated.

Summary of the Effects of Roundoff Errors

There are some significant differences between the pure analog (infinite precision) and digital (finite precision) forms of the LMS algorithm. From pure analog considerations, a relatively large step-size parameter μ is needed to accelerate initial convergence, while a small step-size parameter is needed to reduce the excess mean squared error. When, however, the LMS algorithm is implemented digitally, a decrease in the step-size paramter μ can actually degrade performance. In particular, whenever any component of the product $\mu e'(n)\mathbf{u}'(n)$ in the time-update of the tap-weight vector in Eq. (5.144) is less in magnitude than half a tap-weight quantizing interval, the quantized value of that component is set equal to zero and the corresponding tap weight remains unchanged. That is, the algorithm virtually stops making any further adjustment to the kth tap-weight $w_k'(n)$ if the following condition is satisfied

$$|\mu e'(n)\, u'(n - k + 1)| < 2^{-B_w - 1}, \qquad k = 1, 2, \ldots, M \qquad (5.149)$$

where $u'(n - k + 1)$ is the kth element of the quantized tap-input vector $\mathbf{u}'(n)$. A surprising result is that when adaptation is terminated by quantization effects, the total output mean-squared error can be decreased by *increasing* the step-size parameter (Gitlin, Mazo and Taylor, 1973).

To allow the LMS algorithm to converge completely, the step-size parameter μ has to be chosen large enough to make every component of the product term $ue'(n)\, \mathbf{u}'(n)$ *greater* in magnitude than half a tap-weight quantizing interval for all n. Under this condition, we may use Eq. (5.147) to evaluate the average mean-square value of the total output error. Previously, we indicated that the two terms on the right side of this equation are uncorrelated. To carry out a mean-square error analysis, we may thus consider the two terms separately.

The first term, $e(n)/a$, represents the estimation error assuming infinite precision. We may therefore use results of the analysis in Section 5.11. In particular, the steady-state mean-square value of this error equals the minimum mean-squared error J_{min} plus the steady-state value of the excess mean squared error, $J_{ex}(\infty)$.

The second term, $(1/a) [\delta\mathbf{u}^T(n)\mathbf{w}(n) + \mathbf{u}^T(n)\delta\mathbf{w}(n) + \beta(n)]$, represents the (accumulated) arithmetic error incurred by using fixed-point arithmetic to implement the LMS algorithm. Recognizing that the three components that constitute this term are uncorrelated, we may evaluate its mean-square value by considering the three components, one at a time. The important point that emerges from the evaluation is that the mean-square value of the arithmetic error is, approximately, inversely proportioned to the step-size parameter μ (Caraiscos and Liu, 1984). Accordingly, if we were to use a very small value for μ, in order to reduce the excess mean-squared error, it may result in a considerable arithmetic error. The excess mean-squared error is larger than the mean-square value of the arithmetic error provided that μ is assigned a value larger than that for which adaptation is terminated due to quantization effects. A practical solution for combatting the arithmetic error is to use more bits for the tap weights than for the data (Caraiscos and Liu, 1984; Gitlin and Weinstein, 1979).

PROBLEMS

1. Start with the formula for the estimation error:

$$e(n) = d(n) - \mathbf{w}^H(n)\mathbf{u}(n)$$

where $d(n)$ is the desired response, $\mathbf{u}(n)$ is the tap-input vector, and $\mathbf{w}(n)$ is the tap-weight vector in the transversal filter. Hence, show that the gradient of the instantaneous squared error $|e(n)|^2$ equals

$$\hat{\mathbf{V}}(n) = -2\mathbf{u}(n)d^*(n) + 2\mathbf{u}(n)\mathbf{u}^H(n)\mathbf{w}(n)$$

2. In this problem we explore another way of deriving the steepest-descent algorithm of Eq. (5.9) used to adjust the tap-weight vector in a transversal filter. The inverse of a positive-definite matrix may be expanded in a series as follows:

$$\mathbf{R}^{-1} = \mu \sum_{k=0}^{\infty} (\mathbf{I} - \mu\mathbf{R})^k$$

where $\mathbf{I}$ is the identity matrix, and μ is a positive constant. To ensure convergence of the series, the constant μ must lie inside the range

$$0 < \mu < \frac{2}{\lambda_{max}}$$

where λ_{max} is the largest eigenvalue of the matrix $\mathbf{R}$. By using this series expansion for the inverse of the correlation matrix in the normal equation, develop the recursion

$$\mathbf{w}(n + 1) = \mathbf{w}(n) + \mu[\mathbf{p} - \mathbf{R}\mathbf{w}(n)]$$

where $\mathbf{w}(n)$ is the approximation to the Wiener solution for the tap-weight vector:

$$\mathbf{w}(n) = \mu \sum_{k=0}^{n-1} (\mathbf{I} - \mu \mathbf{R})^k \mathbf{p}$$

3. The zero-mean output $d(n)$ of an unknown real-valued system is represented by the *multiple linear regression model*

$$d(n) = \mathbf{w}_o^T \mathbf{u}(n) + v(n)$$

where $\mathbf{w}_o$ is the (unknown) parameter vector of the model, $\mathbf{u}(n)$ is the input vector, and $v(n)$ is the sample value of an immeasurable white-noise process of zero mean and variance σ_v^2. The block diagram of Fig. P5.1 shows the adaptive modeling of the unknown system, in which the adaptive transversal filter is controlled by a *modified* version of the LMS algorithm. In particular, the tap-weight vector $\mathbf{w}(n)$ of the transversal filter is chosen to minimize the index of performance

$$J(\mathbf{w}, K) = E[e^{2K}(n)]$$

for $K = 1, 2, 3, \ldots$.

(a) By using the instantaneous gradient vector, show that the new adaptation rule for the corresponding estimate of the tap-weight vector is

$$\hat{\mathbf{w}}(n + 1) = \hat{\mathbf{w}}(n) + \mu K \mathbf{u}(n)e(n)$$

where μ is the step-size parameter, and $e(n)$ is the estimation error

$$e(n) = d(n) - \hat{\mathbf{w}}^T(n)\, \mathbf{u}(n)$$

(b) Assume that the weight-error vector

$$\boldsymbol{\varepsilon}(n) = \hat{\mathbf{w}}(n) - \mathbf{w}_o$$

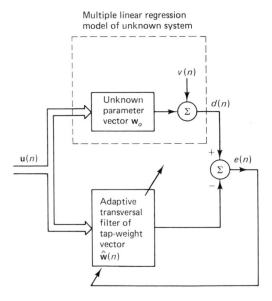

Multiple linear regression
model of unknown system

$v(n)$

Unknown parameter vector $\mathbf{w}_o$

$d(n)$

$u(n)$

$e(n)$

Adaptive transversal filter of tap-weight vector $\hat{\mathbf{w}}(n)$

Figure P5.1

is close to zero, and that $v(n)$ is independent of $\mathbf{u}(n)$. Hence, show that

$$E[\boldsymbol{\varepsilon}(n + 1)] = \left(\mathbf{I} - \mu K(2K - 1) E[v^{2K-2}(n)]\mathbf{R}\right) E[\boldsymbol{\varepsilon}(n)]$$

where $\mathbf{R}$ is the correlation matrix of the input vector $\mathbf{u}(n)$.

(c) Show that the modified LMS algorithm described in part (a) converges in the mean if the step-size parameter μ satisfies the condition

$$0 < \mu < \frac{1}{K(2K - 1)E[v^{2(K-1)}(n)]\lambda_{max}}$$

where λ_{max} is the largest eigenvalue of matrix $\mathbf{R}$.

(d) For $K = 1$, show that the results given in parts (a), (b), and (c) reduce to those in the conventional LMS algorithm.

4. Show that the misadjustment produced by LMS algorithm is approximately given by

$$\mathcal{M} \simeq \frac{M}{4(\tau_{mse})_{av}}$$

where M is the number of taps in the transversal filter, and $(\tau_{mse})_{av}$ is the time constant of a single decaying exponential that approximates the learning curve of the LMS algorithm. Assume that the step-size parameter μ is small.

5. *The Leaky LMS Algorithm:* The time-update for the tap-weight vector $\hat{\mathbf{w}}(n)$ in the *leaky LMS algorithm* is defined by

$$\hat{\mathbf{w}}(n + 1) = \beta\hat{\mathbf{w}}(n) + \mu\mathbf{u}(n) \, e^*(n)$$

where $\mathbf{u}(n)$ is the tap-input vector and μ is the step-size parameter. The estimation error $e(n)$, as in the conventional LMS algorithm, is defined by

$$e(n) = d(n) - \hat{\mathbf{w}}^H(n)\mathbf{u}(n)$$

where $d(n)$ is the desired response The parameter β is a positive constant, limited in value as follows

$$0 < \beta \le 1$$

(a) Using the independence theory, show that

$$\lim_{n\to\infty} E[\hat{\mathbf{w}}(n)] = \left(\mathbf{R} + \frac{1 - \beta}{\mu} \mathbf{I}\right)^{-1}\mathbf{p}$$

where $\mathbf{R}$ is the correlation matrix of the tap inputs and $\mathbf{p}$ is the cross-correlation vector between the tap inputs and the desired response. What is the condition for the algorithm to converge in the mean?

(b) How would you modify the tap-input vector in the conventional LMS algorithm to get the equivalent result described in part (a)?

6. The *normalized LMS algorithm* is described by the following recursion for the tap-weight vector:

$$\hat{\mathbf{w}}(n + 1) = \hat{\mathbf{w}}(n) + \frac{\alpha}{\|\mathbf{u}(n)\|^2} \mathbf{u}(n) \, e^*(n)$$

where α is a positive constant, and $\|\mathbf{u}(n)\|$ is the *norm* of the tap-input vector. The estimation error $e(n)$ is defined by

$$e(n) = d(n) - \mathbf{w}^H(n)\mathbf{u}(n)$$

where $d(n)$ is the desired response.

Using results of the independence theory, show that the necessary and sufficient condition for the normalized LMS algorithm to be convergent in the mean square is $0 < \alpha < 2$.

Hint: You may use the following approximation:

$$E\left[\frac{\mathbf{u}(n)\mathbf{u}^H(n)}{\|\mathbf{u}(n)\|^2}\right] \approx \frac{E[\mathbf{u}(n)\mathbf{u}^H(n)]}{E[\|\mathbf{u}(n)\|^2]}$$

7. **(a)** Let $\mathbf{K}(n)$ denote the correlation matrix of the weight-error vector $\boldsymbol{\varepsilon}(n)$. Show that the trace of $\mathbf{K}(n)$ equals the expected value of the squared norm of $\boldsymbol{\varepsilon}(n)$.

   (b) Using the result of part (a) and Eq. (5.74) that is based on the independence theory, develop a difference equation that describes the time evolution of $E[\|\boldsymbol{\varepsilon}(n)\|^2]$. Hence, show that:

   (i) The condition for $E[\|\boldsymbol{\varepsilon}(n)\|^2]$ to be convergent is the same as that for the convergence of the average mean-squared error, $E[J(n)]$.

   (ii) The presence of the minimum mean-squared error, $J_{\min}$, prevents the estimate of the tap-weight vector, $\hat{\mathbf{w}}(n)$, from converging to the optimum Wiener solution $\mathbf{w}_o$ in the mean square.

   (c) The convergence ratio, $\mathscr{C}(n)$, of an adaptive algorithm is defined by

   $$\mathscr{C}(n) = \frac{E[\|\boldsymbol{\varepsilon}(n+1)\|^2]}{E[\|\boldsymbol{\varepsilon}(n)\|^2]}$$

   Show that for small n, the convergence ratio of the LMS algorithm for stationary inputs approximately equals

   $$\mathscr{C}(n) \approx 1 - 2\mu\sigma_u^2 + \mu^2 M\sigma_u^2, \qquad n \text{ small}$$

   where μ is the step-size parameter, σ_u^2 is the variance of the tap input, and M is the number of taps in the filter. Hence, find the value of μ for which this convergence ratio is optimum.

8. *Gradient adaptive lattice algorithm:* In the *gradient adaptive lattice* (GAL) algorithm the estimate of the reflection coefficient $\Gamma_m(n)$ for stage m of a lattice predictor, operating on stationary inputs, is updated as follows

$$\hat{\Gamma}_m(n+1) = \hat{\Gamma}_m(n) - \frac{1}{2}\mu_m(n)\hat{\nabla}_m(n)$$

where $\mu_m(n)$ is a time-varying step-size parameter. The gradient estimate $\hat{\nabla}_m(n)$ is obtained by substituting instantaneous values for expectations in the gradient

$$\nabla_m(n) = \frac{\partial J_m(n)}{\partial \Gamma_m(n)}$$

where the index of performance $J_m(n)$ is defined by Eq. (4.188). To maintain the same adaptive time constant and misadjustment for each stage of the lattice predictor, the step-size parameter $\mu_m(n)$ is chosen as

$$\mu_m(n) = \frac{1}{\sum_{i=1}^{n} |f_{m-1}(i)|^2 + |b_{m-1}(i-1)|^2}$$

(a) Show that

$$\hat{\Gamma}_m(n+1) = \hat{\Gamma}_m(n) - \mu_m(n)\{\hat{\Gamma}_m(n)[|f_{m-1}(n)|^2 + |b_{m-1}(n-1)|^2] + 2b_{m-1}(n-1)f^*_{m-1}(n)\}$$

where $f_{m-1}(n)$ is the forward prediction error and $b_{m-1}(n-1)$ is the delayed backward prediction error at the input of stage m.

(b) Show that the GAL algorithm is inherently stable.

9. In this problem we continue with the example of time-varying system identification considered in Section 5.17. Assume that the time evolution of the target value $\mathbf{w}_o(n)$ of the tap-weight vector is governed by the difference equation

$$\mathbf{w}_o(n) = a\mathbf{w}_o(n-1) + \mathbf{v}(n)$$

where a is a constant, and the vector $\mathbf{v}(n)$ consists of independent white noise excitations each of which has zero mean and variance σ^2. For the case when the step-size parameter μ is small and the constant a is close to 1, show that the expectation of the Hermitian form $\varepsilon_2^H(n)\mathbf{R}\varepsilon_2(n)$ is approximately equal to $M\sigma^2/2\mu$, wherre M is the number of taps in the filter. The matrix $\mathbf{R}$ is the correlation matrix of the tap-input vector and the weight vector lag $\varepsilon_2(n)$ is defined by

$$\varepsilon_2(n) = \hat{\mathbf{w}}(n) - \mathbf{w}_o(n)$$

Hints: (1) In the absence of gradient vector noise, the tap-weight adaptation in the LMS algorithm takes on the same form as that in the method of steepest descent.

(2) Use z-transform notation to define the z-transform of the kth element of $\mathbf{Q}^H\varepsilon_2(n)$ in terms of the z-transform of the kth element of $\mathbf{Q}^H\mathbf{v}(n)$, where $\mathbf{Q}$ has for its columns the eigenvectors of $\mathbf{R}$.

10. Consider the digital implementation of the LMS algorithm using fixed-point arithmetic, as discussed in Section 5.18. Show that the M-by-1 error vector $\delta\mathbf{w}(n)$, incurred by quantizing the tap-weight vector, may be updated as follows:

$$\delta\mathbf{w}(n+1) = \mathbf{F}(n)\,\delta\mathbf{w}(n) + \mathbf{t}(n), \qquad n = 0, 1, 2, \ldots$$

where $\mathbf{F}(n)$ is an M-by-M matrix and $\mathbf{t}(n)$ is an M-by-1 vector. Hence, define $\mathbf{F}(n)$ and $\mathbf{t}(n)$. Base your analysis on real-valued data.

Computer-oriented Problems

11. An autoregressive (AR) process $\{u(n)\}$ of order 1 is described by the difference equation

$$u(n) + a_1 u(n-1) = v(n)$$

where $\{v(n)\}$ is a white-noise process of zero mean and variance σ_v^2. The AR parameter

a_1 and variance σ_v^2 are assigned one of the following four sets of values:

$$\text{(i)} \quad a_1 = -0.1000, \qquad \sigma_v^2 = 0.9900$$

$$\text{(ii)} \quad a_1 = -0.5000, \qquad \sigma_v^2 = 0.7500$$

$$\text{(iii)} \quad a_1 = -0.8182, \qquad \sigma_v^2 = 0.3305$$

$$\text{(iv)} \quad a_1 = -0.9802, \qquad \sigma_v^2 = 0.0392$$

The AR process $\{u(n)\}$ is applied to a two-tap linear predictor.

(a) Calculate the eigenvalues of the 2-by-2 correlation matrix of the tap inputs of the predictor for each parameter set (i) through (iv). Hence, determine the corresponding permissible ranges for the step-size parameter μ in the steepest-descent algorithm.

(b) Using the two elements of the transformed tap-weight error vector [i.e., $v_1(n)$ and $v_2(n)$] as the variables in the steepest-descent algorithm, construct loci for constant values of mean-squared error for each of the parameter sets (i), (ii), and (iii). Superimpose on each of the three figures the trajectory that describes the change in the coordinates $v_1(n)$ and $v_2(n)$ with time n, assuming that the step-size parameter $\mu = 0.3$ and that the initial values of the two tap weights in the predictor both equal zero.

(c) Repeat the computations in part (b), using the tap weights $w_1(n)$ and $w_2(n)$ themselves as the variables in the steepest-descent algorithm.

(d) Plot the learning curve [i.e., the mean-squared error $J(n)$ versus time n] for each of the four parameters sets (i) through (iv). In each case, repeat the computations for the following step-size parameters:

$$\mu = 0.02$$

$$\mu = 0.05$$

$$\mu = 0.2$$

(e) Discuss the implications of your results.

12. An autoregressive (AR) process $\{u(n)\}$ of order 2 is described by the difference equation

$$u(n) + a_1 u(n-1) + a_2 u(n-2) = v(n)$$

where $\{v(n)\}$ is a white-noise process of zero mean and variance σ_v^2. The parameters a_1 and a_2 and the variance σ_v^2 are assigned one of the following four sets of values:

$$\text{(i)} \quad a_1 = -0.10025, \qquad a_2 = 0.0025, \qquad \sigma_v^2 = 0.9900$$

$$\text{(ii)} \quad a_1 = -0.5359, \qquad a_2 = 0.0718, \qquad \sigma_v^2 = 0.7461$$

$$\text{(iii)} \quad a_1 = -1.0390, \qquad a_2 = 0.2699, \qquad \sigma_v^2 = 0.3065$$

$$\text{(iv)} \quad a_1 = -1.6364, \qquad a_2 = 0.6694, \qquad \sigma_v^2 = 0.0216$$

The AR process $\{u(n)\}$ is applied to a two-tap linear predictor. Repeat the computations made in parts (a) to (d) of Problem 11, and discuss the implications of your results.

13. *Computer experiment on adaptive two-tap predictor using the LMS algorithm, with autoregressive input of order 1:* An autoregressive process $u(n)$ of order 1 is described by the difference equation

$$u(n) + a_1 u(n - 1) = v(n)$$

where $\{v(n)\}$ is a white-noise process of zero mean and variance σ_v^2. The AR parameter a_1 and the variance σ_v^2 are assigned one of the following two sets of values:

(i) $a_1 = -0.8182, \qquad \sigma_v^2 = 0.3305$

(ii) $a_1 = -0.9802, \qquad \sigma_v^2 = 0.0392$

The AR process $\{u(n)\}$ is applied to the input of an adaptive two-tap predictor that uses the LMS algorithm.

(a) What is the permissible range of values for the step-size parameter μ for the LMS algorithm to be convergent in the mean?

(b) What is the permissible range of values for the step-size parameter μ for the LMS algorithm to be convergent in the mean square?

(c) Use the computer to generate a 256-sample sequence representing the auto-regressive process $\{u(n)\}$ for parameter set (i). You may do this by using a random-number subroutine for generating the white noise $\{v(n)\}$. Hence, by averaging over 200 independent trials of the experiement, plot the learning curves of the LMS algorithm for the following two values of step-size parameter: $\mu = 0.05$ and $\mu = 0.005$. Estiamte the corresponding values of the misadjustment by time-averaging over the last 200 iterations of the ensemble-averaged learning curves. Compare the values thus estimated with theory.

(d) For parameter set (i), also estimate mean values for the tap weights $\hat{w}_1(\infty)$ and $\hat{w}_2(\infty)$ that result from the LMS algorithm for two values fo the step-size parameter: $\mu = 0.05$ and $\mu = 0.005$. You may do this by averaging the steady-state values of the tap weights (obtained for the last iteration) over 200 independent trials of the experiment. Compare the mean values thus obtained with theory.

(e) Repeat experiments (c) and (d) for parameter set (ii).

(f) Discuss the implications of your results.

14. *Computer experiment on adaptive two-tap predictor using the LMS algorithm, with autoregressive input of order 2:* An autoregressive (AR) process $\{u(n)\}$ of order 2 is described by the difference equation

$$u(n) + a_1 u(n - 1) + a_2 u(n - 2) = v(n)$$

where $\{v(n)\}$ is a white-noise process of zero mean and variance σ_v^2. The AR parameters, a_1 and a_2 and the variance σ_v^2 are assigned one of the following three sets of values:

(i) $a_1 = -0.10, \qquad a_2 = -0.8, \qquad \sigma_v^2 = 0.2700$

(ii) $a_1 = -0.1636, \qquad a_2 = -0.8, \qquad \sigma_v^2 = 0.1190$

(iii) $a_1 = -0.1960, \qquad a_2 = -0.8, \qquad \sigma_v^2 = 0.0140$

The AR process $\{u(n)\}$ is applied to an adaptive two-tap predictor whose tap weights are adjusted in accordance with the LMS algorithm.

Repeat the computations under parts (a) and (b) and the computer experiments under parts (c) and (d) of Problem 13, and discuss the implications of your results.

15. *Computer experiment on adaptive 2-tap predictor using the clipped LMS algorithm, with autoregressive input of order two*: In the *clipped LMS algorithm* (for *real data*), the tap-input vector $\mathbf{u}(n)$ in the correction term of the update for the tap-weight $\hat{\mathbf{w}}(n)$ is replaced by sgn[$\mathbf{u}(n)$]. The update is thus written as

$$\hat{\mathbf{w}}(n+1) = \hat{\mathbf{w}}(n) + \mu e(n)\ sgn[\mathbf{u}(n)]$$

As before, μ is the step-size parameter and the estimation error

$$e(n) = d(n) - \mathbf{u}^T(n)\ \hat{\mathbf{w}}(n)$$

where $d(n)$ is the desired response. The signum function is defined as

$$sgn[x] = \begin{cases} 1, & \text{if } x > 0 \\ -1, & \text{if } x < 0 \end{cases}$$

The purpose of the clipping is to simplify the implementation of the LMS algorithm without seriously affecting its performance.

Repeat the computer experiments under part (d) of Problems 13 and 14 that deal with parameter set (ii) and parameter set (iii) for the first-order and second-order AR processes, respectively. Hence, compare your results with the corresponding ones obtained for Problems 13 and 14, and comment.

16. *Computer experiment on adaptive antenna using the LMS algorithm, with a single source of noncoherent interference*: Consider an array antenna tht consists of $2N + 1$ uniformly spaced elements. Two plane waves are incident upon the array, one representing a target signal and the other representing a source of interference that is noncoherent with the target signal. The target signal and the interference arrive along directions θ_1 and θ_2 measured with respect to the normal to the array, respectively. The element signals of the array antenna are expressed in baseband form as follows

$$u(n) = A_1 e^{jn\phi_1} + A_2 e^{j(n\phi_2 + \psi)} + v(n),$$

$$n = 0, \pm 1, \ldots, \pm N$$

where A_1 is fixed amplitude of the target signal, and A_2 is fixed amplitude of the interference. The electrical angles ϕ_1 and ϕ_2 are related to the individual angles of arrival θ_1 and θ_2, respectively, by

$$\phi_i = \frac{2\pi}{\lambda}\ d\ \sin\theta_i, \qquad i = 1, 2$$

where d is the element-to-element spacing and λ is the wavelength. The angle ψ represents the phase difference between the target signal and the interference, measured at the center of the array. When the target signal and the interference are noncoherent, the phase difference ψ is a uniformly distributed random variable with the probability density function:

$$f_\Psi(\psi) = \begin{cases} \dfrac{1}{2\pi}, & \pi \leq \psi \leq -\pi \\ 0, & \text{otherwise} \end{cases}$$

Lastly, the additive element noise $v(n)$ is a *complex-valued* Gaussian random variable with zero mean and unit variance. The noise contributions of the individual elements are uncorrelated, as shown by

$$E[v(k)\, v^*(n)] = \begin{cases} 1, & k = n \\ 0, & k \neq n \end{cases}$$

The element-to-element spacing d equals one-half wavelength. The array contains a total of $2N + 1 = 5$ elements. The element outputs of the array are applied to a beamforming network that produces a set of four orthogonal beams that correspond to the uniform phase factor $\alpha = \pm\, \pi/5,\ \pm\, 3\pi/5$, as described in Section 1.4. The beam corresponding to $\alpha = \pi/5$ is chosen as the primary beam, and the remaining three are used as the auxiliary beams whose outputs are adaptively adjusted by means of the LMS algorithm.

The target signal is characterized by the parameters:

$$\text{angle of arrival, } \theta_1 = \sin^{-1}(0.2)$$

$$\text{signal-to-noise ratio, SNR} = 10 \text{ dB}$$

The interference is characterized by the parameters:

$$\text{angle of arrival, } \theta_2 = -\sin^{-1}(0.4)$$

$$\text{interference-to-noise ratio, INR} = 40 \text{ dB}$$

The LMS algorithm is started with the initial values of the weights set equal to zero. The step-size parameter $\mu = 4 \times 10^{-7}$. Each iteration of the algorithm corresponds to one snapshot of data, with each snapshot being independent of another. For the various parameters as specified above, do the following:

(a) Justify the admissibility of the value 4×10^{-7} for the step-size parameter μ.

(b) Plot the adapted antenna pattern after 50 iterations of the LMS algorithm.

(c) Plot the adapted antenna pattern after 200 iterations of the LMS algorithm.

Suppose next that the source of interference is reduced in strength, but with its angular position remaining the same as before. To investigate this new effect, carry out the following experiments:

(d) With the interference-to-noise ratio reduced to 30 dB, plot the adapted antenna pattern after 200 iterations of the LMS algorithm.

(e) Repeat the experiment with the interference-to-noise ratio reduced further to 20 dB.

Comment on the results of your experiments.

17. *Computer experiment on adaptive antenna using the LMS algorithm, with two noncoherent sources of interference*: The adaptive antenna used in this experiment has the same configuration as that used in Problem 16. It uses four orthogonal beams corresponding to the uniform phase factor $\alpha = \pm\pi/5,\ \pm 3\pi/5$, with the one represented by $\alpha = \pi/5$ serving as the primary beam and the remaining three as the auxiliary beams. The adaptive antenna uses the LMS algorithm with the step-size parameter $\mu = 4 \times 10^{-7}$. Plot the adapted antenna pattern after 200 iterations of the LMS algorithm using independent snapshots of data for the following environmental conditions:

(a) The environment includes a target signal and two sources of interference, all three of which are noncoherent with each other. The target signal is characterized by

$$\text{angle of arrival, } \theta_1 = \sin^{-1}(0.2)$$

$$\text{signal-to-noise ratio, SNR} = 10 \text{ dB}$$

The first source of interference is characterized by

$$\text{angle of arrival, } \theta_2 = -\sin^{-1}(0.4)$$

$$\text{interference-to-noise ratio, } (\text{INR})_1 = 40 \text{ dB}$$

The second source of interference is characterized by

$$\text{angle of arrival, } \theta_3 = \sin^{-1}(0.8)$$

$$\text{signal-to-interference ratio, } (\text{INR})_2 = 40 \text{ dB}$$

That is, both interferences lie outside the main beam. The additive complex-valued noise at the output of each element in the array is characterized in the same way as in Problem 16 in that it is Gaussian with zero mean and unit variance and it is uncorrelated with the noise at all other elements.

(b) The environmental conditions are the same as in part (a), except that the first source of interference arrives along the angle $\theta_2 = 0°$, that is, inside the main beam of the antenna.

(c) The target signal arrives along the angle $\theta_1 = \sin^{-1}(0.1)$, that is, the adaptive antenna suffers from pointing inaccuracy in that the main beam (represented by $\alpha = \pi/5$) points along a direction slightly different from the target. Otherwise, the environmental conditions are the same as in part (a).

Comment on your results.

18. *Computer experiment on adaptive antenna using the LMS algorithm with two coherent sources*: Here again the adaptive antenna uses the same configuration as that in Problem 16 in that it has four beams represented by $\alpha = \pm \pi/5, \pm 3\pi/5$, with that corresponding to $\alpha = \pi/5$ acting as the primary beam and the remaining three acting as auxiliary beams. In this problem, however, the two external sources responsible for illuminating the array are coherent. In particular, in the formula that defines the element signal $u(n)$ the phase difference ψ equals zero.

Plot the adapted antenna pattern after 200 iterations of the LMS algorithm with step-size parameter $\mu = 4 \times 10^{-7}$ and for the following three different environmental conditions:

(a) The target signal arrives along $\theta_1 = \sin^{-1}(0.2)$ and the coherent interference arrives along $\theta_2 = -\sin^{-1}(0.4)$. Both are of equal strength, with SNR = 30 dB.

(b) The target signal and the coherent interference are positioned symmetrically with respect to the main beam, with $\theta_1 = -\sin^{-1}(0.4)$ and $\theta_2 = \sin^{-1}(0.8)$. They are of equal strength with SNR = 30 dB.

As in Problem 16, the additive complex-valued noise at each element of the array is Gaussian with zero mean and unit variance, and it is noncoherent with the noise at other elements of the array.

Comment on the results of your experiment.

6

Kalman Filter Theory and Its Application to Adaptive Transversal Filters

6.1 INTRODUCTION

A limitation of the LMS algorithm is that it does not make full use of all the information available to it at the time of adaptation, with the result that its rate of convergence is relatively slow. One method of overcoming this limitation is to use *Kalman filter theory*, which provides the solution to a class of recursive minimum mean-square estimation problems (Kalman, 1960). The Kalman filter is formulated using the *state-space approach*, in which a dynamical system is described by a set of variables called the *state*. The state contains all the necessary information about the behavior of the system such that, given the present and future values of the input, we may compute the future state and output of the system (Kailath, 1980). The significance of Kalman filter theory will be fully appreciated when we apply it to formulate adaptive filtering algorithms for stationary or nonstationary environments later in the chapter.

From Chapter 3, we recall that Wiener filter theory requires the inversion of the correlation matrix of the tap-input vector of a transversal filter for its solution. On the other hand, we find that the application of Kalman filter theory results in a set of difference equations, the solutions of which can be obtained *recursively*. In particular, we find that each updated estimate can be computed from the previous estimate and the new input data, so only the previous estimate must be stored. In addition to eliminating the need for storing the entire past observed data, the Kalman filtering algorithm is computationally more efficient than computing the estimates directly from the entire past observed data at each step. The Kalman filter is thus ideally suited for implementation on a digital computer.

We begin the study of recursive minimum mean-square estimation by considering the simple case of scalar random variables. For this development, we use the *innovations approach* (Kailath, 1968, 1970), which utilizes the correlation properties of the *innovations process*. The idea of innovations was perhaps first used by Kolmogorov (1939).

6.2 RECURSIVE MINIMUM MEAN-SQUARE ESTIMATION FOR SCALAR RANDOM VARIABLES

Let us assume that, based on a complete set of observed random variables $y(1)$, $y(2), \ldots, y(n - 1)$, starting with the first observation at time 1 and extending up to and including time $n - 1$, we have found the minimum mean-square estimate $\hat{x}(n - 1|\mathcal{Y}_{n-1})$ of a related random variable $x(n - 1)$. We are assuming that the observation at (or before) $n = 0$ is zero. The space spanned by the observations $y(1), \ldots, y(n - 1)$ is denoted by $\mathcal{Y}_{n-1}$. Suppose we now have an additional observation $y(n)$ at time n, and the requirement is to compute an *updated* estimate $\hat{x}(n|\mathcal{Y}_n)$ of the related random variable $x(n)$, where $\mathcal{Y}_n$ denotes the space spanned by $y(1), \ldots, y(n)$. We may do this computation by storing the *past* observations, $y(1), y(2), \ldots, y(n - 1)$, and then redoing the whole problem with the available data $y(1), y(2), \ldots, y(n - 1), y(n)$, including the new observation. Computationally, however, it is much more efficient to use a *recursive estimation procedure*. In this procedure we *store* the previous estimate $\hat{x}(n - 1|\mathcal{Y}_{n-1})$ and exploit it to compute the updated estimate $\hat{x}(n|\mathcal{Y}_n)$ in the light of the new observation $y(n)$. There are several ways of developing the algorithm to do this recursive estimation. We will use the notion of *innovations* (Kailath, 1968, 1970).

Define the forward prediction error

$$f_{n-1}(n) = y(n) - \hat{y}(n|\mathcal{Y}_{n-1}) \qquad n = 1, 2, \ldots \qquad (6.1)$$

where $\hat{y}(n|\mathcal{Y}_{n-1})$ is the *one-step prediction* of the observed random variable $y(n)$ at time n, using *all* past observations available up to and including time $n - 1$. The past observations used in this estimation are $y(1), y(2), \ldots, y(n - 1)$, so the order of the prediction equals $n - 1$. We may view $f_{n-1}(n)$ as the output of a forward prediction-error filter of order $n - 1$, and with the filter input fed by the time series $y(1), y(2), \ldots, y(n)$. Note that the *prediction order $n - 1$ increases linearly with n*. According to the principle of orthogonality, the prediction error $f_{n-1}(n)$ is orthogonal to all past observations $y(1), y(2), \ldots, y(n - 1)$ and may therefore be regarded as a *measure* of the new information in the random variable $y(n)$ observed at time n; hence, the name "innovation." The fact is that the observation $y(n)$ does not itself convey completely new information, since the predictable part, $\hat{y}(n|\mathcal{Y}_{n-1})$, is already completely determined by the past observations $y(1), y(2), \ldots, y(n-1)$. Rather, the part of the observation $y(n)$ that is really new is contained in the forward prediction error $f_{n-1}(n)$. We may therefore refer to this prediction error as the innovation, and for simplicity of notation write

$$\alpha(n) = f_{n-1}(n), \qquad n = 1, 2, \ldots \tag{6.2}$$

The innovation $\alpha(n)$ has several important properties, as follows:

Property 1. *The innovation $\alpha(n)$, associated with the observed random variable $y(n)$, is orthogonal to the past observations $y(1), y(2), \ldots, y(n-1)$, as shown by*

$$E[\alpha(n)y^*(k)] = 0, \qquad 1 \le k \le n - 1 \tag{6.3}$$

This is simply a restatement of the principle of orthogonality.

Property 2. *The innovations $\alpha(1), \alpha(2), \ldots, \alpha(n)$ are orthogonal to each other, as shown by*

$$E[\alpha(n)\alpha^*(k)] = 0, \qquad 1 \le k \le n - 1 \tag{6.4}$$

This is a restatement of the fact that [see Eq. (4.137)]

$$E[f_{n-1}(n)f^*_{k-1}(k)] = 0, \qquad 1 \le k \le n - 1$$

Property 3. *There is a one-to-one correspondence between the observed data $\{y(1), y(2), \ldots, y(n)\}$ and the innovations $\{\alpha(1), \alpha(2), \ldots, \alpha(n)\}$ in that the one sequence may be obtained from the other without any loss of information. We may thus write*

$$\{y(1), y(2), \ldots, y(n)\} \leftrightarrows \{\alpha(1), \alpha(2), \ldots, \alpha(n)\} \tag{6.5}$$

To prove this property, we use a form of the Gram–Schmidt orthogonalization procedure (described in Chapter 4). The procedure assumes that the observations $y(1), y(2), \ldots, y(n)$ are linearly independent in an algebraic sense. We first put

$$\alpha(1) = y(1) \tag{6.6}$$

where it is assumed that $\hat{y}(1|\mathcal{Y}_0)$ is zero. Next, we put

$$\alpha(2) = y(2) + a_{1,1}y(1) \tag{6.7}$$

The coefficient $a_{1,1}$ is chosen such that the innovations $\alpha(1)$ and $\alpha(2)$ are orthogonal, as shown by

$$E[\alpha(2)\alpha^*(1)] = 0 \tag{6.8}$$

This requirement is satisfied by choosing

$$a_{1,1} = -\frac{E[y(2)y^*(1)]}{E[y(1)y^*(1)]} \tag{6.9}$$

Hence, except for a minus sign, $a_{1,1}$ is a partial correlation coefficient in that it equals the cross-correlation between the observations $y(2)$ and $y(1)$, normalized with respect to the mean-square value of $y(1)$.

Next we put

$$\alpha(3) = y(3) + a_{2,1}y(2) + a_{2,2}y(1) \tag{6.10}$$

where the coefficients $a_{2,1}$ and $a_{2,2}$ are chosen such that $\alpha(3)$ is orthogonal to both $\alpha(1)$ and $\alpha(2)$, and so on. Thus, in general, we may express the transformation of the observed data $y(1)$, $y(2)$, . . . , $y(n)$ into the innovations $\alpha(1)$, $\alpha(2)$, . . . , $\alpha(n)$ by writing

$$\begin{bmatrix} \alpha(1) \\ \alpha(2) \\ \vdots \\ \vdots \\ \alpha(n) \end{bmatrix} = \begin{bmatrix} 1 & 0 & \cdots & 0 \\ a_{1,1} & 1 & \cdots & 0 \\ \vdots & \vdots & \ddots & \vdots \\ \vdots & \vdots & \ddots & \vdots \\ a_{n-1,\,n-1} & a_{n-1,\,n-2} & \cdots & 1 \end{bmatrix} \begin{bmatrix} y(1) \\ y(2) \\ \vdots \\ \vdots \\ y(n) \end{bmatrix} \tag{6.11}$$

The nonzero elements of row k of the *lower triangular transformation matrix* on the right side of Eq. (6.11) equal $a_{k-1,\,k-1}$, $a_{k-1,\,k-2}$, . . . , 1, where $k = 1, 2,$ . . . , n. These elements represent the coefficients of a *backward* prediction-error filter of order $k - 1$. Note that $a_{k,\,0} = 1$ for all k. Accordingly, given the observed data $y(1)$, $y(2)$, . . . , $y(n)$, we may compute the innovations $\alpha(1)$, $\alpha(2)$, . . . , $\alpha(n)$. There is no loss of information in the course of this transformation, since we may recover the original observed data $y(1)$, $y(2)$, . . . , $y(n)$ from the innovations $\alpha(1)$, $\alpha(2)$, . . . , $\alpha(n)$. This we do by premultiplying both sides of Eq. (6.11) by the inverse of the lower triangular transformation matrix. This matrix is nonsingular since its determinant equals 1 for all n. The transformation is therefore reversible.

Using Eq. (6.5), we may thus write:

$$\hat{x}(n|\mathcal{Y}_n) = \text{minimum mean-square estimate of } x(n)$$
$$\text{given the observed data } y(1), y(2), \ldots, y(n)$$

or, equivalently,

$$\hat{x}(n|\mathcal{Y}_n) = \text{minimum mean-square estimate of } x(n)$$
$$\text{given the innovations } \alpha(1), \alpha(2), \ldots, \alpha(n)$$

Define the estimate $\hat{x}(n|\mathcal{Y}_n)$ as a linear combinaton of the innovations $\alpha(1)$, $\alpha(2)$, . . . , $\alpha(n)$:

$$\hat{x}(n|\mathcal{Y}_n) = \sum_{k=1}^{n} b_k \alpha(k) \tag{6.12}$$

where the b_k are to be determined. With the innovations $\alpha(1)$, $\alpha(2)$, . . . , $\alpha(n)$ orthogonal to each other, and the b_k chosen to minimize the mean-squared value of the estimation error $x(n) - \hat{x}(n|\mathcal{Y}_n)$, we find that

$$b_k = \frac{E[x(n)\alpha^*(k)]}{E[\alpha(k)\alpha^*(k)]}, \qquad 1 \le k \le n \tag{6.13}$$

We rewrite Eq. (6.12) in the form

$$\hat{x}(n|\mathcal{Y}_n) = \sum_{k=0}^{n-1} b_k \alpha(k) + b_n \alpha(n) \tag{6.14}$$

where

$$b_n = \frac{E[x(n)\alpha^*(n)]}{E[\alpha(n)\alpha^*(n)]} \tag{6.15}$$

However, by definition, the summation term on the right side of Eq. (6.14) equals the previous estimate $\hat{x}(n - 1|\mathcal{Y}_{n-1})$. We may thus express the desired recursive estimation algorithm as

$$\hat{x}(n|\mathcal{Y}_n) = \hat{x}(n - 1|\mathcal{Y}_{n-1}) + b_n \alpha(n) \tag{6.16}$$

where b_n is defined by Eq. (6.15). Thus, by adding a *correction term* $b_n\alpha(n)$ to the previous estimate $\hat{x}(n - 1|\mathcal{Y}_{n-1})$, with the correction proportional to the innovation $\alpha(n)$, we get the updated estimate $\hat{x}(n|\mathcal{Y}_n)$.

The simple formulas of Eqs. (6.12) and (6.16) are the basis of all recursive estimation schemes. Equipped with these simple and yet basic ideas, we are ready to study the more general Kalman filtering problem.

6.3 STATEMENT OF THE KALMAN FILTERING PROBLEM

Let an *M*-dimensional parameter vector $\mathbf{x}(n)$ denote the *state* of a discrete-time, linear, dynamical system, and let an *N*-dimensional vector $\mathbf{y}(n)$ denote the *observed data* of the system, both measured at time n. In general, the vectors $\mathbf{x}(n)$ and $\mathbf{y}(n)$ consist of vector random variables. Thus, the system *model* is described by two equations represented in the form of a signal-flow graph, as in Fig. 6.1. They are as follows:

1. A *process equation*

$$\mathbf{x}(n + 1) = \mathbf{\Phi}(n + 1, n)\mathbf{x}(n) + \mathbf{v}_1(n) \tag{6.17}$$

where $\mathbf{\Phi}(n + 1, n)$ is a known *M-by-M state transition matrix* relating the states of the system at times $n + 1$ and n. The *M*-by-1 vector $\mathbf{v}_1(n)$ represents *process noise*. The vector $\mathbf{v}_1(n)$ is modeled as zero-mean, white-noise proc-

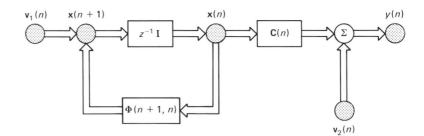

Figure 6.1 Signal-flow graph representation of discrete-time, linear, dynamical system.

esses whose correlation matrix is defined by

$$E[\mathbf{v}_1(n)\mathbf{v}_1^H(k)] = \begin{cases} \mathbf{Q}_1(n), & n = k \\ \mathbf{0}, & n \neq k \end{cases} \tag{6.18}$$

2. A *measurement equation*, describing the observation vector as follows:

$$\mathbf{y}(n) = \mathbf{C}(n)\mathbf{x}(n) + \mathbf{v}_2(n) \tag{6.19}$$

where $\mathbf{C}(n)$ is a known N-by-M *measurement matrix*. The N-by-1 vector $\mathbf{v}_2(n)$ is called *measurement noise*. It is modeled as zero-mean, white-noise processes whose correlation matrix is

$$E[\mathbf{v}_2(n)\mathbf{v}_2^H(k)] = \begin{cases} \mathbf{Q}_2(n), & n = k \\ \mathbf{0}, & n \neq k \end{cases} \tag{6.20}$$

The noise vectors $\mathbf{v}_1(n)$ and $\mathbf{v}_2(n)$ are statistically independent, so we may write

$$E[\mathbf{v}_1(n)\mathbf{v}_2^H(k)] = \mathbf{0}, \qquad \text{for all } n \text{ and } k \tag{6.21}$$

Note that the state transition matrix $\mathbf{\Phi}(n + 1, n)$ and the measurement matrix $\mathbf{C}(n)$ are both assumed *known*. The problem is to use the observed data, consisting of the vectors $\mathbf{y}(1), \mathbf{y}(2), \ldots, \mathbf{y}(n)$, to find for each $n \geq 1$ the minimum mean-square estimates of the components of the state $\mathbf{x}(i)$. The problem is called the *filtering* problem if $i = n$, the *prediction* problem if $i > n$, and the *smoothing* problem if $1 \leq i < n$. We will only be concerned with the filtering and prediction problems, which are closely related.

As remarked earlier in the introduction, we will solve the Kalman filtering problem by using the innovations approach (Kailath, 1968, 1970, 1981; Tretter, 1976).

6.4 THE INNOVATIONS PROCESS

Let the vector $\hat{\mathbf{y}}(n|\mathcal{Y}_{n-1})$ denote the minimum mean-square estimate of the observed data $\mathbf{y}(n)$ at time n, given all the past values of the observed data starting at time $n = 1$ and extending up to and including time $n - 1$. These past values are represented by the vectors $\mathbf{y}(1), \mathbf{y}(2), \ldots, \mathbf{y}(n - 1)$, which span the vector space $\mathcal{Y}_{n-1}$. We define the *innovations process* associated with $\mathbf{y}(n)$ as

$$\boldsymbol{\alpha}(n) = \mathbf{y}(n) - \hat{\mathbf{y}}(n|\mathcal{Y}_{n-1}), \qquad n = 1, 2, \ldots \tag{6.22}$$

The M-by-1 vector $\boldsymbol{\alpha}(n)$ represents the new information in the observed data $\mathbf{y}(n)$.

Generalizing the results of Eqs. (6.3), (6.4) and (6.5), we find that the innovations process $\boldsymbol{\alpha}(n)$ has the following properties:

1. The innovations process $\boldsymbol{\alpha}(n)$, associated with the observed data $\mathbf{y}(n)$ at time n, is orthogonal to all past observations $\mathbf{y}(1), \mathbf{y}(2), \ldots, \mathbf{y}(n - 1)$ as shown by

$$E[\boldsymbol{\alpha}(n)\mathbf{y}^H(k)] = \mathbf{0}, \qquad 1 \le k \le n - 1 \tag{6.23}$$

2. The innovations process consists of a sequence of vector random variables that are orthogonal to each other, as shown by

$$E[\boldsymbol{\alpha}(n)\boldsymbol{\alpha}^H(k)] = \mathbf{0}, \qquad 1 \le k \le n - 1 \tag{6.24}$$

3. There is a one-to-one correspondence between the sequence of vector random variables $\{\mathbf{y}(1), \mathbf{y}(2), \ldots, \mathbf{y}(n)\}$ representing the observed data and the sequence of vector random variables $\{\boldsymbol{\alpha}(1), \boldsymbol{\alpha}(2), \ldots, \boldsymbol{\alpha}(n)\}$ representing the innovations process, in that the one sequence may be obtained from the other by means of linear stable operators without loss of information. Thus, we may state that

$$\{\mathbf{y}(1), \mathbf{y}(2), \ldots, \mathbf{y}(n)\} \rightleftharpoons \{\boldsymbol{\alpha}(1), \boldsymbol{\alpha}(2), \ldots, \boldsymbol{\alpha}(n)\} \tag{6.25}$$

To form the sequence of vector random variables defining the innovations process, we may use a Gram–Schmidt orthogonalization procedure similar to that described in Section 6.2, except that the procedure is now formulated in terms of vectors and matrices (see Problem 1).

Correlation Matrix of the Innovations Process

To determine the correlation matrix of the innovations process $\boldsymbol{\alpha}(n)$, we first solve the state equation (6.17) recursively to obtain

$$\mathbf{x}(k) = \boldsymbol{\Phi}(k, 0)\mathbf{x}(0) + \sum_{i=1}^{k-1} \boldsymbol{\Phi}(k, i + 1)\mathbf{v}_1(i) \tag{6.26}$$

where we have made use of the following assumptions and properties:

1. The initial value of the state vector equals $\mathbf{x}(0)$.
2. As previously assumed, the observed data [and therefore the noise vector $\mathbf{v}_1(n)$] are zero for $n \leq 0$.
3. The state transition matrix has the properties

$$\mathbf{\Phi}(k, k - 1)\mathbf{\Phi}(k - 1, k - 2) \ldots \mathbf{\Phi}(i + 1, i) = \mathbf{\Phi}(k, i)$$

and

$$\mathbf{\Phi}(k, k) = \mathbf{I}$$

where $\mathbf{I}$ is the identity matrix.

Equation (6.26) shows that $\mathbf{x}(k)$ is a linear combination of $\mathbf{x}(0)$ and $\mathbf{v}_1(1)$, $\mathbf{v}_1(2)$, $\ldots$, $\mathbf{v}_1(k-1)$.

By hypothesis, the measurement noise vector $\mathbf{v}_2(n)$ is uncorrelated with both the initial state vector $\mathbf{x}(0)$ and the process noise vector $\mathbf{v}_1(n)$. Accordingly, premultiplying both sides of Eq. (6.26) by $\mathbf{v}_2^H(n)$, and taking expectations, we deduce that

$$E[\mathbf{x}(k)\mathbf{v}_2^H(n)] = \mathbf{0}, \qquad k, n \geq 0 \tag{6.27}$$

Correspondingly, we deduce from the measurement equation (6.19) that

$$E[\mathbf{y}(k)\mathbf{v}_2^H(n)] = \mathbf{0}, \qquad 0 \leq k \leq n - 1 \tag{6.28}$$

Moreover, we may write

$$E[\mathbf{y}(k)\mathbf{v}_1^H(n)] = \mathbf{0}, \qquad 0 \leq k \leq n \tag{6.29}$$

Given the past observations $\mathbf{y}(1), \ldots, \mathbf{y}(n-1)$ that span the space $\mathcal{Y}_{n-1}$, we also find from the measurement equation (6.19) that the minimum mean-square estimate of the present value $\mathbf{y}(n)$ of the observation vector equals

$$\hat{\mathbf{y}}(n|\mathcal{Y}_{n-1}) = \mathbf{C}(n)\,\hat{\mathbf{x}}(n|\mathcal{Y}_{n-1}) + \hat{\mathbf{v}}_2(n|\mathcal{Y}_{n-1})$$

However, the estimate $\hat{\mathbf{v}}_2(n|\mathcal{Y}_{n-1})$ of the measurement noise vector is zero since it is orthogonal to the past observations $\mathbf{y}(1), \ldots, \mathbf{y}(n-1)$; see Eq. (6.28). Hence, we may simply write

$$\hat{\mathbf{y}}(n|\mathcal{Y}_{n-1}) = \mathbf{C}(n)\hat{\mathbf{x}}(n|\mathcal{Y}_{n-1}) \tag{6.30}$$

Therefore, using Eqs. (6.22) and (6.30), we may express the innovations process in the form

$$\boldsymbol{\alpha}(n) = \mathbf{y}(n) - \mathbf{C}(n)\hat{\mathbf{x}}(n|\mathcal{Y}_{n-1}) \tag{6.31}$$

Substituting the measurement equation (6.19) in (6.31), we get

$$\boldsymbol{\alpha}(n) = \mathbf{C}(n)\boldsymbol{\varepsilon}(n, n - 1) + \mathbf{v}_2(n) \tag{6.32}$$

where $\varepsilon(n, n - 1)$ is the *predicted state-error vector* at time n, using data up to time $n - 1$. That is, $\varepsilon(n, n - 1)$ is the difference between the state vector $\mathbf{x}(n)$ and the one-step prediction vector $\hat{\mathbf{x}}(n|\mathcal{Y}_{n-1})$, as shown by

$$\varepsilon(n, n - 1) = \mathbf{x}(n) - \hat{\mathbf{x}}(n|\mathcal{Y}_{n-1}) \tag{6.33}$$

Note that the predicted state-error vector is orthogonal to both the process noise vector $\mathbf{v}_1(n)$ and the measurement noise vector $\mathbf{v}_2(n)$; see Problem 2.

The correlation matrix of the innovations process $\boldsymbol{\alpha}(n)$ is defined by

$$\boldsymbol{\Sigma}(n) = E[\boldsymbol{\alpha}(n)\boldsymbol{\alpha}^H(n)] \tag{6.34}$$

Therefore, substituting Eq. (6.32) in (6.34), expanding the pertinent terms, and then using the fact that the vectors $\varepsilon(n, n - 1)$ and $\mathbf{v}_2(n)$ are orthogonal, we obtain the desired result:

$$\boldsymbol{\Sigma}(n) = \mathbf{C}(n)\mathbf{K}(n, n - 1)\mathbf{C}^H(n) + \mathbf{Q}_2(n) \tag{6.35}$$

where $\mathbf{Q}_2(n)$ is the correlation matrix of the noise vector $\mathbf{v}_2(n)$. The M-by-M matrix $\mathbf{K}(n, n - 1)$ is called the *predicted state-error correlation matrix*, defined by

$$\mathbf{K}(n, n - 1) = E[\varepsilon(n, n - 1)\varepsilon^H(n, n - 1)] \tag{6.36}$$

where $\varepsilon(n, n - 1)$ is the predicted state–error vector.

6.5 *ESTIMATION OF THE STATE USING THE INNOVATIONS PROCESS*

Consider next the problem of deriving the minimum mean-square estimate of the state $\mathbf{x}(i)$ from the innovations process. From the discussion presented in Section 6.2, we deduce that this estimate may be expressed as a linear combination of the sequence of innovations processes $\boldsymbol{\alpha}(1), \boldsymbol{\alpha}(2), \ldots, \boldsymbol{\alpha}(n)$ [see Eq. (6.12)]:

$$\hat{\mathbf{x}}(i|\mathcal{Y}_n) = \sum_{k=1}^{n} \mathbf{B}_i(k)\boldsymbol{\alpha}(k) \tag{6.37}$$

where $\{\mathbf{B}_i(k)\}$ is a set of M-by-N matrices to be determined. According to the principle of orthogonality, the predicted state-error vector is orthogonal to the innovation process, as shown by

$$\begin{aligned}
E[\varepsilon(i, n)\boldsymbol{\alpha}^H(m)] &= E\{[\mathbf{x}(i) - \hat{\mathbf{x}}(i|\mathcal{Y}_n)]\boldsymbol{\alpha}^H(m)\} \\
&= \mathbf{0}, \qquad m = 1, 2, \ldots, n
\end{aligned} \tag{6.38}$$

Substituting Eq. (6.37) in (6.38) and using the orthogonality property of the innovations process, namely Eq. (6.24), we get

$$\begin{aligned}
E[\mathbf{x}(i)\boldsymbol{\alpha}^H(m)] &= \mathbf{B}_i(m)E[\boldsymbol{\alpha}(m)\boldsymbol{\alpha}^H(m)] \\
&= \mathbf{B}_i(m)\boldsymbol{\Sigma}(m)
\end{aligned} \tag{6.39}$$

Hence, postmultiplying both sides of Eq. (6.39) by $\mathbf{\Sigma}^{-1}(m)$, we find that $\mathbf{B}_i(m)$ is given by

$$\mathbf{B}_i(m) = E[\mathbf{x}(i)\boldsymbol{\alpha}^H(m)]\mathbf{\Sigma}^{-1}(m) \tag{6.40}$$

Finally, substituting Eq. (6.40) in (6.37), we get the desired value for the minimum mean-squares estimate $\hat{\mathbf{x}}(i|\mathcal{Y}_n)$ as follows:

$$\hat{\mathbf{x}}(i|\mathcal{Y}_n) = \sum_{k=1}^{n} E[\mathbf{x}(i)\boldsymbol{\alpha}^H(k)] \, \mathbf{\Sigma}^{-1}(k) \, \boldsymbol{\alpha}(k)$$

$$= \sum_{k=1}^{n-1} E[\mathbf{x}(i)\boldsymbol{\alpha}^H(k)] \, \mathbf{\Sigma}^{-1}(k) \, \boldsymbol{\alpha}(k)$$

$$+ E[\mathbf{x}(i)\boldsymbol{\alpha}^H(n)] \, \mathbf{\Sigma}^{-1}(n)\boldsymbol{\alpha}(n)$$

For $i = n + 1$, we may therefore write

$$\hat{\mathbf{x}}(n + 1|\mathcal{Y}_n) = \sum_{k=1}^{n-1} E[\mathbf{x}(n + 1) \, \boldsymbol{\alpha}^H(k)] \, \mathbf{\Sigma}^{-1}(k)\boldsymbol{\alpha}(k)$$
$$+ E[\mathbf{x}(n + 1) \, \boldsymbol{\alpha}^H(n)] \, \mathbf{\Sigma}^{-1}(n)\boldsymbol{\alpha}(n) \tag{6.41}$$

However, the state $\mathbf{x}(n + 1)$ at time $n + 1$ is related to the state $\mathbf{x}(n)$ at time n by Eq. (6.17). Therefore, using this relation, we may write for $0 \leq k \leq n$:

$$E\{\mathbf{x}(n + 1)\boldsymbol{\alpha}^H(k)\} = E\{[\mathbf{\Phi}(n + 1, n)\mathbf{x}(n) + \mathbf{v}_1(n)]\boldsymbol{\alpha}^H(k)\} \tag{6.42}$$
$$= \mathbf{\Phi}(n + 1, n)E[\mathbf{x}(n)\boldsymbol{\alpha}^H(k)]$$

where we have made use of the fact that $\boldsymbol{\alpha}(k)$ depends only on the observed data $\mathbf{y}(1), \ldots, \mathbf{y}(k)$, and therefore from Eq. (6.29) we see that $\mathbf{v}_1(n)$ and $\boldsymbol{\alpha}(k)$ are orthogonal for $0 \leq k \leq n$. We may thus rewrite the summation term on the right side of Eq. (6.41) as follows:

$$\sum_{k=1}^{n-1} E[\mathbf{x}(n + 1)\boldsymbol{\alpha}^H(k)]\mathbf{\Sigma}^{-1}(k)\boldsymbol{\alpha}(k)$$

$$= \mathbf{\Phi}(n + 1, n) \sum_{k=1}^{n-1} E[\mathbf{x}(n)\boldsymbol{\alpha}^H(k)] \, \mathbf{\Sigma}^{-1}(k)\boldsymbol{\alpha}(k) \tag{6.43}$$

$$= \mathbf{\Phi}(n + 1, n)\hat{\mathbf{x}}(n|\mathcal{Y}_{n-1})$$

Next, define the M-by-N matrix:

$$\mathbf{G}(n) = E[\mathbf{x}(n + 1)\boldsymbol{\alpha}^H(n)]\mathbf{\Sigma}^{-1}(n) \tag{6.44}$$

Then, using this definition and the result of Eq. (6.43), we may rewrite Eq. (6.41) as follows:

$$\hat{\mathbf{x}}(n + 1|\mathcal{Y}_n) = \mathbf{\Phi}(n + 1, n) \, \hat{\mathbf{x}}(n|\mathcal{Y}_{n-1}) + \mathbf{G}(n)\boldsymbol{\alpha}(n) \tag{6.45}$$

Equation (6.45) shows that we may compute the minimum mean-square estimate $\hat{\mathbf{x}}(n + 1|\mathcal{Y}_n)$ of the state of a linear dynamical system by adding to the previous estimate $\hat{\mathbf{x}}(n|\mathcal{Y}_{n-1})$ that is premultiplied by the state transition matrix $\mathbf{\Phi}(n + 1, n)$ a correction term equal to $\mathbf{G}(n)\,\boldsymbol{\alpha}(n)$. The correction term equals the innovations process $\boldsymbol{\alpha}(n)$ premultiplied by the matrix $\mathbf{G}(n)$. Accordingly, and in recognition of the pioneering work by Kalman, the matrix $\mathbf{G}(n)$ is called the *Kalman gain*.

There now remains only the problem of expressing the Kalman gain $\mathbf{G}(n)$ in a form convenient for computation. To do this, we first use Eqs. (6.32) and (6.42) to express the expectation of the product of $\mathbf{x}(n + 1)$ and $\boldsymbol{\alpha}^H(n)$ as follows:

$$
\begin{aligned}
E[\mathbf{x}(n + 1)\boldsymbol{\alpha}^H(n)] &= \mathbf{\Phi}(n + 1, n)E[\mathbf{x}(n)\boldsymbol{\alpha}^H(n)] \\
&= \mathbf{\Phi}(n + 1, n)E[\mathbf{x}(n)\,(\mathbf{C}(n)\boldsymbol{\varepsilon}(n, n - 1) + \mathbf{v}_2(n))^H] \quad (6.46) \\
&= \mathbf{\Phi}(n + 1, n)E[\mathbf{x}(n)\boldsymbol{\varepsilon}^H(n, n - 1)]\mathbf{C}^H(n)
\end{aligned}
$$

where we have used the fact that the state $\mathbf{x}(n)$ and noise vector $\mathbf{v}_2(n)$ are uncorrelated; see Eq. (6.27). We further note that the predicted state–error vector $\boldsymbol{\varepsilon}(n, n - 1)$ is orthogonal to the estimate $\hat{\mathbf{x}}(n|\mathcal{Y}_{n-1})$. Therefore, the expectation of the product of $\hat{\mathbf{x}}(n|\mathcal{Y}_{n-1})$ and $\boldsymbol{\varepsilon}^H(n, n - 1)$ is zero, and so we may rewrite Eq. (6.46) by replacing the multiplying factor $\mathbf{x}(n)$ by the predicted state-error vector $\boldsymbol{\varepsilon}(n, n - 1)$ as follows:

$$
E[\mathbf{x}(n + 1)\boldsymbol{\alpha}^H(n)] = \mathbf{\Phi}(n + 1, n)E[\boldsymbol{\varepsilon}(n, n - 1)\boldsymbol{\varepsilon}^H(n, n - 1)]\mathbf{C}^H(n) \quad (6.47)
$$

From Eq. (6.36), we see that the expectation on the right side of Eq. (6.47) equals the predicted state-error correlation matrix. Hence, we may rewrite Eq. (6.47) as follows:

$$
E[\mathbf{x}(n + 1)\boldsymbol{\alpha}^H(n)] = \mathbf{\Phi}(n + 1, n)\,\mathbf{K}(n, n - 1)\,\mathbf{C}^H(n) \quad (6.48)
$$

We may now redefine the Kalman gain. In particular, substituting Eq. (6.48) in (6.44), we get

$$
\mathbf{G}(n) = \mathbf{\Phi}(n + 1, n)\mathbf{K}(n, n - 1)\mathbf{C}^H(n)\mathbf{\Sigma}^{-1}(n) \quad (6.49)
$$

where the correlation matrix $\mathbf{\Sigma}(n)$ itself is defined in Eq. (6.35).

As it stands, the formula of Eq. (6.49) is not particularly useful for computing the Kalman gain $\mathbf{G}(n)$, since it requires that the predicted state-error correlation matrix $\mathbf{K}(n, n - 1)$ be known. To overcome this difficulty, we derive a formula for the recursive computation of $\mathbf{K}(n, n - 1)$.

The predicted state-error vector $\boldsymbol{\varepsilon}(n + 1, n)$ equals the difference between the state $\mathbf{x}(n + 1)$ and the one-step prediction $\hat{\mathbf{x}}(n + 1|\mathcal{Y}_n)$ [see Eq. (6.33)]:

$$
\boldsymbol{\varepsilon}(n + 1, n) = \mathbf{x}(n + 1) - \hat{\mathbf{x}}(n + 1|\mathcal{Y}_n) \quad (6.50)
$$

Substituting Eqs. (6.17) and (6.45) in (6.50), and using Eq. (6.31) for the innovations process $\boldsymbol{\alpha}(n)$, we get

$$
\begin{aligned}
\boldsymbol{\varepsilon}(n + 1, n) = {} & \boldsymbol{\Phi}(n + 1, n)[\mathbf{x}(n) - \hat{\mathbf{x}}(n|\mathcal{Y}_{n-1})] \\
& - \mathbf{G}(n)\,[\mathbf{y}(n) - \mathbf{C}(n)\hat{\mathbf{x}}(n|\mathcal{Y}_{n-1})] + \mathbf{v}_1(n)
\end{aligned} \tag{6.51}
$$

Next, using the measurement equation (6.19) to eliminate $\mathbf{y}(n)$ in Eq. (6.51), we get the following difference equation for recursive computation of the predicted state-error vector:

$$
\begin{aligned}
\boldsymbol{\varepsilon}(n + 1, n) = {} & [\boldsymbol{\Phi}(n + 1, n) - \mathbf{G}(n)\,\mathbf{C}(n)]\,\boldsymbol{\varepsilon}(n, n - 1) \\
& + \mathbf{v}_1(n) - \mathbf{G}(n)\,\mathbf{v}_2(n)
\end{aligned} \tag{6.52}
$$

The correlation matrix of the predicted state-error vector $\boldsymbol{\varepsilon}(n + 1, n)$ equals [see Eq. (6.36)]

$$
\mathbf{K}(n + 1, n) = E[\boldsymbol{\varepsilon}(n + 1, n)\,\boldsymbol{\varepsilon}^H(n + 1, n)] \tag{6.53}
$$

Substituting Eq. (6.52) in (6.53), and recognizing that the error vector $\boldsymbol{\varepsilon}(n, n - 1)$ and the noise vectors $\mathbf{v}_1(n)$ and $\mathbf{v}_2(n)$ are mutually uncorrelated, we may express the predicted state-error correlation matrix as follows:

$$
\begin{aligned}
\mathbf{K}(n+1, n) = {} & [\boldsymbol{\Phi}(n+1, n) - \mathbf{G}(n)\mathbf{C}(n)]\,\mathbf{K}(n, n-1) \\
& \cdot [\boldsymbol{\Phi}(n+1, n) - \mathbf{G}(n)\mathbf{C}(n)]^H \\
& + \mathbf{Q}_1(n) + \mathbf{G}(n)\,\mathbf{Q}_2(n)\,\mathbf{G}^H(n)
\end{aligned} \tag{6.54}
$$

where $\mathbf{Q}_1(n)$ and $\mathbf{Q}_2(n)$ are the correlation matrices of $\mathbf{v}_1(n)$ and $\mathbf{v}_2(n)$, respectively. By expanding the right side of Eq. (6.54), and then using Eqs. (6.49) and (6.35) for the Kalman gain, we get the *Riccati difference equation* for the recursive computation of the predicted state-error correlation matrix:

$$
\mathbf{K}(n + 1, n) = \boldsymbol{\Phi}(n + 1, n)\,\mathbf{K}(n)\boldsymbol{\Phi}^H(n + 1, n) + \mathbf{Q}_1(n) \tag{6.55}
$$

The M-by-M matrix $\mathbf{K}(n)$ is described by the recursion

$$
\mathbf{K}(n) = \mathbf{K}(n, n - 1) - \boldsymbol{\Phi}(n, n + 1)\,\mathbf{G}(n)\,\mathbf{C}(n)\,\mathbf{K}(n, n - 1) \tag{6.56}
$$

Here we have used the fact that

$$
\boldsymbol{\Phi}(n + 1, n)\,\boldsymbol{\Phi}(n, n + 1) = \mathbf{I} \tag{6.57}
$$

where $\mathbf{I}$ is the identity matrix. This property follows from the definition of the transition matrix. The mathematical significance of the matrix $\mathbf{K}(n)$ in Eq. (6.56) will be explained in Section 6.6.

Equations (6.49), (6.31), (6.45), (6.56), and (6.55), in that order, define Kalman's one-step prediction algorithm.

Comments

The process applied to the input of the Kalman filter consists of the observed data $\mathbf{y}(1)$, $\mathbf{y}(2)$, . . . , $\mathbf{y}(n)$ that span the space $\mathcal{Y}_n$. The resulting filter output equals the predicted state vector $\hat{\mathbf{x}}(n+1|\mathcal{Y}_n)$. Given that the matrices $\mathbf{\Phi}(n+1, n)$, $\mathbf{C}(n)$, $\mathbf{Q}_1(n)$, and $\mathbf{Q}_2(n)$ are all known quantities, we find from Eqs. (6.44), (6.55) and (6.56) that the predicted state-error correlation matrix $\mathbf{K}(n+1, n)$ is actually independent of the input $\mathbf{y}(n)$; see Problem 2.   Therefore, no one set of measurements helps more than any other to eliminate some uncertainty about the state vector $\mathbf{x}(n)$ (Anderson and Moore, 1979).   The Kalman gain $\mathbf{G}(n)$ is also independent of the input $\mathbf{y}(n)$.   Consequently, the predicted state-error correlation matrix $\mathbf{K}(n + 1, n)$ and the Kalman gain $\mathbf{G}(n)$ may be computed before the Kalman filter is actually put into operation.

As already mentioned, the Kalman filter theory assumes knowledge of the matrices $\mathbf{\Phi}(n + 1, n)$, $\mathbf{C}(n)$, $\mathbf{Q}_1(n)$ and $\mathbf{Q}_2(n)$.   However, the theory may be *generalized* to include a situation where one or more of these matrices may assume values that depend on the input $\mathbf{y}(n)$.   In such a situation, we find that although $\hat{\mathbf{x}}(n + 1|\mathcal{Y}_n)$ and $\mathbf{K}(n + 1, n)$ are still given by Eqs. (6.45) and (6.55), respectively, the Kalman gain $\mathbf{G}(n)$ and the predicted state-error correlation matrix $\mathbf{K}(n + 1, n)$ are *not* precomputable (Anderson and Moore, 1979).   Rather, they both now depend on the input $\mathbf{y}(n)$.   This means that $\mathbf{K}(n + 1, n)$ is a *conditional* error correlation matrix, conditional on the input $\mathbf{y}(n)$.

Initial Conditions

To operate the one-step prediction algorithm described, we need to know the pertinent set of *initial conditions*.   We now address this issue.

We first note that the state $\mathbf{x}(n)$ and the noise vector $\mathbf{v}_1(n)$ are independent of each other.   Hence, from the state equation (6.17) we find that the minimum mean-square estimate of the state $\mathbf{x}(n + 1)$ at time $n + 1$, given the observed data up to and including time n [i.e., given $\mathbf{y}(1)$, . . . , $\mathbf{y}(n)$], equals

$$\hat{\mathbf{x}}(n + 1|\mathcal{Y}_n) = \mathbf{\Phi}(n + 1, n)\, \hat{\mathbf{x}}(n|\mathcal{Y}_n) + \hat{\mathbf{v}}_1(n|\mathcal{Y}_n) \tag{6.58}$$

Since the noise vector $\mathbf{v}_1(n)$ is independent of the observed data $\mathbf{y}(1)$, . . . , $\mathbf{y}(n)$, it follows that the corresponding minimum mean-square estimate $\hat{\mathbf{v}}_1(n|\mathcal{Y}_n)$ is zero.   Accordingly, Eq. (6.58) simplifies as

$$\hat{\mathbf{x}}(n + 1|\mathcal{Y}_n) = \mathbf{\Phi}(n + 1, n)\, \hat{\mathbf{x}}(n|\mathcal{Y}_n) \tag{6.59}$$

Based on this relation, we are ready to define the required initial conditions. Putting $n = 0$ in Eq. (6.59), we get

$$\hat{\mathbf{x}}(1|\mathcal{Y}_0) = \mathbf{\Phi}(1, 0)\, \hat{\mathbf{x}}(0|\mathcal{Y}_0)$$

However, $\hat{\mathbf{x}}(0|\mathcal{Y}_0)$ is the estimate of $\mathbf{x}(0)$ given no observed data, since the observed data are assumed to start at time $n = 1$. Hence,

$$\hat{\mathbf{x}}(1|\mathcal{Y}_0) = \mathbf{0} \tag{6.60}$$

Correspondingly, putting $n = 0$ in Eq. (6.53), we find that the initial value of the predicted state-error correlation matrix equals

$$\begin{aligned}
\mathbf{K}(1, 0) &= E[\boldsymbol{\varepsilon}(1, 0) \, \boldsymbol{\varepsilon}^H(1, 0)] \\
&= E[(\mathbf{x}(1) - \hat{\mathbf{x}}(1|\mathcal{Y}_0)) \, (\mathbf{x}(1) - \hat{\mathbf{x}}(1|\mathcal{Y}_0)^H] \\
&= E[\mathbf{x}(1) \, \mathbf{x}^H(1)]
\end{aligned} \tag{6.61}$$

Equations (6.60) and (6.61) define the required initial conditions.

6.6 FILTERING

The next signal-processing operation we wish to consider is that of filtering. In particular, we wish to compute the filtered estimate $\hat{\mathbf{x}}(n|\mathcal{Y}_n)$ by using the one-step prediction algorithm described previously.

To find this estimate, we premultiply both sides of Eq. (6.59) by the inverse of the transition matrix $\boldsymbol{\Phi}(n + 1, n)$, and thus write

$$\hat{\mathbf{x}}(n|\mathcal{Y}_n) = \boldsymbol{\Phi}^{-1}(n + 1, n) \, \hat{\mathbf{x}}(n + 1|\mathcal{Y}_n) \tag{6.62}$$

Using the property of the state transition matrix given in Eq. (6.57), we have

$$\boldsymbol{\Phi}^{-1}(n + 1, n) = \boldsymbol{\Phi}(n, n + 1) \tag{6.63}$$

We may therefore rewrite Eq. (6.62) in the form

$$\hat{\mathbf{x}}(n|\mathcal{Y}_n) = \boldsymbol{\Phi}(n, n + 1) \, \hat{\mathbf{x}}(n + 1|\mathcal{Y}_n) \tag{6.64}$$

This shows that knowing the solution to the one-step prediction problem, that is, the minimum mean-square estimate $\hat{\mathbf{x}}(n + 1|\mathcal{Y}_n)$, we may determine the corresponding filtered estimate $\hat{\mathbf{x}}(n|\mathcal{Y}_n)$ simply by multiplying $\hat{\mathbf{x}}(n + 1|\mathcal{Y}_n)$ by the state transition matrix $\boldsymbol{\Phi}(n, n + 1)$.

Thus, we may reformulate Eq. (6.31) to express the innovations process $\boldsymbol{\alpha}(n)$ in terms of the filtered estimate of the state vector at time $n - 1$ as follows:

$$\boldsymbol{\alpha}(n) = \mathbf{y}(n) - \mathbf{C}(n) \, \boldsymbol{\Phi}(n, n - 1) \, \hat{\mathbf{x}}(n - 1|\mathcal{Y}_{n-1}) \tag{6.65}$$

'Similarly, we may reformulate Eq. (6.45) to express the recursion for time-updating the filtered estimate of the state vector as follows:

$$\hat{\mathbf{x}}(n|\mathcal{Y}_n) = \boldsymbol{\Phi}(n, n - 1) \, \hat{\mathbf{x}}(n - 1|\mathcal{Y}_{n-1}) + \boldsymbol{\Phi}(n, n + 1) \, \mathbf{G}(n) \, \boldsymbol{\alpha}(n) \tag{6.66}$$

Equations (6.49), (6.65), (6.66), (6.56), and (6.55) collectively and in that order represent another way of describing the Kalman filter, based on the filtered estimate of the state vector.

Filtered State-error Correlation Matrix

Earlier we introduced the M-by-M matrix $\mathbf{K}(n)$ in the formulation of the Riccati difference equation (6.55). We conclude our present discussion of the Kalman filter theory by showing that this matrix equals the correlation matrix of the error inherent in the filtered estimate $\hat{\mathbf{x}}(n|\mathcal{Y}_n)$.

Define the *filtered state-error vector* $\boldsymbol{\varepsilon}(n)$ as the difference between the state $\mathbf{x}(n)$ and the filtered estimate $\hat{\mathbf{x}}(n|\mathcal{Y}_n)$, as shown by

$$\boldsymbol{\varepsilon}(n) = \mathbf{x}(n) - \hat{\mathbf{x}}(n|\mathcal{Y}_n) \tag{6.67}$$

Substituting Eqs. (6.45) and (6.64) in (6.67), and recognizing that the product of $\boldsymbol{\Phi}(n, n + 1)$ and $\boldsymbol{\Phi}(n + 1, n)$ equals the identity matrix, we get

$$\begin{aligned}\boldsymbol{\varepsilon}(n) &= \mathbf{x}(n) - \hat{\mathbf{x}}(n|\mathcal{Y}_{n-1}) - \boldsymbol{\Phi}(n, n + 1)\, \mathbf{G}(n)\, \boldsymbol{\alpha}(n) \\ &= \boldsymbol{\varepsilon}(n, n - 1) - \boldsymbol{\Phi}(n, n + 1)\, \mathbf{G}(n)\, \boldsymbol{\alpha}(n)\end{aligned} \tag{6.68}$$

where $\boldsymbol{\varepsilon}(n, n, -1)$ is the predicted state-error vector at time n, using data up to time $n - 1$, and $\boldsymbol{\alpha}(n)$ is the innovations process.

By definition, the correlation matrix of the filtered state-error vector $\boldsymbol{\varepsilon}(n)$ equals the expectation $E[\boldsymbol{\varepsilon}(n)\boldsymbol{\varepsilon}^H(n)]$. Hence, using Eq. (6.68), we may express this expectation as follows:

$$\begin{aligned}E[\boldsymbol{\varepsilon}(n)\boldsymbol{\varepsilon}^H(n)] &= E[\boldsymbol{\varepsilon}(n, n - 1)\, \boldsymbol{\varepsilon}^H(n, n - 1)] \\ &\quad + \boldsymbol{\Phi}(n, n + 1)\, \mathbf{G}(n)\, E[\boldsymbol{\alpha}(n)\boldsymbol{\alpha}^H(n)]\, \mathbf{G}^H(n)\, \boldsymbol{\Phi}^H(n, n + 1) \\ &\quad - 2E[\boldsymbol{\varepsilon}(n, n - 1)\boldsymbol{\alpha}^H(n)]\, \mathbf{G}^H(n)\, \boldsymbol{\Phi}^H(n, n + 1)\end{aligned} \tag{6.69}$$

Examining the right side of Eq. (6.69), we find that the three expectations contained in it may be interpreted individually as follows:

1. The first expectation equals the predicted state-error correlation matrix:

$$\mathbf{K}(n, n - 1) = E[\boldsymbol{\varepsilon}(n, n - 1)\, \boldsymbol{\varepsilon}^H(n, n - 1)]$$

2. The expectation in the second term equals the correlation matrix of the innovations process $\boldsymbol{\alpha}(n)$:

$$\boldsymbol{\Sigma}(n) = E[\boldsymbol{\alpha}(n)\boldsymbol{\alpha}^H(n)]$$

3. The expectation in the third term may be expressed as follows:

$$\begin{aligned}E[\boldsymbol{\varepsilon}(n, n - 1)\, \boldsymbol{\alpha}^H(n)] &= E[(\mathbf{x}(n) - \hat{\mathbf{x}}(n|\mathcal{Y}_{n-1}))\boldsymbol{\alpha}^H(n)] \\ &= E[\mathbf{x}(n)\, \boldsymbol{\alpha}^H(n)]\end{aligned}$$

where, in the last line, we have used the fact that the estimate $\hat{\mathbf{x}}(n|\mathcal{Y}_{n-1})$ is orthogonal to the innovations process $\boldsymbol{\alpha}(n)$, acting as input. Next, from Eq. (6.42) we see, by putting $k = n$ and premultiplying both sides by the inverse matrix $\mathbf{\Phi}^{-1}(n + 1, n) = \mathbf{\Phi}(n, n + 1)$, that

$$E[\mathbf{x}(n)\,\boldsymbol{\alpha}^H(n)] = \mathbf{\Phi}(n, n + 1)\,E[\mathbf{x}(n + 1)\boldsymbol{\alpha}^H(n)]$$

$$= \mathbf{\Phi}(n, n + 1)\,\mathbf{G}(n)\,\mathbf{\Sigma}(n)$$

where, in the last line, we have made use of Eq. (6.44). Hence,

$$E[\boldsymbol{\varepsilon}(n, n - 1)\,\boldsymbol{\alpha}^H(n)] = \mathbf{\Phi}(n, n + 1)\,\mathbf{G}(n)\mathbf{\Sigma}(n)$$

We may now use these results in Eq. (6.69), and so obtain

$$E[\boldsymbol{\varepsilon}(n)\boldsymbol{\varepsilon}^H(n)] = \mathbf{K}(n, n - 1)$$

$$- \mathbf{\Phi}(n, n + 1)\,\mathbf{G}(n)\,\mathbf{\Sigma}(n)\mathbf{G}^H(n)\,\mathbf{\Phi}^H(n, n + 1) \qquad (6.70)$$

We may further simplify this result by noting that [see Eq. (6.49)]

$$\mathbf{G}(n)\,\mathbf{\Sigma}(n) = \mathbf{\Phi}(n + 1, n)\,\mathbf{K}(n, n - 1)\,\mathbf{C}^H(n) \qquad (6.71)$$

Accordingly, using Eqs. (6.70) and (6.71), and recognizing that the product of $\mathbf{\Phi}(n, n + 1)$ and $\mathbf{\Phi}(n + 1, n)$ equals the identity matrix, we get the desired result for the filtered state-error correlation matrix:

$$E[\boldsymbol{\varepsilon}(n)\boldsymbol{\varepsilon}^H(n)] = \mathbf{K}(n, n - 1)$$

$$- \mathbf{K}(n, n - 1)\mathbf{C}^H(n)\,\mathbf{G}^H(n)\,\mathbf{\Phi}^H(n, n + 1) \qquad (6.72)$$

Equivalently, using the Hermitian property of $E[\boldsymbol{\varepsilon}(n)\boldsymbol{\varepsilon}^H(n)]$ and that of $\mathbf{K}(n, n - 1)$, we may write

$$E[\boldsymbol{\varepsilon}(n)\boldsymbol{\varepsilon}^H(n)] = \mathbf{K}(n, n - 1) - \mathbf{\Phi}(n, n + 1)\,\mathbf{G}(n)\,\mathbf{C}(n)\,\mathbf{K}(n, n - 1) \qquad (6.73)$$

Comparing Eq. (6.73) with (6.56), we see that

$$E[\boldsymbol{\varepsilon}(n)\,\boldsymbol{\varepsilon}^H(n)] = \mathbf{K}(n)$$

This shows that the matrix $\mathbf{K}(n)$ used in the Riccati difference equation is in fact the filtered state-error correlation matrix.

Initial Conditions

From Eqs. (6.60) and (6.64), we deduce that the initial value of the filtered estimate $\hat{\mathbf{x}}(n|\mathcal{Y}_n)$ equals

$$\hat{\mathbf{x}}(0|\mathcal{Y}_0) = \mathbf{\Phi}(0, 1)\,\hat{\mathbf{x}}(1|\mathcal{Y}_0)$$

$$= \mathbf{0}$$

As before, for the initial condition of the predicted state-error correlation matrix $\boldsymbol{\alpha}(n + 1, n)$, we use the value defined by Eq. (6.61).

6.7 *SUMMARY AND DISCUSSION*

The Kalman filter is a *linear, discrete-time, finite-dimensional system*, the implementation of which is well suited for a digital computer. The input of the filter is the vector process $\{y(n)\}$, represented by the vector space $\mathcal{Y}_n$, and the output is the filtered estimate $\hat{x}(n|\mathcal{Y}_n)$ of the state vector. In Table 6.1, we present a summary of the Kalman filter (including initial conditions) based on the one-step prediction algorithm.

A key property of the Kalman filter is that it leads to minimization of the trace of the filtered state-error correlation matrix $\mathbf{K}(n)$. This means that the Kalman filter is the *linear minimum variance estimator* of the state vector $x(n)$ (Goodwin and Sin, 1984; Anderson and Moore, 1979).

The input–output relation of the Kalman filter is depicted in Fig. 6.2(a), representing a signal-flow graph interpretation of Eqs. (6.65) and (6.66) and the fact that $\hat{x}(n - 1|\mathcal{Y}_{n-1})$ is the delayed version of $\hat{x}(n|\mathcal{Y}_n)$. In this model, $y(n)$ is the input and $\hat{x}(n|\mathcal{Y}_n)$ is the output.

We may rearrange Eq. (6.65) in the form

$$\mathbf{y}(n) = \boldsymbol{\alpha}(n) + \mathbf{C}(n)\,\boldsymbol{\Phi}(n, n - 1)\,\hat{x}(n - 1|\mathcal{Y}_{n-1}) \tag{6.74}$$

Accordingly, we may represent Eqs. (6.66) and (6.74) by the signal-flow graph shown in Fig. 6.2(b). Here again we have included a branch in the graph to represent the fact that $\hat{x}(n - 1|\mathcal{Y}_{n-1})$ is the delayed version of $\hat{x}(n|\mathcal{Y}_n)$. We may view Fig. 6.2(b) as a model for generating the process $\{y(n)\}$ by driving it with the innovations process $\{\boldsymbol{\alpha}(n)\}$. The model of Fig. 6.2(b) is known as the *inverse*

TABLE 6.1 SUMMARY OF THE KALMAN FILTER BASED ON ONE-STEP PREDICTION

Input Vector Process
    Observations = $\{\mathbf{y}(1), \mathbf{y}(2), \ldots, \mathbf{y}(n)\}$
Known Parameters
    State transition matrix = $\boldsymbol{\Phi}(n + 1, n)$
    Measurement matrix = $\mathbf{C}(n)$
    Correlation matrix of process noise vector = $\mathbf{Q}_1(n)$
    Correlation matrix of measurement noise vector = $\mathbf{Q}_2(n)$
Computation: $n = 1, 2, 3, \ldots$
    $\mathbf{G}(n) = \boldsymbol{\Phi}(n + 1, n)\mathbf{K}(n, n - 1)\mathbf{C}^H(n)[\mathbf{C}(n)\mathbf{K}(n, n - 1)\mathbf{C}^H(n) + \mathbf{Q}_2(n)]^{-1}$
    $\boldsymbol{\alpha}(n) = \mathbf{y}(n) - \mathbf{C}(n)\hat{x}(n|\mathcal{Y}_{n-1})$
    $\hat{x}(n + 1|\mathcal{Y}_n) = \boldsymbol{\Phi}(n + 1, n)\hat{x}(n|\mathcal{Y}_{n-1}) + \mathbf{G}(n)\boldsymbol{\alpha}(n)$
    $\hat{x}(n|\mathbf{y}_n) = \boldsymbol{\Phi}(n, n + 1)\hat{x}(n + 1|\mathcal{Y}_n)$
    $\mathbf{K}(n) = \mathbf{K}(n, n - 1) - \boldsymbol{\Phi}(n, n + 1)\mathbf{G}(n)\mathbf{C}(n)\mathbf{K}(n, n - 1)$
    $\mathbf{K}(n + 1, n) = \boldsymbol{\Phi}(n + 1, n)\mathbf{K}(n)\boldsymbol{\Phi}^H(n + 1, n) + \mathbf{Q}_1(n)$
Initial Conditions
    $\hat{x}(1|\mathcal{Y}_0) = \mathbf{0}$
    $\mathbf{K}(1, 0) = E[\mathbf{x}(1)\mathbf{x}^H(1)]$

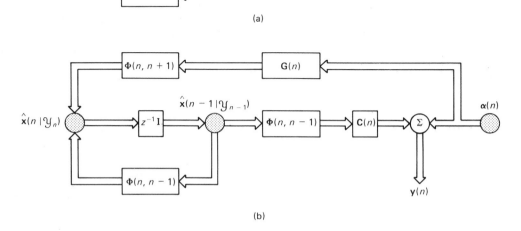

Figure 6.2 Signal-flow graph representations of (a) the Kalman filter, and (b) the inverse model.

model. In this model, the innovations process $\{\boldsymbol{\alpha}(n)\}$ acts as the input, and the process $\{\mathbf{y}(n)\}$ is the output of interest.

We may develop a third model by viewing the observations process $\{\mathbf{y}(n)\}$ as the input and the innovations process $\{\boldsymbol{\alpha}(n)\}$ as the output. The signal-flow graph representation of this model is the same as that shown in Fig. 6.2(a), except that now $\boldsymbol{\alpha}(n)$ represents the desired output at time n. The resulting model is known as the *whitening filter*.

We conclude therefore that the Kalman filter, the inverse model, and the whitening filter merely represent three different, and yet equivalent, ways of viewing the optimum linear filter.

The Kalman filter has been successfully applied to solve many real-world problems as can be seen in the literature on control systems. Invariably, assumptions are made so as to manipulate the problem of interest into a form amenable to the application of the Kalman filter theory, the aim being to produce a near-optimum and yet workable solution.

Our interest in the Kalman filter theory in this book is that it provides a general framework for the development of various adaptive filtering algorithms with a fast rate of convergence. The application of Kalman filter theory to adaptive filtering was apparently first considered by Lawrence and Kaufman (1971). This was followed by Godard (1974), who used a different approach. In particular, Godard formulated the adaptive filtering problem (using a transversal structure) as the estimation of a state vector in Gaussian noise, a classical Kalman filtering problem. The important feature of the adaptive filtering algorithm derived by Godard is that, when all the random variables pertaining to the problem have joint Gaussian distributions, the algorithm yields the fastest possible rate of convergence. Although these assumptions are not always satisfied, nevertheless, in the majority of environments encountered in practice, we find that the algorithm derived by Godard offers a much faster rate of convergence than is attainable by the simple LMS algorithm. The price that has to be paid for this improvement, however, is increased algorithm complexity.

In the next two sections we follow Godard's approach to derive two adaptive transversal filtering algorithms as special cases of Kalman filter theory. The first algorithm, presented in Section 6.8, assumes a stationary environment. The second algorithm, presented in Section 6.9, assumes a nonstationary environment.

6.8 THE KALMAN ALGORITHM APPLIED TO ADAPTIVE TRANSVERSAL FILTERS WITH STATIONARY INPUTS

Consider a linear transversal filter operating in a stationary environment. Suppose that the M-by-1 tap-weight vector of the filter at time n is set equal to the optimum Wiener value $\mathbf{w}_o(n)$, determined in accordance with the normal equation (see Chapter 2). In a stationary environment, the error-performance surface of the transversal filter has fixed shape and fixed orientation, with the result that the optimum tap-weight vector $\mathbf{w}_o(n)$ has a constant value for all time. Under this condition, we may write

$$\mathbf{w}_o^*(n + 1) = \mathbf{w}_o^*(n) \qquad (6.75)$$

Let $\mathbf{u}(n)$ denote the M-by-1 *tap-input vector* applied to this filter at time n. The resulting response of the filter equals the inner product $\mathbf{u}^T(n)\,\mathbf{w}_o^*(n)$. Let $e_o(n)$ denote the optimum value of the *estimation error*, measured with respect to the *desired response* $d(n)$. We may thus model the desired response as

$$d(n) = \mathbf{u}^T(n)\,\mathbf{w}_o^*(n) + e_o(n) \qquad (6.76)$$

Equations (6.75) and (6.76) describe the optimum Wiener condition of the transversal filter when operating in a stationary environment. The tap-weight vector on the right sides of these two equations appears in exactly the same form; hence, the use of complex conjugation in (6.75).

In the context of Kalman filter theory, we may view Eq. (6.75) as the process equation and (6.76) as the measurement equation. We may thus state the filtering problem as follows: *Identify the optimum transversal filter characterized by the process equation* (6.75), *knowing that its output signal satisfies the measurement equation* (6.76).

The interpretation of Eq. (6.75) as the process equation is straightforward. In particular, the following observations suggest themselves:

1. The state vector $\mathbf{x}(n)$ equals the complex conjugate of the optimum tap-weight vector $\mathbf{w}_o(n)$.
2. The state transition matrix $\mathbf{\Phi}(n + 1, n)$ equals the identity matrix.
3. The process noise vector $\mathbf{v}_1(n)$ is zero.

On the other hand, the interpretation of Eq. (6.76) as the measurement equation requires some special considerations:

1. With the complex conjugate of the optimum tap-weight vector $\mathbf{w}_o(n)$ viewed as the state vector, it would seem logical to say that the measurement matrix $\mathbf{C}(n)$ equals the transposed tap-input vector $\mathbf{u}^T(n)$. However, this identification deviates from the assumption made in the original formulation of the Kalman filter. In particular, in the measurement equation (6.19) of the Kalman filter, the measurement matrix $\mathbf{C}(n)$ is assumed to be *known*. On the other hand, in the measurement equation (6.76) of the optimum transversal filter, the measurement matrix identified as $\mathbf{u}^T(n)$ is *random* since the tap-input vector $\mathbf{u}(n)$ is itself random. Nevertheless, as remarked earlier, we may still apply the Kalman filter theory, in particular, Eqs. (6.45) and (6.55) to compute estimates of the state vector and the predicted state error correlation matrix. However, because of the dependence of the measurement matrix $\mathbf{C}(n)$ on the tap-input vector $\mathbf{u}(n)$, we now find that the Kalman gain $\mathbf{G}(n)$ and the predicted state error correlation matrix $\mathbf{K}(n + 1, n)$ cannot be precomputed. Consequently, $\mathbf{K}(n + 1, n)$ loses its interpretation as an unconditional error correlation matrix of the estimator. That is, $\mathbf{K}(n + 1, n)$ represents a *conditional* error correlation matrix, conditional on the tap-input vector $\mathbf{u}(n)$.
2. It would also seem logical to identify the measurement noise vector $\mathbf{v}_2(n)$ as the optimum estimation error $e_o(n)$. Accordingly, to apply the Kalman filter approach, we require that the process $\{e_o(n)\}$ be assumed white with zero mean and variance $J_{\min}$. This is a reasonable approximation, because the optimum mean-squared error is usually quite small, so the correlation between successive samples of the error process $\{e_o(n)\}$ may be neglected.

We thus find that the algorithm for adjusting the tap-weight vector in the adaptive transversal filter (toward the optimum Wiener solution), which results from the use of Eq. (6.75) as the process equation and Eq. (6.76) as the measurement equation, is a *nonlinear approximation to the linear Kalman filter*.

TABLE 6.2

State model of the Kalman filter	Constant-state model of the optimum transversal filter
$\mathbf{x}(n)$	$\mathbf{w}_o^*(n)$
$\mathbf{\Phi}(n + 1, n)$	$\mathbf{I}$
$\mathbf{v}_1(n)$	$\mathbf{0}$
$\mathbf{Q}_1(n)$	$\mathbf{0}$
$\mathbf{y}(n)$	$d(n)$
$\mathbf{C}(n)$	$\mathbf{u}^T(n)$
$\mathbf{v}_2(n)$	$e_o(n)$
$\mathbf{Q}_2(n)$	$J_{\min}$

In Table 6.2, we present a summary of the correspondences between the terms defining the process equation and the measurement equation for the Kalman filter approach and those defining the corresponding equations for the optimum transversal filter.

Since the state transition matrix $\mathbf{\Phi}(n, n+1)$ equals the identity matrix, the predicted estimate $\hat{x}(n+1|\mathcal{Y}_n)$ and the filtered estimate $\hat{x}(n|\mathcal{Y}_n)$ of the state vector assume the same value. In addition, the predicted weight-error correlation matrix $\mathbf{K}(n, n-1)$ and the filtered weight-error correlation matrix $\mathbf{K}(n-1)$ assume the same value. In any event, through the use of the identifications listed in Table 6.2 we may make the following deductions regarding the adaptive algorithm for adjusting the tap weights of the transversal filter in a stationary environment:

1. Let $\hat{\mathbf{w}}(n) = \hat{\mathbf{w}}(n|\mathcal{D}_n)$ denote the estimate of the M-by-1 tap-weight vector at time n, based on the samples $d(1), d(2), \ldots, d(n)$ of the desired response that span the space $\mathcal{D}_n$. The application of Eqs. (6.66) and (6.65) yields the time-update recursion:

$$\hat{\mathbf{w}}(n) = \hat{\mathbf{w}}(n-1) + \mathbf{g}(n)\alpha^*(n) \qquad (6.77)$$

where

$$\alpha(n) = d(n) - \mathbf{u}^T(n)\hat{\mathbf{w}}^*(n-1) \qquad (6.78)$$

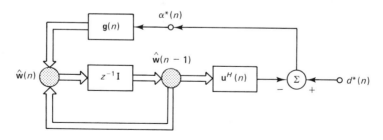

Figure 6.3 Signal-flow graph representation of the Kalman algorithm applied to an adaptive transversal filter operating in a stationary environment.

Note that in this case the Kalman gain $\mathbf{g}(n)$ is an M-by-1 vector; hence, the use of a boldfaced lowercase letter. Note also that innovation $\alpha(n)$ is a scalar. Figure 6.3 shows a signal-flow graph representation of Eqs. (6.77) and (6.78). This figure also includes a branch to represent the fact that $\hat{\mathbf{w}}(n-1)$ represents the delayed version of the tap-weight vector estimate $\hat{\mathbf{w}}(n)$.

2. The application of Eq. (6.49) yields the following formula for computing the gain vector:

$$\mathbf{g}(n) = \mathbf{K}(n-1)\,\mathbf{u}(n)\,[\mathbf{u}^H(n)\mathbf{K}(n-1)\mathbf{u}(n) + J_{\min}]^{-1} \tag{6.79}$$

where $J_{\min}$ is the minimum mean-squared error, and $\mathbf{K}(n-1)$ is the correlation matrix of the weight-error vector:[1]

$$\boldsymbol{\varepsilon}(n-1) = \hat{\mathbf{w}}(n-1) - \mathbf{w}_o \tag{6.80}$$

That is,

$$\mathbf{K}(n-1) = E[\boldsymbol{\varepsilon}(n-1)\boldsymbol{\varepsilon}^H(n-1)] \tag{6.81}$$

Note that the use of Eq. (6.79) to compute the Kalman gain $\mathbf{g}(n)$ requires *prior* knowledge of the minimum mean-squared error, $J_{\min}$. Clearly, when the adaptive transversal filter operates in an unknown environment, it is impossible to meet such a requirement exactly. However, in most cases of practical interest, we find that after adaptation is complete the signal-to-noise ratio at the output of the filter is quite high (typically, between 20 and 30 dB). Thus, to compute the Kalman gain $\mathbf{g}(n)$, we may use a value between 0.01 and 0.001 times the variance of the desired response $d(n)$ as a *reasonable guess* for $J_{\min}$. Later in the section, we show that $J_{\min}$ has a minor influence on the operation of the algorithm.

3. The application of Eq. (6.56) yields the recursion

$$\mathbf{K}(n) = \mathbf{K}(n-1) - \mathbf{g}(n)\mathbf{u}^H(n)\mathbf{K}(n-1) \tag{6.82}$$

for time-updating the correlation matrix of the weight–error vector.

4. In addition to the initial guess at the value of $J_{\min}$, we need the initial values $\hat{\mathbf{w}}(0)$ and $\mathbf{K}(0)$ to start the adaptive process. For the tap-weight vector, it is customary to set $\hat{\mathbf{w}}(0) = \mathbf{0}$. As for $\mathbf{K}(0)$, we may set

$$\begin{aligned}\mathbf{K}(0) &= E[\mathbf{w}_o\mathbf{w}^H_o] \\ &= c\mathbf{I}\end{aligned} \tag{6.83}$$

[1]In Eq. (6.81), the weight-error correlation matrix is defined with respect to the optimum tap-weight vector $\mathbf{w}_o$ rather than the state vector $\mathbf{w}_o^*$. Accordingly, it plays a complex-conjugate role to the state-error correlation matrix of the Kalman filter theory in Sections 6.6 and 6.7. Likewise, the Kalman gain vector $\mathbf{g}(n)$ of Eq. (6.79) plays a complex-conjugate role to the Kalman gain matrix $\mathbf{G}(n)$ of Sections 6.6 and 6.7. Note also that the weight-error vector $\boldsymbol{\varepsilon}(n-1)$ in Eq. (6.80) is defined as the difference between the estimate $\hat{\mathbf{w}}(n-1)$ and the optimum value $\mathbf{w}_o$, so that we are consistent with the notation used in Chapter 5. Such a definition, however, is the opposite of that used in the Kalman filter theory. Clearly, the error correlation matrix is unaffected by such a reversal in algebraic sign.

TABLE 6.3 SUMMARY OF KALMAN ALGORITHM FOR ADAPTIVE TRANSVERSAL FILTERS
ASSUMING A CONSTANT-STATE MODEL (STATIONARY ENVIRONMENT)

Input Variables
  Tap-input vectors: $\mathbf{u}(1), \mathbf{u}(2), \ldots, \mathbf{u}(n)$
  Desired response: $d(1), d(2), \ldots, d(n)$
Computation: $n = 1, 2, 3, \ldots$
  $\mathbf{g}(n) = \mathbf{K}(n-1)\,\mathbf{u}(n)[\mathbf{u}^H(n)\mathbf{K}(n-1)\mathbf{u}(n) + J_{\min}]^{-1}$
  $\alpha(n) = d(n) - \hat{\mathbf{w}}^H(n-1)\mathbf{u}(n)$
  $\hat{\mathbf{w}}(n) = \hat{\mathbf{w}}(n-1) + \mathbf{g}(n)\alpha^*(n)$
  $\mathbf{K}(n) = \mathbf{K}(n-1) - \mathbf{g}(n)\mathbf{u}^H(n)\mathbf{K}(n-1)$
Assumption:
  $J_{\min} = 0.001$ to 0.01 times the variance of desired response $d(n)$.
Initial Conditions
  $\hat{\mathbf{w}}(0) = \mathbf{0}$
  $\mathbf{K}(0) = c\mathbf{I}, \quad c > 0$

where $\mathbf{I}$ is the M-by-M identity matrix, and c is a positive constant. The off-diagonal terms of $\mathbf{K}(0)$ in Eq. (6.83) are all zero, reflecting the assumed statistical independence of the elements of $\mathbf{w}_o$.

In Table 6.3, we present a summary of the Kalman algorithm, assuming a constant-state model (i.e., stationary environment).

Another Form of the Algorithm

Multiplying both sides of Eq. (6.79) by the scalar

$$\mathbf{u}^H(n)\mathbf{K}(n-1)\mathbf{u}(n) + J_{\min}$$

and rearranging terms so as to factor out $\mathbf{K}(n-1)\mathbf{u}(n)$, we get

$$[\mathbf{I} - \mathbf{g}(n)\mathbf{u}^H(n)]\mathbf{K}(n-1)\mathbf{u}(n) = J_{\min}\,\mathbf{g}(n) \tag{6.84}$$

where $\mathbf{I}$ is the M-by-M identity matrix. Next, postmultiplying both sides of Eq. (6.82) by $\mathbf{u}(n)$, we get

$$\mathbf{K}(n)\mathbf{u}(n) = [\mathbf{I} - \mathbf{g}(n)\mathbf{u}^H(n)]\mathbf{K}(n-1)\mathbf{u}(n) \tag{6.85}$$

Substituting Eq. (6.84) in (6.85) and solving for the Kalman gain $\mathbf{g}(n)$, we obtain

$$\mathbf{g}(n) = \frac{1}{J_{\min}}\,\mathbf{K}(n)\mathbf{u}(n) \tag{6.86}$$

This result may be used as a definition of the Kalman gain for the case of an adaptive transversal filter operating in a stationary environment.

By using Eq. (6.86) in (6.77), we may now rewrite the time update relation for the estimate of the tap-weight vector as follows:

$$\hat{\mathbf{w}}(n) = \hat{\mathbf{w}}(n-1) + \mathbf{S}(n)\mathbf{u}(n)\alpha^*(n) \tag{6.87}$$

where the M-by-M matrix $\mathbf{S}(n)$ is obtained by normalizing the weight-error correlation matrix $\mathbf{K}(n)$ with respect to the minimum mean-squared error $J_{\min}$; that is,

$$\mathbf{S}(n) = \frac{1}{J_{\min}} \mathbf{K}(n) \tag{6.88}$$

The correction term in the recursion of Eq. (6.87) consists of the quantity $\mathbf{u}(n)\alpha^*(n)$, representing an estimate of the gradient vector (based on instantaneous values of the tap-input vector and the innovation), premultiplied by the matrix $\mathbf{S}(n)$. This new ingredient in the Kalman algorithm, namely, premultiplication by $\mathbf{S}(n)$, has a profound effect on the convergence properties of the algorithm. Indeed, it is responsible for the much faster rate of convergence that is attained with the Kalman algorithm, compared to the simple LMS algorithm.

Eliminating $\mathbf{g}(n)$ between Eqs. (6.82) and (6.86) and using Eq. (6.88), we get

$$\mathbf{S}(n) = \mathbf{S}(n-1) - \mathbf{S}(n)\mathbf{u}(n)\mathbf{u}^H(n)\mathbf{S}(n-1) \tag{6.89}$$

We note that since $\mathbf{K}(n)$, and therefore $\mathbf{S}(n)$, is a correlation matrix, it is positive definite and has an inverse, and so does $\mathbf{S}(n - 1)$. Hence, premultiplying both sides of Eq. (6.89) by the inverse matrix $\mathbf{S}^{-1}(n)$, postmultiplying them by the second inverse matrix $\mathbf{S}^{-1}(n-1)$, and rearranging terms, we get

$$\mathbf{S}^{-1}(n) = \mathbf{S}^{-1}(n-1) + \mathbf{u}(n)\mathbf{u}^H(n) \tag{6.90}$$

This equation represents a difference equation (of order 1) in the inverse of the normalized weight-error correlation matrix. Thus, starting with the initial condition

$$\mathbf{S}(0) = \frac{1}{J_{\min}} \mathbf{K}(0) \tag{6.91}$$

we derive the following solution:

$$\mathbf{S}^{-1}(n) = \mathbf{S}^{-1}(0) + \sum_{i=1}^{n} \mathbf{u}(i)\mathbf{u}^H(i)$$

Equivalently, we may write

$$\mathbf{S}(n) = \left[\mathbf{S}^{-1}(0) + \sum_{i=1}^{n} \mathbf{u}(i)\mathbf{u}^H(i) \right]^{-1} \tag{6.92}$$

For sufficiently large n, such as $n \geq M$, where M is the number of taps, we usually find that the diagonal matrix $\mathbf{S}^{-1}(0) = J_{\min}[\mathbf{K}^{-1}(0)]$ is small enough to be neglected in comparison with the matrix

$$\sum_{i=1}^{n} \mathbf{u}(i)\mathbf{u}^H(i)$$

Then we may approximate Eq. (6.92) as follows:

$$\mathbf{S}(n) \simeq \left[\sum_{i=1}^{n} \mathbf{u}(i)\mathbf{u}^{H}(i) \right]^{-1} \tag{6.93}$$

Let $\hat{\mathbf{R}}(n)$ denote the *sample mean estimate* of the correlation matrix of the tap inputs, as shown by

$$\hat{\mathbf{R}}(n) = \frac{1}{n} \sum_{i=1}^{n} \mathbf{u}(i)\mathbf{u}^{H}(i) \tag{6.94}$$

The estimate $\hat{\mathbf{R}}(n)$ is unbiased in that its expected value equals the true value of the ensemble averaged correlation matrix $\mathbf{R}$. Accordingly, the matrix $\mathbf{S}(n)$ converges rapidly to a form that is proportional to the sample mean estimate of the inverse matrix $\mathbf{R}^{-1}$, as shown by

$$\mathbf{S}(n) = \frac{1}{n} \hat{\mathbf{R}}^{-1}(n) \tag{6.95}$$

We see therefore that the M-by-M matrix $\mathbf{S}(n)$, by which the instantaneous estimate of the gradient vector is premultiplied, converges rapidly to $1/n$ times the sample mean estimate of the inverse matrix $\mathbf{R}^{-1}$.

Tap-weight Vector Estimate

The innovation $\alpha(n)$ is defined by Eq. (6.78). Hence, using this definition in Eq. (6.87) and rearranging terms, we get

$$\hat{\mathbf{w}}(n) = [\mathbf{I} - \mathbf{S}(n)\mathbf{u}(n)\,\mathbf{u}^{H}(n)]\hat{\mathbf{w}}(n-1) + \mathbf{S}(n)\mathbf{u}(n)d^{*}(n)$$

Premultiplying both sides of this equation by the inverse matrix $\mathbf{S}^{-1}(n)$, we get

$$\mathbf{S}^{-1}(n)\hat{\mathbf{w}}(n) = [\mathbf{S}^{-1}(n) - \mathbf{u}(n)\mathbf{u}^{H}(n)]\hat{\mathbf{w}}(n-1) + \mathbf{u}(n)d^{*}(n) \tag{6.96}$$

However, from Eq. (6.90), we observe that $\mathbf{S}^{-1}(n) - \mathbf{u}(n)\mathbf{u}^{H}(n)$ equals the inverse matrix $\mathbf{S}^{-1}(n-1)$. We may rewrite Eq. (6.96) in the form

$$\mathbf{S}^{-1}(n)\hat{\mathbf{w}}(n) = \mathbf{S}^{-1}(n-1)\hat{\mathbf{w}}(n-1) + \mathbf{u}(n)d^{*}(n) \tag{6.97}$$

Equation (6.97) represents a difference equation in $\mathbf{S}^{-1}(n)\hat{\mathbf{w}}(n)$. Therefore, starting with the initial condition $\hat{\mathbf{w}}(0) = \mathbf{0}$, we find that the solution of this equation is

$$\hat{\mathbf{w}}(n) = \mathbf{S}(n) \sum_{i=1}^{n} \mathbf{u}(i)d^{*}(i) \tag{6.98}$$

where $\mathbf{S}(n)$ is itself defined by Eq. (6.92). Hence, substituting the definition of $\mathbf{S}(n)$ in Eq. (6.98), we get

$$\hat{\mathbf{w}}(n) = \left[\mathbf{S}^{-1}(0) + \sum_{i=1}^{n} \mathbf{u}(i)\mathbf{u}^{H}(i) \right]^{-1} \sum_{i=1}^{n} \mathbf{u}(i)d^{*}(i) \tag{6.99}$$

Earlier, we remarked that after a sufficiently large number of iterations, such as $n \geq M$, where M is the number of taps, we may neglect the diagonal matrix $\mathbf{S}^{-1}(0)$ compared to the sum

$$\sum_{i=1}^{n} \mathbf{u}(i)\mathbf{u}^{H}(i)$$

Under this condition, we may therefore approximate the tap-weight vector at the nth iteration as follows:

$$\hat{\mathbf{w}}(n) = \left[\sum_{i=1}^{n} \mathbf{u}(i)\mathbf{u}^{H}(i)\right]^{-1} \sum_{i=1}^{n} \mathbf{u}(i)d^{*}(i) \tag{6.100}$$

We assume that the processes $\{d(n)\}$ and $\{u(n)\}$ are jointly ergodic, so that we may interchange ensemble averages and time averages, as shown by

$$\mathbf{R} = \lim_{n \to \infty} \frac{1}{n} \sum_{i=1}^{n} \mathbf{u}(i)\mathbf{u}^{H}(i) \tag{6.101}$$

and

$$\mathbf{p} = \lim_{n \to \infty} \frac{1}{n} \sum_{i=1}^{n} \mathbf{u}(i)d^{*}(i) \tag{6.102}$$

where $\mathbf{R}$ is the ensemble-averaged correlation matrix of the tap inputs, and $\mathbf{p}$ is the cross-correlation vector between the tap inputs and the desired response.

Accordingly, we deduce from Eq. (6.100) that

$$\lim_{n \to \infty} \hat{\mathbf{w}}(n) = \mathbf{R}^{-1}\mathbf{p}$$
$$= \mathbf{w}_{o} \tag{6.103}$$

That is, as the number of iterations approaches infinity, the estimate of the tap-weight vector produced by the Kalman algorithm approaches the optimum Wiener solution, $\mathbf{w}_{o}$.

Average Mean-squared Error

The average mean-squared error $E[J(n)]$ may be expressed as follows [see Eqs. (5.79) and (5.80)]:

$$E[J(n)] = J_{\min} + \text{tr}[\mathbf{R}\mathbf{K}(n)] \tag{6.104}$$
$$= J_{\min} + J_{\min}\,\text{tr}[\mathbf{R}\mathbf{S}(n)]$$

where in the second line we have used the definition of Eq. (6.88). For $n \geq M$, we may approximate $\mathbf{S}(n)$ by Eq. (6.95). Correspondingly, we may approximate Eq. (6.104) as follows:

$$E[J(n)] \simeq J_{\min} + \frac{1}{n} J_{\min}\,\text{tr}[\mathbf{R}\hat{\mathbf{R}}^{-1}(n)]$$

Putting $\hat{\mathbf{R}}(n) \simeq \mathbf{R}$, which is justified to a first order of approximation for large n, and recognizing that the trace of the M-by-M identity matrix equals M, the number of taps, we have

$$E[J(n)] \simeq J_{\min} \left(1 + \frac{M}{n} \right) \tag{6.105}$$

This result shows that convergence of the Kalman algorithm, applied to an adaptive transversal filter operating in a stationary environment, can theoretically be obtained within $2M$ iterations (Godard, 1974). The average mean-squared error will have then dropped approximately to $1.5\,J_{\min}$, which is close enough to the optimum condition. Note also that when the number of iterations approaches infinity the average excess mean-squared error approaches zero and, correspondingly, the average mean-squared error approaches the optimum value $J_{\min}$.

Discussion

The special form of the Kalman algorithm just described for the operation of an adaptive transveral filter with stationary inputs differs from the LMS algorithm in the following ways:

1. Typically, the Kalman algorithm requires about $2M$ iterations to converge, whereas the LMS algorithm requires about $20M$ iterations to converge, where M is the number of taps in the transversal filter. Thus, the Kalman algorithm provides a much faster rate of convergence than the LMS algorithm. The reason for this is that, in the Kalman algorithm, all the information available from the start of the adaptive process up to the current time is exploited in the update procedure for the tap-weight vector. On the other hand, the LMS algorithm relies mainly on currently available information.

2. As the number of iterations approaches infinity, the tap-weight vector estimate produced by the Kalman algorithm approaches the optimum Wiener value, provided that the stochastic processes supplying the desired response and the tap inputs are jointly ergodic. On the other hand, the LMS algorithm is convergent in the mean in that, in this case, as the number of iterations approaches infinity, the ensemble-averaged value, and *not* the sample value, of the tap-weight vector estimate approaches the optimum Wiener value.

3. The average excess mean-square error produced by the Kalman algorithm approaches zero as the number of iterations approaches infinity. That is, the Kalman algorithm exhibits zero misadjustment. On other hand, the LMS algorithm produces a misadjustment that is proportional to the number of taps. This misadjustment can be made negligibly small, however, by using a small enough step-size parameter μ.

4. The Kalman algorithm is inherently stable (in theory). On the other hand, for the LMS algorithm to be stable, the step-size parameter used in the algorithm has to be chosen within certain bounds.

5. The Kalman algorithm is more complex in its implementation than the LMS algorithm. Thus, the advantages of the Kalman algorithm summarized here are obtained at the expense of increased computational complexity.

6.9 *THE KALMAN ALGORITHM APPLIED TO ADAPTIVE TRANSVERSAL FILTERS WITH NONSTATIONARY INPUTS*

Consider next the application of the Kalman filter approach to an adaptive transversal filter that operates in a nonstationary environment. In this case, the error-performance surface of the filter changes its position randomly. To account for this random motion, we express the process equation of the filter as follows:

$$\mathbf{w}_o^*(n+1) = \mathbf{w}_o^*(n) + \mathbf{v}(n) \tag{6.106}$$

It is assumed that:

1. The state transition matrix $\mathbf{\Phi}(n+1, n)$ equals the identity matrix
2. The process noise vector, $\mathbf{v}(n)$, has zero mean and correlation matrix $\mathbf{Q}(n)$; hence, the tap-weight vector varies randomly about a mean value.

It is the second assumption that distinguishes the nonstationary case considered here from the stationary case considered in the previous section. For the measurement equation, we write (as for the stationary case)

$$d(n) = \mathbf{u}^T(n)\mathbf{w}_o^*(n) + e_o(n) \tag{6.107}$$

where $\mathbf{u}(n)$ is the tap-input vector, and the estimation error $e_o(n)$ has zero mean and variance J_{min}.

Comparing Eqs. (6.17) and (6.19) for the Kalman filter with Eqs. (6.106) and (6.107) for an optimum transversal filter with nonstationary inputs, we may make the identifications listed in Table 6.4. The presence of the process noise vector $\mathbf{v}(n)$ makes the predicted weight-error correlation matrix and the filtered-weight-

TABLE 6.4

State model of the Kalman filter	Random-walk state model of the optimum transversal filter
$\mathbf{x}(n)$	$\mathbf{w}_o^*(n)$
$\mathbf{\Phi}(n + 1, n)$	$\mathbf{I}$
$\mathbf{v}_1(n)$	$\mathbf{v}(n)$
$\mathbf{Q}_1(n)$	$\mathbf{Q}(n)$
$\mathbf{y}(n)$	$d(n)$
$\mathbf{C}(n)$	$\mathbf{u}^T(n)$
$\mathbf{v}_2(n)$	$e_o(n)$
$\mathbf{Q}_2(n)$	J_{min}

error correlation matrix assume different values. We define the predicted weight-error correlation matrix as

$$\mathbf{K}(n, n - 1) = E[\boldsymbol{\varepsilon}(n, n - 1)\boldsymbol{\varepsilon}^H(n, n - 1)] \tag{6.108}$$

where

$$\boldsymbol{\varepsilon}(n, n - 1) = \hat{\mathbf{w}}(n - 1) - \mathbf{w}_o(n) \tag{6.109}$$

We define the filtered weight-error correlation matrix as

$$\mathbf{K}(n) = E[\boldsymbol{\varepsilon}(n)\boldsymbol{\varepsilon}^H(n)] \tag{6.110}$$

where

$$\boldsymbol{\varepsilon}(n) = \hat{\mathbf{w}}(n) - \mathbf{w}_o(n) \tag{6.111}$$

For the nonstationary case, we find that, in general, $\mathbf{K}(n, n-1) \neq \mathbf{K}(n - 1)$.

Thus, in view of the identifications listed in Table 6.4, the Kalman algorithm takes on the following special form for adaptive transversal filters modeled by Eqs. (6.106) and (6.107):[2]

$$\mathbf{g}(n) = \mathbf{K}(n, n-1)\mathbf{u}(n)[\mathbf{u}^H(n)\mathbf{K}(n, n-1)\mathbf{u}(n) + J_{\min}]^{-1} \tag{6.112}$$

$$\alpha(n) = d(n) - \mathbf{u}^T(n)\hat{\mathbf{w}}^*(n-1) \tag{6.113}$$

$$\hat{\mathbf{w}}(n) = \hat{\mathbf{w}}(n-1) + \mathbf{g}(n)\alpha^*(n) \tag{6.114}$$

$$\mathbf{K}(n) = \mathbf{K}(n, n-1) - \mathbf{g}(n)\mathbf{u}^H(n)\mathbf{K}(n, n-1) \tag{6.115}$$

$$\mathbf{K}(n+1, n) = \mathbf{K}(n) + \mathbf{Q}(n) \tag{6.116}$$

In a stationary environment, the matrix $\mathbf{Q}(n)$ is identically zero, in which case this algorithm reduces to the special form given in Section 6.8.

The application of this algorithm requires knowledge of the correlation matrix $\mathbf{Q}(n)$. We consider two possibilities:

1. The correlation matrix of the process noise vector $\mathbf{v}(n)$ equals

$$\mathbf{Q}(n) = E[\mathbf{v}(n)\mathbf{v}^H(n)] = q\mathbf{I} \tag{6.117}$$

   where q is a scalar and $\mathbf{I}$ is the identity matrix. This, in effect, assumes that the elements of $\mathbf{v}(n)$ constitute a set of independent white-noise processes, each having zero mean and variance q. Such a model is commonly referred to as a *random-walk state model* (Godard, 1974; Zhang and Haykin, 1983).

---

[2]As in the case of stationary environment considered in Section 6.8, the predicted and filtered versions of the weight-error correlation matrix are both defined with respect to the optimum weight vector $\mathbf{w}_o(n)$ rather than the state vector $\mathbf{w}_o^*(n)$. Accordingly, $\mathbf{K}(n, n - 1)$ and $\mathbf{K}(n)$ play conjugate roles to their counterparts in the Kalman filter theory. Likewise, the Kalman gain vector $\mathbf{g}(n)$ and the correlation matrix $\mathbf{Q}(n)$ of the noise vector $\mathbf{v}(n)$ play conjugate roles to the Kalman gain $\mathbf{G}(n)$ and the correlation matrix $\mathbf{Q}_1(n)$ of the process noise vector in the Kalman filter theory, respectively.

2. The correlation matrix of the process noise vector $\mathbf{v}(n)$ equals a scaled version of the filtered weight-error correlation matrix $\mathbf{K}(n)$, as shown by

$$\mathbf{Q}(n) = E\left[\mathbf{v}(n)\mathbf{v}^H(n)\right] = q\mathbf{K}(n) \tag{6.118}$$

where q is the scaling factor. In this case, the Kalman algorithm described by Eqs. (6.112) to (6.116) simplifies as follows (Goodwin and Payne, 1977):

$$\mathbf{g}(n) = \mathbf{K}(n-1)\mathbf{u}(n)[\mathbf{u}^H(n)\mathbf{K}(n-1)\mathbf{u}(n) + \xi_{\min}]^{-1} \tag{6.119}$$

$$\alpha(n) = d(n) - \mathbf{u}^T(n)\hat{\mathbf{w}}^*(n-1) \tag{6.120}$$

$$\hat{\mathbf{w}}(n) = \hat{\mathbf{w}}(n-1) + \mathbf{g}(n)\alpha^*(n) \tag{6.121}$$

$$\mathbf{K}(n) = (1+q)[\mathbf{K}(n-1) - \mathbf{g}(n)\mathbf{u}^H(n)\mathbf{K}(n-1)] \tag{6.122}$$

$$\xi_{\min} = \frac{J_{\min}}{1+q} \tag{6.123}$$

It is of interest to note that the formula for the Kalman gain defined by Eq. (6.119) has the same form as that of Eq. (6.79) for the case of stationary inputs, except for the fact that $J_{\min}$ is replaced by $\xi_{\min}$. Also, the formula for updating the filtered weight-error correlation matrix, defined by Eq. (6.122), has the same form as the corresponding formula of Eq. (6.82) for the case of stationary inputs, except for the scaling factor $(1+q)$. The rate of convergence of the algorithm described by Eqs. (6.119) through (6.122) is controlled by the parameter $\xi_{\min}$. Specifically, a large value of $\xi_{\min}$ (equivalently, a small value of q) results in a slow rate of convergence. On the other hand, a small value of $\xi_{\min}$ produces a fast rate of convergence but a larger variance of estimation error.

Example 1: Dynamic Autoregressive Model

Consider an autoregressive (AR) process $\{u(n)\}$ of order M, which is described by the difference equation

$$\sum_{k=0}^{M} a_k^*(n)u(n-k) = e_o(n) \tag{6.124}$$

where $a_0 = 1$, and $a_1(n), \ldots, a_M(n)$ are time-varying AR parameters, and $\{e_o(n)\}$ is a stationary noise process of zero mean and variance σ^2. The dependence of the AR parameters on time n signifies the dynamic nature of the process. We may rewrite Eq. (6.124) in the form

$$u(n) = \mathbf{u}^T(n-1)\,\mathbf{w}_o^*(n) + e_o(n) \tag{6.125}$$

where the M-by-1 vector $\mathbf{u}(n-1)$ has $u(n-1), \ldots, u(n-M)$ as elements, and the kth element of the M-by-1 vector $\mathbf{w}_o(n)$ equals $w_k(n) = -a_k(n)$, $k = 1, 2, \ldots, M$. We may view Eq. (6.125) as the measurement equation of the AR process.

Assuming a random-walk model for the process equation of the AR model, we write

$$\mathbf{w}_o^*(n+1) = \mathbf{w}_o^*(n) + \mathbf{v}^*(n) \tag{6.126}$$

where $\{v(n)\}$ is a stationary noise process of zero mean and correlation matrix $\mathbf{Q}(n)$ $= q\mathbf{I}$. The processes $\{e_o(n)\}$ and $\{v(n)\}$ are statistically independent.

In Eq. (6.125), we see that the measurement matrix $\mathbf{C}(n)$ of the AR process equals $\mathbf{u}^T(n-1)$. On the other hand, in the Kalman algorithm of Eqs. (6.112) to (6.116), pertaining to an adaptive transversal filter described by the random-walk state model, the measurement matrix $\mathbf{C}(n)$ equals $\mathbf{u}^T(n)$. Accordingly, the use of Eqs. (6.112) to (6.116), with $\mathbf{u}(n-1)$ in place of $\mathbf{u}(n)$ and $u(n)$ as the desired response $d(n)$, yields

$$\mathbf{g}(n) = \mathbf{K}(n, n-1)\mathbf{u}(n-1)[\mathbf{u}^H(n-1)\mathbf{K}(n, n-1)\mathbf{u}(n-1) + \sigma^2]^{-1} \tag{6.127}$$

$$\alpha(n) = u(n) - \mathbf{u}^T(n-1)\hat{\mathbf{w}}^*(n-1) \tag{6.128}$$

$$\hat{\mathbf{w}}(n) = \hat{\mathbf{w}}(n-1) + \mathbf{g}(n)\alpha^*(n) \tag{6.129}$$

$$\mathbf{K}(n) = \mathbf{K}(n, n-1) - \mathbf{g}(n)\mathbf{u}^H(n-1)\mathbf{K}(n, n-1) \tag{6.130}$$

$$\mathbf{K}(n+1, n) = \mathbf{K}(n) + q\mathbf{I} \tag{6.131}$$

Thus, starting with the initial condition

$$\hat{\mathbf{w}}(0) = \mathbf{0}$$

$$\mathbf{K}(1,0) = c\mathbf{I}$$

we may use this algorithm to model a nonstationary process $\{u(n)\}$ as an adaptive AR process, the parameters of which vary randomly about some mean values (Nau and Oliver, 1979). The use of this algorithm assumes prior knowledge of (1) the variance σ^2 of the residual $e_o(n)$ contained in Eq. (6.124) describing the AR equation of dynamic behavior, and (2) the variance q of each element of the noise vector $\mathbf{v}(n)$ that governs the random walk of the AR parameters with time.

Discussion

A nonstationary environment may be well-described by the state-space approach if we know the transition matrix $\mathbf{\Phi}(n+1, n)$. Unfortunately, in practice, we have no a priori knowledge of the transition matrix. As an alternative, the use of a random walk state model offers a simple and yet adequate description of a nonstationary environment, especially when the environment is *slowly varying*.

Equations (6.112) through (6.116) describe a special form of the Kalman algorithm that uses the random-walk state model. Basically, the effect of nonstationarity on the behavior of the algorithm is summed up exclusively in the makeup of the correlation matrix $\mathbf{Q}(n)$ that determines the time-updating of the predicted weight-error correlation matrix $\mathbf{K}(n+1, n)$ [see Eq. (6.116)]. In practice, the degree of success achieved in the application of this algorithm depends entirely on the manner in which the correlation matrix $\mathbf{Q}(n)$ is formulated. In the material presented, we have discussed two special forms of the matrix $\mathbf{Q}(n)$, defined by Eqs. (6.117) and (6.118). Although neither of these two models describes a true situation exactly, nevertheless, it appears that their use could produce good results in the adaptive equalization of slowly time-varying communication channels (Godard, 1974; Hsu, 1983) and system identification (Ljung and Söderström, 1984).

It is also of interest to note that the Kalman algorithm described by Eqs. (6.119) through (6.122) has the same mathematical form as the recursive algorithm that would result from minimization of the sum of exponentially weighted error squares

$$\mathscr{E}(n) = \sum_{i=1}^{n} \lambda^{n-i} |d(i) - \mathbf{u}^T(i)\mathbf{w}^*(i)|^2 \tag{6.132}$$

Specifically, Eqs. (6.119) through (6.122) are the solution to this minimization problem if we choose $\lambda = (1+q)^{-1}$ and $\lambda = \xi_{min}$, where ξ_{min} is defined in Eq. (6.123). The exponentially weighted error criterion has intuitive appeal in that the smaller we make λ the more rapidly the distant past is forgotten. Clearly, this is a desirable property in tracking the time variations of a nonstationary environment. We will discuss this issue in greater detail in Chapter 8.

6.10 SQUARE-ROOT KALMAN FILTERING ALGORITHM

The basic form of the Kalman filter described in Section 6.6 suffers from a *numerical instability problem* that originates from the recursive formula used to compute the filtered state-error correlation matrix $\mathbf{K}(n)$ [see Eq. (6.56)]. We see that in this equation and in all the special forms of it the correlation matrix $\mathbf{K}(n)$ is computed as the difference between two nonnegative definite matrices. Hence, unless the numerical accuracy used at every iteration is high enough, the matrix $\mathbf{K}(n)$ resulting from this computation may *not* be nonnegative definite, as required. Another factor that contributes to numerical instability of the Kalman filter is the integrated effect of roundoff errors, especially when the dimension M of the state vector is high. As a result of the numerical degeneration of the classical Kalman filter, the matrix $\mathbf{K}(n)$ may assume an indefinite form, having both positive and negative eigenvalues. Clearly, this is unacceptable.

To overcome this problem, Bierman (1977) has developed a numerically favorable version of the Kalman filter for *real-valued data* by factoring the filtered state-error correlation matrix $\mathbf{K}(n)$ into an upper triangular matrix $\mathbf{U}(n)$ with 1's along its main diagonal and a diagonal matrix $\mathbf{D}(n)$ as shown by

$$\mathbf{K}(n) = \mathbf{U}(n)\mathbf{D}(n)\mathbf{U}^T(n)$$

For obvious reasons, the factorization is known as the *real UD-factorization*. (Note that this factorization has a similar mathematical form to that discussed in Section

4.17.) Equivalently, the factorization may be written as

$$\mathbf{K}(n) = \left(\mathbf{U}(n)\mathbf{D}^{1/2}(n)\right)\left(\mathbf{U}(n)\mathbf{D}^{1/2}(n)\right)^{T}$$

where $\mathbf{D}^{1/2}(n)$ is the square root of $\mathbf{D}(n)$. Accordingly, the adaptive filtering algorithm that results from this modification is known as the *square-root Kalman filter* even though the computation of square roots is not required. In any event, the nonnegative definiteness of the computed matrix $\mathbf{K}(n)$ is guaranteed, which is achieved by updating the factors $\mathbf{U}(n)$ and $\mathbf{D}(n)$ instead of $\mathbf{K}(n)$ itself. The price paid for this improvement, however, is an algorithm that consists of a somewhat lengthy set of expressions; for details, the reader is referred to Bierman (1977).

In the context of matters of interest in this chapter, Bierman's square-root Kalman algorithm is of limited applicability. Specifically, in the random-walk state model considered in Section 6.9 for nonstationary environment, it does not include the noise correlation matrix $\mathbf{Q}(n)$ that appears in the update relations for the filtered weight-error correlation matrix $\mathbf{K}(n)$ [see Eqs. (6.115) and (6.116)]. Accordingly, it cannot be used to improve the numerical stability of the Kalman algorithm for the random-walk state model. To get around this limitation, Hsu have developed another square-root Kalman algorithm for *complex-valued data* that is based on the random-walk state model with $\mathbf{Q}(n) = q\mathbf{I}$ (Hsu, 1982). The derivation of the algorithm involves a *complex UD-factorization* of the filtered weight-error correlation matrix, as shown by

$$\mathbf{K}(n) = \mathbf{U}(n)\mathbf{D}(n)\mathbf{U}^{H}(n)$$

Efficient and numerically stable updating recursions are achieved for the upper triangular matrix $\mathbf{U}(n)$ and the diagonal matrix $\mathbf{D}(n)$. Moreover, the correlation matrix $\mathbf{Q}(n)$ is computed implicitly in each iteration. For details of the algorithm, the reader is referred to the paper by Hsu (1982).

PROBLEMS

1. The Gram–Schmidt orthogonalization procedure enables the set of observation vectors $\mathbf{y}(1), \mathbf{y}(2), \ldots, \mathbf{y}(n)$ to be transformed into the set of innovations processes $\boldsymbol{\alpha}(1), \boldsymbol{\alpha}(2), \ldots, \boldsymbol{\alpha}(n)$ without loss of information, and vice versa. Illustrate this procedure for $n = 2$, and comment on the procedure for $n > 2$.

2. The predicted state-error vector is defined by

$$\boldsymbol{\varepsilon}(n, n-1) = \mathbf{x}(n) - \hat{\mathbf{x}}(n|\mathcal{Y}_{n-1})$$

where $\hat{\mathbf{x}}(n|\mathcal{Y}_{n-1})$ is the minimum mean-square estimate of the state $\mathbf{x}(n)$, given the space $\mathcal{Y}_{n-1}$ that is spanned by the observed data $\mathbf{y}(1), \ldots, \mathbf{y}(n-1)$. Let $\mathbf{v}_1(n)$ and $\mathbf{v}_2(n)$ denote the process noise and measurement noise vectors, respectively. Show that

$\varepsilon(n, n - 1)$ is orthogonal to both $\mathbf{v}_1(n)$ and $\mathbf{v}_2(n)$; that is,

$$E[\varepsilon(n, n - 1)\mathbf{v}_1^H(n)] = \mathbf{0}$$

and

$$E[\varepsilon(n, n - 1)\mathbf{v}_2^H(n)] = \mathbf{0}$$

3. Consider a set of scalar observations $\{y(n)\}$ of zero mean, which is transformed into the corresponding set of innovations $\{\alpha(n)\}$ of zero mean and variance $\sigma_\alpha^2(n)$. Let the estimate of the state vector $\mathbf{x}(i)$, given this set of data, be expressed as

$$\hat{\mathbf{x}}(i|\mathcal{Y}_n) = \sum_{k=1}^{n} \mathbf{b}_i(k)\alpha(k)$$

where $\mathcal{Y}_n$ is the space spanned by $y(1), \ldots, y(n)$, and $\{\mathbf{b}_i(k)\}$ is a set of vectors to be determined. The requirement is to choose the $\mathbf{b}_i(k)$ so as to minimize the expected value of the squared norm of the estimated state-error vector

$$\varepsilon(i|\mathcal{Y}_n) = \mathbf{x}(i) - \hat{\mathbf{x}}(i|\mathcal{Y}_n)$$

show that this minimization yields the result

$$\hat{\mathbf{x}}(i|\mathcal{Y}_n) = \sum_{k=1}^{n} E[\mathbf{x}(i)\phi^*(k)]\phi(k)$$

where $\phi(k)$ is the normalized innovation

$$\phi(k) = \frac{\alpha(k)}{\sigma_\alpha(k)}$$

This result may be viewed as a special case of Eqs. (6.37) and (6.40).

4. In many cases the predicted state-error correlation matrix $\mathbf{K}(n + 1, n)$ converges to the steady-state value $\mathbf{K}$ as time n approaches infinity. Show that the limiting value $\mathbf{K}$ satisfies the *algebraic Riccati equation*

$$\mathbf{K}\mathbf{C}^H(\mathbf{C}\mathbf{K}\mathbf{C}^H + \mathbf{Q}_2)^{-1}\mathbf{C}\mathbf{K} - \mathbf{Q}_1 = \mathbf{0}$$

where it is assumed that the state transition matrix equals the identity matrix, and the matrices $\mathbf{C}$, $\mathbf{Q}_1$ and $\mathbf{Q}_2$ are the limiting values of $\mathbf{C}(n)$, $\mathbf{Q}_1(n)$ and $\mathbf{Q}_2(n)$, respectively.

5. Equations (6.49) and (6.55) define the Kalman gain $\mathbf{G}(n)$ and the predicted state-error correlation matrix $\mathbf{K}(n + 1, n)$. Using these relations, show that both of these matrices are independent of the input $\mathbf{y}(n)$. Assume that the state transition matrix $\mathbf{\Phi}(n + 1, n)$, the measurement matrix $\mathbf{C}(n)$, and the noise correlation matrices $\mathbf{Q}_1(n)$ and $\mathbf{Q}_2(n)$ are all independent of the space $\mathcal{Y}_n$ that is spanned by the input data $\mathbf{y}(1), \ldots, \mathbf{y}(n)$.

6. Equations (6.112) through (6.116) define the Kalman filtering algorithm for an adaptive transversal filter that assumes a random-walk state model.

   (a) Show that the estimate for the tap-weight vector produced by this algorithm at time n may be expressed in the form

   $$\hat{\mathbf{w}}(n) = \mathcal{R}^{-1}(n)\not{p}(n)$$

where

$$\mathcal{R}(n) = \mathbf{u}(n)\mathbf{u}^H(n) + \mathbf{B}(n-1)\mathbf{u}(n-1)\mathbf{u}^H(n-1) + \cdots + \mathbf{B}(1)\mathbf{u}(1)\mathbf{u}^H(1)$$

$$\rho(n) = \mathbf{u}(n)d^*(n) + \mathbf{B}(n-1)\mathbf{u}(n-1)d^*(n-1) + \cdots + \mathbf{B}(1)\mathbf{u}(1)d^*(1)$$

$$\mathbf{B}(l) = \mathbf{D}(n-1)\mathbf{D}(n-2) \ldots \mathbf{D}(l)$$

$$\mathbf{D}(n) = [\mathbf{I} + \mathbf{K}^{-1}(n)\mathbf{Q}(n)]^{-1}$$

The matrix $\mathbf{K}(n)$ is the filtered weight-error correlation matrix, and $\mathbf{Q}(n)$ is the correlation matrix of the process noise vector that defines the random-walk state model.

(b) Show that the algorithm has the ability to forget data in the distant past if the process noise correlation matrix $\mathbf{Q}(n)$ has the value

$$\mathbf{Q}(n) = q\mathbf{I}, \quad q > 0$$

or

$$\mathbf{Q}(n) = q\mathbf{K}(n), \quad q > 0$$

7. Consider the Kalman algorithm for the operation of an adaptive transversal filter in a stationary environment, as described in Section 6.8.

(a) Show that the estimation error e(n) and the innovation $\alpha(n)$ are related by

$$e(n) = \frac{J_{min}\alpha(n)}{J_{min} + \mathbf{u}^H(n)\mathbf{K}(n-1)\mathbf{u}(n)}$$

where $\mathbf{u}(n)$ is the tap-input vector, $\mathbf{K}(n-1)$ the correlation matrix of the weight-error vector, and J_{min} the minimum mean-squared error.

(b) Show that the mean-square value of the innovation $\alpha(n)$ equals

$$E[|\alpha(n)|^2] = J_{min} + \mathbf{u}^H(n)\mathbf{K}(n-1)\mathbf{u}(n)$$

Hint: Make use of Eqs. (6.34) and (6.35).

(c) How are the relations in parts (a) and (b) affected by adopting the random-walk state model for the environment in which the filter operates?

8. (a) Consider the detection problem described by the model

$$\text{hypothesis } H_1': \quad u(n) = s(n) + v(n), \quad n = 1, 2, \ldots, N$$

$$\text{hypothesis } H_0': \quad u(n) = v(n), \quad n = 1, 2, \ldots, N$$

where $\{u(n)\}$ is the received signal, $\{s(n)\}$ is the signal process, and $\{v(n)\}$ is a zero-mean white Gaussian noise process of variance σ_v^2. Assume that the sample $u(n)$ is complex valued. By transforming the received signal $\{u(n)\}$ into the associated innovations process $\{\alpha(n)\}$, show that the log-likelihood ratio for this detection problem may be expressed as follows:

$$\ln \Lambda_{H_1', H_0'} = \frac{1}{2} \sum_{k=1}^{N} \ln\left(\frac{\sigma_v^2}{\sigma^2(k|H_1')}\right) + \frac{|u(k)|^2}{\sigma_v^2} - \frac{|\alpha(k|H_1')|^2}{\sigma^2(k|H_1')}.$$

where $\{\alpha(k|H_1')\}$ is the innovations process, given that hypothesis H_1' is true and $\sigma^2(k|H_1')$ is its variance.

(b) Consider next the detection problem described by the model

$$\text{hypothesis } H_2: \quad u(n) = s(n) + c(n) + v(n), \quad n = 1, 2, \ldots, N$$

$$\text{hypothesis } H_1: \quad u(n) = c(n) + v(n), \quad n = 1, 2, \ldots, N$$

where $\{u(n)\}$ is the received signal, $\{s(n)\}$ is the signal process, $\{c(n)\}$ is an unknown colored noise process, and $\{v(n)\}$ is a zero-mean white Gaussian noise process of variance σ_v^2. With the composite process $\{s(n) + c(n)\}$ assumed to have finite energy, we may apply the *chain rule of likelihood ratios* to this second detection problem through the use of a dummy hypothesis:

$$\text{hypothesis } H_0: \quad u(n) = v(n), \quad n = 1, 2, \ldots, N$$

The two coupled detection problems are described by

$$\text{hypothesis } H_2: \quad u(n) = s(n) + c(n) + v(n), \quad n = 1, 2, \ldots, N$$

$$\text{hypothesis } H_0: \quad u(n) = v(n), \quad n = 1, 2, \ldots, N$$

and

$$\text{hypothesis } H_1: \quad u(n) = c(n) + v(n), \quad n = 1, 2, \ldots, N)$$

$$\text{hypothesis } H_0: \quad u(n) = v(n), \quad n = 1, 2, \ldots, N$$

Accordingly, we may express the likelihood ratio for the original detection problem as

$$\Lambda_{H_2,H_1} = \frac{\Lambda_{H_2,H_0}}{\Lambda_{H_1,H_0}}$$

where Λ_{H_2,H_1} and Λ_{H_1,H_0} are the likelihood ratios for the two coupled detection problems, respectively. Hence, using this relation and the result given in part (a), show that the log-likelihood ratio for the original detection problem equals

$$\ln \Lambda_{H_2,H_1} = \frac{1}{2} \sum_{k=1}^{N} \ln\left(\frac{\sigma^2(k|H_1)}{\sigma^2(k|H_2)}\right) + \frac{|\alpha(k|H_1)|^2}{\sigma^2(k|H_1)} - \frac{|\alpha(k|H_2)|^2}{\sigma^2(k|H_2)}$$

where $\{\alpha(k|H_i)\}$ is the innovations process and $\sigma^2(k|H_i)$ is its variance, both conditional on hypothesis H_i being true, where $i = 1, 2$.

(c) Discuss the use of the Kalman algorithm as a method for implementing the detection algorithm in part (b) in an adaptive manner. You may assume that under hypothesis H_i the received signal may be modeled as an autoregressive process of order M_i, $i = 1, 2$. You may also use the constant-state version of the Kalman algorithm.

9. Consider a time-varying system that is modeled as a transversal filter whose tap-weight vector $\mathbf{w}_o(n)$ is modeled as a first-order Markov process; that is

$$\mathbf{w}_o^*(n) = a\mathbf{w}_o^*(n - 1) + \mathbf{v}(n)$$

where a is a constant, and the vector $\mathbf{v}(n)$ has its elements drawn from a white noise process of zero mean and variance σ^2. For the time evolution of $\mathbf{w}_o(n)$ to be stable and for it to be approximated by a random-walk state model, the constant a must satisfy the following condition:

$$1 - \delta \le a \le 1$$

where δ is a small positive constant.   Justify the validity of this condition.

Computer-oriented Problem

10. *Computer experiment on the adapative autoregressive modeling of a nonstationary process*: Consider a real-valued nonstationary environment described by the model shown in Fig. P6.1.   The weights of the unknown system vary according to a *first-order Markov process*:

$$\mathbf{w}_o(n) = a\mathbf{w}_o(n - 1) + \mathbf{v}(n)$$

where $\mathbf{v}(n)$ is a white-noise vector of zero mean and correlation matrix

$$E[\mathbf{v}(n)\mathbf{v}^T(n)] = \mathbf{I}$$

The constant a controls the degree of nonstationarity of the environment.   In particular, we define

$$\text{Nonstationarity constant} = \frac{1}{1 - a}$$

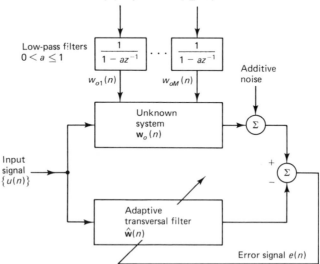

Independent noise sequences

Figure P6.1

Use a computer to simulate the system of Fig. P6.1 for the following conditions:

(i) The input process $\{u(n)\}$ is white noise with zero mean and unit variance.

(ii) The coefficient $a = 0.95$, or nonstationarity constant $= 20$.

(iii) The adaptive transversal filter has the same order as the unknown system described by $\mathbf{w}_o(n)$. In particular, assume the order to equal 5.

(iv) The additive noise at the output of the unknown system has zero mean and variance of 0.2.

(v) The minimum mean-squared error $J_{min} = 0.3$.

(vi) In the random-walk model for the state equation, assume that $q = 0.1$.

(vii) For the initial conditions, set

$$\mathbf{K}(0) = 0.5\mathbf{I}$$

and

$$\hat{\mathbf{w}}(0) = \mathbf{0}$$

(a) Plot a single realization of the first (unknown) tap weight $w_{o,1}(n)$ and the corresponding value $\hat{w}_1(n)$ in the adaptive transversal filter versus time n over the interval $0 \le n \le 120$.

(b) By averaging your simulation results over 120 independent trials of the experiment, plot the ensemble-averaged value of the first (unknown) tap weight $w_{o,1}(n)$ and the corresponding value $\hat{w}_1(n)$ in the adaptive transversal filter over the interval $0 \le n \le 120$.

7

Method of Least Squares

7.1 INTRODUCTION

In this chapter, we use the *method of least squares* to solve the linear filtering problem, without invoking assumptions on the statistics of the inputs applied to the filter (Lawson and Hanson, 1974). To illustrate the basic idea of least squares, suppose we have a set of real-valued measurements $u(1)$, $u(2)$, . . . , $u(N)$, made at times t_1, t_2, . . . , t_N, respectively, and the requirement is to construct a curve that is used to *fit* these points in some optimum fashion. Let the time dependence of this curve be denoted by $f(t_i)$. According to the method of least squares, the "best" fit is obtained by *minimizing the sum of squares of difference* between $f(t_i)$ and $u(i)$ for $i = 1, 2, . . . , N$; hence, the name of the method.

As mentioned in Chapter 1, the method of least squares is perhaps the oldest estimation procedure, going back to Gauss (1809). It was already being used in the 1920s to solve curving-fitting problems mathematically. Nowadays, it is used extensively in statistics (Weisberg, 1980); in system identification (Ljung and Söderström, 1983); in spectrum estimation (Tufts and Kumaresan, 1982); in speech processing (Markel and Gray 1976); and in econometrics (Dhrymus, 1970).

307

The method of least squares may be viewed as the deterministic counterpart of Wiener filter theory. Basically, Wiener filters are derived from *ensemble averages* with the result that one filter (optimum in a probabilistic sense) is obtained for all realizations of the operational environment, assumed to be wide-sense stationary. On the other hand, the method of least squares yields a different filter for each collection of input data.

We begin our study by deriving the *deterministic form of the normal equation*. This equation defines the tap weights of a linear transversal filter that produces an estimate of some desired response due to a set of inputs, which is optimum in the least-squares sense. We refer to the resulting filter as the *least-squares filter*. We study some important properties of this filter and consider its application to the linear prediction problem and the estimation of complex sinusoids in additive noise.

7.2 STATEMENT OF THE LINEAR LEAST-SQUARES ESTIMATION PROBLEM

Consider a stochastic phenomenon that is characterized by two sets of variables, $\{d(i)\}$ and $\{u(i)\}$. The variable $d(i)$ is observed at time i in *response* to the subset of variables $u(i)$, $u(i - 1)$, . . . , $u(i - M + 1)$ applied as *inputs*. That is, $d(i)$ is a function of the inputs $u(i)$, $u(i - 1)$, . . . , $u(i - M + 1)$. This functional relationship is hypothesized to be *linear*. We may thus express the response $d(i)$ as

$$d(i) = \sum_{k=1}^{M} w_{ok}^* u(i - k + 1) + e_o(i) \tag{7.1}$$

where the w_{ok} are *unknown parameters* of the *model*, and $e_o(i)$ represents the *measurement error* to which the statistical nature of the phenomenon is ascribed; each term in the summation represents an inner product. In effect, the model of Eq. (7.1) says that the variable $d(i)$ may be determined as a linear combination of the input variables $u(i)$, $u(i - 1)$, . . . , $(u(i - M + 1)$, except for the error $e_o(i)$. This model, represented by the signal-flow graph shown in Fig. 7.1, is called a *multiple linear regression model*.

The measurement error $e_o(i)$ is an *unobservable* random variable that is introduced into the model to account for its inaccuracy. It is customary to assume that the measurement error process $\{e_o(i)\}$ is white with zero mean and variance σ^2. That is,

$$E[e_o(i)] = 0, \qquad \text{for all } i$$

and

$$E[e_o(i)e_o^*(k)] = \begin{cases} \sigma^2, & i = k \\ 0, & i \neq k \end{cases}$$

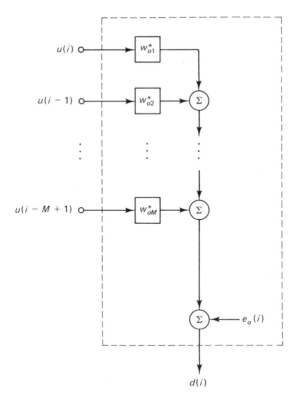

Figure 7.1 Multiple linear regression model.

The implication of this assumption is that we may rewrite Eq. (7.1) in the equivalent form

$$E[d(i)] = \sum_{k=1}^{M} w_{ok}^* u(i - k + 1)$$

where the values of $u(i)$, $u(i - 1)$, ..., $u(i - M + 1)$ are *known*. Hence, the mean of the response $d(i)$, in theory, is uniquely determined by the model.

The problem we have to solve is to *estimate* the unknown parameters of the multiple linear regression model of Fig. 7.1, the w_{ok}, given the two *observable* sets of variables: $\{u(i)\}$ and $\{d(i)\}$, $i = 1, 2, ..., N$. To do this, we postulate the linear transversal filter of Fig. 7.2 as the model of interest. By forming inner products of the *tap inputs* $u(i)$, $u(i - 1)$, ..., $u(i - M + 1)$ and the corresponding *tap weights* w_1, w_2, ..., w_M, respectively, and by utilizing $d(i)$ as the *desired response*, we define the *estimation error* or *residual* $e(i)$ as the difference between the desired response $d(i)$ and the *filter output*, as shown by

$$e(i) = d(i) - \sum_{k=1}^{M} w_k^* u(i - k + 1) \tag{7.2}$$

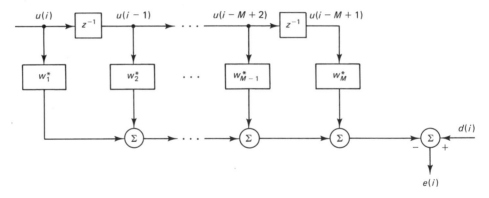

Figure 7.2   Linear transversal filter model.

where the convolution sum represents the filter output.   In the method of least squares, we choose the tap weights of the transversal filter, the w_k, so as to *minimize* an index of performance that consists of the *sum of error squares*:

$$\mathcal{E}(w_1, \ldots, w_M) = \sum_{i=i_1}^{i_2} |e(i)|^2 \tag{7.3}$$

where i_1 and i_2 define the index limits at which the error minimization occurs.   The values assigned to these limits depend on the type of data *windowing* employed, as discussed in section 7.3.   Basically, the problem we have to solve is to substitute Eq. (7.2) into (7.3) and then minimize the objective function $\mathcal{E}(w_1, \ldots, w_M)$ with respect to the tap weights of the transversal filter in Fig. 7.2.   The tap weights of the filter, $w_1, w_2, \ldots, w_M$, are held constant during the interval $i_1 < i < i_2$.

7.3 WINDOWING OF THE DATA

Given M as the number of tap weights used in the transversal filter model of Fig. 7.2, the rectangular matrix constructed from the input data, $u(1), u(2), \ldots, u(N)$, may assume different forms, depending on the values assigned to the limits i_1 and i_2 in Eq. (7.3).   In particular, we may distinguish four different methods of *windowing* the input data:

1. *Covariance method*, which makes no assumptions about the data outside the interval $(1, N)$.   Thus, by defining the limits of interest as $i_1 = M$ and $i_2 = N$, the input data may be arranged in the matrix form:

$$\begin{bmatrix} u(M) & u(M+1) & \cdots & u(N) \\ u(M-1) & u(M) & \cdots & u(N-1) \\ \cdot & \cdot & \cdot & \cdot \\ \cdot & \cdot & \cdot & \cdot \\ \cdot & \cdot & \cdot & \cdot \\ u(1) & u(2) & \cdots & u(N-M+1) \end{bmatrix}$$

2. *Autocorrelation method*, which makes the assumption that the data prior to time $i = 1$ and the data after $i = N$ are zero. Thus, by using $i_1 = 1$ and $i_2 = N + M - 1$, the matrix of input data takes on the form

$$\begin{bmatrix} u(1) & u(2) & \cdots & u(M) & u(M+1) & \cdots & u(N) & 0 & \cdots & 0 \\ 0 & u(1) & \cdots & u(M-1) & u(M) & \cdots & u(N-1) & u(N) & \cdots & 0 \\ \cdot & \cdot & \cdot & \cdot & \cdot & & \cdot & & & \\ \cdot & \cdot & \cdot & \cdot & \cdot & & \cdot & & & \cdot \\ \cdot & \cdot & \cdot & \cdot & \cdot & & \cdot & & & \\ 0 & 0 & \cdots & u(1) & u(2) & \cdots & u(N-M+1) & u(N-M) & \cdots & u(N) \end{bmatrix}$$

3. *Prewindowing method*, which makes the assumption that the input data prior to $i = 1$ are zero, but makes no assumption about the data after $i = N$. Thus, by using $i_1 = 1$ and $i_2 = N$, the matrix of input data assumes the form

$$\begin{bmatrix} u(1) & u(2) & \cdots & u(M) & u(M+1) & \cdots & u(N) \\ 0 & u(1) & \cdots & u(M-1) & u(M) & \cdots & u(N-1) \\ \cdot & \cdot & \cdot & \cdot & \cdot & & \cdot \\ \cdot & \cdot & \cdot & \cdot & \cdot & & \cdot \\ \cdot & \cdot & \cdot & \cdot & \cdot & & \cdot \\ 0 & 0 & \cdots & u(1) & u(2) & \cdots & u(N-M+1) \end{bmatrix}$$

4. *Postwindowing method*, which makes no assumption about the data prior to time $i = 1$, but makes the assumption that the data after $i = N$ are zero. Thus, by using $i_1 = M$ and $i_2 = N + M - 1$, the matrix of input data takes on the form

$$\begin{bmatrix} u(M) & u(M+1) & \cdots & u(N) & 0 & \cdots & 0 \\ u(M-1) & u(M) & \cdots & u(N-1) & u(N) & \cdots & 0 \\ \cdot & \cdot & \cdot & \cdot & \cdot & & \cdot \\ \cdot & \cdot & \cdot & \cdot & \cdot & & \cdot \\ \cdot & \cdot & \cdot & \cdot & \cdot & & \cdot \\ u(1) & u(2) & \cdots & u(N-M+1) & u(N-M) & \cdots & u(N) \end{bmatrix}$$

The terms "covariance method" and "autocorrelation method" are commonly used in speech-processing literature (Makhoul, 1975; Markel and Gray, 1976). It should, however, be emphasized that the use of these two terms is *not* based on the standard definition of the covariance function as the correlation function with the means removed. Rather, these two terms derive their names from the way we interpret the meaning of the *known parameters* contained in the deterministic form of the normal equation that results from minimizing the index of performance of Eq. (7.3). The covariance method derives its name from control theory literature where, with zero-mean tap inputs, these known parameters represent the elements of a *covariance matrix*; hence, the name of the method. The autocorrelation method, on the other hand, derives its name from the fact that, for the conditions stated, these known parameters represent the *short-term autocorrelation function* of the tap inputs; hence, the name of the second method.

In the remainder of this chapter, except for Problem 7.4, which deals with the autocorrelation method, we will be exclusively concerned with the covariance method. The prewindowing method is considered in Chapters 8 through 10.

7.4 THE DETERMINISTIC NORMAL EQUATION

In the covariance method, the sum of error squares is defined by

$$\mathscr{E}(w_1, \ldots, w_M) = \sum_{i=M}^{N} |e(i)|^2 \tag{7.4}$$

By choosing the limits on the time index i in this way, in effect, we make sure that for each value of i, all the M tap inputs of the transversal filter in Fig. 7.2 have nonzero values. As mentioned previously, the problem we have to solve is to determine the tap weights of the transversal filter of Fig. 7.2 for which the sum of error squares is minimum. The approach we will take to solve this problem is based on the use of matrix notation.

Let the M-by-1 *tap-weight vector* $\mathbf{w}$ of the transversal filter be defined by

$$\mathbf{w}^T = [w_1, w_2, \ldots, w_M] \tag{7.5}$$

Correspondingly, let the M-by-1 *input vector* $\mathbf{u}(i)$ denote the tap inputs of the transversal filter, defined by

$$\mathbf{u}^T(i) = [u(i), u(i - 1), \ldots, u(i - M + 1)], \quad M \leq i \leq N \tag{7.6}$$

Then we may express the estimation error $e(i)$ in the form [see Eq. (7.2)]

$$e(i) = d(i) - \mathbf{w}^H \mathbf{u}(i), \quad M \leq i \leq N \tag{7.7}$$

We may view the $e(i)$, $M \leq i \leq N$, as elements of an $(N - M + 1)$-by-1 vector $\boldsymbol{\varepsilon}$, called the *estimation error vector* or *residual vector*; that is

$$\boldsymbol{\varepsilon}^H = [e(M), e(M + 1), \ldots, e(N)] \tag{7.8}$$

Let the $d(i)$, $M \leq i \leq N$, define the elements of $(N - M + 1)$-by-1 vector $\mathbf{b}$, called the *desired response vector*; that is

$$\mathbf{b}^H = [d(M), d(M + 1), \ldots, d(N)] \tag{7.9}$$

We may then rewrite Eq. (7.7) in the matrix form

$$\boldsymbol{\varepsilon}^H = \mathbf{b}^H - \mathbf{w}^H[\mathbf{u}(M), \mathbf{u}(M + 1), \ldots, \mathbf{u}(N)]$$
$$= \mathbf{b}^H - \mathbf{w}^H \mathbf{A}^H \tag{7.10}$$

where   (7.11)

$$\mathbf{A}^H = [\mathbf{u}(M), \mathbf{u}(M + 1), \ldots, \mathbf{u}(N)]$$

By taking the Hermitian transpose of both sides of Eq. (7.10), we may also express the estimation error vector in the form

$$\boldsymbol{\varepsilon} = \mathbf{b} - \mathbf{Aw} \tag{7.12}$$

The *data matrix* $\mathbf{A}$ is an $(N - M + 1)$ by-M matrix that is completely defined by the set of observations $u(1), u(2), \ldots, u(N)$. This is shown in the expanded form

$$\mathbf{A}^H = \begin{bmatrix} u(M) & u(M + 1) & \cdots & u(N) \\ u(M - 1) & u(M) & \cdots & u(N - 1) \\ \cdot & \cdot & \cdot & \cdot \\ \cdot & \cdot & \cdot & \cdot \\ u(1) & u(2) & \cdots & u(N - M + 1) \end{bmatrix} \tag{7.13}$$

We ordinarily have $N \geq M$, where M is the number of taps in the transversal filter model of Fig. 7.2. Note that the t, kth element of data matrix $\mathbf{A}$ equals $u^*(M + t - k)$, where $0 \leq t \leq N - M$ and $0 \leq k \leq M - 1$.

The elements of the estimation error vector $\boldsymbol{\varepsilon}$, the desired response vector $\mathbf{b}$, and the data matrix $\mathbf{A}$ are all defined as *complex conjugates* of the pertinent variables. The reason for this will become apparent in Section 7.12.

Returning to Eq. (7.4), which defines the sum of error squares, we may rewrite this expression in terms of the estimation error vector $\boldsymbol{\varepsilon}$ as

$$\mathscr{E}(\mathbf{w}) = \boldsymbol{\varepsilon}^H \boldsymbol{\varepsilon} \tag{7.14}$$

Substituting Eqs. (7.10) and (7.12) in (7.14), and then expanding, we may express the dependence of the sum of error squares on the tap-weight vector $\mathbf{w}$ explicitly as follows

$$\mathscr{E}(\mathbf{w}) = \mathbf{b}^H \mathbf{b} - \mathbf{b}^H \mathbf{AW} - \mathbf{w}^H \mathbf{A}^H \mathbf{b} + \mathbf{w}^H \mathbf{A}^H \mathbf{Aw} \tag{7.15}$$

Differentiating $\mathscr{E}(\mathbf{w})$ with respect to $\mathbf{w}$, we get the gradient vector

$$\frac{\partial \mathscr{E}}{\partial \mathbf{w}} = -2\mathbf{A}^H \mathbf{b} + 2\mathbf{A}^H \mathbf{Aw} \tag{7.16}$$

The gradient vector may also be expressed in terms of the estimation error vector as follows

$$\frac{\partial \mathscr{E}}{\partial \mathbf{w}} = -2\mathbf{A}^H \boldsymbol{\varepsilon} \tag{7.17}$$

Let $\hat{\mathbf{w}}$ denote the value of the tap-weight vector $\mathbf{w}$ for which the sum of error squares $\mathscr{E}$ is minimum or, equivalently, the gradient vector $\partial \mathscr{E}/\partial \mathbf{w}$ is zero. Then, from Eq. (7.16), we immediately deduce that

$$\mathbf{A}^H \mathbf{A}\hat{\mathbf{w}} = \mathbf{A}^H \mathbf{b} \qquad (7.18)$$

The matrix equation (7.18) is called the *deterministic normal equation* for the linear least-squares problem; the reason for the name is given in the next section.

7.5 THE PRINCIPLE OF ORTHOGONALITY

Let $\boldsymbol{\varepsilon}_{\min}$ denote the *minimum* estimation error vector that results when the tap-weight vector equals $\hat{\mathbf{w}}$; using Eq. (7.12), we may thus write

$$\boldsymbol{\varepsilon}_{\min} = \mathbf{b} - \mathbf{A}\hat{\mathbf{w}} \qquad (7.19)$$

Accordingly, we may rewrite Eq. (7.18) in the equivalent form

$$\mathbf{A}^H \boldsymbol{\varepsilon}_{\min} = \mathbf{0} \qquad (7.20)$$

This equation states that the minimum estimation error vector $\boldsymbol{\varepsilon}_{\min}$ is *orthogonal* to each column vector of the data matrix $\mathbf{A}$. Examining the composition of the matrix $\mathbf{A}$ given in Eq. (7.13), we see that the elements of the kth column of this matrix represent the time series $\{u^*(i - k + 1)\}$ observed at the $(M - k + 1)$th tap of the transversal filter, $k = 1, 2, \ldots, M$, over the time interval $M \leq i \leq N$. Also, the elements of the error vector $\boldsymbol{\varepsilon}_{\min}$ represent the time series $\{e^*_{\min}(i)\}$. Accordingly, we may state the *principle of orthogonality*:

The minimum error time series $\{e_{\min}(i)\}$ is orthogonal to the time series applied to each tap of the transversal filter, as shown by

$$\sum_{i=M}^{N} u(i - k + 1) \, e^*_{\min}(i) = 0, \qquad k = 1, 2, \ldots, M \qquad (7.21)$$

This principle provides the basis of a simple *test* that we can carry out in practice to check whether or not the transversal filter is operating in its *least-squares condition*. We merely have to determine the deterministic cross-correlation between the estimation error and the time series applied to *each* tap input of the filter. It is *only* when *all* these M deterministic cross-correlation functions are identically zero that we find the tap-weight vector of the filter equal to the least-squares estimate $\hat{\mathbf{w}}$.

Let $\hat{\mathbf{b}} = \mathbf{A}\hat{\mathbf{w}}$ denote the *least-squares estimate of the desired response vector* $\mathbf{b}$. Premultiply both sides of Eq. (7.20) by $\hat{\mathbf{w}}^H$, and substitute this estimate. We thus obtain the corollary to the principle of orthogonality, described by

$$\hat{\mathbf{b}}^H \boldsymbol{\varepsilon}_{\min} = \mathbf{0} \qquad (7.22)$$

Equation (7.22) states that *the minimum estimation error vector $\boldsymbol{\varepsilon}_{\min}$ and the least-squares estimate $\hat{\mathbf{b}}$ of the desired response vector are orthogonal to each other*.

We may use this new result to provide a geometric interpretation of the least-

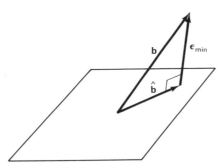

Figure 7.3 Geometric interpretation of the principle of orthogonality.

squares solution, which provides invaluable insight into the problem. Figure 7.3 depicts such an interpretation. The figure shows that the estimation error vector $\boldsymbol{\varepsilon}$ attains its *minimum* length only when it is perpendicular (i.e., normal) to the estimate $\hat{\mathbf{b}}$; hence, the name "normal" equation. For any tap-weight vector $\mathbf{w}$ other than the least-squares estimate $\hat{\mathbf{w}}$, the estimation error vector $\boldsymbol{\varepsilon}$ has a length greater than that of $\boldsymbol{\varepsilon}_{\text{min}}$.

7.6 *UNIQUENESS THEOREM*

The least-squares problem of minimizing the sum of error squares, $\mathscr{E}$, *always* has a solution. That is, for given values of the data matrix $\mathbf{A}$ and the desired response vector $\mathbf{b}$, we can always find a vector $\hat{\mathbf{w}}$ that satisfies the normal equation (7.18). It is therefore important that we know if and when the solution is *unique*. This requirement is covered by the following *uniqueness theorem* (Stewart, 1973):

The least-squares estimate $\hat{\mathbf{w}}$ is unique if and only if the nullity of the data matrix $\mathbf{A}$ equals zero.

Let $\mathbf{A}$ be an L-by-M matrix; in the case of the data matrix $\mathbf{A}$ defined in Eq. (7.13), we have $L = N - M + 1$. We define the *null space* of matrix $\mathbf{A}$, written as $\mathcal{N}(\mathbf{A})$, as the space of all vectors $\mathbf{x}$ such that $\mathbf{A}\mathbf{x} = 0$. We define the *nullity* of matrix $\mathbf{A}$, written as null($\mathbf{A}$), as the dimension of the null space $\mathcal{N}(\mathbf{A})$. In general, we find that null($\mathbf{A}$) $\neq$ null($\mathbf{A}^H$).

In the light of the uniqueness theorem, which is intuitively satisfying, we may expect a unique solution to the least-squares problem *only* when the data matrix $\mathbf{A}$ has *linearly independent columns*. This implies that the matrix $\mathbf{A}$ has at least as many rows as columns; that is, $(N - M + 1) \geq M$. This latter condition means that the system of equations represented by $\mathbf{A}\hat{\mathbf{w}} = \mathbf{b}$ used in the minimization is *overdetermined* in that it has more equations than unknowns. Thus, provided that the data matrix $\mathbf{A}$ has linearly independent columns, the M-by-M matrix $\mathbf{A}^H\mathbf{A}$ is *nonsingular*, and the least-squares estimate has the unique value

$$\hat{\mathbf{w}} = (\mathbf{A}^H\mathbf{A})^{-1}\mathbf{A}^H\mathbf{b} \qquad (7.23)$$

When, however, the matrix **A** has *linearly dependent columns*, the nullity of the matrix **A** is nonzero, and the result is that an infinite number of solutions can be found for minimizing the sum of error squares. In such a situation, the least-squares problem becomes quite involved in that we now have the new problem of deciding which particular solution to adopt. We defer discussion of this issue until Section 7.12. In the meantime, we assume that the data matrix **A** has linearly independent columns so that the least-squares estimate $\hat{\mathbf{w}}$ has the unique value defined by Eq. (7.23).

7.7 MINIMUM SUM OF ERROR SQUARES

A transversal filter designed with the least-squares estimate $\hat{\mathbf{w}}$ as the tap-weight vector is called the *deterministic least-squares filter*. For this filter, the sum of error squares takes on the *minimum value*

$$
\begin{aligned}
\mathscr{E}_{\min} &= \boldsymbol{\varepsilon}_{\min}^{H} \boldsymbol{\varepsilon}_{\min} \\
&= (\mathbf{b}^{H} - \hat{\mathbf{w}}^{H}\mathbf{A}^{H})(\mathbf{b} - \mathbf{A}\hat{\mathbf{w}}) \\
&= \mathbf{b}^{H}\mathbf{b} - \hat{\mathbf{w}}^{H}\mathbf{A}^{H}\mathbf{b} - \mathbf{b}^{H}\mathbf{A}\hat{\mathbf{w}} + \hat{\mathbf{w}}^{H}\mathbf{A}^{H}\mathbf{A}\hat{\mathbf{w}}
\end{aligned}
\tag{7.24}
$$

Substituting Eq. (7.18) into (7.24), the expression for the minimum sum of error squares simplifies as

$$
\begin{aligned}
\mathscr{E}_{\min} &= \mathbf{b}^{H}\mathbf{b} - \mathbf{b}^{H}\mathbf{A}\hat{\mathbf{w}} \\
&= \mathbf{b}^{H}\mathbf{b} - \mathbf{b}^{H}\mathbf{A}(\mathbf{A}^{H}\mathbf{A})^{-1}\mathbf{A}^{H}\mathbf{b}
\end{aligned}
\tag{7.25}
$$

The first term represents the *energy of the desired response*

$$
\mathscr{E}_{d} = \mathbf{b}^{H}\mathbf{b}
\tag{7.26}
$$

or, in expanded form,

$$
\mathscr{E}_{d} = \sum_{i=M}^{N} |d(i)|^{2}
\tag{7.27}
$$

Since $\mathscr{E}_{\min}$ is nonnegative, it follows that the second term on the right side of Eq. (7.25) can never exceed $\mathscr{E}_{d}$. Indeed, it reaches the value of $\mathscr{E}_{d}$ when the measurement error $e_{o}(i)$ in the multiple linear regression model of Fig. 7.1 is zero, a practical impossibility.

Another case for which $\mathscr{E}_{\min}$ equals $\mathscr{E}_{d}$ occurs when the least-squares problem is *underdetermined*. Such a situation arises when there are fewer data points than parameters. This situation is considered in Section 7.12.

7.8 REFORMULATION OF THE DETERMINISTIC NORMAL EQUATION IN TERMS OF CORRELATION FUNCTIONS

The matrix products $\mathbf{A}^H \mathbf{b}$ and $\mathbf{A}^H \mathbf{A}$ in the deterministic normal equation (7.18) have special meanings of their own, as follows:

1. The product $\mathbf{A}^H \mathbf{b}$ is an M-by-1 vector, representing the *deterministic cross-correlation vector* between the inputs of the transversal filter and the desired response. Let $\boldsymbol{\theta}$ denote this cross-correlation vector, as shown by

$$\boldsymbol{\theta} = \mathbf{A}^H \mathbf{b} \qquad (7.28)$$

It may also be expressed in the expanded form

$$\boldsymbol{\theta} = \sum_{i=M}^{M} \mathbf{u}(i) d^*(i) \qquad (7.29)$$

where the tap-input vector $\mathbf{u}(i)$ is defined by Eq. (7.7). The tth element of the vector $\boldsymbol{\theta}$ equals

$$\theta(t) = \sum_{i=M}^{N} u(i - t) d^*(i), \qquad 0 \le t \le M - 1 \qquad (7.30)$$

2. The product $\mathbf{A}^H \mathbf{A}$ is an M-by-M matrix, representing the *deterministic correlation matrix* of the tap inputs. Let $\boldsymbol{\Phi}$ denote this correlation matrix:

$$\boldsymbol{\Phi} = \mathbf{A}^H \mathbf{A} \qquad (7.31)$$

It may also be expressed in the equivalent form

$$\boldsymbol{\Phi} = \sum_{i=M}^{N} \mathbf{u}(i) \mathbf{u}^H(i) \qquad (7.32)$$

The t, kth element of the correlation matrix $\boldsymbol{\Phi}$ equals

$$\phi(t, k) = \sum_{i=M}^{N} u(i - t) u^*(i - k), \qquad 0 \le t, k \le M - 1 \qquad (7.33)$$

Thus, using the definitions of Eqs. (7.28) and (7.31), we may rewrite the deterministic normal equation (7.18) in the new form

$$\boldsymbol{\Phi} \hat{\mathbf{w}} = \boldsymbol{\theta} \qquad (7.34)$$

which may also be written in the expanded forms

$$\sum_{i=M}^{N} \hat{w}_k \phi(t, k) = \theta(t), \qquad t = 0, 1, \ldots, M - 1 \qquad (7.35)$$

Equations (7.34) and (7.35) bear a direct resemblance to the ensemble-averaged normal equation (3.35) or (3.38) that were derived for the Wiener filter for stationary inputs by minimizing the mean-squared error.

Also, we may write the expression for the minimum sum of error squares, defined by Eq. (7.25), as

$$\mathscr{E}_{\min} = \mathscr{E}_d - \boldsymbol{\theta}^H \hat{\mathbf{w}}$$
$$= \mathscr{E}_d - \boldsymbol{\theta}^H \boldsymbol{\Phi}^{-1} \boldsymbol{\theta} \qquad (7.36)$$

Properties of the Deterministic Correlation Matrix

The deterministic correlation matrix $\boldsymbol{\Phi}$ has the following properties:

Property 1. *The correlation matrix $\boldsymbol{\Phi}$ is Hermitian; that is*

$$\boldsymbol{\Phi}^H = \boldsymbol{\Phi}$$

This property follows directly from Eq. (7.31).

Property 2. *The correlation matrix $\boldsymbol{\Phi}$ is nonnegative definite; that is,*

$$\mathbf{x}^H \boldsymbol{\Phi} \mathbf{x} \geq 0$$

for any M-by-1 vector $\mathbf{x}$.

Using the definition of Eq. (7.32), we may write

$$\mathbf{x}^H \boldsymbol{\Phi} \mathbf{x} = \sum_{i=M}^{N} \mathbf{x}^H \mathbf{u}(i) \mathbf{u}^H(i) \mathbf{x}$$

$$= \sum_{i=M}^{N} [\mathbf{x}^H \mathbf{u}(i)] [\mathbf{x}^H \mathbf{u}(i)]^H$$

$$= \sum_{i=M}^{N} |\mathbf{x}^H \mathbf{u}(i)|^2 \geq 0$$

which is the desired result. The fact that the deterministic correlation matrix $\boldsymbol{\Phi}$ is nonnegative definite means that its determinant and all principal minors are nonnegative. When the above condition is satisfied with the inequality sign, the determinant of $\boldsymbol{\Phi}$ and its principal minors are likewise nonzero. In the latter case, $\boldsymbol{\Phi}$ is nonsingular, and the inverse $\boldsymbol{\Phi}^{-1}$ exists.

Property 3. *The eigenvalues of the correlation matrix $\boldsymbol{\Phi}$ are all real and nonnegative.*

The real requirement of the eigenvalues of $\boldsymbol{\Phi}$ follows from Property 1. The fact that all these eigenvalues are also nonnegative follows from Property 2.

Property 4.   *The deterministic correlation matrix is the product of two rectangular Toeplitz matrices.*

The deterministic correlation matrix $\boldsymbol{\Phi}$ is non-Toeplitz, which is clearly seen by examining the expanded form of the deterministic correlation matrix:

$$
\boldsymbol{\Phi} = \begin{bmatrix} \phi(0,0) & \phi(0,1) & \cdots & \phi(0,M-1) \\ \phi(1,0) & \phi(1,1) & \cdots & \phi(1,M-1) \\ \cdot & \cdot & \cdot & \cdot \\ \cdot & \cdot & \cdot & \cdot \\ \cdot & \cdot & \cdot & \cdot \\ \phi(M-1,0) & \phi(M-1,1) & \cdots & \phi(M-1,M-1) \end{bmatrix}
$$

The elements on the main diagonal, $\phi(0,0)$, $\phi(1,1)$, . . . , $\phi(M-1, M-1)$, have different values and similarly for any other diagonal above or below the main diagonal.   However, the matrix $\boldsymbol{\Phi}$ has a special structure in that it is the product of two Toeplitz rectangular matrices.   This follows directly from the definition of Eq. (7.31), which shows that $\boldsymbol{\Phi}$ is the product of the data matrix $\mathbf{A}$ and its Hermitian transpose $\mathbf{A}^H$.   The M-by-$(N-M+1)$ matrix $\mathbf{A}^H$ is a rectangular Toeplitz matrix, which is seen by examining its composition shown in Eq. (7.13).   The data matrix $\mathbf{A}$ itself is likewise a rectangular Toeplitz matrix.

7.9 PROPERTIES OF LEAST-SQUARES ESTIMATES

The method of least squares has a strong intuitive feel that is reinforced by several outstanding properties of the method.   These properties are described next (Miller, 1974; Goodwin and Payne, 1977):

Property 1.   *The least-squares estimate $\hat{\mathbf{w}}$ is unbiased, provided that the measurement error process $\{e_o(i)\}$ has zero mean.*

From the multiple linear regression model of Fig. 7.1, we have

$$
\mathbf{b} = \mathbf{A}\mathbf{w}_o + \boldsymbol{\varepsilon}_o \tag{7.37}
$$

Hence, substituting Eq. (7.37) into (7.23), we may express the least-squares estimate $\hat{\mathbf{w}}$ as

$$
\begin{aligned} \hat{\mathbf{w}} &= (\mathbf{A}^H\mathbf{A})^{-1}\,\mathbf{A}^H\mathbf{A}\mathbf{w}_o + (\mathbf{A}^H\mathbf{A})^{-1}\,\mathbf{A}^H\boldsymbol{\varepsilon}_o \\ &= \mathbf{w}_o + (\mathbf{A}^H\mathbf{A})^{-1}\,\mathbf{A}^H\boldsymbol{\varepsilon}_o \end{aligned} \tag{7.38}
$$

The matrix product $(\mathbf{A}^H\mathbf{A})^{-1}\mathbf{A}$ is a known quantity, since the data matrix $\mathbf{A}$ is completely defined by the set of known observations $u(1)$, $u(2)$, . . . , $u(N)$; see Eq. (7.13).   Hence, if the measurement error process $\{e_o(i)\}$ or, equivalently, the vector $\boldsymbol{\varepsilon}_o$ has zero mean, we find by taking the expectation of both sides of Eq. (7.38) that the least-squares estimate $\hat{\mathbf{w}}$ is *unbiased*; that is,

$$E[\hat{\mathbf{w}}] = \mathbf{w}_o \tag{7.39}$$

Property 2.   *When the measurement error process $\{e_o(i)\}$ is white with zero mean and variance σ^2, the covariance matrix of the least-squares estimate $\hat{\mathbf{w}}$ equals $\sigma^2 \boldsymbol{\Phi}^{-1}$.*

Using the relation of Eq. (7.38), we find that the covariance matrix of the least-squares estimate $\hat{\mathbf{w}}$ equals

$$
\begin{aligned}
\text{cov}[\hat{\mathbf{w}}] &= E[(\hat{\mathbf{w}} - \mathbf{w}_o)(\hat{\mathbf{w}} - \mathbf{w}_o)^H] \\
&= E[(\mathbf{A}^H\mathbf{A})^{-1}\mathbf{A}^H\boldsymbol{\varepsilon}_o\boldsymbol{\varepsilon}_o^H\mathbf{A}(\mathbf{A}^H\mathbf{A})^{-1}] \\
&= (\mathbf{A}^H\mathbf{A})^{-1}\mathbf{A}^H E[\boldsymbol{\varepsilon}_o\boldsymbol{\varepsilon}_o^H]\mathbf{A}(\mathbf{A}^H\mathbf{A})^{-1}
\end{aligned}
\tag{7.40}
$$

With the measurement error process $\{e_o(i)\}$ assumed to be white with zero mean and variance σ^2, we have

$$E[\boldsymbol{\varepsilon}_o\boldsymbol{\varepsilon}_o^H] = \sigma^2\mathbf{I} \tag{7.41}$$

where $\mathbf{I}$ is the identity matrix.   Hence, Eq. (7.40) simplifies as follows:

$$
\begin{aligned}
\text{cov}[\hat{\mathbf{w}}] &= \sigma^2(\mathbf{A}^H\mathbf{A})^{-1}\mathbf{A}^H\mathbf{A}(\mathbf{A}^H\mathbf{A})^{-1} \\
&= \sigma^2(\mathbf{A}^H\mathbf{A})^{-1} \\
&= \sigma^2\boldsymbol{\Phi}^{-1}
\end{aligned}
\tag{7.42}
$$

which is the desired result.

Property 3.   *When the measurement error process $\{e_o(i)\}$ is white with zero mean, the least-squares estimate $\hat{\mathbf{w}}$ is the best linear unbiased estimate.*

Consider any linear unbiased estimator $\tilde{\mathbf{w}}$ that is defined by

$$\tilde{\mathbf{w}} = \mathbf{Bb} \tag{7.43}$$

where $\mathbf{B}$ is an M-by-$(N-M+1)$ matrix.   Substituting Eq. (7.37) into (7.43), we get

$$\tilde{\mathbf{w}} = \mathbf{BA}\mathbf{w}_o + \mathbf{B}\boldsymbol{\varepsilon}_o \tag{7.44}$$

With the measurement error vector $\boldsymbol{\varepsilon}_o$ assumed to have zero mean, we find that the expected value of $\tilde{\mathbf{w}}$ equals

$$E[\tilde{\mathbf{w}}] = \mathbf{BA}\mathbf{w}_o$$

For the linear estimator $\tilde{\mathbf{w}}$ to be unbiased, we therefore require that the matrix $\mathbf{B}$ satisfy the condition

$$\mathbf{BA} = \mathbf{I}$$

Accordingly, we may rewrite Eq. (7.44) as follows:

$$\tilde{\mathbf{w}} = \mathbf{w}_o + \mathbf{B}\boldsymbol{\varepsilon}_o$$

The covariance matrix of $\tilde{\mathbf{w}}$ equals

$$
\begin{aligned}
\text{cov}[\tilde{\mathbf{w}}] &= E[(\tilde{\mathbf{w}} - \mathbf{w}_o)(\tilde{\mathbf{w}} - \mathbf{w}_o)^H] \\
&= E[\mathbf{B}\boldsymbol{\varepsilon}_o\boldsymbol{\varepsilon}_o^H\mathbf{B}^H] \qquad\qquad (7.45) \\
&= \sigma^2\mathbf{BB}^H
\end{aligned}
$$

Here, we have made use of Eq. (7.41), which describes the assumption that the elements of the measurement error vector $\boldsymbol{\varepsilon}_o$ are uncorrelated and have the common variance σ^2; that is, the measurement error process $\{e_o(i)\}$ is white. We next define a new matrix $\boldsymbol{\Psi}$ in terms of $\mathbf{B}$ as

$$\boldsymbol{\Psi} = \mathbf{B} - (\mathbf{A}^H\mathbf{A})^{-1}\mathbf{A}^H \qquad\qquad (7.46)$$

Now we form the matrix product $\boldsymbol{\Psi}\boldsymbol{\Psi}^H$ as

$$
\begin{aligned}
\boldsymbol{\Psi}\boldsymbol{\Psi}^H &= [\mathbf{B} - (\mathbf{A}^H\mathbf{A})^{-1}\mathbf{A}^H][\mathbf{B}^H - \mathbf{A}(\mathbf{A}^H\mathbf{A})^{-1}] \\
&= \mathbf{BB}^H - \mathbf{BA}(\mathbf{A}^H\mathbf{A})^{-1} - (\mathbf{A}^H\mathbf{A})^{-1}\mathbf{A}^H\mathbf{B}^H + (\mathbf{A}^H\mathbf{A})^{-1} \\
&= \mathbf{BB}^H - (\mathbf{A}^H\mathbf{A})^{-1}
\end{aligned}
$$

Since we always have $\boldsymbol{\Psi}\boldsymbol{\Psi}^H \geq 0$, it follows that

$$\mathbf{BB}^H \geq (\mathbf{A}^H\mathbf{A})^{-1}$$

Equivalently, we may write

$$\sigma^2\mathbf{BB}^H \geq \sigma^2(\mathbf{A}^H\mathbf{A})^{-1} \qquad\qquad (7.47)$$

The term $\sigma^2\mathbf{BB}^H$ equals the covariance matrix of the linear estimate $\tilde{\mathbf{w}}$, as in Eq. (7.45). From Property 2, we also know that the term $\sigma^2(\mathbf{A}^H\mathbf{A})^{-1}$ equals the covariance matrix of the least-squares estimate $\hat{\mathbf{w}}$. Thus, Eq. (7.47) shows that within the class of linear unbiased estimates the least-squares estimate $\hat{\mathbf{w}}$ is the "best" estimate of the unknown parameter vector $\mathbf{w}_o$ of the multiple linear regression model. Accordingly, when the measurement error process $\{e_o\}$ contained in this model is white with zero mean, the least-squares estimate $\hat{\mathbf{w}}$ is the *best linear unbiased estimate* (BLUE).

Thus far we have not made any assumption about the statistical distribution of the measurement error process $\{e_o(i)\}$ other than that it is a zero-mean white-noise process. We next make the further assumption that the process $\{e_o(i)\}$ is Gaussian distributed, which yields the following stronger result on the optimality of the least-squares estimate $\hat{\mathbf{w}}$.

Property 4. *When the measurement error process $\{e_o(i)\}$ is white, Gaussian, and has zero mean, the least-squares estimate $\hat{\mathbf{w}}$ achieves the Cramér–Rao lower bound for unbiased estimates.*

Let $f_{\mathbf{E}}(\boldsymbol{\varepsilon}_o)$ denote the joint probability density function of the measurement error vector $\boldsymbol{\varepsilon}_o$. Let $\tilde{\mathbf{w}}$ denote any unbiased estimate of the unknown parameter vector $\mathbf{w}_o$ of the multiple linear regression model. Then the covariance matrix of $\tilde{\mathbf{w}}$ satisfies the inequality[1]

$$\text{cov}[\tilde{\mathbf{w}}] \geq \mathbf{J}^{-1} \tag{7.48}$$

where

$$\text{cov}[\tilde{\mathbf{w}}] = E[(\tilde{\mathbf{w}} - \mathbf{w}_o)(\tilde{\mathbf{w}} - \mathbf{w}_o)^H]$$

The matrix $\mathbf{J}$ is called *Fisher's information matrix*; it is defined by

$$\mathbf{J} = E\left\{ \left[\frac{\partial \ln f_{\mathbf{E}}(\boldsymbol{\varepsilon}_o)}{\partial \mathbf{w}_o} \right] \left[\frac{\partial \ln f_{\mathbf{E}}(\boldsymbol{\varepsilon}_o)}{\partial \mathbf{w}_o} \right]^H \right\} \tag{7.49}$$

Since the measurement error process $\{e_o(i)\}$ is white, the elements of the vector $\boldsymbol{\varepsilon}_o$ are uncorrelated. Furthermore, since the process $\{e_o(i)\}$ is Gaussian, the elements of $\boldsymbol{\varepsilon}_o$ are statistically independent. With $e_o(i)$ assumed to have zero mean and variance σ^2, we have

$$f_E(e_o(i)) = \frac{1}{\sqrt{2\pi}\sigma} \exp\left(-\frac{|e_o(i)|^2}{2\sigma^2} \right)$$

Hence, the joint-probability density function of $\boldsymbol{\varepsilon}_o$ equals

$$f_{\mathbf{E}}(\boldsymbol{\varepsilon}_o) = \prod_{i=M}^{N} f_E(e_o(i))$$

$$= \frac{1}{(2\pi\sigma^2)^{(N-M+1)/2}} \exp\left[-\frac{1}{2\sigma^2} \sum_{i=M}^{N} |e_o(i)|^2 \right]$$

Taking the natural logarithm,

$$\ln f_{\mathbf{E}}(\boldsymbol{\varepsilon}_o) = F - \frac{1}{2\sigma^2} \sum_{i=M}^{N} |e_o(i)|^2$$

$$= F - \frac{1}{2\sigma^2} \boldsymbol{\varepsilon}_o^H \boldsymbol{\varepsilon}_o \tag{7.50}$$

---

[1] For discussion of the Cramér–Rao bound, see Appendix A.

where

$$F = -\frac{(N-M+1)}{2}\ln(2\pi\sigma^2)$$

Differentiating Eq. (7.50) with respect to $\mathbf{w}_o$, we therefore get

$$\frac{\partial \ln f_{\mathbf{E}}(\boldsymbol{\varepsilon}_o)}{\partial \mathbf{w}_o} = -\frac{1}{\sigma^2}\frac{\partial \boldsymbol{\varepsilon}_o^H}{\partial \mathbf{w}_o}\boldsymbol{\varepsilon}_o$$

From Eq. (7.12), we deduce that

$$\frac{\partial \boldsymbol{\varepsilon}_o^H}{\partial \mathbf{w}_o} = -\mathbf{A}^H$$

Hence

$$\frac{\partial \ln f(\boldsymbol{\varepsilon}_o)}{\partial \mathbf{w}_o} = \frac{1}{\sigma^2}\mathbf{A}^H\boldsymbol{\varepsilon}_o \qquad (7.51)$$

Thus, substituting Eq. (7.51) into (7.49) yields Fisher's information matrix for the problem at hand as

$$\begin{aligned}
\mathbf{J} &= \frac{1}{\sigma^4}E[\mathbf{A}^H\boldsymbol{\varepsilon}_o\boldsymbol{\varepsilon}_o^H\mathbf{A}] \\
&= \frac{1}{\sigma^4}\mathbf{A}^HE[\boldsymbol{\varepsilon}_o\boldsymbol{\varepsilon}_o^H]\mathbf{A} \\
&= \frac{1}{\sigma^2}\mathbf{A}^H\mathbf{A} \\
&= \frac{1}{\sigma^2}\boldsymbol{\Phi}
\end{aligned} \qquad (7.52)$$

where, in the third line, we have made use of Eq. (7.41) describing the assumption that the measurement error process $\{e_o(i)\}$ is white. Accordingly, the use of Eq. (7.48) shows that the covariance matrix of the unbiased estimate $\tilde{\mathbf{w}}$ satisfies the inequality

$$\text{cov}[\tilde{\mathbf{w}}] \geq \sigma^2\boldsymbol{\Phi}^{-1}$$

However, from Property 2, we know that $\sigma^2\boldsymbol{\Phi}^{-1}$ equals the covariance matrix of the least-squares estimate $\hat{\mathbf{w}}$. Accordingly, $\hat{\mathbf{w}}$ achieves the Cramer–Rao lower bound. Using Property 1, we conclude therfore that when the measurement error process $\{e_o(i)\}$ is a zero-mean white Gaussian noise process, the least-squares estimate $\hat{\mathbf{w}}$ is a *minimum variance unbiased estimate* (MVUE).

7.10 *THE LINEAR PREDICTION PROBLEM*

The problem of *linear prediction* may be viewed as a special type of parameter estimation that is well suited for the method of least squares. In this application, the least-squares transversal filter is used as the *linear predictor*.

In *forward linear prediction,* we use the set of inputs $u(i-1), u(i-2), \ldots,$ $u(i-M)$ to make a linear prediction of $u(i)$. According to this notation, the prediction is made at time $(i-1)$, *one step into the future.* We define the *forward linear prediction error* as the difference between the *desired response $u(i)$* and the *predictor output* produced by the tap inputs $u(i-1), u(i-2), \ldots, u(i-M)$. When the method of least squares is used to design the predictor, we choose its tap weights so as to minimize the *sum of forward prediction-error squares* or the *forward prediction-error energy.* We refer to this method of designing a predictor as the *forward linear prediction (FLP) method.*

From our study of the linear prediction problem, however, we recall that there is a second form of prediction, backward prediction, that also deserves attention. In *backward linear prediction,* we use the set of inputs $u(i-M+1), \ldots,$ $u(i-1), u(i)$ to make a linear prediction of $u(i-M)$. According to the notation, the prediction is made as time $(i-M+1)$, *one step into the past.* We define the *backward prediction error* as the difference between the desired response $u(i-M)$ and the predictor output produced by the tap inputs $u(i-M+1), \ldots, u(i-1)$, $u(i)$. When the method of least squares is used to design the predictor, we choose its tap weights so as to minimize the *sum of backward prediction-error squares* or the *backward prediction-error energy.* We refer to this second method of designing a predictor as the *backward linear prediction (BLP) method.*

For a given set of input data $u(1), u(2), \ldots, u(N)$, we ordinarily find that the FLP and BLP methods yield entirely different results. It is only natural for this to be so, because, after all, the two methods use entirely different optimization criteria.

We may develop yet another method for the design of a predictor by using an optimization criterion that combines the FLP and BLP methods. In particular, we choose the tap weights of the linear predictor so as to minimize a new objective function that consists of the *forward prediction-error energy plus the backward prediction-error energy.* Accordingly, this third method of designing a predictor is called the *forward–backward linear prediction (FBLP) method.* A predictor designed in this way offers some useful properties that have made the FBLP method a useful design tool, particularly in modeling and spectrum estimation.

The idea of combining forward prediction and backward prediction in the design of predictors was originated by Burg, (1968, 1975). He developed it in the context of lattice predictors. The first application of the FBLP method to the design of a linear predictor that has a transversal filter structure, in accordance with the method of least squares, was developed independently by Ulrych and Clayton (1976) and Nuttall (1976).

In the next section, we develop the theory of the FBLP method for the design of transversal filters used as linear predictors. We do this by considering the FBLP method as a special case of the method of least squares. The FLP and BLP methods are presented to the reader as Problems 5 and 6, respectively.

7.11 THE FORWARD–BACKWARD LINEAR PREDICTION METHOD

Consider the *forward linear predictor,* shown in Fig. 7.4(a). The tap weights of the predictor are denoted by w_1, w_2, . . . , w_M and the tap inputs by $u(i-1)$, $u(i-2)$, . . . , $u(i-M)$, respectively. The forward prediction error, denoted by $f_M(i)$, equals

$$f_M(i) = u(i) - \sum_{k=1}^{M} w_k^* u(i-k) \tag{7.53}$$

The first term, $u(i)$, represents the desired response. The convolution sum, constituting the second term, represents the predictor output; it consists of the sum of inner products. Using matrix notation, we may also express the forward prediction error as

$$f_M(i) = u(i) - \mathbf{u}^T(i-1)\hat{\mathbf{w}}^* \tag{7.54}$$

where $\hat{\mathbf{w}}$ is the M-by-1 tap-weight vector of the predictor:

$$\hat{w}^T = [\hat{w}_1, \hat{w}_2, \ldots, \hat{w}_M]$$

and $\mathbf{u}(i-1)$ is the corresponding tap-input vector:

$$\mathbf{u}^T(i-1) = [u(i-1), u(i-2), \ldots, u(i-M)]$$

Consider next Fig. 7.4(b), which depicts the reconfiguration of the predictor so that it performs backward linear prediction. We have *purposely* retained $\hat{w}_1$, $\hat{w}_2$, . . . , $\hat{w}_M$ as the tap weights of the predictor. The change in the format of the tap inputs is inspired by the discussion presented in Section 4.4 on backward linear prediction and its relation with forward linear prediction for the case of stationary inputs. In particular, the tap inputs in the predictor of Fig. 7.4(b) differ from those of the forward linear predictor of Fig. 7.4(a) in two respects:

1. The tap inputs in Fig. 7.4(b) are *time reversed* in that they appear from right to left, whereas in Fig. 7.4(a) they appear from left to right.
2. With $u(i)$, $u(i-1)$, . . . , $u(i-M+1)$ used as tap inputs, the structure of Fig. 7.4(b) produces a linear prediction of $u(i-M)$. In other words, it performs backward linear prediction. Denoting the backward prediction error by $b_M(i)$, we may thus express it as

$$b_M(i) = u(i-M) - \sum_{k=1}^{M} \hat{w}_k u(i-M+k) \tag{7.55}$$

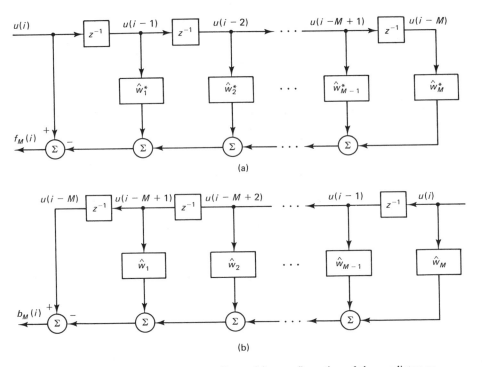

Figure 7.4 (a) Forward linear predictor; (b) reconfiguration of the predictor so as to perform backward linear prediction.

where the first term represents the desired response and the second term is the predictor output. Equivalently, in terms of matrix notation, we may write

$$b_M(i) = u(i-M) - \mathbf{u}^{BT}(i)\hat{\mathbf{w}} \qquad (7.56)$$

where $\mathbf{u}^B(i)$ is the *time-reversed tap-input vector*:

$$\mathbf{u}^{BT}(i) = [u(i-M+1), \ldots, u(i-1), u(i)]$$

Let $\mathscr{E}_M$ denote the *minimum value of the forward–backward prediction-error energy*. In accordance with the method of least squares, we may therefore write

$$\mathscr{E}_M = \sum_{i=M+1}^{N} \left[|f_M(i)|^2 + |b_M(i)|^2\right] \qquad (7.57)$$

where the subscript M signifies the order of the predictor. The lower limit on the time index i equals $M+1$ so as to ensure that the forward and backward prediction errors are formed only when all the tap inputs of interest assume nonzero values. In particular, we may make two observations:

1. The variable $u(i-M)$, representing the last tap input in the forward prediction of Fig. 7.4(a), assumes a nonzero value for the first time when $i=M+1$.

2. The variable $u(i-M)$, playing the role of desired response in the backward predictor of Fig. 7.4(b), also assumes a nonzero value for the first time when $i = M + 1$.

Thus, by choosing $(M+1)$ as the lower limit on i and N as the upper limit, as in Eq. (7.60), we make no assumptions about the data outside the interval $(1,N)$, as required by the covariance method.

Let $\mathbf{A}$ denote the $2(N-M)$-by-M *data matrix* that is defined by

$$
\mathbf{A}^H = \begin{bmatrix}
u(M) & \cdots & u(N-1) & u^*(2) & \cdots & u^*(N-M+1) \\
u(M-1) & \cdots & u(N-2) & u^*(3) & \cdots & u^*(N-M+2) \\
\vdots & & \vdots & \vdots & & \vdots \\
u(1) & \cdots & u(N-M) & u^*(M+1) & \cdots & u^*(N)
\end{bmatrix}
\tag{7.58}
$$

The elements constituting the left half of matrix $\mathbf{A}^H$ represent the various sets of tap inputs used to make a total of $(N-M)$ forward linear predictions. The complex-conjugated elements constituting the right half of matrix $\mathbf{A}^H$ represent the corresponding sets of tap inputs used to make a total of $(N-M)$ backward linear predictions.

Let $\mathbf{b}$ denote the $2(N-M)$-by-1 *desired response vector* that is defined by

$$
\mathbf{b}^H = [u(M+1), \ldots, u(N), u^*(1), \ldots, u^*(N-M)]
\tag{7.59}
$$

Each element in the left half of the vector $\mathbf{b}^H$ represents a desired response for forward linear prediction. Each complex-conjugated element in the right half represents a desired response for backward linear prediction.

Thus, from Eq. (7.18), we find that the deterministic normal equation for the FBLP method is as follows:

$$
\mathbf{A}^H \mathbf{A} \hat{\mathbf{w}} = \mathbf{A}^H \mathbf{b}
\tag{7.60}
$$

From the first line of Eq. (7.25), we find that the minimum value of the forward–backward prediction error energy equals

$$
\mathscr{E}_M = \mathbf{b}^H \mathbf{b} - \mathbf{b}^H \mathbf{A} \hat{\mathbf{w}}
\tag{7.61}
$$

Augmented Normal Equation

We may combine Eqs. (7.60) and (7.61) into a single matrix relation, as shown by

$$
\begin{bmatrix}
\mathbf{b}^H \mathbf{b} & \mathbf{b}^H \mathbf{A} \\
\mathbf{A}^H \mathbf{b} & \mathbf{A}^H \mathbf{A}
\end{bmatrix}
\begin{bmatrix}
1 \\
-\hat{\mathbf{w}}
\end{bmatrix}
=
\begin{bmatrix}
\mathscr{E}_M \\
\mathbf{0}
\end{bmatrix}
\tag{7.62}
$$

where $\mathbf{0}$ is the M-by-1 null vector. Equation (7.62) is the deterministic form of the *augmented normal equation for the FBLP method*. Define the $(M+1)$-by-

$(M+1)$ deterministic correlation matrix:

$$\mathbf{C} = \begin{bmatrix} \mathbf{b}^H\mathbf{b} & \mathbf{b}^H\mathbf{A} \\ \mathbf{A}^H\mathbf{b} & \mathbf{A}^H\mathbf{A} \end{bmatrix} \tag{7.63}$$

Define the $(M+1)$-by-tap-weight vector of the *prediction-error filter of order M:*

$$\mathbf{a}_M = \begin{bmatrix} 1 \\ -\hat{\mathbf{w}} \end{bmatrix} \tag{7.64}$$

Figure 7.5 shows the structure of the prediction-error filter, where $a_{M,0}, a_{M,1}, \ldots,$ $a_{M,M}$ denote the tap weights and $a_{M,0} = 1$.   Then

$$\mathbf{C}\mathbf{a}_M = \begin{bmatrix} \mathscr{E}_M \\ \mathbf{0} \end{bmatrix} \tag{7.65}$$

Note that we may also express the deterministic correlation matrix $\mathbf{C}$ as

$$\mathbf{C} = \mathbf{T}^H\mathbf{T} \tag{7.66}$$

where $\mathbf{T}$ is the $2(N-M)$-by-$(M+1)$ *augmented data matrix* that incorporates the elements of both the desired response vector $\mathbf{b}$ and the data matrix $\mathbf{A}$.   That is, the matrix $\mathbf{T}$ is defined by

$$\mathbf{T}^H = \begin{bmatrix} \mathbf{b}^H \\ \mathbf{A}^H \end{bmatrix}$$

$$= \begin{bmatrix} u(M+1) & \cdots & u(N) & u^*(1) & \cdots & u^*(N-M) \\ u(M) & \cdots & u(N-1) & u^*(2) & \cdots & u^*(N-M+1) \\ \cdot & \cdot & \cdot & \cdot & \cdot & \cdot \\ \cdot & \cdot & \cdot & \cdot & \cdot & \cdot \\ \cdot & \cdot & \cdot & \cdot & \cdot & \cdot \\ u(1) & \cdots & u(N-M) & u^*(M+1) & \cdots & u^*(N) \end{bmatrix} \tag{7.67}$$

We may express the deterministic correlation matrix $\mathbf{C}$ in yet another way as follows (Marple, 1980):

$$\mathbf{C} = [\mathbf{B}^H, \mathbf{E}\mathbf{B}^T]\begin{bmatrix} \mathbf{B} \\ \mathbf{B}^*\mathbf{E} \end{bmatrix}$$
$$= \mathbf{B}^H\mathbf{B} + \mathbf{E}\mathbf{B}^T\mathbf{B}^*\mathbf{E} \tag{7.68}$$

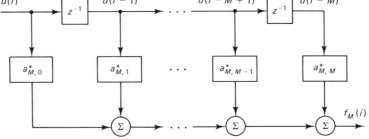

Figure 7.5   Forward prediction-error filter.

where the $(N-M)$-by-$(M+1)$ data matrix $\mathbf{B}$ is defined by

$$\mathbf{B}^H = \begin{bmatrix} u(M+1) & u(M+2) & \cdots & u(N) \\ u(M) & u(M+1) & \cdots & u(N-1) \\ u(M-1) & u(M) & & u(N-2) \\ \cdot & \cdot & \cdot & \cdot \\ \cdot & \cdot & \cdot & \cdot \\ \cdot & \cdot & \cdot & \cdot \\ u(1) & u(2) & \cdots & u(N-M) \end{bmatrix} \tag{7.69}$$

That is, matrix $\mathbf{B}^H$ equals the left half of matrix $\mathbf{T}^H$. The matrix product $\mathbf{EB}^T$ equals the right half of matrix $\mathbf{T}^H$, as shown by

$$\mathbf{EB}^T = \begin{bmatrix} u^*(1) & u^*(2) & \cdots & u^*(N-M) \\ u^*(2) & u^*(3) & \cdots & u^*(N-M+1) \\ u^*(3) & u^*(4) & \cdots & u^*(N-M+2) \\ \cdot & \cdot & \cdot & \cdot \\ \cdot & \cdot & \cdot & \cdot \\ \cdot & \cdot & \cdot & \cdot \\ u^*(M+1) & u^*(M+2) & \cdots & u^*(N) \end{bmatrix} \tag{7.70}$$

From Eqs. (7.69) and (7.70), we deduce that the composition of $(M+1)$-by-$(M+1)$ matrix $\mathbf{E}$ is as follows

$$\mathbf{E} = \begin{bmatrix} 0 & 0 & \cdots & 1 \\ \cdot & \cdot & & \cdot & \cdot \\ \cdot & \cdot & \cdot & \cdot \\ \cdot & \cdot & \cdot & \cdot \\ 0 & 1 & \cdots & 0 \\ 1 & 0 & \cdots & 0 \end{bmatrix} \tag{7.71}$$

The matrix $\mathbf{E}$ is called the *exchange matrix*. Note that the matrix $\mathbf{E}$ is real valued and symmetric.

The deterministic correlation matrix $\mathbf{C}$ has the following properties:

1. The matrix $\mathbf{C}$ is *Hermitian symmetric*; that is,

$$c(k, t) = c^*(t, k)$$

where $c(t,k)$ is the t,kth element of matrix $\mathbf{C}$. Equivalently, we have

$$\mathbf{C}^H = \mathbf{C}$$

This property follows directly from Eq. (7.66) or (7.68). Here we are restating a property that is inherent to every correlation matrix.

2. The matrix $\mathbf{C}$ is *Hermitian persymmetric*; that is,

$$c(M - k, M - t) = c^*(k, t)$$

This property follows directly from the definition of element $c(t,k)$, shown by

$$c(t,k) = \sum_{i=M+1}^{N} [u(i-t)u^*(i-k) + u^*(i-M+t)u(i-M+k)],$$

$$0 \le t,k \le M \tag{7.72}$$

The persymmetric property is unique to a deterministic correlation matrix that is obtained by time-averaging of the input data in the forward as well as backward direction.

3. The matrix $\mathbf{C}$ has a special structure that is composed of the sum of two Toeplitz matrix products. This property is described in Eq. (7.68), where we recognize that $\mathbf{B}$, $\mathbf{B}^H$, $\mathbf{EB}^T$ and $\mathbf{B}^*\mathbf{E}$ are all rectangular Toeplitz matrices; See Eqs. (7.69) and (7.70).

Discussion

The essence of the FBLP method is that it provides a procedure for evaluating the tap weights of a *forward linear predictor of fixed order* by minimizing the combined sum of the forward prediction-error energy and the backward prediction-error energy. The tap-weight vector of the predictor is determined by solving the normal equation (7.60) where the data matrix $\mathbf{A}$ and desired response vector $\mathbf{b}$ represent the known parameters. The resulting minimum value of the forward–backward prediction-error energy $\mathscr{E}_M$ is defined by Eq. (7.61).

Alternatively, we may use the augmented normal equation (7.65) where the deterministic correlation matrix $\mathbf{C}$ represents the known parameters, and the unknowns are represented by the prediction-error filter coefficient $a_{M,1}$, $a_{M,2}$, . . . , $a_{M,M}$ (the coefficient $a_{M,0}$ equals unity) and the minimum forward–backward prediction-error energy $\mathscr{E}_M$. Assuming that the matrix $\mathbf{C}$ is nonsingular, we may solve for these unknowns by using matrix inversion. This direct procedure for solving the augmented normal equation (7.65) requires on the order of M^3 computational operations, where M is the order of the predictor.

The deterministic correlation matrix $\mathbf{C}$ does *not* possess Toeplitz symmetry. Accordingly, we cannot use the Levinson recursion to develop a fast solution of the augmented normal equation (7.65), as was the case with the augmented normal equation based on Wiener filter theory for stationary inputs. However, Marple (1980, 1981) describes fast recursive algorithms for the efficient solution of the augmented normal equation (7.65). Marple exploits the special Toeplitz structure of the deterministic correlation matrix $\mathbf{C}$ described previously. The computational complexity of Marple's fast algorithm is proportional to M^2. When the predictor order M is large, the use of Marple's algorithm results in significant savings in computation.

7.12 SINGULAR-VALUE DECOMPOSITION

To determine the least-squares value of the tap-weight vector of the linear predictor resulting from the application of the FBLP method, we have to solve the normal equation (7.60). However, the solution is unique only when the columns of the data matrix $\mathbf{A}$ are linearly independent; see Section 7.6. When this condition is satisfied, the M-by-M deterministic correlation matrix $\mathbf{\Phi} = \mathbf{A}^H\mathbf{A}$ is nonsingular, and the least-squares solution for the tap-weight vector of the linear predictor is uniquely defined by the formula[2]

$$\hat{\mathbf{w}} = (\mathbf{A}^H\mathbf{A})^{-1}\mathbf{A}^H\mathbf{b} \qquad (7.73)$$

Define the $2(N - M)$-by-M matrix

$$\mathbf{A}^{\#} = (\mathbf{A}^H\mathbf{A})^{-1}\mathbf{A}^H \qquad (7.74)$$

Then we may rewrite Eq. (7.73) simply as

$$\hat{\mathbf{w}} = \mathbf{A}^{\#}\mathbf{b} \qquad (7.75)$$

The matrix $\mathbf{A}^{\#}$ is called the *pseudoinverse* or the *Moore–Penrose generalized inverse* of the matrix $\mathbf{A}$ (Stewart, 1973). Equation (7.75) represents a convenient way of saying that "the vector $\hat{\mathbf{w}}$ solves the least-squares problem." Indeed, it was with the simple format of Eq. (7.75) in mind and also the desire to be consistent with definitions of the correlation matrix and the cross-correlation vector used previously that we defined the data matrix $\mathbf{A}$ and the desired response vector $\mathbf{b}$ in the manner shown in Eqs. (7.58) and (7.59). This remark also applies to the original definitions of the matrix $\mathbf{A}$ and vector $\mathbf{b}$ given in Eqs. (7.13) and (7.9).

However, in practice, we often find that the data matrix $\mathbf{A}$ contains linearly dependent columns. Consequently, we are faced with a new situation where we now have to decide on which of an infinite number of possible solutions to the least-squares problem to work with. The issue can be resolved by using a technique called the *singular-value decomposition* to develop a general definition of a pseudoinverse that solves the least-squares problem even when null ($\mathbf{A}$) $\neq 0$ (Stewart, 1973).

The Singular-value Decomposition Theorem

The singular value decomposition (*SVD*) of a matrix is one of the most elegant algorithms in numerical algebra for providing quantitative information about the structure of a system of linear equations (Klema and Laub, 1980). The system of

---

[2]Equation (7.73) is a restatement of Eq. (7.23). Although the discussion presented in this section relates to a linear forward predictor resulting from the application of the FBLP method, nevertheless, it applies to the solution of any linear least-squares problem.

linear equations that is of specific interest to us is described by

$$\mathbf{A\hat{w}} = \mathbf{b} \tag{7.76}$$

in which $\mathbf{A}$ is an L-by-M matrix, $\mathbf{b}$ is an L-by-1 vector, and $\mathbf{\tilde{w}}$ (representing the unknown parameters) is an M-by-1 vector. Equation (7.76) represents a simplified form of the deterministic normal equation. In particular, premultiplication of both sides of the equation by the vector $\mathbf{A}^H$ yields the deterministic normal equation for the least-squares weight vector $\mathbf{\tilde{w}}$. For example, in the context of the forward-backward linear prediction method, $\mathbf{A}$ represents the data matrix with $L = 2(N-M)$, $\mathbf{b}$ represents the desired response vector, and $\mathbf{\tilde{w}}$ represents the least-squares tap-weight vector of the predictor. In any event, let W denote the rank of matrix $\mathbf{A}$. *Then there are two unitary matrices $\mathbf{X}$ and $\mathbf{Y}$, such that*

$$\mathbf{Y}^H \mathbf{AX} = \begin{bmatrix} \mathbf{\Sigma} & \mathbf{0} \\ \mathbf{0} & \mathbf{0} \end{bmatrix} \tag{7.77}$$

where

$$\mathbf{\Sigma} = \text{diag}(\sigma_1, \sigma_2, \ldots, \sigma_W)$$

and

$$\sigma_1 \geq \sigma_2 \geq \cdots \geq \sigma_W > 0$$

Equation (7.77) is a mathematical statement of the *singular value decomposition theorem*.[3]

The *rank* of matrix $\mathbf{A}$, written as rank $(\mathbf{A})$, is defined as the number of linearly independent columns in the matrix $\mathbf{A}$. Note that we always have rank$(\mathbf{A}^H) = $ rank$(\mathbf{A})$. The rank $W \leq \min (L, M)$. Since it is possible that $L > M$ or $L < M$, there are two distinct cases to be considered. We prove the singular value decomposition theorem by considering both cases, independently of each other. For the case when $L > M$, we have an *overdetermined system* in that we have more equations than unknowns. On the other hand, when $L < M$, we have an *undetermined system* in that we have more unknowns than equations.

For the case of when $L > M$, we form the M-by-M matrix $\mathbf{A}^H\mathbf{A}$ by premultiplying the matrix $\mathbf{A}$ by its Hermitian transpose $\mathbf{A}^H$. Since the matrix $\mathbf{A}^H\mathbf{A}$ is Hermitian and nonnegative definite, its eigenvalues are all real nonnegative numbers. Let these eigenvalues be denoted by $\sigma_1^2, \sigma_2^2, \ldots, \sigma_M^2$, where $\sigma_1 \geq \sigma_2 \geq \ldots \geq \sigma_W \geq 0$, and $\sigma_{W+1} = \sigma_{W+2} = \cdots = \sigma_M = 0$. The matrix $\mathbf{A}^H\mathbf{A}$ has the same rank as $\mathbf{A}$; hence, there are W nonzero eigenvalues. Let $\mathbf{x}_1, \mathbf{x}_2, \ldots, \mathbf{x}_M$ denote a set of orthonormal eigenvectors of $\mathbf{A}^H\mathbf{A}$ that are associated with the eigenvalues $\sigma_1^2, \sigma_2^2, \ldots, \sigma_M^2$, respectively. Also, let $\mathbf{X}$ denote the M-by-M unitary matrix whose columns are made up of the eigenvectors $\mathbf{x}_1, \mathbf{x}_2, \ldots, \mathbf{x}_M$. Thus,

---

[3]Klema and Laub (1980) trace the history of the singular value decomposition to its first development in the 1870's by Betrami and Jordan for real square matrices. For general rectangular matrices, they credit it to Eckart and Young (1939).

using the eigenvalue–eigenvector decomposition of the matrix $\mathbf{A}^H\mathbf{A}$, we may write

$$\mathbf{X}^H\mathbf{A}^H\mathbf{A}\mathbf{X} = \begin{bmatrix} \mathbf{\Sigma}^2 & \mathbf{0} \\ \mathbf{0} & \mathbf{0} \end{bmatrix}$$

Let the unitary matrix $\mathbf{X}$ be partitioned as

$$\mathbf{X} = [\mathbf{X}_1, \mathbf{X}_2]$$

where $\mathbf{X}_1$ is an M-by-W matrix,

$$\mathbf{X}_1 = [\mathbf{x}_1, \mathbf{x}_2, \ldots, \mathbf{x}_W]$$

and $\mathbf{X}_2$ is an M-by-$(M - W)$ matrix,

$$\mathbf{X}_2 = [\mathbf{x}_{W+1}, \mathbf{x}_{W+2}, \ldots, \mathbf{x}_M]$$

with

$$\mathbf{X}_1^H\mathbf{X}_2 = \mathbf{0}$$

We may therefore make two deductions:

1. For matrix $\mathbf{X}_1$, we have

$$\mathbf{X}_1^H\mathbf{A}^H\mathbf{A}\mathbf{X}_1 = \mathbf{\Sigma}^2$$

   Consequently,

$$\mathbf{\Sigma}^{-1}\mathbf{X}_1^H\mathbf{A}^H\mathbf{A}\mathbf{X}_1\mathbf{\Sigma}^{-1} = \mathbf{I} \tag{7.78}$$

2. For matrix $\mathbf{X}_2$, we have

$$\mathbf{X}_2^H\mathbf{A}^H\mathbf{A}\mathbf{X}_2 = \mathbf{0}$$

   Consequently,

$$\mathbf{A}\mathbf{X}_2 = \mathbf{0}$$

We now define a new L-by-W matrix

$$\mathbf{Y}_1 = \mathbf{A}\mathbf{X}_1\mathbf{\Sigma}^{-1} \tag{7.79}$$

Then, from Eq. (7.78), it follows that

$$\mathbf{Y}_1^H\mathbf{Y}_1 = \mathbf{I}$$

which means that the columns of the matrix $\mathbf{Y}_1$ are orthonormal with respect to each other. Next, we choose another L-by-$(L - W)$ matrix $\mathbf{Y}_2$ such that the L-by-L matrix formed from $\mathbf{Y}_1$ and $\mathbf{Y}_2$, namely,

$$\mathbf{Y} = [\mathbf{Y}_1, \mathbf{Y}_2]$$

is a unitary matrix. This means that

$$\mathbf{Y}_1^H\mathbf{Y}_2 = \mathbf{0}$$

Accordingly, we may write

$$
\mathbf{Y}^H \mathbf{A} \mathbf{X} = \begin{bmatrix} \mathbf{Y}_1^H \\ \mathbf{Y}_2^H \end{bmatrix} \mathbf{A} [\mathbf{X}_1, \mathbf{X}_2]
$$

$$
= \begin{bmatrix} \mathbf{Y}_1^H \mathbf{A} \mathbf{X}_1 & \mathbf{Y}_1^H \mathbf{A} \mathbf{X}_2 \\ \mathbf{Y}_2^H \mathbf{A} \mathbf{X}_1 & \mathbf{Y}_2^H \mathbf{A} \mathbf{X}_2 \end{bmatrix}
$$

$$
= \begin{bmatrix} (\mathbf{\Sigma}^{-1} \mathbf{X}_1^H \mathbf{A}^H) \mathbf{A} \mathbf{X}_1 & \mathbf{Y}_1^H(\mathbf{0}) \\ \mathbf{Y}_2^H(\mathbf{Y}_1 \mathbf{\Sigma}) & \mathbf{Y}_2^H(\mathbf{0}) \end{bmatrix}
$$

$$
= \begin{bmatrix} \mathbf{\Sigma} & \mathbf{0} \\ \mathbf{0} & \mathbf{0} \end{bmatrix}
$$

which is the desired result.

Consider next the case when $L < M$. This time we form the L-by-L matrix $\mathbf{A}\mathbf{A}^H$ by postmultiplying the matrix $\mathbf{A}$ by its Hermitian transpose $\mathbf{A}^H$. The matrix $\mathbf{A}\mathbf{A}^H$ is also Hermitian and nonnegative definite, so its eigenvalues are likewise real nonnegative numbers. The nonzero eigenvalues of $\mathbf{A}\mathbf{A}^H$ are the *same* as those of $\mathbf{A}^H\mathbf{A}$. We may therefore denote the eigenvalues of $\mathbf{A}\mathbf{A}^H$ as $\sigma_1^2, \sigma_2^2, \ldots, \sigma_L^2$, where $\sigma_1 \geq \sigma_2 \geq \cdots \geq \sigma_W > 0$, and $\sigma_{W+1} = \sigma_{W+2} = \cdots = \sigma_L = 0$. Let $\mathbf{y}_1, \mathbf{y}_2, \ldots, \mathbf{y}_L$ denote a set of orthonormal eigenvectors of the matrix $\mathbf{A}\mathbf{A}^H$ that are associated with the eigenvalues $\sigma_1^2, \sigma_2^2, \ldots, \sigma_L^2$, respectively. Also, let $\mathbf{Y}$ denote the unitary matrix whose columns are made up of the eigenvectors $\mathbf{y}_1, \mathbf{y}_2, \ldots, \mathbf{y}_L$. Thus, using the eigenvalue–eigenvector decomposition of $\mathbf{A}\mathbf{A}^H$, we may write

$$
\mathbf{Y}^H \mathbf{A} \mathbf{A}^H \mathbf{Y} = \begin{bmatrix} \mathbf{\Sigma}^2 & \mathbf{0} \\ \mathbf{0} & \mathbf{0} \end{bmatrix}
$$

Let the unitary matrix $\mathbf{Y}$ be partitioned as

$$
\mathbf{Y} = [\mathbf{Y}_1, \mathbf{Y}_2]
$$

where

$$
\mathbf{Y}_1 = [\mathbf{y}_1, \mathbf{y}_2, \ldots, \mathbf{y}_W]
$$

$$
\mathbf{Y}_2 = [\mathbf{y}_{W+1}, \mathbf{y}_{W+2}, \ldots, \mathbf{y}_L]
$$

and

$$
\mathbf{Y}_1^H \mathbf{Y}_2 = \mathbf{0}
$$

We may therefore make two deductions:

1. For matrix $\mathbf{Y}_1$, we have

$$
\mathbf{Y}_1^H \mathbf{A} \mathbf{A}^H \mathbf{Y}_1 = \mathbf{\Sigma}^2
$$

Consequently,

$$
\mathbf{\Sigma}^{-1} \mathbf{Y}_1^H \mathbf{A} \mathbf{A}^H \mathbf{Y}_1 \mathbf{\Sigma}^{-1} = \mathbf{I} \tag{7.80}
$$

2. For matrix $\mathbf{Y}_2$, we have

$$\mathbf{Y}_2^H \mathbf{A} \mathbf{A}^H \mathbf{Y}_2 = \mathbf{0}$$

Consequently,

$$\mathbf{A}^H \mathbf{Y}_2 = \mathbf{0}$$

We now define an M-by-W matrix

$$\mathbf{X}_1 = \mathbf{A}^H \mathbf{Y}_1 \mathbf{\Sigma}^{-1} \qquad (7.81)$$

Then from Eq. (7.80), it follows that

$$\mathbf{X}_1^H \mathbf{X}_1 = \mathbf{I}$$

which means that the columns of the matrix $\mathbf{X}_1$ are orthonormal with respect to each other. Next, we choose another M-by-$(M - W)$ matrix $\mathbf{X}_2$ such that the M-by-M matrix formed from $\mathbf{X}_1$ and $\mathbf{X}_2$, namely,

$$\mathbf{X} = [\mathbf{X}_1, \mathbf{X}_2]$$

is a unitary matrix. This means that

$$\mathbf{X}_2^H \mathbf{X}_1 = \mathbf{0}$$

Accordingly, we may write

$$
\begin{aligned}
\mathbf{Y}^H \mathbf{A} \mathbf{X} &= \begin{bmatrix} \mathbf{Y}_1^H \\ \mathbf{Y}_2^H \end{bmatrix} \mathbf{A} [\mathbf{X}_1, \mathbf{X}_2] \\
&= \begin{bmatrix} \mathbf{Y}_1^H \mathbf{A} \mathbf{X}_1 & \mathbf{Y}_1^H \mathbf{A} \mathbf{X}_2 \\ \mathbf{Y}_2^H \mathbf{A} \mathbf{X}_1 & \mathbf{Y}_2^H \mathbf{A} \mathbf{X}_2 \end{bmatrix} \\
&= \begin{bmatrix} \mathbf{Y}_1^H \mathbf{A} (\mathbf{A}^H \mathbf{Y}_1 \mathbf{\Sigma}^{-1}) & (\mathbf{\Sigma} \mathbf{X}_1^H) \mathbf{X}_2 \\ (\mathbf{0}) \mathbf{X}_1 & (\mathbf{0}) \mathbf{X}_2 \end{bmatrix} \\
&= \begin{bmatrix} \mathbf{\Sigma} & \mathbf{0} \\ \mathbf{0} & \mathbf{0} \end{bmatrix}
\end{aligned}
$$

which again is the desired result.

This completes the proof of the singular-value decomposition (SVD) theorem, described by Eq. (7.77).

The numbers $\sigma_1, \sigma_2, \ldots, \sigma_W$, constituting the diagonal matrix $\mathbf{\Sigma}$, are called the *singular values* of the matrix $\mathbf{A}$. The columns of the unitary matrix $\mathbf{X}$, that is, $\mathbf{x}_1, \mathbf{x}_2, \ldots, \mathbf{x}_M$, are the *right singular vectors* of $\mathbf{A}$, and the columns of the second unitary matrix $\mathbf{Y}$, that is, $\mathbf{y}_1, \mathbf{y}_2, \ldots, \mathbf{y}_L$, are the *left singular vectors* of $\mathbf{A}$. We note from the preceding discussion that the right singular vectors, $\mathbf{x}_1, \mathbf{x}_2, \ldots, \mathbf{x}_M$, are eigenvectors of $\mathbf{A}^H \mathbf{A}$, whereas the left singular vectors, $\mathbf{y}_1, \mathbf{y}_2, \ldots, \mathbf{y}_L$, are eigenvectors of $\mathbf{A} \mathbf{A}^H$.

Since $\mathbf{Y}\mathbf{Y}^H$ equals the identity matrix, we find from Eq. (7.77) that

$$\mathbf{A}\mathbf{X} = \mathbf{Y}\begin{bmatrix} \boldsymbol{\Sigma} & \mathbf{0} \\ \mathbf{0} & \mathbf{0} \end{bmatrix}$$

It follows therefore that

$$\begin{aligned} \mathbf{A}\mathbf{x}_i &= \sigma_i\mathbf{y}_i, & i &= 1, 2, \ldots, W \\ \mathbf{A}\mathbf{x}_i &= \mathbf{0} & i &= W+1, \ldots, L \end{aligned} \tag{7.82}$$

Since $\mathbf{X}\mathbf{X}^H$ equals the identity matrix, we also find from Eq. (7.77) that

$$\mathbf{Y}^H\mathbf{A} = \begin{bmatrix} \boldsymbol{\Sigma} & \mathbf{0} \\ \mathbf{0} & \mathbf{0} \end{bmatrix}\mathbf{X}^H$$

or, equivalently,

$$\mathbf{A}^H\mathbf{Y} = \mathbf{X}\begin{bmatrix} \boldsymbol{\Sigma} & \mathbf{0} \\ \mathbf{0} & \mathbf{0} \end{bmatrix}$$

It follows therefore that

$$\begin{aligned} \mathbf{A}^H\mathbf{y}_i &= \sigma_i\mathbf{x}_i, & i &= 1, 2, \ldots, W \\ \mathbf{A}^H\mathbf{y}_i &= \mathbf{0}, & i &= W+1, \ldots, M \end{aligned} \tag{7.83}$$

The singular values of a matrix have many appealing analogies with the eigenvalues of a Hermitian matrix. Indeed, if the matrix $\mathbf{A}$ is Hermitian, then the singular values of $\mathbf{A}$ are just the absolute values of the eigenvalues of $\mathbf{A}$.

General Definition of Pseudoinverse

Our interest in the singular-value decomposition is to formulate a general definition of pseudoinverse. Let $\mathbf{A}$ denote an L-by-M matrix that has the singular-value decomposition described in Eq. (7.77). We define the pseudoinverse of the matrix $\mathbf{A}$ as (Stewart, 1973)

$$\mathbf{A}^\# = \mathbf{X}\begin{bmatrix} \boldsymbol{\Sigma}^{-1} & \mathbf{0} \\ \mathbf{0} & \mathbf{0} \end{bmatrix}\mathbf{Y}^H \tag{7.84}$$

where

$$\boldsymbol{\Sigma}^{-1} = \text{diag}(\sigma_1^{-1}, \sigma_2^{-1}, \ldots, \sigma_W^{-1})$$

and W is the rank of the matrix $\mathbf{A}$.

We may identify the following two special cases of interest that can arise when the matrix $\mathbf{A}$ is of full rank; that is, $W = \min(L, M)$:

1. $L > M$ and $W = M$: In this case, the pseudoinverse of matrix $\mathbf{A}$ equals

$$\mathbf{A}^\# = (\mathbf{A}^H\mathbf{A})^{-1}\mathbf{A}^H \tag{7.85}$$

To show the validity of this special formula, we note from Eqs. (7.78) and (7.79) that

$$(\mathbf{A}^H\mathbf{A})^{-1} = \mathbf{X}_1\boldsymbol{\Sigma}^{-2}\mathbf{X}_1^H$$

and

$$\mathbf{A}^H = \mathbf{X}_1\boldsymbol{\Sigma}\mathbf{Y}_1^H$$

Therefore,

$$(\mathbf{A}^H\mathbf{A})^{-1}\mathbf{A}^H = (\mathbf{X}_1\boldsymbol{\Sigma}^{-2}\mathbf{X}_1^H)(\mathbf{X}_1\boldsymbol{\Sigma}\mathbf{Y}_1^H)$$

$$= \mathbf{X}_1\boldsymbol{\Sigma}^{-1}\mathbf{Y}_1^H$$

$$= \mathbf{X}\begin{bmatrix} \boldsymbol{\Sigma}^{-1} & \mathbf{0} \\ \mathbf{0} & \mathbf{0} \end{bmatrix}\mathbf{Y}^H$$

$$= \mathbf{A}^\#$$

which is the desired result. Note that the definition of Eq. (7.85) coincides with that given in Eq. (7.74).

2. $M > L$ and $W = L$: In this case, the pseudoinverse of matrix $\mathbf{A}$ equals

$$\mathbf{A}^\# = \mathbf{A}^H(\mathbf{A}\mathbf{A}^H)^{-1} \tag{7.86}$$

To show the validity of this second special formula, we note from Eqs. (7.80) and (7.81) that

$$(\mathbf{A}\mathbf{A}^H)^{-1} = \mathbf{Y}_1\boldsymbol{\Sigma}^{-2}\mathbf{Y}_1^H$$

and

$$\mathbf{A}^H = \mathbf{X}_1\boldsymbol{\Sigma}\mathbf{Y}_1^H$$

Therefore,

$$\mathbf{A}^H(\mathbf{A}\mathbf{A}^H)^{-1} = \mathbf{X}_1\boldsymbol{\Sigma}\mathbf{Y}_1^H(\mathbf{Y}_1\boldsymbol{\Sigma}^{-2}\mathbf{Y}_1^H)$$

$$= \mathbf{X}_1\boldsymbol{\Sigma}^{-1}\mathbf{Y}_1^H$$

$$= \mathbf{X}\begin{bmatrix} \boldsymbol{\Sigma}^{-1} & \mathbf{0} \\ \mathbf{0} & \mathbf{0} \end{bmatrix}\mathbf{Y}^H$$

$$= \mathbf{A}^\#$$

which is the desired result.

The Minimum Norm Solution to the Linear Least-Squares Problem

Having equipped ourselves with the general definition of the pseudoinverse of the matrix $\mathbf{A}$ in terms of its singular-value decomposition, we are now ready to tackle the solution to the linear least-squares problem even when null($\mathbf{A}$) $\neq 0$. In particular, we define the solution to the least-squares problem as in Eq. (7.75), reproduced here for convenience,

$$\hat{\mathbf{w}} = \mathbf{A}^{\#}\mathbf{b} \tag{7.87}$$

The pseudoinverse matrix $\mathbf{A}^{\#}$ is itself defined by Eq. (7.74). We thus find that, out of the many vectors that solve the least-squares problem when null($\mathbf{A}$) $\neq 0$, the one defined by Eq. (7.87) is *unique* in that it has the *shortest length possible in the Euclidean sense* (Stewart, 1973).

We prove this important result by manipulating Eq. (7.61) that defines the minimum value of the forward–backward prediction error energy $\mathscr{E}_M$. Since both matrix products $\mathbf{X}\mathbf{X}^H$ and $\mathbf{Y}\mathbf{Y}^H$ equal identity matrices, we may write, starting from Eq. (7.61),

$$
\begin{aligned}
\mathscr{E}_M &= \mathbf{b}^H\mathbf{b} - \mathbf{b}^H\mathbf{A}\hat{\mathbf{w}} \\
&= \mathbf{b}^H(\mathbf{b} - \mathbf{A}\hat{\mathbf{w}}) \\
&= \mathbf{b}^H\mathbf{Y}\mathbf{Y}^H(\mathbf{b} - \mathbf{A}\mathbf{X}\mathbf{X}^H\hat{\mathbf{w}}) \\
&= \mathbf{b}^H\mathbf{Y}(\mathbf{Y}^H\mathbf{b} - \mathbf{Y}^H\mathbf{A}\mathbf{X}\mathbf{X}^H\hat{\mathbf{w}})
\end{aligned} \tag{7.88}
$$

Let

$$
\begin{aligned}
\mathbf{X}^H\hat{\mathbf{w}} &= \mathbf{z} \\
&= \begin{bmatrix} \mathbf{z}_1 \\ \mathbf{z}_2 \end{bmatrix}
\end{aligned} \tag{7.89}
$$

and

$$
\begin{aligned}
\mathbf{Y}^H\mathbf{b} &= \mathbf{c} \\
&= \begin{bmatrix} \mathbf{c}_1 \\ \mathbf{c}_2 \end{bmatrix}
\end{aligned} \tag{7.90}
$$

where $\mathbf{z}_1$ and $\mathbf{c}_1$ are W-by-1 vectors, and $\mathbf{z}_2$ and $\mathbf{c}_2$ are two other vectors. Thus, substituting Eqs. (7.77), (7.89), and (7.90) in (7.88), we get

$$
\begin{aligned}
\mathscr{E}_M &= \mathbf{b}^H\mathbf{Y}\left(\begin{bmatrix} \mathbf{c}_1 \\ \mathbf{c}_2 \end{bmatrix} - \begin{bmatrix} \mathbf{\Sigma} & \mathbf{0} \\ \mathbf{0} & \mathbf{0} \end{bmatrix}\begin{bmatrix} \mathbf{z}_1 \\ \mathbf{z}_2 \end{bmatrix}\right) \\
&= \mathbf{b}^H\mathbf{Y}\begin{bmatrix} \mathbf{c}_1 - \mathbf{\Sigma}\mathbf{z}_1 \\ \mathbf{c}_2 \end{bmatrix}
\end{aligned}
$$

For $\mathscr{E}_M$ to be minimum, we require that

$$\mathbf{c}_1 = \mathbf{\Sigma}\mathbf{z}_1$$

or, equivalently,

$$\mathbf{z}_1 = \mathbf{\Sigma}^{-1}\mathbf{c}_1$$

We observe that $\mathscr{E}_M$ is independent of z_2. Hence, the value of z_2 is arbitrary. However, if we let $z_2 = 0$, we get the special result

$$\hat{\mathbf{w}} = \mathbf{X}\mathbf{z}$$
$$= \mathbf{X}\begin{bmatrix} \mathbf{\Sigma}^{-1}\mathbf{c}_1 \\ \mathbf{0} \end{bmatrix} \tag{7.91}$$

We may also express $\hat{\mathbf{w}}$ in the equivalent form:

$$\hat{\mathbf{w}} = \mathbf{X}\begin{bmatrix} \mathbf{\Sigma}^{-1} & \mathbf{0} \\ \mathbf{0} & \mathbf{0} \end{bmatrix}\begin{bmatrix} \mathbf{c}_1 \\ \mathbf{c}_2 \end{bmatrix}$$
$$= \mathbf{X}\begin{bmatrix} \mathbf{\Sigma}^{-1} & \mathbf{0} \\ \mathbf{0} & \mathbf{0} \end{bmatrix}\mathbf{Y}^H\mathbf{b}$$
$$= \mathbf{A}^{\#}\mathbf{b}$$

This coincides exactly with the value defined by Eq. (7.87), where the pseudoinverse $\mathbf{A}^{\#}$ is defined by Eq. (7.74). In effect, we have shown that this value of $\hat{\mathbf{w}}$ does indeed solve the least-squares problem.

Moreover, the vector $\hat{\mathbf{w}}$ so defined is *unique* in that it has the minimum Euclidean norm possible. In particular, since $\mathbf{X}\mathbf{X}^H = \mathbf{I}$, we find from Eq. (7.91) that the Euclidean norm of $\hat{\mathbf{w}}$ equals

$$\|\hat{\mathbf{w}}\|^2 = \|\mathbf{\Sigma}^{-1}\mathbf{c}_1\|^2$$

Consider now another solution to the least-squares problem that is denoted by

$$\mathbf{w}' = \mathbf{X}\begin{bmatrix} \mathbf{\Sigma}^{-1}\mathbf{c}_1 \\ \mathbf{z}_2 \end{bmatrix}, \qquad \mathbf{z}_2 \neq \mathbf{0}$$

The squared Euclidean norm of $\mathbf{w}'$ equals

$$\|\mathbf{w}'\|^2 = \|\mathbf{\Sigma}^{-1}\mathbf{c}_1\|^2 + \|\mathbf{z}_2\|^2$$

For any $\mathbf{z}_2 \neq \mathbf{0}$, we see therefore that

$$\|\mathbf{w}\| \leq \|\mathbf{w}'\|$$

In summary, the tap-weight $\hat{\mathbf{w}}$ of the linear predictor defined by Eq. (7.87) is a unique solution to the forward–backward linear prediction problem, even when null$(\mathbf{A}) \neq 0$. *The vector $\hat{\mathbf{w}}$ is unique in that it is the only tap-weight vector that simultaneously satisfies two requirements: (1) it produces the minimum forward–backward prediction error energy, and (2) it has the smallest Euclidean norm possible.* This special value of the tap-weight vector $\hat{\mathbf{w}}$ is called the *minimum norm solution.*

Another Formulation of the Minimum Norm Solution

We may develop another formulation of the minimum norm solution by substituting Eq. (7.84) in (7.87), and then using the partioned forms of $\mathbf{X}$ and $\mathbf{Y}$; we may thus write

$$
\begin{aligned}
\hat{\mathbf{w}} &= \mathbf{X} \begin{bmatrix} \mathbf{\Sigma}^{-1} & \mathbf{0} \\ \mathbf{0} & \mathbf{0} \end{bmatrix} \mathbf{Y}^H \mathbf{b} \\
&= [\mathbf{X}_1, \mathbf{X}_2] \begin{bmatrix} \mathbf{\Sigma}^{-1} & \mathbf{0} \\ \mathbf{0} & \mathbf{0} \end{bmatrix} \begin{bmatrix} \mathbf{Y}_1^H \\ \mathbf{Y}_2^H \end{bmatrix} \mathbf{b} \\
&= \mathbf{X}_1 \mathbf{\Sigma}^{-1} \mathbf{Y}_1^H \mathbf{b}
\end{aligned}
\tag{7.92}
$$

Next, substituting Eqs. (7.79) in (7.92), we get (assuming that $L > M$)

$$
\begin{aligned}
\hat{\mathbf{w}} &= \mathbf{X}_1 \mathbf{\Sigma}^{-2} \mathbf{X}_1^H \mathbf{A}^H \mathbf{b} \\
&= \sum_{i=1}^{W} \frac{\mathbf{x}_i}{\sigma_i^2} \mathbf{x}_i^H \mathbf{A}^H \mathbf{b}
\end{aligned}
\tag{7.93}
$$

We next use two facts:

1. The matrix product $\mathbf{A}^H \mathbf{b}$ equals the deterministic cross-correlation vector $\boldsymbol{\theta}$.
2. The vector $\mathbf{x}_i$ is the eigenvector of the deterministic correlation matrix $\mathbf{\Phi} = \mathbf{A}^H \mathbf{A}$ associated with the eigenvalue $\lambda_i = \sigma_i^2$.

Accordingly, we may rewrite Eq. (7.93) as

$$
\hat{\mathbf{w}} = \sum_{i=1}^{W} \frac{\mathbf{x}_i}{\lambda_i} (\mathbf{x}_i^H \boldsymbol{\theta})
\tag{7.94}
$$

Equation (7.94) defines the minimum norm solution of the FBLP method in terms of the cross-correlation vector $\boldsymbol{\theta}$ and the nonzero eigenvalues of the deterministic correlation matrix $\mathbf{\Phi}$ and their associated eigenvectors.

Summary and Discussion

In general, it is not a good idea to use the formula of Eq. (7.94) to compute the minimum-norm value of the tap-weight vector $\hat{\mathbf{w}}$ by first finding the eigenvalues of the deterministic correlation matrix $\mathbf{\Phi}$, tempting as such a procedure may be. The preferred method is to use the SVD–based procedure, which has the advantage in that it is numerically stable.[4] An algorithm is said to be *numerically stable* if it does not introduce any more sensitivity to perturbation than that which is inherently present in the problem (Klema and Laub, 1980). Thus, to compute the minimum-norm value $\hat{\mathbf{w}}$ of the tap-weight vector, we proceed as follows:

---

[4]For a discussion of the issues involved in SVD computation, see Klema and Laub (1980).

1 We compute the singular value decomposition of the data matrix $\mathbf{A}$, and thereby find the singular values $\sigma_1, \sigma_2, \ldots, \sigma_w$ and the associated right singular vectors $\mathbf{x}_1, \mathbf{x}_2, \ldots, \mathbf{x}_w$.

2 We use the formula of Eq. (7.93) to compute $\hat{\mathbf{w}}$, assuming that $L > M$.

For the case when $L < M$, see Problem 9.

7.13 ESTIMATION OF SINE WAVES IN THE PRESENCE OF ADDITIVE NOISE

In this section of the chapter, we will study the use of least squares as the basis for a method for solving the classic problem of estimating the parameters of a set of complex sinusoids in the presence of additive noise. The time series that is representative of such a combination may be expressed as follows:

$$u(i) = \sum_{k=1}^{K} \alpha_k \exp(j\omega_k i) + v(i), \qquad 1 \leq i \leq N \qquad (7.95)$$

where K is the *number of complex sinusoids*, and N is the data length. The $\{\alpha_k\}$ is the set of unknown complex amplitudes, and the $\{\omega_k\}$ is the corresponding set of unknown angular frequencies. Each angular frequency lies in the interval $(0, 2\pi)$. The observation $u(i)$ is a linear function of the α_k, but a nonlinear function of the ω_k. Our primary interest is to estimate the ω_k, $k = 1, 2, \ldots, K$. Once the ω_k have been determined accurately, the α_k, $k = 1, 2, \ldots, K$, can then be determined by a linear least-squares fit to the data.

When the ω_k are well separated from each other, it is well known that the use of the *discrete Fourier transform* (DFT) provides the basis of an efficient and effective method for estimating the angular frequencies of the complex sinusoids (Rife and Boorstyn, 1976). When, however, the ω_k are more closely spaced than the reciprocal of the observation interval or, more generally, when the sinusoids are not orthogonal to each other over the observation interval, we find that the DFT processing is ineffective, and the use of a *high-resolution technique* is the preferred way of solving the problem. Examples of high-resolution techniques include the linear prediction method (Tufts and Kumaresan, 1982), the minimum norm method (Kumaresan and Tufts, 1983) and the MUSIC algorithm (Schmidt, 1981). The first of these methods is considered in this section, and the other two in the next section.

The forward–backward linear prediction method (FBLP) described in Section 7.11 is particularly effective for processing sinusoidal signals (Ulrych and Clayton, 1976; Nutall, 1976). Starting with the time series $\{u(i)\}$, $1 \leq i \leq N$, the FBLP method is used to estimate the tap-weight vector $\hat{\mathbf{w}}$ of the forward linear predictor or the tap-weight vector $\mathbf{a}_M$ of the prediction-error filter. This computation is equivalent to fitting an autoregressive (AR) model to the time series with the model parameters equal to $a_{M,1}, a_{M,2}, \ldots, a_{M,M}$, where these coefficients are elements of the vector $\mathbf{a}_M$; the element $a_{M,0}$ equals unity. The AR power spectrum is next computed by using the formula (see Chapter 2):

$$S_{AR}(\omega) = \frac{\mathscr{E}_M}{|1 + \sum_{m=1}^{M} a_{M,m} \exp(j\omega_m)|^2} \tag{7.96}$$

The unknown frequencies of the complex sinusoids contained in the time series are determined by locating the values of ω for which the AR power spectrum $S_{AR}(\omega)$ attains its *peak* values.

However, the FBLP method suffers from two major limitations (Tufts and Kumaresan, 1982):

1. At high signal-to-noise ratios, the FBLP method is a few decibles poorer than the *Cramér–Rao bound*, which is attained by the maximum-likelihood method.

2. The FBLP method exhibits a *threshold* that appears to be reached at relatively high signal-to-noise ratios, above which the performance of the method deteriorates rapidly. This threshold effect is caused by one of two factors, depending on the predictor order. When the predictor order M is a large fraction of the data length, spurious spectral peaks are induced by the presence of additive noise in the data; the threshold effect is then caused by the spurious spectral peaks overpowering the effect of the signal. On the other hand, when the predictor order M is small, the threshold effect is caused by the merging of closely spaced spectral peaks.

In order to overcome these problems associated with the FBLP mehtod, Tufts and Kumaresan (1982) have developed a two-step procedure for modifying the method. The result is impressive in that the performance of the *modified FBLP method* is close to the Cramér–Rao bound, even for sinusoids of closely spaced angular frequencies at lower values of signal-to-noise ratio than would be possible with other linear prediction methods.

Before proceeding to describe the modified FBLP method, however, we consider first the special case of noiseless data. By so doing, we develop invaluable insight into the operation of the modified FBLP method.

Special Case: Noiseless Data

Let $\mathbf{A}_0$ denote the *noiseless* version of the data matrix in Eq. (7.58). It is easy to verify that the largest size of nonzero determinant that can be formed from the matrix $\mathbf{A}_0$ is K. Hence, the matrix $\mathbf{A}_0$ has rank K. Let $\mathbf{b}_0$ denote the noiseless version of the desired response vector in Eq. (7.62). Thus, we may use Eq. (7.94) to express the minimum norm solution of the FBLP method for noiseless data as follows:

$$\hat{\mathbf{w}}_0 = \sum_{i=1}^{K} \frac{\mathbf{x}_{0i}}{\lambda_{0i}} (\mathbf{x}_{0i}^H \boldsymbol{\theta}_0)$$

where

$$\boldsymbol{\theta}_0 = \mathbf{A}_0^H \mathbf{b}_0$$

and λ_{0i}, $i = 1, 2, \ldots, K$, are the nonzero eigenvalues of the noiseless correlation matrix $\boldsymbol{\Phi}_0 = \mathbf{A}_0^H \mathbf{A}_0$, and $\mathbf{x}_{0i}$, $i = 1, 2, \ldots, K$, are the associated eigenvectors.

Let $\mathbf{a}_{0M}$ and $\mathbf{C}_0$ denote the noiseless versions of the minimum-norm value of the $(M + 1)$-by-1 tap-weight vector of the prediction-error filter $\mathbf{a}_M$ and the $(M + 1)$-by-$(M + 1)$ correlation matrix $\mathbf{C}$, respectively. The tap-weight vector $\mathbf{a}_{0M}$ has a number of important properties, described next:

Property 1.   *The tap-weight vector $\mathbf{a}_{0M}$ is an eigenvector of the deterministic correlation matrix $\mathbf{C}_0$ and the corresponding eigenvalue is zero.*

For noiseless data, the prediction error is zero. Hence, setting the minimum forward–backward prediction error energy $\mathcal{E}_M$ equal to zero, we find from Eq. (7.65) that

$$\mathbf{C}_0 \mathbf{a}_{0M} = \mathbf{0} \qquad (7.97)$$

The correlation matrix $\mathbf{C}_0$ has K nonzero eigenvalues and $(M + 1 - K)$ zero eigenvalues. Thus, Eq. (7.97) is an eigenequation corresponding to a zero eigenvalue, and the tap-weight vector $\mathbf{a}_{0M}$ is a corresponding eigenvector.

Property 2.   *The transfer function of the prediction-error filter has K zeros on the unit circle in the z-plane at angular locations corresponding to the angular frequencies of the complex sinusoids, provided that the predictor order M satisfies the condition*

$$K \le M \le N - \frac{K}{2} \qquad (7.98)$$

where N is the data length.

To prove the property, we first show that the tap-weight vector $\mathbf{a}_{0M}$ of the prediction-error filter is orthogonal to a set of $(M + 1)$-by-1 *sinusoidal vectors* $\mathbf{s}_k$ defined by

$$\mathbf{s}_k^T = [1, \exp(-j\omega_k), \ldots, \exp(-jM\omega_k)], \qquad k = 1, 2, \ldots, K \qquad (7.99)$$

Putting the noise $v(i) = 0$ in Eq. (7.95), we may express the noiseless input data as

$$u(i) = \sum_{k=1}^{K} \alpha_k \exp(j\omega_k i), \qquad i = 1, 2, \ldots, N \qquad (7.100)$$

The (t, k)th element of the correlation matrix $\mathbf{C}_0$ is defined by [see Eq. (7.72)]

$$c_0(t, k) = \sum_{i=M+1}^{N} [u(i - t)u^*(i - k) + u^*(i - M + t)u(i - M + k)],$$

$$0 \le t, k \le M \tag{7.101}$$

Substituting Eq. (7.100) in (7.101) and rearranging terms, we get (after some manipulation)

$$c_0(t, k) = \sum_{l=1}^{K} \sum_{q=1}^{K} s_{tl} p_{lq} s_{qk}^*, \ 0 \le t, k \le M \tag{7.102}$$

where

$$s_{tl} = \exp(-jt\omega_l), \quad \begin{array}{l} 0 \le t \le M \\ 1 \le l \le K \end{array} \tag{7.103}$$

and

$$p_{lq} = \sum_{i=M+1}^{N} \alpha_l \alpha_q^* \exp(ji)(\omega_l - \omega_q) + \alpha_l^* \alpha_q \exp(-j)(i - m)(\omega_l - \omega_q) \tag{7.104}$$

Hence, we may express the noiseless correlation matrix $\mathbf{C}_0$ as the product of three matrices, as shown by

$$\mathbf{C}_0 = \mathbf{SPS}^H \tag{7.105}$$

where $\mathbf{S}$ is an $(M + 1)$-by-K sinusoidal *signal matrix* defined by

$$\mathbf{S} = [\mathbf{s}_1, \mathbf{s}_2, \ldots, \mathbf{s}_K]$$

$$= \begin{bmatrix} 1 & 1 & \cdots & 1 \\ \exp(-j\omega_1) & \exp(-j\omega_2) & \cdots & \exp(-j\omega_K) \\ \exp(-j2\omega_1) & \exp(-j2\omega_2) & \cdots & \exp(-j2\omega_K) \\ \cdot & \cdot & \cdot & \cdot \\ \cdot & \cdot & \cdot & \cdot \\ \cdot & \cdot & \cdot & \cdot \\ \exp(-jM\omega_1) & \exp(-jM\omega_2) & \cdots & \exp(-jM\omega_K) \end{bmatrix} \tag{7.106}$$

and $\mathbf{P}$ is a K-by-K *interaction matrix* defined by

$$\mathbf{P} = \begin{bmatrix} p_{11} & p_{12} & \cdots & p_{1K} \\ p_{21} & p_{22} & \cdots & p_{2K} \\ \cdot & \cdot & \cdot & \cdot \\ \cdot & \cdot & \cdot & \cdot \\ \cdot & \cdot & \cdot & \cdot \\ p_{K1} & p_{k2} & \cdots & p_{KK} \end{bmatrix} \tag{7.107}$$

The matrix $\mathbf{S}$ is a *Vandermonde* matrix (Strang, 1980); each column of the matrix is determined by the angular frequency of a complex sinusoid in the input.

The square matrix $\mathbf{P}$ results from the interaction between the K complex sinusoids. It has the following properties:

1. The matrix $\mathbf{P}$ is Hermitian; that is, $\mathbf{P}^H = \mathbf{P}$.

2. The kth element on the main diagonal of the matrix $\mathbf{P}$ equals

$$p_{kk} = 2(N - M)|\alpha_k|^2$$

where α_k is the amplitude of the kth complex sinusoid

3. Each element of the matrix $\mathbf{P}$ is an inner product of two vectors in the set $\{\mathbf{f}_k\}$, $k = 1, 2, \ldots, K$, where

$$\mathbf{f}_k^T = [\alpha_k^* e^{-j\omega_k(M+1)}, \ldots, \alpha_k^* e^{-j\omega_k N}, \alpha_k e^{j\omega_k}, \ldots, \alpha_k e^{j\omega_k(N-M)}] \qquad (7.108)$$

A matrix, such as $\mathbf{P}$, that has this property is called a Gramian matrix (Tretter, 1976).

4. The matrix $\mathbf{P}$ is nonsingular. This follows from Eq. (4.108), where we see that the vectors $\mathbf{f}_k, k = 1, 2, \ldots, K$, are linearly independent.

Since the tap-weight vector of the prediction-error filter, $\mathbf{a}_{0M}$, is an eigenvector of the correlation matrix $\mathbf{C}_0$ and the associated eigenvalue is zero [see Eq. (7.97)], we may write

$$\mathbf{a}_{0M}^H \mathbf{C}_0 \mathbf{a}_{0M} = 0 \qquad (7.109)$$

Substituting Eq. (7.105) in (7.109), we get

$$\mathbf{a}_{0M}^H \mathbf{S} \mathbf{P} \mathbf{S}^H \mathbf{a}_{0M} = 0 \qquad (7.110)$$

Since the matrix $\mathbf{P}$ is nonsingular, this condition can be satisfied if and only if

$$\mathbf{a}_{0M}^H \mathbf{S} = \mathbf{0} \qquad (7.111)$$

Equivalently, we may use the first line of Eq. (7.106) to write that the tap-weight vector $\mathbf{a}_{0M}$ is orthogonal to the set of $(M + 1)$-by-1 sinusoidal vectors $\{\mathbf{s}_k\}$, $k = 1, 2, \ldots, K$, as shown by

$$\mathbf{a}_{0M}^H \mathbf{s}_k = 1 + \sum_{l=1}^{M} a_{0M,l}^* \exp(-jl\omega_k), \qquad k = 1, 2, \ldots, K$$
$$= 0 \qquad (7.112)$$

where $a_{0M,1}, a_{0M,2}, \ldots, a_{0M,M}$ are the tap weights of the prediction-error filter; the tap weight $a_{0M,0}$ equals unity. The expression on the right side of the first line of Eq. (7.112) is recognized as the transfer function of the prediction-error filter, evaluated on the unit circle at $z = \exp(j\omega_k)$, $k = 1, 2, \ldots, K$. We have thus proved that the transfer function of the prediction-error filter, for noiseless data, has K zeros on the unit circle in the z-plane, which determine the angular frequencies of the K complex sinusoids.

There now only remains the problem of determining the range of values of the predictor order M for which this property holds. First, we note that the minimum value of M is K, the number of complex sinusoids. For input data of fixed length N, as the predictor order M is increased up to the point where $2(N - M) < K$, we find that the vectors $\mathbf{f}_k$, $k = 1, 2, \ldots, K$, cannot be independent, and, consequently, the matrix $\mathbf{P}$ is singular. Then Eq. (7.111) will no longer be true, in general. The maximum value of predictor order M is therefore $N - (K/2)$. Thus, the predictor order should satisfy the condition

$$K \leq M \leq N - \frac{K}{2}$$

as stated previously.

Property 3. *For a predictor order satisfying the condition*

$$K \leq M \leq N - \frac{K}{2}$$

the $M - K$ extraneous zeros of the transfer function of the prediction-error filter are uniformly distributed in angle around the inside of the unit circle in the z-plane.

The proof of this property is quite involved. For details of the proof, the reader is referred to Kumaresan (1983).

We see, therefore, that for noiseless data the zeros of the transfer function of the prediction-error filter take on a special geometry that is uniquely determined by the number of complex sinusoids contained in the input, their angular frequencies, and the predictor order. In particular, these zeros may be divided into two sets. One set consists of K zeros that are located on the unit circle at angular locations uniquely determined by the angular frequencies of the complex sinusoids. The second set consists of $M - K$ extraneous zeros that are uniformly distributed inside the unit circle. Given this geometry, we may determine the exact values of the angular frequencies of the complex sinusoids simply by reading off the angles of the zeros located on the unit circle.

The picture presented is ideal, assuming noiseless input data. The presence of noise, unless very small, destroys this well-defined geometry. However, by modifying the FBLP method in the manner described next, an attempt is made to restore some order into the zeros of the transfer function of the prediction-error filter so that their distribution resembles the well-defined geometry for noiseless data.

Modified FBLP Method

We now describe a *two-step* procedure for modifying the FBLP method (Tufts and Kumaresan, 1982). The need for this modification is motivated by the desire to overcome the ill effects associated with the conventional FBLP method, which were discussed previously.

The two steps of the *modified FBLP method* are as follows:

1. We compute the M eigenvalues of the correlation matrix $\mathbf{\Phi} = \mathbf{A}^H\mathbf{A}$ in the normal equation (7.60) and divide them in two parts: (a) the *signal subspace* spanned by the eigenvectors associated with the K largest eigenvalues, and (b) the *noise subspace* spanned by the remaining $(M - K)$ eigenvectors. The notion that is used here is that the K *principal* eigenvectors of the matrix $\mathbf{\Phi}$ are considerably more *robust* to noisy perturbations than the rest. In the idealized case of noiseless data consisting of K complex sinusoids, the noiseless correlation matrix $\mathbf{\Phi}_0 = \mathbf{A}_0^H\mathbf{A}_0$ has K nonzero eigenvalues and $(M - K)$ zero eigenvalues. Thus, in the practical case of noisy data, as in Eq. (7.95), by retaining the K largest eigenvalues of the correlation matrix $\mathbf{\Phi}$ and ignoring the rest, we are in effect attempting to increase the signal-to-noise ratio by approaching the idealized noiseless case. Indeed, there is experimental evidence to suggest that, for high values of signal-to-noise ratio, the K principal eigenvalues and eigenvectors of the correlation matrix $\mathbf{\Phi}$ remain "close" to their noiseless counterparts (Tufts and Kumaresan, 1982). Let $\lambda_1, \lambda_2, \dots,$ λ_K denote the K largest eigenvalues of the correlation matrix $\mathbf{\Phi}$, and let $\mathbf{x}_1,$ $\mathbf{x}_2, \dots, \mathbf{x}_K$ denote the associated eigenvectors, respectively. As an *improved estimate* of the correlation matrix $\mathbf{\Phi}$, we use

$$\hat{\mathbf{\Phi}} = \sum_{k=1}^{K} \lambda_k \mathbf{x}_k \mathbf{x}_k^H \tag{7.113}$$

Thus, in step 1, we replace the full-rank signal-plus-noise correlation matrix $\mathbf{\Phi}$ by $\hat{\mathbf{\Phi}}$ that is of *lower rank*.

2. Intuitively, the order M should be as large as possible in order to have a large aperture for the predictor. However, in the conventional FBLP method, the use of large values of M gives rise to spurious spectral peaks. For best performance of the FBLP method, Lang and McLellan (1980) suggest the value

$$M \simeq \frac{N}{3}$$

where N is the data length. Now, once we remove the undesirable effects of the noise subspace eigenvectors, we may increase the predictor order M

beyond this limit. This increase in M helps to improve the resolution capability of the predictor since it represents an increase in the aperture or degrees of freedom available in the design of the filter. However, even in using the modified FBLP method, there is a practical limit as to how large a value we may use for the predictor order M. At very large values of M, the effective signal-to-noise ratio improvement decreases, with the result that the resolution capability of the FBLP method decreases despite the increased aperture. This decrease in effective signal-to-noise ratio improvement is due to the fact that, with increased M, fewer product terms [equal to $2(N - M)$ in number] are averaged in computing the elements of the correlation matrix $\boldsymbol{\Phi}$. This, in turn, results in larger perturbations of the eigenvalues λ_k and eigenvectors $\mathbf{x}_k$. We thus find that, on the one hand, increasing the value of M provides the potential for higher resolution, and yet, on the other hand, it also increases the fluctuations in the matrix $\boldsymbol{\Phi}$, and hence in the estimate $\hat{\boldsymbol{\Phi}}$ of Eq. (7.113). We therefore have to use a compromise value of predictor order M, so as to balance the two effects of resolution and stability. Tufts and Kumaresan (1982) have experimentally determined the compromise value to be

$$M \simeq 3N/4.$$

In Table 7.1, we have summarized the modified FBLP method.

Kumaresan–Prony Case

Equation (7.98) defines the permissible range of values for the predictor order M. When M equals the upper limit, $N - (K/2)$, the number of rows contained in the data matrix $\mathbf{A}$, defined in Eq. (7.58) assumes the value

$$2(N - M) = K$$

We typically have $M > K$. Hence, for this special case, the rank W of the data matrix $\mathbf{A}$ equals K. Consequently, the estimate $\hat{\boldsymbol{\Phi}}$ for the M-by-M correlation matrix defined in Eq. (7.113) coincides exactly with the actual value $\boldsymbol{\Phi} = \mathbf{A}\mathbf{A}^H$, thereby making the first step in the modified FBLP method redundant. This special case is called the *Kumaresan–Prony case*.

7.14 COMPUTER EXPERIMENT

Consider the noisy signal

$$u(i) = \exp(j\pi i - j\pi) + \exp(j1.04\pi i - j0.79\pi) + v(i), \qquad 1 \le i \le N$$

where the noise samples $v(i)$ are independent complex Gaussian random variables with zero mean. The data length $N = 25$.

Figure 7.6 shows the geometry of the zeros of the transfer function of the prediction-error filter of order $M = 20$ for the special case of noiseless data; that

TABLE 7.1 SUMMARY OF THE MODIFIED FBLP METHOD

1. Use the singular value decomposition, applied to the data matrix **A**, to compute the sample eigenvalues and eigenvectors by the M-by-M deterministic correlation matrix

$$\mathbf{\Phi} = \mathbf{A}^H\mathbf{A}$$

where **A** is the $2(N - M)$-by-M data matrix, defined by

$$\mathbf{A}^H = \begin{bmatrix} u(M) & u(M+1) & \cdots & u(N-1) & u^*(2) & u^*(3) & \cdots & u^*(N-M+1) \\ u(M-1) & u(M) & \cdots & u(N-2) & u^*(3) & u^*(4) & \cdots & u^*(N-M+2) \\ \cdot & \cdot & \cdot & \cdot & \cdot & \cdot & \cdot & \cdot \\ \cdot & \cdot & \cdot & \cdot & \cdot & \cdot & \cdot & \cdot \\ \cdot & \cdot & \cdot & \cdot & \cdot & \cdot & \cdot & \cdot \\ u(1) & u(2) & \cdots & u(N-M) & u^*(M+1) & u^*(M+2) & \cdots & u^*(N) \end{bmatrix}$$

2. Identify the K largest eigenvalues, where K equals the number of complex sinusoids (assumed known). Let these be denoted as $\lambda_1, \lambda_2, \ldots, \lambda_K$, and their associated eigenvectors as $\mathbf{x}_1, \mathbf{x}_2, \ldots, \mathbf{x}_K$, respectively.

3. Use these eigenvalues and eigenvectors to compute the following estimate of the M-by-1 tap-weight vector of the predictor:

$$\hat{\mathbf{w}} = \sum_{i=1}^{K} \frac{\mathbf{x}_i}{\lambda_i}(\mathbf{x}_i^H\mathbf{\theta})$$

where $\mathbf{\theta}$ is the M-by-1 deterministic cross-correlation vector

$$\mathbf{\theta} = \mathbf{A}^H\mathbf{b}$$

in which the $2(N-M)$-by-1 desired response vector **b** is defined by

$$\mathbf{b}^H = [u(M+1), u(M+2), \ldots, u(M), u^*(1), u^*(2), \ldots, u^*(N-M)]$$

For the prediction order M, use the (empirically derived) optimum value $3N/4$, where N is the data length.

4. Compute the $(M+1)$-by-1 tap-weight vector of the prediction-error filter.

$$\mathbf{a} = \begin{bmatrix} 1 \\ -\hat{\mathbf{w}} \end{bmatrix}$$

5. Determine the angular frequencies of the complex sinusoids as the peaks of the sample spectrum

$$S(\omega) = \frac{1}{|\mathbf{a}^H\mathbf{s}(\omega)|^2}$$

where $\mathbf{s}(\omega)$ is the $(M+1)$-by-1 sinusoidal signal vector:

$$\mathbf{s}^T(\omega) = [1, e^{-j\omega}, \ldots, e^{-jM\omega}]$$

---

is, $v(i) = 0$ for all i. This diagram clearly illustrates Properties 2 and 3 for the noiseless case described previously.

Figure 7.7 shows the locations of the prediction-error filter zeros in the z-plane for the conventional FBLP method, assuming a signal-to-noise ratio of 10 dB. Parts (a) through (h) of this figure show the results obtained for predictor order $M = 4, 8, 12, 16, 18, 20, 22,$ and 24, respectively. This figure shows the

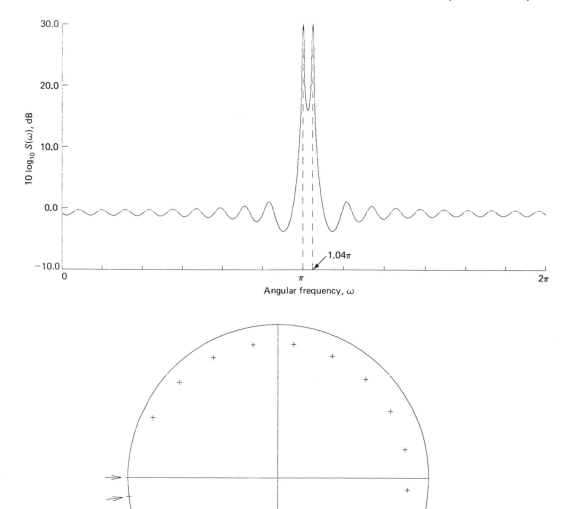

Figure 7.6 Response of forward–backward linear predictor, noiseless data, $M = 20$.

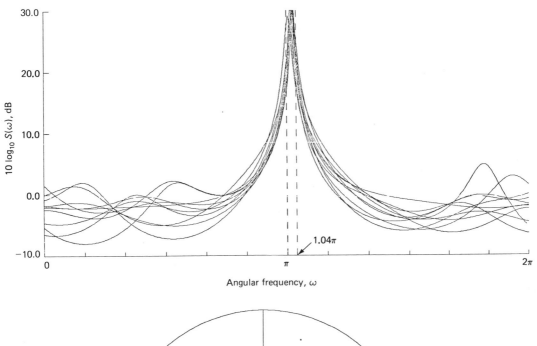

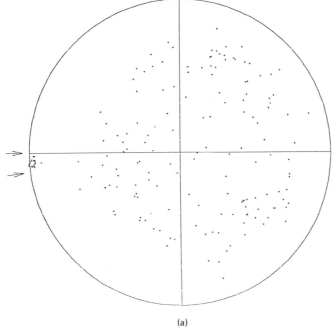

(a)

Figure 7.7   Response of forward–backward linear predictor, SNR = 10 dB, for 50 trials of the experiment: (a) $M = 4$; (b) $M = 8$; (c) $M = 12$; (d) $M = 16$; (e) $M = 18$; (f) $M = 20$; (g) $M = 22$; (h) $M = 24$.

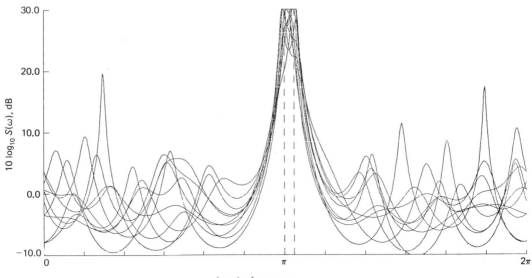

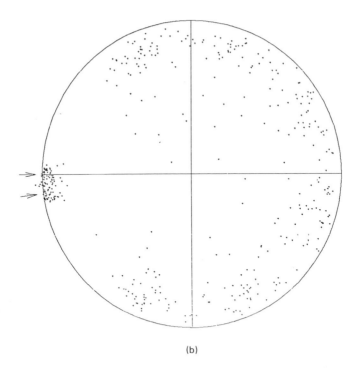

(b)

Figure 7.7 (*cont.*)

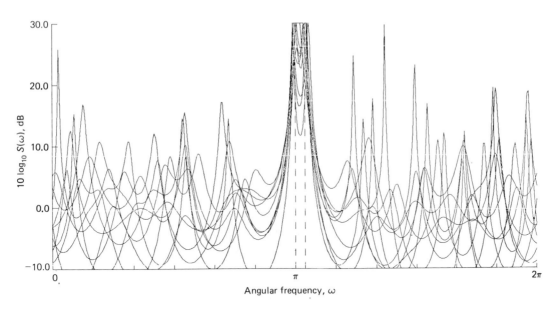

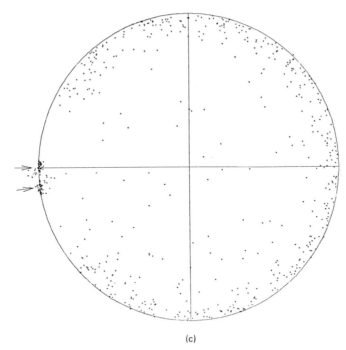

(c)

Figure 7.7   (*cont.*)

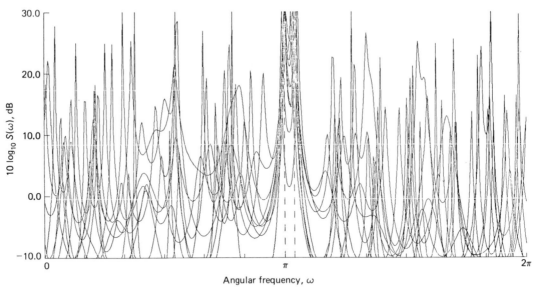

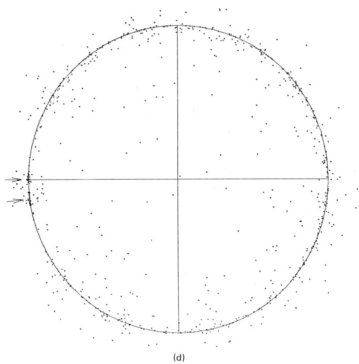

(d)

Figure 7.7 (*cont.*)

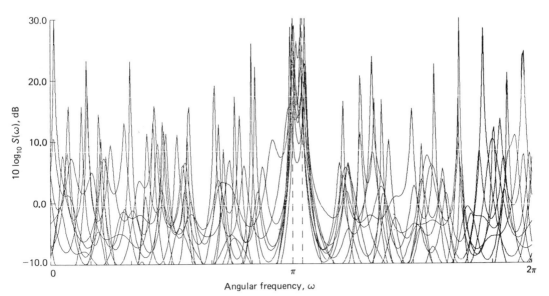

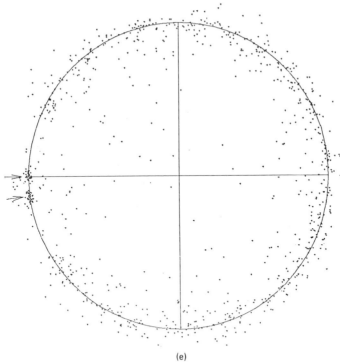

(e)

Figure 7.7 (*cont.*)

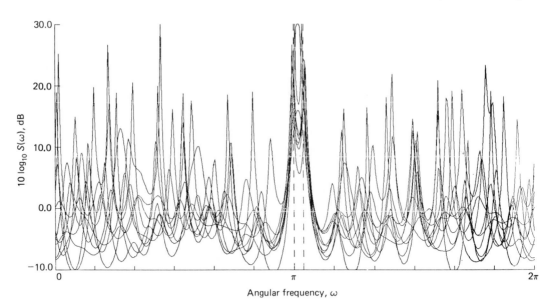

Angular frequency, ω

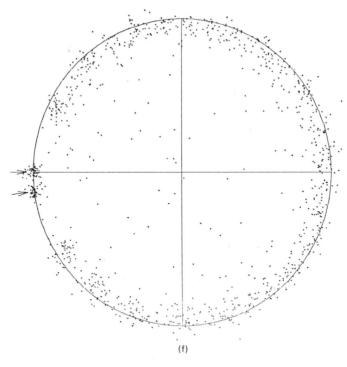

(f)

Figure 7.7 (*cont.*)

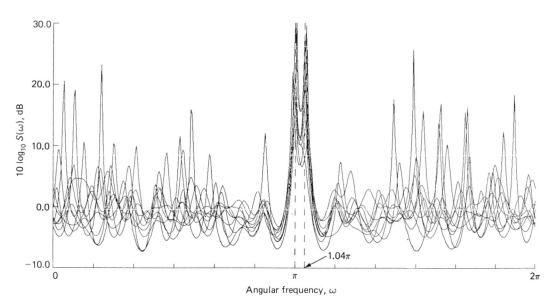

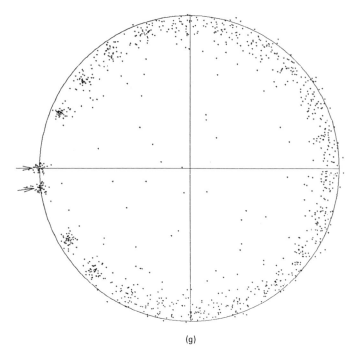

(g)

Figure 7.7   (*cont.*)

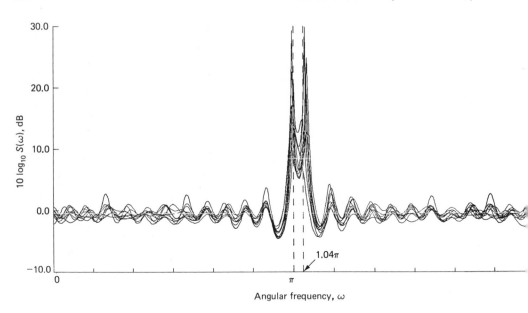

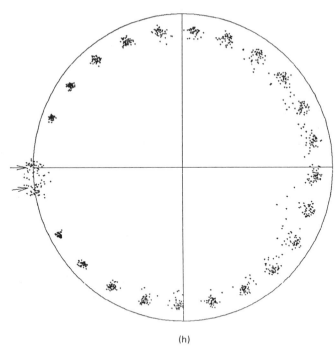

(h)

Figure 7.7 (*cont.*)

combined results of 50 independent trials.   The actual angular locations of the complex sinusoidals are shown by arrows.   For each trial, the two zeros that are close to the angular locations of the complex sinusoids are called *signal zeros*, and the remaining $M - 2$ zeros are called *extraneous zeros*.   Based on the results of Fig. 7.7, we may make the following observations:

1. As the predictor order M is increased, the presence of aditional noise eigenvectors causes considerable fluctuations in the locations of the signal zeros and extraneous zeros.

2. For $M = 16$ and 18, in parts (d) and (e) of the figure, the noise eigenvectors swamp the signal eigenvectors, thereby rendering the prediction-error filter useless.

3. For $M \geq 20$, the situation begins to improve.   This may be related to the smaller number of noise eigenvectors contributing to the predictor computation as M is increased beyond $2N/3$.   For example, for $M = 22$, the rank of the correlation matrix Φ is $2(N - M) = 6$, and so we find that two signal eigenvectors and only four noise eigenvectors enter the predictor computation.   With $M = N - (K/2) = 24$, the situation improves dramatically, with the signal-zero and noise-zero clusters appearing in much more orderly fashion.   They fall into 24 distinct clusters.   This is due to the noise eigenvectors not entering the prediction filter computation at all for this case.   The results presented in Fig. 7.7(h) correspond to the *Kumaresan–Prony case.*

Figure 7.8 shows the zeros of the prediction-error filter, computed by using the modified FBLP method, with the same 50 data sets as before.   Parts (a) through (h) of this figure correspond to the same M values used in Figs. 7.7 (a) through (h).   The following two observations are immediately apparent from these figures:

1. The spreads of the noise zero clusters are considerably less, thereby reducing the chance of spurious frequency estimation.

2. As the predictor order M increases, the spread of the signal zero and also noise zero clusters decreases, reaching a minimum at about $M = 18$ (about $3N/4$), and then increases.   This confirms the choice of $3N/4$ as the optimum value for the predictor order in the modified FBLP method.

In Figs. 7.7 and 7.8, we have also included plots of 10 *sample* spectra for the different sets of parameters for the conventional FBLP and modified FBLP methods, respectively, with each curve corresponding to one trial of the experiment. These spectra are obtained by computing the function

$$S(\omega) = \frac{1}{|H(e^{j\omega})|^2}$$

$$= \frac{1}{|\mathbf{a}^H \mathbf{s}(\omega)|^2}$$

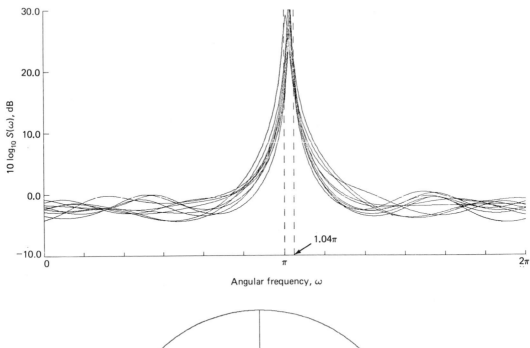

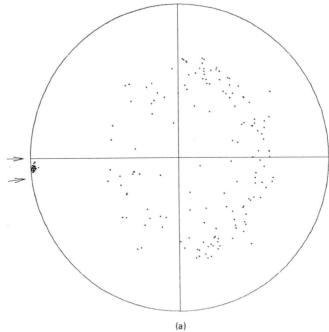

(a)

Figure 7.8 Response of modified FBLP predictor, SNR = 10 dB, for 50 trials of the experiment: (a) $M = 4$; (b) $M = 8$; (c) $M = 12$; (d) $M = 16$; (e) $M = 18$; (f) $M = 20$; (g) $M = 22$; (h) $M = 24$.

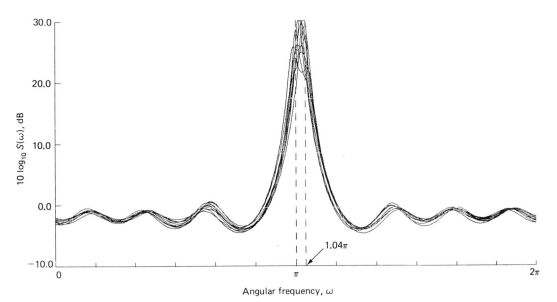

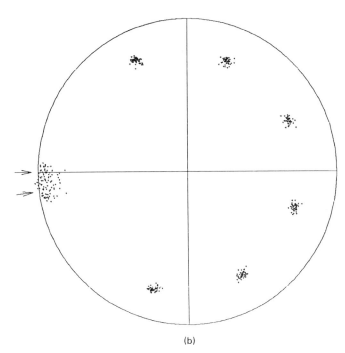

(b)

Figure 7.8 (*cont.*)

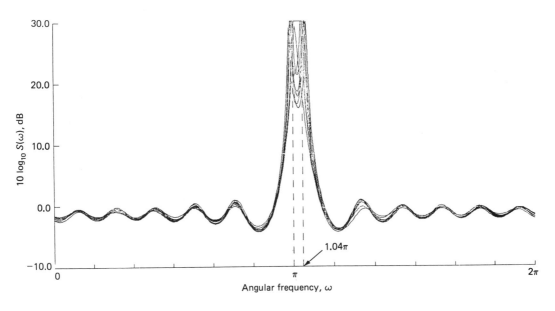

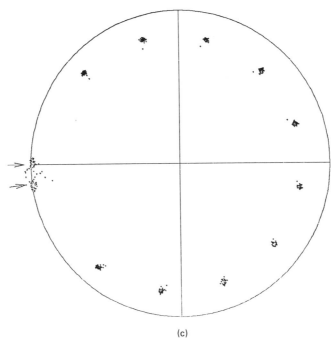

(c)

Figure 7.8 *(cont.)*

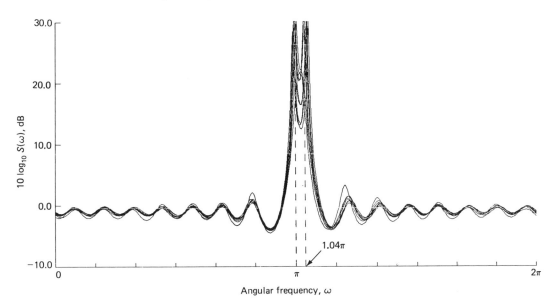

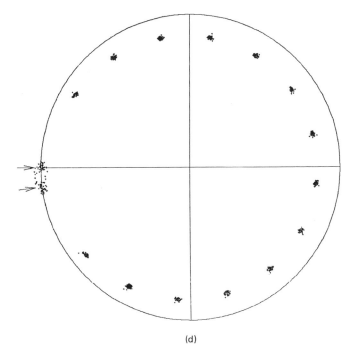

(d)

Figure 7.8   (*cont.*)

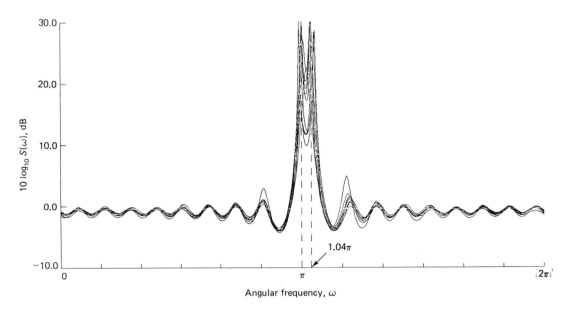

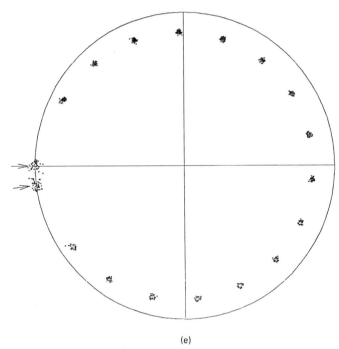

(e)

Figure 7.8 (*cont.*)

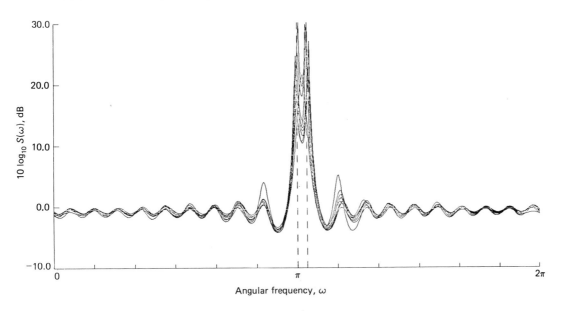

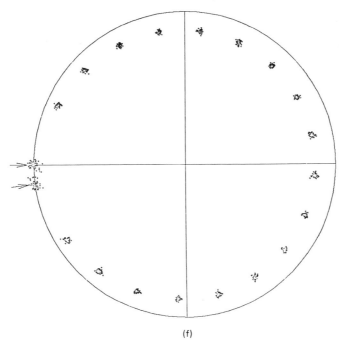

(f)

Figure 7.8   (*cont.*)

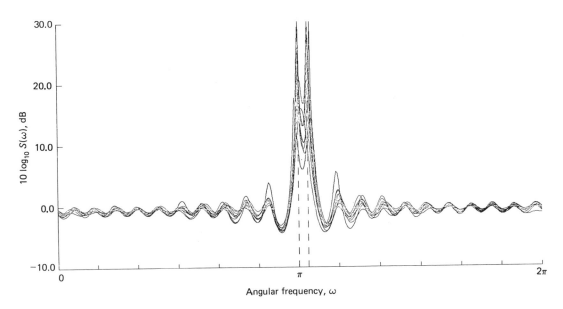

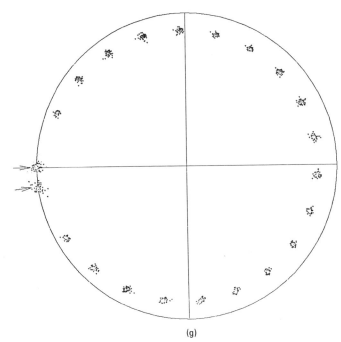

(g)

Figure 7.8 (*cont.*)

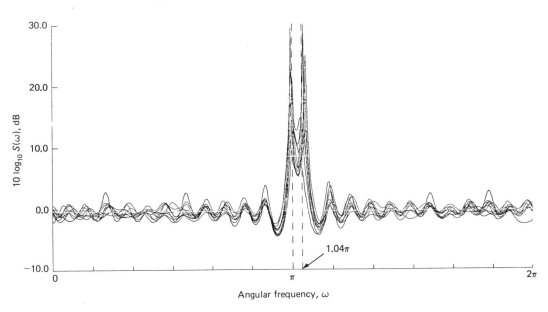

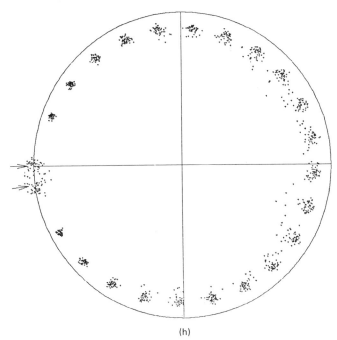

(h)

Figure 7.8   (*cont.*)

where $H(e^{j\omega})$ is the transfer function of the prediction-error filter, **a** is its tap-weight vector, and **s**(ω) is the signal vector whose mth element equals $\exp(-jm\omega)$, $m = 0, 1, \ldots, M$. In the conventional FBLP method, $S(\omega)$ represents the autoregressive (AR) power spectrum of the process. However, this is *not* so in the modified FBLP method. In any event, an ideal spectral model for the case when the process consists of a number of complex sinusoids in additive white noise is one for which $|H(e^{j\omega_k})| = 0$ and $S(\omega_k) = \infty$ for $k = 1, 2, \ldots, K$. The sample spectra shown in Figs. 7.7 and 7.8 thus provide another graphical way of assessing the effect of predictor order on the utility of the conventional FBLP and modified FBLP methods for spectral analysis. In their own way, they substantiate the observations made previously concerning the conventional and modified forms of the FBLP method.

7.15 *TWO OTHER EIGENDECOMPOSITION-BASED METHODS FOR THE ESTIMATION OF SINE WAVES IN NOISE*

A discussion of high-resolution methods for the estimation of sine waves in noise would be incomplete without two other methods, namely, the *MUSIC algorithm* (Schmidt, 1981) and the *minimum-norm method* (Kumaresan and Tufts, 1983). A common feature of both of these methods and the modified FBLP method is the decomposition of a correlation matrix of the input signal into two orthogonal subspaces: signal and noise subspaces. The algorithm for estimating the sine waves is formulated in one or the other of these two subspaces. In this section we briefly describe the MUSIC algorithm and then the minimum-norm method.

The MUSIC Algorithm

To motivate the development of the *mu*ltiple *si*gnal *c*haracterization (MUSIC) algorithm, consider first the $(M + 1)$-by-$(M + 1)$ ensemble-averaged correlation matrix **R** for an input signal that consists of K *uncorrelated* zero-mean complex sinusoids and an additive white noise process of zero mean and variance σ^2. The angular frequencies of the sinusoids are denoted by $\omega_1, \omega_2, \ldots, \omega_K$, and their average powers by $P_1, P_2, \ldots, P_K$. We may thus express the correlation matrix **R** in the form (see Problem 9, Chapter 2).

$$\mathbf{R} = \mathbf{SDS}^H + \sigma^2 \mathbf{I} \tag{7.114}$$

where **S** is the $(M + 1)$-by-K sinusoidal signal matrix defined in Eq. (7.106), **D** is the K-by-K correlation matrix of the sinusoids

$$\mathbf{D} = \text{diag}(P_1, P_2, \ldots, P_K), \tag{7.115}$$

and **I** is the $(M + 1)$-by-$(M + 1)$ identity matrix.

Let $\lambda_1 \geq \lambda_2 \cdots \geq \lambda_{M+1}$ denote the eigenvalues of the correlation matrix **R**, and $v_1 \geq v_2 \cdots v_{M+1}$ denote the eigenvalues of $\mathbf{SDS}^H$, respectively. Then, from the representation shown in Eq. (7.114) we deduce that

$$\lambda_i = v_i + \sigma^2, \quad i = 1, 2, \ldots, M + 1 \quad (7.116)$$

We assume that the signal matrix $\mathbf{S}$ is of full column rank K. This implies that the $(M + 1 - K)$ smallest eigenvalues of the matrix $\mathbf{SDS}^H$ are equal to zero. Correspondingly, the smallest eigenvalue of the correlation matrix $\mathbf{R}$ is equal to σ^2 with multiplicity $(M + 1 - K)$, as shown by

$$\lambda_i = \begin{cases} v_i + \sigma^2, & i = 1, \ldots, K \\ \sigma^2, & i = K + 1, \ldots, M + 1 \end{cases} \quad (7.117)$$

Note that $v_i = P_i$, $i = 1, 2, \ldots, K$.

Let $\mathbf{q}_1, \mathbf{q}_2, \ldots, \mathbf{q}_{M+1}$ denote the eigenvectors of the correlation matrix $\mathbf{R}$. All the $(M + 1 - K)$ eigenvectors associated with the smallest eigenvalues of $\mathbf{R}$ satisfy the relation

$$\mathbf{Rq}_i = \sigma^2 \mathbf{q}_i, \quad i = K + 1, \ldots, M + 1$$

or, equivalently,

$$(\mathbf{R} - \sigma^2\mathbf{I})\mathbf{q}_i = \mathbf{0}, \quad i = K + 1, \ldots, M + 1 \quad (7.118)$$

Using Eq. (7.114), we may rewrite this relation as

$$\mathbf{SDS}^H\mathbf{q}_i = \mathbf{0}, \quad i = K + 1, \ldots, M + 1 \quad (7.119)$$

Since the matrix $\mathbf{S}$ is assumed to be of full column rank K (which is justified if the sinusoidal components have distinct frequencies), and since the matrix $\mathbf{D}$ is diagonal (which is justified if the sinusoidal components are uncorrelated), it follows from Eq. (7.119) that

$$\mathbf{S}^H\mathbf{q}_i = \mathbf{0}, \quad i = K + 1, \ldots, M + 1 \quad (7.120)$$

or more explicitly [from Eq. (7.106)]

$$\mathbf{s}_k^H\mathbf{q}_i = 0, \quad i = K + 1, \ldots, M + 1 \\ k = 1, 2, \ldots, K \quad (7.121)$$

where the vector $\mathbf{s}_k$ constitutes the kth column of matrix $\mathbf{S}$.

A fundamental property of the eigenvectors of a correlation matrix is that they are orthogonal to each other. Hence, the eigenvectors $\mathbf{q}_1, \ldots, \mathbf{q}_K$ span the orthogonal complement of the space spanned by the eigenvectors $q_{K+1}, \ldots, \mathbf{q}_{M+1}$. Accordingly, we deduce from Eq. (7.121) that

$$\text{span } \{\mathbf{s}_1, \ldots, \mathbf{s}_K\} = \text{span } \{\mathbf{q}_1, \ldots, \mathbf{q}_K\} \quad (7.122)$$

Thus, the eigenvalue decomposition of the $(M + 1)$-by-$(M + 1)$ correlation matrix $\mathbf{R}$ of superimposed complex sinusoids in noise suggests the following two observations:

1. The space spanned by the eigenvectors of $\mathbf{R}$ consists of two disjoint subspaces: one (called the *signal subspace*) that is spanned by the eigenvectors associated with the K *largest* eigenvalues of $\mathbf{R}$, and its orthogonal complement (called the *noise subspace*) that is spanned by the eigenvectors associated with the $(M + 1 - K)$ *smallest* eigenvalues of $\mathbf{R}$. This follows from Eq. (7.122).
2. Given the eigenvectors of $\mathbf{R}$, we may determine the frequencies of the complex sinusoids in the input signal by searching for those sinusoidal signal vectors $\mathbf{s}_k$ that are orthogonal to the noise subspace. This follows from Eq. (7.121).

In practice, however, we have to base our analysis on some estimate of the ensemble-averaged correlation matrix $\mathbf{R}$. For example, we may use a sample average that equals the scaled version of the deterministic correlation matrix $\mathbf{C}$. The computation of $\mathbf{C}$ involves time averaging in both the forward and backward directions. Thus, we may use the estimate:

$$\hat{\mathbf{R}} = \frac{1}{2(N - M)} \mathbf{C}$$

where $\mathbf{C}$ is defined in Eq. (7.66), and the scaling factor $1/2(N - M)$ accounts for the fact that the time-averaging is performed over $2(N - M)$ data points. In any event, let $\mathbf{x}_1, \mathbf{x}_2, \dots, \mathbf{x}_{M+1}$ denote the eigenvectors of this estimate. In accordance with point (1), we define a sample signal subspace spanned by the eigenvectors $\mathbf{x}_1, \dots, \mathbf{x}_K$ that are associated with the K largest eigenvalues of the estimate $\hat{\mathbf{R}}$, and a sample noise subspace spanned by eigenvectors $\mathbf{x}_{K+1}, \dots, \mathbf{x}_{M+1}$ associated with the $(M + 1 - K)$ smallest eigenvalues of $\hat{\mathbf{R}}$. Let $\mathbf{X}_N$ denote the $(M+1)$-by-$(M+1-K)$ matrix constructed with the sample noise subspace eigenvectors:

$$\mathbf{X}_N = [\mathbf{x}_{K+1}, \dots, \mathbf{x}_{M+1}] \tag{7.123}$$

and let $\mathbf{X}_S$ denote the $(M + 1)$-by-K matrix constructed with the sample signal subspace eigenvectors:

$$\mathbf{X}_S = [\mathbf{x}_1, \dots, \mathbf{x}_K] \tag{7.124}$$

Naturally, the sample signal and noise subspaces are orthogonal to each other; that is,

$$\mathbf{X}_N^H \mathbf{X}_S = \mathbf{0} \tag{7.125}$$

Owing to the presence of noise in the eigenvector estimates that span these two sample subspaces, however, the orthogonality relations of Eq. (7.121) that are the basis of point (2) no longer hold. The best that we can do is to search for the signal vectors that are most closely orthogonal to the noise subspace. Accordingly, in the MUSIC algorithm it is proposed to estimate the angular frequencies of the complex sinusoids in the input signal as the peaks of the following sample spectrum

(Schmidt, 1979, 1981):

$$S(\omega) = \frac{1}{\displaystyle\sum_{i=K+1}^{M+1} |\mathbf{s}^H(\omega)\mathbf{x}_i|^2}$$

(7.126)

$$= \frac{1}{\mathbf{s}^H(\omega)\mathbf{X}_N\mathbf{X}_N^H\,\mathbf{s}(\omega)}$$

where the signal vector $\mathbf{s}(\omega)$ is defined by

$$\mathbf{s}^T(\omega) = [1, e^{-j\omega}, \ldots, e^{-j\omega(M-K)}]$$

(7.127)

and the matrix $\mathbf{X}_N$ is defined by Eq. (7.123). Equation (7.126) is the formula used in the MUSIC algorithm for estimating the frequencies of complex sinusoids that are corrupted by additive white noise. Note that in Eq. (7.126) all the eigenvectors that constitute the noise subspace are weighted equally. This has the effect of "whitening" those portions of the spectrum that do not belong to sinusoidal signals.

A summary of the MUSIC algorithm is presented in Table 7.2.

TABLE 7.2 SUMMARY OF THE MUSIC ALGORITHM

1. Use the singular value decomposition of the data matrix to compute the sample eigenvalues and eigenvectors of the estimate of the $(M+1)$-by-$(M+1)$ correlation matrix:

$$\hat{\mathbf{R}} = \frac{1}{2(N-M)} \sum_{n=M+1}^{N} [\mathbf{u}(n)\,\mathbf{u}^H(n) + \mathbf{u}^{B*}(n)\,\mathbf{u}^{BT}(n)]$$

   where $\mathbf{u}(n)$ is the $(M+1)$-by-1 input vector, defined by

$$\mathbf{u}^T(n) = [u(n), u(n-1), \ldots, u(n-M)]$$

   and $\mathbf{u}^B(n)$ is the $(M+1)$-by-1 reversed input vector, defined by

$$\mathbf{u}^{BT}(n) = [u(n-M), u(n-M+1), \ldots, u(n)]$$

2. Given that there are K (uncorrelated) complex sinusoids in the input signal, with $K \le M$, classify the eigenvalues into two groups: one consisting of the K largest eigenvalues and the other consisting of the $(M+1-K)$ smallest eigenvalues. Use the eigenvectors associated with the second group to construct the $(M+1)$-by-$(M+1-K)$ matrix $\mathbf{X}_N$ whose elements span the sample noise subspace.

3. Determine the angular frequencies of the complex sinusoids as the peaks of the sample spectrum

$$S(\omega) = \frac{1}{\mathbf{s}^H(\omega)\,\mathbf{X}_N\mathbf{X}_N^H\,\mathbf{s}(\omega)}$$

   where $\mathbf{s}(\omega)$ is the $(M+1)$-by-1 sinusoidal signal vector, defined by

$$\mathbf{s}^T(\omega) = [1, e^{-j\omega}, \ldots, e^{-j\omega(M-K)}]$$

The Minimum-norm Method

In the *minimum-norm method*, as with the MUSIC algorithm, the sample signal space spanned by the eigenvectors of the correlation matrix estimate $\hat{\mathbf{R}}$ is divided into two disjoint subspaces: the sample signal subspace represented by matrix $\mathbf{X}_S$ and the sample noise subspace represented by matrix $\mathbf{X}_N$. Let these two matrices be partitioned as follows:

$$\mathbf{X}_S = \begin{bmatrix} \mathbf{g}_S^T \\ \hline \mathbf{G}_S \end{bmatrix} \tag{7.128}$$

$$\mathbf{X}_N = \begin{bmatrix} \mathbf{g}_N^T \\ \hline \mathbf{G}_N \end{bmatrix} \tag{7.129}$$

where the 1-by-K row vector $\mathbf{g}_S^T$ and the 1-by-$(M + 1 - K)$ row vector $\mathbf{g}_N^T$ have the first elements of the signal and noise subspace eigenvectors, respectively, and the M-by-K matrix $\mathbf{G}_S$ and the M-by-$(M + 1 - K)$ matrix $\mathbf{G}_N$ have the same elements as $\mathbf{X}_S$ and $\mathbf{X}_N$, respectively, except for their first rows.

In the minimum-norm method, the aim is to find an $(M + 1)$-by-1 vector $\mathbf{a}$ that satisfies three requirements (Kumaresan and Tufts, 1983):

1. The vector $\mathbf{a}^*$ lies in the range of $\mathbf{X}_N$, so that it is orthogonal to the columns of $\mathbf{X}_S$, as shown by

$$\mathbf{X}_S^T \mathbf{a} = \mathbf{0} \tag{7.130}$$

2. The first element of $\mathbf{a}$ is set equal to unity.
3. The Euclidean norm of $\mathbf{a}$ is minimum; hence, the name of the method.

The second requirement suggests that we may view the vector $\mathbf{a}$ as the tap-weight vector of a prediction-error filter of order M. Indeed, it was with this interpretation in mind and the desire to be consistent with the notation used previously that we chose the vector $\mathbf{a}^*$ to lie in the range of $\mathbf{X}_N$.

Let the vector $\mathbf{a}$ be partitioned in the form

$$\mathbf{a} = \begin{bmatrix} 1 \\ \hline -\hat{\mathbf{w}} \end{bmatrix} \tag{7.131}$$

where $\hat{\mathbf{w}}$ may be interpreted as the tap-weight vector of a predictor of order M. Then the use of Eqs. (7.128) and (7.131) in (7.130) yields

$$\mathbf{G}_S^T \hat{\mathbf{w}} = \mathbf{g}_S \tag{7.132}$$

This matrix relation represents a linear system of K simultaneous equations in M unknowns (i.e., the elements of vector $\hat{\mathbf{w}}$). Thus, with $K \leq M$, this system of equations is *undetermined*, and therefore has no unique solution. It is in anticipation of this difficulty that the vector **a** is required to be of minimum norm. Clearly, when the vector **a** has minimum norm, so will the vector $\hat{\mathbf{w}}$. Accordingly, the problem we have to solve is to find the vector $\hat{\mathbf{w}}$ that satisfies Eq. (7.132), subject to the condition that its Euclidean norm $\| \hat{\mathbf{w}} \|$ is minimized. This problem is similar to that we encountered in Section 7.14, when dealing with the modified FBLP method. In particular, by using the idea of pseudoinverse of a matrix, we may express the minimum-norm solution to Eq. (7.132) as follows

$$\hat{\mathbf{w}} = (1 - \mathbf{g}_S^H \mathbf{g}_S)^{-1} \mathbf{G}_S^* \mathbf{g}_S \tag{7.133}$$

For the derivation of this formula, the reader is referred to Problem 10.

Correspondingly, the minimum-norm solution for the vector **a** in terms of the signal subspace eigenvectors equals

$$\mathbf{a} = \begin{bmatrix} 1 \\ \hline -(1 - \mathbf{g}_S^H \mathbf{g}_S)^{-1} \mathbf{G}_S^* \mathbf{g}_S \end{bmatrix} \tag{7.134}$$

We may also express the minimum-norm solution for the vector **a** in terms of the noise subspace eigenvectors as follows. Since, by definition, we have

$$\mathbf{X}\mathbf{X}^H = \mathbf{I}_{M+1}$$

where $\mathbf{I}_{M+1}$ is the $(M + 1)$-by-$(M + 1)$ identity matrix, we may use Eqs. (7.128) and (7.129) to write

$$\begin{bmatrix} \mathbf{g}_N^T & \vdots & \mathbf{g}_S^T \\ \hline \mathbf{G}_N & \vdots & \mathbf{G}_S \end{bmatrix} \begin{bmatrix} \mathbf{g}_N^* & \vdots & \mathbf{g}_N^H \\ \hline \mathbf{g}_S^* & \vdots & \mathbf{G}_S^H \end{bmatrix} = \mathbf{I}_{M+1} \tag{7.135}$$

Hence, we deduce the following relations

$$\mathbf{g}_N^T \mathbf{g}_N^* + \mathbf{g}_S^T \mathbf{g}_S^* = 1$$

$$\mathbf{G}_N \mathbf{g}_N^* + \mathbf{G}_S \mathbf{g}_S^* = \mathbf{0} \tag{7.136}$$

$$\mathbf{G}_N \mathbf{G}_N^H + \mathbf{G}_S \mathbf{G}_S^H = \mathbf{I}_M$$

Accordingly, we may rewrite the minimum-norm solution for the vector **a** in terms of the noise subspace eigenvectors as follows

$$\mathbf{a} = \begin{bmatrix} 1 \\ \hline (\mathbf{g}_N^H \mathbf{g}_N)^{-1} \mathbf{G}_N^* \mathbf{g}_N \end{bmatrix} \tag{7.137}$$

We may thus use Eq. (7.134), based on the signal subspace, or Eq. (7.137), based on the noise subspace, for computing the vector **a**. Having computed this vector, we may then determine the angular frequencies of the complex sinusoids as the peaks of the sample spectrum

$$S(\omega) = \frac{1}{|\mathbf{s}^H(\omega)\mathbf{a}|^2} \qquad (7.138)$$

where $\mathbf{s}(\omega)$ is the sinusoidal signal vector defined by

$$\mathbf{s}^T(\omega) = [1, e^{-j\omega}, \ldots, e^{-jM\omega}] \qquad (7.139)$$

A summary of the minimum-norm method is presented in Table 7.3. The essence of the minimum-norm method had been published by Reddy (1979), prior to the work of Kumaresan and Tufts (1983). In particular, he constructed a primary polynomial (in the unit-delay operator z^{-1}) from the columns of matrix $\mathbf{X}_S$ in an ad hoc manner without intentionally constraining the norm of its coefficient vector. Yet, the coefficient vector $\mathbf{a}$ that he obtained for this polynomial turned out to be identical to that in Eq. (7.133). Basically, then, Reddy's work was reformulated by Kumaresan and Tufts, giving it the minimum-norm significance.

TABLE 7.3 SUMMARY OF THE MINIMUM-NORM METHOD

1. Compute the sample eigenvalues and eigenvectors of the estimate of the $(M+1)$-by-$(M+1)$ correlation matrix $\mathbf{R}$ (see Table 7.2 for its definition).
2. Given that these are K (uncorrelated) complex sinusoids in the input signal, with $K \leq M$, classify the eigenvalues into two groups: one consisting of the K largest eigenvalues and the other consisting of the $(M+1-K)$ smallest eigenvalues. Use the eigenvectors associated with the first group to construct the $(M+1)$-by-K matrix $\mathbf{X}_S$ whose elements span the sample signal subspace.
3. Partition the matrix $\mathbf{X}_S$ as

$$\mathbf{X}_S = \begin{bmatrix} \mathbf{g}_S^T \\ \cdots \\ \mathbf{G}_S \end{bmatrix}$$

   where $\mathbf{g}_S^T$ has the first elements of the signal subspace eigenvectors, and the M-by-K matrix $\mathbf{G}_S$ has the rest of the elements of $\mathbf{X}_S$.
4. Compute the minimum-norm value of the $(M+1)$-by-1 vector:

$$\mathbf{a} = \begin{bmatrix} 1 \\ \cdots\cdots\cdots \\ -(1-\mathbf{g}_S^H\mathbf{g}_S)^{-1}\,\mathbf{G}_S^*\mathbf{g}_S \end{bmatrix}$$

5. Determine the angular frequencies of the complex sinusoids as the peaks of the sample spectrum

$$S(\omega) = \frac{1}{|\mathbf{a}^H\mathbf{s}(\omega)|^2}$$

   where $\mathbf{s}(\omega)$ is the $(M+1)$-by-1 sinusoidal vector defined by

$$\mathbf{s}^T(\omega) = [1, e^{-j\omega}, \ldots, e^{-jM\omega}]$$

7.16 *DISCUSSION*

The modified FBLP method, the minimum-norm method and the MUSIC algorithm belong to an eigendecomposition-based class of high-resolution methods for the estimation of superimposed signals in noise (Johnson, 1982; Wax, 1985). The gneral aim of these methods is to exploit the eigenvalue decomposition of a correlation matrix in a way that the estimation of sine waves in noise is improved. This is achieved by partitioning the space spanned by the eigenvectors of a sample correlation matrix into two subspaces: a signal subspace that is spanned by the *principal* eigenvectors associated with the K largest eigenvalues, where *K* equals the number of complex sinewaves, and a noise subspace that is spanned by the eigenvectors associated with the remaining eigenvalues. The rationale that is used here is that the principal eigenvectors are considerably more robust to noise perturbations than the rest. It is noteworthy that the idea of a finite-dimensional signal subspace and its orthogonal noise subspace was first described (in a somewhat different setting) by McDonough and Huggins (1968).

As pointed out previously, the angular frequencies of the complex sinusoids may be determined by locating the *peaks* of the sample spectrum $S(\omega)$. Alternatively, we may work with the reciprocal of this function, namely, the *sample eigen-spectrum* $D(\omega) = S^{-1}(\omega)$, in which case the angular frequencies of the complex sinusoids are determined by locating the *nulls* of the function $D(\omega)$. Accordingly, $D(\omega)$ is also referred to as the *null spectrum*. It should, however, be emphasized that as a result of the processing based on the eigenvalue decomposition the function $S(\omega)$, in general, loses its meaning as the power spectrum of the process.

Another way of estimating the angular frequencies of the complex sinusoids is to solve for the roots of the denominator polynomial of the function $S(\omega)$ with $e^{j\omega}$ replaced by the complex variable z. Those K roots that lie on the unit circle in the z-plane or are closest to it determine the desired angular frequencies. This procedure bypasses the need for scanning the function $S(\omega)$ across the complete frequency interval $-\pi \leq \omega \leq \pi$.

Although in our discussion we have focussed attention only on complex sinusoids in noise, nevertheless, the theory also applies to the problem of estimating the *angles of arrival* of superimposed plane waves that are incident on a linear array of sensors. In such an application, the vector **s** takes on a new meaning, namely, that of a *steering vector* with the angular frequency ω replaced by the electrical angle ϕ (see Section 1.4h). In this context, mention should be made of the fact that the MUSIC algorithm (with appropriate provision) may be used to deal with an array of arbitrary geometry (Schmidt, 1981, 1986).

Kaveh and Barabell (1985) present an asymptotic statistical analysis of the sample eigen-spectra of the MUSIC algorithm and the minimum-norm method for resolving independent closely-spaced plane waves in noise. The results of their analysis, supported by computer simulations, indicate a smaller bias in the minimum-norm eigen-spectrum, at a source angle, than in the MUSIC eigen-spectrum.

This indicates a resolution threshold that is at a lower signal-to-noise ratio for the minimum-norm method than for the MUSIC algorithm. Above and below these thresholds, computer simulations appear to show that the two methods behave similarly in terms of resolution probability, bias and variance of the estimated angles of arrival.

Another statistical result of interest has been reported by Sharman and others (1985). In this paper, it is shown that when the input signal consists of multiple sine waves in noise and the observations are Gaussian distributed, the signal parameter estimates, obtained by locating the nulls in the eigen-spectrum of the MUSIC algorithm, are *asymptotically* zero-mean Gaussian random variables. This observation (that is only strictly valid as the data length becomes large) is made by exploiting previous results published by Anderson (1963). Hence, confidence regions for the signal parameters can be established in an asymptotic sense.

Detection of the Number of Signals

Throughout our discussion, we have assumed that we know the number of sources that are responsible for the generation of sine waves or incident plane waves contained in the received signal. In many practical situations, however, we may not have this prior knowledge. For example, in harmonic retrieval, some sonar and radar array processing applications, the received signal may be modeled as the superposition of a finite number of complex sinusoids or plane waves that are embedded in a background of additive white noise. In these applications, a key issue involved in the development of a suitable *model* for the received signal is the *detection* of the number of signals contained in the model.

There is no unique approach to the solution of this detection problem. As one possible approach, we may use the observation that the number of signals can be determined from the eigenvalues of the correlation matrix of the received signal, and thus use hypothesis testing to solve the problem (Schmidt, 1981). Another approach, based on the application of information-theoretic criteria for model selection, is described by Wax and Kailath (1985). The two criteria considered in this paper are the AIC introduced by Akaike and the MDL introduced by Rissanen; these two criteria were briefly discussed in Section 2.10. The advantage of this second approach is that *no* subjective judgment is required in the decision process in that the number of signals is determined naturally as the value for which the AIC or the MDL criterion is *minimized*.

Wax and Kailath show that the MDL criterion yields, in theory, a *consistent* estimate of the number of signals while the AIC yields an *inconsistent estimate* that tends to overestimate the number of signals asympotically (Wax and Kailath, 1985). A related result has also been reported by Wax (1985); by casting the superimposed signals problem as a model selection problem and then applying the MDL criterion, he has derived an optimal solution for jointly estimating the number of signals, their parameters and waveforms.

PROBLEMS

1. Consider a linear array consisting of M uniformly spaced sensors. The output of sensor k observed at time i is denoted by $u(k,i)$ where $k = 1, 2, \ldots, M$ and $i = 1, 2, \ldots, n$. In effect, the observations $u(1, i), u(2, i), \ldots, u(M,i)$ define snapshot i. Let $\mathbf{A}$ denote the n-by-M-data matrix that is defined by

$$
\mathbf{A}^H = \begin{bmatrix}
u(1,1) & u(1,2) & \cdots & u(1,n) \\
u(2,1) & u(2,2) & \cdots & u(2,n) \\
\vdots & \vdots & & \vdots \\
u(M,1) & u(M,2) & \cdots & u(M,n)
\end{bmatrix}
$$

where the number of columns equals the number of snapshots, and the number of rows equals the number of sensors in the array. Demonstrate the following interpretations:
   (a) The M-by-M matrix $\mathbf{A}^H\mathbf{A}$ is the *spatial* (deterministic) correlation matrix with temporal averaging. This form of averaging assumes that the environment is temporally stationary.
   (b) The n-by-n matrix $\mathbf{A}\mathbf{A}^H$ is the *temporal* (deterministic) correlation matrix with spatial averaging. This form of averaging assumes that the environment is spatially stationary.

2. Assume that the elements of the measurement error vector $\boldsymbol{\varepsilon}_o$ in formulating the method of least squares are statistically independent and have the same Gaussian distribution. Hence, show that the method of least squares yields the same estimate for the tap-weight vector as the principle of maximum likelihood. See Appendix A for details of the principle of maximum likelihood.

3. We say that the least-squares estimate $\hat{\mathbf{w}}$ is *consistent* if, in the long run, the difference between $\hat{\mathbf{w}}$ and the unknown parameter vector $\mathbf{w}_o$ of the multiple linear regression model becomes negligibly small in the mean-square sense. Hence, show that the least-squares estimate $\hat{\mathbf{w}}$ is consistent if the error vector $\boldsymbol{\varepsilon}_o$ has zero mean and its elements are uncorrelated and if the trace of the inverse matrix Φ^{-1} approaches zero as the number of observations, N, approaches infinity.

4. In the autocorrelation method of linear prediction, we choose the tap-weight vector of a transversal predictor to minimize the error energy

$$
\mathscr{E}_f = \sum_{n=1}^{\infty} |f(n)|^2
$$

where $f(n)$ is the prediction error. Show that the transfer function $H_f(z)$ of the (forward) prediction-error filter is minimum phase in that its roots must lie strictly within the unit circle.
   Hints: 1. Express the transfer function $H_f(z)$ of order M (say) as the product of a sample zero factor $(1 - z_i z^{-1})$ and a function $H_f'(z)$. Hence, minimize the prediction-error energy with respect to the magnitude of zero z_i

2. You may also use the Cauchy–Schwartz inequality:

$$\text{Re}\left[\sum_{n=1}^{\infty} e^{j\theta}g(n-1)g^*(n)\right] \leq \left[\sum_{n=1}^{\infty} |g(n)|^2\right]^{1/2}\left[\sum_{n=1}^{\infty} |e^{j\theta}g(n-1)|^2\right]^{1/2}$$

The equality holds if and only if $g(n) = e^{j\theta}g(n-1)$ for $n = 1, 2, \ldots, \infty$.

5. Figure 7.4(a) shows a *forward linear predictor* using a transversal structure, with the tap inputs $u(i-1), u(i-2), \ldots, u(i-M)$ used to make a linear prediction of $u(i)$. The problem is to find the tap-weight vector $\hat{\mathbf{w}}$ that minimizes the sum of forward prediction-error squares:

$$\mathscr{E}_f = \sum_{i=M+1}^{N} |f_M(i)|^2$$

where $f_M(i)$ is the forward prediction error. Find the following parameters:

(a) The M-by-M deterministic correlation matrix of the tap inputs of the predictor.

(b) The M-by-1 deterministic cross-correlation vector between the tap inputs of the predictor and the desired response $u(i)$.

(c) The minimum value of $\mathscr{E}_f$.

6. Figure 7.4(b) shows a *backward linear predictor* using a transversal structure, with the tap inputs $u(i-M+1), \ldots, u(i-1), u(i)$ used to make a linear prediction of the input $u(i-M)$. The problem is to find the tap-weight vector $\hat{\mathbf{w}}$ that minimizes the sum of backward prediction-error squares

$$\mathscr{E}_b = \sum_{i=M+1}^{N} |b_M(i)|^2$$

where $b_M(i)$ is the backward prediction error. Find the following parameters:

(a) The M-by-M deterministic correlation matrix of the tap inputs.

(b) The M-by-1 deterministic correlation vector between the tap inputs and the desired response $u(i-M)$.

(c) The minimum value of $\mathscr{E}_b$.

7. Consider a linear array of M uniformly spaced sensors. The output of sensor k observed at time i is denoted by $u(k, i)$ where $k = 1, 2, \ldots, M$ and $i = 1, 2, \ldots, n$. The element outputs are individually weighted and then summed to produce the output

$$e(i) = \mathbf{u}^T(i)\mathbf{w}^* = \mathbf{w}^H\mathbf{u}(i)$$

where $\mathbf{u}(i)$ is the M-by-1 elemental output vector at time i, defined by

$$\mathbf{u}^T(i) = [u(1, i), u(2, i), \ldots, u(M, i)]$$

and $\mathbf{w}$ is the M-by-1 weight vector, defined by

$$\mathbf{w}^T = [w_1, w_2, \ldots, w_M]$$

The weight vector $\mathbf{w}$ is held constant for the total observation interval $1 \leq i \leq n$. The choice of $\mathbf{w}$ is constrained to fulfill the condition

$$\mathbf{w}^H\mathbf{s} = 1$$

where $\mathbf{s}$ is the M-by-1 *steering vector*. The elements of the vector $\mathbf{s}$ are determined by the look direction of interest. Using the method of Lagrange multipliers, show that the optimum weight vector $\hat{\mathbf{w}}$ that minimizes the sum of weighted error squares, subject to this constraint, equals

$$\hat{\mathbf{w}} = \frac{\mathbf{\Phi}^{-1}\mathbf{s}}{\mathbf{s}^H\mathbf{\Phi}^{-1}\mathbf{s}}$$

For details of the method of Lagrange multipliers, see Appendix D.

8. Show that the formula for the AR power spectrum given in Eq. (7.96) may be rewritten in the form

$$S_{\mathrm{AR}}(\omega) = \frac{\mathbf{\delta}^T\mathbf{C}^{-1}\mathbf{\delta}}{|\mathbf{\delta}^T\mathbf{C}^{-1}\mathbf{s}|^2}$$

where $\mathbf{\delta}$ is the $(M + 1)$-by-1 first unit vector with its first element equal to 1 and its remaining elements equal to zero, $\mathbf{C}$ is the $(M + 1)$-by-$(M + 1)$ deterministic correlation matrix of the input signal, and $\mathbf{s}$ is the $(M + 1)$-by-1 sinusoidal signal vector.

9. Equation (7.93) defines the minimum-norm value $\hat{\mathbf{w}}$ of the tap-weight vector, assuming that the data matrix $\mathbf{A}$ has more rows than columns; that is, $L > M$. For the case when the reverse is true (i.e., $L < M$), show that

$$\hat{\mathbf{w}} = \sum_{i=1}^{W} \frac{\mathbf{y}_i^H \mathbf{b}}{\sigma_i^2} \mathbf{A}^H \mathbf{y}_i$$

where the $\mathbf{y}_i$ are the left singular vectors and the σ_i are the singular values of $\mathbf{A}$; W is the rank of $\mathbf{A}$.

10. Equation (7.133) defines the vector $\hat{\mathbf{w}}$ in accordance with the minimum-norm method. Prove the validity of this formula by using the following two relations:

   (1) The formula for the pseudoinverse of a matrix of coefficients in a system of linear equations.

   (2) The matrix inversion lemma that states if the set of matrices $\mathbf{A}$, $\mathbf{B}$, $\mathbf{C}$ and $\mathbf{D}$, with appropriate dimensions, are related by

$$\mathbf{A} = \mathbf{B}^{-1} + \mathbf{C}\mathbf{D}^{-1}\mathbf{C}^H$$

then

$$\mathbf{A}^{-1} = \mathbf{B} - \mathbf{B}\mathbf{C}(\mathbf{D} + \mathbf{C}^H\mathbf{B}\mathbf{C})^{-1}\mathbf{C}^H\mathbf{B}$$

(For discussion of this lemma, see Section 8.3.)

Computer-oriented Problem

11. *Computer experiment on angle of arrival estimation*: The response of a linear array of antenna elements to two *coherent incident plane waves* is described by

$$u(n) = \exp(jn\phi) + \rho \exp(-jn\phi + j\psi) + v(n)$$

$$n = 0, \pm 1, \pm 2, \ldots, \pm\frac{N-1}{2}$$

The parameter ϕ is related to the angle of arrival θ of the incident plane wave (measured with respect to the normal to the array) by

$$\phi = \frac{2\pi d}{\lambda} \sin \theta$$

where d is element-to-element spacing of the array and λ is the radar wavelength. The parameter ρ is the normalized amplitude of the incident plane wave represented by the second exponential term. The parameter ψ denotes the *phase* difference between the two incident waves measured at the center of the array. The process $\{v(n)\}$ is a white-noise process of zero mean and variance σ_v^2.

The configuration described represents a model of a low-angle tracking radar environment in which the target lies in close proximity to a reflecting sea surface. The first exponential term represents the target echo, and the second exponential term arises because of the "image" of the target. The angle ψ accounts for the difference between the lengths of the two paths. The parameter ρ represents the reflection coefficient of the sea surface. The parameters have the following values:

$$\rho = 0.9$$

Angular separation, $2\phi = 0.5$ beamwidth

Number of elements, $N = 25$

Variance of noise, $\sigma_v^2 = 0.1$

$$\text{Spacing, } d = \frac{\lambda}{3}$$

A *standard beamwidth* is defined as the angular separation between the peak of the main lobe and first null in the array pattern.

(a) For prediction order $M = 18$, plot the zeros of the transfer function of the prediction-error filter designed by using the FBLP method for 100 independent trials of the experiment for each of the following values of phase difference: $\psi = 0$, $\pm\pi/4$, $\pm\pi/2$, $\pm 3\pi/4$, π.

(b) Repeat the experiment using the modified FBLP method by retaining the largest two eigenvalues of the correlation matrix.

Comment on your results.

8

Adaptive Transversal Filters Using Recursive Least Squares

8.1 INTRODUCTION

In this chapter we extend the use of the method of least squares to develop a recursive algorithm for the design of adaptive transversal filters such that, given the least-squares estimate of the tap-weight vector of the filter at time $n - 1$, we may compute the updated estimate of this vector at time n upon the arrival of new data. We refer to the resulting algorithm as the *recursive least-squares (RLS) algorithm*.

As mentioned in Chapter 7, the origin of the method of least squares can be traced back to Gauss (1809). It appears that the original reference for the RLS algorithm is Plackett (1950). However, the algorithm has been rederived independently by several other authors; see, for example, Hastings-James and Sage (1969).

The RLS algorithm may be viewed as the *deterministic* counterpart of the Kalman filter theory. Kalman filters are derived from ensemble averages, resulting in a recursive filter that is optimum in the mean-square sense. On the other hand,

381

the RLS algorithm yields a data-adaptive filter in that for each set of input data there exists a different filter.

We begin the development of the RLS algorithm by reviewing some basic relations that pertain to the method of least squares. Then, by exploiting a relation in matrix algebra known as the *matrix inversion lemma*, we develop the RLS algorithm. An important feature of the RLS algorithm is that it utilizes all the information contained in the input data, extending back to the instant of time when the algorithm is initiated. The resulting rate of convergence is therefore typically an order of magnitude faster than the simple LMS algorithm. This improvement in performance, however, is achieved at the expense of a large increase in computational complexity.

Later in the chapter, we describe the *fast transversal filters* (FTF) implementation of the RLS algorithm, which exploits the *shifting property* of serialized data, thereby resulting in substantial reduction in computational complexity (Cioffi and Kailath, 1984). The FTF algorithm uses a combination of four transversal filters, always realizing the true solution of the RLS algorithm. Another fast realization of the RLS algorithm, with the same computational requirement as the FTF algorithm, has been developed independently by Carayannis, Manolakis, and Kalouptsidis (1983). The deriviations of both of these algorithms were inspired by pioneering work by (1) Morf (1974) on shift-low-rank properties of deterministic correlation matrices, (2) Ljung, Morf, and Falconer (1978) on the fast calculation of a gain vector, and (3) Falconer and Ljung (1978) on the development of the *fast RLS algorithm*.

8.2 SOME PRELIMINARIES

In *recursive* implementations of the method of least squares, we start the computation with *known initial conditions* and use the information contained in new data samples to *update* the old estimates. We therefore find that the length of observable data is variable. Accordingly, we express the *index of performance* to be minimized as $\mathscr{E}(n)$, where n is the variable length of the observable data. Also, it is customary to introduce a *weighting factor* or *forgetting factor* into the definition of the performance index $\mathscr{E}(n)$. We thus write

$$\mathscr{E}(n) = \sum_{i=1}^{n} \beta(n, i)|e(i)|^2 \tag{8.1}$$

where $e(i)$ is the difference between *desired response* $d(i)$ and the *output* $y(i)$ produced by a transversal filter whose tap inputs (at time i) equal $u(i)$, $u(i - 1)$, . . . , $u(i - M + 1)$, as in Fig. 8.1. That is, $e(i)$ is defined by

$$\begin{aligned} e(i) &= d(i) - y(i) \\ &= d(i) - \mathbf{w}^H(n)\mathbf{u}(i) \end{aligned} \tag{8.2}$$

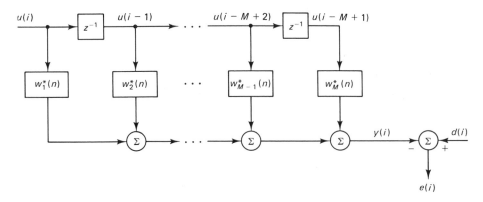

Figure 8.1 Transversal filter.

where $\mathbf{u}(i)$ is the *tap-input vector at time i,* defined by

$$\mathbf{u}^T(i) = [u(i), u(i - 1), \ldots, u(i - M + 1)] \tag{8.3}$$

and $\mathbf{w}(n)$ is the *tap-weight vector at time n* defined by

$$\mathbf{w}^T(n) = [w_1(n), w_2(n), \ldots, w_M(n)] \tag{8.4}$$

Note that the tap weights of the transversal filter remain fixed during the observation interval $1 \leq i \leq n$.

The weighting factor $\beta(n, i)$ in Eq. (8.1) has the property that

$$0 < \beta(n, i) \leq 1, \qquad i = 1, 2, \ldots, n \tag{8.5}$$

The use of the weighting factor $\beta(n, i)$ is intended to ensure that data in the distant past are "forgotten" in order to afford the possibility of following the statistical variations of the observable data when the filter operates in a nonstationary environment. One such form of weighting that is commonly used is the *exponential weighting factor* defined by

$$\beta(n, i) = \lambda^{n-i}, \qquad i = 1, 2, \ldots, n \tag{8.6}$$

where λ is a positive constant close to, but less than, 1. When λ equals 1, we have the ordinary method of least squares. The inverse of $1 - \lambda$ is, roughly speaking, a measure of the *memory* of the algorithm. The special case $\lambda = 1$ corresponds to *infinite memory.* Thus, in the *method of exponentially weighted least squares,* we minimize the index of performance

$$\mathscr{E}(n) = \sum_{i=1}^{n} \lambda^{n-i} |e(i)|^2 \tag{8.7}$$

The optimum value of the tap-weight vector, $\hat{\mathbf{w}}(n)$, for which the performance index $\mathscr{E}(n)$ of Eq. (8.7) attains its minimum value is defined by the *normal equation*:

$$\boldsymbol{\Phi}(n)\hat{\mathbf{w}}(n) = \boldsymbol{\theta}(n) \tag{8.8}$$

The M-by-M correlation matrix $\mathbf{\Phi}(n)$ is defined by

$$\mathbf{\Phi}(n) = \sum_{i=1}^{n} \lambda^{n-i} \mathbf{u}(i)\mathbf{u}^H(i) \tag{8.9}$$

The M-by-1 cross-correlation $\mathbf{\theta}(n)$ between the tap inputs of the transversal filter and the desired response is defined by

$$\mathbf{\theta}(n) = \sum_{i=1}^{n} \lambda^{n-i} \mathbf{u}(i)d^*(i) \tag{8.10}$$

The correlation matrix $\mathbf{\Phi}(n)$ differs from that of Eq. (7.32) in two respects:

1. The matrix product $\mathbf{u}(i)\mathbf{u}^H(i)$ inside the summation on the right side of Eq. (8.9) is weighted by the exponential factor λ^{n-1}, which arises naturally from the adoption of Eq. (8.7) as the index of performance.
2. The use of *prewindowing* is assumed, according to which the input data prior to time $i = 1$ are equal to zero; hence, the use of $i = 1$ as the lower limit of the summation.

Likewise, the cross-correlation vector $\mathbf{\theta}(n)$ differs from that of Eq. (7.29) by the introduction of the exponential weighting factor λ^{n-i}.

Isolating the term corresponding to $i = n$ from the rest of the summation on the right side of Eq. (8.9), we may write

$$\mathbf{\Phi}(n) = \lambda \left[\sum_{i=1}^{n-1} \lambda^{n-1-i} \mathbf{u}(i)\mathbf{u}^H(i) \right] + \mathbf{u}(n)\mathbf{u}^H(n) \tag{8.11}$$

However, by definition, the expression inside the square brackets on the right side of Eq. (8.11) equals the correlation matrix $\mathbf{\Phi}(n-1)$. Hence, we have the following recursion for updating the value of the deterministic correlation matrix of the tap inputs:

$$\mathbf{\Phi}(n) = \lambda\mathbf{\Phi}(n-1) + \mathbf{u}(n)\mathbf{u}^H(n) \tag{8.12}$$

where the matrix product $\mathbf{u}(n)\mathbf{u}^H(n)$ plays the role of a correction term.

Similarly, we may use Eq. (8.10) to derive the following recursion for updating the deterministic cross-correlation vector between the tap inputs and the desired response:

$$\mathbf{\theta}(n) = \lambda\mathbf{\theta}(n-1) + \mathbf{u}(n)d^*(n) \tag{8.13}$$

To compute the least-square estimate $\hat{\mathbf{w}}(n)$ for the tap-weight vector in accordance with Eq. (8.8), we have to determine the inverse of the correlation matrix $\mathbf{\Phi}(n)$. In practice, however, we usually try to avoid performing such an operation as it can be very time consuming, particularly if the number of tap weights, M, is high. Also, we would like to be able to compute the least-squares estimate $\hat{\mathbf{w}}(n)$ for the tap-weight vector recursively for $n = 1, 2, \ldots, \infty$. We can realize both of these objectives by using a basic result in matrix algebra known as the *matrix inversion lemma*. We assume that the initial conditions have been chosen to ensure the nonsingularity of the correlation matrix $\Phi(n)$; this issue is discussed later in Section 8.4.

8.3 THE MATRIX INVERSION LEMMA

Let $\mathbf{A}$ and $\mathbf{B}$ be two positive-definite, M-by-M matrices related by

$$\mathbf{A} = \mathbf{B}^{-1} + \mathbf{C}\mathbf{D}^{-1}\mathbf{C}^H \tag{8.14}$$

where $\mathbf{D}$ is another positive-definite, N-by-N matrix, and $\mathbf{C}$ is an M-by-N matrix. According to the *matrix inversion lemma*,[1] we may express the inverse of the matrix $\mathbf{A}$ as follows:

$$\mathbf{A}^{-1} = \mathbf{B} - \mathbf{B}\mathbf{C}(\mathbf{D} + \mathbf{C}^H\mathbf{B}\mathbf{C})^{-1}\,\mathbf{C}^H\mathbf{B} \tag{8.15}$$

The proof of this lemma is established by multiplying Eq. (8.14) by (8.15) and recognizing that the product of a square matrix and its inverse is equal to the identity matrix (see Problem 2).

In the next section we show how the matrix inversion lemma can be applied to obtain a recursive equation for computing the least-squares solution $\hat{\mathbf{w}}(n)$ for the tap-weight vector.

8.4 THE EXPONENTIALLY WEIGHTED RECURSIVE LEAST-SQUARES ALGORITHM

With the correlation matrix $\Phi(n)$ assumed to be positive definite and therefore nonsingular, we may apply the matrix inversion lemma to the recursive equation (8.12). We first make the following identifications:

$$\mathbf{A} = \Phi(n)$$

$$\mathbf{B}^{-1} = \lambda\Phi(n - 1)$$

$$\mathbf{C} = \mathbf{u}(n)$$

$$\mathbf{D} = 1$$

---

[1] The exact origin of the matrix inversion lemma is not known. Householder (1964) attrributes it to Woodbury (1950). Nevertheless, application of the matrix inversion lemma in the filtering literature was first made by Kailath who used a form of this lemma to prove the equivalence of the Wiener filter and the maximum-likelihood procedure for estimating the output of a random linear time-invariant channel that is corrupted by additive white Gaussian noise (Kailath, 1960). Early use of the matrix inversion lemma was also made by Ho (1963). Another interesting application of the matrix inversion lemma was made by Brooks and Reed, who used it to prove the equivalence of the Wiener filter, the maximum signal-to-noise ratio filter, and the likelihood ratio processor for detecting a signal in additive white Gaussian noise (Brooks and Reed, 1972); see Problem 7, Chapter 3.

Then, substituting these definitions in the matrix inversion lemma of Eq. (8.15), we obtain the following recursive equation for the inverse of the correlation matrix:

$$\mathbf{\Phi}^{-1}(n) = \lambda^{-1}\mathbf{\Phi}^{-1}(n - 1) - \frac{\lambda^{-2}\mathbf{\Phi}^{-1}(n - 1)\mathbf{u}(n)\mathbf{u}^{H}(n)\mathbf{\Phi}^{-1}(n - 1)}{1 + \lambda^{-1}\mathbf{u}^{H}(n)\mathbf{\Phi}^{-1}(n - 1)\mathbf{u}(n)} \qquad (8.16)$$

For convenience of computation, let

$$\mathbf{P}(n) = \mathbf{\Phi}^{-1}(n) \qquad (8.17)$$

and

$$\mathbf{k}(n) = \frac{\lambda^{-1}\mathbf{P}(n - 1)\mathbf{u}(n)}{1 + \lambda^{-1}\,\mathbf{u}^{H}(n)\mathbf{P}(n - 1)\mathbf{u}(n)} \qquad (8.18)$$

Using these definitions, we may rewrite Eq. (8.16) as follows:

$$\mathbf{P}(n) = \lambda^{-1}\mathbf{P}(n - 1) - \lambda^{-1}\mathbf{k}(n)\mathbf{u}^{H}(n)\mathbf{P}(n - 1) \qquad (8.19)$$

Note that $\mathbf{P}(n)$ has the dimensions of a matrix (M-by-M), whereas $\mathbf{k}(n)$ has the dimensions of a vector (M-by-1).   We refer to $\mathbf{k}(n)$ as the *gain vector* for reasons that will become apparent later in the section.

By rearranging Eq. (8.18), we have

$$\begin{aligned}
\mathbf{k}(n) &= \lambda^{-1}\mathbf{P}(n - 1)\mathbf{u}(n) - \lambda^{-1}\mathbf{k}(n)\,\mathbf{u}^{H}(n)\mathbf{P}(n - 1)\,\mathbf{u}(n) \\
&= [\lambda^{-1}\mathbf{P}(n - 1) - \lambda^{-1}\,\mathbf{k}(n)\,\mathbf{u}^{H}(n)\,\mathbf{P}(n - 1)]\,\mathbf{u}(n)
\end{aligned} \qquad (8.20)$$

We see from Eq. (8.19) that the expression inside the brackets on the right side of Eq. (8.20) equals $\mathbf{P}(n)$.   Hence, we may simplify Eq. (8.20) as follows:

$$\mathbf{k}(n) = \mathbf{P}(n)\mathbf{u}(n) \qquad (8.21)$$

This result, together with $\mathbf{P}(n) = \mathbf{\Phi}^{-1}(n)$, may be used as a definition for the gain vector $\mathbf{k}(n)$.   That is, we have

$$\mathbf{k}(n) = \mathbf{\Phi}^{-1}(n)\,\mathbf{u}(n) \qquad (8.22)$$

Time-update for the Tap-weight Vector

Next, we wish to develop a recursive equation for updating the least-squares estimate $\hat{\mathbf{w}}(n)$ for the tap-weight vector.   To do this, we use Eq. (8.8), (8.13), and (8.17) to express the least-squares estimate $\hat{\mathbf{w}}(n)$ for the tap-weight vector at time n as follows:

$$\begin{aligned}
\hat{\mathbf{w}}(n) &= \mathbf{\Phi}^{-1}(n)\,\mathbf{\theta}(n) \\
&= \mathbf{P}(n)\,\mathbf{\theta}(n) \\
&= \lambda\mathbf{P}(n)\,\mathbf{\theta}(n - 1) + \mathbf{P}(n)\,\mathbf{u}(n)\,d^{*}(n)
\end{aligned} \qquad (8.23)$$

Substituting Eq. (8.19) for $\mathbf{P}(n)$ in the first term only in the right side of Eq. (8.23), we get

$$\hat{\mathbf{w}}(n) = \mathbf{P}(n - 1)\, \boldsymbol{\theta}(n - 1) - \mathbf{k}(n)\, \mathbf{u}^H(n)\, \mathbf{P}(n - 1)\, \boldsymbol{\theta}(n - 1)$$
$$+ \mathbf{P}(n)\, \mathbf{u}(n)\, d^*(n)$$
$$= \boldsymbol{\Phi}^{-1}(n - 1)\, \boldsymbol{\theta}(n - 1) - \mathbf{k}(n)\, \mathbf{u}^H(n)\, \boldsymbol{\Phi}^{-1}(n - 1)\, \boldsymbol{\theta}(n - 1) \quad (8.24)$$
$$+ \mathbf{P}(n)\, \mathbf{u}(n)\, d^*(n)$$
$$= \hat{\mathbf{w}}(n - 1) - \mathbf{k}(n)\, \mathbf{u}^H(n)\, \hat{\mathbf{w}}(n - 1) + \mathbf{P}(n)\mathbf{u}(n)d^*(n)$$

Finally, using the fact that $\mathbf{P}(n)\, \mathbf{u}(n)$ equals the gain vector $\mathbf{k}(n)$, as in Eq. (8.21), we get the desired recursive equation for updating the tap-weight vector:

$$\hat{\mathbf{w}}(n) = \hat{\mathbf{w}}(n - 1) + \mathbf{k}(n)[d^*(n) - \mathbf{u}^H(n)\, \hat{\mathbf{w}}(n - 1)]$$
$$= \hat{\mathbf{w}}(n - 1) + \mathbf{k}(n)\alpha^*(n) \quad (8.25)$$

where $\alpha(n)$ is the *innovation* defined by

$$\alpha(n) = d(n) - \mathbf{u}^T(n)\, \hat{\mathbf{w}}^*(n - 1)$$
$$= d(n) - \hat{\mathbf{w}}^H(n - 1)\mathbf{u}(n) \quad (8.26)$$

The inner product $\hat{\mathbf{w}}^H(n - 1)\, \mathbf{u}(n)$ represents an estimate of the desired response $d(n)$, based on the *old* least-squares estimate of the tap-weight vector that was made at time $n - 1$. Accordingly, we may also refer to $\alpha(n)$ as the *a priori estimation error*.

This error is, in general, different from the *a posteriori estimation error*[2]

$$e(n) = d(n) - \hat{\mathbf{w}}^H(n)\mathbf{u}(n) \quad (8.27)$$

the computation of which involves the *current* least-squares estimate of the tap-weight vector available at time n. Indeed, we may view $\alpha(n)$ as a "tentative" value of $e(n)$ before updating the tap-weight vector. Note, however, in the least-squares optimization that led to the recursive algorithm of Eq. (8.25) for the tap-weight vector, we actually minimized an index of performance based on $e(n)$ and *not* $\alpha(n)$.

From here on, we will refer to $\alpha(n)$ as the a priori estimation error and to $e(n)$ as the a posteriori estimation error. The motivation for doing this is that in this chapter and the succeeding two chapters that deal with recursive solutions to the linear least-squares problem, we will be confronted with similar situations when the problems of forward linear prediction and backward linear prediction are considered.

Equations (8.18), (8.26), (8.25), and (8.19) collectively and in that order constitute the *RLS algorithm*. In particular, Eq. (8.26) describes the filtering operation of the algorithm, whereby the transversal filter is excited to compute the

---

[2]To be precise, we should modify the symbol for the estimation error in Eq. (8.27) to signify the fact that it corresponds to the least-squares estimate for the tap-weight vector. We have not done this simply for convenience of notation.

a priori estimation error $\alpha(n)$. Equation (8.25) describes the adaptive operation of the algorithm, whereby the tap-weight vector is updated by incrementing its old value by an amount equal to the complex conjugate of the a priori estimation error $\alpha(n)$ times the time-varying gain vector $\mathbf{k}(n)$; hence, the name "gain vector." Equations (8.18) and (8.19) enable us to update the value of the gain vector itself. An important feature of the RLS algorithm described by these equations is that the inversion of the correlation matrix $\boldsymbol{\Phi}(n)$ is replaced at each step by a simple scalar division.

Figure 8.2 depicts the signal-flow graph representation of the RLS algorithm.

Initial Conditions

The applicability of the RLS algorithm requires that we initialize the recursion of Eq. (8.19) by choosing a starting value $\mathbf{P}(0)$ that assures the nonsingularity of the correlation matrix $\boldsymbol{\Phi}(n)$. We may do this by evaluating the inverse

$$\left[\sum_{i=-n_0}^{0} \lambda^{-i}\mathbf{u}(i)\mathbf{u}^H(i) \right]^{-1}$$

where the tap-weight vector $\mathbf{u}(i)$ is obtained from an initial block of data for $-n_0 \le i \le 0$. A simpler way, however, is to slightly modify the expression for the correlation matrix $\boldsymbol{\Phi}(n)$ by writing

$$\boldsymbol{\Phi}(n) = \sum_{i=1}^{n} \lambda^{n-i} \mathbf{u}(i)u^H(\mathrm{i}) + \delta\lambda^n\mathbf{I} \tag{8.28}$$

where $\mathbf{I}$ is the M-by-M identity matrix, and δ is a small positive constant. This modification affects the starting value, leaving the recursions in the RLS algorithm intact (see Problem 3). We usually find that, for large data length n, the choice of δ is unimportant.

For $n = 0$, we have

$$\boldsymbol{\Phi}(0) = \delta\mathbf{I} \tag{8.29}$$

Correspondingly, for the initial value of $\mathbf{P}(n)$, equal to the inverse of the correlation

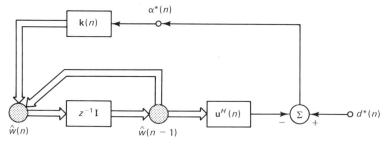

Figure 8.2 Signal-flow graph representation of the RLS algorithm.

matrix $\mathbf{\Phi}(n)$, we set

$$\mathbf{P}(0) = \delta^{-1}\mathbf{I} \tag{8.30}$$

It only remains for us to choose an initial value for the tap-weight vector. It is customary to set

$$\hat{\mathbf{w}}(0) = \mathbf{0} \tag{8.31}$$

where $\mathbf{0}$ is the M-by-1 null vector.

In Table 8.1 we present a summary of the initial conditions and recursions involved in computing the RLS algorithm.

8.5 UPDATE RECURSION FOR THE SUM OF WEIGHTED ERROR SQUARES

The minimum value of the sum of weighted error squares, $\mathscr{E}_{min}(n)$, results when the tap-weight vector is set equal to the least-squares estimate $\hat{\mathbf{w}}(n)$. To compute $\mathscr{E}_{min}(n)$, we may therefore use the relation [see Eq. (7.36)]

$$\mathscr{E}_{min}(n) = \mathscr{E}_d(n) - \mathbf{\theta}^H(n)\,\hat{\mathbf{w}}(n) \tag{8.32}$$

TABLE 8.1 SUMMARY OF THE RLS ALGORITHM

Initialize the algorithm by setting

$\mathbf{P}(0) = \delta^{-1}\mathbf{I},\qquad \delta = $ small positive constant

$\hat{\mathbf{w}}(0) = \mathbf{0}$

For each instant of time, $n = 1, 2, \ldots$, compute[a]

$$\mathbf{k}(n) = \frac{\lambda^{-1}\,\mathbf{P}(n-1)\,\mathbf{u}(n)}{1 + \lambda^{-1}\,\mathbf{u}^H(n)\,\mathbf{P}(n-1)\,\mathbf{u}(n)}$$

$$\alpha(n) = d(n) - \hat{\mathbf{w}}^H(n-1)\,\mathbf{u}(n)$$

$$\hat{\mathbf{w}}(n) = \hat{\mathbf{w}}(n-1) + \mathbf{k}(n)\,\alpha^*(n)$$

$$\mathbf{P}(n) = \lambda^{-1}\mathbf{P}(n-1) - \lambda^{-1}\,\mathbf{k}(n)\,\mathbf{u}^H(n)\,\mathbf{P}(n-1)$$

---

[a] The computational cost of the RLS algorithm may be reduced by defining

$$\mathbf{x}(n) = \lambda^{-1}\,\mathbf{P}(n-1)\,\mathbf{u}(n)$$

We may then rewrite the first and fourth lines as

$$\mathbf{k}(n) = [1 + \mathbf{u}^H(n)\,\mathbf{x}(n)]^{-1}\,\mathbf{x}(n)$$

$$\mathbf{P}(n) = \lambda^{-1}\,\mathbf{P}(n-1) - \mathbf{k}(n)\,\mathbf{x}^H(n)$$

In the recursion for $\mathbf{P}(n)$ we have used the Hermitian property of $\mathbf{P}(n-1)$.

where $\mathscr{E}_d(n)$ is defined by (using the notation of this chapter)

$$\begin{aligned} \mathscr{E}_d(n) &= \sum_{i=1}^{n} \lambda^{n-i} |d(i)|^2 \\ &= \lambda \mathscr{E}_d(n-1) + |d(n)|^2 \end{aligned} \tag{8.33}$$

Therefore, substituting Eqs. (8.13), (8.25), and (8.33) in (8.32), we get

$$\begin{aligned} \mathscr{E}_{\min}(n) &= \lambda[\mathscr{E}_d(n-1) - \boldsymbol{\theta}^H(n-1)\hat{\mathbf{w}}(n-1)] \\ &\quad + d(n)[d^*(n) - \mathbf{u}^H(n)\hat{\mathbf{w}}(n-1)] \\ &\quad - \boldsymbol{\theta}^H(n)\,\mathbf{k}(n)\alpha^*(n) \end{aligned} \tag{8.34}$$

where, in the last term, we have restored $\boldsymbol{\theta}(n)$ to its original form. By definition, the expression inside the first set of brackets on the right side of Eq. (8.34) equals $\mathscr{E}_{\min}(n-1)$. Also, by definition, the expression inside the second set of brackets equals the complex conjugate of the a priori estimation error $\alpha(n)$. For the last term, we use the definition of the gain vector $\mathbf{k}(n)$ to express the inner product $\boldsymbol{\theta}^H(n)\,\mathbf{k}(n)$ as

$$\begin{aligned} \boldsymbol{\theta}^H(n)\,\mathbf{k}(n) &= \boldsymbol{\theta}^H(n)\,\boldsymbol{\Phi}^{-1}(n)\,\mathbf{u}(n) \\ &= [\boldsymbol{\Phi}^{-1}(n)\,\boldsymbol{\theta}(n)]^H\,\mathbf{u}(n) \\ &= \hat{\mathbf{w}}^H(n)\,\mathbf{u}(n) \end{aligned}$$

where (in the second line) we have used the Hermitian property of the correlation matrix $\boldsymbol{\Phi}(n)$, and (in the third line) we have used the fact that $\boldsymbol{\Phi}^{-1}(n)\boldsymbol{\theta}(n)$ equals the least-squares estimate $\hat{\mathbf{w}}(n)$. Accordingly, we may simplify Eq. (8.34) as

$$\begin{aligned} \mathscr{E}_{\min}(n) &= \lambda \mathscr{E}_{\min}(n-1) + d(n)\alpha^*(n) - \hat{\mathbf{w}}^H(n)\mathbf{u}(n)\alpha^*(n) \\ &= \lambda \mathscr{E}_{\min}(n-1) + \alpha^*(n)[d(n) - \hat{\mathbf{w}}^H(n)\mathbf{u}(n)] \\ &= \lambda \mathscr{E}_{\min}(n-1) + \alpha^*(n)e(n) \end{aligned} \tag{8.35}$$

where $e(n)$ is the a posteriori estimation error. Equation (8.35) is the desired recursion for updating the sum of weighted error squares. We thus see that the product of the complex conjugate of $\alpha(n)$ and $e(n)$ represents the correction term in this updating recursion. Note that this product is real valued, which implies that we always have (see Problem 4)

$$\alpha(n)e^*(n) = \alpha^*(n)e(n) \tag{8.36}$$

8.6 CONVERGENCE ANALYSIS OF THE RLS ALGORITHM

In this section we demonstrate the convergence of the RLS algorithm by considering three aspects of the problem: (1) convergence of the estimate $\hat{\mathbf{w}}(n)$ in the mean, (2) convergence of the estimate $\hat{\mathbf{w}}(n)$ in the mean square, and (3) convergence of the average mean-squared value of the a priori estimation error $\alpha(n)$. For this analysis, we assume that the desired response $d(n)$ and the tap-input vector $\mathbf{u}(n)$ are related by the *multiple linear regression model* of Fig. 8.3. In particular, we may write

$$d(n) = e_o(n) + \mathbf{w}_o^H \mathbf{u}(n) \qquad (8.37)$$

where the M-by-1 vector $\mathbf{w}_o$ denotes the *regression parameter vector* of the model, and $e_o(n)$ is the *measurement error*. The measurement error process $\{e_o(n)\}$ is white with zero mean and variance σ^2. The parameter vector $\mathbf{w}_o$ is constant. The latter assumption is equivalent to saying that the adaptive transversal filter operates in a stationary environment. For a stationary environment, the best steady-state results are achieved when the tap weights adapt slowly, a condition that corresponds to the choice of $\lambda = 1$.

Convergence of the Tap-weight vector in the Mean

The modification of the formula for the deterministic correlation matrix $\Phi(n)$ by initializing it with the small term $\delta\mathbf{I}$, as in Eq. (8.28), introduces a *bias* into the estimate $\hat{\mathbf{w}}(n)$ produced by the RLS algorithm. That is, we may express the mean value of $\hat{\mathbf{w}}(n)$ as

$$E[\hat{\mathbf{w}}(n)] = \mathbf{w}_o + \mathbf{b}(n)$$

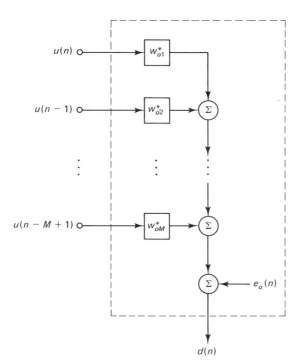

Figure 8.3 Multiple linear regression model.

where $\mathbf{w}_o$ is the parameter vector of the multiple regression model, and $\mathbf{b}(n)$ denotes the bias. The bias itself is defined by (see Problem 6)

$$\mathbf{b}(n) = -\delta\boldsymbol{\Phi}^{-1}(n)\mathbf{w}_o \qquad (8.38)$$

where $\boldsymbol{\Phi}^{-1}(n)$ is the inverse of $\boldsymbol{\Phi}(n)$.

Let $\mathbf{R}$ denote the *M-by-M ensemble-averaged correlation matrix* of the tap-input vector $\mathbf{u}(n)$. Assuming that the stochastic process represented by the tap-input vector $\mathbf{u}(n)$ is *ergodic*, it is well known that for large n we may approximate $\mathbf{R}$ by the time average formula (for $\lambda = 1$)

$$\mathbf{R} \simeq \frac{1}{n}\sum_{i=1}^{n}\mathbf{u}(i)\mathbf{u}^H(i)$$

$$=\frac{1}{n}\boldsymbol{\Phi}(n), \qquad n \text{ large} \qquad (8.39)$$

Hence, Eq. (8.38) yields the bias as

$$\mathbf{b}(n) \simeq -\frac{\delta}{n}\mathbf{R}^{-1}\mathbf{w}_o, \qquad n \text{ large}$$

This result shows that as the number of iterations, n, approaches infinity, the bias $\mathbf{b}(n)$ approaches zero.

We conclude therefore that the RLS algorithm produces an *asymptotically unbiased* estimate of the regression parameter vector $\mathbf{w}_o$. That is, the RLS algorithm is *convergent in the mean*.

Convergence of the Tap-weight Vector in the Mean Square

Let $\boldsymbol{\varepsilon}(n)$ denote the weight-error vector, defined as the difference between the estimate $\hat{\mathbf{w}}(n)$ produced by the RLS algorithm and the parameter vector $\mathbf{w}_o$ of the multiple linear regression model:

$$\boldsymbol{\varepsilon}(n) = \hat{\mathbf{w}}(n) - \mathbf{w}_o \qquad (8.40)$$

From Section 7.9, we also recall that the weight-error correlation matrix equals

$$\mathbf{K}(n) = E[\boldsymbol{\varepsilon}(n)\boldsymbol{\varepsilon}^H(n)]$$

$$= \sigma^2\boldsymbol{\Phi}^{-1}(n) \qquad (8.41)$$

where σ^2 is the variance of the measurement error process $\{e_o(n)\}$, and $\boldsymbol{\Phi}^{-1}(n)$ is the inverse of the time-varying correlation matrix defined in Eq. (8.9). The problem is to show that the weight-error correlation matrix $\mathbf{K}(n)$ approaches zero as the number of iterations, n, approaches infinity.

For large n, we may use the approximate relation of Eq. (8.39) between $\boldsymbol{\Phi}(n)$ and the ensemble-averaged correlation matrix $\mathbf{R}$. Correspondingly, we may approximate the weight-error correlation matrix of Eq. (8.41) as follows:

$$\mathbf{K}(n) \simeq \frac{\sigma^2}{n} \mathbf{R}^{-1}, \qquad \text{large } n$$

Taking the norm of both sides of this equation, we get

$$\|\mathbf{K}(n)\| \simeq \frac{\sigma^2}{n} \|\mathbf{R}^{-1}\|, \qquad \text{large } n \tag{8.42}$$

Let $\lambda_{\min}$ denote the smallest eigenvalue of the correlation matrix $\mathbf{R}$. Hence, we may express the norm of the inverse matrix $\mathbf{R}^{-1}$ as (see Section 2.5)

$$\|\mathbf{R}^{-1}\| = \frac{1}{\lambda_{\min}}$$

Accordingly, we may rewrite Eq. (8.42) as

$$\|\mathbf{K}(n)\| \simeq \frac{\sigma^2}{n\lambda_{\min}}, \qquad \text{large } n \tag{8.43}$$

Based on this result, we may make the following two observations for large n:

1. The norm of the weight-error correlation matrix is *magnified by the inverse of the smallest eigenvalue*. Hence, to a first order of approximation, the sensitivity of the RLS algorithm to eigenvalue spread is determined initially in proportion to the inverse of the smallest eigenvalue.
2. The norm of the weight-error correlation matrix decays almost linearly with time. Hence, the estimate $\hat{\mathbf{w}}(n)$ produced by the RLS algorithm for the tap-weight vector converges in the norm to the parameter vector $\mathbf{w}_o$ of the multiple linear regression model almost *linearly with time*.

Convergence of the RLS Algorithm in the Mean Square

In the RLS algorithm there are two errors, the a priori estimation error $\alpha(n)$ and the a posteriori estimation error $e(n)$, to be considered. Given the initial conditions of Section 8.4, we find that the average mean-square values of these two errors vary differently with time n. At time $n = 1$, the average mean-square value of $\alpha(n)$ attains a *large* value, equal to the mean-square value of the desired response $d(n)$, and then *decays* with increasing n. The average mean-square value of $e(n)$, on the other hand, attains a *small* value at $n = 1$, and then *rises* with increasing n. Accordingly, the choice of $\alpha(n)$ as the error of interest yields a learning curve for the RLS algorithm that has the same general shape as that for the LMS algorithm. By so doing, we can then make a direct graphical comparison between the learning curves of the RLS and LMS algorithms. We will therefore base the convergence analysis of the RLS in the mean square on the a priori estimation error $\alpha(n)$.

Eliminating the desired response $d(n)$ between Eq. (8.26) and (8.37), we may express the a priori estimation error $\alpha(n)$ as

$$
\begin{aligned}
\alpha(n) &= e_o(n) - [\hat{\mathbf{w}}(n-1) - \mathbf{w}_o]^H \mathbf{u}(n) \\
&= e_o(n) - \boldsymbol{\varepsilon}^H(n-1)\mathbf{u}(n)
\end{aligned}
\tag{8.44}
$$

where $\boldsymbol{\varepsilon}(n-1)$ is the weight-error vector at time $n-1$. We assume that the measurement error $e_o(n)$ and the weight-error vector $\boldsymbol{\varepsilon}(n-1)$ are statistically independent, and that $e_o(n)$ has zero mean. Hence, treating the tap-input vector $\mathbf{u}(n)$ as known, we may express the *mean-squared error* as

$$
\begin{aligned}
J'(n) &= E[|\alpha(n)|^2] \\
&= E[|e_o(n)|^2] + \mathbf{u}^H(n)E[\boldsymbol{\varepsilon}(n-1)\boldsymbol{\varepsilon}^H(n-1)]\mathbf{u}(n) \\
&= \sigma^2 + \mathbf{u}^H(n)\mathbf{K}(n-1)\mathbf{u}(n)
\end{aligned}
\tag{8.45}
$$

where σ^2 is the variance of the measurement error $e_o(n)$, and $\mathbf{K}(n-1)$ is the weight-error correlation matrix. The prime in the symbol $J'(n)$ is intended to distinguish the mean-square value of $\alpha(n)$ from that of $e(n)$. Substituting Eq. (8.41) in (8.45), with n replaced by $n-1$, we get

$$
J'(n) = \sigma^2 + \sigma^2 \mathbf{u}^H(n)\mathbf{\Phi}^{-1}(n-1)\mathbf{u}(n)
\tag{8.46}
$$

In a manner similar to that described for the LMS algorithm in Section 5.11,[3] we may average the mean-squared error $J'(n)$ over an ensemble of adaptive transversal filters that operate with the same RLS algorithm, identically initialized, and with the individual tap-input vectors drawn from the same statistical population. Thus, averaging Eq. (8.46) over the tap-input vector $\mathbf{u}(n)$, we get

$$
E[J'(n)] = \sigma^2 + \sigma^2 E[\mathbf{u}^H(n)\mathbf{\Phi}^{-1}(n-1)\mathbf{u}(n)]
\tag{8.47}
$$

Since the trace of a scalar equals the scalar itself, we may rewrite Eq. (8.47) as

$$
E(J'(n)) = \sigma^2 + \sigma^2 \text{ tr } \{E[\mathbf{u}^H(n)\mathbf{\Phi}^{-1}(n-1)\mathbf{u}(n)]\}
$$

Since the operations tr [ ] and $E[\ ]$ are both linear and therefore interchangeable, we may express the average mean-squared error in the equivalent form:

$$
E[J'(n)] = \sigma^2 + \sigma^2 \ E\{\text{tr}[\mathbf{u}^H(n)\mathbf{\Phi}^{-1}(n-1)\mathbf{u}(n)]\}
\tag{8.48}
$$

---

[3]In Section 5.11, the averaging was performed in two stages: (1) the mean-squared error was evaluated by ensemble-averaging over the tap-input vector $\mathbf{u}(n)$ and desired response $d(n)$, assuming that the tap-weight vector $\hat{\mathbf{w}}(n)$ is fixed at time n; (2) the mean-squared error was then ensemble-averaged over the tap-weight vector $\hat{\mathbf{w}}(n)$. In the analysis presented here for the RLS algorithm, this two-stage ensemble-averaging is performed in the reverse order in that we first average over the estimate $\hat{\mathbf{w}}(n)$ for the tap-weight vector, treating the tap-input vector $\mathbf{u}(n)$ as known, and then we average over $\mathbf{u}(n)$.

Now we use the property:

$$\text{tr}[\mathbf{u}^H(n)\mathbf{\Phi}^{-1}(n)\mathbf{u}(n)] = \text{tr}[\mathbf{\Phi}^{-1}(n)\mathbf{u}(n)\mathbf{u}^H(n)]$$

Accordingly, we may rewrite Eq. (8.48) as

$$\begin{aligned}
E[J'(n)] &= \sigma^2 + \sigma^2 E\{\text{tr}[\mathbf{\Phi}^{-1}(n - 1)\mathbf{u}(n)\mathbf{u}^H(n)]\} \\
&= \sigma^2 + \sigma^2\text{tr}\{E[\mathbf{\Phi}^{-1}(n - 1)\mathbf{u}(n)\mathbf{u}^H(n)]\}
\end{aligned} \tag{8.49}$$

For large n, we may approximate the time-variant inverse matrix $\mathbf{\Phi}^{-1}(n - 1)$ as $(n - 1)^{-1}\mathbf{R}^{-1}$, where $\mathbf{R}$ is the ensemble-averaged correlation matrix of the tap-input vector $\mathbf{u}(n)$. Hence, making this substitution in Eq. (8.49), we get

$$E[J'(n)] \simeq \sigma^2 + \frac{\sigma^2}{n - 1}\,\text{tr}\{\mathbf{R}^{-1}E[\mathbf{u}(n)\mathbf{u}^H(n)]\}$$

$$= \sigma^2 + \frac{\sigma^2}{n - 1}\,\text{tr}[\mathbf{R}^{-1}\mathbf{R}]$$

$$= \sigma^2 + \frac{\sigma^2}{n - 1}\,\text{tr}[\mathbf{I}]$$

where $\mathbf{I}$ is the M-by-M identity matrix. Since the trace of a square matrix equals the sum of its diagonal elements, it follows that $\text{tr}[\mathbf{I}]$ equals M, where M is the number of tap weights contained in the transversal filter. Also, for large n, we have $n - 1 \simeq n$. We thus get the following simple formula for the *average mean-squared innovation* that is produced by the RLS algorithm:

$$E[J'(n)] \simeq \sigma^2 + \frac{M\sigma^2}{n} \tag{8.50}$$

Based on this result, we may make the following deductions:

1. The RLS algorithm converges in the mean square in about $2M$ iterations, where M is the number of taps in the transversal filter. This means that the rate of convergence of the RLS algorithm is typically an order of magnitude *faster* than that of the LMS algorithm.

2. Unlike the LMS algorithm, the rate of convergence of the RLS algorithm is essentially insensitive to variations in the eigenvalue spread of the correlation matrix of the tap-input vector $\mathbf{u}(n)$.

3. As the number of iterations, n, approaches infinity, the average mean-squared innovation approaches a final value equal to the variance σ^2 of the measurement error $e_o(n)$. Since the minimum mean-squared error also equals σ^2, it follows that the RLS algorithm, in theory, produces zero excess mean-squared error or, equivalently, zero misadjustment.

It should be emphasized that the above-mentioned improvement in the rate of convergence of the RLS algorithm over the LMS algorithm holds only when the measurement error $e_o(n)$ is small compared to the desired response $d(n)$, that is, when the signal-to-noise ratio is high. Also, the zero misadjustment property of the RLS algorithm assumes that the exponential weighting factor λ equals unity; that is, the algorithm operates with infinite memory.

8.7 COMPUTER EXPERIMENT ON ADAPTIVE PREDICTION

In this section we use the RLS algorithm with the exponential weighting factor $\lambda = 1$ to adaptively estimate the tap weights of the two-tap linear predictor shown in Fig. 8.4. We assume that the tap inputs $u(n - 1)$ and $u(n - 2)$ are drawn from a *real-valued* stationary AR process of order 2. That is, the process $\{u(n)\}$ is described by the second-order difference equation

$$u(n) + a_1 u(n - 1) + a_2 u(n - 2) = v(n)$$

where the sample $v(n)$ is drawn from a white-noise process of zero mean and variance σ_v^2. For specified values of AR parameters a_1 and a_2, the variance σ_v^2 of the white-noise process $\{v(n)\}$ is chosen to make the AR process $\{u(n)\}$ have variance $\sigma_u^2 = 1$. In the experiment, the AR parameters a_1 and a_2 were chosen to satisfy the condition $a_1^2 < 4a_2$, which makes the roots of the characteristic equation of the AR process $\{u(n)\}$ assume complex-conjugate values. The values assigned to a_1 and a_2 also influence the eigenvalue spread of the 2-by-2 correlation matrix of the tap inputs $u(n - 1)$ and $u(n - 2)$. In Table 8.2, we list the three sets of values of a_1, a_2, and σ_v^2, corresponding to the eigenvalue spread $\chi(\mathbf{R}) = 3, 10, 100$, which were used in the experiment. Note that the variance σ_v^2 of $v(n)$ in the AR model describing the generation of the process $\{u(n)\}$ equals the variance σ^2 of the measurement error $e_o(n)$ contained in the multiple linear regression model of Fig. 8.4.

The adaptive version of the predictor in Fig. 8.4 using the LMS algorithm was studied in Section 5.15. We may thus use the predictor as an example to compare the performances of the RLS and LMS algorithms.

We may express the RLS algorithm for the operation of the predictor in Fig. 8.4 (with *real inputs*) as follows

$$\mathbf{k}(n) = \frac{\lambda^{-1}\mathbf{P}(n - 1)\mathbf{u}(n - 1)}{1 + \lambda^{-1}\mathbf{u}^T(n - 1)\mathbf{P}(n - 1)\mathbf{u}(n - 1)}$$

$$\eta(n) = u(n) - \mathbf{u}^T(n - 1)\hat{\mathbf{w}}(n - 1)$$

$$\hat{\mathbf{w}}(n) = \hat{\mathbf{w}}(n - 1) + \eta(n)\,\mathbf{k}(n)$$

$$\mathbf{P}(n) = \lambda^{-1}\mathbf{P}(n - 1) - \lambda^{-1}\,\mathbf{k}(n)\mathbf{u}^T(n - 1)\mathbf{P}(n - 1)$$

where we have used $\eta(n)$ to denote the forward a priori prediction errror and $\hat{\mathbf{w}}(n)$ to denote the tap-weight vector of the predictor. The input $u(n)$ plays the role of desired response. The tap-input vector of the predictor is defined by

$$\mathbf{u}^T(n - 1) = [u(n - 1), u(n - 2)]$$

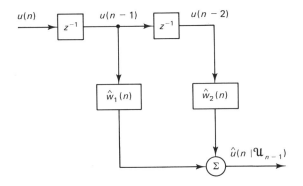

Figure 8.4 Two-tap predictor for real-valued input.

The tap-weight vector of the predictor is defined by

$$\hat{\mathbf{w}}^T(n) = [\hat{w}_1(n), \hat{w}_2(n)]$$

For any one trial of the experiment, 256 samples of a white-noise process $\{v(n)\}$ are obtained from a random-number generator of zero mean and adjustable variance σ_v^2. For each value of n in the range $1 \le n \le 256$, the squared magnitude of the a priori prediction error $\eta(n)$ is computed for the application of the RLS algorithm. The experiment is repeated 200 times, each time using an independent realization of the process $\{v(n)\}$. The mean-squared error is then determined by computing the ensemble average of $\eta^2(n)$ over those 200 independent trials of the experiment. The results of this computation are shown in Fig. 8.5 for the eigenvalue spread $\chi(\mathbf{R}) = 3$, 10, and 100. These results confirm the fast rate of convergence of the RLS algorithm and the fact that it is essentially insensitive to variations in the eigenvalue spread.

In Table 8.3 we have listed (1) the minimum value of the average mean-squared error obtained by time-averaging the last 200 samples of a 256-sample a priori prediction error sequence and then ensemble-averaging over 200 independent trials, and (2) the final values of the tap weights, averaged over 200 trials. Comparing these values with those of Table 8.2, close agreement is observed. In particular, we see that $E[J'(\infty)] \simeq \sigma_v^2$, indicating a negligible misadjustment, and that

$$E[\hat{\mathbf{w}}(\infty)] \simeq -a_i, \qquad \text{where } i = 1, 2$$

TABLE 8.2 PARAMETER VALUES OF THE AR MODEL USED IN THE EXPERIMENT

Eigenvalue spread, $\chi(\mathbf{R})$	σ_v^2	a_1	a_2
3	0.0731	-0.9750	0.95
10	0.0322	-1.5955	0.95
100	0.0038	-1.9114	0.95

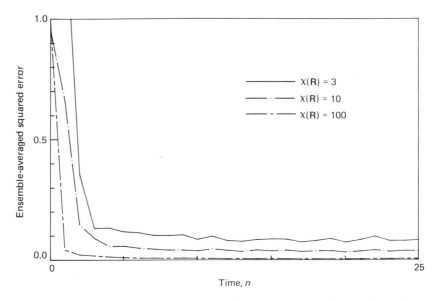

Figure 8.5 Learning curves of RLS algorithm for two-tap predictor with varying eigenvalue spread $\chi(\mathbf{R})$.

Figure 8.6 shows three learning curves, one pertaining to the RLS algorithm and the other two pertaining to two versions of the LMS algorithm with step-size parameter $\mu = 0.005, 0.05$. The eigenvalue spread $\chi(\mathbf{R})$ used in the experiment equals 10. This figure confirms the point we made earlier that the RLS algorithm converges faster than the LMS algorithm by at least an order of magnitude. In particular, we see that the RLS algorithm converges in about 3 iterations, whereas the LMS algorithm for $\mu = 0.05$ requires about 40 iterations.

8.8 COMPUTER EXPERIMENT ON ADAPTIVE EQUALIZATION

For our second experiment, we use the RLS algorithm, with the exponential weighting factor $\lambda = 1$, for the adaptive equalization of a linear dispersive communication channel. The LMS version of this study was presented in Section 5.16. The block diagram of the system used in the study is depicted in Fig. 8.7. Two independent random-number generators are used, one, denoted by $\{a(n)\}$, for probing the channel, and the other, denoted by $\{v(n)\}$, for simulating the effect of additive white noise at the receiver input. The sequence $\{a(n)\}$ is in polar form, with $a(n) = \pm 1$. The second sequence $\{v(n)\}$ has zero mean; its variance σ_v^2 is determined by the desired signal-to-noise ratio. The equalizer has 11 taps. The impulse response

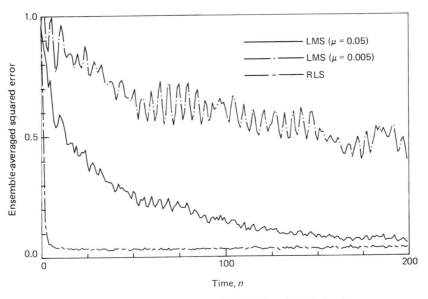

Figure 8.6 Comparison of learning curves of RLS and LMS algorithms.

TABLE 8.3 FINAL PARAMETER VALUES OF THE
ADAPTIVE PREDICTOR USING THE RLS ALGORITHM

Eigenvalue spread, $\chi(\mathbf{R})$	$E[J'(\infty)]$	$E[\hat{w}_1(\infty)]$	$E[\hat{w}_2(\infty)]$
3	0.0749	0.9732	-0.9434
10	0.0331	1.5993	-0.9436
100	0.0039	1.9006	-0.9393

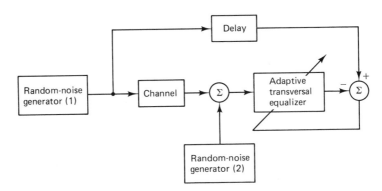

Figure 8.7 Block diagram of adaptive equalizer experiment.

of the channel is defined by

$$h_n = \begin{cases} \dfrac{1}{2}\left[1 + \cos\left(\dfrac{2\pi}{W}(n-2) \right) \right], & n = 1, 2, 3 \\ 0, & \text{otherwise} \end{cases}$$

where W controls the amount of amplitude distortion and therefore the eigenvalue spread produced by the channel. The channel input $\{a(n)\}$, after a delay of seven samples, provides the desired response for the equalizer; see Section 5.16 for details.

The experiment is in two parts: in part 1 the signal-to-noise ratio is high, and in part 2 it is low. In both parts of the experiment, the constant $\delta = 0.004$.

1. Signal-to-noise ratio = 30 dB. Figure 8.8 shows the results of the experiment for fixed signal-to-noise ratio of 30 dB (equivalently, variance $\sigma_v^2 = 0.001$) and varying W or eigenvalue spread $\chi(\mathbf{R})$. The figure is in four parts, for which the parameter W equals 2.9, 3.1, 3.3, and 3.5. The corresponding values

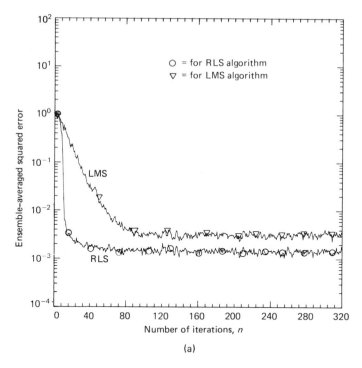

(a)

Figure 8.8 Learning curves for LMS and RLS algorithms: (a) $W = 2.9$, $\delta = 0.004$, $\lambda = 1.0$ for RLS algorithm; step size = 0.075 for LMS algorithm. (b) $W = 3.1$, $\delta = 0.004$, $\lambda = 1.0$ for RLS algorithm; step size = 0.075 for LMS algorithm. (c) $W = 3.3$, $\delta = 0.004$, $\lambda = 1.0$ for RLS algorithm; step size = 0.075 for LMS algorithm. (d) $W = 3.5$, $\delta = 0.004$, $\lambda = 1.0$ for RLS algorithm; step size = 0.075 for LMS algorithm.

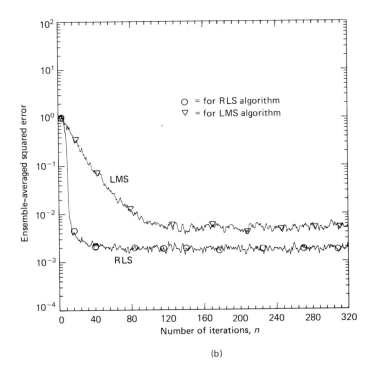

(b)

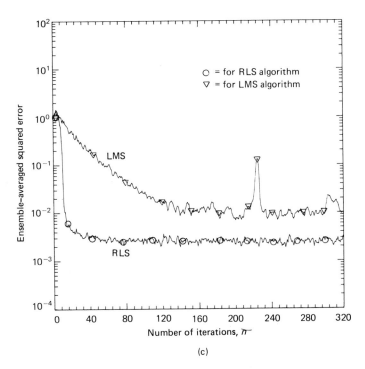

(c)

Figure 8.8 (*cont.*)

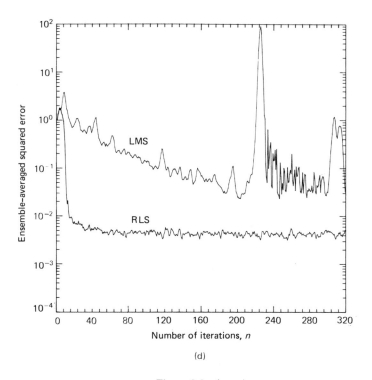

(d)

Figure 8.8 (*cont.*)

of the eigenvalue spread $\chi(\mathbf{R})$ equal 6.0782, 11.1238, 21.7132, and 46.8216, respectively. See Table 5.3 for details. In each part of Fig. 8.8, we have included two learning curves, one for the RLS algorithm obtained by ensemble-averaging (for each iteration n) the squared value of the a priori estimation error $\alpha(n)$, and the other for the LMS algorithm (included for the sake of comparison) obtained by ensemble-averaging the squared value of the a posteriori estimation error $e(n)$. The ensemble-averaging was performed over 200 independent trials of the experiment. For the LMS algorithm, the step-size parameter $\mu = 0.075$ was used. Based on the results shown in Fig. 8.8, we may make the following observations:

(1) Convergence of the RLS algorithm is attained in about 20 iterations, approximately twice the number of taps in the transversal equalizer.

(2) Rate of convergence of the RLS algorithm is essentially independent of the eigenvalue spread $\chi(\mathbf{R})$. This property is clearly illustrated in Fig. 8.9 where we have reproduced the learning curves of the RLS algorithm, corresponding to the four different values of the eigenvalue spread.

(3) The RLS algorithm converges much faster than the LMS algorithm.

(4) The steady-state value of the averaged squared error produced by the RLS algorithm is much smaller than in the case of the LMS algorithm. In both cases, however, it is sensitive to variations in the eigenvalue spread $\chi(\mathbf{R})$.

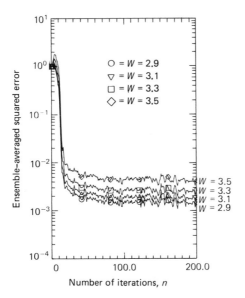

O = W = 2.9
▽ = W = 3.1
□ = W = 3.3
◇ = W = 3.5

Figure 8.9   Learning curves of the RLS algorithm, $\delta = 0.004$, and varying eigenvalue spread.

2. Signal-to-noise ratio = 10 dB.   Figure 8.10 shows the learning curves for the RLS algorithm and the LMS algorithm (with the step-size parameter $\mu = 0.075$) for $W = 3.1$ and signal-to-noise ratio of 10 dB.   Insofar as the rate of convergence is concerned, we now see that the RLS and LMS algorithms perform in roughly the same manner, both requiring about 40 iterations to converge.

The results presented in Figs. 8.8 and 8.10 clearly show that the superior rate of convergence of the RLS over the LMS algorithm is realized only when the signal-to-noise ratio is high.

8.9 OPERATION OF THE RLS ALGORITHM IN A NONSTATIONARY ENVIRONMENT

Throughout the convergence analysis of the RLS algorithm presented in Section 8.6, it was assumed that the exponential weighting factor $\lambda = 1$.   Indeed, the use of $\lambda = 1$ is well-suited for a stationary environment, for which the best steady-state results are obtained.   However, when the RLS algorithm operates in a nonstationary environment, it is customary to use a value of λ less than unity, thereby giving the algorithm only a finite memory.   By so doing, the algorithm attains the capability to track slow statistical variations in the environment in which it operates. Unfortunately, the use of $\lambda < 1$ changes the behavior of the RLS algorithm in a drastic manner.   In this section, we present a summary of the effects produced by using the exponentially windowed RLS algorithm in a nonstationary environment.[4]

[4]For a detailed mathematical analysis of the operation of the RLS algorithm in a nonstationary environment, see Eleftheriou and Falconer (1984).   The summary presented in Section 8.9 is based on this report.

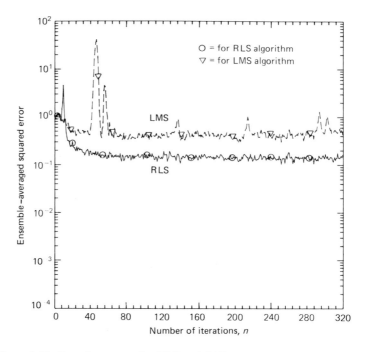

Figure 8.10 Learning curves for RLS and LMS algorithms for $W = 3.1$; $\delta = 0.004$, $\lambda = 1.0$ for RLS; step size $= 0.075$ for LMS; SNR $= 10$ dB.

Weight Vector Noise

When $\lambda < 1$, the exponentially weighted least-squares estimate $\hat{\mathbf{w}}(n)$ of the regression coefficient vector is no longer consistent. This causes noise to appear in the tap weights of the adaptive transversal filter, with the result that (on the average) they become misadjusted from their optimum setting. This type of misadjustment arises whenever λ is less than unity, whether the environment in which the RLS algorithm operates is stationary or not. Following the convention introduced in Section 5.17, we define the *weight vector noise* as

$$\boldsymbol{\varepsilon}_1(n) = \hat{\mathbf{w}}(n) - E[\hat{\mathbf{w}}(n)]$$

where $E[\hat{\mathbf{w}}(n)]$ is the mean value of $\hat{\mathbf{w}}(n)$. The presence of the weight vector noise in the estimate $\hat{\mathbf{w}}(n)$ leads to the *average excess mean squared error* [see Eq. (5.132)]

$$E[J_{\text{est}}(n)] = E[\boldsymbol{\varepsilon}_1^H(n)\mathbf{R}\boldsymbol{\varepsilon}_1(n)]$$

where $\mathbf{R}$ is the ensemble-averaged correlation matrix of the tap inputs. Here we have used the subscript "est" to indicate that the above formula is an expression for the "estimation-noise." For the case when λ is very close to unity, use of this formula yields the following approximate result for the steady-state value of the average excess-mean squared error due to the weight vector noise (Medaugh and Griffiths, 1981; Ling and Proakis, 1984):

$$E[J_{\text{est}}(\infty)] \simeq \frac{1 - \lambda}{1 + \lambda} M\sigma^2 \qquad (8.51)$$

where M is the number of taps in the trasversal filter and σ^2 is the variance of the measurement noise $e_o(n)$ in the multiple linear regression model.

The applicability of this approximate formula is restricted to values of λ very close to unity. In general, calculation of the average excess mean squared error $E(J_{\text{est}}(\infty))]$ depends not only on the exponential windowing but also on some fourth-order statistics of the input signal. The latter effect is noticeable when λ is assigned values as low as 0.9 for fast adaptation (Eleftheriou and Falconer, 1984).

Weight Vector Lag

When the environment in which the RLS algorithm operates is nonstationary, the coefficient vector in the multiple linear regression model takes on a *time-varying* form, as signified by $\mathbf{w}_o(n)$. In such an environment, the RLS algorithm has the task of not only *finding* the regression coefficient vector $\mathbf{w}_o(n)$ but also *tracking* its variation from one iteration to the next. Due to a lag in the adaptive process, we have to account for the *weight vector lag* [see Eq. (5.131)]

$$\boldsymbol{\varepsilon}_2(n) = E[\hat{\mathbf{w}}(n)] - \mathbf{w}_o(n)$$

Provided that the variation in $\mathbf{w}_o(n)$ is slow compared with the memory of the algorithm, $\boldsymbol{\varepsilon}_2(n)$ is approximately defined by the first-order difference equation (Eleftheriou and Falconer, 1984)

$$\boldsymbol{\varepsilon}_2(n) = \lambda\boldsymbol{\varepsilon}_2(n - 1) - \mathbf{w}_o(n) + \mathbf{w}_o(n - 1) \qquad (8.52)$$

From this difference equation we deduce that the weight vector lag $\boldsymbol{\varepsilon}_2(n)$ is a *geometrical* converging vector process. The *time constant* of the process is given by

$$\tau_i = \frac{1}{1 - \lambda}$$

for all coordinates of the process, $1 \le i \le M$. Moreover, it is independent of the eigenvalue spread of the correlation matrix of the tap inputs. This is in direct contrast to the LMS algorithm for which the time constant is different for each tap weight and corresponds to a particular eigenvalue.

Total Value of the Average Excess Mean Squared Error

Consider the example of a multiple linear regression model whose coefficients undergo independent stationary first-order Markov processes. That is, the regression coefficient vector $\mathbf{w}_o(n)$ is described by the real–valued first-order difference equation

$$\mathbf{w}_o(n) = a\mathbf{w}_o(n - 1) + \mathbf{v}(n)$$

The parameter a controls the time constant of nonstationarity, which equals $1/(1 - a)$. The elements of the vector $\mathbf{v}(n)$ consist of independent, white, Gaussian noise processes, each of zero mean and variance σ_v^2. To ensure that the variation in $\mathbf{w}_o(n)$ is slow compared with the memory of the RLS algorithm, the time constant of nonstationarity is chosen to satisfy the condition

$$\frac{1}{1 - a} >> \frac{1}{1 - \lambda}$$

Under these conditions, we find that the average excess-mean squared error $E[J_{\text{lag}}(\infty)]$, due to the weight vector lag $\boldsymbol{\varepsilon}_2(n)$, is inversely proportioned to the factor $(1 - \lambda)$ (Eleftheriou and Falconer, 1984).

By contrast, we see from Eq. (8.51) that for λ very close to unity the average excess-mean squared error $E[J_{\text{est}}(\infty)]$, due to the weight vector noise $\boldsymbol{\varepsilon}_1(n)$, is (for all practical purposes) directly proportioned to $(1 - \lambda)$.

The total value of the average excess-mean squared error, $E[J_{\text{tot}}(\infty)]$, due to the combined action of $\boldsymbol{\varepsilon}_1(n)$ and $\boldsymbol{\varepsilon}_2(n)$, is equal to the sum of $E[J_{\text{est}}(\infty)]$ and $E[J_{\text{lag}}(\infty)]$. At the optimum value λ_{opt}, for which $E[J_{\text{tot}}(\infty)]$ attains its minimum value, the contributions due to the weight vector noise $\boldsymbol{\varepsilon}_1(n)$ and the weight vector lag $\boldsymbol{\varepsilon}_2(n)$ are essentially equal.

From the discussion presented in Section 5.17 where we considered the corresponding behavior of the LMS algorithm operating in a similar nonstationary environment, we recall that the average excess mean-squared error due to the weight vector noise is directly proportional to the step-size parameter μ, and the average excess mean-squared error due to the weight vector lag is inversely proportional to μ.

Thus, comparing the results described for the RLS algorithm with these results for the LMS algorithm, we deduce that the factor $(1 - \lambda)$ plays a role in the RLS algorithm similar to that of the step-size parameter μ in the LMS algorithm.

It is also of interest to note that in system identification applications of time-varying environments, use of the RLS algorithm does not appear to offer a tracking advantage over the LMS algorithm with a judicious choice of the step-size parameter μ (Eleftheriou and Falconer, 1984; Ling and Proakis, 1984).

8.10 RELATIONSHIP BETWEEN THE RLS ALGORITHM AND KALMAN FILTER THEORY

The exponentially weighted RLS algorithm may be viewed as the deterministic counterpart of the Kalman filter theory. Indeed, the equations that define the RLS algorithm have the same basic mathematical structure as those that define a special form of the Kalman algorithm considered in Section 6.9. The special form of the Kalman algorithm that we have in mind assumes a random-walk state model in which the state noise vector has a correlation matrix that consists of the weight-error correlation matrix scaled by the factor q. Thus, comparing Eqs. (8.18), (8.26), (8.25), and (8.19) for the RLS algorithm with Eqs. (6.119), (6.120), (6.121), and (6.122) for the Kalman algorithm, respectively, we may deduce the following correspondences between the two algorithms:

1. The gain vector $\mathbf{k}(n)$ in the RLS algorithm plays the same role as the Kalman gain vector $\mathbf{g}(n)$ in the specialized form of the Kalman algorithm.
2. The matrix $\mathbf{P}(n) = \mathbf{\Phi}^{-1}(n)$ in the RLS algorithm corresponds to $\mathbf{K}(n)/J_{\min}$ in the Kalman algorithm, where $\mathbf{K}(n)$ is the filtered weight-error correlation matrix and $J_{\min}$ is the minimum mean-squared error.
3. The exponential weighting factor λ in the exponentially windowed RLS algorithm corresponds to $(1 + q)^{-1}$ in the Kalman algorithm. In the special case of a stationary environment, $\lambda = 1$ in the RLS algorithm and $q = 0$ in the Kalman algorithm.
4. The constant δ used to initialize the RLS algorithm corresponds to $c^{-1}J_{\min}$ in the initialization of the Kalman algorithm.

8.11 DISCUSSION

The highly superior convergence properties of the RLS algorithm over the LMS algorithm are achieved at the expense of increased computation per iteration. Specifically, to compute the gain vector $\mathbf{k}(n)$ defined by Eq. (8.18), an M-by-M matrix must be adapted and stored once per iteration, where M is the number of variable tap weights contained in the transversal filter. Accordingly, on the order of M^2 arithmetic operations must be performed per iteration of the RLS algorithm. This is in direct contrast to the LMS algorithm, in which on the order of M arithmetic operations per iteration are required.

To overcome this limitation of the RLS algorithm, several computationally efficient modifications of the algorithm have been introduced in the literature. These *fast* algorithms exploit the *shifting property* of serialized input data with the result that they require on the order of M arithmetic operations per iteration, thereby offering the highly improved convergence properties of the RLS algorithm at a computational cost that is competitive with the simple LMS algorithm.

The first major improvement in the computational cost of the RLS algorithm was reported in a series of papers by Morf, Kailath, and Ljung (1976), Ljung, Morf, and Falconer (1978), and Falconer and Ljung (1978). The development was inspired by pioneering work by Morf (1974) on shift-low-rank properties of deterministic correlation matrices. This modification to the RLS algorithm is referred to in the literature as the *fast RLS* (or *fast Kalman*) *algorithm*. The derivation of the fast RLS algorithm uses an approach that parallels that of the Levinson–Durbin recursion in that it relies on the use of forward and backward linear predictions (that are optimum in the least-squares sense) to derive the gain vector $\mathbf{k}(n)$ as a by-product (see Problem 9). This is done without any manipulation or storage of M-by-M matrices as required in the conventional form of the RLS algorithm.

A serious limitation of the fast RLS algorithm, however, is that it has a tendency to become numerically unstable. Mueller (1981a) reports on the results of computer simulation of an adaptive equalization problem that included the use of the fast RLS algorithm with an exponential weighting factor. When floating-point arithmetic (i.e., 24 bits for the mantissa) was used, unstable behavior of the fast RLS algorithm resulted. However, the use of double-precision arithmetic (i.e., 56 bits for the mantissa) eliminated the instability problem.

Cioffi and Kailath (1984) have developed another *fast, fixed-order, least-squares algorithm* for adaptive transversal filter applications, which requires slightly fewer arithmetic operations and exhibits better numerical properties than the fast RLS algorithm. In this algorithm, known as the *fast transversal filters (FTF) algorithm*, a maximum of four transversal filters is used. One filter defines the desired impulse response of the adaptive filter; this component also appears in the LMS algorithm. The three additional filters are required in order to complete the solution whenever the minimum mean-squared error is nonzero. For each of the four filters in the FTF algorithm, two basic computations are performed in a manner similar to that in the LMS algorithm: (1) a filtering operation that involves the excitation of a transversal filter so as to compute an estimation error, and (2) an adaptive operation that involves updating the filters' tap-weight vector. The overall result of these various computations is an adaptive filtering algorithm that always realizes the true solution of the RLS algorithm, and yet computationally it is *almost as fast* as the LMS algorithm.

Carayannis, Manolakis, and Kalouptsidis (1983) have also independently developed a *modified fast RLS* algorithm that requires fewer operations than the conventional form of the fast RLS algorithm. This algorithm has the same computational requirement as the unnormalized version of the FTF algorithm developed by Cioffi and Kailath. The computational efficiency of this third fast algorithm results from the fact that it exploits the relationship between forward and backward linear predictions more efficiently than the fast RLS algorithm.

The fast realizations of the RLS algorithm developed by (1) Falconer and

Ljung, (2) Carayannis, Manolakis and Kalouptsidis, and (3) Cioffi and Kailath exploit the shifting property of serialized input data to perform forward and backward linear predictions adaptively and in the same basic way. Basically, they differ from each other in the manner in which the results of these predictions are used to accomplish fast calculation of the gain vector $\mathbf{k}(n)$ (Cioffi and Kailath, 1984).

In Section 8.12, we develop the background theory for fast algorithms, and in Section 8.13, we derive the FTF algorithm. The reader is referred to Problem 9 for the derivation of the conventional form of the fast RLS algorithm.

8.12 BACKGROUND THEORY FOR FAST ALGORITHMS

The development of fast algorithms relies on certain update relations that arise in the characterization of adaptive forward and backward linear predictions that are optimized in the least-squares sense. In this section, we consider these two forms of linear prediction as special cases of the recursive least-squares problem. The other update relations needed for the development of fast algorithms involve the gain vector $\mathbf{k}(n)$. These are considered later in the section.

Adaptive Forward Linear Prediction

Consider the *forward linear predictor* of order M, depicted in Fig. 8.11(a), whose tap-weight vector $\hat{\mathbf{w}}(n)$ is optimized in the least-squares sense over the observation interval $1 \leq i \leq n$. Let $f_M(i)$ denote the *forward prediction error* produced by the predictor at time i in response to the tap-input vector $\mathbf{u}_M(i - 1)$, as shown by

$$f_M(i) = u(i) - \hat{\mathbf{w}}^H(n)\mathbf{u}_M(i - 1), \qquad 1 \leq i \leq n \tag{8.53}$$

where $u(i)$ plays the role of desired response and

$$\mathbf{u}_M^T(i - 1) = [u(i - 1), u(i - 2), \dots, u(i - M)] \tag{8.54}$$

We refer to $f_M(i)$ as the *a posteriori* prediction error since its computation is based on the *current* value $\hat{\mathbf{w}}(n)$ of the predictor's tap-weight vector.

We may equivalently characterize the forward linear prediction problem by specifying the *prediction-error filter*, as depicted in Fig. 8.11(b). Let $\mathbf{a}_M(n)$ denote the $(M + 1)$-by-1 tap-weight vector of the prediction-error filter of order M. This vector is related to the tap-weight vector of the predictor in Fig. 8.11(a) by

$$\mathbf{a}_M(n) = \begin{bmatrix} 1 \\ -\hat{\mathbf{w}}(n) \end{bmatrix} \tag{8.55}$$

Define the $(M + 1)$-by-1 vector $\mathbf{u}_{M+1}(i)$ that has the desired response $u(i)$ as the leading element and the vector $\mathbf{u}_M(i - 1)$ as the remaining M elements:

$$\mathbf{u}_{M+1}(i) = \begin{bmatrix} u(i) \\ \mathbf{u}_M(i - 1) \end{bmatrix} \tag{8.56}$$

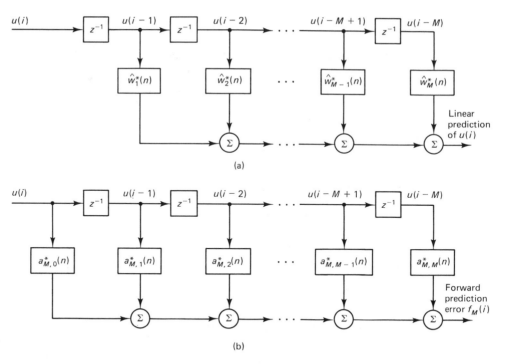

Figure 8.11   (a) Forward predictor of order M; (b) corresponding prediction-error filter.

Then we may redefine the forward a posteriori prediction error $f_M(i)$ as

$$f_M(i) = \mathbf{a}_M^H(n)\, \mathbf{u}_{M+1}(i) \tag{8.57}$$

The tap-weight vector $\hat{\mathbf{w}}(n)$ of the predictor is the solution obtained by minimizing the sum of weighted forward prediction-error squares for $1 \le i \le n$. Equivalently, the tap-weight vector $\mathbf{a}_M(n)$ of the prediction-error filter is the solution to the same minimization problem formulated in terms of the prediction-error filter, subject to the constraint that the first element of $\mathbf{a}_M(n)$ equals unity. In either case, we may express the least-squares solution compactly in terms of the *augmented normal equation for forward linear prediction* (see Chapter 7).

$$\mathbf{\Phi}_{M+1}(n)\, \mathbf{a}_M(n) = \begin{bmatrix} \mathscr{F}_M(n) \\ \mathbf{0} \end{bmatrix} \tag{8.58}$$

where $\mathbf{0}$ is the M-by-1 null vector. The scalar $\mathscr{F}_M(n)$ is the *minimum value of the sum of weighted forward a posteriori prediction-error squares*:

$$\mathscr{F}_M(n) = \sum_{i=1}^{n} \lambda^{n-i} |f_M(i)|^2 \tag{8.59}$$

The $(M + 1)$-by-$(M + 1)$ matrix $\mathbf{\Phi}_{M+1}(n)$ is the *deterministic correlation matrix* of the tap-input vector $\mathbf{u}_{M+1}(i)$ in the prediction-error filter of Fig. 8.11(b):

$$\boldsymbol{\Phi}_{M+1}(n) = \sum_{i=1}^{n} \lambda^{n-i}\mathbf{u}_{M+1}(i)\mathbf{u}_{M+1}^{H}(i) \tag{8.60}$$

Based on the partitioning of the vector $\mathbf{u}_{M+1}(i)$ given in Eq. (8.56), we may also express $\boldsymbol{\Phi}_{M+1}(n)$ in the partitioned form:

$$\boldsymbol{\Phi}_{M+1}(n) = \begin{bmatrix} \mathcal{U}(n) & \boldsymbol{\theta}_1^H(n) \\ \boldsymbol{\theta}_1(n) & \boldsymbol{\Phi}_M(n-1) \end{bmatrix} \tag{8.61}$$

where $\mathcal{U}(n)$ is the sum of weighted squared values of the desired response used for forward prediction,

$$\mathcal{U}(n) = \sum_{i=1}^{n} \lambda^{n-i}|u(i)|^2 \tag{8.62}$$

$\boldsymbol{\theta}_1(n)$ is the M-by-1 cross-correlation vector of the tap-input vector of the predictor in Fig. 8.11(a) and the desired response $u(i)$,

$$\boldsymbol{\theta}_1(n) = \sum_{i=1}^{n} \lambda^{n-1}\mathbf{u}_M(i-1)\, u^*(i) \tag{8.63}$$

and $\boldsymbol{\Phi}_M(n-1)$ is the M-by-M deterministic correlation matrix of the tap-input vector $\mathbf{u}_M(i-1)$ in the predictor of Fig. 8.11(a):

$$\begin{aligned}
\boldsymbol{\Phi}_M(n-1) &= \sum_{i=1}^{n} \lambda^{n-i}\mathbf{u}_M(i-1)\mathbf{u}_M^H(i-1) \\
&= \sum_{i=1}^{n=1} \lambda^{(n-1)-i}\, \mathbf{u}_M(i)\mathbf{u}_M^H(i)
\end{aligned} \tag{8.64}$$

Note that in the last line of Eq. (8.64) we have done two things: (1) we have replaced $i-1$ by i, and (2) we have assumed that the input data are zero prior to time $i = 0$ as prescribed by prewindowing of the data.

We may now turn our attention to the adaptive implementation of the predictor using the RLS algorithm. In Table 8.4, we have listed the correspondences between the quantities characterizing the RLS algorithm and those characterizing the predictor of Fig. 8.11(a). With the aid of this table, it is a straightforward matter to modify the theory of the RLS algorithm developed in Sections 8.4 and 8.5 to write the desired recursions for the forward linear prediction problem.[5] First, from Eqs. (8.22), (8.25), and (8.26), we deduce the following *recursion for updating the tap-weight vector of the predictor*:

$$\hat{\mathbf{w}}(n) = \hat{\mathbf{w}}(n-1) + \mathbf{k}_M(n-1)\, \eta_M^*(n) \tag{8.65}$$

---

[5]We did this modification for the computer experiment in Section 8.7 dealing with the adaptive linear two-tap predictor. The RLS algorithm given there was for real data. In this section, we deal with the general case of complex data.

TABLE 8.4

Quantity	RLS algorithm for linear estimation (General)	RLS algorithm for forward linear prediction of order M	RLS algorithm for backward linear prediction of order M
Tap-input vector	$\mathbf{u}(n)$	$\mathbf{u}_M(n-1)$	$\mathbf{u}_M(n)$
Desired response	$d(n)$	$u(n)$	$u(n-M)$
Tap-weight vector	$\hat{\mathbf{w}}(n)$	$\hat{\mathbf{w}}(n)$	$\mathbf{g}(n)$
A posteriori estimation error	$e(n)$	$f_M(n)$	$b_M(n)$
A priori estimation error	$\alpha(n)$	$\eta_M(n)$	$\psi_M(n)$
Gain vector	$\mathbf{k}(n)$	$\mathbf{k}_M(n-1)$	$\mathbf{k}_M(n)$
Minimum value of sum of weighted errors squares	$\mathcal{E}_{\min}(n)$	$\mathcal{F}_M(n)$	$\mathcal{B}_M(n)$

where $\eta_M(n)$ is the *forward a priori prediction error* (i.e., tentative estimate of the forward prediction error):

$$\eta_M(n) = u(n) - \hat{\mathbf{w}}^H(n-1)\,\mathbf{u}_M(n-1) \qquad (8.66)$$

and $\mathbf{k}_M(n-1)$ is the *gain vector for forward linear prediction of order M*:

$$\mathbf{k}_M(n-1) = \mathbf{\Phi}_M^{-1}(n-1)\,\mathbf{u}_M(n-1) \qquad (8.67)$$

The use of subscript M is intended to signify the order of the predictor. We follow this practice here and in the rest of the chapter since some of the recursions to be developed involve an order update. Correspondingly, we may substitute Eq. (8.65) in (8.55) to write the recursion for updating the tap-weight vector of the prediction-error filter as

$$\mathbf{a}_M(n) = \mathbf{a}_M(n-1) - \begin{bmatrix} 0 \\ \mathbf{k}_M(n-1) \end{bmatrix}\eta_M^*(n) \qquad (8.68)$$

where

$$\begin{aligned}
\eta_M(n) &= [1, \, -\hat{\mathbf{w}}^H(n-1)] \begin{bmatrix} u(n) \\ \mathbf{u}_M(n-1) \end{bmatrix} \\
&= \mathbf{a}_M^H(n-1)\mathbf{u}_{M+1}(n)
\end{aligned} \qquad (8.69)$$

Finally, using Eq. (8.35), we get the following recursion for updating the minimum value of the sum of weighted forward prediction-error squares:

$$\mathcal{F}_M(n) = \lambda\mathcal{F}_M(n-1) + \eta_M(n)f_M^*(n) \qquad (8.70)$$

Note that the correction $\eta_M(n)f_M^*(n)$ in this recursive is always a real-valued scalar; that is,

$$\eta_M(n)f_M^*(n) = \eta_M^*(n)f_M(n)$$

Adaptative Backward Linear Prediction

Consider next the *backward linear predictor of order M*, depicted in Fig. 8.12(a), whose tap-weight vector $\mathbf{g}(n)$ is optimized in the least-squares sense over the observation interval $1 \leq i \leq n$. Let $b_M(i)$ denote the *backward prediction error* produced by this predictor at time i in response to the tap-input vector $\mathbf{u}_M(i)$, as shown by

$$b_M(i) = u(i - M) - \mathbf{u}_M^T(i)\mathbf{g}(n) \tag{8.71}$$

where $u(i - M)$ plays the role of desired response and

$$\mathbf{u}_M^T(i) = [u(i),\, u(i - 1),\, \ldots,\, u(i - M + 1)] \tag{8.72}$$

We refer to $b_M(i)$ as the *backward a posteriori prediction error* since its computation is based on the *current* value $\mathbf{g}(n)$ of the predictor's tap-weight vector.

We may equivalently characterize the backward linear prediction problem by specifying the prediction-error filter, as depicted in Fig. 8.12(b). Let $\mathbf{c}_M(n)$ denote the $(M + 1)$-by-1 tap-weight vector of this prediction-error filter of order M. The vector $\mathbf{c}_M(n)$ is related to the tap-weight vector of the backward predictor in Fig.

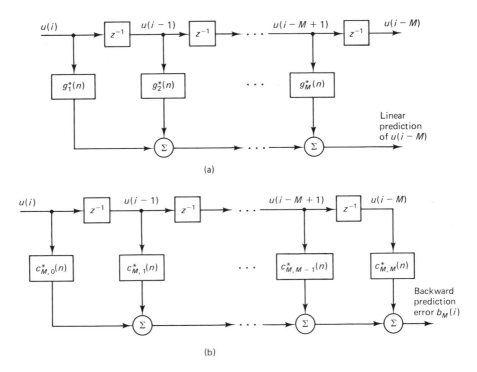

Figure 8.12   (a) Backward predictor of order M; (b) corresponding backward prediction-error filter.

8.12(a) by

$$
\mathbf{c}_M(n) = \begin{bmatrix} -\mathbf{g}(n) \\ 1 \end{bmatrix} \qquad (8.73)
$$

Define the $(M + 1)$-by-1 vector $\mathbf{u}_{M+1}(i)$ that has the desired response $u(i - M)$ as the last element and the vector $\mathbf{u}_M(i)$ as the remaining elements:

$$
\mathbf{u}_{M+1}(i) = \begin{bmatrix} \mathbf{u}_M(i) \\ u(i - M) \end{bmatrix} \qquad (8.74)
$$

Then we may redefine the backward prediction error $b_M(i)$ as

$$
b_M(i) = \mathbf{c}_M^H(n)\, \mathbf{u}_{M+1}(i) \qquad (8.75)
$$

The tap-weight vector $\mathbf{g}(n)$ of the backward linear predictor is the solution that results from minimizing the sum of weighted backward a posteriori prediction-error squares for $1 \leq i \leq n$. Equivalently, the tap-weight vector $\mathbf{c}_M(n)$ of the backward prediction-error filter is the solution of the same minimization problem formulated in terms of the backward prediction-error filter, subject to the constraint that the last element of $\mathbf{c}_M(n)$ equals unity.   In either case, we may express the least-squares solution compactly in terms of the *augmented normal equation for backward linear prediction* (see Chapter 7):

$$
\mathbf{\Phi}_{M+1}(n)\mathbf{c}_M(n) = \begin{bmatrix} \mathbf{0} \\ \mathscr{B}_M(n) \end{bmatrix} \qquad (8.76)
$$

where $\mathbf{0}$ is the M-by-1 null vector.   The scalar $\mathscr{B}_M(n)$ is the *minimum value of the sum of weighted backward a posteriori prediction-error squares*:

$$
\mathscr{B}_M(n) = \sum_{i=1}^{n} \lambda^{n-i}|b_M(i)|^2 \qquad (8.77)
$$

The $(M + 1)$-by-$(M + 1)$ matrix $\mathbf{\Phi}_{M+1}(n)$ is the deterministic correlation matrix of the tap-input vector $\mathbf{u}_{M+1}(i)$ in the prediction-error filter of Fig. 8.12(b).   It is defined by Eq. (8.61).   In a way corresponding to the partitioning of the vector $\mathbf{u}_{M+1}(i)$ shown in Eq. (8.74), we may also express the correlation matrix $\mathbf{\Phi}_{M+1}(n)$ in the partitioned form

$$
\mathbf{\Phi}_{M+1}(n) = \begin{bmatrix} \mathbf{\Phi}_M(n) & \mathbf{\theta}_2(n) \\ \mathbf{\theta}_2^H(n) & \mathscr{U}(n - M) \end{bmatrix} \qquad (8.78)
$$

where $\mathscr{U}(n - M)$ is the sum of weighted squared values of the desired response for backward prediction,

$$
\mathscr{U}(n - M) = \sum_{i=1}^{n} \lambda^{n-i}|u(i - M)|^2
$$
$$
= \sum_{i=1}^{n-M} \lambda^{(n-M)-i}|u(i)|^2 \qquad (8.79)
$$

$\boldsymbol{\theta}_2(n)$ is the M-by-1 cross-correlation vector of the tap-input vector of the predictor in Fig. 8.12(a) and the desired response $u(i - M)$:

$$\boldsymbol{\theta}_2(n) = \sum_{i=1}^{n} \lambda^{n-i} \mathbf{u}_M(i) \, u^*(i-M) \tag{8.80}$$

and $\boldsymbol{\Phi}_M(n)$ is the M-by-M deterministic correlation matrix of the tap-input vector $\mathbf{u}_M(i)$ in the predictor of Fig. 8.12(a):

$$\boldsymbol{\Phi}_M(n) = \sum_{i=1}^{n} \lambda^{n-i} \mathbf{u}_M(i) \, \mathbf{u}_M^H(i) \tag{8.81}$$

To write the recursion for the adaptive implementation of the backward linear predictor using the RLS algorithm, we again modify the theory developed in Sections 8.4 and 8.5 with the aid of the correspondences listed in Table 8.4. Thus, from Eqs. (8.22), (8.25), and (8.26), we deduce the following recursion for updating the tap-weight vector of the backward predictor:

$$\mathbf{g}(n) = \mathbf{g}(n-1) + \mathbf{k}_M(n)\psi_M^*(n) \tag{8.82}$$

where $\psi_M(n)$ is the *backward a priori prediction error* (i.e., tentative estimate of the backward prediction error):

$$\psi_M(n) = u(n-M) - \mathbf{g}^H(n-1)\mathbf{u}_M(n) \tag{8.83}$$

and $\mathbf{k}_M(n)$ is the *gain vector for backward linear prediction of order M*:

$$\mathbf{k}_M(n) = \boldsymbol{\Phi}_M^{-1}(n) \, \mathbf{u}_M(n) \tag{8.84}$$

Correspondingly, we may substitute Eq. (8.82) in (8.73) and so express the recursion for updating the tap-weight vector of the backward prediction-error filter as

$$\mathbf{c}_M(n) = \mathbf{c}_M(n-1) - \begin{bmatrix} \mathbf{k}_M(n) \\ 0 \end{bmatrix} \psi_M^*(n) \tag{8.85}$$

where

$$\begin{aligned} \psi_M(n) &= [-\mathbf{g}^H(n-1), \ 1]\begin{bmatrix} \mathbf{u}_M(n) \\ u(n-M) \end{bmatrix} \\ &= \mathbf{c}_M^H(n-1) \, \mathbf{u}_{M+1}(n) \end{aligned} \tag{8.86}$$

Finally, from Eq. (8.35) we deduce the following expression for updating the minimum value of the sum of weighted backward a posteriori prediction-error squares:

$$\mathcal{B}_M(n) = \lambda \mathcal{B}_M(n-1) + \psi_M(n) \, b_M^*(n) \tag{8.87}$$

where the correction term $\psi_M(n) \, b_M^*(n)$ is always real; that is,

$$\psi_M(n)b_M^*(n) = \psi_M^*(n)b_M(n)$$

Note that the update recursion for the tap-weight vector of the backward linear predictor in Eq. (8.82) or the backward prediction-error filter in Eq. (8.85) requires knowledge of the *current* value $\mathbf{k}_M(n)$ of the gain vector. On the other hand, the update recursion for the tap-weight vector of the forward linear predictor in Eq. (8.65) or the forward prediction-error filter in Eq. (8.68) requires knowledge of the *old* value $\mathbf{k}_M(n-1)$ of the gain vector.

Extended Gain Vector

By analogy with Eq. (8.84), we define the $(M+1)$-by-1 *extended gain vector:*

$$\mathbf{k}_{M+1}(n) = \mathbf{\Phi}_{M+1}^{-1}(n)\mathbf{u}_{M+1}(n) \tag{8.88}$$

The noteworthy feature of the extended gain vector is that, as will be shown later, it incorporates both the gain vector $\mathbf{k}_M(n-1)$ in the adaptive forward linear predictor and the gain vector $\mathbf{k}_M(n)$ in the adaptive backward linear predictor.

The $(M+1)$-by-$(M+1)$ deterministic correlation matrix $\mathbf{\Phi}_{M+1}(n)$ may be partitioned as in Eq. (8.61), which pertains to the forward linear predictor. The inverse of the correlation matrix $\mathbf{\Phi}_{M+1}(n)$ may be expressed as follows (see Problem 10):

$$\mathbf{\Phi}_{M+1}^{-1}(n) = \begin{bmatrix} 0 & \mathbf{0}_M^T \\ \mathbf{0}_M & \mathbf{\Phi}_M^{-1}(n-1) \end{bmatrix} + \frac{1}{\mathscr{F}_M(n)}\, \mathbf{a}_M(n)\, \mathbf{a}_M^H(n) \tag{8.89}$$

when $\mathbf{0}_M$ is the M-by-1 null vector, $\mathbf{a}_M(n)$ is the $(M+1)$-by-1 tap-weight vector of the forward prediction-error filter of order M, and $\mathscr{F}_M(n)$ is the corresponding minimum value of the sum of weighted forward a posteriori prediction-error squares. We also recognize that for forward prediction the $(M+1)$-by-1 tap-input vector $\mathbf{u}_{M+1}(n)$ may be partitioned as

$$\mathbf{u}_{M+1}(n) = \begin{bmatrix} u(n) \\ \mathbf{u}_M(n-1) \end{bmatrix} \tag{8.90}$$

Accordingly, postmultiplying both sides of Eq. (8.89) by $\mathbf{u}_{M+1}(n)$, using the partitioned form of Eq. (8.90) for the first term on the right side of Eq. (8.89), and using in the second term the definition of the forward a posteriori prediction error,

$$f_M(n) = \mathbf{a}_M^H(n)\, \mathbf{u}_{M+1}(n) \tag{8.91}$$

we get the following recursion for the extended gain vector:

$$\mathbf{k}_{M+1}(n) = \begin{bmatrix} 0 \\ \mathbf{k}_M(n-1) \end{bmatrix} + \frac{f_M(n)}{\mathscr{F}_M(n)}\, \mathbf{a}_M(n) \tag{8.92}$$

Consider next the second partitioned form of the $(M+1)$-by-$(M+1)$ deterministic correlation matrix $\mathbf{\Phi}_{M+1}(n)$ shown in Eq. (8.78), which pertains to the backward linear prediction. Using this representation, we find that the inverse of the matrix $\mathbf{\Phi}_{M+1}(n)$ may be expressed as follows (see Problem 10):

$$\mathbf{\Phi}_{M+1}^{-1}(n) = \begin{bmatrix} \mathbf{\Phi}_M^{-1}(n) & \mathbf{0}_M \\ \mathbf{0}_M^T & 0 \end{bmatrix} + \frac{1}{\mathscr{B}_M(n)}\, \mathbf{c}_M(n)\mathbf{c}_M^H(n) \qquad (8.93)$$

where $\mathbf{c}_M(n)$ is the $(M+1)$-by-1 tap-weight vector of the backward prediction-error filter of order M, and $\mathscr{B}_M(n)$ is the corresponding minimum value of the sum of weighted a posteriori backward prediction-error squares. We may also partition the $(M+1)$-by-1 tap-input vector $\mathbf{u}_{M+1}(n)$, in accordance with the requirements of the backward linear prediction, as

$$\mathbf{u}_{M+1}(n) = \begin{bmatrix} \mathbf{u}_M(n) \\ u(n-M) \end{bmatrix} \qquad (8.94)$$

Hence, postmultiplying both sides of Eq. (8.93) by $\mathbf{u}_{M+1}(n)$, using the partitioned form of Eq. (8.94) for the first term in the right side of Eq. (8.93), and using in the second term the definition of the backward a posteriori prediction error

$$b_M(n) = \mathbf{c}_M^H(n)\, \mathbf{u}_{M+1}(n) \qquad (8.95)$$

we get the second recursion for the extended gain vector:

$$\mathbf{k}_{M+1}(n) = \begin{bmatrix} \mathbf{k}_M(n) \\ 0 \end{bmatrix} + \frac{b_M(n)}{\mathscr{B}_M(n)}\, \mathbf{c}_M(n) \qquad (8.96)$$

Another Estimation Error Defined by the Gain Vector

The definition of the M-by-1 gain vector,

$$\mathbf{k}_M(n) = \mathbf{\Phi}_M^{-1}(n)\, \mathbf{u}_M(n)$$

may also be viewed as the solution of a special case of the deterministic normal equation. To be specific, the gain vector $\mathbf{k}_M(n)$ defines the tap-weight vector of a transversal filter that contains M taps and that operates on the input data $u(1)$, $u(2), \ldots, u(n)$ to produce the least-squares estimate of a special desired response that equals

$$d(i) = \begin{cases} 1, & i = n \\ 0, & i = 1, 2, \ldots, n-1 \end{cases} \qquad (8.97)$$

The n-by-1 vector whose elements equal the $d(i)$ of Eq. (8.97) is called the *first unit vector*. This vector has the property that its inner product with any time-dependent vector reproduces the upper or "most recent" element of that vector.

Substituting Eq. (8.97) in (8.10), we find that the M-by-1 deterministic cross-correlation vector $\mathbf{\theta}_M(n)$ between the M tap inputs of the transversal filter and the desired response equals $\mathbf{u}_M(n)$. This therefore confirms the gain vector $\mathbf{k}_M(n)$ as the special solution of the deterministic normal equation that arises when the desired response is defined by Eq. (8.97).

Define the *estimation error*

$$\gamma_M(n) = 1 - \mathbf{k}_M^H(n)\,\mathbf{u}_M(n) \tag{8.98}$$
$$= 1 - \mathbf{u}_M^H(n)\,\boldsymbol{\Phi}_M^{-1}(n)\,\mathbf{u}_M(n)$$

The second term represents the output of a transversal filter whose tap-weight vector equals the gain vector $\mathbf{k}_M(n)$ and which is excited by the tap-input vector $\mathbf{u}_M(n)$, as depicted in Fig. 8.13. Since the filter output has the structure of a Hermitian form, it follows that the estimation error $\gamma_M(n)$ is a real-valued scalar. Moreover, $\gamma_M(n)$ has the important property that it is bounded by zero and one; that is,

$$0 \le \gamma_M(n) \le 1 \tag{8.99}$$

This property is readily proved by substituting the recursion of Eq. (8.16) for the inverse matrix $\boldsymbol{\Phi}_M^{-1}(n)$ in Eq. (8.99), and then simplifying to obtain the result

$$\gamma_M(n) = \frac{1}{1 + \lambda^{-1}\,\mathbf{u}_M^H(n)\,\boldsymbol{\Phi}_M^{-1}(n-1)\,\mathbf{u}_M(n)} \tag{8.100}$$

The Hermitian form $\mathbf{u}_M^H(n)\,\boldsymbol{\Phi}_M^{-1}(n-1)\,\mathbf{u}_M(n) \ge 0$. Consequently, the estimation error $\gamma_M(n)$ is bounded as in Eq. (8.99).

It is noteworthy that $\gamma_M(n)$ also equals the sum of weighted error squares resulting from use of the transversal filter in Fig. (8.13), whose tap-weight vector equals the gain vector $\mathbf{k}_M(n)$, to obtain the least-squares estimate of the first unit vector (see Problem 11).

The new variable $\gamma_M(n)$ has three other useful interpretations:

1. For recursive least-squares estimation, we have

$$\gamma_M(n) = \frac{e_M(n)}{\alpha_M(n)} \tag{8.101}$$

where $e_M(n)$ is the a posteriori estimation error and $\alpha_M(n)$ is the a priori estimation error. This relation is readily proved by postmultiplying the Hermitian transposed

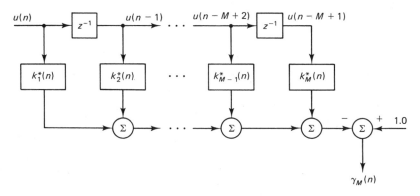

Figure 8.13   Transversal filter for defining the estimation error $\gamma_M(n)$

sides of Eq. (8.25) by $\mathbf{u}_M(n)$, using Eq. (8.26) for the a priori estimation error $\alpha_M(n)$, Eq. (8.27) for the a posteriori estimation error $e_M(n)$, and the first line of Eq. (8.98) for the variable $\gamma_M(n)$. Equation (8.101) states that, given the a priori estimation error $\alpha_M(n)$ as computed in the RLS algorithm, we may determine the corresponding value of the a posteriori estimation error $e_M(n)$ by multiplying $\alpha_M(n)$ by $\gamma_M(n)$. We may therefore view $\alpha_M(n)$ as a tentative value of the estimation error $e_M(n)$ and $\gamma_M(n)$ as the multiplicative correction.

2. For adaptive forward linear prediction, we have

$$\gamma_M(n-1) = \frac{f_M(n)}{\eta_M(n)} \tag{8.102}$$

This relation is readily proved by postmultiplying the Hermitian transposed sides of Eq. (8.68) by $\mathbf{u}_{M+1}(n)$, using the partitioned form of Eq. (8.90) for the post-multiplication of the second term on the right side of Eq. (8.68), and then using the definitions of Eqs. (8.69), (8.91), and (8.98). Equation (8.102) states that, given the foward a priori prediction error $\eta_M(n)$, we may compute the forward a posteriori prediction error $f_M(n)$ by multiplying $\eta_M(n)$ by the delayed estimation error $\gamma_M(n-1)$. We may therefore view $\eta_M(n)$ as a tentative value for the forward a posteriori prediction error $f_M(n)$ and $\gamma_M(n-1)$ as the multiplicative correction.

3. For adaptive backward linear prediction, we have

$$\gamma_M(n) = \frac{b_M(n)}{\psi_M(n)} \tag{8.103}$$

This third relation is readily proved by postmultiplying the Hermitian transposed sides of Eq. (8.85) by $\mathbf{u}_{M+1}(n)$, using the paritioned form of Eq. (8.94) for the postmultiplication of the second term on the right side of Eq. (8.85), and then using the definitions of Eqs. (8.86), (8.95), and (8.98). Equation (8.103) states that, given the backward a priori prediction error $\psi_M(n)$, we may compute the backward a posteriori prediction error $b_M(n)$ by multiplying $\psi_M(n)$ by the estimation error $\gamma_M(n)$. We may therefore view $\psi_M(n)$ as a tentative value for the backward prediction error $b_M(n)$ and $\gamma_M(n)$ as the multiplicative correction.

The above discussion points out the unique role of the variable $\gamma_M(n)$ in that it is the *common* factor (either in its regular or delayed form) in the conversion of an a priori estimation error into the corresponding a posteriori estimation error, be it in the context of ordinary estimation, forward prediction or backward prediction. Accordingly, we may refer to $\gamma_M(n)$ as the *conversion factor*.[6] Indeed, it is remarkable that through the use of this conversion factor we are able to compute the a posteriori errors $e_M(n)$, $f_M(n)$ and $b_M(n)$ at the time n before the tap-weight vectors of the pertinent filters [i.e., $\hat{\mathbf{w}}_M(n)$, $\mathbf{a}_M(n)$, and $\mathbf{c}_M(n)$] that produce them have been actually computed (Carayannis, Manolakis, and Kalouptsidis, 1983).

Next, we develop two recursions for updating of the variable $\gamma_M(n)$ as follows.

---

[6]In the literature, $\gamma_M(n)$ is also referred to as a *likelihood variable*; (Lee, Morf and Friedlander, 1981); for details, see Problem 4, Chapter 9.

First, we postmultiply the Hermitian transposed sides of Eq. (8.92) by $\mathbf{u}_{M+1}(n)$, use the partitioned form of Eq. (8.74) for the postmultiplication of the first term on the right side of Eq. (8.92), and then use the definitions of the forward a posteriori prediction error $f_M(n)$ and the conversion factor $\gamma_M(n)$ given in Eqs. (8.91) and (8.98). We thus obtain the following recursion that involves time as well as order updates of the conversion factor:

$$\gamma_{M+1}(n) = \gamma_M(n-1) - \frac{|f_M(n)|^2}{\mathscr{F}_M(n)} \tag{8.104}$$

For the second recursion, we postmultiply the Hermitian transposed sides of Eq. (8.96) by $\mathbf{u}_{M+1}(n)$, use the partitioned form of Eq. (8.94) for the postmultiplication of the first term on the right side of Eq. (8.96), and then use the definitions of the backward a posteriori prediction error $b_M(n)$ and the conversion factor $\gamma_M(n)$ given in Eqs. (8.95) and (8.98). We thus get the following recursion that only involves an order update:

$$\gamma_{M+1}(n) = \gamma_M(n) - \frac{|b_M(n)|^2}{\mathscr{B}_M(n)} \tag{8.105}$$

Finally, we develop another pair of relations by manipulating these two recursions. First, we use Eqs. (8.70) and (8.102) to eliminate $|f_M(n)|^2$ from (8.104). The result is

$$\gamma_{M+1}(n) = \lambda \frac{\mathscr{F}_M(n-1)}{\mathscr{F}_M(n)} \gamma_M(n-1) \tag{8.106}$$

Similarly, we use Eqs. (8.87) and (8.103) to eliminate $|b_M(n)|^2$ from (8.105), thereby obtaining the relation

$$\gamma_{M+1}(n) = \lambda \frac{\mathscr{B}_M(n-1)}{\mathscr{B}_M(n)} \gamma_M(n) \tag{8.107}$$

We finish off this section by identifying the various recursions and relations that we will use in the derivation of the FTF, a task that we undertake in the next section. For the adaptive forward linear prediction, we use the recursions of Eqs. (8.68) and (8.70) and the definition of Eq. (8.69). For the adaptive backward linear prediction, we use the recursions of Eqs. (8.85) and (8.87) and the definition of Eq. (8.86). For the gain vector, we use the recursions of Eqs. (8.92) and (8.96). Finally, for the estimation error defined by the gain vector, we use the relations of Eqs. (8.102) and (8.103) and the recursions of Eqs. (8.104) through (8.107).

8.13 FAST TRANSVERSAL FILTERS (FTF) ALGORITHM

The FTF algorithm (Cioffi and Kailath, 1984) uses a maximum of four transversal filters that share a common set of tap inputs, as depicted in Fig. 8.14. Each filter performs a distinct task of its own. In particular, filters I and II perform adaptive

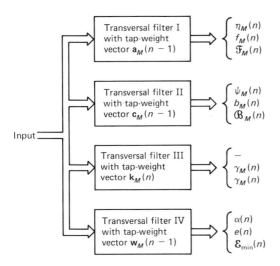

Figure 8.14 Transversal filter computation of RLS variables.

forward and backward linear prediction on the input data, respectively. Filter III defines the gain vector of the RLS algorithm. Filter IV is the adaptive filter whose tap-weight vector is the desired estimate for the problem at hand. The combination of these four filters is designed to produce the *exact solution of the RLS problem at all times*. When the minimum mean-squared is nonzero, as is ordinarily the case, the first three filters are required to complete the solution. We refer to these three filters as *auxiliary filters*. The operation of each filter in Fig. 8.14 involves two basic processes: (1) a filtering process wherein an error signal is produced in response to the excitation applied to the pertinent filter, and (2) an adaptive process wherein the tap-weight vector of that filter is updated by incrementing its old value by an amount equal to a scalar times some vector. The combined result of all these operations is not only an exact solution to the RLS problem but also a computational cost that increases linearly with the number of taps contained in the adaptive filter, as in the LMS algorithm.

We start the development of the FTF algorithm by applying the tap-input vector $\mathbf{u}_{M+1}(n)$ to filter I, the forward prediction-error filter, to produce the forward a priori prediction error [see Eq. (8.69)]:

$$\eta_M(n) = \mathbf{a}_M^H(n-1)\mathbf{u}_{M+1}(n) \tag{8.108}$$

We multiply $\eta_M(n)$ by the delayed estimation error $\gamma_M(n-1)$ to produce the forward a posteriori prediction error [see Eq. (8.102)]:

$$f_M(n) = \gamma_M(n-1)\eta_M(n) \tag{8.109}$$

We compute the updated minimum value of the sum of weighted forward prediction-error squares by using the recursion [see Eq. (8.70)]:

$$\mathscr{F}_M(n) = \lambda\mathscr{F}_M(n-1) + \eta_M(n)f_M^*(n) \tag{8.110}$$

We now use Eq. (8.106) to compute the updated a posteriori estimation error:

$$\gamma_{M+1}(n) = \lambda \frac{\mathcal{F}_M(n-1)}{\mathcal{F}_M(n)} \gamma_M(n-1) \tag{8.111}$$

Next we compute the updated value of the gain vector by using Eq. (8.92) except for a minor modification. The change is intended to make the recursion dependent on the old estimate of the tap-weight vector of the forward prediction-error filter, $\mathbf{a}_M(n-1)$, rather than the current estimate $\mathbf{a}_M(n)$. This we do simply by substituting Eq. (8.68) in (8.92), thereby obtaining (after collecting common terms)

$$\mathbf{k}_{M+1}(n) = \left(1 - \frac{\eta_M^*(n)f_M(n)}{\mathcal{F}_M(n)}\right)\begin{bmatrix} 0 \\ \mathbf{k}_M(n-1) \end{bmatrix} + \frac{f_M(n)}{\mathcal{F}_M(n)}\mathbf{a}_M(n-1) \tag{8.112}$$

We use Eq. (8.70) and then (8.106) to write

$$1 - \frac{\eta_M^*(n)f_M(n)}{\mathcal{F}_M(n)} = \lambda\frac{\mathcal{F}_M(n-1)}{\mathcal{F}_M(n)}$$

$$= \frac{\gamma_{M+1}(n)}{\gamma_M(n-1)} \tag{8.113}$$

We use Eq. (8.102) and then (8.106) to write

$$\frac{f_M(n)}{\mathcal{F}_M(n)} = \frac{\eta_M(n)\gamma_M(n-1)}{\mathcal{F}_M(n)}$$

$$= \lambda^{-1}\frac{\eta_M(n)\gamma_{M+1}(n)}{\mathcal{F}_M(n-1)} \tag{8.114}$$

Therefore, substituting Eqs. (8.113) and (8.114) in (8.112), we get

$$\mathbf{k}_{M+1}(n) = \frac{\gamma_{M+1}(n)}{\gamma_M(n-1)}\begin{bmatrix} 0 \\ \mathbf{k}_M(n-1) \end{bmatrix} + \lambda^{-1}\frac{\eta_M(n)\gamma_{M+1}(n)}{\mathcal{F}_M(n-1)}\mathbf{a}_M(n-1) \tag{8.115}$$

For convenience of presentation, we define the *normalized gain vector*

$$\tilde{\mathbf{k}}_M(n) = \frac{\mathbf{k}_M(n)}{\gamma_M(n)} \tag{8.116}$$

Accordingly, we may simplify the recursion of Eq. (8.115) as

$$\tilde{\mathbf{k}}_{M+1}(n) = \begin{bmatrix} 0 \\ \tilde{\mathbf{k}}_M(n-1) \end{bmatrix} + \lambda^{-1}\frac{\eta_M(n)}{\mathcal{F}_M(n-1)}\mathbf{a}_M(n-1) \tag{8.117}$$

which is the desired recursion for updating the normalized gain vector.

Correspondingly, we use Eq. (8.116) and then (8.102) to redefine the recursion of (8.68) for updating the tap-weight vector of the forward prediction-error filter as

$$\mathbf{a}_M(n) = \mathbf{a}_M(n-1) - \gamma_M(n-1)\,\eta_M^*(n) \begin{bmatrix} 0 \\ \tilde{\mathbf{k}}_M(n-1) \end{bmatrix}$$

$$= \mathbf{a}_M(n-1) - f_M^*(n) \begin{bmatrix} 0 \\ \tilde{\mathbf{k}}_M(n-1) \end{bmatrix} \qquad (8.118)$$

At this stage in our development, we have time-updated the forward prediction-error filter (i.e., filter I) and also updated the gain vector (i.e., filter III) not only in time (as desired) but also order.

Our next task is twofold: (1) to correct for the increase in the order of filter III from M to $M+1$ incurred through the use of Eq. (8.111) to compute $\gamma_{M+1}(n)$, and (2) to perform a time-update on the backward prediction-error filter (i.e., filter II). To do this, we first substitute Eq. (8.85) in (8.96) and collect common terms, obtaining

$$\mathbf{k}_{M+1}(n) = \left(1 - \frac{\psi_M^*(n)b_M(n)}{\mathscr{B}_M(n)}\right) \begin{bmatrix} \mathbf{k}_M(n) \\ 0 \end{bmatrix} + \frac{b_M(n)}{\mathscr{B}_M(n)}\,\mathbf{c}_M(n-1) \qquad (8.119)$$

Let $k_{M+1,M-1}(n)$ denote the *last element* of the $(M+1)$-by-1 gain vector $\mathbf{k}_{M+1}(n)$. We recognize that the last element of the $(M+1)$-by-1 tap-weight vector $\mathbf{c}_{M+1}(n)$, characterizing the backward prediction-error filter, equals unity. Hence, by considering the last elements of the vectors on the left and right sides of Eq. (8.119), we deduce that

$$k_{M+1,M+1}(n) = \frac{b_M(n)}{\mathscr{B}_M(n)} \qquad (8.120)$$

By normalizing with respect to $\gamma_{M+1}(n)$, we may also express the last element of $\hat{\mathbf{k}}_{M+1}(n)$ as

$$\tilde{k}_{M+1,M+1} = \frac{k_{M+1,M+1}(n)}{\gamma_{M+1}(n)}$$

$$= \frac{b_M(n)}{\gamma_{M+1}(n)\mathscr{B}_M(n)} \qquad (8.121)$$

The combined use of Eqs. (8.103) and (8.107) in (8.121) yields

$$\tilde{k}_{M+1,M+1}(n) = \frac{\psi_M(n)}{\lambda\mathscr{B}_M(n-1)} \qquad (8.122)$$

Equivalently, we may express the backward a priori prediction error $\psi_M(n)$ in terms of the last element of the normalized gain vector $\tilde{\mathbf{k}}_{M+1}(n)$ as

$$\psi_M(n) = \lambda\mathscr{B}_M(n-1)\,\tilde{k}_{M+1,M+1}(n) \qquad (8.123)$$

This relation enables us to compute $\psi_M(n)$. Solving Eq. (8.107) for $\gamma_M(n)$, we have

$$\gamma_M(n) = \frac{\mathcal{B}_M(n)}{\lambda \mathcal{B}_M(n-1)} \gamma_{M+1}(n) \tag{8.124}$$

By using Eq. (8.87) and then (8.121), we may write

$$\lambda \frac{\mathcal{B}_M(n-1)}{\mathcal{B}_M(n)} = 1 - \frac{\psi_M(n) b_M^*(n)}{\mathcal{B}_M(n)}$$

$$= 1 - \psi_M(n) \, \gamma_{M+1}(n) \, \bar{k}_{M+1,M+1}^*(n) \tag{8.125}$$

Therefore, substituting Eq. (8.125) in (8.124), we get

$$\gamma_M(n) = [1 - \psi_M(n) \, \gamma_{M+1}(n) \, \bar{k}_{M+1,M+1}^*(n)]^{-1} \, \gamma_{M+1}(n) \tag{8.126}$$

Note that

$$\psi_M(n)\bar{\mathbf{k}}_{M+1,M+1}^*(n) = \psi_M^*(n)\bar{\mathbf{k}}_{M+1,M+1}(n)$$

Equation (8.126) enables us to compute the current value of the conversion factor $\gamma_M(n)$. Thus, this relation corrects for the increase in the order of filter III from M to $M+1$ that results from the use of Eq. (8.111). Having already computed the current values of $\gamma_M(n)$ and $\psi_M(n)$, we may now use Eq. (8.103) to compute the backward a posteriori prediction error:

$$b_M(n) = \gamma_M(n) \, \psi_M(n) \tag{8.127}$$

Furthermore, we may use the recursion of Eq. (8.87) to update the minimum value of the sum of weighted backward a posteriori prediction-error squares:

$$\mathcal{B}_M(n) = \lambda \mathcal{B}_M(n-1) + \psi_M(n) \, b_M^*(n) \tag{8.128}$$

Returning to the recursion of Eq. (8.119), we use the first line of Eq. (8.125), (8.107), (8.116), and (8.121), then rearrange terms, and thereby rewrite this recursion as follows

$$\begin{bmatrix} \bar{k}_M(n) \\ 0 \end{bmatrix} = \bar{\mathbf{k}}_{M+1}(n) - \bar{k}_{M+1,M+1}(n) \, \mathbf{c}_M(n-1) \tag{8.129}$$

We are now ready to update the tap-weight vector that characterizes the backward prediction-error filter [see Eq. (8.85)].

$$\mathbf{c}_M(n) = \mathbf{c}_M(n-1) - \psi_M^*(n) \begin{bmatrix} \mathbf{k}_M(n) \\ 0 \end{bmatrix}$$

$$= \mathbf{c}_M(n-1) - b_M^*(n) \begin{bmatrix} \bar{\mathbf{k}}_M(n) \\ 0 \end{bmatrix} \tag{8.130}$$

where in the last line we have made use of Eqs. (8.103) and (8.116).

We have thus completed the time-updating of the gain vector (i.e., filter III) and the backward prediction-error filter (i.e., filter II). There now only remains the task of updating the tap-weight vector of the adaptive filter (i.e., filter IV). This we do next.

We use Eq. (8.26) to compute the a posteriori estimation error

$$\alpha_M(n) = d(n) - \hat{\mathbf{w}}_M^H(n-1)\,\mathbf{u}_M(n) \tag{8.131}$$

We use Eq. (8.101) to compute the corresponding value of the estimation error:

$$e_M(n) = \gamma_M(n)\alpha_M(n) \tag{8.132}$$

Finally, we use Eq. (8.25) to update the tap-weight vector of the adaptive filter:

$$\begin{aligned} \bar{\mathbf{w}}_M(n) &= \hat{\mathbf{w}}_M(n-1) + \mathbf{k}_M(n)\,\alpha_M^*(n) \\ &= \hat{\mathbf{w}}_M(n-1) + \bar{\mathbf{k}}_M(n)\,e_M^*(n) \end{aligned} \tag{8.133}$$

where in the last line we have made use of Eqs. (8.101) and (8.116).

This completes the updating cycle of the FTF algorithm.

Summary

In Table 8.5, we present a summary of the FTF algorithm by collecting together the recursions derived previously. This table presents a summary of the recursions and relations that are defined by Eqs. (8.108)–(8.111), (8.117), (8.118), (8.123), and (8.126–8.133), in that order.

The FTF algorithm realizes the RLS solution for the tap-weight vector $\hat{\mathbf{w}}_M(n)$ exactly. In contrast to the RLS algorithm, however, its computational complexity increases linearly with M, the number of tap weights contained in the adaptive filter, as in the LMS algorithm. Thus, the FTF algorithm offers the best of two worlds: the desirable convergence properties of the RLS algorithm, and the computational simplicity of the LMS algorithm. However, the simplification in computational complexity is achieved at the expense of a much more complicated program statement, as evidenced by comparing Table 8.5 for the FTF algorithm with the summary presented in Section 5.13 for the LMS algorithm.

8.14 EXACT INITIALIZATION OF THE FTF ALGORITHM USING ZERO INITIAL CONDITION

During the initialization period $1 \le n \le M + 1$, where M is the number of tap weights contained in the adaptive filter, considerable simplification and stabilization of the RLS solution is achieved by special investigation of the FTF algorithm (Cioffi and Kailath, 1984). In this section, we study *exact initialization* of the FTF algorithm for the case when the initial condition is zero. This case is very common, simply because in practice we ordinarily have no prior information about the unknown parameter vector $\mathbf{w}_o(n)$ in the multiple linear regression model of Fig. 8.3.

TABLE 8.5 THE FTF ALGORITHM

$$\eta_M(n) = \mathbf{a}_M^H(n-1)\,\mathbf{u}_{M+1}(n)$$

$$f_M(n) = \gamma_M(n-1)\eta_M(n)$$

$$\mathscr{F}_M(n) = \lambda\mathscr{F}_M(n-1) + \eta_M(n)f_M^*(n)$$

$$\gamma_{M+1}(n) = \lambda\,\frac{\mathscr{F}_M(n-1)}{\mathscr{F}_M(n)}\,\gamma_M(n-1)$$

$$\tilde{\mathbf{k}}_{M+1}(n) = \begin{bmatrix} 0 \\ \tilde{\mathbf{k}}_M(n-1) \end{bmatrix} + \lambda^{-1}\frac{\eta_M(n)}{\mathscr{F}_M(n-1)}\,\mathbf{a}_M(n-1)$$

$$\mathbf{a}_M(n) = \mathbf{a}_M(n-1) - f_M^*(n)\begin{bmatrix} 0 \\ \tilde{k}_M(n-1) \end{bmatrix}$$

$$\psi_M(n) = \lambda\mathscr{B}_M(n-1)\,\tilde{k}_{M+1,M+1}(n)$$

$$\gamma_M(n) = [1 - \psi_M^*(n)\,\gamma_{M+1}(n)\,\tilde{k}_{M+1,M-1}(n)]^{-1}\gamma_{M+1}(n)$$

Rescue[a] variable $= [1 - \psi_M^*(n)\gamma_{M+1}(n)\tilde{k}_{M+1,M+1}(n)]$

$$b_M(n) = \gamma_M(n)\psi_M(n)$$

$$\mathscr{B}_M(n) = \lambda\mathscr{B}_M(n-1) + \psi_M(n)\,b_M^*(n)$$

$$\begin{bmatrix} \tilde{\mathbf{k}}_M(n) \\ 0 \end{bmatrix} = \tilde{\mathbf{k}}_{M+1}(n) - \tilde{k}_{M+1,M+1}(n)\,\mathbf{c}_M(n-1)$$

$$\mathbf{c}_M(n) = \mathbf{c}_M(n-1) - b_M^*(n)\begin{bmatrix} \tilde{k}_M(n) \\ 0 \end{bmatrix}$$

$$\alpha_M(n) = d(n) - \hat{\mathbf{w}}_M^H(n-1)\,\mathbf{u}_M(n)$$

$$e_M(n) = \gamma_M(n)\alpha_M(n)$$

$$\hat{\mathbf{w}}_M(n) = \hat{\mathbf{w}}_M(n-1) + \tilde{\mathbf{k}}_M(n)\quad e_M^*(n)$$

[a] Rescue if variable is negative: Save $\hat{\mathbf{w}}_M(n)$ as initial condition and weight in the manner ζ as discussed in Section 8.15.

Accordingly, by setting the adjustable tap-weight vector $\hat{\mathbf{w}}_M(n)$ of the adaptive filter equal to zero at time $n = 0$, we simplify the initialization of the FTF algorithm.

Before deriving the exact initialization procedure, we discuss some special characteristics of the prewindowed RLS solution during the initialization period, for which $1 \le n \le M + 1$. At time $n = M$, initialization of both the gain vector (filter III) and the adaptive filter (filter IV) is completed. However, the forward and backward prediction-error filters (filters I and II) are both one unit longer; hence, their initialization is completed at time $n = M + 1$. In any event, during the initialization period the length of the input data is less than or equal to the number of adjustable tap weights in the adaptive filter, so the RLS problem is *underdetermined* or *exactly determined,* respectively. Hence, during the initialization period, we may set all the estimation errors, $e_M(n)$, $\gamma_M(n)$, $f_M(n)$, and $b_M(n)$, equal to zero since we have an abundance of adjustable parameters. In such a situation, the unique solution to the RLS problem is defined by the *minimum-norm* values of the tap-weight vectors of the corresponding transversal filters (see Section 7.12).

Another peculiarity of the RLS solution is the special structure assumed by both the gain vector and the tap-weight vector of the backward prediction-error filter at time $n = M$. From the definitions of the gain vector $\mathbf{k}_M(n)$ and the deterministic correlation matrix $\mathbf{\Phi}_M(n)$, we have

$$\left(\sum_{i=1}^{n} \lambda^{n-i} \mathbf{u}_M(i) \mathbf{u}_M^H(i) \right) \mathbf{k}_M(n) = \mathbf{u}_M(n)$$

By separating out the contribution to the deterministic correlation matrix at time $i = n$ from the rest, and using the definition of the estimation error $\gamma_M(n)$ that is real valued, we may equivalently write

$$\lambda \sum_{i=1}^{n-1} \lambda^{n-1-i} \mathbf{u}_M(i) \, \mathbf{u}_M^H(i) \, \mathbf{k}_M(n) = \mathbf{u}_M(n) \, [1 - \mathbf{u}_M^H(n) \mathbf{k}_M(n)] = \mathbf{u}_M(n) \, \gamma_M(n)$$

Since the estimation error $\gamma_M(n)$ is zero at time $n = M$, we thus have

$$\sum_{i=1}^{M-1} \lambda^{M-1-i} \mathbf{u}_M(i) \, \mathbf{u}_M^H(i) \, \mathbf{k}_M(M) = \mathbf{0} \tag{8.134}$$

In the prewindowing method, for $i = 1$ the last $M - 1$ elements of the tap-input vector $\mathbf{u}_M(i)$ are zero, for $i = 2$ the last $M - 2$ elements of $\mathbf{u}_M(i)$ are zero, and so on right up to $i = M - 1$ for which the last element of $\mathbf{u}_M(i)$ is zero. We deduce therefore that if Eq. (8.134) is to hold, subject to the requirement that the value $\mathbf{k}_M(M)$ of the gain vector be of minimum norm, then all the elements of $\mathbf{k}_M(M)$ must be zero except for the last one, as shown by

$$\mathbf{k}_M(M) = \begin{bmatrix} \mathbf{0} \\ k_{M,M}(M) \end{bmatrix} \tag{8.135}$$

where $\mathbf{0}$ is the $(M-1)$-by-1 null vector. We determine the nonzero element $k_{M,M}(M)$ by observing that

$$\gamma_M(M) = 1 - \mathbf{u}_M^H(M) \, \mathbf{k}_M(M)$$

$$= 1 - u^*(1) k_{M,M}(M)$$

Since $\gamma_M(M)$ is zero, we therefore have

$$k_{M,M}(M) = \frac{1}{u^*(1)} \tag{8.136}$$

and

$$\mathbf{k}_M(M) = \begin{bmatrix} \mathbf{0} \\ \dfrac{1}{u^*(1)} \end{bmatrix} \tag{8.137}$$

To determine the corresponding value of the tap-weight vector of the backward prediction-error filer, we observe that the desired response $u(n-M)$ in the backward linear prediction is zero at time $n = M$. Hence, at this time the tap-weight vector of the backward linear predictor is likewise zero, and the minimum-norm value of the $(M+1)$-by-1 tap-weight vector of the backward prediction-error filter equals

$$\mathbf{c}_M(M) = \begin{bmatrix} \mathbf{0} \\ 1 \end{bmatrix} \tag{8.138}$$

where $\mathbf{0}$ is the M-by-1 null vector. We thus see that at $n = M$ the first $M - 1$ elements of $\mathbf{k}_M(M)$ and the first M elements of $\mathbf{c}_M(M)$ are zero.

Consider next the situation at time $n = M+1$. At this time the RLS problem becomes *overdetermined* for the first time in that the length of the input data, n, exceeds the number of adjustable tap weights, M, by 1. Also, at time $n = M+1$ the desired response $u(n-M)$ in the backward linear prediction assumes a nonzero value for the first time. In particular, we see that $u(n-M) = u(1)$ at $n = M+1$. Accordingly, we may express the desired response vector for the backward linear prediction at time $n = M+1$ as

$$\mathbf{b}(M+1) = \begin{bmatrix} u^*(1) \\ \mathbf{0} \end{bmatrix}$$

Hence, the desired response vector in the backward linear prediction at time $n = M + 1$ is exactly the same, except for the scaling factor $u^*(1)$, as the first unit vector. This means that at this time the backward linear predictor and the transversal filter III (that defines the gain vector) make a least-squares estimate of the same desired response, within the multiplicative scalar $u^*(1)$. Both of these filters are of length M, and they operate on the same set of tap inputs, namely, $u(M + 1), u(M), \ldots, u(2)$. Hence, at time $n = M + 1$, the tap-weight vector of the backward linear predictor is equal to $u^*(1)$ times the gain vector $\mathbf{k}_M(M + 1)$. Correspondingly, we may state that the tap-weight vector of the backward prediction-error filter and the gain vector are related by

$$\mathbf{c}_M(M+1) = \begin{bmatrix} -u^*(1)\, \mathbf{k}_M(M+1) \\ 1 \end{bmatrix} \tag{8.139}$$

We see therefore that it is not possible to simultaneously solve Eqs. (8.85) and (8.96) because these two equations are essentially scalar multiples of one another.

We thus have a peculiarity to overcome in the initialization of the FTF algorithm. All the elements of the vectors $\mathbf{c}_M(M)$ and $\mathbf{k}_M(M)$, except for their individual last elements, are zero. In contrast, we usually find in practical applications that all the elements of $\mathbf{c}_M(M+1)$ and $\mathbf{k}_M(M+1)$ are nonzero. To update the zero elements in $\mathbf{c}_M(M)$ and $\mathbf{k}_M(M)$ to the nonzero elements in $\mathbf{c}_M(M+1)$ and $\mathbf{k}_M(M+1)$, we require knowledge of $\mathbf{c}_{M-1}(M)$ and $\mathbf{k}_{M-1}(M)$. This requirement means that we have to perform simultaneous time- and order-updating during the initialization of the FTF algorithm so as to avoid the above-mentioned peculiarity at time $n = M$. This special form of updating continues until time $n = M+1$.

We now derive the simultaneous time- and order-updating that is necessary to complete the exact initialization of the FTF algorithm. We start the derivation by setting $M+1 = n$ or $M = n - 1$ in Eq. (8.68), which yields

$$\mathbf{a}_{n-1}(n) = \mathbf{a}_{n-1}(n-1) - \eta^*_{n-1}(n) \begin{bmatrix} 0 \\ \mathbf{k}_{n-1}(n-1) \end{bmatrix} \qquad (8.140)$$

Figure 8.15(a) depicts the structure of the forward prediction-error filter characterized by the tap-weight vector $\mathbf{a}_{n-1}(n-1)$. We see that the tap input applied to the last element of $\mathbf{a}_{n-1}(n-1)$ is zero. The effect is equivalent to simplifying the structure of the filter as in Fig. 8.15(b). We may therefore express $\mathbf{a}_{n-1}(n-1)$ as

$$\mathbf{a}_{n-1}(n-1) = \begin{bmatrix} \mathbf{a}_{n-2}(n-1) \\ 0 \end{bmatrix} \qquad (8.141)$$

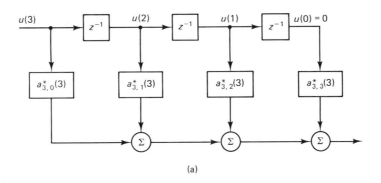

(a)

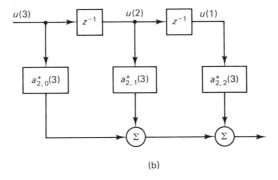

(b)

Figure 8.15 Reduction of $\mathbf{a}_{n-1}(n-1)$ to $\mathbf{a}_{n-2}(n-1)$ for $n = 4$; (a) transversal filter described by $\mathbf{a}_{n-1}(n-1)$; (b) transversal filter described by $\mathbf{a}_{n-2}(n-1)$.

To evaluate the gain vector $\mathbf{k}_{n-1}(n-1)$, we set $M = n-1$ in Eq. (8.137), obtaining

$$\mathbf{k}_{n-1}(n-1) = \begin{bmatrix} \mathbf{0} \\ 1 \\ \hline u^*(1) \end{bmatrix} \tag{8.142}$$

To evaluate the forward a priori prediction error $\eta_{n-1}(n)$, we set $M = n-1$ in Eq. (8.69) and then use Eq. (8.141), thereby obtaining

$$\eta_{n-1}(n) = \mathbf{a}_{n-1}^H(n-1)\mathbf{u}_n(n)$$

$$= [\mathbf{a}_{n-2}^H(n-1),\ 0] \begin{bmatrix} \mathbf{u}_{n-1}(n) \\ u(1) \end{bmatrix}$$

$$= \mathbf{a}_{n-2}^H(n-1)\ \mathbf{u}_{n-1}(n)$$

However, from the definition of the forward a priori prediction error, we have

$$\eta_{n-2}(n) = \mathbf{a}_{n-2}^H(n-1)\mathbf{u}_{n-1}(n)$$

We thus find that during the initialization period

$$\eta_{n-1}(n) = \eta_{n-2}(n) \tag{8.143}$$

Accordingly, the use of Eqs. (8.141), (8.142), and (8.143) in (8.140) yields

$$\mathbf{a}_{n-1}(n) = \begin{bmatrix} \mathbf{a}_{n-2}(n-1) \\ -\dfrac{\eta_{n-2}^*(n)}{u^*(1)} \end{bmatrix} \tag{8.144}$$

Given the forward a priori prediction error $\eta_{n-2}(n)$, we set $M = n-2$ in Eq. (8.102) to compute the corresponding value of the forward a posteriori prediction error:

$$f_{n-2}(n) = \gamma_{n-2}(n-1)\eta_{n-2}(n) \tag{8.145}$$

Next we set $M = n - 2$ in Eq. (8.70), obtaining

$$\mathcal{F}_{n-2}(n) = \lambda\mathcal{F}_{n-2}(n-1) + \eta_{n-2}(n)f_{n-2}^*(n) \tag{8.146}$$

The use of $M = n - 2$ in Eq. (8.106) yields

$$\gamma_{n-1}(n) = \lambda\,\frac{\mathcal{F}_{n-2}(n-1)}{\mathcal{F}_{n-2}(n)}\,\gamma_{n-2}(n-1) \tag{8.147}$$

During the initialization period, however, we may show that the following time-order update holds (see Problem 13)

$$\mathcal{F}_{n-1}(n) = \lambda\mathcal{F}_{n-2}(n - 1) \tag{8.148}$$

Accordingly, we may use this recursion to reformulate Eqs. (8.145) and (8.147) as follows,

$$\mathscr{F}_{n-2}(n) = \mathscr{F}_{n-1}(n) + \eta_{n-2}(n)f_{n-2}^*(n) \tag{8.149}$$

and

$$\gamma_{n-1}(n) = \frac{\mathscr{F}_{n-1}(n)}{\mathscr{F}_{n-2}(n)} \gamma_{n-2}(n-1) \tag{8.150}$$

Equations (8.144), (8.145) and (8.148)–(8.150) define the recursions for the first part of the exact initialization period that pertains to the forward prediction-error filter (i.e., filter I).

Next, using Eq. (8.117) with $M = n-2$, we obtain

$$\bar{\mathbf{k}}_{n-1}(n) = \begin{bmatrix} 0 \\ \bar{\mathbf{k}}_{n-2}(n-1) \end{bmatrix} + \frac{\eta_{n-2}(n)}{\lambda\mathscr{F}_{n-2}(n-1)} \mathbf{a}_{n-2}(n-1) \tag{8.151}$$

This recursion, constituting the second part of the initialization procedure, realizes the time-order update of the gain vector (i.e., filter III).

As explained earlier, the desired response $u(n-M)$ for the backward linear prediction is zero for $1 \le n \le M$. It assumes a nonzero value for the first time at $n = M+1$. Therefore, in the initialization of the FTF algorithm, the computations for the backward prediction-error filter need only be performed when $n = M+1$. In particular, by setting $M = n-1$ in Eq. (8.139), we obtain the following value for the tap-weight vector of the backward prediction-error filter:

$$
\begin{aligned}
\mathbf{c}_{n-1}(n) &= \begin{bmatrix} -u^*(1)\,\mathbf{k}_{n-1}(n) \\ 1 \end{bmatrix} \\
&= \begin{bmatrix} -u^*(1)\gamma_{n-1}(n)\,\bar{\mathbf{k}}_{n-1}(n) \\ 1 \end{bmatrix}
\end{aligned} \tag{8.152}
$$

To compute the corresponding minimum value of the sum of weighted backward a posteriori prediction-error squares, we set $M = n - 1$ in Eq. (8.87), and also use Eq. (8.103), obtaining

$$\mathscr{B}_{n-1}(n) = \lambda\mathscr{B}_{n-1}(n-1) + \gamma_{n-1}(n)\,|\psi_{n-1}(n)|^2 \tag{8.153}$$

However, $\mathscr{B}_{n-1}(n-1)$ is zero, and the backward a priori prediction error $\psi_{n-1}(n)$ equals [see Eq. (8.86)]

$$
\begin{aligned}
\psi_{n-1}(n) &= \mathbf{c}_{n-1}^H(n-1)\,\mathbf{u}_n(n) \\
&= [\mathbf{0}^T,\, 1] \begin{bmatrix} \mathbf{u}_{n-1}(n) \\ u(1) \end{bmatrix} \\
&= u(1)
\end{aligned}
$$

where, in the second line, we have made use of Eq. (8.138). Hence, we may simplify Eq. (8.153) as

$$\mathcal{B}_{n-1}(n) = \gamma_{n-1}(n) \, |u(1)|^2 \tag{8.154}$$

Equations (8.152) and (8.154) define the desired recursions for the third part of the initialization procedure that deals with the backward prediction-error filter (i.e., filter II).

Finally, the updating of the adaptive filter (i.e., filter IV) simplifies considerably during the initialization period $1 \leq n \leq M$. Putting $M = n$ in the first line of Eq. (8.133) yields

$$\hat{\mathbf{w}}_n(n) = \hat{\mathbf{w}}_n(n-1) + \mathbf{k}_n(n) \, \alpha_n^*(n) \tag{8.155}$$

Recognizing that (for reasons similar to those illustrated in Fig. 8.15)

$$\hat{\mathbf{w}}_n(n-1) = \begin{bmatrix} \hat{\mathbf{w}}_{n-1}(n-1) \\ 0 \end{bmatrix}$$

and using Eq. (8.142), we may rewrite Eq. (8.155) as

$$\hat{\mathbf{w}}_n(n) = \begin{bmatrix} \hat{\mathbf{w}}_{n-1}(n-1) \\ \dfrac{\alpha_n^*(n)}{u^*(1)} \end{bmatrix} \tag{8.156}$$

To compute the a priori estimation error $\alpha_n(n)$, we set $M = n$ in Eq. (8.131), obtaining

$$\alpha_n(n) = d(n) - \hat{\mathbf{w}}_n^H(n-1) \, \mathbf{u}_n(n)$$

$$= d(n) - [\hat{\mathbf{w}}_{n-1}^H(n-1), 0] \begin{bmatrix} \mathbf{u}_{n-1}(n) \\ u(1) \end{bmatrix}$$

$$= d(n) - \hat{\mathbf{w}}_{n-1}^H(n-1) \, \mathbf{u}_{n-1}(n)$$

By definition, the final form of this expression equals $\alpha_{n-1}(n)$, the a priori estimation error for order $n - 1$. Hence, we may write

$$\alpha_{n-1}(n) = d(n) - \hat{\mathbf{w}}_{n-1}^H(n-1) \mathbf{u}_{n-1}(n) \tag{8.157}$$

Having computed $\alpha_{n-1}(n)$, we may now determine the corresponding value of the a posteriori estimation error

$$e_{n-1}(n) = \gamma_{n-1}(n) \alpha_{n-1}(n) \tag{8.158}$$

Note that the computation of the tap-weight vector $\hat{\mathbf{w}}_n(n)$ during the initialization period is possible without the use of the forward and backward prediction-error filters and the gain vector. This is possible because during this period the minimum value of the sum of weighted error squares is zero. However, once the initialization period is over (i.e., $n > M$), the minimum sum of weighted error squares assumes a nonzero value, and then these three filters are necessary for the computation of $\hat{\mathbf{w}}(n)$. These three additional filters are updated during the initialization period, so they are correct and available for use for $n > M$. Thus, for $n = M+1$, the use of Eq. (8.133) yields

$$\hat{\mathbf{w}}_{n-1}(n) = \hat{\mathbf{w}}_{n-1}(n-1) + \tilde{\mathbf{k}}_{n-1}(n)\, e_{n-1}^*(n) \tag{8.159}$$

Equations (8.158), (8.156) and (8.159) define the recursions for exact initialization of the adaptive filter (i.e., filter IV).

This completes the exact initialization of the FTF algorithm.

Summary of the Exact Initialization Procedure for Zero Initial Condition

To start off the computation, we use the following set of values at $n = 1$:

$$\mathbf{a}_0(1) = 1$$

$$\mathbf{c}_0(1) = 1$$

$$\tilde{\mathbf{k}}_0(1) = 0 \quad \text{(zero dimension)}$$

$$\hat{\mathbf{w}}_1(1) = \frac{d^*(1)}{u^*(1)} \tag{8.160}$$

$$\gamma_0(1) = 1$$

$$\mathcal{F}_0(1) = |u(1)|^2$$

A summary of the initialization procedure is presented in Table 8.6, where we have collected together the various relations derived previously.

8.15 *EXACT INITIALIZATION OF THE FTF ALGORITHM FOR ARBITRARY INITIAL CONDITION*

Nonzero initial conditions in the form of an initial value $\mathbf{w}_o$ for the M-by-1 tap-weight vector of the adaptive transversal filter component in the FTF algorithm may arise from the prior use of another adaptive algorithm (e.g., the LMS algorithm), or from some additional side information that is made available in a particular situation. In any event, it is desirable to have a method that can exploit this initial condition.

Another reason for developing a method that accommodates the use of an arbitrary initial condition is to provide an effective means to change the tracking capability of the FTF algorithm. In this context, we have to realize that it is not permissible to change the exponential weighting factor λ during the operation of

TABLE 8.6 EXACT INITIALIZATION OF THE FTF ALGORITHM

$n = 1$

$\mathbf{a}_0(1) = \mathbf{c}_0(1) = 1$

$\tilde{\mathbf{k}}_0(1) = 0$ (zero dimension)

$\hat{\mathbf{w}}_1(1) = d^*(1)/u^*(1)$

$\gamma_0(1) = 1$

$\mathcal{F}_0(1) = |u(1)|^2, \qquad u(1) \neq 0$

$2 \leq n \leq M + 1$

$\eta_{n-2}(n) = \mathbf{a}_{n-2}^H(n-1)\,\mathbf{u}_{n-1}(n)$

$\mathbf{a}_{n-1}(n) = \begin{bmatrix} \mathbf{a}_{n-2}(n-1) \\ -\eta_{n-2}^*(n)/u^*(1) \end{bmatrix}$

$f_{n-2}(n) = \gamma_{n-2}(n-1)\eta_{n-2}(n)$

$\mathcal{F}_{n-1}(n) = \lambda\mathcal{F}_{n-2}(n-1)$

$\mathcal{F}_{n-2}(n) = \mathcal{F}_{n-1}(n) + \eta_{n-2}(n)f_{n-2}^*(n)$

$\gamma_{n-1}(n) = \dfrac{\mathcal{F}_{n-1}(n)}{\mathcal{F}_{n-2}(n)}\,\gamma_{n-2}(n-1)$

$\tilde{\mathbf{k}}_{n-1}(n) = \begin{bmatrix} 0 \\ \tilde{\mathbf{k}}_{n-2}(n-1) \end{bmatrix} + \dfrac{\eta_{n-2}(n)}{\lambda\mathcal{F}_{n-2}(n-1)}\,\mathbf{a}_{n-2}(n-1)$

$\mathbf{c}_{n-1}(n) = \begin{bmatrix} -u^*(1)\gamma_{n-1}(n)\,\tilde{\mathbf{k}}_{n-1}(n) \\ 1 \end{bmatrix}$,   compute only when $n = M + 1$

$\mathcal{B}_{n-1}(n) = \gamma_{n-1}(n)|u(1)|^2$,   compute only when $n = M + 1$

$\alpha_{n-1}(n) = d(n) - \hat{\mathbf{w}}_{n-1}^H(n-1)\,\mathbf{u}_{n-1}(n)$

$e_{n-1}(n) = \gamma_{n-1}(n)\alpha_{n-1}(n)$

If $n \leq M$, $\hat{\mathbf{w}}_n(n) = \begin{bmatrix} \hat{\mathbf{w}}_{n-1}(n-1) \\ \alpha_{n-1}^*(n)/u^*(1) \end{bmatrix}$

If $n = M + 1$, $\hat{\mathbf{w}}_{n-1}(n) = \hat{\mathbf{w}}_{n-1}(n-1) + \tilde{\mathbf{k}}_{n-1}(n)e_{n-1}^*(n)$

---

the FTF algorithm since any such variation prevents the exact solution to the RLS problem to be realized.

To accommodate an arbitrary initial condition, we append a *soft constraint* to Eq. (8.7) so that the *augmented* index of performance for the least-squares problem equals

$$\mathcal{E}'(n) = \mu\lambda^n \|\mathbf{w}_M(n) - \mathbf{w}_o\|^2 + \sum_{i=1}^{n} \lambda^{n-i} |e(i)|^2 \tag{8.161}$$

where the factor $\mu > 0$ to ensure a quadratic index of performance, and the M-by-1 vector $\mathbf{w}_o$ denotes the initial condition.   The symbol $\| \; \|$ signifies the norm of the vector enclosed within. For $n \geq 1$, we use the factor μ to weight the effect of the initial condition $\mathbf{w}_o$ on the augmented performance index. For $n \leq 0$, the second term in Eq. (8.161) vanishes, so the minimization of $\mathscr{E}'(n)$ with respect to $\mathbf{w}_M(n)$ yields an optimum value equal to $\mathbf{w}_o$, just as it should.

The application of a nonzero initial condition $\mathbf{w}_o$ is useful only if it arises from previous use of adaptive algorithms, such as the LMS algorithm or other least-squares algorithms, that have already processed a sufficiently large amount of data to average, or diminish, the effect of a relatively large measurement error. Furthermore, the initial condition $\mathbf{w}_o$ must be unbiased so as to maintain the unbiased estimator property of the RLS solution.

As for the factor μ, it allows a degree of flexibility in varying the capability of the FTF algorithm to track statistical variations in the input data (Cioffi and Kailath, 1984). By choosing the factor μ large, we make the FTF algorithm track as "slow" as desired, thereby attaining the least-squares solution with a finer resolution. By contrast, variation of the exponential weighting factor λ, by itself, is usually not as effective for two reasons. First, any variation of λ during the operation of the FTF algorith renders it incorrect; hence, the use of gear-shifting by varying λ is impractical. Second, the upper bound of unity on λ limits the slowest learning rate of the FTF algorithm. A more practical method of attaining slow tracking is to introduce a form of gear-shifting into the FTF algorithm by redefining time $n = 0$ as the time at which the gear-shifting is to occur and then weighting the present tap-weight vector $\mathbf{w}_M(n)$ by an appropriate choice of μ. This procedure is simple and yet robust; it does not increase complexity, nor does it destroy the "exactness" of the solution.

Given the initial condition $\mathbf{w}_o$, we may perform exact initialization of the FTF algorithm by augmenting the time series representing the filter input and that representing the desired response, as illustrated in Fig. 8.16. The augmentation is achieved by adding $(M + 1)$ terms at the beginning of each time series. Thus, the first M iterations of the exact initialization algorithm in Table 8.6 yield the following initial conditions at time $n = 0$:

$$\mathbf{a}_M(0) = \begin{bmatrix} 1 \\ \mathbf{0} \end{bmatrix}$$

$$\mathscr{F}_M(0) = \lambda^M \mu$$

$$\mathbf{c}_M(0) = \begin{bmatrix} \mathbf{0} \\ 1 \end{bmatrix}$$

$$\mathscr{B}_M(0) = \mu \qquad\qquad (8.162)$$

$$\mathbf{k}_M(0) = \mathbf{0}$$

$$\gamma_M(0) = 1$$

$$\hat{\mathbf{w}}_M(0) = \mathbf{w}_o$$

Note that time $n = 0$ coincides with the end of the set of added terms.

$$\overbrace{\hspace{3cm}}^{(M\ +\ 1)\ \text{terms}}$$

Tap inputs $\{u(i)\} \longrightarrow \{\mu^{\frac{1}{2}}, \quad 0, \ldots, 0, 0, u(1), u(2), \ldots, u(n)\}$

Desired response $\{d(i)\} \longrightarrow \{\mu^{\frac{1}{2}}w_{o1}, \mu^{\frac{1}{2}}w_{o2}, \ldots, \mu^{\frac{1}{2}}w_{oM}, 0, d(1), d(2), \ldots, d(n)\}$

$$\uparrow$$

Time, $n = 0$

Figure 8.16   Augmentation of the tap inputs and desired response to accommodate the use of arbitrary initial condition.

With the proper initial conditions established at time $n = 0$, application of the FTF algorithm in Table 8.5 for $n \geq 1$ may now proceed directly, without using Table 8.6 to initialize the algorithm. However, care must be exercised in not assigning too small a value to the factor μ; otherwise, the algorithm can become unstable when $n = M + 1$.

8.16 COMPUTER EXPERIMENT ON SYSTEM IDENTIFICATION

To illustrate the performance of the FTF algorithm, we use it to study the *system identification* problem depicted in Fig. 8.17. The multiple linear regression model shown in the figure involves an *unknown* linear transversal filter with M tap weights.

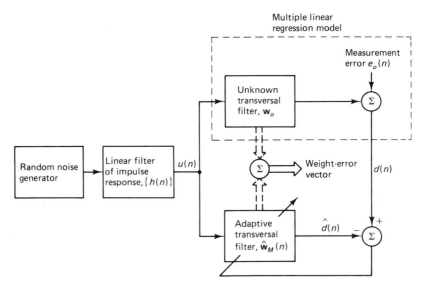

Figure 8.17   System identification.

The requirement is to estimate the tap weights of this filter in the presence of the measurement error $e_o(n)$. The aim of the experiment is to evaluate the effects of varying the number of unknown parameters, M, and the eigenvalue spread of the input $\{u(n)\}$ on the performance of the algorithm.

To vary the eigenvalue spread, the test signal $\{u(n)\}$ is generated by passing a zero-mean white-noise process, independent of the measurement error $\{e_o(n)\}$, through a linear filter whose impulse response equals

$$h_n = \begin{cases} \dfrac{1}{2}\left[1 + \cos\dfrac{2\pi}{W}(n - 2)\right], & n = 1, 2, 3 \\ 0, & \text{otherwise} \end{cases}$$

The variance of the white-noise source is adjusted so as to make the variance of the unknown system output equal unity. For a specified number of unknown parameters, M, the eigenvalue spread is varied by varying W. Table 8.7 shows the eigenvalue spreads used in the experiment.

The measurement error $\{e_o(n)\}$ is modeled as a white Gaussian noise process of zero mean and variance equal to 10^{-4}. This represents a signal-to-noise ratio of 40 dB, measured with respect to the filter input $\{u(n)\}$.

The unknown impulse response of the transversal filter in the multiple linear regression model is chosen to follow a triangular wave form that is symmetric with respect to the center tap point. The exponential weighting factor λ is assumed to equal unity.

Results and Observations

Figure 8.18(a) shows two plots of the estimated impulse response of the adaptive transversal filter (with $M = 31$) after convergence to steady-state conditions has been established. One plot corresponds to parameter $W = 3.1$, and the other two $W = 3.7$. Each of these two plots also includes the theoretical impulse response. Figure 8.18(b) shows the results obtained for $M = 61$ and $W = 3.1$, and 3.7. Figure 8.18(c) shows the results obtained for $M = 99$ and $W = 3.1$, and 3.7. The results shown in these figures were obtained by using the conventional (i.e., unconstrained) form of the FTF algorithm, with the computations performed on a high-precision machine.

TABLE 8.7

Number of unknown parameters, N	Eigenvalue spread	
	$W = 3.1$	$W = 3.7$
31	14.5	212.1
61	21.7	295.5
99	26.35	333.5

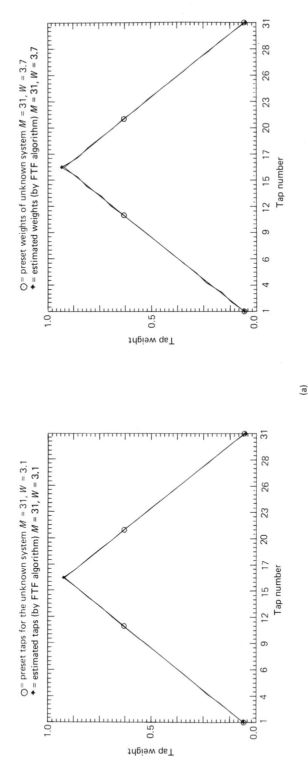

Figure 8.18 System identification experiment. (a) Results of FTF for number of taps $M =$ 31, SNR = 40 dB, $W =$ 3.1, 3.7; (b) comparison of unknown and FTF filter taps for $M =$ 61, SNR = 40 dB, $W =$ 3.1, 3.7; (c) comparison of unknown and FTF filter taps for $M =$ 99, SNR = 40 dB, $N =$ 3.1, 3.7.

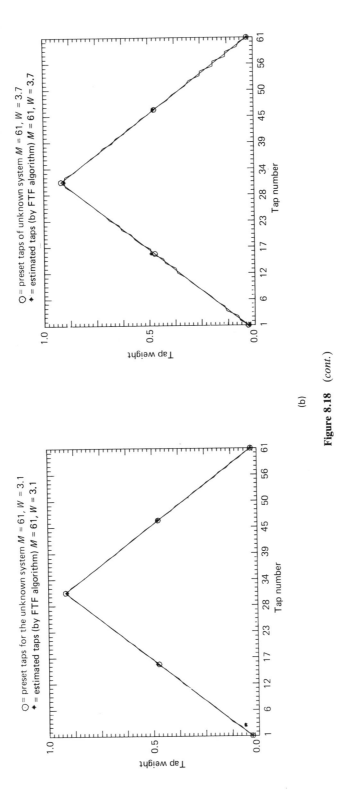

Figure 8.18 (*cont.*)

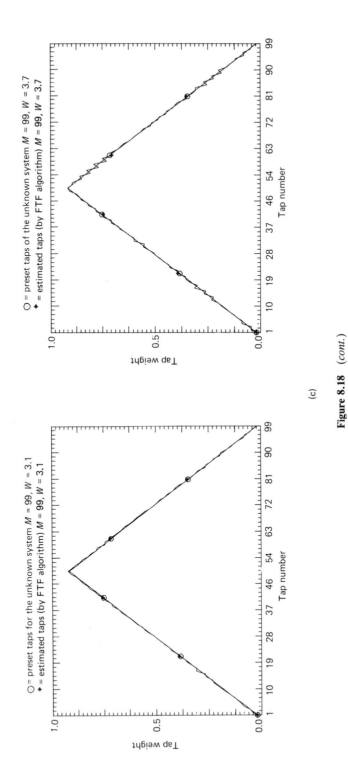

Figure 8.18 (*cont.*)

These results demonstrate the robustness of the FTF algorithm. In particular, the estimate produced by the algorithm for the impulse response of the transversal filter is practically indistinguishable from the theoretical value, even under difficult operating conditions (i.e., when the eigenvalue spread is high and the number of tap weights in the filter is large).

8.17 RESCUE DEVICES

When the FTF algorithm is implemented on a digital computer or special-purpose digital processor with a limited number of bits used to represent the various parameters in the algorithm, *finite-precision errors* result that may cause the algorithm to diverge.[7] Experimentation with the FTF algorithm indicates that a certain, positive variable in the algorithm becomes negative due to the accumulated effect of finite-precision errors just before divergence of the algorithm occurs (Lin, 1984; Cioffi and Kailath, 1984). This variable equals the ratio of two nonnegative quantities, as shown by

$$\zeta_M(n) = \frac{\gamma_{M+1}(n)}{\gamma_M(n)}$$

We refer to $\zeta_M(n)$ as the *rescue variable*. Note that the rescue variable is real valued. Using Eq. (8.107), we may also write

$$\zeta_M(n) = \frac{\gamma \mathcal{B}_M(n-1)}{\mathcal{B}_M(n)} \qquad (8.163)$$

from which we deduce that $0 \leq \zeta_M(n) \leq 1$. We may express the variable $\zeta_M(n)$ in yet another form by using Eq. (8.125) to write

$$\zeta_M(n) = 1 - \psi_M^*(n)\gamma_{M+1}(n)\bar{k}_{M+1,M+1}(n) \qquad (8.164)$$

Suppose that at time n the rescue variable $\zeta_M(n)$ becomes negative (due to precision errors) for the *first time*. In such a situation, we need only save the tap-weight vector $\hat{\mathbf{w}}_M(n)$ pertaining to the adaptive transversal filter component of the

---

[7]Numerical instability due to the use of finite-precision arithmetic poses a potential problem not only for the FTF algorithm but also other *fast* realizations of the RLS algorithm, particularly when the exponential weighting factor λ is chosen less than unity for the purpose of fast adaptation (Mueller, 1981a; Lin, 1984). Indeed, numerical difficulties may also be experienced with the conventional RLS algorithm (Goodwin and Payne, 1977). A numerically favorable version of the RLS algorithm is obtained by using the UD-factorization that is an example of a a square root-type algorithm (Bierman, 1977). According to this factorization, the matrix $\mathbf{P}(n)$ in the RLS algorithm is expressed as

$$\mathbf{P}(n) = \mathbf{U}(n)\mathbf{D}(n)\mathbf{U}^H(n)$$

where $\mathbf{U}(n)$ is an upper triangular matrix with "1"s along the diagonal and $\mathbf{D}(n)$ is a diagonal matrix. Then the factors $\mathbf{U}(n)$ and $\mathbf{D}(n)$ are updated instead of $\mathbf{P}(n)$. A brief discussion of this issue was presented in the context of the Kalman algorithm in Section 6.10.

FTF algorithm. Then, to continue with the algorithm, we may apply a *soft-constrained rescue operation* that consists of using the initialization procedure described in Section 8.15 with the value $\hat{\mathbf{w}}_M(n)$ adopted as the new initial condition. In so doing, we introduce the additional improvement of choosing an appropriate μ to weight the effect of this initial condition. That is, the noise averaging that occurred in the computation of $\hat{\mathbf{w}}_M(n)$ will not have been lost.

Periodic Restart for the FTF Algorithm

Eleftheriou and Falconer describe a *periodic reinitialization procedure* for overcoming the numerical instability problem encountered in the FTF algorithm due to roundoff errors (Eleftheriou and Falconer, 1984). In particular, to avoid the build-up of roundoff errors with time, it is proposed that operation of the FTF algorithm be interrupted and then restarted at periodic intervals, say every N iterations. Immediately following such a restart, a simple LMS algorithm (which has been initialized with the tap-weight vector attained by the FTF algorithm just before the restart was initiated) provides an estimate of the desired response temporarily until the restarted FTF algorithm takes over again after a short time. The *transition* from the FTF algorithm to the LMS algorithm for obtaining the desired estimate is so short in duration (typically, 1 to 1.5 times the memory of the transversal filter) and so smooth that temporary reliance on the slower LMS algorithm causes little or no performance degradation. Figure 8.19 illustrates the timing of operations involved in the periodic reinitialization procedure.

Following a *restart* of the FTF algorithm at time n_0, say, the algorithm minimizes a constrained index of performance for $n \geq n_0$, as described in Section 8.15. Specifically, it minimizes the function [see Eq. (8.161)]

$$
\begin{aligned}
\varepsilon(n) = \ &\mu\lambda^{n-n_0} \|\mathbf{w}_M(n) - \hat{\mathbf{w}}_M(n_0)\|^2 \\
&+ \sum_{i=n_0}^{n} \lambda^{n-i} |d(i) - \hat{\mathbf{w}}_M^H(n)\mathbf{u}_M(i)|^2, \quad n \geq n_0
\end{aligned}
\tag{8.165}
$$

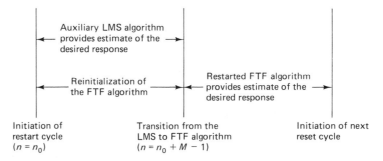

Figure 8.19 Timing of the operations involved in one cycle of the periodic reinitialization procedure of the FTF algorithm.

where $\hat{\mathbf{w}}_M(n_0)$ is the least-squares value of the tap-weight vector updated just prior to restart. The positive constant μ is included to weight the contribution made by this initial condition to the index of performance. All other internal variables of the FTF algorithm are initialized in exactly the same way as during the original start-up procedure at time $n = 0$ (see Section 8.14). For adaptation purposes, the restarted FTF algorithm sees a tap-input vector $\tilde{\mathbf{u}}_M(n)$ whose M elements equal

$$\tilde{u}(n - k) = \begin{cases} u(n - k), & n > n_0 + k \\ 0, & n \le n_0 + k \end{cases} \tag{8.166}$$

$$k = 0, 1, \ldots, M - 1$$

The major feature that distinguishes the periodic reinitialization procedure proposed by Eleftheriou and Falconer from others is the way in which the desired response $d(n)$ is modified so as to mitigate the effect of the truncation of the tap-input vector $\tilde{\mathbf{u}}(n)$ on the tracking ability of the algorithm. The modification involves removing from the desired response those elements of the tap-input vector $\tilde{\mathbf{u}}(n)$ that are not available to the restarted FTF algorithm.

Let $\mathbf{w}'_M(n)$ denote the value of the tap-weight vector that would have resulted if there had been no restart just prior to time $n = n_0$. We may then express the desired response as

$$d(n) = \left(\mathbf{w}'_M(n)\right)^H \mathbf{u}_M(n) + v(n) \tag{8.167}$$

where $v(n)$ is error due to noise and random data. Define the vector

$$\mathbf{u}_M^{(-)}(n) = \mathbf{u}_M(n) - \tilde{\mathbf{u}}_M(n) \tag{8.168}$$

That is $\mathbf{u}_M^{(-)}$ contains only those elements of $\tilde{\mathbf{u}}_M(n)$ that will have occurred prior to $\mathbf{n} = n_0$, as shown by

$$u^{(-)}(n - k) = \begin{cases} 0, & n > n_0 + k \\ u(n - k), & n \le n_0 + k \end{cases} \tag{8.169}$$

$$k = 0, 1, \ldots, M - 1.$$

Then eliminating $\mathbf{u}_M(n)$ between Eqs. (8.167) and (8.168) and rearranging, we get

$$d(n) - \left(\mathbf{w}'_M(n)\right)^H \mathbf{u}_M^{(-)}(n) = \left(\mathbf{w}'_M(n)\right)^H \tilde{\mathbf{u}}_M(n) + v(n) \tag{8.170}$$

We now make two observations:

1. The only tap-input vector available to the restarted FTF algorithm equals $\tilde{\mathbf{u}}_M(n)$.
2. Ideally, the adapted tap-weight vector $\hat{\mathbf{w}}_M(n)$ should approach the value $\mathbf{w}'_M(n)$.

Accordingly, Eq. (8.170) suggests that during the restart period, we put the desired response for the FTF algorithm equal to

$$\tilde{d}(n) = d(n) - \mathbf{w}_{\text{LMS}}^H(n - 1)\mathbf{u}_M^{(-)}(n), \quad n_0 \le n \le n_0 + M - 1 \quad (8.171)$$

where $\mathbf{w}_{\text{LMS}}(n - 1)$ is the current value of the tap-weight vector as obtained from the auxiliary LMS algorithm, and the M elements of the vector $\mathbf{u}_M^{(-)}(n)$ are as defined in Eq. (8.169). Note that when time n exceeds n_0 by a number of sample periods equal to the memory of the transversal filter (i.e., $n > n_0 + M - 1$), the vector $\mathbf{u}_M^{(-)}(n)$ is zero and $\tilde{d}(n) = d(n)$.

The procedure described for the periodic reinitialization of the FTF algorithm appears to provide a remedy for the numerical instability problem in the FTF algorithm, thereby making it viable for use in adaptive equalization or system identification.[8]

PROBLEMS

1. To permit a recursive implementation of the method of least squares, the window or weighting function $\beta(n, i)$ must have a suitable structure. Assume that $\beta(n, i)$ may be expressed as

$$\beta(n, i) = \lambda(i)\beta(n, i - 1), \quad i = 1, \ldots, n$$

where $\beta(n, n) = 1$. Hence, show that

$$\beta(n, i) = \prod_{k=i+1}^{n} \lambda^{-1}(k)$$

What is the form of $\lambda(k)$ for which $\beta(n, i) = \lambda^{n-i}$ is obtained?

2. Establish the validity of the matrix inversion lemma.

[8]This procedure assumes no *reinitialization* of the tap-weight vector $\hat{\mathbf{w}}_M(n)$ of the restarted FTF algorithm at the initiation of each restart cycle, and also assumes use of Eq. (8.170) for the desired response during the restart period. Eleftheriou and Falconer have investigated three other versions of their periodic reinitialization procedure (Eleftheriou and Falconer, 1984):

1. The elements of the tap-weight vector $\hat{\mathbf{w}}_M(n)$ of the restarted FTF algorithm are *set equal to zero* at the initialization of the restart [i.e., $\hat{\mathbf{w}}_M(n_0) = \mathbf{0}$]. Equation (8.170) is used to provide the desired response during the restart period.
2. As in the procedure described in the text, $\hat{\mathbf{w}}_M(n)$ is not reinitialized, but the desired response is redefined as

$$\tilde{d}(n) = \begin{cases} \mathbf{w}_{\text{LMS}}^H(n - 1)\tilde{\mathbf{u}}(n), & n_0 \le n \le n_0 + M - 1 \\ d(n), & n > n_0 + M - 1 \end{cases}$$

where the M elements of vector $\tilde{\mathbf{u}}_M(n)$ are themselves defined by Eq. (8.166). This modification of the desired response is also suggested by Eq. (8.170).
3. The elements of $\hat{\mathbf{w}}_M(n)$ are set equal to zero, as in (1), but the desired response is defined as in (2).

Eleftheriou and Falconer present computer simulation results that show the periodic reinitialization procedure described above having a slightly better performance than the procedure under (1), which in turn is somewhat better than the procedures under (2) and (3).

3. Consider the modified definition of the correlation matrix $\boldsymbol{\Phi}(n)$ given in Eq. (8.28), which is reproduced here for convenience.

$$\boldsymbol{\Phi}(n) = \sum_{i=1}^{n} \lambda^{n-i}\, \mathbf{u}(i)\mathbf{u}^H(i) + \delta\lambda^n \mathbf{I}$$

where $\mathbf{u}(i)$ is the tap-input vector, λ is the exponential weighting factor, and δ is a small positive constant. Show that the use of this new definition for $\boldsymbol{\Phi}(n)$ leaves the equations that define the RLS algorithm completely unchanged.

4. Let $\alpha(n)$ denote the a priori estimation error

$$\alpha(n) = d(n) - \hat{\mathbf{w}}^H(n-1)\mathbf{u}(n)$$

where $d(n)$ is the desired response, $\mathbf{u}(n)$ is the tap-input vector, and $\hat{\mathbf{w}}(n - 1)$ is the old estimate of the tap-weight vector. Let $e(n)$ denote the a posteriori estimation error

$$e(n) = d(n) - \hat{\mathbf{w}}^H(n)\mathbf{u}(n)$$

where $\hat{\mathbf{w}}(n)$ is the current estimate of the tap-weight vector. For complex-valued data, both $\alpha(n)$ and $e(n)$ are likewise complex valued. Show that the product $\alpha(n)e^*(n)$ is always real valued.

5. Given the initial conditions of Section 8.4 for the RLS algorithm, explain the reasons for the fact that the average mean-square value of the a posteriori estimation error $e(n)$ attains a small value at $n = 1$ and then rises with increasing n.

6. Consider the RLS algorithm with the exponential weighting factor $\lambda \leq 1$. Show that the tap-weight vector $\hat{\mathbf{w}}(n)$ produced by the RLS algorithm has a bias that is approximately defined by

$$\mathbf{b}(n) = E[\hat{\mathbf{w}}(n)] - \mathbf{w}_o$$

$$= -\delta\lambda^n \boldsymbol{\Phi}^{-1}(n)\mathbf{w}_o, \qquad n \geq 1$$

where $\boldsymbol{\Phi}^{-1}(n)$ is the inverse of the deterministic correlation matrix $\boldsymbol{\Phi}(n)$,

$$\delta\mathbf{I} = \boldsymbol{\Phi}(0)$$

$$\mathbf{w}_o = \text{coefficient vector of the multiple linear regression model}$$

7. Consider a multiple linear regression model whose output is described by

$$d(i) = \mathbf{w}_o^H\mathbf{u}(i) + e_o(i)$$

where $\mathbf{w}_o$ is the M-by-1 regression vector of the model, and $\mathbf{u}(i)$ is the corresponding input vector. The measurement error process $\{e_o(i)\}$ is white with zero mean and variance σ^2. Let $\hat{\mathbf{w}}(n)$ denote the estimate of $\mathbf{w}_o$ that is produced by application of the RLS algorithm.

  (a) Assuming that the number of iterations n is large enough for the effect of initial conditions to be ignored, show that the weight-error vector equals

$$\boldsymbol{\varepsilon}(n) = \hat{\mathbf{w}}(n) - \mathbf{w}_o$$

$$= \boldsymbol{\Phi}^{-1}(n)\sum_{i=1}^{n}\lambda^{n-i}\mathbf{u}(i)e_o^*(i), \qquad n \geq M$$

where

$$\Phi(n) = \sum_{i=1}^{n} \lambda^{n-i} \mathbf{u}(i) \mathbf{u}^H(i)$$

(b) Using the result obtained in part (a), show that the weight-error correlation matrix equals

$$\mathbf{K}(n) = E[\mathbf{\varepsilon}(n)\mathbf{\varepsilon}^H(n)]$$

$$= \sigma^2 \mathbf{\Phi}_{(n)}^{-1} \left(\sum_{i=1}^{n} \lambda^{2(n-i)} \mathbf{u}(i) \mathbf{u}^H(i) \right) \mathbf{\Phi}^{-1}(n), \qquad n \geq M$$

For the special case of $\gamma = 1$, what does this formula reduce to?

8. From the *law of large numbers*, we have

$$\sum_{i=1}^{n} \lambda^{n-i} \mathbf{u}(i) \mathbf{u}^H(i) \simeq \left(\sum_{i=1}^{n} \lambda^{n-i} \right) (\mathbf{R} + \delta\mathbf{R}), \qquad n \text{ large}$$

where λ is the exponential weighting factor, $\mathbf{R}$ is the ensemble-averaged correlation matrix:

$$\mathbf{R} = E[\mathbf{u}(i)\mathbf{u}^H(i)]$$

and $\delta\mathbf{R}$ is composed of elements that have zero mean and variances that are small compared to the elements of $\mathbf{R}$. You may thus assume that

$$(\mathbf{R} + \delta\mathbf{R})^{-1} \simeq \mathbf{R}^{-1} - \mathbf{R}^{-1}\delta\mathbf{R}\mathbf{R}^{-1}$$

Hence, using these results in the formula for the excess mean-squared error

$$J_{\text{ex}}(n) = \text{tr}\{E[\mathbf{R}\mathbf{K}(n)]\}, \qquad n \text{ large}$$

show that the misadjustment produced by the exponentially weighted RLS algorithm equals

$$\mathcal{M} = \frac{E[J_{\text{ex}}(\infty)]}{\sigma^2}$$

$$= \left(\frac{1 - \lambda}{1 + \lambda} \right) M$$

where M is the number of taps in the transversal filter.

9. Table 8.8 is a summary of the fast RLS algorithm. The development of the algorithm follows the theory presented in Section 8.12 up to (and including) the step that involves computing the extended gain vector $\mathbf{k}_{M+1}(n)$. By partitioning this vector as

$$\mathbf{k}_{M+1}(n) = \left[\begin{array}{c} \mathbf{t}(n) \\ \hline \tau(n) \end{array} \right]$$

where $\mathbf{t}(n)$ is an M-by-1 vector and $\tau(n)$ is a scalar, derive the remaining three steps of the algorithm.

TABLE 8.8 FAST RLS ALGORITHM

$$\eta_M(n) = \mathbf{a}_M^H(n-1)\,\mathbf{u}_{M+1}(n)$$

$$\mathbf{a}_M(n) = \mathbf{a}_M(n-1) - \begin{bmatrix} 0 \\ \mathbf{k}_M(n-1) \end{bmatrix}\eta_M^*(n)$$

$$f_M(n) = \mathbf{a}_M^H(n)\,\mathbf{u}_{M+1}(n)$$

$$\mathscr{F}_M(n) = \lambda\mathscr{F}_M(n-1) + \eta_M(n)\,f_M^*(n)$$

$$\mathbf{k}_{M+1}(n) = \begin{bmatrix} 0 \\ \mathbf{k}_M(n-1) \end{bmatrix} + \frac{f_M(n)}{\mathscr{F}_M(n)}\,\mathbf{a}_M(n)$$

$$\mathbf{k}_{M+1}(n) = \begin{bmatrix} \mathbf{t}(n) \\ \text{-----} \\ \tau(n) \end{bmatrix}$$

$$\psi_M(n) = u(n-M) - \mathbf{g}^H(n-1)\mathbf{u}_M(n)$$

$$\mathbf{g}(n) = [1 - \tau(n)\psi_M^*(n)]^{-1}[\mathbf{g}(n-1) + \mathbf{t}(n)\,\psi_M^*(n)]$$

$$\mathbf{k}_M(n) = \mathbf{t}(n) + \mathbf{g}(n)\tau(n)$$

---

10. (a) Show that the inverse of the deterministic correlation matrix $\boldsymbol{\Phi}_{M+1}(n)$ may be expressed as follows:

$$\boldsymbol{\Phi}_{M+1}^{-1}(n) = \begin{bmatrix} 0 & \mathbf{0}_M^T \\ \mathbf{0}_M & \boldsymbol{\Phi}_M^{-1}(n-1) \end{bmatrix} + \frac{1}{\mathscr{F}_M(n)}\,\mathbf{a}_M(n)\,\mathbf{a}_M^H(n)$$

where $\mathbf{0}_M$ is the M-by-1 null vector, $\mathbf{0}_M^T$ its transpose, $\mathscr{F}_M(n)$ is the minimum sum of weighted forward prediction-error squares, and $\mathbf{a}_M(n)$ is the tap-weight vector of forward prediction-error filter. Both $\mathbf{a}_M(n)$ and $\mathscr{F}_M(n)$ refer to prediction order M.

(b) Show that the inverse of $\boldsymbol{\Phi}_{M+1}(n)$ may also be expressed in the form

$$\boldsymbol{\Phi}_{M+1}^{-1}(n) = \begin{bmatrix} \boldsymbol{\Phi}_M^{-1}(n) & \mathbf{0}_M \\ \mathbf{0}_M^T & 0 \end{bmatrix} + \frac{1}{\mathscr{B}_M(n)}\,\mathbf{c}_M(n)\,\mathbf{c}_M^H(n)$$

where $\mathscr{B}_M(n)$ is the minimum sum of weighted backward a posteriori prediction-error squares, and $\mathbf{c}_M(n)$ is the tap-weight vector of the backward prediction-error filter. Both $\mathscr{F}_M(n)$ and $\mathbf{c}_M(n)$ refer to prediction order M.

Hints: For part (a), use the partitioned form of $\boldsymbol{\Phi}_{M+1}(n)$ given in Eq. (8.61). For part (b), use the second partitioned form of this matrix given in Eq. (8.78).

11. Show that the parameter $\gamma_M(n)$ defined by

$$\gamma_M(n) = 1 - \mathbf{k}_M^H(n)\,\mathbf{u}_M(n)$$

equals the sum of weighted error squares resulting from use of the transversal filter in Fig. 8.13. The tap-weight vector of this filter equals the gain vector $\mathbf{k}_M(n)$, and the tap-input vector equals $\mathbf{u}_M(n)$. The filter is designed to produce the least-squares estimate of a desired response that equals the *first unit vector*.

12. Let $\boldsymbol{\Phi}_M(n)$ denote the deterministic correlation matrix of the tap-input vector $\mathbf{u}_M(n)$ at time n; likewise for $\boldsymbol{\Phi}_M(n-1)$. Show that the conversion factor $\gamma_M(n)$ is related to the determinants of these two matrices as follows

$$\gamma_M(n) = \lambda \frac{\det[\Phi_M(n-1)]}{\det[\Phi_M(n)]}$$

where λ is the exponential weighting factor

Hint: Use the identity

$$\det\ (\mathbf{I}_1 + \mathbf{AB}) = \det\ (\mathbf{I}_2 + \mathbf{BA})$$

where $\mathbf{I}_1$ and $\mathbf{I}_2$ are identity matrices of appropriate dimensions, and $\mathbf{A}$ and $\mathbf{B}$ are matrices of compatible dimensions.

13. During the initialization of the FTF algorithm, show that

$$\mathcal{F}_{n-1}(n) = \lambda \mathcal{F}_{n-2}(n-1)$$

where $\mathcal{F}_{n-1}(n)$ is the sum of weighted forward a posteriori prediction-error squares, and λ is the exponential weighting factor.

Computer-oriented Problems

14. *Computer experiment on adaptive two-tap predictor using the RLS algorithm, with autoregressive input*: An autoregressive (AR) process $\{u(n)\}$ of order 2 is described by the difference equation

$$u(n) + a_1 u(n-1) + a_2 u(n-2) = v(n)$$

where $\{v(n)\}$ is a white-noise process of zero mean and variance σ_v^2. The AR parameters a_1 and a_2 and the variance σ_v^2 are assigned one of the following three sets of values:

(i) $a_1 = 0.100,$ $a_2 = -0.8,$ $\sigma_v^2 = 0.2700$

(ii) $a_1 = 0.1636,$ $a_2 = -0.8,$ $\sigma_v^2 = 0.1190$

(iii) $a_1 = 0.1960,$ $a_2 = -0.8,$ $\sigma_v^2 = 0.0140$

The process $\{u(n)\}$ is applied to an adaptive two-tap predictor that uses the RLS algorithm to adjust the tap weights of the predictor.

(a) Use a computer to generate a 256-sample sequence to represent the AR process $\{u(n)\}$ for parameter set (i). You may do this by using a random generator subroutine for $\{v(n)\}$. Hence, plot an ensemble-averaged learning curve for the predictor by averaging over 200 independent trials of the experiment. By averaging over the least 200 iterations of the ensemble-averaged learning curve, estimate the minimum mean-squared error produced by the RLS algorithm. Compare your result with theory.

(b) For parameter set (i), also compute the mean values of the tap weights $\hat{w}_1(\infty)$ and $\hat{w}_2(\infty)$ by ensemble-averaging the steady-state values of the tap weights produced by the RLS algorithm over 200 independent trials of the experiment. Compare your results with theory.

(c) Repeat computations (a) and (b) for parameter set (ii), and again for parameter set (iii).

(d) Comment on the results of parts (a) to (c).

(e) Suppose that the AR parameter a_1 is made negative; otherwise its magnitude is left as specified under parameter sets (i), (ii) and (iii). How would you assess the effect of this change on the performance of the RLS algorithm? Justify your answer.

15. *Computer Experiment on Adaptive Beamforming*: Repeat the computer experiment of Problem 16, Chapter 5, on the adaptive beamformer, using the RLS algorithm. The environment (in which the beamformer operates) consists of a target signal and a single source of interference. Their angular locations are as follows:

$$\text{target: } \phi = 0.2\,\pi \text{ radians}$$

$$\text{interference: } \phi = -0.4\,\pi \text{ radians}$$

where ϕ is related to the actual angle of arrival θ as in Eq. (1.11). The source of interference is noncoherent with the target. The signal-to-noise ratio (SNR) is fixed at 10 dB.

Initialize the RLS algorithm with $\delta = 0.004$. Hence, perform the following computations:

(a) For interference-to-noise ratio INR = 40, 30, 20 dB, compute the eigenvalue spread of the correlation matrix of the signals at the output ports of the beamforming network.

(b) Compute the adapted pattern of the beamformer after 10, 25, and 50 iterations of the algorithm, with the interference-to-noise ratio INR = 40 dB. Repeat your computations for INR = 30, 20 dB. Comment on your results.

(c) By ensemble-averaging over 200 independent trials of the experiment, plot the learning curves of the RLS algorithm for INR = 40, 30, 20 dB. In each case, plot the learning curve up to 225 iterations of the algorithm.

(d) By time-averaging each learning curve over the last 25 iterations, compute the steady-state value of the average mean-squared error for INR = 40, 30, 20 dB.

(e) Using the minimum mean-squared error from theory, and the results of (d) above, compute the misadjustment produced by the RLS algorithm for INR = 40, 30, 20 dB.

(f) Repeat parts (c), (d) and (e) of the problem by using the LMS algorithm for which the step-size parameter $\mu = 4 \times 10^{-7}$. For the case when INR = 40 dB, also plot the learning curve of the LMS algorithm with $\mu = 4 \times 10^{-8}$, up to 500 iterations of the algorithm; as before, perform the ensemble averaging over 200 independent repetitions of the experiment, and calculate the average mean-squared error by time-averaging the learning curve over the last 25 iterations

(g) Compare the results obtained for the RLS algorithm with those obtained for the LMS algorithm.

16. *Computer experiment on the adaptive autoregressive modeling of a nonstationary environment.* In Problem 10, Chapter 6, we considered a real-valued nonstationary environment described by the model shown in Fig. P6.1. For the adaptation, we used the Kalman algorithm assuming a random-walk state model. Repeat this experiment by using the RLS algorithm with the exponential weighting factor $\lambda = 0.85$. Compare your results with those obtained in that problem.

17. *Computer experiment on system identification using the FTF algorithm (continued)*: In the results presented in Section 8.16 on the use of the FTF algorithm for system identification, the conventional (i.e., unconstrained) form of the algorithm was used. With this form of the algorithm, when fast initialization is used, the norm of the tap-weight

error vector, $\|\hat{\mathbf{w}}_M(n) - \mathbf{w}_o\|$, may grow with time during the initialization period. This growth is eliminated by using the soft-constrained version of the FTF algorithm.

Repeat the computer experiment of Section 8.16 for the number of taps $M = 99$ and with the parameter W (that controls the eigenvalue spread) having the following two values: (1) $W = 3.1$, eigenvalue spread $= 26.4$, and (2) $W = 3.7$, eigenvalue spread $= 400$. Use the FTF algorithm with soft constraint, for which the parameters λ and μ are assigned the values

$$\lambda = 1$$

$$\mu = 0.01$$

Hence, carry out the following computations:

(a) For each value of W, evaluate the performance of the algorithm by computing the norm of the tap-weight error vector, $\|\hat{\mathbf{w}}_M(n) - \mathbf{w}_o\|$, divided by $\sqrt{M}$. Plot your results versus time n for about 1000 iterations of the algorithm. In your plot, include the performance of the algorithm during the initialization period.

(b) For each value of W, plot the impulse response of the adaptive transversal filter versus time n, obtained at the end of the experiment.

In this plot include the theoretical value of the impulse response. Compare the results in parts (a) and (b) with those obtained by using the unconstrained form of the FTF algorithm.

9

Recursive Least-squares Lattice Filters

9.1 INTRODUCTION

The *fast transversal filters (FTF) algorithm* (Cioffi and Kailath, 1984) that we derived in Chapter 8 offers a highly efficient and robust method for obtaining the exact *recursive least-squares (RLS) solution* to the linear filtering problem. The FTF algorithm uses a maximum of four transversal filters with a common input, one of which defines the impulse response of the adaptive filter. In this chapter we describe another class of *exact* least-squares algorithms based on a different structure, the multistage *lattice predictor*, which is *modular* in form. They are known collectively as *least-squares lattice (LSL) algorithms*, involving both *order-update* and *time-update recursions*. These algorithms are as efficient as the FTF algorithm in that they both essentially realize the same rate of convergence at the expense of a computational cost that increases linearly with the number of adjustable tap weights. The class of LSL algorithms is also just as robust as the RLS and FTF algorithms in that they are all essentially insensitive to variations in the eigenvalue spread of the correlation matrix of the input signal.

The lattice predictors considered in this chapter may operate in *nonstationary* environments, though with *time-variant* reflection coefficients. As such, they are basically different from the lattice predictors (with *fixed* reflection coefficients) considered in Chapter 4 that were based on the assumption of an underlying *stationary* environment.

Because of the basic structural differences between the LSL and FTF algorithms, these two algorithms present the relevant information in different ways. The FTF algorithm presents information about the input data in the form of instantaneous values of transversal filter coefficients. By contrast, the LSL algorithm presents the information in the form of a corresponding set of reflection coefficients. Such a difference may influence the choice of one algorithm over the other, depending on the application of interest.

The history of LSL algorithms can be traced back to the pioneering work of Morf (1974) on efficient solutions for least-squares predictors for the case when the correlation matrix of the predictor input is non-Toeplitz but has the special structure described in Section 7.8. The formal derivation of LSL algorithms was presented by Morf, Vieira, and Lee (1977), and Morf and Lee (1978). This important work led to a host of refinements, generalizations, and applications (Lee, Morf, and Friedlander, 1981; Kailath, 1982; Satorius and Pack, 1981; Porat, Friedlander, and Morf, 1982; Porat and Kailath, 1983; Lev-Ari, Kailath, and Cioffi, 1984; Ling, Manolakis, and Proakis, 1985).

The class of recursive LSL algorithms is by no means the only one available for the design of adaptive lattice predictors. Indeed, various adaptive lattice algorithms have been derived in the literature for different signal-processing applications (Friedlander, 1982). A particular class of algorithms that deserves special mention is that of *gradient adaptive lattice* (GAL) algorithms proposed by Griffiths for the recursive computation of the reflection coefficients of multistage lattice predictors (Griffiths, 1977, 1978). The development of these algorithms was inspired by the LMS algorithm.[1] The main virtue of GAL algorithms is their reduced computational complexity compared to recursive LSL algorithms. In general, however, their performance is inferior to that of recursive LSL algorithms (Satorius and Pack, 1981; Friedlander, 1982). This may not be surprising since recursive LSL algorithms provide an *exact* solution to the linear least-squares problem at each time step, whereas GAL algorithms invoke the use of *approximations* in their derivations.

In this chapter, we present a detailed discussion of recursive LSL algorithms by using ideas that were developed in Section 8.12 for serialized data. However, we first review some basic relations and notations that pertain to forward and backward prediction-error filters.

---

[1] For one version of the GAL algorithm that involves the recursive computation of a single reflection coefficient per stage of the lattice predictor, see Problem 8, Chapter 5.

9.2 SOME PRELIMINARIES

Consider a forward prediction-error filter of order m whose *least-squares* tap-weight vector is denoted by $\mathbf{a}_m(n)$. By definition, the first element of $\mathbf{a}_m(n)$ equals unity. Let $\mathbf{u}_{m+1}(i)$ denote the $(m + 1)$-by-1 tap-input vector of the filter measured at time i, where $1 \leq i \leq n$. The forward a posteriori prediction error produced at the output of the filter, in response to the tap-input vector $\mathbf{u}_{m+1}(i)$, equals

$$f_m(i) = \mathbf{a}_m^H(n)\mathbf{u}_{m+1}(i), \qquad 1 \leq i \leq n \tag{9.1}$$

We assume the use of *prewindowing*, so we may put $u(i) = 0$ for $i \leq 0$. The sum of weighted forward a posteriori prediction-error squares equals

$$\mathscr{F}_m(n) = \sum_{i=1}^{n} \lambda^{n-i}|f_m(i)|^2 \tag{9.2}$$

where λ is the exponential weighting factor. The tap-weight vector $\mathbf{a}_m(n)$ is the result of minimizing the index of performance of index $\mathscr{F}_m(n)$, subject to the constraint that the first element of $\mathbf{a}_m(n)$ equals unity. Note that $\mathbf{a}_m(n)$ is treated as a constant vector during this minimization.

Consider next the corresponding backward prediction-error filter of order m whose least-squares tap-weight vector is denoted by $\mathbf{c}_m(n)$. By definition, the last element of $\mathbf{c}_m(n)$ equals unity. The backward a posteriori prediction error produced at the output of the filter, in response to the tap-input vector $\mathbf{u}_{m+1}(n)$, equals

$$b_m(i) = \mathbf{c}_m^H(n)\mathbf{u}_{m+1}(i), \qquad 1 \leq i \leq n \tag{9.3}$$

The sum of weighted backward a posteriori prediction-error squares equals

$$\mathscr{B}_m(n) = \sum_{i=1}^{n} \lambda^{n-i}|b_m(i)|^2 \tag{9.4}$$

The tap-weight vector $\mathbf{c}_m(n)$ is the result of minimizing the second index of performance $\mathscr{B}_m(n)$, subject to the constraint that the last element of $\mathbf{c}_m(n)$ equals unity. Here, again, $\mathbf{c}_m(n)$ is treated as a constant vector during this minimization.

9.3 ORDER-UPDATE RECURSIONS

Let $\mathbf{\Phi}_{m+1}(n)$ denote the $(m + 1)$-by-$(m + 1)$ deterministic correlation matrix of the tap-input vector $\mathbf{u}_{m+1}(i)$ applied to the forward prediction-error filter of order m, where $1 \leq i \leq n$. We may characterize this filter by the augmented normal equation [see Eq. (8.58)]

$$\mathbf{\Phi}_{m+1}(n)\,\mathbf{a}_m(n) = \begin{bmatrix} \mathscr{F}_m(n) \\ \mathbf{0}_m \end{bmatrix} \tag{9.5}$$

where $\mathbf{0}_m$ is the m-by-1 null vector, and $\mathbf{a}_m(n)$ and $\mathscr{F}_m(n)$ are as defined previously.

From the discussion presented in Chapter 8, we recall that the correlation matrix $\Phi_{m+1}(n)$ may be partitioned in two different ways, depending on how we interpret the first or last element of the tap-input vector $\mathbf{u}_{m+1}(i)$. The form of partitioning that we like to use is the one that enables us to relate the tap-weight vector $\mathbf{a}_m(n)$, pertaining to prediction order m, to the tap-weight vector $\mathbf{a}_{m-1}(n)$, pertaining to prediction order $m - 1$. This aim is realized by using the result [see Eq. 8.78)].

$$\Phi_{m+1}(n) = \begin{bmatrix} \Phi_m(n) & \vdots & \theta_2(n) \\ \cdots\cdots & \vdots & \cdots\cdots \\ \theta_2^H(n) & \vdots & \mathcal{U}(n-m) \end{bmatrix} \tag{9.6}$$

where $\Phi_m(n)$ is the m-by-m deterministic correlation matrix of the tap-input vector $\mathbf{u}_m(i)$, $\theta_2(n)$ is the m-by-1 deterministic cross-correlation vector between $\mathbf{u}_m(i)$ and $u(i - m)$, and $\mathcal{U}(n - m)$ is the sum of weighted, squared values of the input $u(i - m)$ for $1 \le i \le n$. Note that $\mathcal{U}(n - m)$ is zero for $n - m \le 0$. We postmultiply both sides of Eq. (9.6) by an $(m + 1)$-by-1 vector whose first m elements are defined by the vector $\mathbf{a}_{m-1}(n)$ and whose last element equals zero. We may thus write

$$\Phi_{m+1}(n) \begin{bmatrix} \mathbf{a}_{m-1}(n) \\ \cdots\cdots \\ 0 \end{bmatrix} = \begin{bmatrix} \Phi_m(n) & \vdots & \theta_2(n) \\ \cdots\cdots & \vdots & \cdots\cdots \\ \theta_2^H(n) & \vdots & \mathcal{U}(n - m) \end{bmatrix} \begin{bmatrix} \mathbf{a}_{m-1}(n) \\ \cdots\cdots \\ 0 \end{bmatrix}$$

$$= \begin{bmatrix} \Phi_m(n)\mathbf{a}_{m-1}(n) \\ \theta_2^H(n)\mathbf{a}_{m-1}(n) \end{bmatrix} \tag{9.7}$$

Both $\Phi_m(n)$ and $\mathbf{a}_{m-1}(n)$ have the same time argument n. Furthermore, in the first line of Eq. (9.7), they are both positioned in such a way that when the matrix multiplication is performed, $\Phi_m(n)$ becomes postmultiplied by $\mathbf{a}_{m-1}(n)$. For a forward prediction-error filter of order $m - 1$, evaluated at time n, the augmented normal equation is

$$\Phi_m(n)\mathbf{a}_{m-1}(n) = \begin{bmatrix} \mathcal{F}_{m-1}(n) \\ \mathbf{0}_{m-1} \end{bmatrix} \tag{9.8}$$

where $\mathcal{F}_{m-1}(n)$ is the sum of weighted forward a posteriori prediction-error squares at the filter output, and $\mathbf{0}_{m-1}$ is the $(m - 1)$-by-1 null vector. Define the scalar

$$\Delta_{m-1}(n) = \theta_2^H(n)\mathbf{a}_{m-1}(n) \tag{9.9}$$

Accordingly, we may rewrite Eq. (9.7) as

$$\Phi_{m+1}(n) \begin{bmatrix} \mathbf{a}_{m-1}(n) \\ 0 \end{bmatrix} = \begin{bmatrix} \mathcal{F}_{m-1}(n) \\ \mathbf{0}_{m-1} \\ \Delta_{m-1}(n) \end{bmatrix} \tag{9.10}$$

Consider next the backward prediction-error filter of order m that is characterized by the augmented normal equation

$$\boldsymbol{\Phi}_{m+1}(n)\mathbf{c}_m(n) = \begin{bmatrix} \mathbf{0}_m \\ \mathcal{B}_m(n) \end{bmatrix} \tag{9.11}$$

where $\boldsymbol{\Phi}_{m+1}(n)$, $\mathbf{c}_m(n)$, and $\mathcal{B}_m(n)$ are as defined previously, and $\mathbf{0}_m$ is the m-by-1 null vector. This time we use the other partitioned form of the deterministic correlation matrix $\boldsymbol{\Phi}_m(n)$, as shown by [see Eq. (8.61)]

$$\boldsymbol{\Phi}_{m+1}(n) = \begin{bmatrix} \mathcal{U}(n) & \boldsymbol{\theta}_1^H(n) \\ \hline \boldsymbol{\theta}_1(n) & \boldsymbol{\Phi}_m(n-1) \end{bmatrix} \tag{9.12}$$

where $\mathcal{U}(n)$ is the sum of weighted, squared values of the input $u(i)$ for the time interval $1 \le i \le n$, $\boldsymbol{\theta}_1(n)$ is the m-by-1 deterministic cross-correlation vector between $u(i)$ and the tap-input vector $\mathbf{u}_m(i-1)$, and $\boldsymbol{\Phi}_m(n-1)$ is the m-by-m deterministic correlation matrix of $\mathbf{u}_m(i-1)$. Correspondingly, we postmultiply $\boldsymbol{\Phi}_{m+1}(n)$ by an $(m+1)$-by-1 vector whose first element is zero and whose m remaining elements are defined by the tap-weight vector $\mathbf{c}_{m-1}(n-1)$ that pertains to a backward prediction-error filter of order $m-1$. We may thus write

$$\boldsymbol{\Phi}_{m+1}(n) \begin{bmatrix} 0 \\ \mathbf{c}_{m-1}(n-1) \end{bmatrix} = \begin{bmatrix} \mathcal{U}(n) & \boldsymbol{\theta}_1^H(n) \\ \boldsymbol{\theta}_1(n) & \boldsymbol{\Phi}_m(n-1) \end{bmatrix} \begin{bmatrix} 0 \\ \mathbf{c}_{m-1}(n-1) \end{bmatrix} \tag{9.13}$$
$$= \begin{bmatrix} \boldsymbol{\theta}_1^H(n)\mathbf{c}_{m-1}(n-1) \\ \boldsymbol{\Phi}_m(n-1)\mathbf{c}_{m-1}(n-1) \end{bmatrix}$$

Both $\boldsymbol{\Phi}_m(n-1)$ and $\mathbf{c}_{m-1}(n-1)$ have the same time argument, $n-1$. Also, they are both positioned in the first line of Eq. (9.13) in such a way that, when the matrix multiplication is performed, $\boldsymbol{\Phi}_{m-1}(n-1)$ become postmultiplied by $\mathbf{c}_{m-1}(n-1)$. For a backward prediction-error filter of order $m-1$, evaluated at time $n-1$, the augmented normal equation is

$$\boldsymbol{\Phi}_m(n-1)\mathbf{c}_{m-1}(n-1) = \begin{bmatrix} \mathbf{0}_{m-1} \\ \mathcal{B}_{m-1}(n-1) \end{bmatrix} \tag{9.14}$$

where $\mathcal{B}_{m-1}(n-1)$ is the sum of weighted backward a posteriori prediction-errors squares produced at the output of the filter at time $n-1$. Define the second scalar

$$\Delta_{m-1}'(n) = \boldsymbol{\theta}_1^H(n)\mathbf{c}_{m-1}(n-1) \tag{9.15}$$

where the prime is intended to distinguish this new parameter from $\Delta_{m-1}(n)$.

Accordingly, we may rewrite Eq. (9.13) as

$$
\mathbf{\Phi}_{m+1}(n)
\begin{bmatrix}
0 \\
\mathbf{c}_{m-1}(n-1)
\end{bmatrix}
=
\begin{bmatrix}
\Delta'_{m-1}(n) \\
\mathbf{0}_{m-1} \\
\mathscr{B}_{m-1}(n-1)
\end{bmatrix}
\tag{9.16}
$$

The next manipulation we wish to perform is to combine Eqs. (9.10) and (9.16) in such a way that the resultant assumes a form identical to the augmented normal equation (9.5) that pertains to a forward prediction-error filter of order m. This we can do by multiplying Eq. (9.16) by the ratio $\Delta_{m-1}(n)/\mathscr{B}_{m-1}(n-1)$ and subtracting the result from Eq. (9.10). We thus get

$$
\mathbf{\Phi}_{m+1}(n)
\left[
\begin{bmatrix}
\mathbf{a}_{m-1}(n) \\
0
\end{bmatrix}
-
\frac{\Delta_{m-1}(n)}{\mathscr{B}_{m-1}(n-1)}
\begin{bmatrix}
0 \\
\mathbf{c}_{m-1}(n-1)
\end{bmatrix}
\right]
$$
$$
\tag{9.17}
$$
$$
=
\begin{bmatrix}
\mathscr{F}_{m-1}(n) - \dfrac{\Delta_{m-1}(n)\Delta'_{m-1}(n)}{\mathscr{B}_{m-1}(n-1)} \\
\mathbf{0}_m
\end{bmatrix}
$$

Comparing Eqs. (9.5) and (9.17), we therefore deduce the following two order-update recursions:

$$
\mathbf{a}_m(n) =
\begin{bmatrix}
\mathbf{a}_{m-1}(n) \\
0
\end{bmatrix}
-
\frac{\Delta_{m-1}(n)}{\mathscr{B}_{m-1}(n-1)}
\begin{bmatrix}
0 \\
\mathbf{c}_{m-1}(n-1)
\end{bmatrix}
\tag{9.18}
$$

and

$$
\mathscr{F}_m(n) = \mathscr{F}_{m-1}(n) - \frac{\Delta_{m-1}(n)\Delta'_{m-1}(n)}{\mathscr{B}_{m-1}(n-1)}
\tag{9.19}
$$

A second manipulation that we may also perform is to combine Eqs. (9.10) and (9.16) in such a way that, this time, the resultant assumes a form identical to the augmented normal equation (9.11) pertaining to the backward prediction-error filter of order m. To achieve this end result, we multiply Eq. (9.10) by the ratio $\Delta'_{m-1}(n)/\mathscr{F}_{m-1}(n)$ and subtract the result from Eq. (9.16). We thus obtain

$$
\mathbf{\Phi}_{m+1}(n)
\left[
\begin{bmatrix}
0 \\
\mathbf{c}_{m-1}(n-1)
\end{bmatrix}
-
\frac{\Delta'_{m-1}(n)}{\mathscr{F}_{m-1}(n)}
\begin{bmatrix}
\mathbf{a}_{m-1}(n) \\
0
\end{bmatrix}
\right]
$$
$$
\tag{9.20}
$$
$$
=
\begin{bmatrix}
\mathbf{0}_m \\
\mathscr{B}_{m-1}(n-1) - \dfrac{\Delta_{m-1}(n)\Delta'_{m-1}(n)}{\mathscr{F}_{m-1}(n)}
\end{bmatrix}
$$

Comparing Eqs. (9.11) and (9.20), we therefore deduce two additional order-update recursions:

$$\mathbf{c}_m(n) = \begin{bmatrix} 0 \\ \mathbf{c}_{m-1}(n-1) \end{bmatrix} - \frac{\Delta'_{m-1}(n)}{\mathscr{F}_{m-1}(n)} \begin{bmatrix} \mathbf{a}_{m-1}(n) \\ 0 \end{bmatrix} \tag{9.21}$$

and

$$\mathscr{B}_m(n) = \mathscr{B}_{m-1}(n-1) - \frac{\Delta_{m-1}(n)\Delta'_{m-1}(n)}{\mathscr{F}_{m-1}(n)} \tag{9.22}$$

Relation between the two Parameters $\Delta_{m-1}(n)$ and $\Delta'_{m-1}(n)$

The parameters $\Delta_{m-1}(n)$ and $\Delta'_{m-1}(n)$, defined by Eqs. (9.9) and (9.15), respectively, are the complex conjugate of one another; that is,

$$\Delta'_{m-1}(n) = \Delta^*_{m-1}(n) \tag{9.23}$$

where $\Delta^*_{m-1}(n)$ is the complex conjugate of $\Delta_{m-1}(n)$. We prove this relation in three stages:

1. We premultiply both sides of Eq. (9.10) by the row vector

$$[(0, \mathbf{c}^H_{m-1}(n-1)],$$

where the superscript H denotes Hermitian transposition. The result of this matrix multiplication is the scalar

$$[0, \mathbf{c}^H_{m-1}(n-1)]\mathbf{\Phi}_{m+1}(n) \begin{bmatrix} \mathbf{a}_{m-1}(n) \\ 0 \end{bmatrix}$$

$$= [0, \mathbf{c}^H_{m-1}(n-1)] \begin{bmatrix} \mathscr{F}_{m-1}(n) \\ \mathbf{0}_{m-1} \\ \Delta_{m-1}(n) \end{bmatrix} = \Delta_{m-1}(n) \tag{9.24}$$

where we have used the fact that the last element of $\mathbf{c}_{m-1}(n-1)$ equals unity.

2. We apply Hermitian transposition to both sides of Eq. (9.16), and use the Hermitian property of the deterministic correlation matrix $\mathbf{\Phi}_{m+1}(n)$, thereby obtaining

$$[0, \mathbf{c}^H_{m-1}(n-1)]\mathbf{\Phi}_{m+1}(n) = [\Delta'^*_{m-1}(n), \mathbf{0}^T_{m-1}, \mathscr{B}_{m-1}(n-1)]$$

where $\Delta'^*_{m-1}(n)$ is the complex conjugate of $\Delta'_{m-1}(n)$, and $\mathscr{B}_{m-1}(n-1)$ is real valued.   Next we use this relation to evaluate the scalar

$$[0, \mathbf{c}_{m-1}^H(n-1)]\mathbf{\Phi}_{m+1}(n) \begin{bmatrix} \mathbf{a}_{m-1}(n) \\ 0 \end{bmatrix}$$

$$= [\Delta_{m-1}'^*(n), \mathbf{0}_{m-1}^T, \mathcal{B}_{m-1}(n-1)] \begin{bmatrix} \mathbf{a}_{m-1}(n) \\ 0 \end{bmatrix} \quad (9.25)$$

$$= \Delta_{m-1}'^*(n)$$

where we have used the fact that the first element of $\mathbf{a}_{m-1}(n)$ equals unity.

3. Comparison of Eqs. (9.24) and (9.25) immediately yields the relation of Eq. (9.23) between the parameters $\Delta_{m-1}(n)$ and $\Delta_{m-1}'(n)$.

Order-update Recursions for the a Posteriori Prediction Errors

Define the *forward reflection coefficient*

$$\Gamma_{f,m}(n) = -\frac{\Delta_{m-1}(n)}{\mathcal{B}_{m-1}(n-1)}, \qquad m = 1, 2, \ldots, M \quad (9.26)$$

where M is the final order of the lattice predictor.

Define the *backward reflection coefficient*

$$\Gamma_{b,m}(n) = -\frac{\Delta_{m-1}'(n)}{\mathcal{F}_{m-1}(n)}$$

$$= -\frac{\Delta_{m-1}^*(n)}{\mathcal{F}_{m-1}(n)}, \qquad m = 1, 2, \ldots, M \quad (9.27)$$

In general, $\mathcal{F}_{m-1}(n)$ and $\mathcal{B}_{m-1}(n-1)$ are unequal so that in the LSL algorithms, unlike Burg's formula discussed in Chapter 4, we have, $\Gamma_{f,m}(n) \neq \Gamma_{b,m}^*(n)$.   Using these definitions, we may rewrite Eqs. (9.18) and (9.21) as follows, respectively:

$$\mathbf{a}_m(n) = \begin{bmatrix} \mathbf{a}_{m-1}(n) \\ 0 \end{bmatrix} + \Gamma_{f,m}(n) \begin{bmatrix} 0 \\ \mathbf{c}_{m-1}(n-1) \end{bmatrix} \quad (9.28)$$

and

$$\mathbf{c}_m(n) = \begin{bmatrix} 0 \\ \mathbf{c}_{m-1}(n-1) \end{bmatrix} + \Gamma_{b,m}(n) \begin{bmatrix} \mathbf{a}_{m-1}(n) \\ 0 \end{bmatrix} \quad (9.29)$$

where $m = 1, 2, \ldots, M$.   Equations (9.28) and (9.29) may be viewed as the deterministic counterpart of the Levinson–Durbin recursion.

The forward a posteriori prediction error $f_m(n)$ equals the output of the forward prediction-error filter of order m that is produced in response to the tap-input vector $\mathbf{u}_{m+1}(n)$, as shown by

$$f_m(n) = \mathbf{a}_m^H(n)\,\mathbf{u}_{m+1}(n) \tag{9.30}$$

The tap-input vector $\mathbf{u}_{m+1}(n)$ may be partitioned as

$$\mathbf{u}_{m+1}(n) = \begin{bmatrix} \mathbf{u}_m(n) \\ u(n-m) \end{bmatrix} \tag{9.31}$$

or, alternatively, as

$$\mathbf{u}_{m+1}(n) = \begin{bmatrix} u(n) \\ \mathbf{u}_m(n-1) \end{bmatrix} \tag{9.32}$$

Hence, substituting Eq. (9.28) in (9.30), and using the partitioned form of Eq. (9.31) for the first term on the right side of Eq. (9.28) and that of Eq. (9.32) for the second term, we get the following *order-update recursion for the forward a posteriori prediction error:*

$$f_m(n) = f_{m-1}(n) + \Gamma_{f,m}^*(n)b_{m-1}(n-1), \qquad m = 1, 2, \ldots, M \tag{9.33}$$

where $f_{m-1}(n)$ is the forward a posteriori prediction error and $b_{m-1}(n-1)$ is the *delayed* version of the backward a posteriori prediction error, both for prediction order $m = 1$.

   To develop the corresponding order update recursion for the backward prediction error, we may follow a procedure similar to that just described.   The backward a posteriori prediction error $b_m(n)$ equals the output of the backward prediction-error filter of order m that is produced in response to the tap-input vector $\mathbf{u}_{m+1}(n)$, as shown by

$$b_m(n) = \mathbf{c}_m^H(n)\,\mathbf{u}_{m+1}(n) \tag{9.34}$$

Substituting Eq. (9.29) in (9.34), and using the partitioned form of Eq. (9.32) for the first term on the right side of Eq. (9.29) and that of Eq. (9.31) for the second term, we get the following *order-update recursion for the backward a posteriori prediction error:*

$$b_m(n) = b_{m-1}(n-1) + \Gamma_{b,m}^*(n)f_{m-1}(n), \qquad m = 1, 2, \ldots, M \tag{9.35}$$

   Equations (9.33) and (9.35) are represented by the signal-flow graph shown in Fig. 9.1(a), which has the appearance of a *lattice.*

   The prediction order m is a *variable* parameter.   It takes on the values 0, 1, . . . , M, where M is the *final value* of the prediction order.   When $m = 0$, there is *no* prediction being performed on the input data.   This elementary condition corresponds to the *initial values*

$$f_0(n) = b_0(n) = u(n) \tag{9.36}$$

where $u(n)$ is the input at time n.   Thus, as we vary the prediction order m from zero all the way up to the final value M, we get the *multistage least-squares lattice*

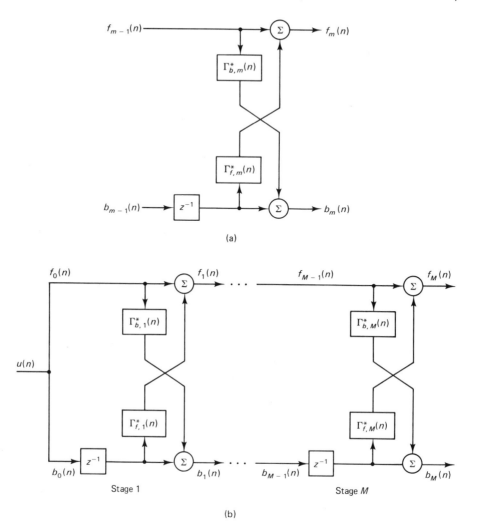

Figure 9.1   (a) Typical least-squares lattice stage; (b) multistage least-squares lattice predictor.

predictor shown in Fig. 9.1(b).   The number of stages contained in this structure equals M, the final order of the predictor.

It is noteworthy that the least-square prediction errors $f_m(n)$ and $b_m(n)$ obey a set of order update recursions that are identical in structure to those we derived in Chapter 4.   The latter recursions are consequences of the Levinson–Durbin recursion that arises basically because of the Toeplitz property of the ensemble-averaged correlation matrix $\mathbf{R}_m$ for stationary inputs.   On the other hand, the deterministic correlation matrix $\mathbf{\Phi}_m(n)$ is, in general, non-Toeplitz and yet we are still able to derive simple order-update recursions for the least-squares prediction errors $f_m(n)$ and $b_m(n)$.   This is indeed a remarkable result.

Order-update Recursions for the Sums of Weighted a posteriori Prediction Error Squares

Equations (9.19) and (9.22) define the order-update recursions for the sum of weighted forward a posteriori prediction-error squares, $\mathcal{F}_m(n)$, and the sum of weighted backward a posteriori prediction-error squares, $\mathcal{B}_m(n)$, respectively.   By using the relation between $\Delta_{m-1}(n)$ and $\Delta'_{m-1}(n)$ shown in Eq. (9.23), we may rewrite these two order-update recursions as follows, respectively:

$$\mathcal{F}_m(n) = \mathcal{F}_{m-1}(n) - \frac{|\Delta_{m-1}(n)|^2}{\mathcal{B}_{m-1}(n-1)} \tag{9.37}$$

and

$$\mathcal{B}_m(n) = \mathcal{B}_{m-1}(n-1) - \frac{|\Delta_{m-1}(n)|^2}{\mathcal{F}_{m-1}(n)} \tag{9.38}$$

where $m = 1, 2, \ldots, M$.

Equations (9.33), (9.35), (9.37), and (9.38), in conjunction with the definitions of (9.26) and (9.27) for the reflection coefficients, constitute the basic order-update recursions for the least-squares lattice predictor.   These recursions generate two sequences of prediction errors: (1) the foward a posteriori prediction errors $f_0(n)$, $f_1(n), \ldots, f_M(n)$, and (2) the backward a posteriori prediction errors $b_0(n)$, $b_1(n)$, $\ldots, b_M(n)$.   These two sequences play key roles in the recursive solution of the linear least-squares problem.

Order-update Recursion for $\gamma_m(n-1)$

To complete the list of order-update recursions required for the formulation of the LSL algorithm, we need one other such recursion for the delayed *estimation error* or *conversion factor* $\gamma_m(n-1)$.   The role of $\gamma_m(n-1)$, in the context of the least-squares lattice predictor, will become apparent in Section 9.4.   The reason for showing the time of interest as $n-1$ is to be consistent with the mathematical development in that section.

The estimation error $\gamma_m(n-1)$ results from applying the tap-input vector $\mathbf{u}_m(n-1)$ to a transversal filter whose tap-weight vector equals the gain vector $\mathbf{k}_m(n-1)$ that enters the derivation of the RLS solution.   From the discussion presented in Section 8.12, we recall that the gain vector $\mathbf{k}_m(n-1)$ is order-updated as follows [see Eq. (8.96)]:

$$\mathbf{k}_m(n-1) = \begin{bmatrix} \mathbf{k}_{m-1}(n-1) \\ 0 \end{bmatrix} + \frac{b_{m-1}(n-1)}{\mathcal{B}_{m-1}(n-1)} \mathbf{c}_{m-1}(n-1),$$
$$m = 1, 2, \ldots, M \tag{9.39}$$

Postmultiplying the Hermitian transposed sides of this equation by $\mathbf{u}_m(n - 1)$, we get

$$\mathbf{k}_m^H(n - 1)\mathbf{u}_m(n - 1) = [\mathbf{k}_{m-1}^H(n - 1), 0]\mathbf{u}_m(n - 1)$$

$$+ \frac{b_{m-1}^*(n - 1)}{\mathcal{B}_{m-1}(n - 1)} \mathbf{c}_{m-1}^H(n - 1)\mathbf{u}_m(n - 1) \qquad (9.40)$$

By definition, we have

$$\mathbf{k}_m^H(n - 1)\mathbf{u}_m(n - 1) = 1 - \gamma_m(n - 1) \qquad (9.41)$$

$$[\mathbf{k}_{m-1}^H(n - 1), 0]\mathbf{u}_m(n - 1) = [\mathbf{k}_{m-1}^H(n - 1), 0] \begin{bmatrix} \mathbf{u}_{m-1}(n - 1) \\ u(n - m) \end{bmatrix}$$

$$= \mathbf{k}_{m-1}^H(n - 1)\mathbf{u}_{m-1}(n - 1) \qquad (9.42)$$

$$= 1 - \gamma_{m-1}(n - 1)$$

and

$$\mathbf{c}_{m-1}^H(n - 1)\mathbf{u}_m(n - 1) = b_{m-1}(n - 1) \qquad (9.43)$$

Therefore, the use of Eqs. (9.41) through (9.43) in (9.40) yields the simple result

$$\gamma_m(n - 1) = \gamma_{m-1}(n - 1) - \frac{|b_{m-1}(n - 1)|^2}{\mathcal{B}_{m-1}(n - 1)}, \qquad m = 1, 2, \ldots, M \qquad (9.44)$$

This is the desired recursion for order-updating $\gamma_m(n - 1)$, which completes the list of desired order-update recursions.

9.4 TIME-UPDATE RECURSION

The recursions derived in Section 9.3 enable us to order-update the forward and backward a posteriori prediction errors as well as the sums of their weighted squares, starting from the elementary case $m = 0$. However, in order to make these recursions adaptive in time, it is necessary to derive a time-update recursion for the parameter $\Delta_{m-1}(n)$ that enters the computation of these order-update recursions. We now address this issue.

Consider the m-by-1 tap-weight vector $\mathbf{a}_{m-1}(n - 1)$ that pertains to forward prediction-error filter of order $m - 1$, evaluated at time $n - 1$. The reason for considering time $n - 1$ will become apparent presently. Since the leading element of the vector $\mathbf{a}_{m-1}(n - 1)$ equals unity, we may express $\Delta_{m-1}(n)$ as follows [see Eqs. (9.23) and (9.25)]:

$$\Delta_{m-1}(n) = [\Delta_{m-1}(n), \mathbf{0}_{m-1}^T, \mathcal{B}_{m-1}(n - 1)] \begin{bmatrix} \mathbf{a}_{m-1}(n - 1) \\ 0 \end{bmatrix} \qquad (9.45)$$

Taking the Hermitian transpose of both sides of Eq. (9.16), recognizing the Hermitian property of $\mathbf{\Phi}_{m+1}(n)$, and using the relation of Eq. (9.23), we get

$$[0, \mathbf{c}_{m-1}^H(n-1)] \, \mathbf{\Phi}_{m+1}(n) = [\mathbf{\Delta}_{m-1}(n), \mathbf{0}_{m-1}^T, \mathscr{B}_{m-1}(n-1)] \qquad (9.46)$$

Hence, substitution of Eq. (9.46) in (9.45) yields

$$\mathbf{\Delta}_{m-1}(n) = [0, \mathbf{c}_{m-1}^H(n-1)] \, \mathbf{\Phi}_{m+1}(n) \begin{bmatrix} \mathbf{a}_{m-1}(n-1) \\ 0 \end{bmatrix} \qquad (9.47)$$

But the deterministic correlation matrix $\mathbf{\Phi}_{m+1}(n)$ may be time-updated as follows [see Eq. (8.12)]:

$$\mathbf{\Phi}_{m+1}(n) = \lambda \mathbf{\Phi}_{m+1}(n-1) + \mathbf{u}_{m+1}(n)\mathbf{u}_{m+1}^H(n) \qquad (9.48)$$

Accordingly, we may use this relation for $\mathbf{\Phi}_{m+1}(n)$ to rewrite Eq. (9.47) as

$$\begin{aligned} \mathbf{\Delta}_{m-1}(n) = {} & \lambda[0, \mathbf{c}_{m-1}^H(n-1)] \, \mathbf{\Phi}_{m+1}(n-1) \begin{bmatrix} \mathbf{a}_{m-1}(n-1) \\ 0 \end{bmatrix} \\ & + [0, \mathbf{c}_{m-1}^H(n-1)] \, \mathbf{u}_{m+1}(n) \, \mathbf{u}_{m+1}^H(n) \begin{bmatrix} \mathbf{a}_{m-1}(n-1) \\ 0 \end{bmatrix} \end{aligned} \qquad (9.49)$$

Next we recognize from the definition of *forward a priori prediction error* that

$$\begin{aligned} \mathbf{u}_{m+1}^H(n) \begin{bmatrix} \mathbf{a}_{m-1}(n-1) \\ 0 \end{bmatrix} &= [\mathbf{u}_m^H(n), u^*(n-m)] \begin{bmatrix} \mathbf{a}_{m-1}(n-1) \\ 0 \end{bmatrix} \\ &= \mathbf{u}_m^H(n) \, \mathbf{a}_{m-1}(n-1) \\ &= \eta_{m-1}^*(n) \end{aligned} \qquad (9.50)$$

and from the definition of the *backward a posteriori prediction error* that

$$\begin{aligned} [0, \mathbf{c}_{m-1}^H(n-1)] \, \mathbf{u}_{m+1}(n) &= [0, \mathbf{c}_{m-1}^H(n-1)] \begin{bmatrix} u(n) \\ \mathbf{u}_m(n-1) \end{bmatrix} \\ &= \mathbf{c}_{m-1}^H(n-1) \, \mathbf{u}_m(n-1) \\ &= b_{m-1}(n-1) \end{aligned} \qquad (9.51)$$

Also, by substituting $n-1$ for n in Eq. (9.10), we have

$$\mathbf{\Phi}_{m+1}(n-1) \begin{bmatrix} \mathbf{a}_{m-1}(n-1) \\ 0 \end{bmatrix} = \begin{bmatrix} \mathscr{F}_{m-1}(n-1) \\ \mathbf{0}_{m-1} \\ \mathbf{\Delta}_{m-1}(n-1) \end{bmatrix} \qquad (9.52)$$

Hence, using this relation and the fact that the last element of the tap-weight vector $\mathbf{c}_{m-1}(n-1)$, pertaining to the backward prediction-error filter, equals unity, we may write the first term on the right side of Eq. (9.49), except for λ, as

$$[0, \mathbf{c}_{m-1}^{H}(n-1)] \, \mathbf{\Phi}_{m+1}(n-1) \begin{bmatrix} \mathbf{a}_{m-1}(n-1) \\ 0 \end{bmatrix}$$

$$= [0, \mathbf{c}_{m-1}^{H}(n-1)] \begin{bmatrix} \mathscr{F}_{m-1}(n-1) \\ \mathbf{0}_{m-1} \\ \Delta_{m-1}(n-1) \end{bmatrix} \tag{9.53}$$

$$= \Delta_{m-1}(n-1)$$

Accordingly, substituting Eqs. (9.50), (9.51), and (9.53) in (9.49), we may express the time-update recursion for $\Delta_{m-1}(n)$ simply as

$$\Delta_{m-1}(n) = \lambda \Delta_{m-1}(n-1) + b_{m-1}(n-1)\eta_{m-1}^{*}(n) \tag{9.54}$$

The forward a priori prediction error $\eta_{m-1}(n)$ is related to the forward a posteriori prediction error $f_{m-1}(n)$ by [see Eq. (8.102)]

$$\eta_{m-1}(n) = \frac{f_{m-1}(n)}{\gamma_{m-1}(n-1)} \tag{9.55}$$

where $\gamma_{m-1}(n-1)$ is the delayed estimation error or conversion factor, defined previously. Thus, the use of Eq. (9.55) in (9.54) yields the time-update recursion

$$\Delta_{m-1}(n) = \lambda \Delta_{m-1}(n-1) + \frac{b_{m-1}(n-1)f_{m-1}^{*}(n)}{\gamma_{m-1}(n-1)} \tag{9.56}$$

The correction term in the time update of Eq. (9.56) is amplified by the reciprocal of the conversion factor $\gamma_{m-1}(n-1)$. This parameter enables the LSL algorithm to adapt rapidly to sudden changes in the input data (Morf and Lee, 1978). It is also of interest to note that, except for the amplification factor $1/\gamma_{m-1}(n-1)$, the time update of the parameter $\Delta_{m-1}(n)$ is, in fact, a *time average of the cross-correlation* between the the delayed backward prediction error $b_{m-1}(n-1)$ and forward prediction error $f_{m-1}(n)$.

In Section 8.12, we showed that the conversion factor $\gamma_{m-1}(n-1)$ is real valued and bounded by zero and one. When $\gamma_{m-1}(n-1)$ is zero, the recursion of Eq. (9.56) will stop, indicating that the deterministic correlation matrix $\mathbf{\Phi}_{m+1}(n)$ is noninvertible.

9.5 SUMMARY OF THE LSL ALGORITHM AND DISCUSSION

The complete list of order- and time-update recursions constituting the recursive LSL algorithm (based on a posteriori estimation errors) is summarized in Table 9.1. This summary includes the recursions and relations of Eqs. (9.56), (9.26), (9.27), (9.33), (9.35), (9.37), (9.38) and (9.44), in that order.

TABLE 9.1 SUMMARY OF THE RECURSIVE LSL ALGORITHM USING A POSTERIORI ESTIMATION ERRORS

Starting with $n = 1$, compute the various order-updates in the following sequences: $m = 1, 2,$ $\ldots, M$, where M is the final order of the least-squares lattice predictor:

$$\Delta_{m-1}(n) = \lambda\Delta_{m-1}(n-1) + \frac{b_{m-1}(n-1)f_{m-1}^*(n)}{\gamma_{m-1}(n-1)}$$

$$\Gamma_{f,m}(n) = -\frac{\Delta_{m-1}(n)}{\mathscr{B}_{m-1}(n-1)}$$

$$\Gamma_{b,m}(n) = -\frac{\Delta_{m-1}^*(n)}{\mathscr{F}_{m-1}(n)}$$

$$f_m(n) = f_{m-1}(n) + \Gamma_{f,m}^*(n)b_{m-1}(n-1)$$

$$b_m(n) = b_{m-1}(n-1) + \Gamma_{b,m}^*(n)f_{m-1}(n)$$

$$\mathscr{F}_m(n) = \mathscr{F}_{m-1}(n) - \frac{|\Delta_{m-1}(n)|^2}{\mathscr{B}_{m-1}(n-1)}$$

$$\mathscr{B}_m(n) = \mathscr{B}_{m-1}(n-1) - \frac{|\Delta_{m-1}(n)|^2}{\mathscr{F}_{m-1}(n)}$$

$$\gamma_m(n-1) = \gamma_{m-1}(n-1) - \frac{|b_{m-1}(n-1)|^2}{\mathscr{B}_{m-1}(n-1)}$$

Joint-process estimation:
Starting with $n = 1$, compute the various order-updates in the following sequence: $m = 0, 1,$ $\ldots, M$

$$\rho_m(n) = \lambda\rho_m(n-1) + \frac{b_m(n)}{\gamma_m(n)}e_m^*(n)$$

$$\kappa_m(n) = \frac{\rho_m(n)}{\mathscr{B}_m(n)}$$

$$e_{m+1}(n) = e_m(n) - \kappa_m^*(n)\,b_m(n)$$

This algorithm performs linear prediction, both in the forward and backward directions, on the input data. The order of prediction determines the number of stages, M, contained in the lattice predictor. An important property of the LSL algorithm is its *modular* structure, as exemplified by the signal-flow graph of Fig. 9.1(b). The implication of this property is that the prediction order can be readily increased without the need to recalculate all previous values. This property is particularly useful when there is no prior knowledge as to what the final value of the prediction order should be.

It is noteworthy that the feature that distinguishes the recursive LSL algorithm from the two-reflection coefficient version of the GAL algorithm (Griffiths, 1978) is the significant role played by the conversion factor $\gamma_{m-1}(n-1)$. Indeed, if we were to nullify this effect by putting $\gamma_{m-1}(n-1)$ equal to 1, the LSL algorithm reduces to the same basic form as this particular GAL algorithm. Consequently, in the latter algorithm, no distinction is made between the a priori and a posteriori values of prediction errors.

9.6 INITIALIZATION OF THE RECURSIVE LSL ALGORITHM

To initialize the recursive LSL algorithm, we start with the elementary case of prediction order $m = 0$, for which we have [see Eqs. (9.30) and (9.34)]

$$f_0(n) = b_0(n) = u(n)$$

where $u(n)$ is the lattice predictor input at time n.

The other set of initial values we need pertain to the sums of weighted a posteriori prediction-error squares for $m = 0$. The sum of weighted forward a posteriori prediction-error squares, $\mathcal{F}_m(n)$, is time-updated as [see Eq. (8.70)]

$$\mathcal{F}_m(n) = \lambda \mathcal{F}_m(n - 1) + f_m(n)\eta_m^*(n) \qquad (9.57)$$

where $f_m(n)$ and $\eta_m(n)$ are the forward a posteriori and a priori prediction errors for prediction order m, respectively. When $m = 0$, we have

$$f_0(n) = \eta_0(n) = u(n)$$

Hence, evaluating the time-update recursion of Eq. (9.57) for $m = 0$, we get

$$\mathcal{F}_0(n) = \lambda \mathcal{F}_0(n - 1) + |u(n)|^2 \qquad (9.58)$$

Similarly, we may show that for $m = 0$ the sum of weighted backward a posteriori prediction-error squares is updated in time as

$$\mathcal{B}_0(n) = \lambda \mathcal{B}_0(n - 1) + |u(n)|^2 \qquad (9.59)$$

With the conversion factor $\gamma_m(n - 1)$ bounded by zero and one, a logical choice for the zeroth-order value of this parameter is

$$\gamma_0(n - 1) = 1 \qquad (9.60)$$

We complete the initialization of the algorithm by setting at $n = 0$ the following conditions:

$$\Delta_{m-1}(0) = 0 \qquad (9.61)$$

and

$$\mathcal{F}_0(0) = \mathcal{B}_0(0) = \delta \qquad (9.62)$$

where δ is a small positive constant. The constant δ is used to ensure nonsingularity of the deterministic correlation matrix $\mathbf{\Phi}_{m+1}(n)$.

Table 9.2 presents a summary of the computations involved in the initialization of the LSL algorithm.

TABLE 9.2 SUMMARY OF THE INITIALIZATION OF THE RECURSIVE LSL ALGORITHM
USING A POSTERIORI ESTIMATION ERRORS

---

1. To initialize the algorithm, at time $n = 0$ set

   $\Delta_{m-1}(0) = 0$

   $\mathcal{F}_{m-1}(0) = \delta, \qquad \delta = $ small positive constant

   $\mathcal{B}_{m-1}(0) = \delta$

2. At each instant $n \geq 1$, generate the various zeroth-order variables as follows:

   $f_0(n) = b_0(n) = u(\hat{n})$

   $\mathcal{F}_0(n) = \mathcal{B}_0(n) = \lambda\mathcal{F}_0(n - 1) + |u(n)|^2$

   $\gamma_0(n - 1) = 1$

3. For joint-process estimation, initialize the algorithm by setting at time $n = 0$

   $\rho_m(0) = 0$

   At each instant $n \geq 1$, generate the zeroth-order variable

   $e_0(n) = d(n)$

Note: For prewindowed data, the input $u(n)$ and desired response $d(n)$ are both zero for $n \leq 0$.

9.7 COMPUTER EXPERIMENT ON ADAPTIVE PREDICTION USING THE RECURSIVE LSL ALGORITHM

In this experiment we use the recursive LSL algorithm to adaptively estimate the tap weights of the linear predictor shown in Fig. 9.2. This predictor was studied in Chapter 5 using the simple LMS algorithm and in Chapter 8 using the conventional RLS algorithm. As before, we assume that (1) the predictor has two tap weights $\hat{w}_1(n)$ and $\hat{w}_2(n)$, and (2) the corresponding tap inputs $u(n - 1)$ and $u(n - 2)$ are drawn from the real-valued autoregressive (AR) process $\{u(n)\}$ described by the second-order difference equation

$$u(n) + a_1u(n - 1) + a_2u(n - 2) = v(n)$$

The sample $v(n)$ itself is drawn from a zero-mean white-noise process whose variance is adjusted so as to make the variance of the AR process $\{u(n)\}$ equal unity.

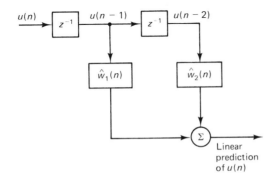

Figure 9.2 Two-tap predictor for real-valued input.

The AR parameters a_1 and a_2 are chosen so as to realize different values for the eigenvalue spread $\chi(\mathbf{R})$ of the correlation matrix of the tap inputs. Their values for three different eigenvalue spreads $\chi(\mathbf{R}) = 3$, 10, and 100 are shown in Table 5.2. The purpose of the experiment is to evaluate the sensitivity of the learning curve of the LSL algorithm to variations in the eigenvalue spread. The experiment will thus enlighten us as to how the LSL algorithm compares in performance to the LMS and RLS algorithms when they all operate under similar stationary environmental conditions.

As before, for any one trial of the experiment, 256 samples of white-noise process $\{v(n)\}$ are obtained from a *random-number generator* of zero mean and adjustable variance. For each value of n in the range $1 \leq n \leq 256$, the squared magnitude of the forward prediction error $f_2(n)$ produced at the output of the two-stage lattice predictor is computed. The experiment is repeated 200 times, each time using an *independent* realization of the white-noise process $\{v(n)\}$. The average squared error is determined by computing the ensemble average of $f_2^2(n)$ over these 200 independent trials of the experiment.

Since the LSL algorithm (summarized in Table 9.1) involves division by updated parameters at some of the steps, care must be taken to ensure that these values are not allowed to become too small. Unless a high-precision computer is used, selection of the constant δ [determining the initial values $\mathcal{F}_0(0)$ and $\mathcal{B}_0(0)$] may have a severe effect on the initial transient performance of the LSL algorithm. Friedlander (1982) suggests using some form of *thresholding*, in that if the divisor (in any computation of the LSL algorithm) is less than this preassigned threshold, the corresponding term involving that divisor is set to be zero. In this experiment, performed on a 16-bit computer, the constant δ was assigned the value 1.0, and the value of the threshold was selected to be 0.5.

The results of the computation are shown in Fig. 9.3 and Table 9.3. The three learning curves in Fig. 9.3 correspond to the eigenvalue spread $\chi(\mathbf{R}) = 3$, 10, 100. The steady-state values of the average mean-squared error $J(\infty)$ shown in Table 9.3 for these three eigenvalue spreads were obtained by time-averaging 200 samples of $f_2^2(n)$, starting from the instant when the tap weights reach essentially constant values, and then ensemble-averaging over 200 independent trials of the experiment.

The results shown in Fig. 9.3 confirm two important properties of the recursive LSL algorithm:

1. The LSL algorithm has a *fast* rate of convergence, approximately equal to the number of stages in the lattice predictor.
2. The LSL algorithm is *robust* in that its rate of convergence is essentially independent of variations in the eigenvalue spread $\chi(\mathbf{R})$.

In Table 9.3, we have also included the theoretical value of the variance σ_v^2 of the white-noise process $\{v(n)\}$ for each of the three eigenvalue spreads used in the experiment. There is very close agreement between the experimental values

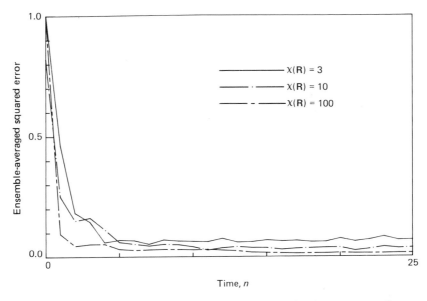

Figure 9.3   Learning curves of the recursive LSL algorithm for two-tap predictor with varying eigenvalue spread $\chi(\mathbf{R})$.

of the average mean-squared error $J(\infty)$ and the corresponding theoretical values of the variance σ_v^2.   This confirms another important property of the recursive LSL algorithm, that the *misadjustment* produced by the LSL algorithm is *negligible* when the measurement noise in the application of interest is low.

TABLE 9.3

Eigenvalue Spread	Variance of White-noise Process $\{v(n)\}$	Average Mean-squared Value of the Forward Prediction Error $f_2(n)$ for $n \simeq \infty$
3	0.0731	0.0737
10	0.0322	0.0323
100	0.0038	0.0053

9.8 JOINT-PROCESS ESTIMATION

For a lattice predictor consisting of m stages, the recursive LSL algorithm produces a sequence of backward prediction errors $b_0(n)$, $b_1(n)$, . . . , $b_m(n)$ that are *uncorrelated with each other in a deterministic sense for all instants of time*.   That is, the time-averaged or deterministic correlation matrix of the backward prediction errors is a *diagonal matrix*.   We refer to this property as the *decoupling property* of the LSL algorithm. Accordingly, by using the backward prediction errors $b_0(n)$, $b_1(n), \ldots, b_{,m}(n)$ as tap inputs that are applied to a corresponding set of *regression coefficients* $\kappa_0, \kappa_1, \ldots, \kappa_m$, respectively, as in Fig. 9.4, we may determine the *least-squares estimate* of some desired response $d(n)$ *exactly* and in a highly *efficient* manner. We refer to the two-channel structure of Fig. 9.4 as a *joint-process estimator*[2] because it solves the problem of estimating one process $\{d(n)\}$ from observations of a related process $\{u(n)\}$ by embedding them into the joint process $\{d(n), u(n)\}$.

Exact Decoupling Property of the LSL Algorithm

Before proceeding to derive the recursive procedure for updating the regression coefficients of the structure in Fig. 9.4, we will first prove the exact decoupling property of the LSL algorithm. As mentioned previously, this property implies that the backward a posteriori prediction errors $b_0(n), b_1(n), \ldots, b_m(n)$ used as inputs are uncorrelated with each other in a time-averaged sense at all instants of time.

Consider a backward prediction-error filter of order m. Let the $(m + 1)$-by-1 tap-weight vector of the filter, optimized in the least-squares sense over the time interval $1 \leq i \leq n$, be denoted by $\mathbf{c}_m(n)$. In expanded form, we have

$$\mathbf{c}_m^T(n) = [c_{m,m}(n), c_{m,m-1}(n), \ldots, 1] \qquad (9.63)$$

Let $b_m(i)$ denote the backward a posteriori prediction error produced at the output of the filter in response to the $(m + 1)$-by-1 input vector $\mathbf{u}_{m+1}(i)$. The expanded form of the input vector is shown by

$$\mathbf{u}_{m+1}^T(i) = [u(i), u(i - 1), \ldots, u(i - m)], \qquad i > m \qquad (9.64)$$

We may thus express the error $b_m(i)$ as

$$
\begin{aligned}
b_m(i) &= \mathbf{c}_m^H(n)\mathbf{u}_{m+1}(i) \\
&= \sum_{k=0}^{m} c_{m,k}^*(n)u(i - m + k), \qquad \begin{matrix} m < i \leq n, \\ m = 0, 1, 2, \ldots \end{matrix}
\end{aligned}
\qquad (9.65)
$$

Let $\mathbf{b}_{m+1}(i)$ denote the $(m + 1)$-by-1 *backward a posteriori prediction-error vector* that is defined by

$$\mathbf{b}_{m+1}^T(i) = [b_0(i), b_1(i), \ldots, b_m(i)], \qquad \begin{matrix} i > m \\ m = 0, 1, 2, \ldots \end{matrix} \qquad (9.66)$$

Substituting Eq. (9.65) in (9.66), we may express the transformation of the input data into the corresponding set of backward a posteriori prediction errors as follows:

$$\mathbf{b}_{m+1}(i) = \mathbf{L}_m(n)\mathbf{u}_{m+1}(i) \qquad (9.67)$$

---

[2]The idea of using a lattice predictor to perform joint-process estimation as in Fig. 9.4 was first proposed by Griffiths (1978). A variation of this idea was proposed independently by Makhoul (1978).

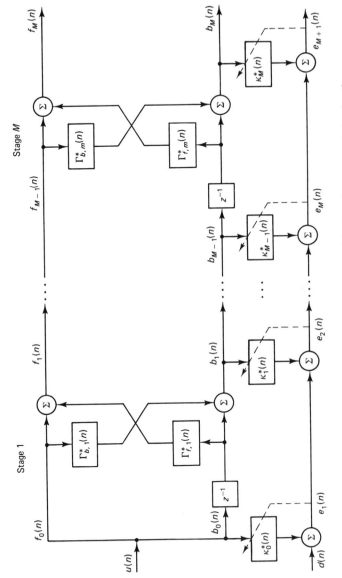

Figure 9.4 Joint-process estimator using the LSL algorithm based on a posteriori estimation errors.

where the $(m + 1)$-by-$(m + 1)$ *transformation matrix* is defined by

$$
\mathbf{L}_m(n) = \begin{bmatrix}
1 & 0 & \cdot & \cdot & \cdot & 0 \\
c^*_{1,1}(n) & 1 & & \cdot & \cdot & 0 \\
\cdot & & \cdot & & & \cdot \\
\cdot & & & \cdot & & \cdot \\
\cdot & & & & \cdot & \cdot \\
c^*_{m,m}(n) & c^*_{m,m-1}(n) & \cdot & \cdot & \cdot & 1
\end{bmatrix} \tag{9.68}
$$

The subscript m in the symbol $\mathbf{L}_m(n)$ refers to the highest order of backward prediction-error filter involved in its constitution. Note also that:

1. The nonzero elements of row l of matrix $\mathbf{L}_m(n)$ are defined by the tap weights of backward prediction-error filter of order $(l - 1)$.
2. The diagonal elements of matrix $\mathbf{L}_m(n)$ equal unity; this follows from the fact that the last tap weight of a backward prediction-error filter equals unity.
3. The determinant of matrix $\mathbf{L}_m(n)$ equals one for all m; hence, the inverse matrix $\mathbf{L}_m^{-1}(n)$ exists.

We define the $(m + 1)$-by-$(m + 1)$ deterministic correlation matrix of the backward a posteriori prediction-error vector as

$$
\mathbf{D}_{m+1}(n) = \sum_{i=1}^{n} \lambda^{n-i} \mathbf{b}_{m+1}(i) \mathbf{b}_{m+1}^H(i), \qquad \begin{array}{l} m < i \leq n, \\ m = 0, 1, 2, \ldots \end{array} \tag{9.69}
$$

Substituting Eq. (9.67) in (9.69), we get

$$
\mathbf{D}_{m+1}(n) = \sum_{i=1}^{n} \lambda^{n-i} \mathbf{L}_m(n) \mathbf{u}_{m+1}(i) \mathbf{u}_{m+1}^H(i) \mathbf{L}_m^H(n)
$$

$$
= \mathbf{L}_m(n) \left[\sum_{i=1}^{n} \lambda^{n-i} \mathbf{u}_{m+1}(i) \mathbf{u}_{m+1}^H(i) \right] \mathbf{L}_m^H(n) \tag{9.70}
$$

The expression inside the square brackets equals the deterministic correlation matrix of the input vector; that is,

$$
\mathbf{\Phi}_{m+1}(n) = \sum_{i=1}^{n} \lambda^{n-i} \mathbf{u}_{m+1}(i) \mathbf{u}_{m+1}^H(i) \tag{9.71}
$$

Accordingly, we may simplify Eq. (9.70) as

$$
\mathbf{D}_{m+1}(n) = \mathbf{L}_m(n) \mathbf{\Phi}_{m+1}(n) \mathbf{L}_m^H(n) \tag{9.72}
$$

Using the formula for the augmented normal equation for backward linear prediction, we may show that the product $\mathbf{\Phi}_{m+1}(n)\mathbf{L}_m^H(n)$ consists of a lower triangular matrix whose diagonal elements equal the various sums of weighted backward a posteriori prediction-error squares, that is, $\mathscr{B}_0(n), \mathscr{B}_1(n), \ldots, \mathscr{B}_m(n)$ (see Problem 1). We thus find that the product $\mathbf{L}_m(n)\,\boldsymbol{\Phi}_{m+1}(n)\,\mathbf{L}_m^H(n)$ is a symmetric product of two lower triangular matrices, which makes $\mathbf{D}_{m+1}(n)$ a diagonal matrix, as shown by

$$\mathbf{D}_{m+1}(n) = \operatorname{diag}\,[\mathscr{B}_0(n), \mathscr{B}_1(n), \ldots, \mathscr{B}_m(n)] \tag{9.73}$$

Equation (9.73) shows that the backward a posteriori prediction errors $b_0(n), b_1(n), \ldots, b_m(n)$ produced by the various stages of the least-squares lattice predictor are uncorrelated (in a time-averaged sense) at all instants of time. This proves another remarkable property of the least-squares lattice predictor that makes it ideally suited for *exact* joint-process estimation.

It is noteworthy that the transformation of the (correlated) input data by the LSL algorithm into a new sequence of (uncorrelated) backward prediction errors may be viewed as a deterministic form of the Gram–Schmidt orthogonalization procedure. For further perusal of this issue, the reader is referred to Problem 2.

Transformation of the RLS Solution

Consider the conventional tapped-delay-line or transversal filter structure shown in Fig. 9.5, where the tap inputs $u(n), u(n-1), \ldots, u(n-m)$ are derived directly from the process $\{u(n)\}$ and the tap weights $\hat{w}_0(n), \hat{w}_1(n), \ldots, \hat{w}_m(n)$ are used to form respective inner products. From Chapter 7, we recall that the least-squares solution for the $(m+1)$-by-1 tap-weight vector $\hat{\mathbf{w}}_m(n)$, consisting of the elements $\hat{w}_0(n), \hat{w}_1(n), \ldots, w_m(n)$, is defined by [see Eq. (7.34)]

$$\boldsymbol{\Phi}_{m+1}(n)\hat{\mathbf{w}}_m(n) = \boldsymbol{\theta}_{m+1}(n) \tag{9.74}$$

where $\boldsymbol{\Phi}_{m+1}(n)$ is the $(m+1)$-by-$(m+1)$ deterministic correlation matrix of the tap inputs, and $\boldsymbol{\theta}_{m+1}(n)$ is the $(m+1)$-by-1 deterministic cross-correlation vector between the tap inputs and desired response. We modify Eq. (9.74) in two ways: (1) we premultiply both sides of the equation by the $(m+1)$-by-1 lower triangular transformation matrix $\mathbf{L}_m(n)$, and (2) we interject the $(m+1)$-by-$(m+1)$ identity matrix $\mathbf{I} = \mathbf{L}_m^H(n)\,\mathbf{L}_m^{-H}(n)$ between the matrix $\boldsymbol{\Phi}_{m+1}(n)$ and the vector $\hat{\mathbf{w}}_m(n)$ on

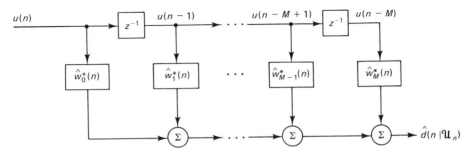

Figure 9.5 Conventional transversal filter.

the left side of the equation. The symbol $\mathbf{L}_m^{-H}(n)$ denotes the Hermitian transpose of the inverse matrix $\mathbf{L}_m^{-1}(n)$. We may thus write

$$\mathbf{L}_m(n)\ \boldsymbol{\Phi}_{m+1}(n)\ \mathbf{L}_m^H(n)\mathbf{L}_m^{-H}(n)\hat{\mathbf{w}}_m(n) = \mathbf{L}_m(n)\ \boldsymbol{\theta}_{m+1}(n) \tag{9.75}$$

From Eq. (9.72), the product $\mathbf{L}_m(n)\ \boldsymbol{\Phi}_{m+1}(n)\ \mathbf{L}_m^H(n)$ on the left side of Eq. (9.75) equals the diagonal matrix $\mathbf{D}_{m+1}(n)$ that represents the deterministic correlation matrix of the backward a posteriori prediction errors used as tap inputs for the regression coefficients of Fig. 9.4. The product $\mathbf{L}_m(n)\ \boldsymbol{\theta}_{m+1}(n)$ on the right side equals the $(m + 1)$-by-1 deterministic cross-correlation vector between these backward prediction errors and the desired response. Let $\mathbf{t}_{m+1}(n)$ denote this cross-correlation vector. By definition, we have

$$\mathbf{t}_{m+1}(n) = \sum_{i=1}^{n} \lambda^{n-i}\mathbf{b}_{m+1}(i)d^*(i) \tag{9.76}$$

where $d(i)$ is the desired response. Substituting Eq. (9.67) in (9.76), we thus get

$$\mathbf{t}_{m+1}(n) = \sum_{i=1}^{n} \lambda^{n-i}\mathbf{L}_m(n)\mathbf{u}_{m+1}(i)d^*(i)$$

$$= \mathbf{L}_m(n) \sum_{i=1}^{n} \lambda^{n-i}\mathbf{u}_{m+1}(i)d^*(i) \tag{9.77}$$

$$= \mathbf{L}_m(n)\boldsymbol{\theta}_{m+1}(n)$$

which is the desired result. Accordingly, the combined use of Eqs. (9.72) and (9.77) in (9.75) yields the *transformed RLS solution*:

$$\mathbf{D}_{m+1}(n)\mathbf{L}_m^{-H}(n)\hat{\mathbf{w}}_m(n) = \mathbf{t}_{m+1}(n) \tag{9.78}$$

Thus far we have considered how the application of lower triangular matrix $\mathbf{L}_m(n)$ transforms the RLS solution for the tap-weight vector of the conventional transversal structure shown in Fig. 9.5. We next wish to consider the RLS solution for the regression coefficient vector $\boldsymbol{\kappa}_m(n)$ in the structure of Fig. 9.4, which is defined by

$$\boldsymbol{\kappa}_m^T(n) = [\kappa_0(n), \kappa_1(n), \ldots, \kappa_m(n)] \tag{9.79}$$

The regression coefficient vector $\boldsymbol{\kappa}_m(n)$ is obtained by minimizing the index of performance

$$\sum_{i=1}^{n} \lambda^{n-i}|d(i) - \mathbf{b}_{m+1}^T(i)\ \boldsymbol{\kappa}_m^*(n)|^2$$

where $\boldsymbol{\kappa}_m(n)$ is held constant for $1 \leq i \leq n$. The resulting solution to this RLS problem may be expressed as (see Problem 3)

$$\mathbf{D}_{m+1}(n)\ \boldsymbol{\kappa}_m(n) = \mathbf{t}_{m+1}(n) \tag{9.80}$$

where, as defined before, $\mathbf{D}_{m+1}(n)$ is the $(m + 1)$-by-$(m + 1)$ deterministic correlation matrix of the backward a posteriori prediction errors used as tap inputs in Fig. 9.4, and $\mathbf{t}_{m+1}(n)$ is the $(m + 1)$-by-1 deterministic cross-correlation between these tap inputs and the desired response.

By comparing the transformed RLS solution of Eq. (9.78) and the RLS solution of Eq. (9.80), we immediately deduce the following simple relationship between the tap-weight vector $\hat{\mathbf{w}}_m(n)$ in the structure of Fig. 9.5 and the corresponding regression coefficient vector $\mathbf{\kappa}_m(n)$ in the structure of Fig. 9.4:

$$\mathbf{\kappa}_m(n) = \mathbf{L}_m^{-H}(n) \, \hat{\mathbf{w}}_m(n) \tag{9.81}$$

or, equivalently,

$$\hat{\mathbf{w}}_m(n) = \mathbf{L}_m^H(n) \, \mathbf{\kappa}_m(n) \tag{9.82}$$

Here, again we see that the lower triangular transformation matrix $\mathbf{L}_m(n)$ represents the connecting link between the RLS solutions for the tap-weight vectors of Figs. 9.4 and 9.5.

Recursive Algorithm for Computing the Regression Coefficient Vector $\kappa_m(n)$

Solving Eqs. (9.77) and (9.80) for the regression coefficient vector $\mathbf{\kappa}_m(n)$, we get

$$\mathbf{\kappa}_m(n) = \mathbf{D}_{m+1}^{-1}(n) \, \mathbf{L}_m(n) \, \mathbf{\theta}_{m+1}(n) \tag{9.83}$$

The exact decoupling property of the LSL algorithm ensures that the deterministic correlation matrix $\mathbf{D}_{m+1}(n)$ of the sequence of backward prediction errors produced by the algorithm is a diagonal matrix for all n, as shown by Eq. (9.73). This means that the inverse matrix $\mathbf{D}_{m+1}^{-1}(n)$ is likewise diagonal:

$$\mathbf{D}_{m+1}^{-1}(n) = \text{diag} \, [\mathcal{B}_0^{-1}(n), \, \mathcal{B}_1^{-1}(n), \, \ldots, \, \mathcal{B}_m^{-1}(n)] \tag{9.84}$$

Hence, only $m + 1$ scalar divisions rather than matrix inversion are needed in the use of Eq. (9.83) to compute the regression coefficient vector $\mathbf{\kappa}_m(n)$. In particular, the use of Eqs. (9.68) and (9.84) in (9.83) yields a system of $(m + 1)$ scalar equations:

$$\mathbf{\kappa}_m(n) = \mathcal{B}_m^{-1}(n) \, \mathbf{c}_m^H(n) \, \mathbf{\theta}_{m+1}(n), \qquad m = 0, 1, \ldots, M \tag{9.85}$$

where M is the final value of the prediction order. To develop a recursion for time-updating the regression coefficient $\kappa_m(n)$, define the scalar

$$\rho_m(n) = \mathbf{c}_m^H(n) \, \mathbf{\theta}_{m+1}(n), \qquad m = 0, 1, 2, \ldots, M \tag{9.86}$$

We may then redefine the regression coefficient $\kappa_m(n)$ of Eq. (9.85) in terms of $\rho_m(n)$ as follows:

$$\mathbf{\kappa}_m(n) = \frac{\rho_m(n)}{\mathcal{B}_m(n)}, \qquad m = 0, 1, \ldots, M \tag{9.87}$$

From Chapter 8, we recall that the tap-weight vector $\mathbf{c}_m(n)$ of the backward prediction-error filter may be time-updated as follows [see Eqs. (8.85) and (8.103)]:

$$\mathbf{c}_m(n) = \mathbf{c}_m(n-1) - \frac{b_m^*(n)}{\gamma_m(n)} \begin{bmatrix} \mathbf{k}_m(n) \\ 0 \end{bmatrix} \tag{9.88}$$

where $\mathbf{k}_m(n)$ is the m-by-1 gain vector. Also, from Chapter 8 we recall that the **(m + 1)-by-1** deterministic cross-correlation vector $\boldsymbol{\theta}_{m+1}(n)$ may be time-updated as follows [see Eq. (8.13)]:

$$\boldsymbol{\theta}_{m+1}(n) = \lambda \boldsymbol{\theta}_{m+1}(n-1) + \mathbf{u}_{m+1}(n)\, d^*(n) \tag{9.89}$$

Hence, substituting Eqs. (9.88) and (9.89) in (9.86), and recognizing that $\gamma_m(n)$ is real valued, we get

$$\rho_m(n) = \lambda \mathbf{c}_m^H(n-1)\, \boldsymbol{\theta}_{m+1}(n-1) + \mathbf{c}_m^H(n-1)\, \mathbf{u}_{m+1}(n)\, d^*(n)$$
$$-\frac{b_m}{\gamma_m(n)}\, \mathbf{k}_m^H(n)\, \boldsymbol{\theta}_m(n) \tag{9.90}$$

In the last element of the right side in Eq. (9.90), we have used the fact that the first m elements of the $(m + 1)$-by-1 vector $\boldsymbol{\theta}_{m+1}(n)$ are contained in the m-by-1 vector $\boldsymbol{\theta}_m(n)$, and since the last element of the $(m + 1)$-by-1 vector on the right side of Eq. (9.88) is zero, only these elements of $\boldsymbol{\theta}_{m+1}(n)$ enter the multiplication of the second term of $\mathbf{c}_m^H(n)$ by $\boldsymbol{\theta}_{m+1}(n)$. We may simplify Eq. (9.90) as follows:

1. From the definition introduced in Eq. (9.86), we deduce that

$$\rho_m(n-1) = \mathbf{c}_m^H(n-1)\boldsymbol{\theta}_{m+1}(n-1)$$

2. From the definition of backward a priori prediction error, we have

$$\psi_m(n) = \mathbf{c}_m^H(n-1)\mathbf{u}_{m+1}(n)$$

Moreover, since $\psi_m(n)$ equals the ratio $b_m(n)/\gamma_m(n)$ [see Eq. (8.103)], it follows therefore that

$$\mathbf{c}_m^H(n-1)\mathbf{u}_{m+1}(n) = \frac{b_m(n)}{\gamma_m(n)}$$

3. Since, by definition, the gain vector $\mathbf{k}_m(n)$ equals $\boldsymbol{\Phi}_m^{-1}(n)\mathbf{u}_m(n)$, and the deterministic correlation matrix $\boldsymbol{\Phi}_m(n)$ is Hermitian, it follows that

$$\mathbf{k}_m^H(n)\, \boldsymbol{\theta}_m(n) = \mathbf{u}_m^H(n)\, \boldsymbol{\Phi}_m^{-1}(n)\, \boldsymbol{\theta}_m(n)$$
$$= \mathbf{u}_m^H(n)\, \hat{\mathbf{w}}_{m-1}(n)$$

where in the last line we have made use of the deterministic normal equation for least-squares estimation. The inner product $\mathbf{u}_m^H(n)\, \hat{\mathbf{w}}_{m-1}(n)$ equals the complex conjugate of the least-squares estimate of the desired response $d(n)$, given the m-by-1 tap-input vector $\mathbf{u}_m(n)$. Let $\hat{d}(n|\mathscr{U}_n)$ denote the value of this estimate, where $\mathcal{U}_n$ is the space spanned by the elements of vector $\mathbf{u}_m(n)$.

Accordingly, we may rewrite Eq. (9.90) simply as

$$\rho_m(n) = \lambda\rho_m(n-1) + \frac{b_m(n)}{\gamma_m(n)}[d^*(n) - \hat{d}^*(n|\mathcal{U}_n)] \tag{9.91}$$

Define the *a posteriori estimation error*, based on the use of m tap inputs, as

$$
\begin{aligned}
e_m(n) &= d(n) - \hat{d}(n|\mathcal{U}_n) \\
&= d(n) - \hat{\mathbf{w}}_{m-1}^H(n)\mathbf{u}_m(n)
\end{aligned}
\tag{9.92}
$$

Hence, we may simplify the time-update recursion of Eq. (9.91) as

$$\rho_m(n) = \lambda\rho_m(n-1) + \frac{b_m(n)}{\gamma_m(n)}e_m^*(n), \qquad m = 0, 1, 2, \ldots, M \tag{9.93}$$

Note that, as in the time-update recursion of Eq. (9.56) for $\Delta_m(n)$, the correction term in the recursion of Eq. (9.93) is amplified by the reciprocal of the conversion factor $\gamma_m(n)$. Also, except for $\gamma_m(n)$, we may interpret $\rho_m(n)$ as the deterministic cross-correlation between $b_m(n)$ and $e_m(n)$.

To complete the recursive procedure for joint-process estimation, we need to accompany the time-update recursion of Eq. (9.93) with an order-update recursion that enables us to compute the updated estimation error $e_{m+1}(n)$, given $e_m(n)$. We may do this by replacing m with $m+1$ in Eq. (9.92) to write

$$e_{m+1}(n) = d(n) - \hat{\mathbf{w}}_m^H(n)\,\mathbf{u}_{m+1}(n)$$

Since the inner product $\hat{\mathbf{w}}_m^H(n)\mathbf{u}_{m+1}(n)$ equals $\mathbf{\kappa}_m^H(n)\,\mathbf{b}_{m+1}(n)$ [see Eqs. (9.67) and (9.82)], we may also express $e_{m+1}(n)$ as follows

$$
\begin{aligned}
e_{m+1}(n) &= d(n) - \mathbf{\kappa}_m^H(n)\,\mathbf{b}_{m+1}(n) \\
&= d(n) - \sum_{l=0}^{m}\kappa_l^*(n)\,b_l(n)
\end{aligned}
\tag{9.94}
$$

Isolating the inner product $\kappa_m^*(n)b_m(n)$ from the rest of the summation on the right side of Eq. (9.94), and recognizing that the remainder equals (by definition)

$$d(n) - \sum_{l=0}^{m-1}\kappa_l^*(n)\,b_l(n) = e_m(n),$$

we may express the order-update recursion for the a posteriori estimation error as

$$e_{m+1}(n) = e_m(n) - \kappa_m^*(n)\,b_m(n), \qquad m = 0, 1, \ldots, M \tag{9.95}$$

Note that the number of regression coefficients [represented by $\kappa_0(n)$, $\kappa_1(n)$, $\ldots$, $\kappa_M(n)$] exceeds the final prediction order M by 1.

We may now sum up the procedure for the extension of the recursive LSL algorithm to perform joint-process estimation. The time-update recursion of Eq. (9.93) enables us to compute the updated value of parameter $\rho_m(n)$. Given this value and the updated value of the sum of weighted backward a posteriori prediction-error squares $\mathcal{B}_m(n)$, obtained from the order-update recursion of Eq. (9.38), we may compute the updated value of $\kappa_m(n)$ by using Eq. (9.87). Hence, we may use the order-update recursion of Eq. (9.95) to compute the updated value of the a posteriori estimation error $e_{m+1}(n)$.

Initial Conditions

For the elementary case of prediction order $m = 0$, we have (see Fig. 9.4)

$$e_0(n) = d(n) \tag{9.96}$$

where $d(n)$ is the desired response. Thus, to initiate the computation, we compute $e_0(n)$ at each instant n.

To complete the initialization of the LSL algorithm for joint-process estimation, at time $n = 0$ we set

$$\rho_m(0) = 0, \qquad m = 0, 1, \ldots, M \tag{9.97}$$

Summary

The back end of the summary of the recursive LSL algorithm presented in Table 9.1 and, likewise, the back end of the summary of the initial conditions presented in Table 9.2 pertain to the extension of the algorithm for joint-process estimation.

9.9 COMPUTER EXPERIMENT ON ADAPTIVE EQUALIZATION USING THE RECURSIVE LSL ALGORITHM

In this second experiment, we study the use of the recursive LSL algorithm for *adaptive equalization* of a linear channel that produces unknown distortion. The parameters of the channel are the same as those used to study the LMS algorithm in Section 5.16 and the RLS algorithm in Section 8.8 for a similar application. The results of the experiment should therefore help us to make a further assessment of the performance of the recursive LSL algorithm, compared to the LMS and the RLS algorithms.

Figure 9.6 shows the superposition of learning curves of the recursive LSL algorithm for the initializing constant $\delta = 0.003$, the exponential weighting factor $\lambda = 1$, and four different values of the eigenvalue spread $\chi(\mathbf{R}) = 6.0782, 11.1238, 21.7132, 46.8216$ that correspond to the channel parameter $W = 2.9, 3.1, 3.3, 3.5$, respectively. Each learning curve was obtained by ensemble-averaging the squared value of the final a priori estimation error $\alpha_{M+1}(n)$ over 200 independent trials of

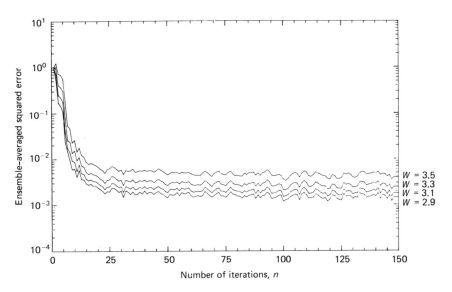

Figure 9.6   Learning curve of the recursive LSL algorithm for adaptive equalizer with $\delta = 0.003$ and varying eigenvalue spread $\chi(\mathbf{R})$.

the experiment, where the final prediction order $M = 10$. To compute the a priori estimation error $\alpha_{M+1}(n)$, we use the relation [see Eq. (8.101)]

$$\alpha_{M+1}(n) = \frac{e_{M+1}(n)}{\gamma_{M+1}(n)}$$

where $e_{M+1}(n)$ is the final a posteriori estimation error, and $\gamma_{M+1}(n)$ is the associated conversion factor. The reason for basing computation of the learning curves on $\alpha_{M+1}(n)$ rather than $e_{M+1}(n)$ is that we would like to make the learning curves of the recursive LSL algorithm consistent with those for the RLS algorithm.

In Fig. 9.7 we show the superposition of four plots of the conversion factor $\gamma_{M+1}(n)$ (for the final stage) versus time n for one trial of the experiment, corresponding to the four different values of the eigenvalue spread $\chi(\mathbf{R})$ as defined above. All the computations in Figs. 9.6 and 9.7, were performed on a high-precision computer.

Based on the results of this experiment and that of Section 8.8 (involving the RLS algorithm), we may make the following observations:

1. The value of $\gamma_{M+1}(n)$ is quite small during the initialization period that lasts for 11 iterations (equal to the number of adjustable parameters in the equalizer). Thereafter, $\gamma_{M+1}(n)$ increases rapidly (toward a final value of 1), reaching a value close to 0.6 in about 22 iterations.

2. The recursive LSL algorithm converges very rapidly, requiring only about 22 iterations (i.e., twice the number of adjustable parameters) to reach a mean

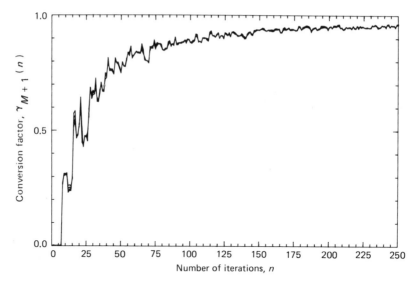

Figure 9.7 Superposition of the curves of $\gamma_{m+1}(n)$ versus n for the adaptive equalizer experiment, pertaining to 4 different values of eigenvalue spread $\chi(\mathbf{R})$.

squared error of no less than 10^{-2}).   This rate of convergence is essentially the same as that of the RLS algorithm.

3. Like the RLS algorithm, the rate of convergence of the recursive LSL algorithm is essentially insensitive to variations in the eigenvalue spread of the correlation matrix of the input signal.   This is well-evidenced in both Figs. 9.6 and 9.7.

9.10 RECURSIVE LSL ALGORITHM USING A PRIORI ESTIMATION ERRORS

The recursive LSL algorithm summarized in Table 9.1 is based on three sets of a posteriori estimation errors: forward prediction errors, backward prediction errors, and joint-process estimation errors.   We now describe another version of the algorithm that uses a priori forms of forward prediction errors, backward prediction errors, and joint-process estimation errors as the variables of interest.

From Section 8.12, we recall that the forward a posteriori prediction error, $f_m(n)$, and the forward a priori prediction error, $\eta_m(n)$, are related by the formula [see Eq. (8.102)]

$$f_m(n) = \gamma_m(n-1)\,\eta_m(n) \tag{9.98}$$

Also, the backward a posteriori prediction error, $b_m(n)$, and the backward a priori prediction error, $\psi_m(n)$, are related by the formula [see Eq. (8.103)]

$$b_m(n) = \gamma_m(n)\,\psi_m(n) \tag{9.99}$$

For the relation between the a posteriori estimation error, $e_m(n)$, and the a priori estimation error $\alpha_m(n)$, we have [see Eq. (8.101)]

$$e_m(n) = \gamma_m(n)\,\alpha_m(n) \tag{9.100}$$

From Section 8.12, we also note that the sum of weighted forward a posteriori prediction-error squares, $\mathscr{F}_{m-1}(n)$, and the sum of weighted backward a posteriori prediction-error squares, $\mathscr{B}_{m-1}(n-1)$, may be updated as follows, respectively

$$\mathscr{F}_{m-1}(n) = \lambda\mathscr{F}_{m-1}(n-1) + \eta_{m-1}(n)f_{m-1}^*(n) \tag{9.101}$$

and

$$\mathscr{B}_{m-1}(n) = \lambda\mathscr{B}_{m-1}(n-1) + \psi_{m-1}(n)b_{m-1}^*(n) \tag{9.102}$$

Using Eqs. (9.26), (9.56), (9.98), (9.99), and (9.102) in (9.33), and then simplifying, we get the following recursion for order-updating the forward a priori prediction error:

$$\eta_m(n) = \eta_{m-1}(n) + \Gamma_{f,m}^*(n-1)\psi_{m-1}(n-1), \qquad m = 1, 2, \ldots, M \tag{9.103}$$

Similarly, using Eqs. (9.27), (9.56), (9.98), (9.99), and (9.101) in (9.35), and then simplifying, we get the corresponding recursion for order-updating the backward a priori prediction error

$$\psi_m(n) = \psi_{m-1}(n-1) + \Gamma_{b,m}^*(n-1)\eta_{m-1}(n), \qquad m = 1, 2, \ldots, M \tag{9.104}$$

Finally, using Eqs. (9.87), (9.93), (9.99), (9.100), and (9.102) in (9.95), and then simplifying, we get the following recursion for order-updating the a priori estimation error:

$$\alpha_{m+1}(n) = \alpha_m(n) - \kappa_m^*(n-1)\psi_m(n), \qquad m = 0, 1, \ldots, M \tag{9.105}$$

Thus, combining the above order-updates with the following recursions: (1) time-update for the deterministic cross-correlation $\Delta_m(n)$ reformulated in terms of the a priori prediction errors $\eta_{m-1}(n)$ and $\psi_{m-1}(n-1)$; (2) time-updates for the sums of weighted a posteriori prediction error squares $\mathscr{F}_{m-1}(n)$ and $\mathscr{B}_{m-1}(n)$ as in Eqs. (9.101) and (9.102); and (3) time-updates for the deterministic cross-correlation $\rho_m(n)$ reformulated in terms of the a priori errors $\alpha_m(n)$ and $\psi_m(n)$, and arranging the various recursions in their proper sequence, we get the second version of the recursive LSL algorithm summarized in Table 9.4 (Ljung and Söderström, 1983; Ling, Manolakis and Proakis, 1985).

In Fig. 9.8, we present the signal-flow graph representation of this second version of the recursive LSL algorithm. Note that in this version, in order to update the variables of interest (namely, the a priori forward prediction, backward prediction, and joint-process estimation errors) at time n, we require knowledge of the values of three basic sets of coefficients (namely, the forward reflection coefficients, the backward reflection coefficients, and the regression coefficients) at time $n-1$. On the other hand, in the first version of the recursive LSL algorithm represented by the signal-flow graph of Fig. 9.4, in order to update the a posteriori

TABLE 9.4 SUMMARY OF THE RECURSIVE LSL ALGORITHM USING A PRIORI ESTIMATION ERRORS

(a) *Lattice predictor*

Starting with $n = 1$, compute the various order-updates in the following sequence $m = 1, 2, \ldots, M$, where M is the final order of the least-squares predictor:

$$\eta_m(n) = \eta_{m-1}(n) + \Gamma^*_{f,m}(n - 1)\,\psi_{m-1}(n - 1)$$

$$\psi_m(n) = \psi_{m-1}(n - 1) + \Gamma^*_{b,m}(n - 1)\,\eta_{m-1}(n)$$

$$\Delta_{m-1}(n) = \lambda\Delta_{m-1}(n - 1) + \gamma_{m-1}(n - 1)\,\psi_{m-1}(n - 1)\,\eta^*_{m-1}(n)$$

$$\mathscr{F}_{m-1}(n) = \lambda\mathscr{F}_{m-1}(n - 1) + \gamma_{m-1}(n - 1)|\eta_{m-1}(n)|^2$$

$$\mathscr{B}_{m-1}(n) = \lambda\mathscr{B}_{m-1}(n - 1) + \gamma_{m-1}(n)|\psi_{m-1}(n)|^2$$

$$\Gamma_{f,m}(n) = -\frac{\Delta_{m-1}(n)}{\mathscr{B}_{m-1}(n - 1)}$$

$$\Gamma_{b,m}(n) = -\frac{\Delta^*_{m-1}(n)}{\mathscr{F}_{m-1}(n)}$$

$$\gamma_m(n) = \gamma_{m-1}(n) - \frac{\gamma^2_{m-1}(n)|\psi_{m-1}(n)|^2}{\mathscr{B}_{m-1}(n)}$$

(b) *Joint-process estimator*

Starting with $n = 1$, compute the various order-updates in the following sequence $m = 0, 1, \ldots, M$:

$$\rho_m(n) = \lambda\rho_m(n - 1) + \gamma_m(n)\,\psi_m(n)\,\alpha^*_m(n)$$

$$\alpha_{m+1}(n) = \alpha_m(n) - \kappa^*_m(n - 1)\,\psi_m(n)$$

$$\kappa_m(n) = \frac{\rho_m(n)}{\mathscr{B}_m(n)}$$

---

forward prediction, backward prediction, and joint-process estimation errors at time n, we require knowledge of these three sets of coefficients at time n.

In Table 9.5, we present a summary of the initialization procedure for the second version of the recursive LSL algorithm based on a priori estimation errors.

9.11 MODIFIED FORMS OF THE RECURSIVE LSL ALGORITHMS

In the two versions of the recursive LSL algorithm summarized in Tables 9.1 and 9.4, we update the forward and backward reflection coefficients of the lattice predictor and the regression coefficients of the joint-process estimator in an *indirect* manner. We first compute the deterministic cross-correlation between forward and delayed backward errors and the deterministic cross-correlation between backward prediction errors and joint-process estimation errors. We next compute the

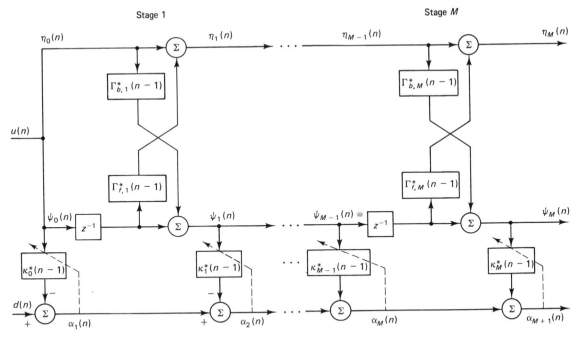

Figure 9.8 Joint-process estimator using LSL algorithm based on a priori estimation errors.

TABLE 9.5 INITIALIZATION OF THE RECURSIVE LSL ALGORITHM USING A PRIORI ESTIMATION ERRORS

1. To initialize the algorithm, at time $n = 0$ set

   $\Delta_{m-1}(0) = 0$

   $\mathcal{F}_{m-1}(0) = \delta, \qquad \delta =$ small positive constant

   $\mathcal{B}_{m-1}(0) = \delta$

2. At each instant $n \geq 1$, generate the zeroth-order variables:

   $\eta_0(n) = \psi_0(n) = u(n)$

   $\mathcal{F}_0(n) = \mathcal{B}_0(n) = \lambda \mathcal{F}_0(n-1) + |u(n)|^2$

   $\gamma_0(n) = 1$

3. For joint-process estimation, initialize the algorithm by setting at time $n = 0$

   $\rho_m(0) = 0$

   At each instant $n \geq 1$, generate the zeroth-order variable

   $\alpha_0(n) = d(n)$

Note: For prewindowed data, the input $u(n)$ and desired response $d(n)$ are both zero for $n \leq 0$.

sum of weighted forward error squares and the sum of weighted backward error squares. We then compute the reflection and regression coefficients by dividing a deterministic cross-correlation by a sum of weighted error squares. Accordingly, the accuracy of the reflection and regression coefficients depends on the accuracy attained in computing the cross-correlations and the sums of weighted error squares. When the recursive LSL algorithm is implemented by using fixed-point arithmetic with a short word-length, these quantities cannot be computed accurately, thereby resulting in numerical inaccuracy in the application of recursive LSL algorithms. We may overcome the cause of this numerical inaccuracy by modifying the recursive LSL algorithms in such a way that we update the reflection and regression coefficients *directly* (Ling, Manolakis and Proakis, 1985).

Below, we describe a modification of the second version of the recursive LSL algorithm (based on a priori estimation errors) that by-passes the need for computing cross-correlations. For the corresponding modification of the first version of the recursive LSL algorithm (based on a posteriori estimation errors), the reader is referred to Problem 10.

Direct Time-updates of Reflection and Regression Coefficients Using a priori Errors

Consider first the direct update of the forward reflection coefficient $\Gamma_{f,m}(n)$. Substituting line 3 into line 6 of the recursive LSL algorithm summarized in Table 9.4, we get

$$
\begin{aligned}
\Gamma_{f,m}(n) &= -\frac{\lambda \Delta_{m-1}(n-1)}{\mathscr{B}_{m-1}(n-1)} - \frac{\gamma_{m-1}(n-1)\psi_{m-1}(n-1)\,\eta^{*}_{m-1}(n)}{\mathscr{B}_{m-1}(n-1)} \\[2mm]
&= -\frac{\Delta_{m-1}(n-1)}{\mathscr{B}_{m-1}(n-2)}\frac{\lambda \mathscr{B}_{m-1}(n-2)}{\mathscr{B}_{m-1}(n-1)} \\[2mm]
&\quad -\frac{\gamma_{m-1}(n-1)\,\psi_{m-1}(n-1)\eta^{*}_{m-1}(n)}{\mathscr{B}_{m-1}(n-1)}
\end{aligned}
\tag{9.106}
$$

We now recognize that, by definition,

$$
\Gamma_{f,m}(n-1) = -\frac{\Delta_{m-1}(n-1)}{\mathscr{B}_{m-1}(n-2)}
$$

Replacing n with $n-1$ in Eq. (9.102) and rearranging, we get

$$
\lambda \mathscr{B}_{m-1}(n-2) = \mathscr{B}_{m-1}(n-1) - \gamma_{m-1}(n-1)|\psi_{m-1}(n-1)|^{2}
$$

Accordingly, we may rewrite Eq. (9.106) as

$$
\Gamma_{f,m}(n) = \Gamma_{f,m}(n-1) - \frac{\gamma_{m-1}(n-1)\psi_{m-1}(n-1)\eta^{*}_{m}(n)}{\mathscr{B}_{m-1}(n-1)},
$$
$$
m = 1, 2, \ldots, M
\tag{9.107}
$$

where we have made use of Eq. (9.103).

Similarly, we may derive the following recursions for directly updating the backward reflection coefficient $\Gamma_{b,m}(n)$ and the regression coefficient $\kappa_m(n)$, respectively:

$$\Gamma_{b,m}(n) = \Gamma_{b,m}(n-1) - \frac{\gamma_{m-1}(n-1)\,\eta_{m-1}(n)\,\psi_m^*(n)}{\mathcal{F}_{m-1}(n)}, \quad m = 1, 2, \ldots, M \qquad (9.108)$$

and

$$\kappa_m(n) = \kappa_m(n-1) + \frac{\gamma_m(n)\,\psi_m(n)\,\alpha_{m+1}^*(n)}{\mathcal{B}_m(n)}, \qquad m = 0, 1, \ldots, M \qquad (9.109)$$

A modified form of the recursive LSL algorithm based on a priori estimation errors is obtained by deleting lines 3 and 9 from Table 9.4 and then replacing lines 6, 7, and 11 by Eqs. (9.107), (9.108), and (9.109), respectively.

A distinctive feature of this new algorithm is that the a priori errors $\eta_m(n)$, $\psi_m(n)$, and $\alpha_{m+1}(n)$ are *fed back* to time-update the forward reflection coefficient $\Gamma_{f,m}(n)$, the backward reflection coefficient $\Gamma_{b,m}(n)$, and the regression coefficient $\kappa_m(n)$, respectively. Accordingly, this method of updating the reflection and regression coefficients is called the *error-feedback formula*.

The error-feedback form of the recursive LSL algorithm (based on a priori estimation errors) does not provide a computational advantage over the conventional form of the algorithm. Rather, it exhibits better numerical accuracy. Ling, Manolakis, and Proakis (1985) have used computer simulation results to compare the numerical behavior of the error-feedback and conventional forms of the algorithm. Fixed-point arithmetic with the word-length varied from 6 to 15 bits was used in the simulation. The results reported therein show that for word lengths less than or equal to 10 bits the error-feedback form of the algorithm produces output mean-squared errors that are orders of magnitude smaller than those produced by the conventional form of the algorithm, and that the error-feedback form of the algorithm using fixed-point arithmetic with a word-length of 12 bits produces an output mean-squared error that is very close to a 22 bit floating-point implementation of the algorithm.

9.12 DISCUSSION

The recursive LSL algorithms considered in this chapter, together with the RLS algorithm and its fast realizations considered in the preceding chapter, represent different recursive procedures for dealing with the same basic issue, namely, on-line solution of the linear least-squares problem. We conclude the study of recursive LSL algorithms by comparing their structure, computational complexity, the physical significance of their parameters, and the effect of roundoff errors on their operation with those of the RLS and FTF algorithms.

Structure

The RLS and FTF algorithms rely on the use of *transversal filters* for their implementation. On the other hand, the recursive LSL algorithms have a *modular* structure in that each one of their various stages consists of 3 *first-order estimators*: forward predictor, backward predictor, and joint-process estimator. This is clearly portrayed in the signal-flow graphs of Figs. 9.4 and 9.8 that represent the two basic forms of the recursive LSL algorithm using a posteriori and a priori estimation errors, respectively. Accordingly, recursive LSL algorithms have an advantage in that it is a straightforward matter to increase the prediction order. This is accomplished simply by adding one or more stages, as required, without affecting previous computations.

Computational Complexity

As in previous chapters, we use the number of operations required to perform one iteration of the algorithm as a measure of computational complexity. By operations we mean the following: multiplications, divisions, and additions/subtractions. In Table 9.6 we present a comparison of the computational complexities of four algorithms: RLS, FTF, conventional form of LSL, and error-feedback form of LSL. In each case, it is assumed that the input data are real-valued and that the order of the estimator equals M [i.e., there are $(M + 1)$ regression coefficients to be estimated]. We may thus make the following observations:

1. Like the FTF algorithm, the computational complexity of an LSL algorithm increases *linearly* with the order M. In the RLS algorithm, on the other hand, it increases with M^2.

2. The computational complexity of the conventional form of an LSL algorithm is slightly larger than that of an FTF algorithm. In particular, it requires more divisions per iteration of the algorithm.

3. The conventional and error-feedback forms of the LSL algorithm have approximately the same computational complexity, with the conventional form of the algorithm having a slight advantage.

TABLE 9.6 COMPARISON OF COMPUTATIONAL COMPLEXITIES OF FOUR LEAST-SQUARES ALGORITHMS FOR REAL DATA

Algorithm	Number of operations per iteration		
	Multiplications	Divisions	Addition/Subtractions
RLS	$2M^2 + 7M + 5$	$M^2 + 4M + 3$	$2M^2 + 6M + 4$
FTF	$7M + 12$	4	$6M + 3$
Conventional LSL	$10M + 3$	$6M + 2$	$8M + 2$
Error-feedback LSL	$18M + 3$	$4M + 1$	$9M + 2$

Physical Significance of Parameters

In both the RLS and FTF algorithms, the instantaneous values of tap weights in a transversal filter represent the information contained in the input data.   This property is exploited in *system identification* (Äström and Eykhoff, 1974; Goodwin and Payne, 1977; Ljung and Söderström, 1983; Goodwin and Sin, 1984) and *parametric spectrum analysis* (Childers, 1978; Haykin, 1983b) where a transversal filter of *fixed* length is used to *model* an unknown system or process.   In such applications, the tap weights of a transversal filter represent the desired information. Furthermore, the instantaneous values of these tap weights often provide useful information about the environment in which an adaptive transversal filter operates. An example of this arises in *adaptive equalization* (Lucky, Salz, and Weldon, 1968; Gersho, 1969; Gitlin and Magee, 1977) where, ideally, the transfer function of the equalizer is equal to the inverse of that of the (unknown) communication channel.

On the other hand, a recursive LSL algorithm presents information about the input data in the form of instantaneous values of reflection coefficients or transformed regression coefficients.   Hence, additional computation is required if the information extracted by the LSL algorithm is to be presented in the form of parameters of a multiple linear regression model as in system identification and parametric spectrum analysis, or the impulse response of a communication channel as in adaptive equalization.

Yet, in another context, the structure of a multistage lattice predictor is closely allied to the modeling of *elastic wave propagation* in a stratified solid medium (Thomson, 1950).   In particular, in *seismic signal processing,* the structural and stratigraphic features of the subsurface are formulated as a lattice filter (Robinson and Treitel, 1980).   Additionally, in *speech processing* the human vocal tract is modeled as a lattice filter, based on a cascade of acoustic tube sections (Flanagan, 1972; Markel and Gray, 1976; Rabiner and Schaefer, 1978).   The significance of the lattice filter model for speech is that the cross-sectional area of the vocal tract specifies certain reflection coefficients in each section of the acoustic tube, corresponding to a particular stage in the model.

The main point of this brief discussion is that neither the use of transversal filters in the RLS and FTF algorithms nor the use of lattice filters in the recursive LSL algorithms can claim monopoly of physical relevance.   Rather, the two classes of adaptive least-squares algorithms are complementary, each portraying a physical significance of its own.

Finite-Precision Problems

Despite the practical importance of finite-precision problems that arise in the digital implementation of adaptive least-squares algorithms, surprisingly little work has been done on the subject.   Nevertheless, special mention should be made of some recent results reported by Mueller (1981a), Lin (1984), Cioffi and Kailath (1984), Eleftheriou and Falconer (1984), Ling, Manolakis, and Proakis (1985), and Ljung and Ljung (1985). Except for the paper by Ljung and Ljung, the first 4 papers (in the context of our present discussion) deal essentially with computer simulation results obtained by using finite-precision arithmetic, and rescue devices for curing the numerical instability problem encountered in fast realizations of the RLS algorithm (i.e., fast RLS and FTF algorithms).[3] Based on the computer simulation results presented therein, it can be said that the fast realizations of the RLS algorithm exhibit the biggest tendency toward numerical instability. This observation is substantiated by the mathematical analysis of error propagation presented in the paper by Ljung and Ljung.

By *error propagation,* we mean the manner in which error introduced at an arbitrary point in an adaptive filtering algorithm propagates to future time instants. Indeed, this property is fundamental to the numerical behavior of the algorithm in the presence of roundoff errors and other numerical disturbances.

Ljung and Ljung take the approach that all adaptive filtering algorithms may be viewed as discrete-time, nonlinear dynamical systems described by the *state equations*

$$\hat{\mathbf{x}}(n) = f\big(\hat{\mathbf{x}}(n-1), \mathbf{z}(n)\big) \tag{9.110}$$

$$\mathbf{y}(n) = \mathbf{g}\big(\hat{\mathbf{x}}(n), \mathbf{z}(n)\big) \tag{9.111}$$

where $\hat{\mathbf{x}}(n)$ is the estimate of the state vector, $\mathbf{z}(n)$ is the input vector, and $\mathbf{y}(n)$ is the output vector. The different adaptive least-squares algorithms correspond to different choices of state vector and associated functions $f(\ )$ and $g(\ )$. The state equations (9.110) and (9.111) are ideal in that they represent the pure analog (infinite precision) form of the algorithms. However, digital (finite precision) implementation of the algorithms will generate round-off errors and representation errors in the computer. Accordingly, the various vectors in the state equations deviate from their exact values. In their study, Ljung and Ljung focus on the propagation of a single error in the state vector occurring at time n_0. Let this error be denoted by $\delta\mathbf{x}(n_0)$. We may thus represent the numerically perturbed form of the algorithms by the modified state equations:

$$\mathbf{x}'(n) = f\big(\mathbf{x}'(n-1), \mathbf{z}(n)\big) \tag{9.112}$$

$$\mathbf{y}'(n) = g\big(\mathbf{x}'(n), \mathbf{z}(n)\big) \tag{9.113}$$

where $\mathbf{x}'(n)$ and $\mathbf{y}'(n)$ denote the values by which the state vector and output vector are represented in the computer, respectively. The state equation (9.112) is subject to the *initial conditions*

$$\mathbf{x}'(n_0) = \hat{\mathbf{x}}(n_0) + \delta\mathbf{x}(n_0) \tag{9.114}$$

---

[3]In the paper by Ling, Manolakis, and Proakis (1985) modifications of the two basic forms of recursive LSL algorithms (based on a posteriori and a priori estimation errors) are presented, which appear to have improved numerical properties. These modifications were discussed in Section 9.11.

To study the effect of the error $\delta\mathbf{x}(n_0)$ on the output vector $\mathbf{y}'(n)$, it is sufficient to consider its effect on the state vector $\mathbf{x}'(n)$, since the function $g(\ )$ in Eq. (9.113) is continuous in $\mathbf{x}'(n)$.

Thus, the problem to be solved is that of a *classical Lyapunov stability problem*[4] for the nonlinear difference equation [see Eq. (9.112)], subject to the initial condition [see Eq. (9.114)]. In particular, the requirement is to determine how the state vector reacts to perturbations in the initial conditions.

Briefly, we say the solution $\mathbf{x}(n)$ [corresponding to $\delta\mathbf{x}(n_0) = \mathbf{0}$] is *stable* if for all $\varepsilon \geq 0$ there exists a δ, such that $\|\delta\mathbf{x}(n_0)\| < \delta$ implies that $\|\mathbf{x}'(n) - \mathbf{x}(n)\| < \varepsilon$ for all $n \leq n_0$. We say the solution is *asymptotically stable* if it is stable and there exists a $\delta > 0$ such that $\|\delta\mathbf{x}(n_0)\| < \delta$ implies that $\|\mathbf{x}'(n) - \mathbf{x}(n)\|$ approaches zero as time n approaches infinity. We say that the solution is *exponentially stable* if it is asymptotically stable and there exists C and $0 < \lambda < 1$, such that

$$\|\mathbf{x}'(n) - \mathbf{x}(n)\| \leq C\lambda^{n-n_0}, \quad n \geq n_0 \text{ if } \|\delta\mathbf{x}(n_0)\| < \delta$$

Exponential stability of a system implies *bounded-input bounded-output stability*. Accordingly, if the error propagation is found to be exponentially stable, then bounded output errors would be guaranteed. If, on the other hand, the error propagation is found to be unstable, so that the effect of a single disturbance tends to infinity, then the algorithm of interest cannot be used in a continuous operation.

The error propagation calculations are relatively straightforward for the RLS and recursive LSL algorithms [Ljung and Ljung (1985)]. However, the fast realizations of the RLS algorithm are considerably more difficult to analyze. To study the problem in this latter case, Ljung and Ljung use a counter example. The results obtained by Ljung and Ljung may be summarized as follows:

1. The RLS and recursive LSL algorithms exhibit an error propagation that is exponentially stable. In particular, the effect of a single disturbance decays exponentially, and the base of the decay is equal to the exponential weighting factor λ.

2. The fast implementations of the RLS algorithm do not have this nice property, and may be unstable with respect to such errors when the exponential weighting factor $\lambda < 1$. Hence, such algorithms are not suited for continuous operation, unless special *rescue devices* are incorporated.

Thus, the paper by Ljung and Ljung presents the first mathematical justification for the numerical superiority of the recursive LSL algorithm over the FTF algorithm, and the necessity of using a reliable rescue device for ensuring continuous operation of the FTF algorithm.

[4]For a discussion of the classical Lyapunov stability problem, see Brogan (1985).

PROBLEMS

1. The deterministic correlation matrix $\Phi_{m+1}(n)$ is postmultiplied by the Hermitian transpose of the lower triangular matrix $\mathbf{L}_m(n)$, where $\mathbf{L}_m(n)$ is defined by Eq. (9.68). Show that the product $\Phi_{m+1}(n)\,\mathbf{L}_m^H(n)$ consists of a lower triangular matrix whose diagonal elements equal the various sums of weighted backward prediction-error squares, $\mathcal{B}_0(n)$, $\mathcal{B}_1(n)$, . . . $\mathcal{B}_m(n)$. Hence, show that the product $\mathbf{L}_m(n)\,\Phi_{m+1}(n)\,\mathbf{L}_m^H(n)$ is a diagonal matrix, as shown by

$$\mathbf{D}_{m+1}(n) = \mathrm{diag}[\mathcal{B}_0(n),\,\mathcal{B}_1(n),\,.\,.\,.\,,\,\mathcal{B}_m(n)]$$

2. Use the deterministic form of the Gram–Schmidt orthogonalization procedure to transform the input sequence $u(n)$, . . . , $u(n-m)$ into a new sequence $b_0(n)$, . . . , $b_m(n)$ whose elements are uncorrelated with each other in a time-averaged sense.

3. Consider the index of performance

$$\mathcal{E}_m(n) = \sum_{i=1}^{n} \lambda^{n-i}|d(i) - \mathbf{b}_{m+1}^T(i)\,\boldsymbol{\kappa}_m^*(n)|^2, \qquad m = 1, 2, .\,.\,.\,, M$$

where λ is the exponential factor, $d(i)$ is the desired response, $\mathbf{b}_{m+1}(i)$ is the vector of backward a posteriori prediction errors, and $\boldsymbol{\kappa}_m(n)$ is the corresponding vector of regression coefficients that are fed by the backward prediction errors (see Fig. 9.4). Hence, show that $\mathcal{E}_m(n)$ is minimized by choosing

$$\mathbf{D}_{m+1}(n)\,\boldsymbol{\kappa}_m(n) = \mathbf{t}_{m+1}(n)$$

where $\mathbf{D}_{m+1}(n)$ is the deterministic correlation matrix of the backward a posteriori prediction errors, and $\mathbf{t}_{m+1}(n)$ is the cross-correlation vector between the backward a posteriori prediction errors and the desired response.

4. Consider the case where the input samples $u(n)$, $u(n-1)$, . . . , $u(n-M)$ have a *joint Gaussian distribution with zero mean*. Assume that, within a scaling factor, the ensemble-averaged correlation matrix $\mathbf{R}_{M+1}$ of the input signal is equal to its deterministic correlation matrix $\Phi_{M+1}(n)$ for time $n \geq M$. Show that the log-likelihood function for this input includes a term equal to the parameter $\gamma_M(n)$ associated with the adaptive LSL algorithm. For this reason, the parameter $\gamma_M(n)$ is sometimes referred to as a *likelihood variable*.

5. In the adaptive lattice equalizer considered in Section 9.9, the impulse response of the channel is symmetric with respect to the mid-point $k = 2$. Would you expect the regression coefficients in this equalizer to be likewise symmetric with respect to its mid-point? Justify your answer.

6. Let $\hat{d}(n|\mathcal{U}_{n-m+1})$ denote the least-squares estimate of the desired response $d(n)$, given the inputs $u(n-m+1)$, . . . , $u(n)$ that span the space $\mathcal{U}_{n-m+1}$. Similarly, let $\hat{d}(n|\mathcal{U}_{n-m})$ denote the least-squares estimate of the desired response, given the inputs $u(n-m)$, $u(n-m+1)$, . . . , $u(n)$ that span the space $\mathcal{U}_{n-m}$. In effect, the latter estimate exploits an additional piece of information represented by the input $u(n-m)$. Show that this new information is represented by the corresponding backward prediction error $b_m(n)$. Also, show that the two estimates are related by the recursion

$$\hat{d}(n|\mathcal{U}_{n-m}) = \hat{d}(n|\mathcal{U}_{n-m+1}) + \kappa_m^*(n)\,b_m(n)$$

where $\kappa_m(n)$ denotes the pertinent regression coefficient in the joint-process estimator. Compare this result with that of Section 6.2 dealing with the concept of innovations.

7. Let $\mathbf{\Phi}(n)$ denote the $(M + 1)$-by-$(M + 1)$ deterministic correlation matrix of the input data $\{u(n)\}$. Using the deterministic theory presented in Section 9.8, show that the change of variables to backward prediction errors brought about by using the lattice predictor achieves exactly the Cholesky decomposition of the matrix $\mathbf{P}(n) = \mathbf{\Phi}^{-1}(n)$.

8. Consider the joint-process estimator of Fig. 9.4. Show that the last transformed regression coefficient $\kappa_M(n)$ equals the last element $\hat{w}_M(n)$ contained in the $(M + 1)$-by-1 least-squares estimate $\hat{\mathbf{w}}_M(n)$ of the coefficient vector in the multiple linear regression model.

9. Expand the joint-process estimator of Fig. 9.4 so as to include (in modular form) the least-squares estimate of the desired response $d(n)$ for increasing prediction order m.

10. In Section 9.11 we discussed a modification of the a priori error LSL algorithm by using a form of error feedback. In this problem we consider the corresponding modified version of the a posteriori LSL algorithm. In particular, show that

$$\Gamma_{f,m}(n) = \frac{\gamma_m(n-1)}{\gamma_{m-1}(n-1)}\left[\Gamma_{f,m}(n-1) - \frac{1}{\lambda}\frac{b_{m-1}(n-1)\,f_{m-1}^*(n)}{\mathcal{B}_{m-1}(n-2)\,\gamma_{m-1}(n-1)}\right]$$

$$\Gamma_{b,m} = \frac{\gamma_m(n)}{\gamma_{m-1}(n-1)}\left[\Gamma_{b,m}(n-1) - \frac{1}{\lambda}\frac{f_{m-1}(n)\,b_{m-1}^*(n-1)}{\mathcal{F}_{m-1}(n-1)\,\gamma_{m-1}(n-1)}\right]$$

11. Table 9.6 includes a summary of the computational requirements of four algorithms: RLS, FTF, conventional LSL, and the error-feedback LSL. Confirm the validity of each entry in this table.

12. Table 9.7 is a summary of the *normalized LSL algorithm*. The normalized parameters are defined by

$$\bar{f}_m(n) = \frac{f_m(n)}{\mathcal{F}_m^{1/2}(n)\gamma_m^{1/2}(n-1)}$$

$$\bar{b}_m(n) = \frac{b_m(n)}{\mathcal{B}_m^{1/2}(n)\,\gamma_m^{1/2}(n)}$$

$$\bar{\Delta}_m(n) = \frac{\Delta_m(n)}{\mathcal{F}_m^{1/2}(n)\,\mathcal{B}_m^{1/2}(n-1)}$$

Hence, derive the steps summarized in Table 9.7.

TABLE 9.7 SUMMARY OF NORMALIZED LSL ALGORITHM

$$\bar{\Delta}_{m-1}(n) = \bar{\Delta}_{m-1}(n-1)[1 - |\bar{f}_{m-1}(n)|^2]^{1/2}$$
$$[1 - |\bar{b}_{m-1}(n-1)|^2]^{1/2} + \bar{b}_{m-1}^*(n-1)\,\bar{f}_{m-1}^*(n)$$

$$\bar{b}_m(n) = \frac{\bar{b}_{m-1}(n-1) - \bar{\Delta}_{m-1}\bar{f}_{m-1}(n)}{[1 - |\bar{\Delta}_{m-1}(n)|^2]^{1/2}[1 - |\bar{f}_{m-1}(n)|^2]^{1/2}}$$

$$\bar{f}_m(n) = \frac{\bar{f}_{m-1}(n) - \bar{\Delta}_{m-1}^*(n)\,\bar{b}_{m-1}(n-1)}{[1 - |\bar{\Delta}_{m-1}(n)|^2]^{1/2}[1 - |\bar{b}_{m-1}(n-1)|^2]^{1/2}}$$

Computer-Oriented Problems

13. *Computer experiment on adaptive two-stage lattice predictor with autoregressive input of order 1:* An autoregressive (AR) process of order 1 is described by the difference equation

$$u(n) + a_1 u(n - 1) = v(n)$$

where $\{v(n)\}$ is a white-noise process of zero mean and variance σ_v^2. The AR parameter a_1 and variance σ_v^2 are assigned the following values:

$$a_1 = -0.8182, \qquad \sigma_v^2 = 0.3305$$

The AR process $\{u(n)\}$ is applied to a two-stage lattice predictor that uses the LSL algorithm to adjust its reflection coefficients. The weighting factor $\lambda = 1.0$.

(a) Use a computer to generate a 256-sample sequence to represent the AR process $\{u(n)\}$. You may use a random-number subroutine for the noise $\{v(n)\}$. Hence, plot the learning curve of the LSL algorithm by averaging over 200 independent trials of the experiment for $\delta = 1$.

(b) Repeat the computer experiment for the adaptive two-tap predictor with parameter δ fixed at 1, and for each of the following two sets of parameter values:

$$\text{(ii)} \quad a_1 = -0.5, \qquad \sigma_v^2 = 0.75$$

$$\text{(iii)} \quad a_1 = -0.9802, \qquad \sigma_v^2 = 0.0392$$

(c) By averaging the ensemble-averaged learning curve for the last 200 iterations, compute the average mean-squared error for parameter sets (i), (ii) and (iii). Compare your results with theory.

(d) By averaging the steady-state values of the two reflection coefficients over 200 independent trials, compute their mean values and compare your results with theory.

14. *Computer experiment on adaptive equalizer:* This problem is a continuation of the computer experiment of Section 9.9 on adaptive equalization using the recursive LSL algorithm (based on a posteriori estimation errors).

Perform the following computations:

(a) To investigate the effect of varying the initialization parameter δ on the performance of the algorithm, plot the learning curves of the algorithm for $\delta = 0.002, 0.004$. In each case, compute the mean-squared error by ensemble-averaging the squared magnitude of the final a priori estimation error $\alpha_{M+1}(n)$, with $M = 10$, over 200 independent trials of the experiment, for the channel parameter $W = 2.9, 3.1, 3.3, 3.5$. Each learning curve should extend up to 200 iterations.

(b) For $\delta = 0.003$ and channel parameter $W = 2.9, 3.1, 3.3, 3.5$, plot the final conversion factor $\gamma_{M+1}(n)$, $M = 10$, versus the number of iterations n (up to 400). For this plot, use ensemble-averaged values of $\gamma_{M+1}(n)$ by averaging your results over 200 independent trials of the experiment. Hence, show that the curve for $\gamma_{M+1}(n)$ is closely approximated by the formula

$$\gamma_{M+1}(n) \approx 1 - \frac{M + 1}{n}, \quad n \geq M + 1$$

Justify the theoretical validity of this formula.

(c) For $W = 2.9$, $\delta = 0.003$, plot the ensemble-averaged value of the squared magnitude of $\alpha_{m+1}(n)$ versus the number of iterations (up to 50), for stage number $m = 0, 1, \ldots, 10$. Perform the averaging over 200 independent trials of the experiment.

(d) For $\delta = 0.003$ and $W = 2.9, 3.1, 3.3, 3.5$, compute the ensemble-averaged square of the final a posteriori estimation error $e_{M+1}(n)$, and plot your results versus the number of interations, n. Do the averaging over 200 independent trials of the experiment. Compare this new set of learning curves with those based on the a priori estimation error $\alpha_{M+1}(n)$, and comment on the differences between them.

(e) For $\delta = 0.003$, compute the transformed regression coefficients of the joint-process estimator for one trial of the experiment, with $W = 2.9, 3.1, 3.3, 3.5$. Do your computations after 10, 20, 30, 40, 50 iterations of the algorithm.

Comment on your results.

10

Recursive Least-squares Estimation Using Systolic Arrays

In Chapter 8, we formulated the recursive solution to the linear least-squares problem by using transversal filters. In Chapter 9, we reformulated the recursive solution to the problem by using the pipelined lattice structure to transform the input data sequence into another sequence of backward prediction errors that are orthogonal to each other in a time-averaged sense. This transformation may be viewed as the deterministic form of Gram–Schmidt orthogonalization. In this final chapter, we complete the study of the linear least-squares problem by using another orthogonal triangularization process based on a unitary transformation known as the *Givens rotation* (Givens, 1958; Ralston, 1965). This new procedure may be implemented efficiently by means of systolic arrays.

The idea of a *systolic array* was developed by Kung and Leiserson (1978). It consists of an array of individual *processing cells* that are arranged in the form of a regular structure. Each cell in the array is provided with local memory of its own, and each cell is connected only to its nearest neighbors. The array is designed such that regular streams of data are clocked through it in a highly rhythmic fashion, much like the pumping action of the human heart; hence, the name "systolic" (Kung, 1982).

The application of systolic arrays to provide *on-the-fly* the least-squares fit to input data up to any time of interest was first described by Gentleman and Kung (1981). The proposed structure consists of a pair of systolic array sections chained together: a *triangular* array to implement the pipelined sequence of Givens rotations and a *linear* array to compute the least-squares value of the weight vector (assuming that the data matrix has full rank). McWhirter (1983) describes a simplification of this structure by eliminating the linear array component. This structural simplification is of practical value in linear least-squares applications (e.g., adaptive noise cancellation, adaptive beamforming) where the main requirement is to compute the estimation error and it is not necessary to know the least-squares value of the weight vector. Schreiber and Kuekes (1985) describe another systolic array structure that is well-suited for the design of an adaptive beamformer with *minimum-variance distortionless response* (MVDR). It is noteworthy that, unlike the FTF and LSL algorithms, the theory of adaptive filters using systolic arrays finds applications in both time- and space-domains.

In this chapter we study the three systolic array structures developed by Kung and Gentleman, McWhirter, and Schreiber and Kuekes. We begin the discussion by reviewing the terminology of the linear least-squares problem.

10.2 SOME PRELIMINARIES

The terminology that we use in this chapter follows a format similar to that in Chapter 7 with some minor modifications in that we assume the use of prewindowing here. Suppose that we are given a time series consisting of the observations $u(1)$, $u(2), \ldots, u(n)$. We use a subset of this time series to form an M-by-1 *input vector*[1] $\mathbf{u}(i)$, defined by

$$\mathbf{u}^T(i) = [u(i), u(i - 1), \ldots, u(i - M + 1)] \qquad (10.1)$$

We use this input vector to make an estimate of some *desired response* $d(i)$ as the inner product $\mathbf{w}^H(n)\mathbf{u}(i)$, where $\mathbf{w}(n)$ is an M-by-1 *weight vector* whose transpose equals

$$\mathbf{w}^T(n) = [w_1(n), w_2(n), \ldots, w_M(n)] \qquad (10.2)$$

We assume that the weight vector $\mathbf{w}(n)$ is held constant during the observation

---

[1] We refer to $\mathbf{u}(i)$ as the input vector rather than the tap-input vector and, likewise, we refer to $\mathbf{w}(n)$ as the weight vector rather than the tap-weight vector simply because we wish to reserve the use of tap-input and tap-weight vectors for the transversal (i.e., tapped-delay-line) filter.

interval $1 \leq i \leq n$. We define the *estimation error* $e(i)$ as

$$e(i) = d(i) - \mathbf{w}^H(n) \, \mathbf{u}(i) \tag{10.3}$$

According to the *exponentially weighted, prewindowed method*, we minimize the index of performance

$$\mathscr{E}(n) = \sum_{i=1}^{n} \lambda^{n-i} |e(i)|^2 \tag{10.4}$$

where λ is the exponential weighting factor. Let the *n*-by-*M* matrix $\mathbf{A}(n)$ denote the *data matrix* that is defined by (using the notation of previous chapters)

$$\mathbf{A}^H(n) = [\mathbf{u}(1), \mathbf{u}(2), \ldots, \mathbf{u}(M), \ldots, \mathbf{u}(n)]$$

$$= \begin{bmatrix} u(1) & u(2) & \cdots & u(M) & \cdots & u(n) \\ 0 & u(1) & \cdots & u(M-1) & \cdots & u(n-1) \\ \cdot & \cdot & \cdot & \cdot & \cdot & \cdot \\ \cdot & \cdot & \cdot & \cdot & \cdot & \cdot \\ \cdot & \cdot & \cdot & \cdot & \cdot & \cdot \\ 0 & 0 & & u(1) & \cdots & u(n-M+1) \end{bmatrix} \tag{10.5}$$

Let the *n*-by-1 vector $\boldsymbol{\varepsilon}(n)$ denote the *error vector*, defined by

$$\boldsymbol{\varepsilon}^H(n) = [e(1), e(2), \ldots, e(n)] \tag{10.6}$$

Let the *n*-by-1 vector $\mathbf{b}(n)$ denote the *desired response vector*, defined by

$$\mathbf{b}^H(n) = [d(1), d(2), \ldots, d(n)] \tag{10.7}$$

We may then redefine the index of performance as

$$\mathscr{E}(n) = \boldsymbol{\varepsilon}^H(n) \, \boldsymbol{\Lambda}(n) \, \boldsymbol{\varepsilon}(n) \tag{10.8}$$

where $\boldsymbol{\Lambda}$ is the *n*-by-*n exponential weighting matrix*

$$\boldsymbol{\Lambda}(n) = \mathrm{diag}[\lambda^{n-1}, \lambda^{n-2}, \ldots, 1] \tag{10.9}$$

The inclusion of the exponential weighting matrix $\boldsymbol{\Lambda}(n)$ has the effect of progressively weighting against the preceding columns of the data matrix $\mathbf{A}^H(n)$ in favor of the last column. The last column of $\mathbf{A}^H(n)$ corresponds to the input vector $\mathbf{u}(n)$ at time n, for which the weighting factor is unity.

We may also express the index of performance $\mathscr{E}(n)$ as the squared Euclidean norm

$$\mathscr{E}(n) = \|\boldsymbol{\Lambda}^{1/2}(n) \, \boldsymbol{\varepsilon}(n)\|^2 \tag{10.10}$$

The problem we have to solve is to find the least-squares value of the weight vector that minimizes the performance index $\mathscr{E}(n)$.

10.3 *GIVENS ROTATION*

In Chapter 7, we reduced the linear least-squares problem to that of solving the deterministic normal equation in which the deterministic (time-averaged) correlation matrix $\mathbf{\Phi}(n)$ and cross-correlation vector $\mathbf{\theta}(n)$ represent the known parameters and the least-squares weight vector $\hat{\mathbf{w}}(n)$ represents the unknown. An alternative approach that is computationally more efficient is to operate on the given data matrix directly by performing an orthogonal triangularization and thereby exploit the special structure of this linear transformation.

The *orthogonal triangularization process* may be carried out by using various techniques. One such technique is the *Gram–Schmidt orthogonalization procedure*[2] that provides the mathematical basis for the pipelined lattice predictor, an approach we studied in Chapter 9. Another powerful technique is the *Givens rotation procedure*,[3] so called as it involves an elementary transformation in the form of *plane rotation* (Givens, 1958). Through successive applications of the Givens rotation, we may develop a very efficient algorithm for solving the linear least-squares problem, whereby the orthogonal triangularization of the data matrix is recursively updated as each new set of data enters the computation (Gentleman and Kung, 1981; McWhirter, 1983).

The matrix structure that arises in such a study of the linear least-squares problem, and for which the Givens rotation procedure is particularly well suited, is *partially triangularized* in that it may be partitioned into an upper triangular matrix, followed by a matrix of zeros, and finally a row of nonzero elements. The requirement is to annihilate all the nonzero elements in the last row of the matrix. This we do by successive applications of the Givens rotation, whereby the elements are annihilated one at a time. Next we present a description of this operation.

Consider an *n*-by-*M* complex-valued matrix $\mathbf{Y}$, where $n > M$ and M is some fixed number. We assume that the first M rows of matrix $\mathbf{Y}$ constitute an upper triangular matrix $\mathbf{R}$, followed by an $(n - M - 1)$-by-$(n - M - 1)$ null matrix, and then finally a row of M nonzero elements. Suppose we wish to annihilate the *m*th element in the last row (i.e., row *n*) of matrix $\mathbf{Y}$ by *rotating* it with row *m* of the matrix, where $m = 1, 2, \ldots, M$. To accomplish this, we premultiply the

---

[2]The Gram–Schmidt orthogonalization procedure for adaptive filtering may also be implemented using a systolic array processor. Gallivan and Leiserson describe such a procedure; in particular, they base their implementation on a modification of the Gram–Schmidt algorithm that avoids the use of division (Gallivan and Leiserson, 1984).

[3]The *Householder transformation* is yet another technique that can be used to carry out the orthogonal triangularization process for solving the linear least-squares problem (Householder, 1958; Golub, 1965; Stewart, 1973). However, in the context of our present discussion, the use of the Givens rotation offers the advantage of being amenable to recursive computation whereby the rotation parameters are updated on a sample–by–sample basis.

matrix $\mathbf{Y}$ by an n-by-n transformation matrix $\mathbf{G}$ the elements of which are defined as follows:[4]

$$
g_{ik} = \begin{cases}
\cos \phi, & i, \, k = m \\
\sin \phi e^{j\theta}, & i = m, \, k = n \\
-\sin \phi \, e^{-j\theta}, & i = n, \, k = m \\
\cos \phi, & i, \, k = n \\
1, & i, \, k = 1, 2, \ldots, n - 1, \text{ excluding } m \\
0, & \text{otherwise}
\end{cases}
\tag{10.11}
$$

We refer to the linear transformation defined by the matrix of Eq. (10.11) as a *plane rotation* because (in the example considered here) it consists of a rotation of the mth and nth coordinates of matrix $\mathbf{Y}$ through angles ϕ and θ. We may express the product $\mathbf{GY}$, highlighting the elements of particular interest to the discussion here, as follows:

$$
\mathbf{GY} = \begin{bmatrix}
& \vdots & & \vdots & \\
\cdots & \cos \phi & \cdots & \sin \phi e^{j\theta} & \\
& \vdots & & \vdots & \\
\cdots & -\sin \phi e^{-j\theta} & \cdots & \cos \phi &
\end{bmatrix}
\begin{bmatrix}
& \vdots & \\
\cdots & y_{mm} & \cdots \\
& \vdots & \\
\cdots & y_{nm} & \cdots
\end{bmatrix}
\tag{10.12}
$$

From this formulation, we see that the mth element in the last row of product $\mathbf{GY}$ is annihilated provided that the angles θ and ϕ satisfy the condition

$$
-\sin\phi \, e^{-j\theta} y_{mm} + \cos\phi \, y_{nm} = 0
\tag{10.13}
$$

where y_{mm} and y_{nm} are the mmth and nmth elements of matrix $\mathbf{Y}$, respectively.

In general, the matrix $\mathbf{Y}$ is complex-valued. For Eq. (10.13) to hold, the real and imaginary parts of the expression on the left side must individually equal zero. Thus, recognizing that

$$
\cos^2\phi + \sin^2\phi = 1
$$

and solving Eq. (10.13) for the parameters of the transformation, we get

$$
\cos\phi = \frac{|y_{mm}|}{\sqrt{|y_{nm}|^2 + |y_{mm}|^2}}
\tag{10.14}
$$

$$
\sin\phi e^{j\theta} = \left(\frac{y_{nm}^*}{y_{mm}^*}\right)\cos\phi
$$

[4]For complex-valued data, the transformation needed to annihilate the element of interest requires two degrees of freedom, as shown in Eq. (10.11). When the matrix $\mathbf{Y}$ is real valued, however, the angle θ in Eq. (10.11) reduces to zero, leaving ϕ as the only degree of freedom. Clearly, the linear transformation for real-valued data is a special case of Eq. (10.11).

When both y_{nm} and y_{mm} are real valued, the angle θ reduces to zero. Note also that

$$\mathbf{G}^H \mathbf{G} = I$$

That is, the Givens rotation operator $\mathbf{G}$ is a *unitary matrix*.

For convenience of notation, we write the parameters of the Givens rotation operator $\mathbf{G}$ as

$$c = \cos\phi \qquad (10.15)$$
$$s = \sin\phi e^{j\theta}$$

where, of course, c is always real and s is complex valued for complex-valued data.

The Givens rotation as described operates on the mth and nth rows of matrix $\mathbf{Y}$ to annihilate the mth element in the last (i.e., nth) row. Both the mth and nth rows of the product $\mathbf{GY}$ will, of course, be modified in the process. In particular, the element y_{mm} is replaced by the new value

$$\cos\phi y_{mm} + \sin\phi e^{j\theta} y_{nm} = \frac{y_{mm}}{|y_{mm}|} \sqrt{|y_{mm}|^2 + |y_{nm}|^2} \qquad (10.16)$$

where we have made use of the definitions for $\cos\phi$ and $\sin\phi \exp(j\theta)$ given in Eq. (10.14). Naturally, all other elements of the product $\mathbf{GY}$ (i.e., those outside its mth and nth rows) are unaffected by the transformation.

We may now describe the procedure for orthogonal triangularization of the partially triangularized matrix $\mathbf{Y}$. We first annihilate the leading element in the last row of matrix $\mathbf{Y}$ by applying the Givens rotation to operate on the first and last rows of matrix $\mathbf{Y}$. We next annihilate the second element in the last row of the transformed matrix by applying the Givens rotation to operate on the second and last rows. We continue in this fashion until all the M elements in the last row have been annihilated. Let $\mathbf{G}_1, \mathbf{G}_2, \ldots, \mathbf{G}_M$ denote the sequence of Givens rotations thus computed, which transforms the matrix $\mathbf{Y}$ into upper triangular form, as shown by

$$\mathbf{G}_M \cdots \mathbf{G}_2 \mathbf{G}_1 \mathbf{Y} = \begin{bmatrix} \mathbf{R} \\ \mathbf{O} \\ \mathbf{0}^T \end{bmatrix} \qquad (10.17)$$

where $\mathbf{R}$ is an upper triangular M-by-M matrix and $\mathbf{O}$ is the $(n-M-1)$-by-$(n-M-1)$ null matrix, and $\mathbf{0}^T$ is the M-by-1 null row vector. Equation (10.17) describes the desired orthogonal triangularization process.

In the next section, we exploit the orthogonal triangularization of the pre-windowed data matrix to develop a new recursive solution to the linear least-squares problem.

10.4 SOLUTION OF THE LINEAR LEAST-SQUARES PROBLEM USING QR-DECOMPOSITION

The essence of the linear least-squares estimation problem is to choose the weight vector $\mathbf{w}(n)$ [for successive values of time $n = M, M + 1, \ldots$, for which the data matrix $\mathbf{A}(n)$ has full rank] so as to minimize the index of performance $\mathscr{E}(n)$, defined in Eq. (10.10) as the squared norm of the weighted error vector $\mathbf{\Lambda}^{1/2}(n)\mathbf{\varepsilon}(n)$. We solve the problem by using the method of orthogonal triangularization that offers good numerical stability.

Since the norm of a vector is unaffected by premultiplication by a unitary matrix (see Problem 1), we may also express the index of performance $\mathscr{E}(n)$ as

$$\mathscr{E}(n) = \|\mathbf{Q}(n)\mathbf{\Lambda}^{1/2}(n)\mathbf{\varepsilon}(n)\|^2 \tag{10.18}$$

where $\mathbf{Q}(n)$ is an n-by-n unitary matrix. The exact role of this unitary matrix will be defined presently. The error vector equals

$$\mathbf{\varepsilon}(n) = \mathbf{b}(n) - \mathbf{A}(n)\mathbf{w}(n) \tag{10.19}$$

Hence, by using this expression for the error vector $\mathbf{\varepsilon}(n)$, we may express the vector $\mathbf{Q}(n)\mathbf{\Lambda}^{1/2}(n)\mathbf{\varepsilon}(n)$ as

$$\mathbf{Q}(n)\mathbf{\Lambda}^{1/2}(n)\mathbf{\varepsilon}(n) = \mathbf{Q}(n)\mathbf{\Lambda}^{1/2}(n)\mathbf{b}(n) - \mathbf{Q}(n)\mathbf{\Lambda}^{1/2}(n)\mathbf{A}(n)\mathbf{w}(n) \tag{10.20}$$

According to Eq. (10.20), the n-by-n unitary matrix $\mathbf{Q}(n)$ operates on the exponentially weighted versions of $\mathbf{A}(n)$ and $\mathbf{b}(n)$.

For any value of $n \geq M$, we assume that the unitary matrix $\mathbf{Q}(n)$ is generated in such a way that it applies an orthogonal triangularization to the weighted data matrix $\mathbf{\Lambda}^{1/2}(n)\mathbf{A}(n)$, as shown by

$$\mathbf{Q}(n)\mathbf{\Lambda}^{1/2}(n)\mathbf{A}(n) = \begin{bmatrix} \mathbf{R}(n) \\ \mathbf{O} \end{bmatrix}, \qquad n \geq M \tag{10.21}$$

where $\mathbf{R}(n)$ is an M-by-M *upper triangular matrix*, and $\mathbf{O}$ is the $(n - M)$-by-$(n - M)$ null matrix. According to Eq. (10.21), the n-by-M matrix $\mathbf{\Lambda}^{1/2}(n)\mathbf{A}(n)$ may be expressed as a product of an orthogonal matrix and an upper triangular matrix; such a factorization is referred to in the literature as the *QR-decomposition* (Stewart, 1973). The Givens rotation procedure described in Section 10.3 provides one method for computing the QR-decomposition.[5]

In any event, through the use of the QR-decomposition of Eq. (10.21), we

---

[5]A distinction between the decompositions given in Eqs. (10.17) and (10.21) should, however, be noted. The decomposition of Eq. (10.17), based on the Givens rotation, applies to a partially triangularized matrix $\mathbf{Y}$. The decomposition of Eq. (10.21), on the other hand, applies to the weighted data matrix $\mathbf{\Lambda}^{1/2}(n)\, \mathbf{A}(n)$ that is not, in general, partially triangularized in the same way as $\mathbf{Y}$. The role of Givens rotations will become apparent in Section 10.6 when we deal with the recursive formulation of the linear least-squares problem.

may express the second term on the right side of Eq. (10.20) as

$$\mathbf{Q}(n)\mathbf{\Lambda}^{1/2}(n)\mathbf{A}(n)\mathbf{w}(n) = \begin{bmatrix} \mathbf{R}(n) \\ \mathbf{O} \end{bmatrix} \mathbf{w}(n)$$

$$= \begin{bmatrix} \mathbf{R}(n)\mathbf{w}(n) \\ \cdots\cdots\cdots \\ \mathbf{0}_{n-M} \end{bmatrix}, \quad n \geq M \tag{10.22}$$

where $\mathbf{0}_{n-M}$ is the $(n - M)$-by-1 null vector.

Suppose, next, the n-by-n unitary matrix $\mathbf{Q}(n)$ is partitioned as

$$\mathbf{Q}(n) = \begin{bmatrix} \mathbf{F}(n) \\ \cdots\cdots \\ \mathbf{S}(n) \end{bmatrix}, \quad n \geq M \tag{10.23}$$

where $\mathbf{F}(n)$ is an M-by-n matrix consisting of the first M rows of $\mathbf{Q}(n)$, and $\mathbf{S}(n)$ is an $(n - M)$-by-n matrix consisting of the remaining rows. Accordingly, we may express the first term on the right side of Eq. (10.20) as

$$\mathbf{Q}(n)\,\mathbf{\Lambda}^{1/2}(n)\,\mathbf{b}(n) = \begin{bmatrix} \mathbf{F}(n) \\ \cdots\cdots \\ \mathbf{S}(n) \end{bmatrix} \mathbf{\Lambda}^{1/2}(n)\,\mathbf{b}(n)$$

$$= \begin{bmatrix} \mathbf{p}(n) \\ \cdots\cdots \\ \mathbf{v}(n) \end{bmatrix}, \quad n \geq M \tag{10.24}$$

where $\mathbf{p}(n)$ is an M-by-1 vector defined by

$$\mathbf{p}(n) = \mathbf{F}(n)\,\mathbf{\Lambda}^{1/2}(n)\,\mathbf{b}(n), \quad n \geq M \tag{10.25}$$

and $\mathbf{v}(n)$ is an $(n - M)$-by-1 vector defined by

$$\mathbf{v}(n) = \mathbf{S}(n)\,\mathbf{\Lambda}^{1/2}(n)\,\mathbf{b}(n), \quad n \geq M \tag{10.26}$$

Note that the dimension of the vector $\mathbf{p}(n)$ is fixed at the value M for all n. On the other hand, the dimension of the vector $\mathbf{v}(n)$ is variable in that it increases with n linearly. Thus, using Eqs. (10.22) and (10.24), we may rewrite Eq. (10.20) as

$$\mathbf{Q}(n)\mathbf{\Lambda}^{1/2}(n)\mathbf{\varepsilon}(n) = \begin{bmatrix} \mathbf{p}(n) \\ \cdots\cdots \\ \mathbf{v}(n) \end{bmatrix} - \begin{bmatrix} \mathbf{R}(n)\,\mathbf{w}(n) \\ \cdots\cdots\cdots \\ \mathbf{0}_{n-M} \end{bmatrix}$$

$$= \begin{bmatrix} \mathbf{p}(n) - \mathbf{R}(n)\,\mathbf{w}(n) \\ \cdots\cdots\cdots\cdots \\ \mathbf{v}(n) \end{bmatrix} \tag{10.27}$$

To solve the linear least-squares problem, we have to choose the weight vector $\mathbf{w}(n)$ so as to minimize the performance index $\mathscr{E}(n)$ or, equivalently, the squared norm of the n-by-1 vector $\mathbf{Q}(n)\,\boldsymbol{\Lambda}^{1/2}(n)\,\boldsymbol{\varepsilon}(n)$. Let $\hat{\mathbf{w}}(n)$ denote the *optimum* value of the weight vector that solves this minimization problem. From Eq. (10.27), we immediately deduce that the squared norm of $\mathbf{Q}(n)\,\boldsymbol{\Lambda}^{1/2}(n)\,\boldsymbol{\varepsilon}(n)$ is minimum when $\hat{\mathbf{w}}(n)$ satisfies the condition

$$\mathbf{R}(n)\hat{\mathbf{w}}(n) = \mathbf{p}(n), \qquad n \geq M \tag{10.28}$$

Correspondingly, the minimum value of the performance index equals

$$\mathscr{E}_{\min}(n) = \|\mathbf{v}(n)\|^2, \qquad n \geq M \tag{10.29}$$

Since the M-by-M matrix $\mathbf{R}(n)$ is upper triangular and therefore invertible, Eq. (10.28) may readily be solved for the optimum weight vector $\hat{\mathbf{w}}(n)$ by a process of back substitution. Moreover, we may use the weight vector thus computed to determine the least-squares value of the estimation error $e(i)$ over the observation interval $1 \leq i \leq n$.

10.5 SOLUTION OF THE RECURSIVE LEAST-SQUARES PROBLEM USING A SEQUENCE OF GIVENS ROTATIONS

The next issue we wish to consider is to make the procedure described previously for solving the linear least-squares problem recursively. Suppose that, at time $n - 1$, the $(n - 1)$-by-M weighted data matrix $\boldsymbol{\Lambda}^{1/2}(n - 1)\,\mathbf{A}(n - 1)$ has been already reduced to an orthogonal triangularized form, as shown by the QR-decomposition:

$$\mathbf{Q}(n - 1)\boldsymbol{\Lambda}^{1/2}(n - 1)\mathbf{A}(n - 1) = \begin{bmatrix} \mathbf{R}(n - 1) \\ \cdots\cdots\cdots \\ \mathbf{O} \end{bmatrix} \tag{10.30}$$

where $\mathbf{R}(n - 1)$ is an M-by-M upper triangular matrix and $\mathbf{O}$ is the $(n - M - 1)$-by-M null matrix. The $(n - 1)$-by-M data matrix $\mathbf{A}(n - 1)$ represents all the prewindowed data collected up to and including time $n - 1$.

At time n, the new input $u(n)$ becomes available for computation. This addition increases the number of rows contained in the updated data matrix $\mathbf{A}(n)$ by 1, as shown by

$$\mathbf{A}(n) = \begin{bmatrix} \mathbf{A}(n - 1) \\ \cdots\cdots \\ \mathbf{u}^H(n) \end{bmatrix} \tag{10.31}$$

where the input vector $\mathbf{u}(n)$ is defined by

$$\mathbf{u}^T(n) = [u(n),\, u(n - 1),\, \ldots,\, u(n - M + 1)]$$

To compute the QR-decomposition of the updated data matrix $\mathbf{A}(n)$, given the QR-decomposition of the old data matrix $\mathbf{A}(n - 1)$ and the new input vector $\mathbf{u}(n)$, we clearly need a *recursion* that relates the updated value of the unitary matrix,

$\mathbf{Q}(n)$, to the old value of the unitary matrix, $\mathbf{Q}(n - 1)$. We now derive this recursion by using the Givens rotation operation.

First, we define a new n-by-n unitary matrix $\overline{\mathbf{Q}}(n - 1)$ related to the old value $\mathbf{Q}(n - 1)$ as follows:

$$\overline{\mathbf{Q}}(n - 1) = \begin{bmatrix} \mathbf{Q}(n - 1) & \vdots & \mathbf{0}_{n-1} \\ \cdots\cdots\cdots\cdots & \vdots & \cdots\cdots \\ \mathbf{0}_{n-1}^T & \vdots & 1 \end{bmatrix} \tag{10.32}$$

where $\mathbf{0}_{n-1}$ is the $(n - 1)$-by-1 vector. We use the matrix $\overline{\mathbf{Q}}(n - 1)$ to apply a linear transformation to the weighted (new) data matrix $\mathbf{\Lambda}^{1/2}(n)\mathbf{A}(n)$. The data matrix $\mathbf{A}(n)$ is related to the old value $\mathbf{A}(n - 1)$ by Eq. (10.31). Correspondingly, the n-by-n exponential weighting matrix $\mathbf{\Lambda}(n)$ is related to the old value $\mathbf{\Lambda}(n - 1)$ as follows:

$$\mathbf{\Lambda}(n) = \begin{bmatrix} \lambda\mathbf{\Lambda}(n - 1) & \vdots & \mathbf{0}_{n-1} \\ \cdots\cdots\cdots\cdots & \vdots & \cdots\cdots \\ \mathbf{0}_{n-1}^T & \vdots & 1 \end{bmatrix} \tag{10.33}$$

This follows directly from the definition of the exponential weighting matrix given in Eq. (10.9). Hence, premultiplying the weighted data matrix $\mathbf{\Lambda}^{1/2}(n)\mathbf{A}(n)$ by the unitary matrix $\overline{\mathbf{Q}}(n - 1)$, and using Eqs. (10.31)–(10.33), we get

$$\overline{\mathbf{Q}}(n - 1)\mathbf{\Lambda}^{1/2}(n)\mathbf{A}(n) = \begin{bmatrix} \lambda^{1/2}\mathbf{Q}(n - 1)\mathbf{\Lambda}^{1/2}(n - 1)\mathbf{A}(n - 1) \\ \cdots\cdots\cdots\cdots\cdots\cdots\cdots\cdots\cdots\cdots \\ \mathbf{u}^H(n) \end{bmatrix} \tag{10.34}$$

However, earlier we assumed that the unitary matrix $\mathbf{Q}(n - 1)$ performs orthogonal transformation on $\mathbf{\Lambda}^{1/2}(n - 1)\mathbf{A}(n - 1)$, as shown in Eq. (10.30). Hence, we may simplify Eq. (10.34) as

$$\overline{\mathbf{Q}}(n - 1)\mathbf{\Lambda}^{1/2}(n)\mathbf{A}(n) = \begin{bmatrix} \lambda^{1/2}\mathbf{R}(n - 1) \\ \cdots\cdots\cdots\cdots \\ \mathbf{O} \\ \cdots\cdots\cdots\cdots \\ \mathbf{u}^H(n) \end{bmatrix} \tag{10.35}$$

The n-by-M matrix on the right side of Eq. (10.35) is partially triangularized in that the last row of the matrix consists of nonzero elements. We may transform this matrix into an orthogonal triangularized form through successive applications of the Givens rotation by following the procedure described in Section 10.3. Let $\mathbf{G}_1(n), \mathbf{G}_2(n), \ldots, \mathbf{G}_M(n)$ denote the sequence of Givens rotations that are applied to annihilate all M elements contained in the last row of the matrix, one by one. Let the unitary matrix $\mathbf{T}(n)$ denote the combined effect of this sequence of Givens rotations:

$$\mathbf{T}(n) = \mathbf{G}_M(n) \ldots \mathbf{G}_2(n) \, \mathbf{G}_1(n) \tag{10.36}$$

Accordingly, we may triangularize Eq. (10.35) as

$$\mathbf{T}(n)\overline{\mathbf{Q}}(n-1)\mathbf{\Lambda}^{1/2}(n)\mathbf{A}(n) = \begin{bmatrix} \mathbf{R}(n) \\ \cdots \\ \mathbf{O} \\ \cdots \\ \mathbf{0}_M^T \end{bmatrix} \tag{10.37}$$

where $\mathbf{O}$ is the $(n - M - 1)$-by-M null matrix, and $\mathbf{0}_M^T$ is the 1-by-M null row vector.

Substituting Eq. (10.35) in (10.37), we get

$$\begin{bmatrix} \mathbf{R}(n) \\ \cdots \\ \mathbf{O} \end{bmatrix} = \mathbf{T}(n) \begin{bmatrix} \lambda^{1/2}\mathbf{R}(n-1) \\ \cdots\cdots\cdots \\ \mathbf{O} \\ \mathbf{u}^H(n) \end{bmatrix} \tag{10.38}$$

Hence, given the unitary matrix $\mathbf{T}(n)$ and the new value $\mathbf{u}(n)$ of the input vector, we may use this recursion to time update the M-by-M upper triangular matrix $\mathbf{R}(n)$.

From Eq. (10.37), we also deduce that the *time-order updated* value of the unitary matrix in the QR-decomposition of the weighted data matrix at time n equals

$$\mathbf{Q}(n) = \mathbf{T}(n)\, \overline{\mathbf{Q}}(n-1)$$

$$= \mathbf{T}(n) \begin{bmatrix} \mathbf{Q}(n-1) & \vdots & \mathbf{0}_{n-1} \\ \cdots\cdots\cdots & \vdots & \cdots\cdots \\ \mathbf{0}_{n-1}^T & \vdots & 1 \end{bmatrix} \tag{10.39}$$

Equation (10.39) shows that, given the $(n - 1)$-by-$(n - 1)$ unitary matrix $\mathbf{Q}(n - 1)$ that performs the QR-decomposition of the weighted data matrix at time $n - 1$ and given the sequence of Givens transformations represented by the unitary matrix $\mathbf{T}(n)$, we may compute the updated unitary matrix $\mathbf{Q}(n)$.

Let the desired response vector $\mathbf{b}(n)$ be partitioned as follows

$$\mathbf{b}(n) = \begin{bmatrix} \mathbf{b}(n-1) \\ \cdots\cdots \\ d^*(n) \end{bmatrix} \tag{10.40}$$

where $d(n)$ is the value of the desired response that becomes available at time n. Then, using Eqs. (10.33), (10.39) and (10.40), we get

$$\mathbf{Q}(n)\, \mathbf{\Lambda}^{1/2}(n)\, \mathbf{b}(n) = \mathbf{T}(n) \begin{bmatrix} \lambda^{1/2}\mathbf{Q}(n-1)\, \mathbf{\Lambda}^{1/2}(n-1)\, \mathbf{b}(n-1) \\ \cdots\cdots\cdots\cdots\cdots\cdots\cdots\cdots\cdots \\ d^*(n) \end{bmatrix} \tag{10.41}$$

But, by definition, we have [see Eq. (10.24)]

$$\mathbf{Q}(n - 1) \, \mathbf{\Lambda}^{1/2}(n - 1) \, \mathbf{b}(n - 1) = \begin{bmatrix} \mathbf{p}(n - 1) \\ \mathbf{v}(n - 1) \end{bmatrix}$$

Hence, we may simplify Eq. (10.41) as

$$\begin{bmatrix} \mathbf{p}(n) \\ \cdots \\ \mathbf{v}(n) \end{bmatrix} = \mathbf{T}(n) \begin{bmatrix} \lambda^{1/2}\mathbf{p}(n - 1) \\ \cdots \\ \lambda^{1/2}\mathbf{v}(n - 1) \\ d^*(n) \end{bmatrix} \tag{10.42}$$

Equation (10.42) shows that both $\mathbf{p}(n)$ and $\mathbf{v}(n)$ may be updated by using the same sequence of Givens rotations, $\mathbf{T}(n)$, that is used to update $\mathbf{R}(n)$.

Having computed the updated M-by-M matrix $\mathbf{R}(n)$ from Eq. (10.38), and the updated M-by-1 vector $\mathbf{p}(n)$ from Eq. (10.42), we may readily use *backward substitution* in Eq. (10.28) to compute the corresponding updated value $\hat{\mathbf{w}}(n)$ of the least-squares weight vector. We refer to the algorithm defined by Eqs. (10.28), (10.38), and (10.42) as the *recursive QR decomposition-least squares (QRD-LS) algorithm*. A summary of the algorithm is presented in Table 10.1; a diagrammatic summary is shown in Fig. 10.1

Exact Initialization of the Algorithm

The solution obtained by this algorithm for the weight vector $\hat{\mathbf{w}}(n)$ is defined only for time $n \geq M$ for which the data matrix $\mathbf{A}(n)$ is of full rank. Nevertheless, we may *initialize* the orthogonal triangularization procedure by setting $\mathbf{R}(0) = \mathbf{O}$ and $\mathbf{p}(0) = \mathbf{0}$. The exact initialization procedure for the algorithm occupies the period $0 \leq n \leq M$, for which the index of performance $\mathscr{E}_{\min}(n)$ is zero. During this period, the data matrix $\mathbf{A}(n)$ assumes a lower triangular form [see Eq. (10.5)]. It is this special form of the data matrix, resulting from the use of prewindowing, that facilitates the application of the initialization procedure described here for *temporal* data. In particular, by using a sequence of Givens rotations, as illustrated next by way of an example, the data matrix $\mathbf{A}(n)$ is linearly transformed into upper

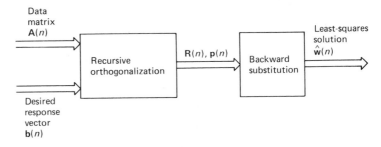

Figure 10.1 Block diagramatic representation of the recursive QRD–LS algorithm

TABLE 10.1 SUMMARY OF THE RECURSIVE QRD-LS ALGORITHM

1. Initialize the orthogonal triangularization procedure by setting $\mathbf{R}(0) = \mathbf{O}$ and $\mathbf{p}(0) = \mathbf{0}$. The exact initialization procedure occupies the period $0 \le n \le M$.
2. For $n > M$, perform the following updates:
    (a) Update the M-by-M upper triangular matrix $\mathbf{R}(n)$ by using the recursion

$$\begin{bmatrix} \mathbf{R}(n) \\ \mathbf{O} \end{bmatrix} = \mathbf{T}(n) \begin{bmatrix} \lambda^{1/2}\mathbf{R}(n-1) \\ \cdots\cdots\cdots \\ \mathbf{O} \\ \mathbf{u}^{H}(n) \end{bmatrix}$$

where $\mathbf{T}(n)$ is the n-by-n unitary matrix denoting the combined effect of the sequence of Givens rotations, and $\mathbf{u}(n)$ is the input vector at time n.
    (b) Update the M-by-1 vector $\mathbf{p}(n)$ and the $(n-M)$-by-1 vector $\mathbf{v}(n)$ by using the recursion

$$\begin{bmatrix} \mathbf{p}(n) \\ \cdots \\ \mathbf{v}(n) \end{bmatrix} = \mathbf{T}(n) \begin{bmatrix} \lambda^{1/2}\mathbf{p}(n-1) \\ \cdots\cdots\cdots \\ \lambda^{1/2}\mathbf{v}(n-1) \\ d^{*}(n) \end{bmatrix}$$

where $d(n)$ is the desired response at time n.
    (c) Compute the least-squares weight vector $\hat{\mathbf{w}}(n)$ on-the-fly:

$$\hat{\mathbf{w}}(n) = \mathbf{R}^{-1}(n)\,\mathbf{p}(n)$$

and the minimum value of the sum of weighted error squares:
$$\varepsilon_{\min}(n) = \|\mathbf{v}(n)\|^2$$

---

triangular form. The successive application of these rotations has the combined effect of order updating and time updating in so far as the upper triangular matrix $\mathbf{R}(n)$ and the vector $\mathbf{p}(n)$ are concerned. At $n = M$, the initialization period is completed, whereafter application of the recursive QRD-LS algorithm may proceed.

Example 10.1 Initialization of the Orthogonal Triangularization Process

To illustrate the exact initialization procedure for the algorithm, we consider the example of a least-squares problem assuming that the data are *real valued* and the weight vector consists of three elements (i.e., $M = 3$). We consider first initialization of the upper triangular matrix $\mathbf{R}(n)$, and then the vector $\mathbf{p}(n)$.

Initialization of matrix R(n). The algorithm is initialized by setting $\mathbf{R}(0)$ equals to zero. This is equivalent to setting all the inputs equal to zero at $n = 0$, in accordance with the prewindowing method. Thus, for $M = 3$, we may express the data matrix for the initialization period, extending from $n = 0$ to $n = 3$, both inclusive, as follows:

$$
\mathbf{A}(3) =
\begin{bmatrix}
0 & & 0 & & 0 \\
\hline
u(1) & & 0 & & 0 \\
\hline
u(2) & u(1) & & & 0 \\
\hline
u(3) & u(2) & & u(1)
\end{bmatrix}
$$

The partitions shown in the composition of this matrix are intended to separate out the elements of interest at time $n = 0, 1, 2, 3$.

At time $n = 1$, we may write [see the right side of Eq. (10.35)]

$$
\begin{bmatrix} \lambda^{1/2}\mathbf{R}(0) \\ \cdots\cdots \\ u(1) \end{bmatrix} =
\begin{bmatrix} 0 \\ \cdots \\ u(1) \end{bmatrix}
$$

The linear transformation of this two-element vector into the desired form is elementary in that we merely require interchanging positions of the two elements. This we accomplish by using the special form of the Givens rotation:

$$
\mathbf{G}_1(1) =
\begin{bmatrix} c_1(1) & s_1(1) \\ -s_1(1) & c_1(1) \end{bmatrix}
$$

$$
c_1(1) = 0
$$

$$
s_1(1) = 1
$$

We thus get

$$
\mathbf{G}_1(1)
\begin{bmatrix} \lambda^{1/2}\mathbf{R}(0) \\ \cdots\cdots\cdots \\ u(1) \end{bmatrix} =
\begin{bmatrix} \mathbf{R}(1) \\ \cdots\cdots \\ 0 \end{bmatrix}
$$

$$
= \begin{bmatrix} r_{11}(1) \\ \cdots\cdots\cdots \\ 0 \end{bmatrix}
$$

where

$$
r_{11}(1) = u(1)
$$

With this transformation in place, we next write, at time $n = 2$,

$$
\begin{bmatrix} \lambda^{1/2}\mathbf{R}(1) & 0 \\ 0 & 0 \\ \cdots\cdots\cdots & \cdots \\ u(2) & u(1) \end{bmatrix} =
\begin{bmatrix} \lambda^{1/2}r_{11}(1) & 0 \\ 0 & 0 \\ \cdots\cdots\cdots & \cdots \\ u(2) & u(1) \end{bmatrix}
$$

To triangularize this matrix, we use a sequence of two Givens rotations, $G_1(2)$ and $G_2(2)$. The first Givens rotation, $G_1(2)$, is defined by

$$
G_1(2) = \begin{bmatrix} c_1(2) & 0 & s_1(2) \\ 0 & 1 & 0 \\ -s_1(2) & 0 & c_1(2) \end{bmatrix}
$$

$$
c_1(2) = \frac{\lambda^{1/2} r_{11}(1)}{r_{11}(2)}
$$

(10.43)

$$
s_1(2) = \frac{u(2)}{r_{11}(2)}
$$

$$
r_{11}(2) = [\lambda r_{11}^2(1) + u^2(2)]^{1/2}
$$

The second Givens rotation, $G_2(2)$, is defined by

$$
G_2(2) = \begin{bmatrix} 1 & 0 & 0 \\ 0 & c_2(2) & s_2(2) \\ 0 & -s_2(2) & c_2(2) \end{bmatrix}
$$

$$
c_2(2) = 0
$$

$$
s_2(2) = 1
$$

Hence, we may write

$$
G_2(2)\, G_1(2) \begin{bmatrix} \lambda^{1/2} r_{11}(1) & 0 \\ 0 & 0 \\ \cdots\cdots\cdots \\ u(2) & u(1) \end{bmatrix} = \begin{bmatrix} R(2) \\ O \end{bmatrix}
$$

$$
= \begin{bmatrix} r_{11}(2) & r_{12}(2) \\ 0 & r_{11}(2) \\ \cdots\cdots\cdots\cdots \\ 0 & 0 \end{bmatrix}
$$

where $r_{11}(2)$ is defined in Eq. (10.43) and

$$
r_{12}(2) = s_1(2) u(1)
$$

$$
r_{22}(2) = c_1(2) u(1)
$$

At time $n = 3$, the end of the initialization period in this example, we have three nonzero inputs. Thus, with the aforementioned transformations in place, we write

$$
\begin{bmatrix}
\lambda^{1/2}\mathbf{R}(2) & \vdots & 0 \\
 & \vdots & 0 \\
\mathbf{O} & \vdots & 0 \\
\cdots\cdots\cdots\cdots & \vdots & \cdots\cdots \\
u(3), \quad u(2), & \vdots & u(1)
\end{bmatrix}
=
\begin{bmatrix}
\lambda^{1/2}r_{11}(2) & \lambda^{1/2}r_{12}(2) & \vdots & 0 \\
0 & \lambda^{1/2}r_{22}(2) & \vdots & 0 \\
0 & 0 & \vdots & 0 \\
\cdots\cdots\cdots & \cdots\cdots\cdots & \vdots & \cdots\cdots \\
u(3) & u(2) & \vdots & u(1)
\end{bmatrix}
$$

To triangularize this matrix, we apply a sequence of three Givens rotations $\mathbf{G}_1(3)$, $\mathbf{G}_2(3)$, and $\mathbf{G}_3(3)$. The first Givens rotation, $\mathbf{G}_1(3)$, is defined by

$$
\mathbf{G}_1(3) =
\begin{bmatrix}
c_1(3) & 0 & 0 & s_1(3) \\
0 & 1 & 0 & 0 \\
0 & 0 & 1 & 0 \\
-s_1(3) & 0 & 0 & c_1(3)
\end{bmatrix}
$$

$$
c_1(3) = \frac{\lambda^{1/2}r_{11}(2)}{r_{11}(3)} \tag{10.44}
$$

$$
s_1(3) = \frac{u(3)}{r_{11}(3)}
$$

$$
r_{11}(3) = [\lambda r_{11}^2(2) + u^2(3)]^{1/2}
$$

The second Givens rotation, $\mathbf{G}_2(3)$, is defined by

$$
\mathbf{G}_2(3) =
\begin{bmatrix}
1 & 0 & 0 & 0 \\
0 & c_2(3) & 0 & s_2(3) \\
0 & 0 & 1 & 0 \\
0 & -s_2(3) & 0 & c_2(3)
\end{bmatrix}
$$

$$
c_2(3) = \frac{\lambda^{1/2}r_{22}(2)}{r_{22}(3)} \tag{10.45}
$$

$$
s_2(3) = \frac{c_1(3)u(2) - \lambda^{1/2}s_1(3)r_{12}(2)}{r_{22}(3)}
$$

$$
r_{22}(3) = [\lambda r_{22}^2(2) + \left(c_1(3)u(2) - \lambda^{1/2}s_1(3)r_{12}(2)\right)^2]^{1/2}
$$

The third Givens rotation, $\mathbf{G}_3(3)$, is defined by

$$
\mathbf{G}_3(3) =
\begin{bmatrix}
1 & 0 & 0 & 0 \\
0 & 1 & 0 & 0 \\
0 & 0 & c_3(3) & s_3(3) \\
0 & 0 & -s_3(3) & c_3(3)
\end{bmatrix}
$$

$$
c_3(3) = 0
$$

$$
s_3(3) = 1
$$

Hence, we may write

$$
G_3(3) \; G_2(3) \; G_1(3)
\begin{bmatrix}
\lambda^{1/2}R(2) & \vdots & 0 \\
 & \vdots & 0 \\
O & \vdots & 0 \\
\cdots\cdots\cdots & \cdots\cdots & \\
u(3), & u(2), & u(1)
\end{bmatrix}
=
\begin{bmatrix}
R(3) \\
\cdots\cdots \\
O
\end{bmatrix}
$$

$$
=
\begin{bmatrix}
r_{11}(3) & r_{12}(3) & r_{13}(3) \\
0 & r_{22}(3) & r_{23}(3) \\
0 & 0 & r_{33}(3) \\
\cdots\cdots\cdots\cdots\cdots \\
0 & 0 & 0
\end{bmatrix}
$$

where $r_{11}(3)$ and $r_{22}(3)$ are defined by Eqs. (10.44) and (10.45), respectively, and

$$r_{12}(3) = c_1(3)\lambda^{1/2}r_{12}(2) + s_1(3)u(2)$$

$$r_{13}(3) = s_1(3)u(1)$$

$$r_{23}(3) = s_2(2)c_1(3)u(1)$$

$$r_{33}(3) = c_2(3)c_1(3)u(1)$$

This completes the initialization of upper triangular matrix $R(n)$.

Initialization of vector p(n).   As mentioned previously, initialization of the recursive QRD-LS algorithm also involves setting $p(0)$ equal to zero. Using this initial value and recognizing that the desired response equals $d(1)$ at time $n = 1$, we determine the updated value $p(1)$ by considering the linearly transformed vector [see the right side of the recursion of Eq. (10.42)]:

$$
G_1(1)
\begin{bmatrix}
\lambda^{1/2}p(0) \\
d(1)
\end{bmatrix}
=
\begin{bmatrix}
0 & 1 \\
-1 & 0
\end{bmatrix}
\begin{bmatrix}
0 \\
d(1)
\end{bmatrix}
$$

$$
=
\begin{bmatrix}
d(1) \\
0
\end{bmatrix}
$$

We therefore deduce that

$$p(1) = p_1(1)$$

where

$$p_1(1) = d(1)$$

and that

$$v(1) = 0$$

This latter result implies (as expected) that the corresponding minimum value $\mathscr{E}_{min}(1)$ of the performance index is zero.

At time $n = 2$, the desired response equals $d(2)$. We determine the updated vector $\mathbf{p}(2)$ by using the right side of the recursion of Eq. (10.42) to write

$$\mathbf{G}_2(2)\,\mathbf{G}_1(2) \begin{bmatrix} \lambda^{1/2}\mathbf{p}(1) \\ 0 \\ d(2) \end{bmatrix} = \begin{bmatrix} p_1(2) \\ p_2(2) \\ 0 \end{bmatrix}$$

where   (10.46)

$$p_1(2) = p_1(1)\lambda^{1/2}c_1(2) + s_1(2)d(2)$$

$$p_2(2) = -p_1(1)\lambda^{1/2}s_1(2) + c_1(2)d(2)$$

Since the minimum value $\mathscr{E}_{min}(2)$ of the performance index at time $n = 2$ is zero, we have $\mathbf{v}(2) = 0$. We therefore deduce from Eq. (10.46) that

$$\mathbf{p}(2) = \begin{bmatrix} p_1(2) \\ p_2(2) \end{bmatrix}$$

The initialization period terminates at $n = 3$. At this time, the desired response equals $d(3)$. We determine the updated vector $p(3)$ by writing

$$\mathbf{G}_3(3)\,\mathbf{G}_2(3)\,\mathbf{G}_1(3) \begin{bmatrix} \lambda^{1/2}\mathbf{p}(2) \\ 0 \\ d(3) \end{bmatrix} = \begin{bmatrix} p_1(3) \\ p_2(3) \\ p_3(3) \\ 0 \end{bmatrix}$$ (10.47)

where

$$p_1(3) = c_1(3)\lambda^{1/2}p_2(1) + s_1(3)d(3)$$

$$p_1(3) = -s_2(3)s_1(3)\lambda^{1/2}p_2(1) + c_2(3)\lambda^{1/2}p_2(2) + s_2(3)c_1(3)d(3)$$

$$p_1(3) = -c_2(3)s_1(3)\lambda^{1/2}p_2(1) - s_2(3)\lambda^{1/2}p_2(2) + c_2(3)c_1(3)d(3)$$

Here, again, since the minimum value $\mathscr{E}_{min}(3)$ of the performance index at time $n = 3$ is zero, we have $\mathbf{v}(3) = 0$. Accordingly, we deduce from Eq. (10.47) that the updated vector $\mathbf{p}(3)$ equals

$$\mathbf{p}(3) = \begin{bmatrix} p_1(3) \\ p_2(3) \\ p_3(3) \end{bmatrix}$$

This completes the initialization process.

Note that, as stated at the beginning of the example, the input data are all real valued. The results developed in this example only apply to real data.

10.6 SYSTOLIC ARRAY IMPLEMENTATION I

Figure 10.2 shows a *systolic array* structure for implementing the recursive QRD-LS algorithm described by Eqs. (10.28), (10.38), and (10.42) for the case when the weight vector has three elements (i.e., $M = 3$). The structure is arranged with two specific points in mind. First, data flow through it from left to right, consistent with all other adaptive filters considered in previous chapters. Second, the systolic array operates directly on the input data that are represented by the matrix $\mathbf{A}^H(n)$ and the row vector $\mathbf{b}^H(n)$. Accordingly, the solution produced at the output of the systolic array will appear as $\hat{\mathbf{w}}^H(n)$. Thus, to produce an estimate of the desired response $d(n)$, we form the inner product $\hat{\mathbf{w}}^H(n)\,\mathbf{u}(n)$ simply by postmultiplying the output of the systolic array processor by the input vector $\mathbf{u}(n)$.

The structure of Fig. 10.2 consists of two distinct sections: a *triangular systolic*

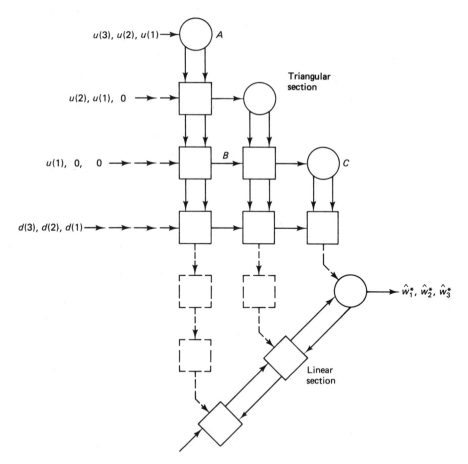

Figure 10.2 Systolic array implementation of the recursive QRD–LS algorithm for $M = 3$.

array and a *linear systolic array* (Gentleman and Kung, 1981). The entire systolic array is controlled by a single *clock*. Each section of the array consists of two types of processing cells: *internal cells* (represented by squares) and *boundary cells* (represented by circles). The specific arithmetic functions of these cells are defined later. Each cell receives its input data from the directions indicated for one clock cycle, performs the specified arithmetic functions, and then, on the next clock cycle, delivers the resulting output values to neighboring cells as indicated. A distinctive feature of systolic arrays is that each processing cell is always kept active as data flow across the array. The triangular systolic array section implements the Givens rotation part of the recursive QRD-LS algorithm, whereas the linear systolic array section computes the least-squares weight vector *at the end of the entire recursion.*

Consider first the operation of the triangular systolic array section labeled *ABC* in Fig. 10.2. The boundary and internal cells of this section are given in Fig. 10.3. Basically, the internal cells perform only additions and multiplications, as illustrated in Fig. 10.3(b). The boundary cells, on the other hand, are considerably more complex in that they compute square roots and reciprocals, as in Fig. 10.3(a). The operations shown in this figure follow directly from the discussion presented in Section 10.3. Each cell of the triangular systolic array section (depending on its location) stores a particular element of the lower triangular matrix $\mathbf{R}^H(n)$, which, at the outset of the least-squares recursion, is initialized to zero and

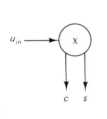

If $u_{in} = 0$, then

$$c \leftarrow 1$$
$$s \leftarrow 0$$

otherwise

$$x' \leftarrow \frac{x}{|x|}\sqrt{\lambda |x|^2 + |u_{in}|^2}$$
$$c \leftarrow \frac{\lambda^{1/2} |x|}{|x'|}$$
$$s \leftarrow \frac{x |u_{in}|^2/|x| u_{in}}{|x'|}$$
$$x \leftarrow x'$$

(a)

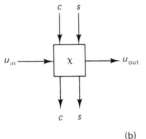

$$u_{out} \leftarrow cu_{in} - s\lambda^{1/2} x$$
$$x \leftarrow su_{in} + c\lambda^{1/2} x$$

Figure 10.3 Cells for the recursive QRD–LS algorithm: (a) boundary cell; (b) internal cell.

(b)

thereafter updated every clock cycle. The function of each column of processing cells in the triangular systolic array section is to *rotate* one column of the stored triangular matrix with a vector of data received from the left in such a way that the leading *element of the received data vector is annihilated.* The reduced data vector is then passed to the right on to the next column of cells. The boundary cell in each column of the section computes the pertinent rotation parameters and then passes them downward on the next clock cycle. The internal cells subsequently apply the same rotation to all other elements in the received data vector. Since a delay of one clock cycle per cell is incurred in passing the rotation parameters downward along a column, it is necessary that the input data vectors enter the triangular systolic array in a *skewed order,* as illustrated in Fig. 10.2 for the case of $M = 3$. This arrangement of the input data ensures that as each column vector $\mathbf{u}(n)$ of the data matrix $\mathbf{A}^H(n)$ propagates through the array, it interacts with the previously stored triangular matrix $\mathbf{R}^H(n - 1)$ and thereby undergoes the sequence of Givens rotations $\mathbf{T}(n)$, as required. Accordingly, all the elements of the column vector $\mathbf{u}(n)$ are annihilated, one by one, and an updated lower triangular matrix $\mathbf{R}^H(n)$ is produced and stored in the process.

The systolic array operates in a highly pipelined manner whereby, as (time-skewed) input data vectors enter the array, we find that in effect each such vector defines a processing *wave front* that moves across the array. It should therefore be appreciated that, on any particular clock cycle, elements of the pertinent lower triangular matrix $\mathbf{R}^H(n)$ only exist along the corresponding wave front. This phenomenon is illustrated in Example 10.2.

At the same time as the orthogonal triangularization process is being performed by the triangular systolic array section labeled *ABC* in Fig. 10.2, the row vector $\mathbf{p}^H(n)$ is computed by the appended bottom row of internal cells. In effect, this computation is made by treating the desired response vector $\mathbf{b}^H(n)$ as an additional row that is appended to the data matrix $\mathbf{A}^H(n)$ at its bottom end.

When the entire orthogonal triangularization process is completed, each particular row of the lower triangular M-by-M matrix $\mathbf{R}^H(n)$ or the associated 1-by-M vector $\mathbf{p}^H(n)$ along the corresponding wave front is *clocked out* for subsequent processing by the linear systolic array section. This section of the processor computes the Hermitian transposed least-squares weight vector, $\hat{\mathbf{w}}^H(n)$. In particular, the elements of the vector $\hat{\mathbf{w}}^H(n)$ are computed by using the *method of backward substitution* (Kung and Leiserson, 1978):

$$z_i^{(M)} = 0$$

$$z_i^{(k-1)} = z_i^{(k)} + r_{ik}^*(n)\hat{w}_k^*(n)$$

$$\hat{w}_i^*(n) = \frac{p_i^*(n) - z_i^{(i)}}{r_{ii}(n)}$$

(10.48)

$$i, k = M - 1, \ldots, 1$$

where the $z_i^{(k)}$ are intermediate variables, the $r_{ik}(n)$ are elements of upper triangular matrix $\mathbf{R}(n)$, the $p_i(n)$ are elements of the vector $\mathbf{p}(n)$, and the $\hat{w}_k(n)$ are elements of the least-squares weight vector $\hat{\mathbf{w}}(n)$.

The linear systolic array section consists of one boundary cell and $(M - 1)$ internal cells that perform the arithmetic functions defined in Fig. 10.4, in accordance with Eq. (10.48). The boundary cell performs subtraction and division, whereas the internal cells perform additions and multiplications. The elements of the complex-conjugated least-squares weight vector appear at the output of the boundary cell at different clock cycles, with $\hat{w}_M^*(n)$ leaving this cell first, followed by $\hat{w}_{M-1}^*(n)$, and so on right up to $\hat{w}_1^*(n)$. In effect, the elements of the weight vector $\hat{\mathbf{w}}^H(n)$ are read out *backward*.

Thus, by chaining the linear and triangular systolic array sections together, as shown in Fig. 10.2, we produce a powerful device that is capable of solving on-the-fly the exact least-squares problem recursively.

Note that, initially, zeros are stored in all boundary and internal cells. Also, the parameters of the Givens rotation at the output of each boundary cell (and therefore every other cell in the triangular systolic array section) are initially set at the values $c_{\text{out}} = 1$ and $s_{\text{out}} = 0$.

Example 10.2

In this example, we illustrate the operation of the systolic array structure of Fig. 10.2 for the case when the input data are *real valued* and $M = 3$. We consider first the orthogonal triangularization process and computation of the vector $\mathbf{p}(n)$ that are performed by the triangular systolic array section. We then consider computation of the elements of the least-squares weight vector $\hat{\mathbf{w}}(n)$ by the linear systolic array section.

Triangular systolic array section. The inputs and states of the triangular systolic array for $M = 3$ are summarized in Table 10.2. These results are determined by applying the update formulas given in Fig. 10.3 in response to two factors: (1) the input entering the boundary or internal cell of the array, and (2) the content of the cell, both at the time in question. The different parts of the table refer to time $n = 1, 2, \ldots, 6$. In each part of the table, we show the pertinent inputs

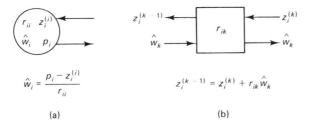

(a)

(b)

Figure 10.4 Cells for linear systolic array; (a) boundary cell; (b) inner cell.

TABLE 10.2 INPUTS AND STATES OF TRIANGULAR SYSTOLIC ARRAY AT DIFFERENT TIMES

Inputs
$u(1)$

$r_{11}(1)$		
0	0	
0	0	0
0	0	0

(a) Time $n = 1$

Inputs
$u(2)$

$r_{11}(2)$		
0	0	
0	0	0
0	0	0

(b) Time $n = 2$

Inputs
$u(3)$
$u(1)$

$r_{11}(3)$		
$r_{12}^*(2)$	$r_{22}(2)$	
0	0	0
0	0	0

(c) Time $n = 3$

Inputs
$u(4)$
$u(2)$

$d(1)$

$r_{11}(4)$		
$r_{12}^*(3)$	$r_{22}(3)$	
0	0	0
$p_1^*(1)$	0	0

(d) Time $n = 4$

Inputs
$u(5)$
$u(3)$
$u(1)$
$d(2)$

$r_{11}(5)$		
$r_{12}^*(4)$	$r_{22}(4)$	
$r_{13}^*(3)$	$r_{23}^*(3)$	$r_{33}(3)$
$p_1^*(2)$	$p_2^*(2)$	0

(e) Time $n = 5$

Inputs
$u(6)$
$u(4)$
$u(2)$
$d(3)$

$r_{11}(6)$		
$r_{12}^*(5)$	$r_{22}(5)$	
$r_{13}^*(4)$	$r_{23}^*(4)$	$r_{33}(4)$
$p_1^*(3)$	$p_2^*(3)$	$p_3^*(3)$

(f) Time $n = 6$

entering the array and the data stored in the various boundary and internal cells of the array. Some of the entries in this table were defined in Example 10.1.

Based on Table 10.2, we may make the following statements for time n (including the initialization period) in the context of complex valued data:

1. The elements of row i of the lower triangular M-by-M matrix $\mathbf{R}^H(n)$ are stored in the cells of row i of the triangular systolic array section at time $n + i - 1$, where $i = 1, 2, \ldots, M$. Note that the diagonal elements of matrix $\mathbf{R}(n)$ are all *real valued*; they appear as the respective contents of the boundary cells of the triangular systolic array at appropriate times.

2. The elements of the 1-by-M vector $\mathbf{p}^H(n)$ are stored in the row of internal cells appended to the triangular systolic array section (at its bottom end) at time $n + M$.

3. The initialization period occupies $2M$ clock cycles of the triangular systolic array section. This is twice as long as the basic initialization period of the QRD–LS algorithm. The reason for the discrepancy is the temporal skew imposed on the input data vectors.

Linear systolic array section. The orthogonal triangularization process ends at time $n + 3$ (for $M = 3$). The elements of the upper triangular 3-by-3 matrix $\mathbf{R}^H(n)$ and the 1-by-3 vector $\mathbf{p}^H(n)$ are stored in the pertinent cells of the triangular systolic array section, which are clocked out as the recursion nears its end. The contents of all the cells in the linear systolic array are initially set equal to zero. Thus, by ordering the elements of matrix $\mathbf{R}^H(n)$ and vector $\mathbf{p}^H(n)$ and using them as inputs into the linear systolic array section as indicated in Fig. 10.5, and using the input–output relations defined in Fig. 10.4 for the boundary cell and the internal cells of this section, we generate the elements $\hat{w}_3^*(n)$, $\hat{w}_2^*(n)$ and $\hat{w}_1^*(n)$ of the least-squares weight vector $\hat{\mathbf{w}}^H(n)$, in that order, at the output of the boundary cell. The operations involved in the generation of these elements at the individual clock cycles of the linear systolic array section are shown in Fig. 10.5.

10.7 COMPUTER EXPERIMENT ON ADAPTIVE EQUALIZATION

In this experiment we study the use of the recursive QRD-LS algorithm for *adaptive equalization* of a linear communication channel that produces unknown distortion. The parameters of the channel are the same as those used to study the LMS algorithm in Section 5.16. The number of taps $M = 11$. The exponential factor $\lambda = 1$. The data are all *real valued*.

The aims of the experiment are as follows:

1. To investigate the evolution of the least-squares weight vector $\hat{\mathbf{w}}(n)$ for one trial of the experiment and after the exact initialization of the triangular systolic array section has been completed.

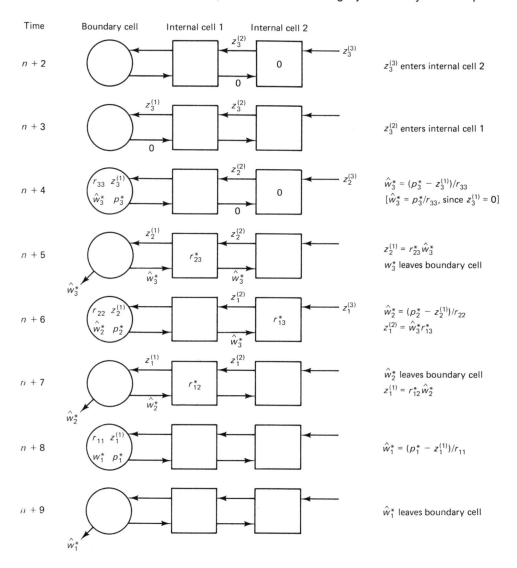

Figure 10.5   Operation of the linear systolic array for $M = 3$.

2. To evaluate the effect of variations in the eigenvalue spread of the M-by-M correlation matrix of the input signal.

3. To compare the values of $\hat{\mathbf{w}}(n)$ computed above with the optimum Wiener solution.

   The input signal $\{u(n)\}$ and the desired response $\{d(n)\}$ are generated in the manner described in Section 5.16.   The elements of $\{u(n)\}$ and $\{d(n)\}$ are fed into the triangular systolic array in a skewed manner (as illustrated in Fig. 10.2).   With the number of taps $M = 11$, exact initialization of the triangular systolic array takes $2M = 22$ clock cycles.

The four parts of Table 10.3 show the results of the experiment for four different values of the eigenvalue spread: $\chi(\mathbf{R}) = 6.0782, 11.1238, 21.7132, 46.8216$, corresponding to the channel parameter $W = 2.9, 3.1, 3.3, 3.5$, respectively. Column 2 of each part shows the optimum Wiener solution for the weight vector. The last four columns show the corresponding values of the least-squares weight vector $\hat{\mathbf{w}}(n)$ after $2(M + 1), 4(M + 1), 6(M + 1), 8(M + 1)$ clock cycles, measured

TABLE 10.3 COMPARISON OF LEAST-SQUARES SOLUTION $\hat{\mathbf{w}}(n)$ OBTAINED BY SYSTOLIC ARRAY *I*, FOR VARYING ITERATIONS, WITH WIENER SOLUTION

Value of tap weight	Wiener solution $\mathbf{w}_0$	Least-squares solution $\hat{\mathbf{w}}(n)$ for varying number of clock cycles			
		$8(M + 1)$	$6(M + 1)$	$4(M + 1)$	$2(M + 1)$
w_1	-0.0006	-0.0130	-0.0491	-0.1217	-0.0101
w_2	0.0031	0.0371	0.0368	0.1636	-0.1393
w_3	-0.0136	-0.0610	-0.0019	-0.1565	0.0620
w_4	0.0590	0.0738	0.0246	0.1973	-0.1237
w_5	-0.2561	-0.2485	-0.2050	-0.3444	-0.3936
w_6	1.1110	1.1096	1.0728	1.1746	1.2371
w_7	-0.2561	-0.2530	-0.2391	-0.3270	-0.3025
w_8	0.0590	0.0637	0.0494	0.1488	0.0697
w_9	-0.0136	-0.0254	-0.0080	-0.1412	0.0459
w_{10}	0.0031	0.0074	-0.0106	0.1063	-0.0693
w_{11}	-0.0006	0.0008	0.0172	-0.0326	-0.0219

(a) $w = 2.9$

Value of tap weight	Wiener solution $\mathbf{w}_0$	Least-squares solution $\hat{\mathbf{w}}(n)$ for varying number of clock cycles			
		$8(M + 1)$	$6(M + 1)$	$4(M + 1)$	$2(M + 1)$
w_1	-0.0026	-0.0226	-0.0531	-0.1222	-0.0050
w_2	0.0101	0.0547	0.0462	0.1860	-0.1396
w_3	-0.0339	-0.0919	-0.0174	-0.2036	0.0585
w_4	0.1118	0.1322	0.0606	0.2734	0.1617
w_5	-0.3671	-0.3570	-0.2914	-0.4781	-0.5180
w_6	1.2037	1.1950	1.1393	1.2863	1.3608
w_7	-0.3671	-0.3589	-0.3282	-0.4514	-0.4516
w_8	0.1118	0.1134	0.0877	0.2164	0.1483
w_9	-0.0339	-0.0442	-0.0201	-0.1758	0.0213
w_{10}	0.0101	0.0138	-0.0074	0.1322	-0.0742
w_{11}	-0.0026	-0.0001	0.0173	-0.0517	-0.0015

(b) $w = 3.1$

TABLE 10.3 CONTINUED

Value of tap weight	Wiener solution $\mathbf{w}_0$	Least-squares solution $\hat{\mathbf{w}}(n)$ for varying number of clock cycles			
		$8(M + 1)$	$6(M + 1)$	$4(M + 1)$	$2(M + 1)$
w_1	-0.0081	-0.0366	-0.0590	-0.1268	-0.0021
w_2	0.0275	0.0837	0.0608	0.2194	-0.1328
w_3	-0.0756	-0.1436	-0.0432	-0.2740	0.0400
w_4	0.1988	0.2211	0.1137	0.3883	0.2292
w_5	-0.5182	-0.4980	-0.3953	-0.6585	-0.6813
w_6	1.3462	1.3195	1.2307	1.4555	1.5435
w_7	-0.5182	-0.4933	-0.4347	-0.6235	-0.6579
w_8	0.1988	0.1871	0.1425	0.3245	0.2793
w_9	-0.0756	-0.0765	-0.0421	-0.2362	-0.0381
w_{10}	0.0275	0.0255	0.0004	0.1712	-0.0622
w_{11}	-0.0081	-0.0024	0.0155	-0.0747	0.0116

(c) $w = 3.3$

Value of tap weight	Wiener solution $\mathbf{w}_0$	Least-squares solution $\hat{\mathbf{w}}(n)$ for varying number of clock cycles			
		$8(M + 1)$	$6(M + 1)$	$4(M + 1)$	$2(M + 1)$
w_1	-0.0222	-0.0567	-0.0671	-0.1389	-0.0062
w_2	0.0683	0.1293	0.0820	0.2730	-0.1044
w_3	-0.1602	-0.2254	-0.0816	-0.3839	-0.0178
w_4	0.3493	0.3523	0.1862	0.5670	0.3561
w_5	-0.7446	-0.6851	-0.5197	-0.9193	-0.9216
w_6	1.5748	1.4950	1.3488	1.7200	1.8312
w_7	-0.7446	-0.6668	-0.5610	-0.8815	-0.9727
w_8	0.3493	0.2915	0.2159	0.5050	0.5147
w_9	-0.1602	-0.1262	-0.0762	-0.3449	-0.1793
w_{10}	0.0683	0.0438	0.0145	0.2353	0.0006
w_{11}	-0.0222	-0.0058	0.0114	-0.1055	0.0033

(d) $w = 3.5$

from the end of the initialization period.   The results shown in these four columns were computed using the linear systolic array section.   They were all obtained for a single trial of the experiment.

Based on these results, we see that after $8(M + 1)$ clock cycles (from the end of the initialization period) the least-squares weight vector takes on a value close to the Wiener solution.

To further emphasize this point, we show in Table 10.4 two sets of results: (a) the convolution of the impulse response of the channel with the impulse response of the Wiener filter, and (b) the convolution of the impulse response of the channel

TABLE 10.4 ANOTHER COMPARISON OF SYSTOLIC ARRAY *I* WITH WIENER FILTER, BASED ON THE CONVOLUTION $y(n) = h_n \otimes u(n)$ WHERE h_n = IMPULSE RESPONSE AND $u(n)$ = INPUT, AND $\otimes$ DENOTES CONVOLUTION

	$y(n)$			
n	$W = 2.9$	$W = 3.1$	$W = 3.3$	$W = 3.5$
-6	-0.0001	-0.0007	-0.0027	-0.0086
-5	0.0001	0.0002	0.0012	0.0043
-4	-0.00001	-0.0001	-0.0007	-0.0026
-3	0.00002	0.0002	0.0005	0.0021
-2	-0.0002	-0.0004	-0.0010	-0.0024
-1	0.0006	0.0010	0.0017	0.0033
0	0.9986	0.9983	0.9975	0.9959
1	0.0006	0.0010	0.0017	0.0033
2	-0.0002	-0.0004	-0.0010	-0.0024
3	0.00002	0.0002	0.0005	0.0021
4	-0.00001	-0.0001	-0.0007	-0.0026
5	0.0001	0.0002	0.0012	0.0043
6	-0.0001	-0.0007	-0.0027	-0.0086

(a)   Wiener Filter

	$y(n)$			
n	$W = 2.9$	$W = 3.1$	$W = 3.3$	$W = 3.5$
-6	-0.0029	-0.0063	-0.0123	-0.0220
-5	-0.0049	-0.0073	-0.0084	-0.0064
-4	0.0209	0.0227	0.0231	0.0196
-3	-0.0367	-0.0396	-0.0410	-0.0382
-2	0.0059	0.0066	0.0052	-0.0016
-1	0.0111	0.0143	0.0204	0.0329
0	0.9996	0.9947	0.9859	0.9695
1	0.0044	0.0072	0.0137	0.0276
2	0.0026	0.0006	-0.0046	-0.0167
3	-0.0098	-0.0086	-0.0050	0.0041
4	0.0020	0.0014	-0.0010	-0.0075
5	0.0024	0.0038	0.0062	0.0112
6	0.0002	-0.00003	-0.0008	-0.0022

(b)   Systolic array after $8(M + 1)$ clock cycles

with the impulse response of the least-squares equalizer attained after $8(M + 1)$ clock cycles from the end of the initialization period. In each case, the eigenvalue spread $\chi(\mathbf{R})$ takes on the four values given above. Ideally, the transfer function of the equalizer should equal the inverse of the transfer function of the channel, so that the convolution of their respective impulse responses would be simply a delta function. In the example considered here, the delta function should be located at the mid-point of the equalizer. Table 10.4 shows that this ideal result is very closely realized by the Wiener solution and to a slightly lesser extent by the systolic array solution.

10.8 RECURSIVE LEAST-SQUARES MINIMIZATION WITHOUT EXPLICIT COMPUTATION OF THE WEIGHT VECTOR

In Section 10.5, we described a procedure, based on the QR-decomposition, for solving the recursive least-squares problem. This procedure also provides a means for computing the optimum weight vector $\hat{\mathbf{w}}(n)$ that achieves the least-squares minimization. However, in some applications (e.g., prediction-error filters, adaptive noise cancellers, adaptive beamformers), it is not necessary that we compute the least-squares weight vector $\hat{\mathbf{w}}(n)$ explicitly. In the remainder of this section, we present a modification of the recursive QRD-LS algorithm that avoids the need to solve Eq. (10.28) at any stage of the process (McWhirter, 1983).

Since the last element of the diagonal matrix $\Lambda(n)$ equals unity [see Eq. (10.9)], it follows that the complex conjugate of the estimation error $e(n)$ is equal to the last element of the n-by-1 vector:

$$\Lambda^{1/2}(n)\varepsilon(n) \;=\; \Lambda^{1/2}(n)\mathbf{b}(n) \;-\; \Lambda^{1/2}(n)\mathbf{A}(n)\hat{\mathbf{w}}(n) \tag{10.49}$$

In the right side of Eq. (10.49), we have used the definition of Eq. (10.19) for the error vector $\varepsilon(n)$. Premultiplying both sides of Eq. (10.21) by $\mathbf{Q}^{-1}(n)$, the inverse of the unitary matrix $\mathbf{Q}(n)$, and recognizing (from the definition of a unitary matrix) that

$$\mathbf{Q}^{-1}(n) \;=\; \mathbf{Q}^H(n)$$

we may express the matrix $\Lambda^{1/2}(n)\mathbf{A}(n)$ as

$$
\begin{aligned}
\Lambda^{1/2}(n)\mathbf{A}(n) &= \mathbf{Q}^H(n)
\begin{bmatrix}
\mathbf{R}(n) \\
\cdots\cdots \\
\mathbf{O}
\end{bmatrix} \\[2mm]
&=
\begin{bmatrix}
& \vdots & \\
\mathbf{F}^H(n & \vdots & \mathbf{S}^H(n) \\
& \vdots &
\end{bmatrix}
\begin{bmatrix}
\mathbf{R}(n) \\
\cdots\cdots \\
\mathbf{O}
\end{bmatrix} \\[2mm]
&= \mathbf{F}^H(n)\,\mathbf{R}(n)
\end{aligned}
\tag{10.50}
$$

In the second line of Eq. (10.50), we have used the partitioned representation of Eq. (10.23) for the unitary matrix $\mathbf{Q}(n)$. The substitution of Eq. (10.50) in (10.49) thus yields

$$\Lambda^{1/2}(n)\varepsilon(n) \;=\; \Lambda^{1/2}(n)\mathbf{b}(n) \;-\; \mathbf{F}^H(n)\,\mathbf{R}(n)\,\hat{\mathbf{w}}(n) \tag{10.51}$$

However, the least-squares weight vector $\hat{\mathbf{w}}(n)$ satisfies Eq. (10.28). We may therefore replace the product $\mathbf{R}(n)\hat{\mathbf{w}}(n)$ by the vector $\mathbf{p}(n)$, and so rewrite Eq.

(10.51) as

$$\mathbf{\Lambda}^{1/2}(n)\boldsymbol{\varepsilon}(n) = \mathbf{\Lambda}^{1/2}(n)\mathbf{b}(n) - \mathbf{F}^H(n)\,\mathbf{p}(n) \tag{10.52}$$

As desired, this result does not depend explicitly on the weight vector $\hat{\mathbf{w}}(n)$. Furthermore, premultiplying both sides of Eq. (10.24) by the inverse matrix $\mathbf{Q}^{-1}(n) = \mathbf{Q}^H(n)$, we may write

$$\mathbf{\Lambda}^{1/2}(n)\,\mathbf{b}(n) = \mathbf{Q}^H(n) \begin{bmatrix} \mathbf{p}(n) \\ \cdots\cdots \\ \mathbf{v}(n) \end{bmatrix} \tag{10.53}$$

$$= \mathbf{F}^H(n)\,\mathbf{p}(n) + \mathbf{S}^H(n)\,\mathbf{v}(n)$$

Hence, using Eq. (10.53) in (10.52) and simplifying, we get

$$\mathbf{\Lambda}^{1/2}(n)\boldsymbol{\varepsilon}(n) = \mathbf{S}^H(n)\mathbf{v}(n) \tag{10.54}$$

Earlier we remarked that the complex conjugate of the estimation error $e(n)$ equals the last element of the vector $\mathbf{\Lambda}^{1/2}(n)\,\boldsymbol{\varepsilon}(n)$. Equation (10.54) shows that, equivalently, $e^*(n)$ equals the last element of the vector $\mathbf{S}^H(n)\,\mathbf{v}(n)$. By using the latter result, we now show how the estimation error $e(n)$ may be obtained directly from the orthogonal triangularization process. To do this, we first examine the structure of the sequence of unitary matrices that are generated during the application of the Givens rotation procedure. From Eq. (10.36), supported by the results of Example 10.1, we deduce that the unitary matrix $\mathbf{T}(n)$ representing the sequence of Givens rotation operations must take the form

$$\mathbf{T}(n) = \gamma^{1/2}(n) \begin{bmatrix} \mathbf{Z}(n) & \vdots & \mathbf{O} & \vdots & -\mathbf{Z}(n)\boldsymbol{\beta}(n) \\ \cdots & & \cdots & & \cdots \\ \mathbf{O} & \vdots & \gamma^{-1/2}(n)\mathbf{I} & \vdots & \mathbf{0} \\ \cdots & & \cdots & & \cdots \\ \boldsymbol{\beta}^H(n) & \vdots & \mathbf{0} & \vdots & 1 \end{bmatrix} \tag{10.55}$$

where $\mathbf{Z}(n)$ is an M-by-M matrix, and $\boldsymbol{\beta}(n)$ is an M-by-1 vector, both of which may be complex-valued, $\mathbf{I}$ is the $(n - M - 1)$-by-$(n - M - 1)$ identity matrix, and $\gamma^{1/2}(n)$ is a real-valued scaling factor. The reason for including the power $\frac{1}{2}$ in the symbol for the scaling factor will become apparent later in the section. Hence, using Eq. (10.55) in (10.42), we get

$$\begin{bmatrix} \mathbf{p}(n) \\ \cdots \\ \mathbf{v}(n) \end{bmatrix} = \gamma^{1/2}(n) \begin{bmatrix} \lambda^{1/2}\mathbf{Z}(n)\,\mathbf{p}(n-1) - d^*(n)\,\mathbf{Z}(n)\boldsymbol{\beta}(n) \\ \cdots\cdots\cdots\cdots\cdots\cdots\cdots\cdots \\ \gamma^{-1/2}(n)\lambda^{1/2}\mathbf{v}(n-1) \\ \lambda^{1/2}\boldsymbol{\beta}^H(n)\mathbf{p}(n-1) + d^*(n) \end{bmatrix} \tag{10.56}$$

From this relation, we deduce that the $(n - M)$-by-1 vector $\mathbf{v}(n)$ equals

$$
\mathbf{v}(n) = \begin{bmatrix} \lambda^{1/2}\mathbf{v}(n - 1) \\ \cdots\cdots\cdots \\ \gamma^{1/2}(n)\alpha^*(n) \end{bmatrix} \tag{10.57}
$$

where the complex-valued variable $\alpha(n)$ is defined by

$$
\alpha(n) = \lambda^{1/2}\mathbf{p}^H(n - 1)\boldsymbol{\beta}(n) + d(n) \tag{10.58}
$$

Similarly, from Eqs. (10.39), (10.55), and the definition [see Eq. (10.23)]

$$
\mathbf{Q}(n - 1) = \begin{bmatrix} \mathbf{F}(n - 1) \\ \mathbf{S}(n - 1) \end{bmatrix} \tag{10.59}
$$

we may express the unitary matrix $\mathbf{Q}(n)$ as

$$
\mathbf{Q}(n) = \mathbf{T}(n) \begin{bmatrix} \mathbf{F}(n - 1) & \mathbf{0} \\ \mathbf{S}(n - 1) & \mathbf{0} \\ \mathbf{0}^T & 1 \end{bmatrix}
$$

$$
= \gamma^{1/2}(n) \begin{bmatrix} \mathbf{Z}(n)\,\mathbf{F}(n - 1) & -\mathbf{Z}(n)\boldsymbol{\beta}(n) \\ \gamma^{-1/2}(n)\mathbf{S}(n - 1) & \mathbf{0} \\ \boldsymbol{\beta}^H(n)\,\mathbf{F}(n - 1) & 1 \end{bmatrix} \tag{10.60}
$$

With $\mathbf{S}(n)$ defined as the last $(n - M)$ rows of the unitary matrix $\mathbf{Q}(n)$, we deduce from Eq. (10.60) that

$$
\mathbf{S}(n) = \gamma^{1/2}(n) \begin{bmatrix} \gamma^{-1/2}(n)\mathbf{S}(n - 1) & \mathbf{0} \\ \boldsymbol{\beta}^H(n)\,\mathbf{F}(n - 1) & 1 \end{bmatrix} \tag{10.61}
$$

Hence, we may use Eqs. (10.57) and (10.61) to express the vector $\mathbf{S}^H(n)\,\mathbf{v}(n)$ as

$$
\mathbf{S}^H(n)\,\mathbf{v}(n) = \lambda^{1/2} \begin{bmatrix} \mathbf{S}^H(n - 1)\,\mathbf{v}(n - 1) \\ \cdots\cdots\cdots\cdots\cdots \\ 0 \end{bmatrix}
$$

$$
+ \alpha^*(n)\,\gamma(n) \begin{bmatrix} \mathbf{F}^H(n - 1)\,\boldsymbol{\beta}(n) \\ \cdots\cdots\cdots\cdots \\ 1 \end{bmatrix} \tag{10.62}
$$

Thus, with the complex conjugate of the estimation error $e(n)$ equal to the last element of the vector $\mathbf{S}^H(n)\mathbf{v}(n)$, we finally obtain the result

$$e(n) = \alpha(n)\gamma(n) \tag{10.63}$$

From the definition of $\gamma^{1/2}(n)$ as the nnth element of the unitary matrix $\mathbf{Q}(n)$ [see Eq. (10.60)] and from the structure of matrix $\mathbf{T}(n)$, defining the sequence of Givens rotations as exemplified by Example 10.1, we deduce that its value is given by

$$\gamma^{1/2}(n) = \prod_{i=1}^{n} c_i(n), \qquad n \geq M \tag{10.64}$$

where $c_i(n)$ is the cosine parameter associated with the ith Givens rotation in the sequence of such operations represented by the transformation matrix $\mathbf{T}(n)$. The parameter $\gamma^{1/2}(n)$ may therefore be computed quite readily during the orthogonal triangularization process. It is of interest to note that this parameter may also be interpreted as simply the result obtained by applying the sequence of Givens rotations (constituting the orthogonal triangularization process) to rotate a unit input with each element of the M-by-1 null vector.

The product $\gamma^{1/2}(n)\alpha(n)$ may be interpreted simply as the result obtained when the desired response $d(n)$ is rotated with each element in the M-by-1 vector $\mathbf{p}(n - 1)$ by means of the same sequence of Givens rotations. We therefore find that this product is also computed quite naturally during the orthogonal triangularization process.

Another Interpretation of the Variable $\alpha(n)$ and Parameter $\gamma(n)$

The product term $\lambda^{1/2}\mathbf{p}^H(n - 1)\, \boldsymbol{\beta}(n)$ in Eq. (10.58) represents (except for a minus sign) the least-squares estimate of the desired response $d(n)$, computed at time n by using the prior value of the least-squares weight vector, namely, $\hat{\mathbf{w}}(n - 1)$. Let $\mathcal{U}_{n-1}$ denote the space spanned by the inputs $u(n - 1), \ldots, u(n - M)$. We may then write (see Problem 4)

$$\lambda^{1/2}\mathbf{p}^H(n - 1)\, \boldsymbol{\beta}(n) = -\hat{d}(n|\mathcal{U}_{n-1}) \tag{10.65}$$
$$= -\hat{\mathbf{w}}^H(n - 1)\mathbf{u}(n)$$

Accordingly, the variable $\alpha(n)$ may be interpreted as the *a priori estimation error*.

The estimation error $e(n)$ is computed at time n by using the value of the weight vector $\hat{\mathbf{w}}(n)$ at time n. Hence, $e(n)$ in Eq. (10.63) is the *a posteriori estimation error*.

Now, we recall from the theory presented in Section 8.12 that the a priori estimation error $\alpha(n)$ and the a posteriori estimation error $e(n)$ are related by

$$e(n) = \gamma(n)\, \alpha(n)$$

where $\gamma(n)$ is the *conversion factor* defined as

$$\gamma(n) = 1 - \mathbf{u}^H(n)\,\boldsymbol{\Phi}^{-1}(n)\mathbf{u}(n) \tag{10.66}$$

We thus deduce that the real-valued scaling factor $\gamma^{1/2}(n)$ first introduced in Eq. (10.55) is the same as the square root of the conversion factor defined in Eq. (10.66). Indeed, it was with this relation in mind that we denoted the scaling factor in Eq. (10.55) as $\gamma^{1/2}(n)$.

10.9 SYSTOLIC ARRAY IMPLEMENTATION II

Figure 10.6 shows the systolic array implementation of the *modified QRD–LS algorithm* (McWhirter, 1983). Apart from the introduction of an extra parameter $\gamma^{1/2}$ into the boundary cell, the boundary and internal cells in the structure of Fig. 10.6 are identical to those shown in the triangular systolic array section of Fig. 10.2. Moreover, the linear systolic array section has been omitted from the structure of Fig. 10.6, as the modified QRD–LS algorithm eliminates the need for computing the least-squares weight vector $\hat{\mathbf{w}}(n)$ explicitly. However, the structure of Fig. 10.6 includes a new element referred to as the *final processing cell* (indicated by a small circle). This cell is intended to compute the a posteriori estimation error $e(n)$. The specific functions of the boundary cells, the internal cells, and final processing cell are given in Fig. 10.7.

The parameter $\gamma^{1/2}(n)$, defined by Eq. (10.64), is also generated as a result of the orthogonal triangularization process. This follows from the fact that the

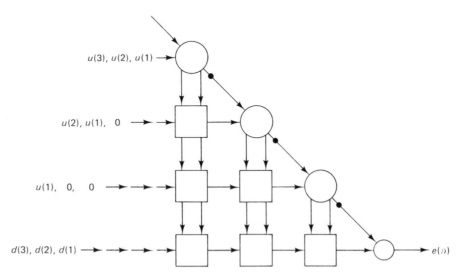

Figure 10.6 Systolic array implementation of the modified recursive QRD–LS algorithm.

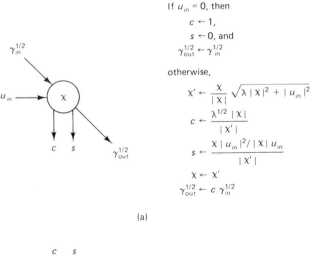

If $u_{in} = 0$, then

$$c \leftarrow 1,$$
$$s \leftarrow 0, \text{ and}$$
$$\gamma_{out}^{1/2} \leftarrow \gamma_{in}^{1/2}$$

otherwise,

$$\chi' \leftarrow \frac{\chi}{|\chi|} \sqrt{\lambda |\chi|^2 + |u_{in}|^2}$$

$$c \leftarrow \frac{\lambda^{1/2} |\chi|}{|\chi'|}$$

$$s \leftarrow \frac{\chi |u_{in}|^2 / |\chi| u_{in}}{|\chi'|}$$

$$\chi \leftarrow \chi'$$
$$\gamma_{out}^{1/2} \leftarrow c \, \gamma_{in}^{1/2}$$

(a)

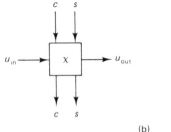

$$u_{out} \leftarrow c u_{in} - s \lambda^{1/2} \chi$$
$$\chi \leftarrow s u_{in} + c \lambda^{1/2} \chi$$

(b)

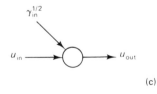

$$u_{out} \leftarrow \gamma_{in}^{1/2} u_{in}$$

(c)

Figure 10.7 Cells for the modified recursive QRD–LS algorithm: (a) boundary cell; (b) internal cell; (c) final cell.

product of the cosine parameters, the $c_i(n)$, is updated recursively by the additional parameter $\gamma^{1/2}(n)$ in each boundary cell and then passed on to the boundary cell in the next column exactly two clock cycles later. The extra delay, which is a direct consequence of the temporal skew imposed on the input data vectors, may be achieved by using an additional storage element that is indicated by means of a black dot in Fig. 10.6. This processing delay may be incorporated within the boundary cell.

As each element $d(n)$ of the desired response vector $\mathbf{b}^H(n)$ moves across the last row of the triangular systolic array in Fig. 10.6, it undergoes the same sequence of Givens rotations, interacting with the previously stored 1-by-M vector $\mathbf{p}^H(n-1)$. An updated vector $\mathbf{p}^H(n)$ is thereby generated in the process. As a result, the scaled version of the a priori estimation error, $\gamma^{1/2}(n) \, \alpha(n)$, emerges from the cell on the extreme right of the bottom row of the systolic array. The final processing cell simply multiplies $\gamma^{1/2}(n)\ \alpha(n)$ thus produced by the output $\gamma^{1/2}(n)$ from the last boundary cell to produce the a posteriori estimation error $e(n)$.

As the time-skewed input data vectors enter the systolic array of Fig. 10.6, we find that updated estimation errors are produced at the output of the array at the rate of one every clock cycle. The estimation error produced on a given clock cycle corresponds, of course, to the particular element of the desired response vector $d(n)$ that entered the array M clock cycles previously.

It is noteworthy that the a priori estimation error $\alpha(n)$ may be obtained by *dividing* the output that emerges from the cell on the extreme right of the bottom row of the systolic array by the output from the last boundary cell. Also, the conversion factor $\gamma(n)$ may be obtained simply by squaring the output that emerges from the last boundary cell.

10.10 COMPUTER EXPERIMENT ON ADAPTIVE EQUALIZATION (CONTINUED)

To illustrate the operation of systolic array II for the solution of the linear least-squares problem, we continue with the computer experiment on adaptive equalization considered in Section 10.7. In particular, we use this form of implementation to study the learning process of the recursive QRD-LS algorithm.

Figure 10.8 shows the learning curves of the algorithm obtained by ensemble-averaging the squared magnitude of the a priori estimation error $\alpha(n)$ over 200

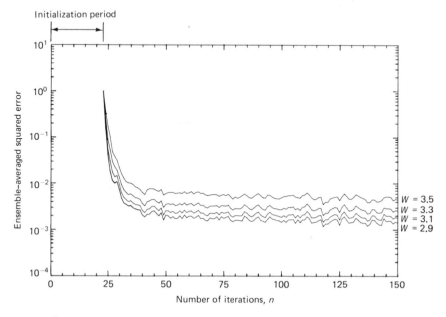

Figure 10.8 Learning curves of the recursive QRD–LS algorithm for adaptive equalization for varying eigenvalue spread $\chi(\mathbf{R})$.

independent trials of the experiment. The learning curves are plotted for 150 iterations of the algorithms and four different values of the eigenvalue spread: $\chi(\mathbf{R})$ = 6.0782, 11.1238, 21.7132, 46.8216. As mentioned in Section 10.9, the error $\alpha(n)$ may be computed by dividing the output emerging from the cell on the extreme right of the bottom row of the systolic array by the output from the last boundary cell.

Fig. 10.9 shows the superposition of four curves obtained by computing (for one trial of the experiment) the conversion factor $\gamma(n)$ versus time n for the four different values of $x(\mathbf{R})$. To compute $\gamma(n)$, we square the output from the last boundary cell.

Comparing these results with those presented in Section 9.9 for the recursive LSL algorithm used to study the same adaptive equalization problem, we may make the following observations:

1. Excluding the initialization periods of the recursive QRD-LS and LSL algorithms, the "$\gamma(n)$ versus n" curve has the same shape for both algorithms. This curve is practically independent of the eigenvalue spread $\chi(\mathbf{R})$.

2. Like the recursive LSL algorithm, the learning curve of the recursive QRD-LS algorithm is insensitive to variations in the eigenvalue spread $\chi(\mathbf{R})$.

3. Once the initialization period is completed, the recursive QRD-LS algorithm converges faster than the recursive LSL algorithm. However, for a prescribed

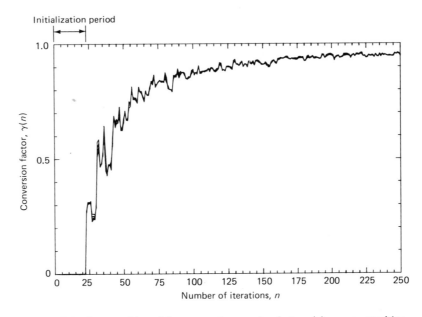

Figure 10.9   Superposition of the curves of conversion factor $\gamma(n)$ versus n resulting from the use of the recursive QRD–LS algorithm for adaptive equalizer for 4 different values of eigenvalue spread $\chi(\mathbf{R})$.

number of elements in the weight vector, initialization of the recursive QRD-LS algorithm takes twice as long as the recursive LSL algorithm.

10.11 APPLICATION OF SYSTOLIC ARRAY II TO ADAPTIVE BEAMFORMING

The objective of an *adaptive beamformer* is to minimize the average power of the beamformer output while, at the same time, maintaining a *fixed* gain along the direction of a target signal. In such an application, explicit knowledge of the least-squares weight vector $\hat{\mathbf{w}}(n)$ is of no direct interest. This makes the systolic array structure II well-suited for use as an adaptive beamformer (Ward and others, 1984; Ward and Hargrave, 1985).

To carry out such an implementation, consider first a linear array that consists of M uniformly spaced sensors, $(M - 1)$ of which serve as *auxiliary* elements and the remaining one serves as the *primary* element. The outputs of the auxiliary sensors constitute the $(M - 1)$-by-n data matrix $\mathbf{A}^{H}(n)$, with the $(M - 1)$ elements of each column defining one *snapshop* of data; n denotes the number of snapshots. The corresponding output of the primary sensor defines the 1-by-n desired response vector $\mathbf{b}^{H}(n)$. Figure 10.10 illustrates the systolic array implementation of an adaptive beamformer for the case of $M = 4$. The individual elements of the data matrix $\mathbf{A}^{H}(n)$ and the desired response vector $\mathbf{b}^{H}(n)$ are fed into the triangular systolic array in a skewed manner. The elements of snapshot k (pertaining to the auxiliary sensors) are denoted by $u(k, i)$, where $i = 1, 2, \ldots, M - 1$. The corresponding output of the primary sensor is denoted by $d(k)$.

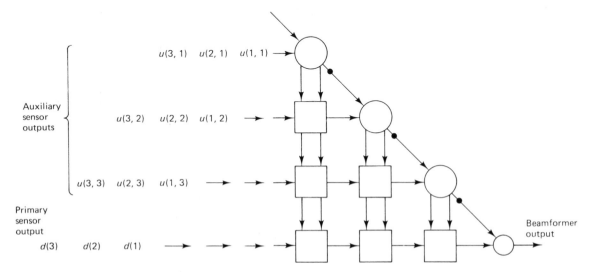

Figure 10.10 Systolic array II used for adaptive beamforming.

For radar applications, the primary sensor consists of an antenna element that is designed to have a relatively narrow main beam and relatively low sidelobes. On the other hand the auxiliary sensors consist of low-gain, omnidirectional dipoles. By excluding the primary antenna from the adaptation process, a fixed gain is maintained along the direction of its main beam.   The outputs of the auxiliary antenna elements are adaptively weighted so as to cancel interferences in the sidelobe region of the primary antenna by placing deep nulls in their directions. Although there are no such adjustable weights visible in the structure of Fig. 10.10, nevertheless, their influence is fully accounted for in the beamformer output in an implicit manner.   This effect is illustrated in Problem 11.

10.12 ANOTHER SYSTOLIC ARRAY FOR IMPLEMENTING THE MINIMUM-VARIANCE DISTORTIONLESS RESPONSE

We now describe another systolic array structure for implementing an adaptive beamformer with *minimum-variance distortionless response* (MVDR) in that the variance (i.e., average power) of the output is minimized while a distortionless response is maintained along the direction of a target signal.

Consider again a linear array of M uniformly spaced sensors whose outputs are individually weighted and then summed to produce the overall response

$$e(i) = \sum_{k=1}^{M} w_k^*(n)\, u(i,\, k) \tag{10.67}$$

where $u(i,\, k)$ is the output of sensor k at time i, and $w_k(n)$ is the associated weight. The requirement is to minimize the index of performance

$$\mathscr{E}(n) = \sum_{i=1}^{n} \lambda^{n-i}\, |e(i)|^2$$

subject to the *constraint*

$$\sum_{k=1}^{M} w_k^*(n)\, s(k,\, \phi) = 1, \quad \text{for all } n \tag{10.68}$$

where $s(1,\, \phi)$, $s(2,\, \phi)$, . . . , $s(M,\, \phi)$ are the elements of a prescribed *steering vector* $\mathbf{s}(\phi)$.   In particular, the element $s(k,\, \phi)$ is the output of sensor k of the array under the condition that there is no signal other than that due to a source of interest, the direction of which determines the angle ϕ (see Section 1.4h).   The solution of this constrained minimization problem is given by (see Problem 2.7)

$$\hat{\mathbf{w}}(n) = \frac{\mathbf{\Phi}^{-1}(n)\, \mathbf{s}(\phi)}{\mathbf{s}^H(\phi)\, \mathbf{\Phi}^{-1}(n)\, \mathbf{s}(\phi)} \tag{10.69}$$

where $\mathbf{\Phi}(n)$ is the M-by-M deterministic correlation matrix of the sensor outputs, averaged over n snapshots, and $\mathbf{s}(\phi)$ is the steering vector.

By applying the QR-decomposition to the data matrix $\mathbf{A}(n)$ [see Eq. (10.21)]

$$\mathbf{Q}(n)\ \mathbf{\Lambda}^{1/2}(n)\ \mathbf{A}(n)\ =\ \begin{bmatrix} \mathbf{R}(n) \\ \mathbf{O} \end{bmatrix}$$

where $\mathbf{\Lambda}(n)$ is the weighting matrix defined in Eq. (10.9), and recognizing that the transformation matrix $\mathbf{Q}(n)$ is a unitary matrix:

$$\mathbf{Q}^H(n)\ \mathbf{Q}(n)\ =\ \mathbf{I},$$

and that $\mathbf{\Phi}(n)$ is related to the data matrix as:

$$\mathbf{\Phi}(n)\ =\ \mathbf{A}^H(n)\ \mathbf{\Lambda}(n)\ \mathbf{A}(n),$$

we may transform the solution of Eq. (10.69) as

$$\hat{\mathbf{w}}(n)\ =\ \frac{\mathbf{R}^{-1}(n)\ \mathbf{R}^{-H}(n)\ \mathbf{s}(\phi)}{\mathbf{s}^H(\phi)\ \mathbf{R}^{-1}(n)\ \mathbf{R}^{-H}(n)\ \mathbf{s}(\phi)} \qquad (10.70)$$

where $\mathbf{R}^{-H}(n)$ is the Hermitian transpose of the inverse of matrix $\mathbf{R}(n)$. In Table 10.5, we present a summary of the algorithm, based on Eq. (10.70), for computing the *constrained* least-squares estimate of the weight vector (Schreiber and Kuekes, 1985). Note that the denominator of Eq. (10.70) consists of an inner product; hence, it equals a real-valued scalar.

We may now describe a systolic array implementation of the algorithm for solving the MVDR problem. The implementation consists of three sections, each

TABLE 10.5 SUMMARY OF THE ALGORITHM FOR COMPUTING THE SOLUTION FOR MINIMUM-VARIANCE DISTORTIONLESS RESPONSE

---

1. Given the n-by-M data matrix $\mathbf{A}(n)$, factor it by using the QR-decomposition as

$$\mathbf{Q}(n)\ \mathbf{\Lambda}^{1/2}(n)\ \mathbf{A}(n)\ =\ \begin{bmatrix} \mathbf{R}(n) \\ \mathbf{O} \end{bmatrix}$$

where $\mathbf{Q}(n)$ is an n-by-n unitary matrix, $\mathbf{\Lambda}(n)$ is the n-by-n weighting matrix, and $\mathbf{R}(n)$ is an M-by-M upper triangular matrix. Hence, determine $\mathbf{R}(n)$.

2. For the look direction of interest, represented by the steering $\mathbf{s}(\phi)$, perform the following computations:

(a) Use *forward* substitution to solve the system of M linear equations:

$$\mathbf{R}^H(n)\ \mathbf{a}(n)\ =\ \mathbf{s}(\phi)$$

for the M-by-1 auxiliary vector $\mathbf{a}(n)$. (Note that $\mathbf{R}^H(n)$ is a lower triangular matrix.)

(b) Use *backward* substitution to solve the associated system of M linear equations:

$$\mathbf{R}(n)\ \hat{\mathbf{w}}(n)\ =\ [\mathbf{a}^H(n)\mathbf{a}(n)]^{-1}\ \mathbf{a}(n)$$

for the optimum weight vector $\hat{\mathbf{w}}(\mathbf{n})$.

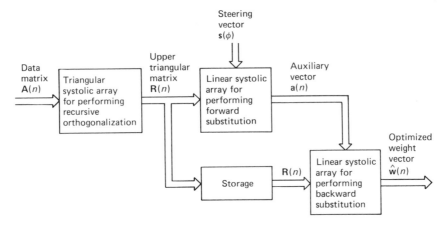

Figure 10.11   Block diagram of systolic array processor for adaptive MVDR beam-former.

of which is designed to perform a specific operation in the algorithm (see Fig. 10.11):

1. A triangular systolic array operates on the input data to perform the QR-decomposition through a recursive sequence of Givens rotations.

2. (a) A linear systolic array performs forward substitution by operating on the lower triangular matrix $\mathbf{R}^H(n)$ and the steering vector $\mathbf{s}(\phi)$ to produce the *auxiliary vector* $\mathbf{a}$. The auxiliary vector $\mathbf{a}(n)$ is itself defined by

$$\mathbf{a}(n) = \mathbf{R}^{-H}(n)\mathbf{s}(\phi) \qquad (10.71)$$

   (b) A second systolic array performs backward substitution by operating on the upper triangular matrix $\mathbf{R}(n)$ and a scaled version of the auxiliary vector $\mathbf{a}(n)$ to produce the optimum weight vector $\hat{\mathbf{w}}(n)$.

For the triangular systolic array, we may use an arrangement similar to that illustrated in Fig. 10.10 for $M = 4$. The content of the boundary cell at the top corner of this array at time n provides the leading element $r_{11}(n)$ of matrix $\mathbf{R}^H(n)$. The contents of the next row of processing cells at time $n + 1$ provide the elements $r_{12}^*(n)$ and $r_{22}(n)$ of matrix $\mathbf{R}^H(n)$, and so on.

For the linear systolic array to perform backward substitution, we may use an arrangement similar to that illustrated in Fig. 10.2. A minor rearrangement of this linear section may be used to perform the forward substitution (see Problem 6).[6]

---

[6]Schreiber and Kuekes (1985) describe a triangular array that solves parts (a) and (b) of step 2.

10.13 *COMPUTER EXPERIMENT ON ADAPTIVE BEAMFORMING*

In this section we illustrate the performance of the systolic array implementation of an adaptive MVDR beamformer by considering a linear array of 5 uniformly spaced sensors. The spacing d between adjacent elements equals $\lambda/2$, where λ is the received wavelength. The array operates in an environment that consists of a target signal and a single source of interference, which are noncoherent with each other. The exponential weighting factor $\lambda = 1$.

The aims of the experiment are two-fold:

1. To examine the evolution of the adapted spatial response (pattern) of the beamformer with time.
2. To evaluate the effect of varying the interference-to-target ratio on the interference-nulling capability of the beamformer.

The directions of the target and source of interference are as follows:

Excitation	Angle of incidence, θ, measured with respect to normal to the array (radians)
Target	$\sin^{-1}(0.2)$
Interference	$-\sin^{-1}(0.4),\ 0$

That is, the interference may originate from any one of 2 possible directions.

The steering vector is defined by

$$\mathbf{s}^H(\phi) = [1,\ e^{j\phi},\ e^{j2\phi},\ e^{j3\phi},\ e^{j4\phi}]$$

where $\phi = \pi\sin\theta$.

Figure 10.12 shows the individual effects of the parameters of interest on the adapted response of the beamformer. The initialization period occupies a total of 10 snapshots. The response is obtained by plotting $20\log_{10}|\hat{\mathbf{w}}^H(n)\mathbf{s}(\phi)|$ versus $\sin\theta = \phi/\pi$. The weight vector $\hat{\mathbf{w}}(n)$ is computed by using the theory described in Section 10.12. The target-to-noise ratio is held constant at 10 dB. In the first part of Fig. 10.12(a), pertaining to a data base of 20 snapshots after the initialization has been completed, the 3 curves correspond to interference-to-noise ratios = 40, 30, 20 dB. In the second part, the data base is increased to 100 snapshots, and in the last part it is increased to 200 snapshots, again after the initialization has been completed. Part (a) of Fig. 10.12 corresponds to an angle of arrival $\theta = -\sin^{-1}(0.4)$ radians for the interference. Part (b) of the figure corresponds $\theta = 0$ for the interference.

Based on these results, we may make the following observations:

1. The response of the beamformer is held fixed at a value of one under all conditions, as required.

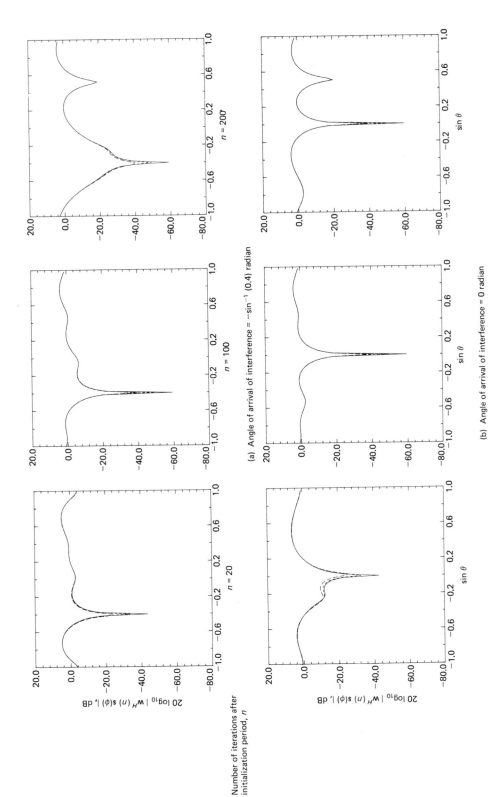

Figure 10.12 Response of adaptive MVDR beamformer, based on systolic array III, for varying signal-to-interference ratio, for 2 difference directions of the interference, and for varying numbers of iteration of the algorithm. The angle ϕ is defined by $\sin \theta = \phi/\pi$.

535

2. With as few as 30 snapshots, including initialization, the beamformer exhibits a reasonably effective nulling capability, which continually improves as the beamformer processes more snapshots.

3. The response of the beamformer is relatively insensitive to variations in the interference-to-target ratio.

10.14 DISCUSSION

In this chapter, we have studied three different systolic array structures: two for solving the linear least-squares problem and the other for solving a constrained version of the problem. In this concluding section, we sum up the common features of these structures, their basic differences, and their areas of application.

To begin with, the three systolic array structures shown in Figs. 10.2, 10.6 and 10.12 share the following *common* features:

1. They all exploit the QR-decomposition method that involves successive applications of a unitary transformation (i.e., the Givens rotation) directly to the data matrix $\mathbf{A}(n)$. The QR-decomposition method is known to be *numerically stable*, with the result that these structures are *less* susceptible to the effects of finite-precision arithmetic than would be the case if direct matrix inversion were to be applied to the deterministic correlation matrix $\mathbf{\Phi}(n)$ in order to solve the normal equation $\mathbf{\Phi}(n)\ \hat{\mathbf{w}}(n)\ =\ \mathbf{\theta}(n)$. Also, the Givens rotations are updated recursively as new data arrive.

2. They are examples of an *open-loop data-adaptive system*, data adaptive in that the values assigned to the parameters of the sequence of Givens rotations vary in accordance with the input data. The "open-loop" feature is exemplified by the absence of feedback in all three structures. This feature distinguishes them from the other closed-loop adaptive filter structures considered in previous chapters of the book. Accordingly, the systolic array structures of Figs. 10.2, 10.6, and 10.12, have the advantage in that they require only *limited* input data to describe accurately their external environment.

3. All three structures are relatively easy to design because of their highly pipelined operation, regular nature, and layout. These features, and the fact that only nearest-neighbor interconnections are involved between individual processing cells in the arrays, make them particularly well suited for implementation in VLSI form.[7]

---

[7]With *very large-scale integration* (*VLSI*), it is possible to build complex array processors that consist of a large number of processing cells, which cooperate with each other to perform extremely fast matrix computations. Kung (1985) presents an overview of VLSI array processors, with particular reference to systolic arrays. A major problem that arises from the use of this technology, however, is that a single flaw in the chip or wafer can render an entire VLSI array processor useless. It is therefore desirable to have a VLSI array processor that achieves high performance, tolerates physical failures, and yet produces correct results. Jou and Abraham (1984) describe a general and complete technique of system–level fault tolerance for matrix operations. This technique can be used to tolerate faults in systolic array processors.

4. The triangular systolic array that performs the sequence of Givens rotations consists of boundary cells, inner cells, and (in the case of Fig. 10.6 only) a final processing cell. The number of boundary cells (representing the most complicated node algorithm in that square roots and divisions as well as multiplications and additions are performed there) equals $N - 1$, where N is the number of data points entering the array.[8] The number of inner cells (where multiplications and additions only are performed), on the other hand, equals $(N^2 - N)/2$. For the structures of Figs. 10.2 and 10.6, we have $N = M + 1$, where M is the dimension of the input vector $\mathbf{u}(n)$ and the 1 accounts for the desired response $d(n)$. In Fig. 10.12 we have $N = M$. In addition to the triangular systolic array, the structure of Fig. 10.2 includes a linear systolic array for computing the M-by-1 least-squares weight vector $\hat{\mathbf{w}}(n)$; this section consists of one boundary cell (where addition, multiplication and division are performed) and $M - 1$ internal cells (where addition and multiplication are performed). The structure of Fig. 10.6 includes a final processing cell in the triangular systolic array, where a single multiplication is performed; this cell is used to compute the (a posteriori) estimation error $e(n)$. The structure of Fig. 10.12 includes two linear systolic arrays (each with one boundary cell and $M - 1$ internal cells), one of which performs a forward substitution to solve for an M-by-1 auxiliary vector, and the other performs a backward substitution to solve for the constrained least-squares weight vector $\hat{\mathbf{w}}(n)$.

Next we sum up the *differences* between these structures, particularly, as they relate to their areas of application:

1. The structure of Fig. 10.2 solves the linear least-squares problem by producing the individual elements of the least-squares weight vector $\hat{\mathbf{w}}(n)$ at the output of the linear systolic array section. Such a structure may therefore find use in applications (e.g., adaptive equalization, system identification) where explicit knowledge of the least-squares weight vector is required.

2. The structure of Fig. 10.6 solves the linear least-squares problem by producing the (a posteriori) estimation error $e(n)$ at the output of the final processing cell without computing the weight vector $\hat{\mathbf{w}}(n)$ explicitly. The second structure may therefore find use in applications (e.g., adaptive prediction-error filters, adaptive noise cancellers, adaptive beamformers) where explicit knowledge of the weight vector $\hat{\mathbf{w}}(n)$ is not required. This structure achieves the highest possible throughput rate with the minimum amount of computation and circuitry. The structure of Fig. 10.6 may also have a numerical

---

[8] Since all the cells in the systolic array must operate at the same throughput rate, boundary cells may constitute a *bottleneck* for the overall performance. It is therefore desirable to reduce the complexity of boundary cells in the triangular systolic array so that it becomes close to that of internal cells. Gentleman (1973) describes methods for performing Givens rotations *without square roots* for the case of real-valued data. See Problem 8.

advantage over that of Fig. 10.2 since the former structure avoids the need to solve a triangular linear system; such a system of equations may sometimes be ill conditioned.

3. The addition of a linear systolic array section to the structure of Fig. 10.6 suggests itself. In this way, the unique features of the structures in Figs. 10.2 and 10.6 are combined. The new structure provides a means for computing the least-squares weight vector (at the end of the entire recursions or at regular intervals spaced $2M$ clock cycles apart) and also for *continuous* monitoring of the estimation error $e(n)$ at the output of the final processing cell.

4. The structure of Fig. 10.6 may be used for adaptive beamforming in one of two ways: (a) the output of a primary sensor (with a beamsteering capability) provides the desired response while the outputs from a set of auxiliary sensors provide the input vector for nulling interferences that impinge on the array from unknown directions; (b) the outputs from a set of similar sensors are applied to a multiple-beamforming network with one of them selected as the desired response (depending on the look direction of interest) and the other auxiliary beams used for interference nulling. This second configuration has a computational disadvantage in that every time a new look direction is selected, there is a corresponding change in the set of auxiliary beams used for interference nulling, with the result that a new triangular systolic array is required for the orthogonal triangularization.

5. When the MVDR structure of Fig. 10.12 is used for adaptive beamforming, the structure automatically applies a constraint on the way in which the outputs from a set of similar sensors are weighted before summing. This constraint is determined by the choice of the steering vector $s(\phi)$. The structure of the triangular systolic array section remains invariant to this choice. However, for every desired steering vector $s(\phi)$, we have to use a corresponding pair of linear systolic array sections to perform the pertinent forward and backward computations.

In conclusion, each of the systolic array structures considered in this chapter has unique features of its own. The decision as to which one to use can only be determined by the requirements of the problem of interest.

PROBLEMS

1. Consider the squared Euclidean norm

$$\mathcal{E}(n) = \|\Lambda^{1/2}(n)\,\varepsilon(n)\|^2$$

where $\varepsilon(n)$ is the estimation error vector and $\Lambda(n)$ is the diagonal weighting matrix. These two parameters are defined by Eqs. (10.6) and (10.9), respectively. Show that $\mathcal{E}(n)$ is unaffected by postmultiplying the vector $\Lambda^{1/2}(n)\varepsilon(n)$ by a unitary matrix $Q(n)$.

2. Start with the deterministic normal equation

$$\boldsymbol{\Phi}(n)\ \hat{\mathbf{w}}(n)\ =\ \boldsymbol{\theta}(n)$$

where $\boldsymbol{\Phi}(n)$ is the deterministic correlation matrix of the input vector $\mathbf{u}(n)$, $\boldsymbol{\theta}(n)$ is the deterministic cross-correlation vector between $\mathbf{u}(n)$ and the desired response $d(n)$, and $\hat{\mathbf{w}}(n)$ is the least-squares weight vector. Hence, using the QR-decomposition

$$\mathbf{Q}(n)\boldsymbol{\Lambda}^{1/2}(n)\mathbf{A}(n)\ =\ \begin{bmatrix} \mathbf{R}(n) \\ \mathbf{O} \end{bmatrix}$$

where $\mathbf{A}(n)$ is the data matrix, $\boldsymbol{\Lambda}(n)$ is the exponential weighting matrix, $\mathbf{Q}(n)$ is a unitary matrix, and $\mathbf{R}(n)$ is an upper triangular matrix, transform the normal equation into the form

$$\mathbf{R}(n)\ \hat{\mathbf{w}}(n)\ =\ \mathbf{p}(n)$$

How are the matrices $\mathbf{R}(n)$ and $\boldsymbol{\Phi}(n)$, and the vectors $\mathbf{p}(n)$ and $\boldsymbol{\theta}(n)$ related?

3. Equation (10.55) describes a method of partitioning the unitary matrix $\mathbf{T}(n)$ that represents the sequence of Givens rotations. Show that the scaling factor $\gamma^{1/2}(n)$ introduced in this equation equals the square root of the real-valued scalar $1 - \mathbf{u}^H(n)\,\boldsymbol{\Phi}^{-1}(n)\,\mathbf{u}(n)$, where $\mathbf{u}(n)$ is the input vector and $\boldsymbol{\Phi}^{-1}(n)$ is the inverse of the associated correlation matrix.

4. Show that the variable $\alpha(n)$ defined in Eq. (10.58) equals the a priori estimation error, as used in Chapters 8 and 9.

5. Explain the way in which the systolic array structure of Fig. 10.6 may be used to operate as a prediction-error filter.

6. Equation (10.55) defines the unitary matrix $\mathbf{T}(n)$ that represents the sequence of Givens rotations. Find the conditions which the matrix $\mathbf{Z}(n)$ and vector $\boldsymbol{\beta}(n)$ have to satisfy in order to ensure that

$$\mathbf{T}^{-1}(n)\ =\ \mathbf{T}^H(n)$$

7. In Section 10.6 we described a linear systolic array, based on backward substitution, for solving the equation $\mathbf{R}(n)\ \hat{\mathbf{w}}(n)\ =\ \mathbf{p}(n)$ for $\hat{\mathbf{w}}(n)$. Discuss the use of such a section, based on forward substitution, for solving the equation $\mathbf{R}^H(n)\ \mathbf{a}(n)\ =\ \mathbf{s}$ for the vector $\mathbf{a}(n)$; the matrix $\mathbf{R}$ is an M-by-M upper triangular matrix and $\mathbf{s}$ is an M-by-1 vector.

8. The QRD-LS decomposition algorithm may be reformulated by using the *square-root-free* version of the Givens rotation. Such a change has obvious computational advantages. In particular, when the square-root-free Givens rotation is used to perform orthogonal triangularization of a matrix, the upper triangular matrix $\mathbf{R}$ is represented in terms of a *diagonal matrix* $\mathbf{D}$ and a *unit upper triangular matrix* $\overline{\mathbf{R}}$ (with its diagonal elements equal to 1) as follows:

$$\mathbf{R}\ =\ \mathbf{D}^{1/2}\overline{\mathbf{R}}$$

   (a) Using this new approach, reformulate the recursive QRD-LS algorithm.
   (b) Describe the operation of cells in the systolic array required for implementing this new algorithm.

Computer-oriented Problems

9. *Computer experiment on adaptive equalization using systolic array structure I (continued)*: In Section 10.7 we used the systolic array structure in Fig. 10.2 to compute the least-squares weight vector of an adaptive equalizer with $M = 11$ taps. The values of the weight vector were computed after $2(M + 1)$, $4(M + 1)$, $6(M + 1)$, $8(M + 1)$ clock cycles from the end of the exact initialization period. Details of the environment in which the equalizer operates were given in Section 5.16.

   Continue this experiment by carrying out the following computations:
   (a) Compute the values of the weight vector for one trial of the experiment and after $16(M + 1)$, $32(M + 1)$ clock cycles from the end of the initialization period.
   (b) Convolve the impulse response of the channel with that of the systolic array equalizer computed after $32(M + 1)$ clock cycles from the end of the initialization period.
   Carry out the computations in (a) and (b) for each of the four eigenvalue spreads specified in Section 10.7. Compare your results with the Wiener solution.

10. *Computer experiment on adaptive equalization using systolic array structure II*: In Section 10.10 we used the systolic array structure II to study the learning process of an adaptive equalizer (the same one considered in Problem 9). In particular, the learning curves of the equalizer were plotted for different eigenvalue spreads by basing the analysis on the a priori estimation error $\alpha(n)$.

   Continue this experiment by basing your analysis on the a posteriori estimation error $e(n)$. In particular, plot the ensemble-averaged squared value of $e(n)$ versus n by averaging your results over 200 independent trials of the experiment for the four different values of channel parameter $W = 2.9, 3.1, 3.3, 3.5$ that control the eigenvalue spread of the correlation matrix of the input vector.

11. *Computer experiment on adaptive beamforming using systolic array II*: Consider an environment that contains a target signal and a single source of interference that are noncoherent with each other. The environmental parameters are as follows:
   signal-to-noise ratio (SNR) = 10 dB
   angle of arrival of target = $\sin^{-1}(0.2)$
   interference-to-noise ratio (INR) = 40, 30, 20 dB
   angle of arrival of interference = $-\sin^{-1}(0.4)$
   The angles of arrival are measured with respect to the normal to the linear array of sensors used to obtain the spatial data. The number of sensors $M = 5$. The sensor outputs are applied to a beamforming network to produce 4 orthogonal beams directed at $\phi = \pm 0.2\pi, \pm 0.6\pi$. The beam pointing along $\phi = 0.2\pi$ is used as the desired response, and the remaining three beams are used as auxiliary inputs for the adaptive nulling of the interference.

   Perform the following computations:
   (a) Compute the antenna pattern (i.e., $10 \log_{10}|e(n, \phi)|^2$) after $n = 8$ clock cycles, measured from the end of the initialization period, for each of the three prescribed values of INR. Note that the processor output $e(n, \phi)$ is a function of angle ϕ.
   (b) Repeat the computations in (a) for $n = 16, 24, 32, 64$ clock cycles, measured from the end of the initialization period.
   Comment on your results.

12. *Computer experiment on adaptive beamforming using systolic array III*: Consider an environment that contains a target signal and two sources of interference, all of which

are noncoherent with each other. The signal-to-noise ratio SNR = 10 dB. The two interferences are equal in strength with INR = 40, 30, 20 dB. The pertinent angles of arrival are as described below:

Excitation	Angle of arrival, θ, measured with respect to the normal to the array
Target	$\sin^{-1}(0.2)$
Interference I	$-\sin^{-1}(0.4)$, $\sin^{-1}(0.1)$
Interference II	$\sin^{-1}(0.8)$

That is, the directions of the target and interference II are fixed, while interference I may originate from one of two possible directions. The array itself consists of 5 similar sensors with uniform spacing $d = \lambda/2$, where λ is the wavelength.

Perform the following computations:

(a) Compute the adapted response of the beamformer after 20 snapshots from the end of the initialization period for the case when interference I arrives along the direction $\theta = -\sin^{-1}(0.4)$, and for each of the prescribed values of INR. Repeat your computations for 100, 200 snapshots.

(b) Repeat your computations in (a) for the case when interference I has the direction $\theta = \sin^{-1}(0.1)$.

A

Maximum-Likelihood Estimation

A.1 INTRODUCTION

Estimation theory is a branch of probability and statistics that deals with the problem of deriving information about properties of random variables and stochastic processes, given a set of observed samples. This problem arises frequently in the study of communications and control systems. *Maximum likelihood* is by far the most general and powerful method of estimation. It was first used by Fisher in 1906. In principle, the method of maximum likelihood may be applied to any estimation problem with the proviso that we formulate the joint probability density function of the available set of observed data. As such, the method yields almost all the well-known estimates as special cases.

A.2 *LIKELIHOOD FUNCTION*

The method of maximum likelihood is based on a relatively simple idea: different populations generate different data samples and any given data sample is more *likely* to have come from some population than from others (Kmenta, 1971).

Let $f_U(u|\theta)$ denote the *conditional joint probability density function* of the *random vector* U represented by the observed *sample* vector u, where the sample vector u has $u_1, u_2, \ldots, u_M$ for its elements, and θ is a *parameter vector* with θ_1, $\theta_2, \ldots, \theta_K$ as elements. The method of maximum likelihood is based on the principle that we should estimate the parameter vector θ by its most *plausible values*, given the observed sample vector u. In other words, the maximum-likelihood estimators of $\theta_1, \theta_2, \ldots, \theta_K$ are those values of the parameter vector for which the conditional joint probability density function $f_U(u|\theta)$ is at maximum.

The name *likelihood function*, denoted by $l(\theta)$, is given to the conditional joint probability density function $f_U(u|\theta)$, viewed as a function of the parameter vector θ. We thus write

$$l(\theta) = \mathbf{f}_U(u|\theta) \tag{A.1}$$

Although the conditional joint probability density function and the likelihood function have exactly the same formula, nevertheless, it is vital that we appreciate the physical distinction between them. In the case of the conditional joint probability density function, the parameter vector θ is fixed and the observation vector u is variable. On the other hand, in the case of the likelihood function, the parameter vector θ is variable and the observation vector u is fixed.

In many cases, it turns out to be more convenient to work with the logarithm of the likelihood function rather than with the likelihood function itself. Thus, using $L(\theta)$ to denote the *log-likelihood function*, we write

$$\begin{aligned} L(\theta) &= \ln[l(\theta)] \\ &= \ln[f_U(u|\theta)] \end{aligned} \tag{A.2}$$

The logarithm of $l(\theta)$ is a *monotonic transformation* of $l(\theta)$. This means that whenever $l(\theta)$ decreases its logarithm $L(\theta)$ also decreases. Since $l(\theta)$, being a formula for conditional joint probability density function, can never become negative, it follows that there is no problem in evaluating its logarithm $L(\theta)$. We conclude therefore that the parameter vector for which the likelihood function $l(\theta)$ is at maximum is exactly the same as the parameter vector for which the log-likelihood function $L(\theta)$ is at its maximum.

To obtain the *i*th element of the maximum-likelihood estimate of the parameter vector θ, we differentiate the log-likelihood function with respect to θ_i and set the result equal to zero. We thus get a set of first-order conditions:

$$\frac{\partial L}{\partial \theta_i} = 0, \qquad i = 1, 2, \ldots, K \tag{A.3}$$

The first derivative of the log-likelihood function with respect to parameter θ_i is called the *score* for that parameter. The vector of such parameters is known as the *scores vector* (i.e., the gradient vector). The scores vector is identically zero at the maximum-likelihood estimates of the parameters, that is, at the values of θ that result from the solutions of Eq. (A.3).

To find how effective the method of maximum likelihood is, we can compute the *bias* and *variance* for the estimate of each parameter. However, this is frequently difficult to do. Rather than approach the computation directly, we may derive a *lower bound* on the variance of any *unbiased* estimate. We say an estimate is unbiased if the average value of the estimate equals the parameter we are trying to estimate. Later we show how the variance of the maximum-likelihood estimate compares with this lower bound.

A.3 CRAMÉR–RAO INEQUALITY

Let U be a random vector with conditional joint probability density function $f_U(u|\theta)$, where **u** is the observed sample vector with elements $u_1, u_2, \ldots, u_M$ and θ is the parameter vector with elements $\theta_1, \theta_2, \ldots, \theta_K$. Using the definition of Eq. (A.2) for the log-likelihood function $L(\theta)$ in terms of the conditional joint probability density function $f(u|\theta)$, we form the K-by-K matrix:

$$
\mathbf{J} = -
\begin{bmatrix}
E\left[\dfrac{\partial^2 L}{\partial \theta_1^2}\right] & E\left[\dfrac{\partial^2 L}{\partial \theta_1 \partial \theta_2}\right] & \cdots & E\left[\dfrac{\partial^2 L}{\partial \theta_1 \partial \theta_K}\right] \\[2ex]
E\left[\dfrac{\partial^2 L}{\partial \theta_2 \partial \theta_1}\right] & E\left[\dfrac{\partial^2 L}{\partial \theta_2^2}\right] & \cdots & E\left[\dfrac{\partial^2 L}{\partial \theta_2 \partial \theta_K}\right] \\[2ex]
\vdots & \vdots & & \vdots \\[2ex]
E\left[\dfrac{\partial^2 L}{\partial \theta_K \partial \theta_1}\right] & E\left[\dfrac{\partial^2 L}{\partial \theta_K \partial \theta_2}\right] & \cdots & E\left[\dfrac{\partial^2 L}{\partial \theta_K^2}\right]
\end{bmatrix}
\qquad (A.4)
$$

The matrix **J** is called *Fisher's information matrix*.

Let I denote the inverse of Fisher's information matrix **J**. Let I_{ii} denote the ith diagonal element (i.e., the element in the ith row and ith column) of the inverse matrix **I**. Let $\hat{\theta}_i$ be *any* unbiased estimate of the parameter θ_i, based on the observed sample vector **u**. We may then write (Van Trees, 1968; Nahi, 1969)

$$
\text{var}[\hat{\theta}_i] \geq I_{ii}, \qquad i = 1, 2, \ldots, K \qquad (A.5)
$$

Equation (A.5) is called the *Cramér–Rao inequality*. This theorem enables us to construct a lower limit (greater than zero) for the variance of any unbiased estimator, provided, of course, that we know the functional form of the log-likelihood function. The lower limit specified in the theorem is called the *Cramér–Rao lower bound*.

If we can find an unbiased estimator whose variance equals the Cramér–Rao lower bound, then according to the theorem of Eq. (A.5) there is no other unbiased estimator with a smaller variance. Such an estimator is said to be *efficient*.

A.4 PROPERTIES OF MAXIMUM-LIKELIHOOD ESTIMATORS

Not only is the method of maximum likelihood based on an intuitively appealing idea (that of choosing those parameters from which the actually observed sample vector is most likely to have come), but also the resulting estimates have some desirable properties. Indeed, under quite general conditions, the following *asymptotic* properties may be proved (Kmenta, 1971):

1. Maximum-likelihood estimators are *consistent*. That is, the value of θ_i for which the score $\partial L/\partial \theta_1$ is identically zero *converges in probability* to the true value of the parameter θ_i, $i = 1, 2, \ldots, K$, as the *sample size M* approaches infinity.

2. Maximum-likelihood estimators are *asymptotically efficient*; that is,

$$\lim_{M \to \infty} \left\{ \frac{\text{var}[\theta_{i,ml} - \theta_i]}{I_{ii}} \right\} = 1, \qquad i = 1, 2, \ldots, K$$

   where $\theta_{i,ml}$ is the maximum-likelihood estimate of parameter θ_i, and I_{ii} is the ith diagonal element of the inverse of Fisher's information matrix.

3. Maximum-likelihood estimators are *asymptotical Gaussian*.

In practice, we find that the large-sample (asymptotic) properties of maximum-likelihood estimators hold rather well for sample size $M \geq 50$.

B

Maximum-entropy Method

B.1 INTRODUCTION

The *maximum-entropy method* (MEM) was originally devised by Burg in 1967 to overcome fundamental limitations of Fourier-based methods for estimating the power spectrum of a stationary stochastic process (Burg 1967, 1975). The basic idea of the MEM is to choose the particular spectrum that corresponds to the most *random* or the most *unpredictable* time series whose autocorrelation function agrees with a set of known values. This condition is equivalent to an extrapolation of the autocorrelation function of the available time series by *maximizing* the *entropy* of the process; hence, the name of the method. Entropy is a measure of the average information content of the process (Shannon, 1948). Thus, the MEM bypasses the problems that arise from the use of window functions, a feature that is common to all Fourier-based methods of spectrum analysis. In particular, the MEM avoids the use of a periodic extension of the data (as in the method based on smoothing the periodogram and its computation using the fast Fourier transform algorithm) or of the assumption that the data outside the available record length are zero (as in the Blackman–Tukey method based on the sample autocorrelation function). An important feature of the spectrum estimate resulting from the use of the MEM is that it is *nonnegative at all frequencies*, which is precisely the way it should be.

B.2 MAXIMUM-ENTROPY SPECTRUM ESTIMATOR

Suppose that we are given $2M + 1$ values of the autocorrelation function of a stationary stochastic process $\{u(n)\}$ of zero mean. We wish to obtain an estimate for the power spectrum of the process that corresponds to the most random time series whose autocorrelation function is consistent with the set of $2M + 1$ known values. In terms of information theory, this statement corresponds to the *principle of maximum entropy* (Jaynes, 1982).

In the case of a set of Gaussian-distributed random variables of zero mean, the entropy is given by (see Section 4.16)

$$H = \tfrac{1}{2}\ln[\det(\mathbf{R})] \tag{B.1}$$

where $\mathbf{R}$ is the correlation matrix of the process. When the process is of infinite duration, however, we find that the entropy H diverges, and so we cannot use it as a measure of information content. To overcome the divergence problem, we may use the *entropy rate* defined by (Middletown, 1960)

$$\begin{aligned} h &= \lim_{M\to\infty} \frac{H}{M + 1} \\ &= \lim_{M\to\infty} \tfrac{1}{2}\ln[\det(\mathbf{R})]^{1/(M+1)} \end{aligned} \tag{B.2}$$

Let $S(\omega)$ denote the power spectrum of the process $\{u(n)\}$. The limiting form of the determinant of the correlation matrix $\mathbf{R}$ is related to the power spectrum $S(\omega)$ as follows (see Problem 17, Chapter 2)

$$\lim_{M\to\infty} [\det(\mathbf{R})]^{1/(M+1)} = \exp\left\{ \frac{1}{2\pi} \int_{-\pi}^{\pi} \ln S(\omega)\, d\omega \right\} \tag{B.3}$$

Hence, substituting Eq. (B.3) in (B.2), we get

$$h = \frac{1}{4\pi} \int_{-\pi}^{\pi} \ln[S(\omega)]\, d\omega \tag{B.4}$$

Although this relation was derived on the assumption that the process $\{u(n)\}$ is Gaussian, nevertheless, the form of the relation is valid for any stationary process.

We may now restate the power spectrum estimation problem in terms of the entropy rate. We wish to find a real positive-valued power spectrum estimate $\hat{S}(\omega)$ characterized by entropy rate h, satisfying two simultaneous requirements:

1. The entropy rate h is *stationary* with respect to the *unknown* values of the autocorrelation function of the process.
2. The power spectrum estimate $\hat{S}(\omega)$ is *consistent* with respect to the *known* values of the autocorrelation function of the process.

We will address these two requirements in turn.

Since the autocorrelation sequence $\{r(m)\}$ and power spectrum $S(\omega)$ of a stationary process $\{u(n)\}$ form a discrete-time Fourier transform pair, we write

$$S(\omega) = \sum_{m=-\infty}^{\infty} r(m) \exp(-jm\omega) \tag{B.5}$$

Equation (B.5) assumes that the sampling period of the discrete-time process $\{u(n)\}$ is normalized to unity. Substituting Eq. (B.5) in (B.4), we get

$$h = \frac{1}{4\pi} \int_{-\pi}^{\pi} \ln\left[\sum_{m=-\infty}^{\infty} r(m) \exp(-jm\omega) \right] d\omega \tag{B.6}$$

We extrapolate the autocorrelation sequence $\{r(m)\}$ outside the range of known values, $-M \le m \le M$, by choosing the unknown values of the autocorrelation function in such a way that no information or entropy is added to the process. That is, we impose the condition

$$\frac{\partial h}{\partial r(m)} = 0, \qquad |m| \ge M + 1 \tag{B.7}$$

Hence, differentiating Eq. (B.6) with respect to $r(m)$ and setting the result equal to zero, we find that the conditions for *maximum entropy* are as follows:

$$\int_{-\pi}^{\pi} \frac{\exp(-jm\omega)}{\hat{S}(\omega)} d\omega = 0, \qquad |m| \ge M + 1 \tag{B.8}$$

Equation (B.8) implies that the power spectrum estimate $\hat{S}(\omega)$ is expressible in the form of a truncated Fourier series:

$$\frac{1}{\hat{S}(\omega)} = \sum_{k=-M}^{M} c_k \exp(-jk\omega) \tag{B.9}$$

The complex Fourier coefficient c_k of the expansion satisfies the Hermitian condition

$$c_k^* = c_{-k} \tag{B.10}$$

so as to ensure that the estimate $\hat{S}(\omega)$ is real.

The next requirement is to make the power spectrum estimate $\hat{S}(\omega)$ consistent with the set of known values of the autocorrelation function $r(m)$ for the interval $-M \le m \le M$. Since $r(m)$ is a Hermitian function, we need only concern ourselves with $0 \leq m \leq M$. Accordingly, $r(m)$ must euqal the inverse discrete-time Fourier transform of $S(\omega)$ for $0 \leq m \leq M$, as shown by

$$r(m) = \frac{1}{2\pi} \int_{-\pi}^{\pi} \hat{S}(\omega) \exp(jm\omega)\, d\omega, \qquad 0 \leq m \leq M \qquad \text{(B.11)}$$

Therefore, substituting Eq. (B.9) in (B.11), we get

$$r(m) = \int_{-\pi}^{\pi} \frac{\exp(jm\omega)}{\displaystyle\sum_{k=-M}^{M} c_k \exp(-jk\omega)}\, d\omega, \qquad 0 \leq m \leq M \qquad \text{(B.12)}$$

Clearly, in the set of complex Fourier coefficients $\{c_k\}$, we have the available degrees of freedom needed to satisfy the conditions of Eq. (B.12).

To proceed with the analysis, however, we find it convenient to use z-transform notation by changing from the variable ω to z. Define

$$z = \exp(j\omega) \qquad \text{(B.13)}$$

Hence,

$$d\omega = \frac{1}{j} \frac{dz}{z}$$

and so we rewrite Eq. (B.12) in terms of the variable z as the contour integral:

$$r(m) = \frac{1}{j2\pi} \oint \frac{z^{m-1}}{\displaystyle\sum_{k=-M}^{M} c_k z^{-k}}\, dz, \qquad 0 \leq m \leq M \qquad \text{(B.14)}$$

The contour integration in Eq. (B.14) is performed on the unit circle in the z-plane in a counterclockwise direction. Since the complex Fourier coefficient c_k satisfies the Hermitian condition of Eq. (B.10), we may express the summation in the denominator of the integral in Eq. (B.14) as the product of two polynomials, as follows:

$$\sum_{k=-M}^{M} c_k z^{-k} = G(z)\, G^*\!\left(\frac{1}{z^*}\right) \qquad \text{(B.15)}$$

where

$$G(z) = \sum_{k=0}^{M} g_k z^{-k} \qquad \text{(B.16)}$$

and

$$G^*\!\left(\frac{1}{z^*}\right) = \sum_{k=0}^{M} g_k^* z^{k} \qquad \text{(B.17)}$$

We choose the first polynomial $G(z)$ to be minimum phase in that its zeros are all located inside the unit circle in the z-plane. Correspondingly, we choose the second polynomial $G^*(1/z^*)$ to be maximum phase in that all its zeros are located outside the unit circle in the z-plane. Moreover, the zeros of these two polynomials are the inverse of each other with respect to the unit circle. Thus, substituting Eq. (B.15) in (B.14), we get

$$r(m) = \frac{1}{j2\pi} \oint \frac{z^{m-1}}{G(z)G^*(1/z^*)}\, dz, \qquad 0 \le m \le M \tag{B.18}$$

We next form the summation

$$\sum_{k=0}^{M} g_k r(m - k) = \frac{1}{j2\pi} \oint \frac{z^{m-1}\sum\limits_{k=0}^{M} g_k z^{-k}}{G(z)G^*(1/z^*)}\, dz$$

$$\tag{B.19}$$

$$= \frac{1}{j2\pi} \oint \frac{z^{m-1}}{G^*(1/z^*)}\, dz, \qquad 0 \le m \le M$$

where in the first line we have used Eq. (B.18), and in the second line we have used Eq. (B.16).

To evaluate the contour integral of Eq. (B.19), we use *Cauchy's residue theorem* of complex variable theory (Levinson and Redheffer, 1970). According to this theorem, the contour integral equals $2\pi j$ times the sum of *residues* of the poles of the integral $z^{m-1}/G^*(1/z^*)$ that lie inside the unit circle, the contour of integration. Since the polynomial $G^*(1/z^*)$ is chosen to have no zeros inside the unit circle, it follows that the integral in Eq. (B.19) is analytic on and inside the unit circle for $m \ge 1$. For $m = 0$ the integral has a simple pole at $z = 0$ with a *residue* equal to $1/g_0^*$. Hence, application of Cauchy's residue theorem yields

$$\oint \frac{z^{m-1}}{G^*(1/z^*)}\, dz = \begin{cases} \dfrac{2\pi j}{g_0^*} & m = 0 \\[2mm] 0 & m = 1, 2, \ldots, M \end{cases} \tag{B.20}$$

Thus, substituting Eq. (B.20) in (B.19), we get

$$\sum_{k=0}^{M} g_k r(m - k) = \begin{cases} \dfrac{1}{g_0^*}, & m = 0 \\[2mm] 0 & m = 1, 2, \ldots, M \end{cases} \tag{B.21}$$

We recognize that the set of $(M + 1)$ equations in (B.21) have a mathematical form similar to that of the augmented normal equation for forward prediction of order M (see Section 4.3). In particular, by comparing Eqs. (B.21) and (4.19), we deduce that

$$g_k^* = \frac{1}{g_0 P_M} a_{M,k}, \qquad 0 \le k \le M \tag{B.22}$$

where $\{a_{M,k}\}$ are the coefficients of a prediction-error filter of order M, and P_M is the average output power of the filter. Since $a_{M,0} = 1$ for all M, we find from Eq. (B.22) that, for $k = 0$;

$$|g_0|^2 = \frac{1}{P_M} \tag{B.23}$$

Finally, substituting Eqs. (B.15), (B.22), (B.23) in (B.9) with $z = \exp(j\omega)$, we get the desired expression for the power spectrum estimate:

$$S(\omega) = \frac{P_M}{\left| 1 + \sum_{k=1}^{M} a_{M,k} \exp(jk\omega) \right|^2} \tag{B.24}$$

We refer to the formula of Eq. (B.24) as the *maximum-entropy spectrum estimator* (MESE).

C
z-Transform

The *z-transform* plays an important role in the analysis of *discrete-time, linear, time-invariant systems*. In this appendix, we define the *z*-transform of a sequence of numbers and briefly discuss the interrelationships between properties of the *z*-transform and those of the original sequence (Oppenheim and Schafer, 1975).

C.1 DEFINITIONS

Consider the sequence $\{u(n)\}$, $n = 0, \pm 1, \pm 2, \ldots$. We define the *z*-transform $U(z)$ of this sequence as

$$U(z) = \sum_{n=-\infty}^{\infty} u(n)z^{-n} \tag{C.1}$$

where z is a *complex variable*. We refer to the z-transform of Eq. (C.1) as the *two-sided z-transform*.

We may also define the *one-sided z-transform* as

$$U(z) = \sum_{n=0}^{\infty} u(n)z^{-n} \tag{C.2}$$

When $u(n)$ is zero for $n < 0$, it is clear that the one- and two-sided z-transforms of the sequence are equivalent. Otherwise, they have different values.

For $z = e^{j\omega}$, that is, points on the unit circle in the z-plane, we have, from Eq. (C.1),

$$U(e^{j\omega}) = \sum_{n=-\infty}^{\infty} u(n)e^{-jn\omega} \tag{C.3}$$

which is recognized as the discrete-time Fourier transform of $\{u(n)\}$.

C.2 PROPERTIES OF THE z-TRANSFORM

In this section we briefly discuss some properties of the z-transform as defined in Eq. (C.1). For proofs of these properties, the reader is referred to Oppenheim and Schafer (1975), Rabiner and Gold (1975).

(1) Region of Convergence. For a given sequence $\{u(n)\}$, the set of values of z for which the z-transform $U(z)$ converges is called the *region of convergence*. In general, we find that the power series of Eq. (C.1) converges in an annulus centered at the origin of the z-plane and defined by

$$R_{u^-} < |z| < R_{u^+} \tag{C.4}$$

where the lower limit R_{u^-} may be as small as zero and the upper limit R_{u^+} may be as large as infinity.

In an important class of z-transforms, $U(z)$ is a rational function; that is, $U(z)$ consists of a ratio of two polynomials in z. We refer to the roots of the numerator polynomial [i.e., those values of z for which $U(z) = 0$] as the *zeros* of $U(z)$. We refer to the roots of the denominator polynomial [i.e., those values of z for which $U(z) = \infty$] as the *poles* of $U(z)$.

It is clear that, in the case of rational z-transforms, no poles of $U(z)$ can occur inside the region of convergence simply because the z-transform cannot converge at a pole. Furthermore, the region of convergence is bounded by poles or by zero or infinity.

(2) Linearity. Let $U_1(z)$ and $U_2(z)$ denote the z-transforms of two sequences $\{u_1(n)\}$ and $\{u_2(n)\}$, respectively. Then the z-transform of a new sequence defined by $\{au_1(n) + bu_2(n)\}$ equals $aU_1(z) + bU_2(z)$, where a and b are constants.

(3) Shift of a Sequence.   Let $U(z)$ denote the z-transform of the sequence $\{u(n)\}$. Let the sequence be shifted by k units of time, yielding the new sequence $\{u(n - k)\}$. If k is positive, the shift is to the right, and if k is negative, the shift is to the left. The z-transform of the shifted sequence $\{u(n - k)\}$ equals $z^{-k}U(z)$.

(4) Conjugation of a Complex Sequence.   Let $U(z)$ denote the z-transform of a complex-valued sequence $\{u(n)\}$. The z-transform of the sequence $\{u^*(n)\}$ equals $U^*(z^*)$, where the asterisk denotes complex conjugation.

(5) Initial-value Theorem.   Consider a one-sided sequence $\{u(n)\}$ for which $u(n) = 0$ for $n < 0$. Let $U(z)$ denote the z-transform of the sequence. Then

$$u(0) = \lim_{z \to \infty} U(z) \qquad\qquad\qquad (C.5)$$

(6) Convolution.   Let $U(z)$ and $H(z)$ denote the z-transforms of the sequences $\{u(n)\}$ and $\{h_n\}$, respectively. The *convolution* of sequences $\{u(n)\}$ and $\{h_n\}$ is defined by

$$y(n) = \sum_{k=-\infty}^{\infty} h_k u(n - k), \qquad n = 0, \pm 1, \pm 2, \ldots \qquad (C.6)$$

Then the z-transform of the sequence $\{y(n)\}$ equals the product $H(z)U(z)$.

The convolution of Eq. (C.6) may be viewed as a linear filtering operation with $\{h_n\}$ representing the *impulse response* of the filter and $\{u(n)\}$ representing the *excitation* applied to the filter input. Accordingly, $\{y(n)\}$ represents the *response* produced at the filter output. The z-transform $H(z)$ represents the *transfer function* of the filter; it equals the z-transform of the response divided by the z-transform of the excitation.

(7) Inverse z-Transform.   Let $U(z)$ denote the z-transform of the sequence $\{u(n)\}$. The *inverse z-transform* relation is given by the *contour integral*

$$u(n) = \frac{1}{2\pi j} \oint_C U(z)z^{n-1}\, dz, \qquad n = 0, \pm 1, \pm 2, \ldots \qquad (C.7)$$

where C is a closed contour in the region of convergence of $U(z)$, encircling the origin of the z-plane and traversed in the *counterclockwise* direction.

For rational z-transforms, the contour integral of Eq. (C.7) is conveniently evaluated by using the *residue theorem* of complex variable theory, as shown by

$$u(n) = \text{sum of residues of } U(z)z^{n-1} \text{ at the poles inside } C \qquad (C.8)$$

(8) Parseval's Theorem.   Let $U(z)$ and $X(z)$ denote the z-transforms of two complex-valued sequences $\{u(n)\}$ and $\{x(n)\}$, respectively. Parseval's theorem states that

$$\sum_{n=-\infty}^{\infty} u(n)x^*(n) = \frac{1}{2\pi j} \oint_C U(v)X^*\left(\frac{1}{v^*}\right) v^{-1} \, dv \tag{C.9}$$

where the contour C is taken in the overlap of the regions of convergence of $U(v)$ and $X^*(1/v^*)$.

When both $U(z)$ and $X(z)$ converge on the unit circle, we may choose

$$v = e^{j\omega}$$

Hence, Eq. (C.9) becomes

$$\sum_{n=-\infty}^{\infty} u(n)x^*(n) = \frac{1}{2\pi} \int_{-\pi}^{\pi} U(e^{j\omega})X^*(e^{j\omega}) \, d\omega \tag{C.10}$$

As a special case of this result, we have

$$\sum_{n=-\infty}^{\infty} |u(n)|^2 = \frac{1}{2\pi} \int_{-\pi}^{\pi} |U(e^{j\omega})|^2 \, d\omega \tag{C.11}$$

where the sum on the left side represents the *energy* of the sequence $\{u(n)\}$.

D

Method of Lagrange Multipliers

D.1 INTRODUCTION

Optimization consists of determining the values of some specified variables that minimize or maximize an *index of performance* or *objective function*, which combines important properties of a system into a single real-valued number. The optimization may be *constrained* or *unconstrained*, depending on whether the variables are also required to satisfy side equations or not. Needless to say, the additional requirement to satisfy one or more side equations complicates the issue of constrained optimization. In this appendix, we derive the classical *method of Lagrange multipliers* for solving the constrained optimization problem (Dorney, 1975). The notation used in the derivation is influenced by the nature of applications that are of interest to us. We consider first the case when the problem involves a single side equation, followed by the more general case of multiple side equations.

D.2 OPTIMIZATION INVOLVING A SINGLE EQUALITY CONSTRAINT

Consider the optimization of a real-valued function $f(\mathbf{w})$ that is a quadratic function of an M-by-1 vector $\mathbf{w}$, subject to the *constraint*

$$\mathbf{w}^H \mathbf{s} = 1 \tag{D.1}$$

or equivalently

$$\mathbf{s}^H \mathbf{w} = 1 \tag{D.2}$$

where $\mathbf{s}$ is a prescribed vector of compatible dimension. We may redefine the constraint by introducing a new function $g(\mathbf{w})$ that is linear in $\mathbf{w}$, as shown by

$$\begin{aligned} g(\mathbf{w}) &= \mathbf{w}^H \mathbf{s} + \mathbf{s}^H \mathbf{w} - 2 \\ &= 0 \end{aligned} \tag{D.3}$$

In effect, this equation combines (D.1) and (D.2) into a single statement. For example, in a beamforming application the vector $\mathbf{w}$ represents a set of weights applied to the individual sensor outputs, and $\mathbf{s}$ represents a steering vector whose elements are defined by a prescribed "look" direction; the function $f(\mathbf{w})$ to be minimized represents the mean-square value of the overall beamformer output. In a harmonic retrieval application, $\mathbf{w}$ represents the tap-weight vector of a transversal filter, and $\mathbf{s}$ represents a sinusoidal vector whose elements are determined by the angular frequency of a complex sinusoid contained in the filter input; the function $f(\mathbf{w})$ represents the mean-square value of the filter output. In any event, assuming that the issue is one of minimization, we may state the constrained optimization problem as follows:

$$\textit{Minimize } f(\mathbf{w}) \textit{ subject to } g(\mathbf{w}) = 0 \tag{D.4}$$

If the function $f(\mathbf{w})$ has a constrained minimum at a particular point $\mathbf{w}$, then a very small excursion $\delta\mathbf{w}$ about $\mathbf{w}$ that does not violate the constraint should not affect the value of the function f. Accordingly, it is reasonable to replace the functions f and g by their tangent functions (Dorney, 1975). That is, we may reformulate the problem as one of picking the excursion $\delta\mathbf{w}$ so as to minimize the function

$$f(\mathbf{w} + \delta\mathbf{w}) \simeq f(\mathbf{w}) + \delta f(\mathbf{w}, \delta\mathbf{w})$$

subject to the constraint

$$\begin{aligned} g(\mathbf{w} + \delta\mathbf{w}) &\simeq g(\mathbf{w}) + \delta g(\mathbf{w}, \delta\mathbf{w}) \\ &= \delta g(\mathbf{w}, \delta\mathbf{w}) \\ &= 0 \end{aligned}$$

where, in the second line, we have made use of Eq. (D.3).   Thus, excursions $\delta\mathbf{w}$ that lie along the tangent to the constraint

$$\delta g(\mathbf{w}, \delta\mathbf{w}) = 0$$

must not cause $f(\mathbf{w}) + \delta f(\mathbf{w}, \delta\mathbf{w})$ to deviate from the value $f(\mathbf{w})$.   Consequently, if $\mathbf{w}$ is to be the solution to Eq. (D.4), it is necessary that we have

$$\delta f(\mathbf{w}, \delta\mathbf{w}) = 0$$

for all $\delta\mathbf{w}$ in such a way that the constraint

$$\delta g(\mathbf{w}, \delta\mathbf{w}) = 0$$

is satisfied.   The *differential* δf is defined as the inner product of the excursion $\delta\mathbf{w}$ and the gradient vector $\partial f/\partial\mathbf{w}$, as shown by

$$\delta f(\mathbf{w}, \delta\mathbf{w}) = (\delta\mathbf{w})^H\left(\frac{\partial f}{\partial\mathbf{w}}\right) \tag{D.5}$$

Similarly, we write

$$\delta g(\mathbf{w}, \delta\mathbf{w}) = (\delta\mathbf{w})^H\left(\frac{\partial g}{\partial\mathbf{w}}\right) \tag{D.6}$$

Multiplying Eq. (D.6) by a scalar λ, to be determined, and then subtracting the result from Eq. (D.5), we get

$$(\delta\mathbf{w})^H\left(\frac{\partial f}{\partial\mathbf{w}} - \lambda\frac{\partial g}{\partial\mathbf{w}}\right) = 0 \tag{D.7}$$

Provided that the difference vector inside the parentheses in Eq. (D.7) vanishes, then for any (small) excursion $\delta\mathbf{w}$, we will have a horizontal tangent.   Thus, the system of $(M + 1)$ simultaneous equations, consisting of M equations in matrix form, namely,

$$\frac{\partial f}{\partial\mathbf{w}} - \lambda\frac{\partial g}{\partial\mathbf{w}} = \mathbf{0} \tag{D.8}$$

and the original constraint

$$g(\mathbf{w}) = 0 \tag{D.9}$$

define the optimum solutions for the M-by-1 vector $\mathbf{w}$ and the scalar λ.

We call λ the *Lagrange multiplier*, Eq. (D.8) the *adjoint equation*, and Eq. (D.9) the *primal equation* (Dorney, 1975).   Note that the sign on λ is arbitrary; a reversal in the sign convention with respect to that used in Eq. (D.8) merely results in a change in the sign of the Lagrange multiplier.

D.3 OPTIMIZATION INVOLVING MULTIPLE EQUALITY CONSTRAINTS

Consider next the optimization of the real-valued function $f(\mathbf{w})$ that is a quadratic function of the vector $\mathbf{w}$, subject to a set of *multiple constraints*

$$\mathbf{w}^H \mathbf{s}_k = c_k, \qquad k = 1, 2, \ldots, K \tag{D.10}$$

or equivalently

$$\mathbf{s}_k^H \mathbf{w} = c_k, \qquad k = 1, 2, \ldots, K \tag{D.11}$$

where the number of constraints, K, is less than M, the dimension of vector $\mathbf{w}$. Note that the c_k are real valued constants. Here, again, we combine Eqs. (D.10) and (D.11), thereby redefining the set of multiple constraints as

$$\begin{aligned} g_k(\mathbf{w}) &= \mathbf{w}^H \mathbf{s}_k + \mathbf{s}_k^H \mathbf{w} - 2c_k \\ &= 0, \qquad\qquad\qquad k = 1, 2, \ldots, K \end{aligned} \tag{D.12}$$

We may thus state the multiple-constrained optimization problem as follows

$$\textit{Minimize } f(\mathbf{w}) \textit{ subject to } g_k(\mathbf{w}) = 0, \qquad k = 1, 2, \ldots, K \tag{D.13}$$

The solution to this optimization problem is readily obtained by generalizing the results of Section D.2, particularly, Eqs. (D.8) and (D.9). Thus, the system of $(M + K)$ simultaneous equations, consisting of the adjoint equation

$$\frac{\partial f}{\partial \mathbf{w}} - \sum_{k=1}^{K} \lambda_k \frac{\partial g_k}{\partial \mathbf{w}} = \mathbf{0} \tag{D.14}$$

and the primal equation

$$g_k(\mathbf{w}) = 0, \qquad k = 1, 2, \ldots, K \tag{D.15}$$

define the optimum solutions for the M-by-1 vector $\mathbf{w}$ and the Lagrange multipliers $\lambda_1, \lambda_2, \ldots, \lambda_K$.

Note that the Lagrange multipliers are all real valued, and so we may use the first line of Eq. (D.12) to write

$$\sum_{k=1}^{K} \lambda_k g_k(\mathbf{w}) = \mathbf{w}^H \mathbf{S} \boldsymbol{\lambda} + \boldsymbol{\lambda}^T \mathbf{S}^H \mathbf{w} - 2 \boldsymbol{\lambda}^T \mathbf{c} \tag{D.16}$$

where

$$\mathbf{S} = [\mathbf{s}_1, \mathbf{s}_2, \ldots, \mathbf{s}_K]$$

$$\mathbf{c}^T = [c_1, c_2, \ldots, c_K]$$

and

$$\boldsymbol{\lambda}^T = [\lambda_1, \lambda_2, \ldots, \lambda_K]$$

The summation term on the left-side of the adjoint equation (D.14) may thus be replaced by the derivative of the expression on the right-side of Eq. (D.16) with respect to $\mathbf{w}$.

Glossary

TEXT CONVENTIONS

1. Boldfaced lowercase letters are used to denote column vectors. Boldfaced uppercase letters are used to denote matrices.
2. The estimate of a scalar, vector, or matrix is designated by the use of a hat (ˆ) placed over the pertinent symbol.
3. The symbol | | denotes the magnitude or absolute value of the scalar enclosed within.
4. The symbol ‖ ‖ denotes the norm of the vector or matrix enclosed within.
5. The symbol det( ) denotes the determinant of the square matrix enclosed within.
6. The inverse of nonsingular (square) matrix $\mathbf{A}$ is denoted by $\mathbf{A}^{-1}$.
7. The pseudoinverse of matrix $\mathbf{A}$ (not necessarily square) is denoted by $\mathbf{A}^{\#}$.
8. Complex conjugation of a scalar, vector, or matrix is denoted by the use of

an asterisk as superscript. Transposition of a vector or matrix is denoted by superscript T. Hermitian transposition (i.e., complex conjugation and transposition combined) of a vector or matrix is denoted by superscript H. Backward rearrangement of the elements of a vector is denoted by superscript B.

9. The symbol $\mathbf{A}^{-H}$ denotes the Hermitian transpose of the inverse of a nonsingular (square) matrix $\mathbf{A}$.

10. The square root of a square matrix $\mathbf{A}$ is denoted by $\mathbf{A}^{1/2}$.

11. The symbol $\text{diag}(\lambda_1, \lambda_2, \ldots, \lambda_M)$ denotes a diagonal matrix whose elements on the main diagonal equal $\lambda_1, \lambda_2, \ldots, \lambda_M$.

12. The symbol $fx(\ )$ denotes the fixed-point arithmetic representation of the quantity enclosed within. The use of a primed symbol denotes a quantity of finite precision.

13. The order of linear predictor or the order of autoregressive model is signified by a subscript added to the pertinent scalar or vector parameter.

14. The expectation operator is denoted by $E[\ ]$, where the quantity enclosed is the random variable or random vector of interest. The variance of a random variable is denoted by $\text{var}[\ ]$, where the quantity enclosed is the random variable.

15. The conditional probability density function of random variable U, given that hypothesis H_i is true, is denoted by $f_U(u|H_i)$, where u is the sample value of random variable U.

16. The inner product of two vectors $\mathbf{x}$ and $\mathbf{y}$ is defined as $\mathbf{x}^H\mathbf{y} = \mathbf{y}^T\mathbf{x}^*$. Another possible inner product is $\mathbf{y}^H\mathbf{x} = \mathbf{x}^T\mathbf{y}^*$. In general, these two inner products are different. The outer product of the vectors $\mathbf{x}$ and $\mathbf{y}$ is defined as $\mathbf{x}\mathbf{y}^H$. The inner product is a scalar, whereas the outer product is a matrix.

17. The autocorrelation function of stationary discrete-time stochastic process $\{u(n)\}$ is defined by

$$r(k) = E[u(n)u^*(n - k)]$$

The cross-correlation function between two jointly stationary discrete-time stochastic process $\{u(n)\}$ and $\{d(n)\}$ is defined by

$$p(k) = E[u(n)d^*(n - k)]$$

18. The ensemble-averaged correlation matrix of a random vector $\mathbf{u}(n)$ is defined by

$$\mathbf{R} = E[\mathbf{u}(n)\mathbf{u}^H(n)]$$

19. The ensemble-averaged cross-correlation vector between a random vector $\mathbf{u}(n)$ and a random variable $d(n)$ is defined by

$$\mathbf{p} = E[\mathbf{u}(n)d^*(n)]$$

20. The deterministic correlation matrix of a vector $\mathbf{u}(i)$ over the observation interval $1 \leq i \leq n$ is defined by

$$\mathbf{\Phi}(n) = \sum_{i=1}^{n} \mathbf{u}(i)\mathbf{u}^{H}(i)$$

21. The deterministic cross-correlation vector between a vector $\mathbf{u}(i)$ and a scalar $d(i)$ over the observation interval $1 \leq i \leq n$ is defined by

$$\mathbf{\theta} = \sum_{i=1}^{n} \mathbf{u}(i)d^{*}(i)$$

22. In constructing block diagrams, the following symbols are used. The symbol

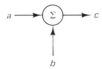

denotes an adder with $c = a + b$. The same symbol with algebraic signs added as in the following

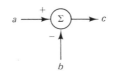

denotes a substractor with $c = a - b$. The symbol

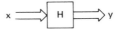

denotes a multiplier with $y = hx$.

23. In constructing signal-flow graphs, the following symbols are used. The symbol

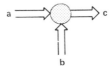

denotes a summing node with $\mathbf{c} = \mathbf{a} + \mathbf{b}$. The symbol

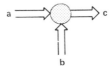

denotes a branch having transmittance **H**, with $\mathbf{y} = \mathbf{Hx}$. The unit-sample operator is denoted by the symbol

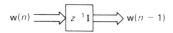

$$\mathsf{w}(n) \Longrightarrow \boxed{z^{-1}\mathbf{I}} \Longrightarrow \mathsf{w}(n-1)$$

ABBREVIATIONS

ALE: Adapative line enhancer
APC: Adaptive predictive coding
AR: Autoregressive
ARMA: Autoregressive–moving average
b/s: Bit per second
DPCM: Differential pulse-code modulation
FBLP: Forward-backward linear prediction
FTF: Fast transversal filters
Hz: Hertz
INR: Interference-to-noise ratio
kb/s: Kilobit per second
kHz: Kilohertz
LMS: Least-mean square
LSL: Least-squares lattice
LPC: Linear predictive coding
MA: Moving average
MEM: Maximum-entropy method
MESE: Maximum-entropy spectrum estimator
MUSIC: Multiple signal characterization
MVDR: Minimum-variance distortionless response
PARCOR: Partial correlation
PCM: Pulse-code modulation
PN: Pseudonoise
QRD-LS: QR decomposition-least squares
RLS: Recursive least squares
ROC: Rate of convergence
s: Second
SNR: Signal-to-noise ratio
SVD: Singular value decomposition

PRINCIPAL SYMBOLS

$a_{M,k}(n)$ kth tap weight of forward prediction-error filter of order M (at time n), with $k = 0, 1, \ldots, M$; note that $a_{m,0}(n) = 1$

$\mathbf{a}_M(n)$ Tap-weight vector of forward prediction-error filter of order M (at time n)

A	Data matrix in the covariance method
$\mathbf{A}(n)$	Data matrix in the prewindowing method
$b_M(n)$	Backward (a posteriori) prediction error produced at time n by prediction-error filter of order M
$\mathbf{b}(n)$	Backward (a posteriori) prediction-error vector representing sequence of errors produced by backward prediction-error filters of orders $0, 1, \ldots, M$
$\mathbf{b}$	Desired response vector in the covariance method
$\mathbf{b}(n)$	Desired response vector in the prewindowing method
$\mathcal{B}_M(n)$	Sum of weighted backward prediction error squares produced by backward prediction-error filter of order M
$c(n)$	Colored Gaussian noise
$c_{M,k}(n)$	kth tap weight of backward prediction-error filter of order M (at time n), with $k = 0, 1, \ldots, M$; note that $c_{M,M}(n) = 1$
$\mathbf{c}_M(n)$	Tap-weight vector of backward prediction-error filter of order M (at time n)
$\mathbf{c}(n)$	Weight-error vector in steepest-descent algorithm
$\mathcal{C}(n)$	Convergence ratio
$d(n)$	Desired response
D	Unit-delay operator
$\mathbf{D}_{m+1}(n)$	Correlation matrix of backward prediction errors
$e(n)$	A posteriori estimation error
$e_{m+1}(n)$	A posteriori estimation error at the output of stage m in the joint-process estimator using the recursive LSL algorithm
e	Base of natural logarithm
exp	Exponential
E	Expectation operator
$\mathcal{E}(w)$	Index of performance defined as the sum of error squares, expressed as a function of the tap-weight vector
$\mathcal{E}_{\min}$	Minimum value of $\mathcal{E}(\mathbf{w})$
$\mathcal{E}(n)$	Index of performance defined as the sum of weighted error squares, expressed as a function of time n
$f_M(n)$	Forward (a posteriori) prediction error produced at time n by forward prediction-error filter of order M
$\mathbf{f}(n)$	Forward (a posteriori) prediction error vector representing sequence of errors produced by forward prediction-error filters of orders $0, 1, \ldots, M$
$f_U(u)$	Probability density function of random variable U whose sample value equals u
$f_{\mathbf{U}}(\mathbf{u})$	Joint probability density function of random vector $\mathbf{U}$ whose sample value equals $\mathbf{u}$
$\mathbf{F}(n)$	Matrix used in formulating the recursive QRD-LS algorithm
$F_M(z)$	z-transform of sequence of forward prediction errors produced by forward prediction-error filter of order M
$\mathcal{F}_M(n)$	Weighted sum of forward prediction-error squares produced by forward prediction-error filter of order M

g_k	kth tap weight of backward linear predictor
$\mathbf{g}_s$	Tap-weight vector of backward linear predictor
$\mathbf{g}_N$	Vector in the minimum-norm method
$\mathbf{g}_s$	Another vector in the minimum-norm method.
$\mathbf{g}(n)$	Kalman gain vector used in the application of Kalman filter theory to adaptive transversal filters
$\mathbf{G}(n)$	Kalman gain
$\mathbf{G}(n)$	Matrix denoting Givens rotation
$\mathbf{G}_N$	Matrix in the minimum-norm method
$\mathbf{G}_s$	Another matrix in the minimum-norm method
H_i	ith hypothesis
$H(z)$	Transfer function of discrete-time linear filter
$\mathbf{I}$	Identity matrix
$\mathbf{I}$	Inverse of Fisher's information matrix $\mathbf{J}$
j	Square root of -1
$J(\mathbf{w})$	Index of performance used to formulate the Wiener filtering problem, expressed as a function of the tap-weight vector $\mathbf{w}$
$\mathbf{J}$	Fisher's information matrix
$\mathbf{k}(n)$	Gain vector in the RLS algorithm
K	Number of complex sine waves
K	Final order of moving-average model
$\mathbf{K}(n)$	Correlation matrix of weight-error vector
$\ln$	Natural logarithm
$\mathbf{L}(n)$	Transformation matrix in the form of lower triangular matrix
m	Variable order of linear predictor or autoregressive model
M	Final order of linear predictor or autoregressive model
M, K	Final order of autoregressive–moving average model
$\mathcal{M}$	Misadjustment
n	Discrete-time or number of iterations applied to recursive algorithm
N	Data length
$p(k)$	Element of cross-correlation vector $\mathbf{p}$ for lag k
$\mathbf{p}$	Cross-correlation vector between tap-input vector $\mathbf{u}(n)$ and desired response $d(n)$
$\mathbf{p}$	Vector used in formulating the recursive QRD-LS algorithm
P_M	Average value of (forward or backward) prediction-error power for prediction order M
$\mathbf{P}(n)$	Matrix equal to the inverse of the deterministic correlation matrix $\boldsymbol{\Phi}(n)$ used in formulating the RLS algorithm
q_{ki}	ith element of kth eigenvector
$\mathbf{q}_k$	kth eigenvector
$\mathbf{Q}$	Unitary matrix that consists of normalized eigenvectors in the set $\{\mathbf{q}_k\}$ used as columns

$\mathbf{Q}$	Unitary matrix used in formulating the recursive QRD-LS algorithm
$r(k)$	Element of (ensemble-averaged) correlation matrix $\mathbf{R}$ for lag k
$\mathbf{R}$	Ensembled-average correlation matrix of stationary discrete-time process $\{u(n)\}$
$\mathbf{R}$	Matrix used in formulating the recursive QRD-LS algorithm
$\mathbf{s}$	Signal vector; steering vector
$S(\omega)$	Power spectral density
$\mathbf{S}(n)$	Tap-weight error correlation matrix $\mathbf{K}(n)$ divided by $J_{\min}$; this matrix is used in the convergence analysis of the Kalman algorithm
$\mathbf{S}(n)$	Matrix used in the development of the recursive QRD-LS algorithm
t	Time
$\mathbf{t}(n)$	Vector used in the fast RLS algorithm
$\mathbf{T}(n)$	Matrix consisting of the sequence of Givens rotations used in formulating the recursive QRD-LS algorithm
$u(n)$	Sample value of tap input in transversal filter at time n
$\mathbf{u}(n)$	Tap-input vector consisting of $u(n)$, $u(n-1)$, . . . , as elements
U	Random variable whose sample value equals u
$\mathbf{U}$	Random vector whose sample value equals $\mathbf{u}$
$\mathcal{U}_n$	Space spanned by tap inputs $u(n)$, $u(n-1)$, . . .
$\mathcal{U}(n)$	Sum of weighted squared values of tap inputs $u(i)$, $i = 1, 2, . . ., n$
$v(n)$	Sample value of white-noise process of zero mean
$\mathbf{v}_1(n)$	Process noise vector
$\mathbf{v}_2(n)$	Measurement noise vector
$\mathbf{v}(n)$	Process noise vector in random-walk state model
$\mathbf{v}(n)$	Vector used in formulating the recursive QRD-LS algorithm
$w_k(n)$	kth tap weight of transversal filter at time n
$\mathbf{w}(n)$	Tap-weight vector of transversal filter at time n
$\mathbf{x}(n)$	State vector
$\mathbf{X}$	Unitary matrix used in singular-value decomposition
$\mathbf{X}_N$	Matrix constructed from noise subspace eigenvectors
$\mathbf{X}_S$	Matrix constructed from signal subspace eigenvectors
$\mathbf{y}(n)$	Observation vector used in formulating Kalman filter theory
$\mathbf{Y}$	Matrix used in singular-value decomposition
$\mathcal{Y}_n$	Vector space spanned by $\mathbf{y}(n)$, $\mathbf{y}(n-1)$, . . .
z^{-1}	Unit-sample delay used in defining the z-transform of a sequence
$\alpha(n)$	Innovation or a priori estimation error
$\boldsymbol{\alpha}(n)$	Innovations vector
$\chi(\mathbf{R})$	Eigenvalue spread (i.e., ratio of maximum eigenvalue to minimum eigenvalue) of correlation matrix $\mathbf{R}$
δ	Constant used in the initialization of RLS and recursive LSL algorithms
$\boldsymbol{\delta}$	First unit vector
Δ_m	Cross-correlation between forward prediction error $f_m(n)$ and delayed backward prediction error $b_m(n-1)$

$\Delta_m(n)$	Parameter in recursive LSL algorithm
$\boldsymbol{\varepsilon}(n)$	Weight-error vector
$\boldsymbol{\varepsilon}$	Estimation error vector in the covariance method
$\boldsymbol{\varepsilon}(n)$	Estimation error vector in the prewindowed method
$\eta(n)$	Forward (a priori) prediction error
$\gamma(n)$	Conversion factor used in the FTF algorithm, recursive LSL algorithm, or recursive QRD-LS algorithm
Γ_m	Reflection coefficient in mth stage of lattice predictor for stationary inputs
$\Gamma_{b,m}(n)$	Backward reflection coefficient in mth stage of least-squares lattice predictor at time n
$\Gamma_{f,m}(n)$	Forward reflection coefficient in mth stage of least-squares lattice predictor at time n
$\boldsymbol{\kappa}$	Regression coefficient vector in joint-estimator using lattice predictor
λ	Exponential weighting vector in RLS, LSL, and QRD-LS algorithms
λ_k	kth eigenvalue of correlation matrix $\mathbf{R}$
$\lambda_{\max}$	Maximum eigenvalue of correlation matrix $\mathbf{R}$
$\lambda_{\min}$	Minimum eigenvalue of correlation matrix $\mathbf{R}$
$\lambda(n)$	Threshold
Λ	Likelihood ratio
$\ln\Lambda$	Log-likelihood ratio
$\Lambda(n)$	Diagonal matrix of exponential weighting factors
μ	Mean value
μ	Step-size parameter in steepest-descent algorithm or LMS algorithm
μ	Constant used in the FTF algorithm with soft constraint
$\boldsymbol{v}(n)$	Normalized weight-error vector in steepest-descent algorithm
$v_k(n)$	kth element of $\boldsymbol{v}(n)$
ω	Normalized angular frequency; $0 \le \omega \le 2\pi$
$\phi(t, k)$	t, kth element of deterministic correlation matrix $\boldsymbol{\Phi}$
$\boldsymbol{\Phi}$	Deterministic correlation matrix
$\boldsymbol{\Phi}(n + 1, n)$	State transition matrix
$\psi(n)$	Backward (a priori) prediction error
ρ_m	Correlation coefficient or normalized value of autocorrelation function for log m
ρ_m	Parameter in recursive LSL algorithm used for joint-process estimation
σ^2	Variance
$\boldsymbol{\Sigma}$	Correlation matrix of innovations process $\{\boldsymbol{\alpha}(n)\}$
$\boldsymbol{\theta}$	Deterministic cross-correlation vector between tap-input vector $\{\mathbf{u}(i)\}$ and desired response $\{d(i)\}$
$\boldsymbol{\theta}$	Parameter vector
τ_k	Time constant of kth natural mode of steepest-descent algorithm
τ_{ms}	Average time constant of LMS algorithm

References and Bibliography

AKAIKE, H. (1973). "Maximum likelihood identification of Gaussian autoregressive moving average models," *Biometrika*, vol. 60, pp. 255–265.

AKAIKE, H. (1974). "A new look at the statistical model identification," *IEEE Trans. Automatic Control*, vol. AC-19, pp. 716–723.

AKAIKE, H. (1977). "An entropy maximisation principle," in P. Krishnaiah, ed., *Proceedings of Symposium on Applied Statistics*, North-Holland, Amsterdam.

ANDERSON, B. D. O., and J. B. MOORE (1979). *Linear Optimal Control*, Prentice-Hall, Englewood Cliffs, N.J.

ANDERSON, T. W. (1963). "Asymptotic theory for principal component analysis," *Annals of Mathematical Statistics*, vol. 34, pp. 122–148.

APPLEBAUM, S. P. (1966). "Adaptive arrays," *Syracuse University Research Corporation*, Rep. SPL TR 66-1. This report is reproduced in *IEEE Trans. Antennas and Propagation*, Special Issue on Adaptive Antennas, vol. AP-24, pp. 585–598, 1976.

APPLEBAUM, S. P., and D. J. CHAPMAN (1976). "Adaptive arrays with main beam constraints," *IEEE Trans. Antennas and Propagation*, vol. AP-24, pp. 650–662.

ÄSTRÖM, K. J., and P. EYKHOFF (1971). "System identification—A survey," *Automatica*, vol. 7, pp. 123–162.

ATAL, B. S. (1970). "Speech analysis and synthesis by linear prediction of the speech wave," *J. Acoust. Soc. Amer.*, vol. 47, p. 65.

ATAL, B. S., and S. L. HANAUER (1971). "Speech analysis and synthesis by linear prediction of the speech wave," *J. Acoust. Soc. Amer.*, vol. 50, pp. 637–655.

ATAL, B. S., and M. R. SCHROEDER (1967). "Predictive coding of speech signals," *Proc. 1967 Conf. Commun. and Processes*, pp. 360–361.

ATAL, B. S., and M. R. SCHROEDER (1970). "Adaptive predictive coding of speech signals," *Bell Syst. Tech. J.*, vol. 49, pp. 1973–1986.

AUSTIN, M. E. (1967). "Decision-feedback equalization for digital communication over dispersive channels," *MIT Lincoln Laboratory Tech., Rep. 437.*

BASSEVILLE, M., and A. BENVENISTE (1983). "Sequential detection of abrupt changes in spectral characteristics of digital signals," *IEEE Trans. Information Theory*, vol. IT-29, pp. 709–724.

BELFIORE, C. A., and J. H. PARK, JR. (1979). "Decision feedback equalization," *Proc. IEEE*, vol. 67, pp. 1143–1156.

BELLMAN, R. (1960). *Introduction to Matrix Analysis*, McGraw-Hill, New York.

BERKHOUT, A. J., and P. R. ZAANEN (1976). "A comparison between Wiener filtering, Kalman filtering, and deterministic least squares estimation," *Geophysical Prospecting*, vol. 24, pp. 141–197.

BIENVENUE, G., and L. KOPP (1980). "Adaptivity to background noise spatial coherence for high resolution passive methods," *Proceedings ICASSP 80*, Denver, Colo., pp. 307–310.

BIENVENUE, G., and H. MERMOZ (1985). "Principles of high-resolution array processing," in the book entitled *VLSI and Modern Signal Processing*, edited by S. Y. Kung, H. J. Whitehouse, and T. Kailath, pp. 83–105, Prentice-Hall, Englewood Cliffs, N.J.

BIERMAN, G. J. (1977). *Factorization Methods for Discrete Sequential Estimation*, Academic Press, New York.

BODE, H. W., and C. E. SHANNON (1950). "A simplified derivation of linear least square smoothing and predicition theory," *Proc. IRE*, vol. 38, pp. 417–425.

BOWYER, D. E., P. K. RAJASEKARAN, and W. W. GEBHART (1979). "Adaptive clutter filtering using autoregressive spectral estimation," *IEEE Trans. Aerospace and Electronic Syst.*, vol. AES-15, pp. 538–546.

BOX, G. E. P., and G. M. JENKINS (1976). *Time Series Analysis: Forecasting and Control*, Holden-Day, San Francisco.

BRACEWELL, R. N. (1978). *The Fourier Transform and its Applications*, 2nd ed., McGraw-Hill, New York.

BRADY, D. M. (1970). "An adaptive coherent diversity receiver for data transmission through dispersive media," *Conference Record*, ICC 70, pp. 21-35 to 21-40.

BROGAN, W. L. (1985). *Modern Control Theory*, 2nd ed., Prentice-Hall, Englewood Cliffs, N.J.

BROMLEY, K., J. M. SPEISER, and H. J. WHITEHOUSE (1985). Course notes on *Advanced Signal Processing*, Evolving Technology Institute, Los Angeles, Calif.

BROOKS, L. W., and I. S. REED (1972). "Equivalence of the likelihood ratio processor, the maximum signal-to-noise ratio filter, and the Wiener filter," *IEEE Trans. Aerospace and Electronic Syst.*, vol. AES-8, pp. 690–692.

BURG, J. P. (1967). "Maximum entropy spectral analysis," *37th Ann. Intern. Meeting, Soc. Explor. Geophysics*, Oklahoma City, Okla.

BURG, J. P. (1968). "A new analysis technique for time series data," NATO Advanced Study Institute on Signal Processing, Enschede, Holland.

BURG, J. P. (1975). "Maximum Entropy Spectral Analysis," Ph.D. dissertation, Stanford University, Stanford, Calif.

CAPON, J. (1969). "High-resolution frequency-wavenumber spectrum analysis," *Proc. IEEE*, vol. 57, pp. 1408–1418.

CARAISCOS, C. and B. LIU (1984). "A roundoff error analysis of the LMS adaptive algorithm," *IEEE Trans. Acoust., Speech, and Signal Processing*, vol. ASSP-32, pp. 34–41.

CARAYANNIS, G., D. G. MANOLAKIS, and N. KALOUPTSIDIS (1983). "A fast sequential algorithm for least-squares filtering and prediction," *IEEE Trans. Acoust., Speech, and Signal Processing*, vol. ASSP-31, pp. 1394–1402.

CHILDERS, D. G., ed. (1978). *Modern Spectrum Analysis*, IEEE Press, New York.

CIOFFI, J. M., and T. KAILATH (1984). "Fast, recursive-least-squares transversal filters for adaptive filtering," *IEEE Trans. Acoust., Speech, and Signal Processing*, vol. ASSP-32, pp. 304–337.

CIOFFI, J. M., and T. KAILATH, (1984). "A classification of fast fixed-order RLS algorithms," *ASSP Digital Signal Processing Workshop*, Chatham, Mass.

COHN, A. (1922). "Über die anzahl der wurzeln einer algebraischen gleichung in einem kreise," *Mathematische Zeitschrift*, vol. 14, pp. 110–148.

COMPTON, R. T., JR., *Adaptive Antennas: Concepts and Performance*, to be published.

COWAN, C. F. N., and P. M. GRANT (1985). *Adaptive Filters*, Prentice-Hall, Englewood Cliffs, N.J.

CUTLER, C. C. (1952). *Differential Quantization for Communication Signals*, U.S. Patent 2,605,361.

DEUTSCH, R. (1965). *Estimation Theory*, Prentice-Hall, Englewood Cliffs, N.J.

DEWILDE, P. (1969). "*Cascade Scattering Matrix Synthesis*," Ph.D. dissertation, Stanford University, Stanford, Calif.

DEWILDE, P., A. C. VIEIRA, and T. KAILATH (1978). "On a generalized Szegö–Levinson realization algorithm for optimal linear predictors based on a network synthesis approach," *IEEE Trans. Circuits and Syst.*, vol. CAS-25, pp. 663–675.

DHRYMUS, P. J. (1970). *Econometrics: Statistical Foundations and Applications*, Harper & Row, New York.

DITORO, M. J. (1965). "A new method for high speed adaptive signal communication through any time variable and dispersive transmission medium," *1st IEEE Ann. Commun. Conf.*, pp. 763–767.

DONGARRA, J. J., and others (1979). *LINPACK User's Guide*, Society for Industrial and Applied Mathematics, Philadelphia, PA.

DOOB, L. J. (1953). *Stochastic Processes*, Wiley, New York.

DORNY, C. N. (1975). *A Vector Space Approach to Models and Optimization*, Wiley-Interscience, New York.

DURBIN, J., (1960). "The fitting of time series models," *Rev. Intern. Statist. Inst.*, vol. 28, pp. 233–244.

DUTTWEILER, D. L., and Y. S. CHEN (1980). "A single-chip VLSI echo canceler," *Bell Syst. Tech. J.*, vol. 59, pp. 149–160.

ECKART, C., and G. YOUNG (1939). "A principal axis transformation for non-Hermitian matrices," *Bull. Amer. Math. Soc.*, vol. 45, pp. 118–121.

EDWARDS, A. W. F. (1972). *Likelihood*, Cambridge University Press, New York.

ELEFTHERIOU, E., and D. D. FALCONER (1984). *Tracking Properties and Steady State Performance of RLS Adaptive Filter Algorithms*, Report SCE-84-14, Department of Systems and Computer Engineering, Carleton University, Ottawa, Canada.

ER, M. H., and A. CANTONI (1986). "A new set of linear constraints for broadband time domain element space processors," *IEEE Trans. Antennas and Propagation*, vol. AP-34, March.

EVANS, J. E., J. R. JOHNSON, and D. F. SUN (1982). *Applications of Advanced Signal Processing Techniques to Angle of Arrival Estimation in ATC Navigation and Surveillance Systems*, MIT Lincoln Laboratory, Lexington, MA, Tech. Rept. 582.

EVANS, J. E. (1979). "Aperture sampling techniques for precision direction finding," *IEEE Trans. Aerospace and Electronic Systems*, vol. AES-15, pp. 891–895.

FALCONER, D. D., and L. LJUNG (1978). "Application of fast Kalman estimation to adaptive equalization," *IEEE Trans. Commun.*, vol. COM-26, pp. 1439–1446.

FLANAGAN, J. L. (1972). *Speech Analysis, Synthesis and Perception*, 2nd ed., Springer-Verlag, New York.

FLANAGAN, J. L., and others (1979). "Speech coding," *IEEE Trans. Commun.*, vol. COM-27, pp. 710–737.

FORNEY, G. D. (1972). "Maximum-likelihood sequence estimation of digital sequence in the presence of intersymbol interference," *IEEE Trans. Information Theory*, vol. IT-18, pp. 363–378.

FRANKS, L. E., ed. (1974). *Data Communication: Fundamentals of Baseband Transmission*, Benchmark Papers in Electrical Engineering and Computer Science, Dowden, Hutchinson, and Ross, Stroudsburg, Pa.

FRIEDLANDER, B. (1982). "Lattice filters for adaptive processing," *Proc. IEEE*, vol. 70, pp. 829–867.

GABOR, D., W. P. L. WILBY, and R. WOODCOCK (1960). "A universal non-linear filter, predictor and simulator which optimizes itself by a learning process," *Proc. IEEE (London)*, vol. 108, pp. 422–438.

GABRIEL, W. F. (1976). "Adaptive arrays: an introduction," *Proc. IEEE*, vol. 64, pp. 239–272.

GABRIEL, W. F. (1980). "Spectral analysis and adaptive array superresolution techniques," *Proc. IEEE*, vol. 68, pp. 654–666.

GALLIVAN, K. A., and C. E. LEISERSON (1984). "High-performance architectures for adaptive filtering based on the Gram–Schmidt algorithm," *Proc. SPIE*, vol. 495, Real Time Signal Processing VII, pp. 30–38.

GARBOW, B. S., and others (1977). *Matrix Eigensystem Routines—EISPACK Guide Extension*, Lecture Notes in Computer Science, vol. 51, Springer–Verlag, New York.

GAUSS, C. F. (1809). *Theoria Motus Corporum Coelestium in Sectionibus Conicus Solem Ambientum*, Hamburg (Translation: Dover, New York, 1963).

GENTLEMAN, W. M. (1973). "Least squares computations by Givens transformations without square-roots," *J. Inst. Maths. Applics.*, vol. 12, pp. 329–336.

GENTLEMAN, W. M., and H. T. KUNG (1981). "Matrix triangularization by systolic arrays," *Proc. SPIE*, vol. 298, Real Time Signal Processing IV, pp. 298–303.

GERSHO, A., (1968). "Adaptation in a quantized parameter space," *Allerton Conference*, pp. 646–653.

GERSHO, A. (1969). "Adaptive equalization of highly dispersive channels for data transmission," *Bell Syst. Tech. J.*, vol. 48, pp. 55–70.

GERSHO, A., B. GOPINATH, and A. M. OLDYZKO, (1979). "Coefficient inaccuracy in transversal filtering," *Bell Syst. Tech. J.*, vol. 58, pp. 2301–2316.

GIBSON, J. D. (1980). "Adaptive prediction in speech differential encoding systems," *Proc. IEEE*, vol. 68, pp. 488–525.

GIBSON, C., S. HAYKIN, and S. B. KESLER (1979). "Maximum entropy (adaptive) filtering applied to radar clutter," *Proc. IEEE Intern. Conf. Acoust., Speech, and Signal Processing*, Washington, D.C., pp. 166–169.

GITLIN, R. D., and F. R. MAGEE, JR. (1977). "Self-orthogonalizing adaptive equalization algorithms," *IEEE Trans. Commun.*, vol. COM-25, pp. 666–672.

GITLIN, R. D., and S. B. WEINSTEIN (1979). "On the required tap-weight precision for digitally implemented mean-squared equalizers," *Bell Syst. Tech. J.*, vol. 58, pp. 301–321.

GITLIN, R. D., and S. B. WEINSTEIN (1981). "Fractionally spaced equalization: an improved digital transversal equalizer," *Bell Syst. Tech. J.*, vol. 60, pp. 275–296.

GITLIN, R. D., J. E. MAZO, and M. G. TAYLOR (1973). "On the design of gradient algorithms for digitally implemented adaptive filters," *IEEE Trans. Circuit Theory*, vol. CT-20, pp. 125–136.

GIVENS, W. (1958). "Computation of plane unitary rotations transforming a general matrix to triangular form," *J. Soc. Industr. Appl. Math*, vol. 6, pp. 26–50.

GLASER, E. M. (1961). "Signal detection by adaptive filters," *IRE Trans. Information Theory*, vol. IT-7, pp. 87–98.

GODARA, L. C., and A. CANTONI (1986). "Analysis of constrained LMS algorithm with application to adaptive beamforming using perturbation sequences," *IEEE Trans. Antennas and Propagation*, vol. AP-34, March.

GODARD, D. (1974). "Channel equalization using a Kalman filter for fast data transmission," *IBM. J. Res. Develop.*, vol. 18, pp. 267–273.

GOLD, B. (1977). "Digital speech networks," *Proc. IEEE*, vol. 65, pp. 1636–1658.

GOLUB, G. H. (1965). "Numerical methods for solving linear least squares problems," *Numer. Math.*, vol. 7, pp. 206–216.

GOLUB, G. H., and C. REINSCH (1970). "Singular value decomposition and least squares problems," *Numer. Math.*, vol. 14, pp. 403–420.

GOLUB, G. H., and C. F. VAN LOAN (1983). *Matrix Computations*, The Johns Hopkins University Press, Baltimore, MD.

GOODWIN, G. C., and R. L. PAYNE (1977). *Dynamic System Identification: Experiment Design and Data Analysis*, Academic Press, New York.

GOODWIN, G. C., and K. S. SIN (1984). *Adaptive Filtering, Prediction and Control*, Prentice-Hall, Englewood Cliffs, N.J.

GRAY, A. H., JR., and J. D. MARKEL (1975). "A normalized digital filter structure," *IEEE Trans. Acoust., Speech, and Signal Processing*, vol. ASSP-23, pp. 268–277.

GRENANDER, U., and G. SZEGÖ (1958). *Toeplitz Forms and Their Applications*, University of California Press, Berkeley, Calif.

GRIFFITHS, L. J. (1975). "Rapid measurement of digital instantaneous frequency," *IEEE Trans. Acoust., Speech, and Signal Processing*, vol. ASSP-23, pp. 207–222.

GRIFFITHS, L. J. (1977). "A continuously-adaptive filter implemented as a lattice structure," *Proc. IEEE Intern. Conf. Acoust., Speech, and Signal Processing*, Hartford, Conn., pp. 683–686.

GRIFFITHS, L. J., and R. PRIETO-DIAZ (1977). "Spectral analysis of natural seismic events using autoregressive techniques," *IEEE Trans. Geoscience Electronics*, vol. GE-15, pp. 13–25.

GRIFFITHS, L. J. (1978). "An adaptive lattice structure for noise-cancelling applications," *Proc. IEEE Intern. Conf. Acoust., Speech, and Signal Processing*, Tulsa, Okla., pp. 87–90.

GUPTA, I. J., and A. A. KSIENSKI (1986). "Adaptive antenna arrays for weak interfering signals," *IEEE Trans. Antennas and Propagation*, vol. AP-34, March.

GUTOWSKI, P. R., E. A. ROBINSON, and S. TREITEL (1978). "Spectral estimation: fact or fiction," *IEEE Trans. Geosci. Electronics*, vol. GE-16, pp. 80–84.

HANSEN, D. S. (1984). "AR spectral modeling of simulated incoherent Doppler sonar data," *MPL Technical Memorandum 367*, Marine Physical Laboratory, Scripps Institution of Oceanography, San Diego, Calif.

HASTINGS-JAMES, R., and M. W. SAGE (1969). "Recursive generalized-least-squares procedure for online identification of process parameters," *Proc. IEE (London)*, vol. 116, pp. 2057–2062.

HAYKIN, S., B. W. CURRIE, and S. B. KESLER (1982). "Maximum-entropy spectral analysis of radar clutter," *Proc. IEEE*, vol. 70, 1982, pp. 953–962.

HAYKIN, S. (1983a). *Communication Systems*, 2nd ed., Wiley, New York.

HAYKIN, S., ed. (1983b). *Nonlinear Methods of Spectral Analysis*, 2nd ed., Springer-Verlag, New York.

HAYKIN, S. (1984). *Introduction to Adaptive Filters*, Macmillan, New York.

HAYKIN, S., ed., (1984). *Array Signal Processing*, Prentice-Hall, Englewood Cliffs, N.J.

HAYKIN, S., T. GREENLAY, and J. LITVA (1985). "Performance evaluation of the modified FBLP method for angle of arrival estimation using real radar multipath data," *Proc. IEE (London)*, vol. 132, pt. F, pp. 159–174.

HAYKIN, S. (1985). "Radar signal processing," *IEEE ASSP Magazine*, vol. 2, no. 2, pp. 2–18.

HO, Y. C. (1963). "On the stochastic approximation method and optimal filter theory," *J. Math. Anal. Appl.*, vol. 6, pp. 152–154.

HODGKISS, W. S., JR., and J. A. PRESLEY, JR. (1981). "Adaptive tracking of multiple sinusoids whose power levels are widely separated," *IEEE Trans. on Circuits and Systems*, vol. CAS-28, pp. 550–561.

HODGKISS, W. S., and D. ALEXANDROU (1983). "Applications of adaptive least-squares lattice structures to problems in underwater acoustics," *Proceedings of SPIE*, vol. 431, Real Time Signal Processing VI, pp. 48–54, San Diego, California.

HONIG, M. L., and D. G. MESSERSCHMITT (1984). *Adaptive Filters: Structures, Algorithms and Applications*, Kluwer Academic Publishers, Boston.

HOUSEHOLDER, A. S. (1958). "The approximate solution of matrix problems." *J. Assoc. Comput. Mach.*, vol. 5, pp. 204–243.

HOUSEHOLDER, A. S. (1964). *The Theory of Matrices in Numerical Analysis*, Blaisdell, Waltham, Mass.

HOWELLS, P. W. (1965). *Intermediate Frequency Sidelobe Canceller*, U.S. Patent 3,202,990, August 24.

HOWELLS, P. W. (1976). "Explorations in fixed and adaptive resolution at GE and SURC," *IEEE Trans. Antennas and Propagation*, vol. AP-24, Special Issue on Adaptive Antennas, pp. 575–584.

HSIA, T. C. (1983). "Convergence analysis of LMS and NLMS adaptive algorithms," *Proc. Intern. Conf. Acoust., Speech, and Signal Processing*, Boston, pp. 667–670.

HSU, F. M. (1982). "Square root Kalman filtering for high-speed data received over fading dispersive HF channels," *IEEE Trans. Information Theory*, vol. IT-28, pp. 753–763.

HUDSON, J. E. (1981). *Adaptive Array Principles*, Peter Peregrinus Ltd., London.

ITAKURA, F., and S. SAITO (1971). "Digital filtering techniques for speech analysis and synthesis," *Proc. 7th Intern. Conf. Acoust.*, Budapest, vol. 25-C-1, pp. 261–264.

ITAKURA, F., and S. SAITO (1972). "On the optimum quantization of feature parameters in the PARCOR speech synthesizer," *IEEE 1972 Conf. Speech Communications and Processing*, New York, pp. 434–437.

JABLON, N.K. (1986). "Steady state analysis of the generalized sidelobe canceller by adaptive noise canceling techniques," *IEEE Trans. Antennas and Propagation*, vol. AP-34, March.

JAYANT, N. S., and P. NOLL (1984). *Digital Coding of Waveforms*, Prentice-Hall, Englewood Cliffs, N.J.

JAYNES, E.T., (1982). "On the rationale of maximum-entropy methods," *Proc. IEEE*, vol. 70, pp. 939–952.

JAZWINSKI, A. H. (1969). "Adaptive filtering," *Automatica*, vol. 5., pp. 475–485.

JOHNSON, D. H. (1982). "The application of spectral estimation methods to bearing estimation problems," *Proc. IEEE*, vol. 70, pp. 1018–1028.

JOHNSON, C. R., JR. (1984). "Adaptive IIR filtering: Current results and open issues," *IEEE Trans. Information Theory*, vol. IT-30, Special Issue on Linear Adaptive Filtering, pp. 237–250.

JONES, S. K., R. K. CAVIN III, and W. M. REED (1982). "Amalysis of error-gradient adaptive linear equalizers for a class of stationary-dependent processes," *IEEE Trans. Information Theory*, vol. IT-28, pp. 318–329.

JOU, J-Y., and J. A. ABRAHAM (1984). "Fault-tolerant matrix operations on multiple processor systems using weighted checksums," *Proc. SPIE*, vol. 495, Real Time Signal Processing VII, pp. 94–101.

JURKEVICS, A. J., and T. J. ULRYCH (1978). "Representing and simulating strong ground motion," *B.S.S.A.*, vol. 68, pp. 781–801.

KAILATH, T. (1960). "Estimating filters for linear time-invariant channels," *MIT Research Laboratory for Electronics, Quarterly Progress Report* No. 58, pp. 185–197.

KAILATH, T. (1968). "An innovations approach to least-squares estimation—Part I: Linear filtering in additive white noise," *IEEE Trans. Automatic Control*, vol. AC-13, pp. 646–655.

KAILATH, T. (1969). "A generalized likelihood ratio formula for random signals in Gaussian noise," *IEEE Trans. on Information Theory*, vol. IT-15, pp. 350–361.

KAILATH, T., and P. A. FROST (1968). "An innovations approach to least-squares estimation, Part 2: Linear smoothing in additive white noise," *IEEE Trans. on Automatic Control*, vol. AC-13, pp. 655–660.

KAILATH, T., and R. A. GEESEY (1973). "In innovations approach to least-squares estimation; Part 5: Innovation representations and recursive estimation in colored noise," *IEEE Trans. on Automatic Control*, vol. AC-18, pp. 435–453.

KAILATH, T. (1970). "The innovations approach to detection and estimation theory," *Proc. IEEE*, vol. 58, pp. 680–695.

KAILATH, T. (1974). "A view of three decades of linear filtering theory," *IEEE Trans. Information Theory*, vol. IT-20, pp. 146–181.

KAILATH, T., ed. (1977). *Linear Least-Squares Estimation*, Benchmark Papers in Electrical Engineering and Computer Science, Dowden, Hutchinson and Ross, Stroudsburg, Pa.

KAILATH, T. (1980). *Linear Systems*, Prentice-Hall, Englewood Cliffs, N.J.

KAILATH, T. (1981). *Lectures on Linear Least-squares Estimation*, Springer-Verlag, New York.

KAILATH, T. (1982). "Time-variant and time-invariant lattice filters for nonstationary processes" in the book entitled *Outils et Modeles Mathematique pour L'Automatique, L'Analyse de Systems et le Traitement du Signal*, vol. 2, edited by I. Landau, France, CNRS, pp. 417–464.

KALLMANN, H. J. (1940). "Transversal filters," *Proc. IRE*, vol. 28, pp. 302–310.

KALMAN, R. E. (1960). "A new approach to linear filtering and prediction problems," *Trans. ASME J. Basic Engineering*, vol. 82, pp. 35–45.

KALMAN, R. E., and R. S. BUCY (1961). "New results in linear filtering and prediction theory," *Trans. ASME J. Basic Engineering*, vol. 83, pp. 95–108.

KAVEH, M., and A. BARABELL (1985). Private Communication.

KAY, S. M., and S. L. MARPLE, JR. (1981). "Spectrum analysis—A modern perspective," *Proc. IEEE*, vol. 69, pp. 1380–1419

KAY, S. M. (1983). "Recursive maximum likelihood estimation of autoregressive processes," *IEEE Trans. Acoustics, Speech, and Signal Processing*, vol. ASSP-31, pp. 56–65.

KAY, S. M. (1983). "Asymptotically optimal detection in unknown colored noise via autoregressive modeling," *IEEE Trans. Acoust., Speech, and Signal Processing*, vol. ASSP-31, pp. 927–940.

KELLY, J. L., JR., and R. F. LOGAN (1970). *Self-Adaptive Echo Canceller*, U.S. Patent 3,500,000, March 10.

KLEMA, V. C., and A. J. LAUB (1980). "The singular value decomposition: its computation and some applications," *IEEE Trans Automatic Control*, vol. AC-25, pp. 164–176.

KMENTA, J. (1971). *Elements of Econometrics*, Macmillan, New York.

KNIGHT, W. C., R. G. PRIDHAM, and S. M. KAY (1981). "Digital signal processing for sonar," *Proc. IEEE*, vol. 69, pp. 1451–1506.

KOLMOGOROV, A. N. (1939). "Sur l'interpolation et extrapolation des suites stationaires," *Comptes Rendus de l'Acad. Sci.*, Paris, vol. 208, pp. 2043–2045. [English translation reprinted in the book edited by Kailath (1977)].

KOLMOGOROV, A. N. (1968). "Three approaches to the quantitative definition of information," *Probl. Inf. Transmission*, vol. 1, pp. 1–7.

KREIN, M. G. (1945). "On a problem of extrapolation of A. N. Kolmogorov," *C.R. (Dokl.) Akad. Nauk SSSR*, vol. 46, pp. 306–309. [This paper is reproduced in the book edited by Kailath (1977)].

KUMARESAN, R. (1983). "On the zeros of the linear prediction-error filter for deterministic signals," *IEEE Trans. Acoust., Speech, and Signal Processing*, vol. ASSP-31, pp. 217–220.

KUMARESAN, R., and D. W. TUFTS (1983). "Estimating the angles of arrival of multiple plane waves," *IEEE Trans. Aerospace and Electronic Systems*, vol. AES-19, pp. 134–139.

KUNG, H. T. (1982). "Why systolic architectures?," *IEEE Computer*, vol. 15, pp. 37–46.

KUNG, H. T., and C. E. LEISERSON (1978). "Systolic arrays (for VLSI)," *Sparse Matrix Proc. 1978, Soc. Industrial and Appl. Math.*, 1979, pp. 256–282. [A version of this paper is reproduced in Mead and Conway (1980)].

KUNG, S. Y., H. J. WHITEHOUSE, and T. KAILATH, eds. (1985). *VLSI and Modern Signal Processing*, Prentice-Hall, Englewood Cliffs, N.J.

KUNG, S.Y. (1985). "VLSI array processors," *IEEE ASSP Magazine*, vol. 2, no. 3, pp. 4–22.

LACROSS, R. T. (1971). "Data adaptive spectral analysis methods," *Geophysics*, vol. 36, pp. 661–675.

LANDAU, I. D. (1984). "A feedback system approach to adaptive filtering," *IEEE Trans. Information Theory*, vol. IT-30, Special Issue on Linear Adaptive Filtering, pp. 251–262.

LANG, S. W., and J. H. McCLELLAN (1979). "A simple proof of stability for all-pole linear prediction models," *Proc. IEEE*, vol. 67, pp. 860–861.

LANG, S. W., and J. H. McCLELLAN (1980). "Frequency estimation with maximum entropy spectral estimators," *IEEE Trans. Acoust., Speech, and Signal Processing*, vol. ASSP-28, pp. 716–724.

LAWRENCE, R. E., and H. KAUFMAN (1971). "The Kalman filter for the equalization of a digital communication channel," *IEEE Trans. Commun. Tech.*, vol. COM-19, p. 1137.

LAWSON, C. L., and R. J. HANSON (1974). *Solving Least Squares Problems*, Prentice-Hall, Englewood Cliffs, N.J.

LEE, D. T. L. (1980). *Canonical Ladder Form Realizations and Fast Estimation Algorithms*, Ph.D. dissertation, Stanford University, Stanford, Calif.

LEE, D. T. L., M. MORF, and B. FRIEDLANDER (1981). "Recursive least-squares ladder estimation algorithms," *IEEE Trans. Circuits and Systems*, vol. CAS-28, pp. 467–481.

LEGENDRE, A. M. (1810). "Methode des moindres quarre's, pour trouver le milieu le plus probable entre les résultats de differentes observations," *Mem. Inst. France*, pp. 149–154.

LEHMER, D. H. (1961). "A machine method for solving polynomial equations," *J. Assoc. Comput. Mach.*, vol. 8, pp. 151–162.

LEV-ARI, H., T. KAILATH, and J. CIOFFI (1984). "Least-squares adaptive lattice and transversal filters: A unified geometric theory," *IEEE Trans. Information Theory*, vol. IT-30, pp. 222–236.

LEVINSON, N. (1947). "The Wiener RMS (root-mean-square) error criterion in filter design and prediction," *J. Math Phys.*, vol. 25, pp. 261–278.

LEVINSON, N., and R. M. REDHEFFER (1970). *Complex Variables*, Holden-Day, San Francisco.

LILES, W. C., J. W. DEMMEL, and L. E. BRENNAN (1980). *Gram–Schmidt Adaptive Algorithms*, Tech. Rept. RADC-TR-79-319, RADC, GRIFFISS AIR FORCE BASE, NY.

LIN, D. W. (1984). "On digital implementation of the fast Kalman algorithm," *IEEE Trans. Acoust. Speech, and Signal Processing*, vol. ASSP-32, pp. 998–1005.

LING, F., D. MANOLAKIS, and J. G. PROAKIS (1985). "New forms of LS lattice algorithms and an analysis of their round-off error characteristics," *Proceedings ICASSP 85*, Tampa, Fla., pp. 1739–1742.

LING, F., and J. G. PROAKIS (1984). "Numerical accuracy and stability: Two problems of adaptive estimation algorithms caused by round-off error," *Proceedings ICASSP 84*, San Diego, Calif., pp. 30.3.1 to 30.3.4.

LING, F., and J. G. PROAKIS (1984). "Nonstationary learning characteristics of least squares adaptive estimation algorithims," *Proceedings* ICASSP84, San Diego, Calif., pp. 3.7.1 to 3.7.4.

LJUNG, L. (1977). "Analysis of recursive stochastic algorithms," *IEEE Trans. Automatic Control*, vol. AC-22, pp. 551–575.

LJUNG, L. (1984). "Analysis of stochastic gradient algorithms for linear regression problems," *IEEE Trans. Information Theory*, vol. IT-30, Special Issue on Linear Adaptive Filtering, pp. 151–160.

LJUNG, L., and T. SÖDERSTRÖM (1983). *Theory and Practice of Recursive Identification*, MIT Press, Cambridge, Mass.

LJUNG, L., M. MORF, and D. Falconer (1978). "Fast calculation of gain matrices for recursive estimation schemes," *Intern. J. Control*, vol. 27, pp. 1–19.

LJUNG, S., and L. LJUNG (1985). "Error propagation properties of recursive least-squares adaptation algorithms," *Autometica*, vol. 21, pp. 157–167.

LUCKY, R. W. (1965). "Automatic equalization for digital communication," *Bell Syst. Tech. J.*, vol. 44, pp. 547–588.

LUCKY, R. W. (1966). "Techniques for adaptive equalization of digital communication systems," *Bell Syst. Tech. J.*, vol. 45, pp. 255–286.

LUCKY, R. W. (1973). "A survey of the communication theory literature: 1968–1973," *IEEE Trans. Information Theory*, vol. IT-19, pp. 725–739.

LUCKY, R. W., J. SALZ, and E. J. WELDON, JR. (1968). *Principles of Data Communication*, McGraw-Hill, New York.

MACCHI, O., and E. EWEDA (1984). "Convergence analysis of self-adaptive equalizers," *IEEE Trans. Information Theory*, vol. IT-30, Special Issue on Linear Adaptive Filtering, pp. 161–176.

MAKHOUL, J. (1975). "Linear prediction: A tutorial review," *Proc. IEEE*, vol. 63, pp. 561–580.

MAKHOUL, J. (1977). "Stable and efficient lattice methods for linear prediction," *IEEE Trans. Acoust., Speech, and Signal Processing*, vol. ASSP-25, pp. 423–428.

MAKHOUL, J. (1978). "A class of all-zero lattice digital filters: properties and applications," *IEEE Trans. Acoust., Speech, and Signal Processing*, vol. ASSP-26, pp. 304–314.

MAKHOUL, J., and L. K. COSSELL (1981). "Adaptive lattice analysis of speech," *IEEE Trans. Circuits and Systems*, vol. CAS-28, pp. 494–499.

MARDEN, M. (1949). "The geometry of the zeros of a polynomial in a complex variable," *Amer. Math. Soc. Surveys* (New York), no. 3, chapter 10.

MARKEL, J. D., and A. H. GRAY, JR. (1976). *Linear Prediction of Speech*, Springer-Verlag, New York.

MARPLE, L. (1980). "A new autoregressive spectrum analysis algorithm," *IEEE Trans. Acoust., Speech, and Signal Processing*, vol. ASSP-28, pp. 441–454.

MARPLE, S. L., JR. (1981). "Efficient least squares FIR system identification," *IEEE Trans. Acoust,. Speech, and Signal Processing*, vol. ASSP-29, pp. 62–73.

MASON, S. J. (1956). "Feedback theory: further properties of signal flow graphs," *Proc. IRE*, vol. 44, pp. 920–926.

MAZO, J. E. (1979). "On the independence theory of equalizer convergence," *Bell Syst. Tech. J.*, vol. 58, pp. 963–993.

MCCOOL, J. M., and others (1980). *Adaptive Line Enhancer*, U.S. Patent 4,238,746, December 9.

MCCOOL, J. M., and others (1981). *An Adaptive Detector*, U.S. Patent 4,243,935, January 6.

MCDONALD, R. A. (1966). "Signal-to-noise performance and idle channel performance of differential pulse code modulation systems with particular applications to voice signals," *Bell Syst. Tech. J.*, vol. 45, pp. 1123–1151.

MCDONOUGH, R. N., and W. H. HUGGINS (1968). "Best least-squares representation of signals by exponentials," *IEEE Trans. Automatic Control*, vol. AC-13, pp. 408–412.

MCWHIRTER, J. G. (1983). "Recursive least-squares minimization using a systolic array," *Proc. SPIE*, vol. 431, Real-Time Signal Porcessing VI, pp. 105–112.

MEAD, C., and L. CONWAY (1980). *Introduction to VLSI Systems*, Addison-Wesley, Reading, Mass.

MEDAUGH, R. S., and L. J. GRIFFITHS (1981). "A comparison of two linear predictors," *Proceedings ICASSP 81*, Atlanta, Ga., pp. 293–296.

MERMOZ, H. F. (1981). "Spatial processing beyond adaptive beamforming," *J. Acoust. Soc. Amer.*, vol. 70, pp. 74–79.

METFORD, P. A. S., and S. HAYKIN (1985). "Experimental analysis of an innovations-based detection algorithm for surveillance radar," *Proc. IEE (London)*, vol. 132, pp. 18–26.

METFORD, P. A. S., S. HAYKIN, and D. P. TAYLOR (1982). "An innovations approach to discrete-time detection theory," *IEEE Trans. Information Theory*, vol. IT-28, pp. 376–380.

MIDDLETON, D. (1960). *An Introduction to Statistical Communication Theory*, McGraw-Hill, New York.

MILLER, K. S. (1974). *Complex Stochastic Processes: An Introduction to Theory and Application*, Addison-Wesley, Reading, Mass.

MONSEN, P. (1971). "Feedback equalization for fading dispersive channels," *IEEE Trans. Information Theory*, vol. IT-17, pp. 56–64.

MONZINGO, R. A., and T. W. MILLER (1980). *Introduction to Adaptive Arrays*, Wiley-Interscience, New York.

MORF, M. (1974). "*Fast Algorithms for Multivariable Systems*," Ph.D. dissertation, Stanford University, Stanford, Calif.

MORF, M., and D. T. LEE (1978). "Recursive least squares ladder forms for fast parameter tracking," *Proc. 1978 Conf. Decision and Control*, San Diego, Calif., pp. 1362–1367.

MORF, M., T. KAILATH, and L. LJUNG (1976). "Fast algorithms for recursive identification," *Proc. 1976 Conf. Decision and Control,* Clearwater Beach, Fla., pp. 916–921.

MORF, M., A. VIEIRA, and D. T. LEE (1977). "Ladder forms for identification and speech processing," *Proc. 1977 IEEE Conf. Decision and Control*, New Orleans, pp. 1074–1078.

MOSCHNER, J. L. (1970). *Adaptive Filter with Clipped Input Data*, Stanford University Center for Systems Research, Tech. Rept. 6796-1, Stanford, Calif.

MUELLER, M. S. (1981a). Least-squares algorithms for adaptive equalizers," *Bell Syst. Tech. J.*, vol. 60, pp. 1905–1925.

MUELLER, M. S. (1981b). "On the rapid initial convergence of least-squares equalizer adjustment algorithms," *Bell Syst. Tech. J.*, vol. 60, pp. 2345–2358.

MURRAY, W., ed. (1972). *Numerical Methods for Unconstrained Optimization*, Academic Press, New York.

NAHI, N. E. (1969). *Estimation Theory and Applications*, Wiley, New York.

NAU, R. F., and R. M. OLIVER (1979). "Adaptive filtering revisted," *J. Operational Res. Soc.*, vol. 30, pp. 825–831.

NUTTALL, A. H. (1976). *Spectral Analysis of a Univariate Process with Bad Data Points via Maximum Entropy and Linear Predictive Techniques*, Naval Underwater System Center (NUSC) Scientific and Engineering Studies, Spectral Estimation, NUSC, New London, Conn.

OWSLEY, N. L. (1978). "Data adaptive orthonormalization," *Proceedings ICASSP 78*, Tulsa, OK, pp. 109–112.

OWSLEY, N. L. (1973). "A recent trend in adaptive spatial processing for sensor arrays: constrained adaptation," in *Signal Processing*, eds. J W. R. Griffiths and others, Academic Press, New York, pp. 591–604.

OWSLEY, N. L. (1984). "Sonar array processing," in S. Haykin, ed., *Array Signal Processing*, pp. 115–193, Prentice-Hall, Englewood Cliffs, N.J.

OPPENHEIM, A. V., and R. W. SCHAFER (1975). *Digital Signal Processing*, Prentice-Hall, Englewood Cliffs, N.J.

PAPOULIS, A. (1984). *Probability, Random Variables, and Stochastic Processes*, 2nd ed., McGraw-Hill, New York.

PLACKETT, R. L. (1950). "Some theorems in least squares," *Biometrika*, vol. 37, p. 149.

PORAT, B., B. FRIEDLANDER, and M. MORF (1982). "Square root covariance ladder algorithms," *IEEE Trans. Automatic Control*, vol. AC-27, pp. 813–829.

PORAT, B., and T. KAILATH (1983). "Normalized lattice algorithms for least-squares FIR system identification," *IEEE Trans. Acoust., Speech, and Signal Processing*, vol. ASSP-31, pp. 122–128.

PRIESTLEY, M. B. (1981). *Spectral Analysis and Time Series*, vols. 1 and 2, Academic Press, New York.

PROAKIS, J. G. (1975). "Advances in equalization for intersymbol interference," in *Advances in Communication Systems*, vol. 4, pp. 123–198, Academic Press, New York.

PROAKIS, J. G. (1983). *Digital Communications*, McGraw-Hill, New York.

PROAKIS, J. G., and J. H. MILLER (1969). "An adaptive receiver for digital signaling through channels with intersymbol interference," *IEEE Trans. Information Theory*, vol. IT-15, pp. 484–497.

PRONY, R. (1795). "Essai experimental et analytique, etc.," *L'ecole Polytechnique, Paris*, vol. 1, no. 2, pp. 24–76.

QURESHI, S. (1982). "Adaptive equalization," *IEEE Commun. Magazine*, vol. 20, pp. 9–16.

RABINER, L. R., and B. GOLD (1975). *Theory and Application of Digital Signal Processing*, Prentice-Hall, Englewood Cliffs, N.J.

RABINER, L. R., and R. W. SCHAFER (1978). *Digital Processing of Speech Signals*, Prentice-Hall, Englewood Cliffs, N.J.

RALSTON, A. (1965). *A First Course in Numerical Analysis*, McGraw-Hill, New York.

REDDI, S. S. (1979). "Multiple source location—A digital approach," *IEEE Trans. on Aerospace and Electronic Systems*, vol. AES-15, pp. 95–105.

REEVES, A. H. (1975). "The past, present, and future of PCM," *IEEE Spectrum*, vol. 12, pp. 58–63.

RIFE, D. C., and R. R. BOORSTYN (1976). "Multiple tone parameter estimation from discrete time observations," *Bell Syst. Tech. J.*, vol. 55, pp. 1389–1410.

RISSANEN, J. (1978). "Modelling by shortest data description," *Automatica*, vol. 14, pp. 465–471.

ROBBINS, H., and S. MONRO (1951). "A stochastic approximation method," *Ann. Math. Statis.*, vol. 22, pp. 400–407.

ROBINSON, E. A. (1967). *Multichannel Time-Series Analysis with Digital Computer Programs*, Holden-Day, San Francisco.

ROBINSON, E. A. (1982). "A historical perspective of spectrum estimation," *Proc. IEEE*, Vol. 70, Special Issue on Spectral Estimation, pp. 885–907.

ROBINSON, E. A., and S. TREITEL (1980). *Geophysical Signal Analysis*, Prentice-Hall, Englewood Cliffs, N.J.

SAITO, S., and F. ITAKURA (1966). *The Theoretical Consideration of Statistically Optimum Methods for Speech Spectral Density*, Report No. 3107, Electrical Communication Laboratory, N.T.T., Tokyo (in Japanese).

SAMSON, C. (1982). "A unified treatment of fast Kalman algorithms for identification," *Intern. J. Control*, vol. 35, pp. 909–934.

SATORIUS, E. H., and S. T. ALEXANDER (1979). "Channel equalization using adaptive lattice algorithms," *IEEE Trans. Commun.*, vol. COM-27, pp. 899–905.

SATORIUS, E. H., and J. D. PACK (1981). "Application of least squares lattice algorithms to adaptive equalization," *IEEE Trans. Commun.*, vol. COM-29, pp. 136–142.

SCHMIDT, R. O. (1981). *A Signal Subspace Approach to Multiple Emitter Location and Spectral Estimation*, Ph.D. dissertation, Stanford University, Stanford, Calif.

SCHMIDT, R. (1979). "Multiple emitter location and signal parameter estimation," *RADC Spectrum Estimation Workshop*, pp. 243–258. This paper is reproduced in the 1986 Special Issue of IEEE Transactions on Antennas and Propagation devoted to Adaptive Processing Antenna Systems.

SCHMIDT, R., and R. FRANKS (1986). "Multiple source DF signal processing: an experimental system," *IEEE Trans. Antennas and Propagation*, vol. AP-34, March.

SCHREIBER, R. J., and P. J. KUEKES (1985). "Systolic linear algebra machines in digital signal processing," in *VLSI and Modern Signal Processing*, edited by S.Y. Kung, H. J. Whitehouse, and T. Kailath, Prentice-Hall, Englewood Cliffs, N.J., pp. 389–405.

SCHROEDER, M. R. (1966). "Vocoders: Analysis and synthesis of speech," *Proc. IEEE*, vol. 54, pp. 720–734.

SCHROEDER, M. R. (1985). "Linear predictive coding of speech: review and current directions," *IEEE Communications Magazine,* vol. 23, pp. 54–61.

SCHUR, I. (1917). "Über potenzreihen, die im innern des einheitskreises beschränkt sind," *Journal für die Reine und Angewandte Mathematik*, vol. 147, pp. 205–232, Berlin. See also vol. 148, pp. 122–145.

SCHUSTER, A. (1898). "On the investigation of hidden periodicities with applications to a supposed 26-day period of meterological phenomena," *Terr. Mag. Atmos. Elect.*, vol. 3, pp. 13–41.

SCHWARTZ, G. (1978). "Estimating the dimension of a model," *Ann. Statis.*, vol. 6, pp. 461–464.

SENNE, K. D. (1968). *Adaptive Linear Discrete-Time Estimation*, Stanford University Center for Systems Research, Tech. Rept., 6778–5, Stanford, Calif.

SHAN, T-J., and T. KAILATH (1985). "Adaptive beamforming for coherent signals and interference," *IEEE Trans. Acoust., Speech, and Signal Processing*, vol. ASSP-33, pp. 527–536.

SHANNON, C. E. (1948). "The mathematical theory of communication," *Bell Syst. Tech. J.*, vol. 27, pp. 379–423, 623–656.

SHARPE, S. M., and L. W. NOLTE (1981). "Adaptive MSE estimation," *Proc. IEEE Intern. Conf. Acoust., Speech, and Signal Processing*, Atlanta, Ga., pp. 518–521.

SKOLNIK, M. I. (1982). *Introduction to Radar Systems*, McGraw-Hill, New York.

SHICHOR, E. (1982). "Fast recursive estimation using the lattice structure," *Bell Syst. Tech. J.*, vol. 61, pp. 97–115.

SONDHI, M. M., (1967). "An adaptive echo canceller," *Bell Syst. Tech. J.*, vol. 46, pp. 497–511.

SONDHI, M. M., (1970). *Closed Loop Adaptive Echo Canceller Using Generalized Filter Networks*, U.S. Patent, 3,499,999, March 10.

SORENSON, H. W. (1970). "Least-squares estimation: from Gauss to Kalman," *IEEE Spectrum*, vol. 7, pp. 63–68.

Special Issue on System Identification and Time-series Analysis (1974). *IEEE Trans. Automatic Control*, vol. AC-19, pp. 638–951.

Special Issue on Adaptive Systems (1976). *Proc. IEEE*, vol. 64, pp. 1123–1240.

Special Issue on Adaptive Antennas (1976). *IEEE Trans. Antennas and Propagation*, vol. AP-24, September.

Special Issue on Adaptive Signal Processing (1981). *IEEE Trans. Circuits and Systems*, vol. CAS-28, pp. 465–602.

Special Issue on Spectral Estimation (1982). *Proc. IEEE*, vol. 70, pp. 883–1125.

Special Issue on Adaptive Arrays (1983). *IEE Proceedings on Communications, Radar and Signal Processing*, London, vol. 130, pp. 1–151.

Special Issue on Linear Adaptive Filtering (1984). *IEEE Trans. Information Theory*, vol. IT-30, pp. 131–295.

Special Issue on Adaptive Processing Antenna Systems (1986). *IEEE Trans. Antennas and Propagation*, vol. AP-34, March.

SPEISER, J. M., and H. J. WHITEHOUSE (1983). "A review of signal processing with systolic arrays, "*Proceedings of SPIE*, vol. 431, Real Time Signal Processing VI, pp. 2–6, San Diego, Calif.

STEWART, G. W., (1973). *Introduction to Matrix Computations*, Academic Press, New York.

STRANG, G. (1980). *Linear Algebra and Its Applications*, 2nd ed., Academic Press, N.Y.

SWERLING, P. (1958). *A Proposed Stagewise Differential Correction Procedure for Satellite Tracking and Prediction*, RAND Corp. Rep. P-1292.

SWERLING, P. (1963). "Comment on 'A statistical optimizing navigation procedure for space flight,' " *AIAA J.*, vol. 1, p. 1968.

SZEGÖ, G. (1939). "Orthogonal polynomials," *Colloquium Publications*, no. 23, American Mathematical Society, Providence, R.I., 4th ed., 1975.

TANG, T., and others (1985). "An alternative to Fourier transform spectral analysis with improved resolution," *Magnetic Resonance*, vol. 62, pp. 167–171.

THOMAS, P. H. (1984). "*ARMA Modelling of Baseband Radar Signals*," M.Eng. thesis, McMaster University, Hamilton, Ontario.

THOMSON, W. T. (1950). "Transmission of elastic waves through a stratified solid medium," *J. Applied Phys.*, vol. 21, pp. 89–93.

TRETTER, S. A. (1976). *Introduction to Discrete-Time Signal Processing*, Wiley, New York.

TUFTS, D. W., and R. KUMARESAN (1982). "Estimation of frequencies of multiple sinusoids: Making linear prediction perform like maximum likelihood," *Proc. IEEE*, Vol. 70, pp. 975–989.

ULRYCH, T. J., and R. W. CLAYTON (1976). "Time series modelling and maximum entropy," *Phys. Earth Planet. Interiors*, vol. 12, pp. 188–200.

ULRYCH, T. J., and M. OOE (1983), "Autoregressive and mixed autoregressive-moving average models and spectra," in *Nonlinear Methods of Spectral Analysis* (S. Haykin, ed.), Springer-Verlag, New York.

UNGERBOECK, G. (1972). "Theory on the speed of convergence in adaptive equalizers for digital communication," *IBM J. Res. Devel.*, vol. 16, pp. 546–555.

UNGERBOECK, G. (1976). "Fractional tap-spacing equalizer and consequences for clock recovery in data modems," *IEEE Trans. Commun.*, vol. COM-24, pp. 856–864.

VAN DEN BOS, A. (1971). "Alternative interpretation of maximum entropy spectral analysis," *IEEE Trans. Information Theory*, vol. IT-17, pp. 493–494.

VAN TREES, H. L. (1968). *Detection, Estimation and Modulation Theory*, Part I, Wiley, New York.

WAKITA, H. (1973). "Direct estimation of the vocal tract shape by inverse filtering of acoustic speech waveforms," *IEEE Trans. Audio and Electroacoustics*, vol. AU-21, pp. 417–427.

WALACH, E., and B. WIDROW (1984). "The least mean fourth (LMF) adaptive algorithm and its family," *IEEE Trans. Information Theory*, vol. IT-30, Special Issue on Linear Adaptive Filtering, pp. 275–283.

WALKER, G. (1931). "On periodicity in series of related terms," *Proc. Royal Soc.*, vol. A131, pp. 518–532.

WARD, C. R., and others (1984). "Application of a systolic array to adaptive beamforming," *Proc. IEE (London)*, vol. 131, pt. F, pp. 638–645.

WARD, C. R., and P. J. HARGRAVE (1985). "A systolic array for high performance adaptive beamforming," *AGARD Conference Preprint No. 380*, Lisbon, Portugal.

WAX, M. (1985). *Detection and Estimation of Superimposed Signals*, Ph.D. dissertation, Stanford University, Stanford, Calif.

WAX, M., and T. KAILATH (1985). "Detection of Signals by Information Theoretic Criteria," *IEEE Trans. Acoustics, Speech, and Signal Processing*, vol. ASSP-33, pp. 387–392.

WEISBERG, S. (1980). *Applied Linear Regression*, Wiley, New York.

WHITE, W. D. (1979). "Angular spectra in radar applications," *IEEE Trans. Aerospace and Electronic Systems*, vol. AES-15, pp. 895–899.

WHITTLE, P. (1963). "On the fitting of multivariate autoregressions and the approximate canonical factorization of a spectral density matrix," *Biometrika*, vol. 50, pp. 129–134.

WIDROW, B. (1966). *Adaptive Filters I: Fundamentals*, Rept. SEL-66-126 (TR 6764-6), Stanford Electronics Laboratories, Stanford, Calif.

WIDROW, B. (1970). "Adaptive filters," in R. E. Kalman and N. DeClaris, eds. *Aspects of Network and System Theory*, Holt, Rinehart and Winston, New York.

WIDROW, B., and M. E. HOFF, JR. (1960). "Adaptive switching circuits," *IRE WESCON Conv. Rec.*, Part 4, pp. 96–104.

WIDROW, B., and E. WALACH (1984). "On the statistical efficiency of the LMS algorithm with nonstationary inputs," *IEEE Trans. Information Theory*, vol. IT-30, Special Issue on Linear Adaptive Filtering, pp. 211–221.

WIDROW, B., J. McCOOL, and M. BALL (1975). "The complex LMS algorithm," *Proc. IEEE*, vol. 63, pp. 719–720.

WIDROW, B., and others (1967). "Adaptive antenna systems," *Proc. IEEE*, vol. 55, pp. 2143–2159.

WIDROW, B., and others (1975). "Adaptive noise cancelling: Principles and applications," *Proc. IEEE*, vol. 63, pp. 1692–1716.

WIDROW, B., and others (1976). "Stationary and nonstationary learning characteristics of the LMS adaptive filter," *Proc. IEEE*, vol. 64, pp. 1151–1162.

WIDROW, B., K. M. DUVALL, R. P. GOOCH, and W. C. NEWMAN (1982). "Signal cancellation phenomena in adaptive antennas: Causes and cures," *IEEE Trans. on Antennas and Propagation*, vol. AP-30, pp. 469–478.

WIDROW, B., and S. D. STEARNS (1985). *Adaptive Signal Processing*, Prentice-Hall, Englewood Cliffs, N.J.

WIENER, N. (1949). *Extrapolation, Interpolation, and Smoothing of Stationary Time Series, with Engineering Applications*, MIT Press, Cambridge, Mass. (This was originally issued as a classified National Defense Research Report in February 1942.)

WIENER, N., and E. HOPF (1931). "On a class of singular integral equations," *Proc. Prussian Acad. Math-Phys. Ser.*, p. 696.

WILKINSON, J. H. (1965). *The Algebraic Eigenvalue Problem*, Oxford University Press, England.

WILSKY, A. S. (1979). *Digital Signal Processing and Control and Estimation Theory: Points of Tangency, Areas of Intersection, and Parallel Directions*, MIT Press, Cambridge, Mass.

WOLD, H. (1938). *A Study in the Analysis of Stationary Time Series*, Almqvist and Wicksells, Uppsala, Sweden.

WOODBURY, M. (1950). *Inverting Modified Matrices*, Memorandum Rep. 42, Statistical Research Group, Princeton University, Princeton, N.J.

WOZENCRAFT, J. M., and I. M. JACOBS (1965). *Principles of Communication Engineering*, Wiley, New York.

YULE, G. U. (1927). "On a method of investigating periodicities in disturbed series, with special reference to Wölfer's sunspot numbers. *Phil. Trans. Royal Soc. (London)*, vol. A226, pp. 267–298.

ZHANG, QI-TU, and S. HAYKIN (1983). "Tracking characteristics of the Kalman filter in a nonstationary environment for adaptive filter applications," *Proc. IEEE Intern. Conf. Acoust., Speech, and Signal Processing*, Boston, pp. 671–674.

ZHANG, QI-TU,, and S. HAYKIN (1984). "Radar clutter suppression schemes," *Electronics Letters*, vol. 20, pp. 1007–1008.

ZHANG, QI-TU, and S. HAYKIN (1985). "Adaptive radar detection," *Electronics Letters*, vol. 21, pp. 808–810.

Index

Adaptive backward linear prediction, 413
  a posteriori error, 413
  a priori error, 415
  augmented normal equation, 414
  gain vector, 415
  minimum sum of error squares, 415
Adaptive beamformer, 24, 42
  with minimum-variance distortionless response, 26, 43
  spatial smoothing, 43
Adaptive beamformer using LMS algorithm, 266, 267, 268
  RLS algorithm, 449
  systolic array, 530, 534, 540
Adaptive combiner with fixed beams, 27
Adaptive differential pulse-code modulation, 17
Adaptive equalizer, 10, 35
  decision feedback, 36
  fractionally spaced, 36
  linear synchronous, 36
  tracking mode, 12
  training mode, 11
Adaptive equalizer using LMS algorithm, 244
  LSL algorithm, 478, 492
  RLS algorithm, 398
  systolic arrays, 517, 528, 540
Adaptive estimation of sine waves in noise, 341
  using minimum-norm method, 341, 372
  using modified FBLP method, 349, 379
  using MUSIC algorithm, 341, 368
Adaptive filter (general), 3, 34
  adaptive process, 194

  applications, 7
  classification, 8
  computational requirements, 4
  filtering process, 194
  finite-impulse response, 31, 69
  infinite-impulse response, 31, 69
  misadjustment, 3
  numerical properties, 4
  rate of convergence, 3
  robustness, 4
  structure, 4
  viewed as linear device, 3
  viewed as nonlinear device, 3
Adaptive filter theory (general):
  approach based on Kalman filter, 5
  approach based on method of least squares, 6
  approach based on Wiener filter, 4
Adaptive forward linear prediction, 409
  a posteriori error, 409
  a priori error, 412
  augmented normal equation, 410
  gain vector, 412
  minimum sum of error squares, 412
Adaptive line enhancer, 24
Adaptive linear prediction using LMS algorithm, 240, 264
  LSL algorithm, 467, 492
  RLS algorithm, 398, 448
Adaptive noise canceller, 40
Adaptive sidelobe canceller, 42
Adaptive signal detection, 18, 40

Adaptive systolic (MVDR) beamformer, 531, 540
  computational complexity, 537
  summary, 532
Adjoint equation, 559
Algebraic Riccati equation, 302
All-pass filter, 155
All-pole filter, 69, 159
All-zero filter, 69, 159
An information-theoretic criterion, 88, 376
Array pattern, 28
Augmented index of performance, 434
Augmented normal equation:
  backward linear prediction, 136, 189
  forward linear prediction, 128, 188
Autocorrelation function, 45
Autocorrelation method, 37, 97, 377
Autocovariance function, 46
Autoregressive-moving average process, 86
Autoregressive process, 67
  asymptotic stationarity, 73
  autocorrelation function, 74
  maximum-likelihood estimates, 97
  order one, 98
  order two, 76
  parameters, 17, 67
  power spectrum, 17, 38, 76, 379
  process analyzer, 69
  process generator, 69
  Yule-Walker equations, 75

Backward linear prediction, 131, 378
  relation to forward linear

prediction, 134
Backward linear prediction (BLP) method, 324, 378
Backward method, 192
Backward prediction error, 131
   relation to forward prediction error, 154
Backward prediction-error filter, 135
   maximum-phase property, 153
   relation to forward prediction-error filter, 152
   transfer function, 152
Backward prediction-error power, 134
Backward substitution, 514, 533
Baseband, 8
Best linear unbiased estimate, 321
Block diagram, conventions, 563
Burg formula, 185, 192
   block estimation, 188

Cauchy's residue theorem, 551
Cauchy-Schwartz inequality (see Schwartz inequality)
Cholesky decomposition, 181
Clipped LMS algorithm, 266
Clutter, 20
Coding of speech, 37
Condition number, 61
   sensitivity to parameter variations, 96
Constant-state model (see Kalman algorithm, for transversal filters with stationary inputs)
Convergence ratio, 262
Conversion factor, 419
   viewed as estimation error, 418
   viewed as likelihood variable, 490
   viewed as sum of weighted error squares, 447
Convolution theorem, 555
Correlation matrix (deterministic):
   definition, 317
   properties, 318
Correlation matrix (ensemble averaged), 47, 104
   backward rearrangement of observables, 50
   bounds on eigenvalues, 65
   characteristic equation, 53
   eigenvalues, 53
   eigenvectors, 54

Hermitian symmetric property, 48
   partitioning, 51
   positive definiteness property, 49, 180
   Toeplitz property, 48
Cramér-Rao inequality, 322, 545
Cross-correlation vector (deterministic), 317
Cross-correlation vector (ensemble averaged), 103

Data matrix:
   autocorrelation method, 311
   covariance method, 310, 313
   postwindowing method, 311
   prewindowing method, 311, 496
Delay-and-sum beamformer, 25
Deterministic gradient algorithm (see Method of steepest descent)
Differential pulse-code modulation, 16, 38
Discrete-time Fourier transform, 63
   inverse, 63
Discrete-time stochastic process, 44
   partial characterization, 45

Echo canceller, 22
Eigen-spectrum, 375
Eigenvalue problem, 53
Eigenvalue spread, 5, 62
Eigenvectors, 54
   orthogonal property, 57
Energy, 556
Entropy, 178, 548
   of backward prediction-error vector, 178
Entropy rate, 548
Ergodicity, 47
Error feedback least-squares lattice algorithm, 485
   computational complexity, 486
Error performance surface, 5, 102
   canonical representation, 117
Error propagation, 488
Estimation error, 196
Excess mean-squared error, 227
Exchange matrix, 329
Exponential weighting matrix, 496
Extended gain vector, 416

Fast algorithms, 6, 382, 408

background theory, 409
Fast recursive least-squares (RLS) algorithm, 6, 408, 446
   summary, 447
Fast transversal filters (FTF) algorithm, 6, 240
   computational complexity, 486
   exact initialization for arbitrary initial condition, 433
   exact initialization for zero initial condition, 425
   finite-precision arithmetic effects, 441
   normalized gain vector, 422
   periodic restart procedures, 442
   rescue devices, 441
   rescue variable, 441
   summary, 426
Filter, 1
   linear, 2
Filtered state-error correlation matrix, 283
Filtering, 1
First unit vector, 161, 417
Fisher's information matrix, 322, 545
Forward-backward linear prediction (FBLP) method, 39, 325, 341
   augmented data matrix, 328
   augmented normal equation, 327
   data matrix, 327
   noiseless data, 342
Forward linear prediction, 123, 378
   relation to backward linear prediction, 134, 152
Forward linear prediction (FLP) method, 324
Forward method, 192
Forward prediction error, 124
   relation to backward prediction error, 154
Forward prediction-error filter, 127
   minimum-phase property, 150, 180
   relation to backward prediction-error filter, 152
   transfer function, 149
Forward prediction-error power, 125
Forward substitution, 533, 539
Frequency response, 63

Gershgorin's theorem, 60
Givens rotation, 497

Givens rotation (*cont.*)
  without square roots, 537, 539
Godard algorithm (*see* Kalman algorithm)
Gradient adaptive lattice (GAL) algorithm:
  one-coefficient version, 262, 452
  two-coefficient version, 465
Gradient vector, 197
Gramian matrix, 345
Gram-Schmidt orthogonalization procedure, 173, 301, 497

Harmonic mean method (*see* Burg formula)
Hermitian form, 49
Hermitian persymmetric matrix, 329
Hermitian symmetric matrix, 48
High-resolution spectrum estimation, 341
Householder transformation, 497

Impulse response, 102, 555
Index of performance, 102, 557
  differentiation with respect to weight vector, 105
Inner product, definition, 562
Innovations, 33, 90, 270
  scalar random variables, 271
Innovations-based detection algorithm, 20, 41, 304
Innovations process:
  correlation matrix, 277
  vector random variables, 275
Interference:
  coherent, 43
  noncoherent, 43
Inverse discrete-time Fourier transform, 63
Inverse filter, 92
Inverse recursion, 145
  relation to Schur-Cohn test, 157, 189
Inverse *z*-transform, 555

Jacobian, 179
Joint-process estimation (stationary inputs), 182
Joint-process estimation (adaptive), using recursive LSL algorithm, 469

Kalman algorithm, 34

convergence, 295
  for transversal filters with nonstationary inputs, 296
  for transversal filters with stationary inputs, 287
  signal-flow graph representation, 289
  summary, 291
Kalman-Bucy filter (*see* Kalman filter)
Kalman filter, 2, 269
  based on filtered estimate of state, 282
  based on one-step prediction of state, 280
  initial conditions, 281, 284
  signal-flow graph representation, 286
  summary, 285
Kalman filtering problem, 273
  measurement equation, 274
  process equation, 273
Kalman gain, 279
Kullback-Leibler mean information, 88
Kumaresan-Prony case, special case of modified FBLP method, 348

Lagrange multiplier, 559
Lattice inverse filter, 184, 191
Lattice predictor (stationary inputs), 162
  correlation properties, 167
  order-update recursions, 163
  normalized version, 190
Law of large numbers, 446
Leaky least-mean square (LMS) algorithm, 261
Learning curve, 203, 227
Least mean-square (LMS) algorithm, 5, 34, 216
  average mean-squared-error, 227
  average tap-weight vector, 219
  clipped, 266
  convergence in the mean, 221
  convergence in the mean square, 233
  digital implementation, 255
  fundamental assumption, 218
  instantaneous estimates, 216
  learning curve, 227
  memory, 238
  misadjustment, 236
  normalized, 261
  operation in nonstationary environment, 251

signal-flow graph representation, 217
  summary, 237
  weight-error correlation matrix, 221
Least-squares estimator (filter), 314
  properties, 319, 377
Lehmer-Schur test (*see* Schur-Cohn test)
Levinson-Durbin recursion, 32, 137, 143, 145
Likelihood function, 544
Likelihood ratio, 18
  chain rule, 304
Likelihood ratio processor, 121
  equivalence with Wiener filter, 121
Likelihood variable (*see* conversion factor)
Linear estimation theory, 31
  least-squares, 308
Linear prediction, 122
  least-squares, 324
Linear predictive coding (LPC), 14, 38
Log-likelihood function, 544
Log-likelihood ratio, 303

Markov process, first order, 98, 252
Matrix-inversion lemma, 385, 444
Maximum-entropy method, 547
Maximum-entropy spectrum estimator, 548
Maximum-likelihood estimator, 89, 543
  properties, 546
Maximum signal-to-noise ratio (SNR) filter, 120
  equivalence with Wiener filter, 121
Mean-square value, 46
Mean-value function, 45
Method of exponentially weighted least squares, (*see* Recursive least-squares algorithm)
Method of Lagrange multipliers, 557
Method of least squares, 6, 307
  desired response vector, 312
  estimation error vector, 312
  minimum sum of error squares, 316
  uniqueness theorem, 315
Method of steepest descent, 5, 197, 259

eigen-decomposition, 199
fastest rate of convergence, 203
feedback model, 198
learning curve, 203
overdamped response, 215
slowest rate of convergence, 202
stability, 199
time constant, 201
underdamped response, 215
Minimum-description-length (MDL) criterion, 89, 376
Minimum mean-squared estimation, 32
Minimum-norm method, 40, 341, 372
summary, 374
Minimum-variance distortionless response (MVDR) filter:
multiple constraints, 120
single constraint, 119
Minimum-variance unbiased estimate, 323
Misadjustment, 3, 236
Model, stochastic, 67
Modified forward-backward linear prediction (FBLP) method, 40, 347
Kumaresan-Prony case, 348
noise subspace, 347
signal subspace, 347
summary, 349
Modified QR decomposition-least squares (QRD-LS) algorithm, 522
computational complexity, 537
conversion factor, 526, 539
systolic array implementation, 526
Moore-Penrose generalized inverse (*see* pseudoinverse)
Moving average (MA) process, 86
Multiple linear regression model, 260, 308
Multiple side-lobe canceller, 27
Multiple signal characterization (MUSIC) algorithm, 40, 341, 368
noise subspace, 370
signal subspace, 370
summary, 371

Nonsingular matrix, 50
Normal equation, 32, 107
geometric interpretation, 110

Normal equation, deterministic, 312
geometric interpretation, 315
Normalized least mean-square (LMS) algorithm, 261
Norm of matrix, 61
Norm of vector, 62
Nullity of matrix, 316
Null space of matrix, 315
Null spectrum, 375

Objective function (*see* Index of performance)
Open-loop data-adaptive filter, 536
Optimization theory, constrained, 557
multiple side equations, 560
single side equation, 558
Optimum filtering problem, 101
Orthogonal multiple-beam-forming network, 27
Outer product, 562

Parametric spectrum analysis (*see* High-resolution spectrum estimation)
Parseval's theorem, 555
Partial correlation (PARCOR) coefficient, 143
Perron-Fronbenius theorem (*see* Perron theorem)
Perron theorem, 235
Phoneme, 13
Pole, 70
Power spectral density, 62
properties, 63
Power spectrum (*see* Power spectral density)
Predicted state-error correlation matrix, 277
Predicted state-error vector, 277
Prediction, 2
Prediction-error filter, 127
augmented normal equation, 129
eigenvector representation, 161
relation to linear prediction, 128
viewed as autoregressive (AR) process analyzer, 159
whitening property, 158
Primal equation, 559
Principle of maximum entropy, 548

Principle of orthogonality, deterministic, 314
geometric interpretation, 315
Principle of orthogonality, stationary inputs, 109
geometric interpretation, 110
Pseudoinverse, 331, 336
Pulse-amplitude modulation, 10
Pulse-code modulation, 15, 38

QR-decomposition, 500

Radar, surveillance, 19
Random process (*see* Stochastic process)
Random-walk state model (*see* Kalman algorithm, for transversal filters with nonstationary inputs)
Rank of matrix, 94, 332
Rayleigh quotient, 56
Recursive least-squares (RLS) algorithm, 6, 389
a posteriori error, 387
a priori error, 387
bias, 445
computational complexity, 486
convergence analysis, 390
convergence in the mean, 391
convergence in the mean square, 392, 393
gain vector, 386
initial conditions, 388
memory, 406
misadjustment, 404, 446
operation in nonstationary environment, 403
physical significance of parameters, 487
prewindowing, 384
relation to the Kalman algorithm, 407
structure, 486
summary, 389
transformed solution, 473
update of sum of weighted error squares, 389
update of tap-weight vector, 387
weighting factor, 382, 444
weight vector lag, 404
Recursive least-squares lattice (LSL) algorithm, 7, 451
a posteriori version, 465
a priori version, 482
backward reflection coefficient, 458

Recursive least-squares lattice (LSL) algorithm (*cont.*)
computational complexity, 486
error feedback formula, 485
exact decoupling property, 470
forward reflection coefficient, 458
initialization, 466, 467
joint-process estimation, 469
normalized, 491
order update for conversion factor, 461
order update for estimation error, 477
order update recursions, 453
physical significance of parameters, 487
prewindowing, 453
structure, 486
summary, 465
time-update recursion, 462
Recursive minimum mean-square estimation:
scalar random variables, 270
vector random variables, 277
Recursive QR decomposition-least squares (QRD-LS) algorithm, 505
a posteriori estimation error, 525
a priori estimation error, 525
computational complexity, 537
exact initialization, 505
summary, 506
systolic array implementation, 512
Reflection coefficients, stationary inputs:
definition, 142
relation to autocorrelation sequence, 147
relation to minimum-phase property, 150
Regression coefficients, 183
Reverberation, 20
Riccati difference equation, 280
Rouché's theorem, 150

Sampling theorem, 2
Schur-Cohn test, 156
relation to inverse recursion, 157, 189
Schwartz inequality, 61, 378
Scores vector, 545
Signal-flow graph, 198
conventions, 563

Singular value decomposition (SVD), 59, 331
Singular values, 335
relation to eigenvalues, 336
Singular vectors:
left, 335
right, 335
Smoothing, 1
Sonar, active, 19
Spectral norm, 61
Spectral theorem, 59
Speech:
linear predictive coding (LPC), 14, 38
production model, 13
unvoiced, 13
voiced, 13
wave-form coding, 15
Square-root Kalman filter, 300
Standard beamwidth, 380
State-transition matrix, 273
State vector, 273
Steering vector, 27
Step-size parameter, 198
Stochastic approximation, 34
Stochastic gradient algorithm (*see* Least mean-square (LMS) algorithm)
Stochastic model, 67
order selection, 87
Stochastic process, 44
discrete-time, 45
innovations representation, 90
strictly stationary, 45
wide-sense stationary, 46
Symmetric matrix, 48
System identification, 8
using FTF algorithm, 436, 449
using Kalman algorithm, 305
using RLS algorithm, 449
Systolic arrays, 7, 494
fault-tolerant, 536
linear section, 514
triangular section, 512
Szegö's theorem, 96

Time constant of nonstationarity, 252
Time constant of steepest descent algorithm, 201
Toeplitz matrix, 32, 48
rectangular, 319
Total input power, 235
Transfer function, 555
relation to impulse response, 555

Transversal filter, 4
Triangle inequality, 61

UD factorization:
complex, 301, 441
real, 300
Uniform phase factor, 28
Uniqueness theorem, 315
Unitary matrix, 58
Unitary similarity transformation, 59

Vandermonde matrix, 55, 345
Variance, 46
Very large-scale integration (VLSI), 4, 536

Wave-form coding, 15
Weight-error vector, 219
Weight vector lag, 252
Weight vector noise, 251
Whitening filter, 19, 91, 158
Wiener filter, 2
equivalence with likelihood ratio processor, 121
equivalence with maximum signal-to-noise ratio (SNR) filter, 120
minimum mean-squared error, 110
theory, 100
Wiener-Hopf equation, 32
Wiener-Khintchine relations, 63
Windowing of data:
autocorrelation method, 311
covariance method, 310
postwindowing method, 311
prewindowing method, 311
Wold's decomposition theorem, 71

Yule-Walker equations, 67, 127

Zero, 69
Zero-forcing algorithm, 35
z-transform, 553
complex-conjugation property, 555
convolution property, 555
inverse, 555
linearity property, 554
one-sided, 554
Parseval's theorem, 555
region of convergence, 554
shifting property, 555
two-sided, 554